THE MATHEMATICAL HERITAGE OF
C. F. GAUSS

THE MATHEMATICAL HERITAGE OF
C. F. GAUSS

A Collection of Papers in Memory of C. F. Gauss

Editor

GEORGE M. RASSIAS

Department of Mathematics
The American College of Greece
Athens

World Scientific
Singapore • New Jersey • London • Hong Kong

Published by

World Scientific Publishing Co. Pte. Ltd.
5 Toh Tuck Link, Singapore 596224
USA office: 27 Warren Street, Suite 401-402, Hackensack, NJ 07601
UK office: 57 Shelton Street, Covent Garden, London WC2H 9HE

British Library Cataloguing-in-Publication Data
A catalogue record for this book is available from the British Library.

THE MATHEMATICAL HERITAGE OF C F GAUSS

ISBN-13 978-981-02-0201-9
ISBN-10 981-02-0201-6
ISBN-13 978-981-02-3797-4 (pbk)
ISBN-10 981-02-3797-9 (pbk)

Preface

The Mathematical Heritage of C. F. Gauss is a volume dedicated to the memory of Carl Friedrich Gauss (1777–1855) — one of the greatest mathematicians of all time.

It contains a series of papers written by eminent scientists from the international community on the occasion of the 135th anniversary (1855–1990) of C. F. Gauss's death.

This volume hopes to provide an insight and analysis on various research problems and new theories in the fields of mathematics and its applications in which Gauss had made many fundamental discoveries. It also provides the opportunity for publication of new aspects of scientific research problems related to Gauss's investigations that will be of the greatest value to all scientists currently working within these areas of research. The presentation of concepts and methods of the volume makes it an invaluable reference for advanced graduate students and research mathematicians interested in introducing themselves to these areas of research as well as for teachers and other professionals in mathematics interested in these scientific fields.

It is a pleasure to express my wholehearted thanks to all of the scientists who so willingly responded to my invitation and enthusiastically participated in this international tribute of honor to Carl Friedrich Gauss. In addition, I would like to express my sincere appreciation and gratitude to the staff of World Scientific Publishing Company, Singapore, for their care and superb assistance in the production of this volume.

Athens, Greece   George M. Rassias
January 30, 1991

Contents

Contents   xiii

THE MATH. HERITAGE OF C.F. GAUSS (pp. 1-11)
edited by George M. Rassias
©1991 World Scientific Publ. Co. Singapore

A SUMMARY OF C. F. GAUSS'S LIFE
AND MATHEMATICAL WORK

George M. Rassias

Carl (or Karl) Friedrich Gauss (1777–1855) was a German mathematician, physicist and astronomer. He was the son of poor, uneducated parents born in a humble cottage at Braunshweig (Brunswick), Germany, on April 30, 1777. His full baptismal name was Johann Friedrich Carl Gauss. In later life he signed his masterpieces simply Carl Friedrich Gauss.

His father was a farmer who wished that his young son to follow one of the family trades and become a bricklayer or a gardener. But at a very early age it was clear that his son had unusual talents. He is said to have corrected an error to his father's payroll accounts at the age of three. At elementary school, at the age of eight, he added up all the numbers from one to a hundred in his first lesson. Recognising his precocious talent, the teacher persuaded his father that Gauss should be encouraged to train to pursuit a profession rather than learn a trade. When he went to high school he proved to be just as good at philology as at mathematics.

His remarkable abilities attracted the attention of the Duke of Brunswick. The Duke was impressed with the boy and undertook to support his further education, first at the Collegium Carolynum of Brunswick (1792–1795) and later at the University of Göttingen (1795–1798).

While as a student at the Collegium Carolynum at the age of fourteen of fifteen he discovered the prime number theorem. As a teenager he discovered (published in 1795) the method of least squares for minimizing the effect of errors inherent in any set of statistical data.

In his nineteenth year he made an exciting mathematical discovery by showing that the 17-sided regular polygon is constructible using ruler and compass only. This represented the first discovery in Euclidean geometry after 2000 years. It was this discovery, announced on 1 June 1796, but made on 30 March 1796, which induced the young Gauss to choose mathematics instead of philology as his life work. In 1798 he transferred to the University of Helmstädt and there obtained his Ph.D. in 1799 under the supervision of Johann Friedrich Pfaff (1765–1825). In his dissertation he gave the first rigorous proof of the now known as the fundamental theorem of Algebra which states that every algebraic equation with real coefficients has at least one root and hence has n roots.

D'Alembert had tried to prove it in 1746. Two more proofs were given by Gauss and a fourth one in which he returned to his first proof. After finishing his doctorate he returned to Brunswick where he wrote some of his most famous papers.

In 1801, Gauss published his monumental treatise *Disquisitiones Arithmeticae* (Arithmetical Researches) which is usually considered as the beginning of modern number theory. In it he summarized all the work that had been carried out up to that time, and formulated concepts and questions that are still relevant today. In this book Gauss provided a proof of the fundamental theorem of Arithmetic. The book was finished while Gauss was still a student and he dedicated it to his patron, the Duke of Brunswick, who had taken charge of his education.

In the years between 1800 and 1810 he concentrated on *astronomy*. On the first night of the 19th century (January 1, 1801) Piazzi in Palermo discovered the first of the so called small planets (planetoids or asteroids), which was given the name of Ceres. Since only a few observations of the new platenoid could be made to find it again, its position had to be determined with a great deal of precision from a small number of observations. Gauss was attracted by the challenge and making use of his method of least squares and with his virtuosity at numerical computation he solved the problem and told the astronomers where to look with their telescopes and the new planet Ceres was found again at precisely the place predicted by the computations of Gauss.

On March 28, 1802 a second planetoid, Pallas, was discovered by Wilhelm Olbers (1758–1840) at his private observatory in Bremen.

Certain pecularities of the motion of Pallas namely its perturbation by Jupiter presented exceptional difficulties. Gauss began to take interest in that problem that led him to the *Theoria motus corporum coelistium*

(1809) translated as "Theory of the Motion of the Heavenly Bodies moving about the Sun in Conic Sections", to his paper on the attraction of general ellipsoids (1813), to his work on mechanical quadrature (1814), and to his study of seculiar perturbations of planets (1818). The 1001st planetoid to be discovered was named *Gaussia* in his honour.

In 1807 he became Professor of Astronomy and Director of the Göttingen observatory, a position he held for some forty years until his death. Gauss thought of astronomy as his profession and pure mathematics as his recreation. Except for one visit to Berlin to attend a scientific meeting he remained at Göttingen for the rest of his life.

Around 1820 the government of Hanover asked him to supervise a geodetic survey of the Kingdom. This work provided the stimulus that led him to the ideas of his paper *Disquisitiones generales circa superficies curvas* (1827), in which he founded the intrinsic geometry of a surface, in which he used curvilinear coordinates to express the linear element ds in a quadratic differential form

$$ds^2 = Edu^2 + 2Fdudv + Gdv^2 \; ;$$

he formulated the concepts of *Gaussian curvature* and integral curvature. There was also a climax, the *theorema egregium*, which states that the Gaussian curvature of the surface depends only on E, F and G and their derivatives, and is therefore a bending invariant, and the *Gauss-Bonnet theorem* on integral curvature for the case of a geodesic triangle which in its general form is the most essential fact of modern global differential geometry.

His research in geodecy led him to the development of the concept of *conformal mapping*. Lagrange had treated the conformal mapping of a surface of revolution into the plane. In 1822 Gauss won a prize offered by the Danish Royal Society of Sciences for a paper on the problem of finding the analytic condition for transforming any surface conformally onto any other surface.

Until Gauss, surfaces had been studied as figures in three-dimensional Euclidean space but he showed that the geometry of a surface could be studied concentrating on the surface itself. Gauss's work in differential geometry is a landmark in itself. But its implications were far deeper than he himself appreciated.

In 1831 his great work on *biquadratic residues* appeared. It was an extension of some of early discoveries in number theory with the aid of a new method, the theory of complex numbers. He defined complex integers (now called Gaussian integers) as complex numbers $a + ib$ where a and b are integers and introduced a new theory of prime numbers in which 3 remains a prime but $5 = (1 + 2i)(1 - 2i)$ does not.

In this paper Gauss clarified the confusion which still surrounded the subject of *complex numbers* by representing them as points in a plane. He coined the term "complex number" and introduced the notation i for $\sqrt{-1}$. He also proved that the ring $Z[i]$ is a UFD and a Euclidean domain.

His work in the field of geodesy was followed by a prolific work in *physics*. In the period 1833–34 he worked experimentally on geomagnetism and in collaboration with his friend and almost 30 years younger colleague Wilhelm Weber (1804–1891) they built and operated a new magnetic observatory free of all metal that might affect magnetic forces and they organized the Magnetische Verein, which united a worldwide network of observatories. In 1834 there were already twenty-three magnetic observatories in Europe.

J. C. Maxwell says in his *Electricity and Magnetism* that Gauss's studies of magnetism reconstructed the whole science, the instruments used, the methods of observation, and the calculation of results. Gauss's papers on terrestrial magnetism are models of physical research and supplied the best method of measuring the earth's magnetic field. His work on astronomy and magnetism opened up a new and brilliant period of alliance between mathematics and physics.

Gauss and Weber also invented and constructed *the first electromagnetic telegraph* in 1833, with which they made possible for the Gauss's Observatory to communicate with Weber's Institute of Physics. Early in 1833 the first words were sent and then whole sentences. Gauss devised logical sets of units for magnetic phenomena as well as the unit of magnetic flux density which is called *the "gauss"* after him. Steinheil in Germany (in 1837) and Morse in U.S.A. (in 1838) independently invented more efficient telegraphs.

At Göttingen he invented *the heliotrope*, an instrument which enabled more precise trigonometric determinations of the shape of the Earth to be made. At Göttingen there is a statue of Gauss and Weber in the process of inventing the electric telegraph. James Clerk Maxwell refered several times to Gauss's work in his famous Treatise on Electricity and Magnetism (1873). In 1839 Gauss published his fundamental paper on the general theory of inverse square forces, *Allgemeine Lehrsätze* that was the beginning of *potential theory* as a coherent branch of mathematics and it led to certain minimal principles concerning space integrals which later became known as "Dirichlet's principle" which was eventually proved by D. Hilbert.

Among his discoveries were also *the divergence theorem* (known as Gauss's theorem) of modern vector analysis and the basic mean value theorem for harmonic functions.

In 1811 Gauss proposed himself the problem of investigating what are called *analytic functions* of a complex variable today which as it was mentioned before he represented as points on the plane. In his letter to his friend F. Bessel in 1811 he stated what amounts to be the fundamental theorem in the theory of analytic functions which was one of the most active fields of mathematics in the nineteenth century. This theorem that was rediscovered independently by A.-L. Cauchy with another similar theorem gave rise to many significant results in Complex Analysis.

Around 1810, N. H. Abel, J. Fourier, C. F. Gauss and B. Bolzano began the exact handling of infinite series. In his 1812 paper *Disquisitiones Generales Circa Seriem Infinitam* (General Investigations of Infinite series), Gauss made the first important and stricly rigorous investigation of convergence of infinite series wherein he studied *the hypergeometric series $F(\alpha,\beta,\gamma,x)$*. In most of his work he called a series convergent if the terms from a certain one onwards decrease to zero. But in his 1812 paper he noted that this is not the correct concept. Because the hypergeometric series can represent many functions for different choices of α,β, and γ it seemed desirable to him to develop an exact criterion for convergence for this series.

Gauss showed that the hypergeometric series converges for real and complex x if $|x| < 1$ and diverges if $|x| > 1$. For $x = 1$, the series converges if and only if $\alpha + \beta < \gamma$ and for $x = -1$ the series converges if and only if $\alpha + \beta < \gamma + 1$.

In his astronomical and geodetic work, Gauss, like the eighteenth-century men, followed the practice of using a finite number of terms of an infinite series and neglecting the rest. He stopped including terms when he saw that the succeeding terms were numerically small and of course did not estimate the error. In unpublished papers he recorded his innovative work in two major fields: the *elliptic functions* and *non-Euclidean geometry*.

From Euclid's time to Gauss's boyhood the postulates of Euclidean geometry were taken for granted as logical statements (axioms). However for two thousand years mathematicians were puzzled with the question of whether Euclid's parallel postulate is not independent of the others and tried to prove it as a theorem but without success.

Gauss was the first man who believed in the independence of Euclid's parallel postulate, which implied that another choice of axiom could produce other geometries.

In 1820 he possessed the main theorems of *non-Euclidean geometry* (a term that Gauss coined) but never published his results because as he wrote in a letter to Bessel of January 27, 1829, he feared ridicule, or, as he put it, he feared the "shouts of the Boetians" that would be heard if his secrets were not kept.

The unsucccessful attempts of mathematicians to reduce the parallel postulate to a theorem led to a full appreciation of Euclid's wisdom in taking it as an axiom as well as to the birth of other, so called non-Euclidean geometries. In 1829 and 1832 the Russian mathematician Nikolai Ivanovitch Lobachevsky (1792–1856) and the Hungarian mathematician Janos Bolyai (1802–1860), son of Wolfgang, published their own independent work on the non-Euclidean geometry.

Similarly Gauss had published nothing on the subject of *elliptic functions*, a rich field of analysis, and claimed nothing from the works of Abel (1827) and Jacobi (1828–29), so the world of mathematicians was astonished when it gradually became known that he had discovered many of the results before Abel and Jacobi were born.

After having visited Gauss in 1840, Jacobi wrote to his brother, "Mathematics would be in a very different position if practical astronomy had not diverted this colossal genius from his glorious career."

Gauss also was a pioneer in Topology. In his dissertation of 1799 he made the prediction that *analysis situs* would become one of the chief concerns of mathematics. He was probably the first to recognize the possibility of studying *knots* mathematically and he included an analytic formulation of *linking number*, a basic concept to knot theory and other fields of topology, in his investigations of electrodynamics (1833) *Zur mathematischen Theorie der electrodynamischen Wirkungen*. Werke, Koniglichen Gesellschaft der Wissinchaften Zu Göttingen, vol. 5, 602–629." Although Gauss did not publish papers on topology he made a beginning as was found out in his posthumous papers. His prediction of the great importance of topology has been fulfilled in our own generation.

In 1848 Johann B. Listing (1806–1882), a student of Gauss in 1834, and later professor of physics at Göttingen, published *Vorstudien zur Topologie* (Introductory Studies in Topology), in which he discussed what he preferred to call the geometry of position but, since this term was used for projective geometry by von Staudt, he used the term *topology*. J. B. Listing sought qualitative laws for geometrical figures. Thus he attempted to generalize the Euler relation $V - E + F = 2$.

It was A. Möbius however, who was an assistant to Gauss in 1813, that first formulated properly the nature of topological investigations in his *Theorie der elementaren Verwandschaft* (Theory of elementary Relationships), in 1863. Möbius and Listing independently discovered one-sided surfaces in 1858, of which the Möbius strip is best known.

The greatest impetus to topological investigations came from the work of another student of Gauss, namely Riemann, in his thesis of 1851 on complex functions and in his study of Abelian functions where he stressed that to work with functions some theorems of analysis situs were indispensable. In these investigations he found it necessary to introduce the *connectivity* of Riemann surfaces. He classified surfaces according to their connectivity and, as he himself realized, had introduced *a topological property*.

Gauss also worked in *optics*, which had been neglected since Euler's days, and his investigations of 1838–41 gave a totally new basis for the handling of optical problems. He also made fundamental contributions in *capillarity, crystallography, mechanics* and there is rarely any mathematical, physical or astronomical field in which Gauss did not work. Among his students were Dedekind, Dirichlet, Eisenstein, Kummer. Gauss had little enthusiasm for teaching, which he regarded as a waste of his time and as essentially useless (for different reasons) for both talented and untalented students. However, when teaching was unavoidable he apparently did it superbly. According to R. Dedekind who was one of his students, Gauss's lectures after the passage of 50 years remained "unforgettable in memory as among the finest which I have ever heard."

Gauss enjoyed social life, was married twice, and raised a family (he had six children, two of whom emigrated to America). Apart from science and his family, his main interests were history and world literature, international politics, and public finance. Perhaps stimulated by his actuarial work, he fell into the habit of collecting all sorts of statistics from the newspapers, books, and daily observations. Undoubtedly some of these data helped him with financial speculations shrewd enough to create an estate equal to nearly 200 times his annual salary. The "star gazer", as his father called him, had, as an afterthought, achieved the financial status denied his more "practical" relatives. Luxury never attracted the Prince of

Mathematicians whose life had been unaffectedly dedicated to science long before he was twenty. As his friend Sartorius von Waltershausen writes, "As he was in his youth, so he remained through his old age to his dying day, the unaffectedly simple Gauss. A small study, a little work table with a green cover, a standing-desk painted white, a narrow sopha and, after his seventieth year, an arm chair, a shaded lamp, an unheated bedroom, plain food, a dressing gown and a velvet cap, these were so becomingly all his needs." He owned a large library of about 6000 volumes in many languages, including Greek, Latin, English, French, Russian, Danish, and of course German.

Reports from several of his students who attended his lectures when he was an old man show that Gauss now enjoyed teaching and lecturing much more than in his early years. One of the reasons for his change of attitude may have been that students were now much better prepared and more interested. Among Gauss's last students were M. Cantor, who is known as the author of a voluminous history of mathematics, and R. Dedekind, famous for his work in number theory and algebra. We quote from an account which Dedekind gave in 1901. "... usually he sat in a comfortable attitude, looking down, slightly stooped, with hands folded above his lap. He spoke quite freely, very clearly, simply and plainly; but when he wanted to emphasize a new viewpoint, in which he used an especially characteristic word, then he suddenly lifted his head, turned to one of those sitting next to him, and gazed at him with his beautiful, penetrating blue eyes during the emphatic speech ... If he proceeded from an explanation of principles to the development of mathematical formulas, then he got up, and in stately, very upright posture he wrote on a blackboard beside him in his peculiarly beautiful handwriting; he always succeeded through economy and deliberate arrangement in making do with the rather small space. For numerical examples, on whose careful completion he placed special value, he brought along the requisite data on little slips of paper."

From the early 1840's the intensity of Gauss's activity gradually decreased. Further publications were either variations on old themes, reviews, reports, or solutions of minor problems. However he continued astronomical observing. He served several times as dean of the Göttingen faculty. He was busy during the 1840's in finishing many old projects, such as the last calculations on the Hanover survey. In 1847 he eloquently praised number theory and G. Eisenstein in the preface to the collected works of this ill-fated young man who had been one of the few to tell Gauss anything he did not already know. He spent several years putting the university widows' fund on a sound actuarial basis, calculating the necessary tables. He learned to read and speak Russian fluently. The fiftieth anniversary celebration of his doctorate in 1849 brought him many messages and formal honors. The paper that Gauss delivered was his fourth proof of the fundamental theorem of algebra, appropriately a variation of the first in his thesis of 1799. After this celebration, Gauss continued his interests at a slower pace and became more than ever a legendary figure unapproachable by those outside his personal circle.

In January 1854, Gauss had a thorough physical examination. The diagnosis was dilatation of the heart and did not leave any hope that he might live much longer. Subsequently, his health improved again; it was during this period (June 1854) that Riemann was given the opportunity to deliver his great *Habilitationsschrift* on the foundations of geometry, a topic chosen by Gauss. Riemann gave his lecure "Über die Hypothesen, Welche der Geometrie Zu Grunde liegen" in fulfillment of the requirements for attaining the permission to lecture, the *venia legendi* at the University of Göttingen. A few days later he left Göttingen for the last time to observe construction of the railway from Kassel. By the beginning of August, his health had deteriorated again and he was no longer able to leave his house. By autumn his illness was much worse. Although gradually more bedridden, he kept up his reading, correspondence, and trading in securities. On December 7, the end seemed to have arrived, but Gauss fought it off once more. He died in his sleep early in the morning of February 23, 1855, after steadily losing strength since mid-January. His watch is reported to have stopped minutes after his death. Gauss is buried in Göttingen.

Gauss's contemporaries appreciated his genius and by the time of his death he was widely venerated and called *the Prince of Mathematicians.*

In 1880 a memorial whose base is a 17-sided polygon was erected to Gauss at Brunswick, his birthplace. In his honor, the reigning King of Hanover ordered a medal which hailed him as the Prince of Mathematicians. Felix Klein, in his lectures on the development of mathematics in the 19th century at Göttingen from 1914 to 1919 included the following splendid tribute to Gauss: If we now inquire what was the unusual and unique quality of this man, the answer must be: a combination of the greatest individual achievement in every chosen field with the greatest versatility; a complete balance between mathematical creativity, the strength for pursuing its development, and the practical sense for its application, including faultless observation and measurement; and finally, the presentation of this great self-created wealth in the most polished form.

"He had only two peers, Archimedes and Newton, who were equally gifted. In common with both, Gauss had the unusually long life span which makes possible a full development of personality. Archimedes personifies the scientific achievements of classical antiquity, Newton is the initiator of higher mathematics, while Gauss represents the emergence of a new mathematical era."

Gauss's seal bore the motto, *pauca sed matura* – few, but ripe — and he was so full of original ideas that he did not have time to see that all of them ripened to the point of perfection on which he insisted before publication. He was a perfectionist and never published his work unless he was certain that he had the complete answer in the most elegant and precise form. Many of his results including his work on the non-Euclidean geometry, were not published in his lifetime. All of Gauss's publications (including his fine reviews of his own papers) are reprinted in the *Werke*, published in 12 volumes by the Königliche Gesellschaft der Wissenschaften zu Göttingen (Leipzig-Berlin, 1863–1933). The *Werke* contains also a generous selection of his unpublished notes and papers, related correspondence, commentaries, and extensive analyses of his work in each field.

Acknowledgement

To cover the infomation of this biography I have relied on many sources that I have cited in the references. Anyone interested in Gauss's life and work can enrich much more his knowledge from these works than I have extracted. I am deeply grateful to the Departments of Mathematics of the University of California at Berkeley and of Stanford University, California for my visit at those two Universities in August, 1990 that played a significant role for completing the preparation of this volume.

References

1. D. Abbott, (ed.), *Mathematicians*, The Biographical Dictionary of Scientists, Blond Educational, Great Britain, 1985.
2. E. T. Bell, *The Prince of Mathematicians*, The World of Mathematics, Simon and Schuster, New York, 1956.

3. C. B. Boyer, *A History of Mathematics*, Princeton Univ. Press, 1985.

4. W. K. Bühler, *Gauss, A Biographical Study*, Springer Verlag, New York Inc., 1981.

5. D. M. Campbell and J. C. Higgins, (eds), *Mathematics, People. Problems. Results*, Wadsworth, Inc. 1984.

6. J. A. Gallian, *Contemporary Abstract Algebra*, D. C. Heath Company, Lexington, Massachusetts, and Toronto, 1986.

7. M. Kline, *Mathematical Thought from Ancient to Modern Times*, Oxford University Press, Inc., New York, 1972.

8. K. O. May, *Carl Friedrich Gauss*, Dictionary of Scientific Biography (edited by C. C. Gillispie), Charles Scribner's Sons, New York, 1972.

9. G. F. Simmons, *Differential Equations with Applications and Historical Notes*, McGraw-Hill Book Company, 1972.

10. D. J. Struik, *A Concise History of Mathematics*, Dover Publications, Inc., New York, 1967.

11. H. Tietze, *Famous Problems of Mathematics*, Graylock Press, Baltimore, 1965.

George M. Rassias
The American College of Greece
Department of Mathematics
Aghia Paraskevi Attikis
GR 15342 Athens
Greece

THE MATH. HERITAGE OF C.F. GAUSS (pp. 12-31)
edited by George M. Rassias
©1991 World Scientific Publ. Co. Singapore

ON GAUSS' DIFFERENTIAL EQUATION AND ITS TWENTY-FOUR KUMMER'S SOLUTIONS

M. A. Al-Bassam

In a previous article ([12], pp. 1–20) this author has studied Gauss' differential equation as a particular case of Riemann-Papperitz second order differential equation. The existence of Kummer's twenty-four solutions has been established through finding solutions of the equivalent operator (transform) equations. In this paper the author develops ways and means by which any second order differential equations of this type may be represented by its equivalent operator equations whose solutions represent those of their equivalent differential equations. As a particular case the author establishes the derivation of Gauss's equation, its representation by operator equations and the existence of Kummer's twenty-four solutions. A new approach has been used in finding these solutions. It has been shown that the number of solutions of any differential equations of Riemann-Papperitz type of order n with $(n+1)$ singularities including $(-\infty)$ equals $2n^2(n+1)$. Since Gauss' equation of order two with three singularities (including $-\infty$) and there are four transformations leaving the integrals unaltered, therefore the number of solutions is $(2\times4\times3=24)$. These solutions have been derived by using the generalized (fractional) calculus and the properties of the integro— differential operator of generalized order, together with using the singular points as lower limits for the integrals involved. Also, the extension of Riemannian' P-Function related to equations of higher order has been mentioned. In fact the author in other articles, have developed and extended Riemann P-Function to that of nth order differential equations of Riemann-Papperitz type.

1. Introduction

In a previous article [12] the author has shown that the operator equation

$$\mathop{I}_{a}^{x-w}\prod_{i=1}^{m}(a_i+x)^{\alpha_i}\mathop{I}_{a}^{x-1}\prod_{i=1}^{m}(a_i+x)^{1-\alpha_i}\mathop{I}_{a}^{x-w-n+1}y(x) = 0\,, \qquad \text{(E)}$$

where α_i, a_i are numbers, $\mathrm{Re}\,(\lambda - w) > 0, (\lambda = 1, 2, \ldots), y$ is a function of class $C^{(n)}$ on $[a, b]$ and m, n are positive integers, is equivalent to the integro-differential equation

$$\frac{d^n y}{dx^n} + \sum_{p=-1}^{m-2} \varphi_p \sum_{1 \le i_1 < i_2 < \ldots < i_{p+2} \le m} \frac{w - \sum\limits_{k=1}^{p+2} \alpha_{i_k} + 1}{\prod\limits_{k=1}^{p+2} (a_{i_k} + x)} \, \underset{a}{\overset{x}{\mathrm{I}}}{}^{p-n+2} y(x) = 0 \,, \quad (\overline{\mathrm{E}})$$

where $\varphi_{-1} = 1, \varphi_p = w(w-1)\ldots(w-p), (p = 0, 1, 2, \ldots, m-2)$. These equations may be classified as follows:

(i) For $m = n$, the equation (E) represents nth differential equation of Fuchsian class with singularities at $x = a_i (i = 1, 2, \ldots, n)$. ([5], pp. 62–79).

(ii) For $m > n, (m - n = \nu)$, equation (E) is an integro-differential equation of Fuchs-Volterra type of order (ν, n) where ν is the highest integral order and n is the highest derivative order of y. The integrals appearing in the terms of $(\overline{\mathrm{E}})$ are of the forms

$$\underset{a}{\overset{x}{\mathrm{I}}}{}^q y = \frac{1}{\Gamma(q)} \int_a^x (x - t)^{q-1} y(t)dt \,,$$

$(q = 1, 2, \ldots, \nu)$.

(iii) If $m < n(n - m = \nu)$ then it is a differential equation of nth order with the last term containing a derivative of y of order μ.

(iv) For $m = n + 1$ equation (E) represents the integro-differential equation $(1, n)$ and it is reduced to the differential equation of order $(0, n)$ indicated by (i) if $w - \sum\limits_{k=1}^{n+1} \alpha_{i_k} + 1 = 0$. Differential equations of Riemann type are derived from $(\overline{\mathrm{E}})$ as it will be seen from our study. Furthermore the Gauss' hypergeometric differential equation is obtained from these equations and its Kummer's twenty-four solutions will be found through new approach and even it is different from that used in [12]. In addition, the extension of Riemann's P-Function associated with differential equations of order n will be discussed.

2. Some Cases of (E)

Let in (E) $m = 4, n = 3$, then the operator equation

$$\underset{a}{\overset{x}{\mathrm{I}}}{}^{-w} \prod_{i=1}^{4} (a_i + x)^{\alpha_i} \underset{a}{\overset{x}{\mathrm{I}}}{}^{-1} \prod_{i=1}^{4} (a_i + x)^{1-\alpha_i} \underset{a}{\overset{x}{\mathrm{I}}}{}^{w-2} y(x) = 0 \qquad (\mathrm{E}_1)$$

14   *M. A. Al-Bassam*

is equivalent to the integro-differential equation of Riemann type:

$$y''' + \sum_{i=1}^{4} \frac{w - \alpha_i + 1}{a_i + x} y'' + w \sum_{1 \leq j < j \leq 4} \frac{w - \alpha_i - \alpha_j + 1}{(a_i + x)(a_j + x)} y'$$

$$+ w(w-1) \sum_{1 \leq i < j < k \leq 4} \frac{w - \alpha_i - \alpha_j - \alpha_k + 1}{(a_i + x)(a_j + x)(a_k + x)} y \qquad (\overline{\text{E}}_1)$$

$$+ w(w-1)(w-2) \frac{w - \sum_{i=1}^{4} \alpha_i + 1}{\prod_{i=1}^{4}(a_i + x)} \underset{a}{\overset{x}{\text{I}}} y = 0 \,.$$

When $m = n = 3$ in (E) we have the operator equation:

$$\underset{a}{\overset{x}{\text{I}}}^{-w} \prod_{i=1}^{3}(a_i + x)^{\alpha_i} \underset{a}{\overset{x}{\text{I}}}^{-1} \prod_{i=1}^{3}(a_i + x)^{1-\alpha_i} \underset{a}{\overset{x}{\text{I}}}^{w-2} y(x) = 0 \qquad (\text{E}_2)$$

which is equivalent to the third order differential equation

$$y''' + \sum_{i=1}^{3} \frac{w - \alpha_i + 1}{a_i + x} y'' + w \sum_{1 \leq i < j \leq 3} \frac{w - \alpha_i - \alpha_j + 1}{(a_i + x)(a_j + x)} y'$$

$$+ w(w-1) \frac{w - \sum_{i=1}^{4} \alpha_i + 1}{\prod_{i=1}^{3}(a_i + x)} y = 0 \,. \qquad (\overline{\text{E}}_2)$$

From (E_1) and (E_2) which are derived from (E) we may conclude the following remarks:

(i) (E_1) is reduced to (E_2) and their equivalent differential equations $(\overline{\text{E}}_1)$ is reduced to $(\overline{\text{E}}_2)$ when $w - \sum_{i=1}^{4} \alpha_i + 1 = 0$ and $a_4 \to \pm\infty$.

(ii) $(\overline{\text{E}}_1)$ is an integro-differential equation of order (1,3) and it becomes a fourth order differential equation, i.e., of order (0,4) in Y if $Y = \underset{a}{\overset{x}{\text{I}}} y = \int_a^x y(t)dt$.

(iii) This fourth order differential equation is equivalent to the operator equation (E_1) if $(w - 2)$ is replaced by $(w - 3)$ in the equation.

These are just examples of particular cases of (E). In fact in some other papers this writer has treated and studied cases with order up to (0,5), ([28],

[30], [31]). In this article we are going to study Gauss' equations and cases related to it.

3. Gauss' Equation

Let in (E) $m = 3, n = 2$, then the operator equation:

$$\underset{a}{\overset{x}{I}}{}^{-w} \prod_{i=1}^{3} (a_i + x)^{\alpha_i} \underset{a}{\overset{x}{I}}{}^{-1} \prod_{i=1}^{3} (a_i + x)^{1-\alpha_i} \underset{a}{\overset{x}{I}}{}^{w-1} y(x) = 0 \qquad (E_2)$$

is equivalent to the integro-differential equation of Riemann type of order $(1,2)$:

$$y'' + \sum_{i=1}^{3} \frac{w - \alpha_i + 1}{a_i + x} y' + w \sum_{1 \le i < j \le 3} \frac{w - \alpha_1 - \alpha_j + 1}{(a_i + x)(a_j + x)} y$$

$$+ w(w-1) \frac{w - \overset{3}{\underset{i=1}{\Sigma}} \alpha_i + 1}{\prod\limits_{i=1}^{3} (a_i + x)} \underset{a}{\overset{x}{I}} y = 0 . \qquad (\overline{E}_2)$$

From (E_2) and $(\overline{E}_2)$ we find that:

(i) (E_2) represents a linear differential equation of Riemann type if and only if

$$w - \sum_{i=1}^{3} \alpha_i + 1 = 0 . \qquad (3.1)$$

(ii) The condition (3.1) is satisfied by the values of w, and α_i which may be given by the matrix equation

$$\begin{bmatrix} w \\ \alpha_1 \\ \alpha_2 \\ \alpha_3 \end{bmatrix} = U \begin{bmatrix} \alpha + \beta + \gamma \\ \alpha' + \beta + \gamma \\ \alpha + \beta' + \gamma \\ \alpha + \beta + \gamma' \end{bmatrix} \qquad (3.2)$$

where U is the unit matrix of order (4×4), and $\alpha, \beta, \gamma, \alpha', \beta', \gamma', (\Sigma \alpha + \alpha' = \alpha + \beta + \gamma + \alpha' + \beta' + \gamma' = 1)$, are the indices of the Riemann P-Function.

$$P \left\{ \begin{array}{ccc} \overline{a}_1 & \overline{a}_2 & \overline{a}_3 \\ \alpha & \beta & \gamma \; x \\ \alpha' & \beta' & \gamma' \end{array} \right\} , \; \overline{a}_i = -a_i (i = 1, 2, 3) \qquad (3.2.1)$$

16   *M. A. Al-Bassam*

which is associated with Riemann-Papperitz differential equation ([10], p.291).

(iii) If (3.1) holds and the values (3.2) are assumed by the indices of (E_2) or $(\overline{E}_2)$, then (E_2) represents the Riemann second order differential equation

$$y'' + \sum_{i=1}^{3} \frac{1 + \alpha + \alpha'}{x + a_i} y' + \frac{\Sigma\alpha[(1 + \Sigma\alpha)x - \Sigma a_i(\alpha + \beta' + \gamma' - 1)]}{\prod_{i=1}^{3}(x + a_i)} y = 0 \quad (E_3)$$

where $\Sigma a_i(\alpha + \beta' + \gamma' - 1) = a_1(\alpha + \beta' + \gamma' - 1) + a_2(\alpha' + \beta + \gamma' - 1) + a_3(\alpha' + \beta' + \gamma - 1)$.

(iv) If (i) condition (3.1) holds

(ii) w and $\alpha_i(i = 1, 2, 3)$ take the values in (3.2)

(iii) the transformation

$$y(x) = (x + a_1)^{-\alpha}(x + a_2)^{-\beta}(x + a_3)^{-\gamma}u(x) \tag{3.3}$$

is applied to (E_2), then we would get the Riemann-Papperitz differential equation

$$u'' + \Sigma\frac{1 - \alpha - \alpha'}{x + a_i}u' + \left\{ \frac{\alpha\alpha'(a_1 - a_2)(a_1 - a_3)}{x + a_1} + \frac{\beta\beta'(a_2 - a_3)(a_2 - a_1)}{x + a_3} \right.$$

$$\left. + \frac{\gamma\gamma'(a_3 - a_1)(a_3 - a_2)}{x + a_3} \right\} \frac{u}{\prod_{i=1}^{3}(x + a_i)} = 0. \tag{E_4}$$

(v) (E_2) represents differential equations of Gauss' types with singularities at $-a_1, -\infty, -a_2$ if for fixed i (say $i = 2$), $a_3 \to -\infty$ or $+\infty$. In particular this equation becomes Gauss' hypergeometric differential equation when $a_1 = -1$ and $a_2 = 0$ and

$$\left.\begin{array}{l} w = a \\ \alpha_1 = c - b \\ \alpha_2 = a - c + 1, \end{array}\right\} \quad \text{or} \quad \left\{\begin{array}{l} w = b \\ \alpha_1 = c - a \\ \alpha_2 = b - c + 1 \end{array}\right.$$

This particular case can be obtained from $(\overline{E}_2)$ which, as $a_3 \to \pm\infty$, takes the form

$$y'' + \left[\frac{w - \alpha_1 + 1}{x + a_1} + \frac{w - \alpha_2 + 1}{x + a_2} \right] y' + \frac{w(w - \alpha_1 - \alpha_2 + 1)}{\prod_{i=1}^{2}(x + a_i)} y = 0. \tag{E_5}$$

4. Equivalence Properties for Equations of Gauss-Riemann Types

Let in (E) $m = n = 2$. Then the operator equation

$$\underset{a}{\overset{x}{\mathrm{I}}}{}^{-w} \prod_{i=1}^{2} (a_i + x)^{\alpha_i} \underset{a}{\overset{x}{\mathrm{I}}}{}^{-1} \prod_{i=1}^{2} (a_i + x)^{1-\alpha_i} \underset{a}{\overset{x}{\mathrm{I}}}{}^{w-1} y(x) = 0 , \qquad (\overline{\mathrm{E}}_5)$$

is equivalent to the differential equation

$$\prod_{i=1}^{2} (x + a_i) y'' + P(x) y' + Q y = 0 , \qquad (\mathrm{E}_5)$$

where

$$P(x) = \left[2(w + 1) - \sum_{i=1}^{2} \alpha_i \right] x + (w + 1) \sum_{i=1}^{2} a_i - (a_1 \alpha_2 + a_2 \alpha_1)$$

$$Q = w \left(w - \sum_{i=1}^{2} \alpha_i + 1 \right) . \qquad (4.1)$$

This differential equation is the same as (E_5).

Equivalent Operator Equation

It may be concluded from Eqs. (E_5) and $(\overline{\mathrm{E}}_5)$ that any second order differential equation of the form

$$(x + a_1)(x + a_2) y'' + (Ax + B) y' + S y = 0 \qquad (4.2)$$

can be represented by an operator equation of the form $(\overline{\mathrm{E}}_5)$ if and only if we have

$$\left. \begin{array}{l} A = 2(w + 1) - \displaystyle\sum_{i=1}^{2} \alpha_i \\[2mm] B = (w + 1) \displaystyle\sum_{i=1}^{2} a_i - (\alpha_1 a_2 + \alpha_2 a_1) \\[2mm] S = w \left(w - \displaystyle\sum_{i=1}^{2} \alpha_i + 1 \right) \end{array} \right\} \qquad (4.3)$$

Suppose that we are given $a_1 = -1, a_2 = 0, w = \alpha, \alpha_1 = \gamma - \beta, \alpha_2 = \alpha - \gamma + 1$, then from (4.3) we have

$$A = \alpha + \beta + 1$$
$$B = -\gamma$$
$$S = \alpha \beta$$

and consequently the operator equation

$$\mathop{I}_{a}^{x}{}^{-\alpha}x^{\alpha-\gamma+1}(1-x)^{\gamma-\beta}\mathop{I}_{a}^{x}{}^{-1}x^{\gamma-\alpha}(1-x)^{\beta-\gamma+1}\mathop{I}_{a}^{x}{}^{\alpha-1}y = 0 \qquad (4.4)$$

is equivalent to the Gauss' equation

$$x(1-x)y'' + \{\gamma - (\alpha+\beta+1)\}y' - \alpha\beta y = 0 \,. \qquad (4.4)$$

If the Gauss' equation

$$x(1-x)y'' + \{c - (a+b+1)x\} - aby = 0 \qquad (4.5)$$

is given, then from (4.2) we have $a_1 = -1, a_2 = 0, A = -(a+b+1), B = c$ and $S = -ab$. Also from (4.3) we have:

$$a + b + 1 = 2(w+1) - \alpha_1 - \alpha_2$$
$$-c = -(w+1) + \alpha_2 \qquad (4.3.1)$$
$$ab = w(w - \alpha_1 - \alpha_2 + 1) \,.$$

From the first and second equations of (4.3.1)

$$\alpha_2 = w + 1 - c$$
$$\alpha_1 = c - a - b + w \,.$$

Combining these equations with the third equation (4.3.1) we find that

$$w^2 - (a+b)w + ab = 0 \,,$$

which yields:

$$w = a \quad \text{for} \quad w = b \,.$$

Thus we have the two solutions: $w = a, \alpha_1 = c - b, \alpha_2 = a - c + 1$ and $w = b, \alpha_1 = c - a, \alpha_2 = b - c + 1$ and consequently Eq. (4.5) is equivalent to any one of the two operator equations:

$$\mathop{I}_{x_0}^{x}{}^{-a}x^{a-c+1}(1-x)^{c-b}\mathop{I}_{x_0}^{x}{}^{-1}x^{c-a}(1-x)^{b-c+1}\mathop{I}_{x_0}^{x}{}^{a-1}y = 0 \qquad (4.5.1)$$

or

$$\mathop{I}_{x_0}^{x}{}^{-b}x^{b-c+1}(1-x)^{c-a}\mathop{I}_{x_0}^{x}{}^{-1}x^{c-b}(1-x)^{a-c+1}\mathop{I}_{x_0}^{x}{}^{b-1}y = 0 \,. \qquad (4.5.2)$$

Once the equivalent operator equation is found, then the differential equation can be easily solved by finding solutions of the first through the use of the Fractional Calculus and the properties of the integro-differential operator of generalized order as it will be discussed later in this article.

5. Some Definitions and Hypergeometric Identities

Some definitions ad properties of integro-differential operators of generalized order, transformations and hypergeometric functions and identities, together with their references, will be presented here as they may be referred to later in the work. It may be pointed out that the hypergeometric identities have been recently developed by the author.

Definition 1. If $f(x)$ is a real-valued functions of class $C^{(n)}$ on $a \leq t, x \leq b$ and $\operatorname{Re} \alpha + n > 0$, then

$$\mathop{I}_{a}^{x}{}^{\alpha} f = \frac{1}{\Gamma(\alpha + n)} D_x^n \int_a^x (x - t)^{\alpha + n - 1} f(t) dt, \quad (n = 0, 1, 2, \dots). \qquad (5.1)$$

When $a \to -\infty$

$$\mathop{I}_{-\infty}^{x}{}^{\alpha} f = \lim_{x = -\infty} \mathop{I}_{a}^{x}{}^{\alpha} f = D_x^n \mathop{I}_{-\infty}^{x}{}^{\alpha + n} f, \qquad (5.2)$$

where $D_x^n = \frac{d^n}{dx^n}$ and Γ is the Gamma function.

Details and properties of this operator are given in [11], [12].

Definition 2. If $\mathop{I}_{(a,-\infty)}^{x}{}^{-\alpha} f = 0, \operatorname{Re} \alpha - n \geq 0, (n = 1, 2, \dots)$ and

(1) If $f \in C^{(n)}$ on $[a, b]$, then $\mathop{I}_{a}^{x}{}^{\alpha} 0 = f(x) = \sum_{k=1}^{n} \frac{C_k (x-a)^{\alpha-k}}{\Gamma(\alpha-k+1)}$ (5.3)

(2) $f \in C^{(n)}$ on $(-\infty, b]$, then $\mathop{I}_{-\infty}^{x}{}^{\alpha}(0) = f(x) = \sum_{i=1}^{n} \frac{C_i x^{i-1}}{\Gamma(i)}$,   (5.4)

where C_i and C_k are arbitrary constants ([12], pp. 5–6).

Remark. When $\mathop{I}_{a}^{x}{}^{-\alpha} F = 0$, then $D_x^2 \mathop{I}_{a}^{x}{}^{2-\alpha} = 0$ which implies that $\mathop{I}_{a}^{x}{}^{2-\alpha} F = C_1(x - a) + C_2$. Thus

$$F = C_1 \frac{(x - a)^{\alpha - 1}}{\Gamma(\alpha)} + C_2 \frac{(x - a)^{\alpha - 2}}{\Gamma(\alpha - 1)}, \qquad (5.5)$$

C_1 and C_2 are arbitrary constants. These results may also be obtained from (5.3) if C_k (the arbitrary constants) are chosen that $C_k = 0$ for $(k = 3, 4, \ldots, n)$.

Definition 3. The hypergeometric functions in the variables $x_i, (i = 1, 2, \ldots, n)$ may be defined by

$$F_D^{(n)}(A, B_1, B_2, \ldots, B_n \, ; C \, ; x_1, x_2, \ldots, x_n) = \frac{\Gamma(C)}{\Gamma(A)\Gamma(C-A)}$$
$$\int_0^1 u^{A-1}(1-u)^{C-A-1} \prod_{i=1}^{n}(1-ux_i)^{-B_i} du \, , \tag{5.6}$$

where $A, B_i, C, (i = 1, 2 \ldots, n)$ are numbers and $\operatorname{Re} C > \operatorname{Re} A > 0$, $|\operatorname{Arg}(1 - x_i)| < \pi$. Also,

$$F_D^{(n)}(A, B_1, B_2, \ldots, B_n \, ; C \, ; x_1, x_2, \ldots, x_n) =$$
$$\sum_{k_1, \ldots, k_n = 0} \frac{(A)_{k_1 + \ldots + k_n}(B)_{k_1} \ldots (B_n)_{k_n}}{(C)_{k_1 + k_2 + \ldots + k_n}} \frac{x_1^{k_1} x_2^{k_2} \ldots x_n^{k_n}}{\Gamma(k_1 + 1) \ldots \Gamma(k_n + 1)} \, , \tag{5.7}$$

$(|x_i| < 1, i = 1, 2, \ldots, n)$, where

$$(B)_r = B(B+1)\ldots(B+r-1), r \geq 1, (B)_0 = 1, B \neq 0 \, .$$

From the above it may be concluded the following:-

(a) the integrals (5.6) remain unaltered under the $2(n+1)$ transformations

$$
\begin{aligned}
&\text{(i)} && u = v \, , \\
&\text{(ii)} && u = 1 - v \, , \\
&\text{(iii)} && u = \frac{v}{1 - x_k + vx_k} \, , \\
&\text{(iv)} && u = \frac{1 - v}{1 - vx_k} \, , (k = 1, 2, \ldots, n) \, .
\end{aligned}
\tag{5.8}
$$

(b) (i) is clearly the identity transformations.

(c) In applying (ii) to (5.6) we find that

$$F_D^{(n)}(A, B_1, B_2, \ldots, B_n \, ; C \, ; x_1, x_2, \ldots, x_n) = \prod_{i=1}^{n}(1 - x_i)^{-B_i} \times$$
$$F_D^{(n)}\left[C - A, B_1, \ldots, B_n \, ; C \, ; \frac{x_1}{x_1 - 1}, \frac{x_2}{x_2 - 1}, \ldots, \frac{x_n}{x_n - 1}\right] \tag{5.9}$$

(d) By applying (iii), $\left[u = \frac{v}{1-x_i+vx_i}, i = 1, \ldots, n\right]$ to the integrals (5.6) we obtain

$$F_D^{(n)}(A, B_1, \ldots, B_n\,;\,C\,;\,x_1, \ldots, x_n) = (1 - x_i)^{-A}\times$$
$$F_D^{(n)}\left[A, B_1, B_2, \ldots, B_{i-1}, C - \sum_{k=1}^{n} B_k, B_{i+1}, \ldots, B_n\,;\right.$$
$$\left. C\,;\,\frac{x_1 - x_i}{1 - x_i}, \ldots, \frac{x_{i-1} - x_i}{1 - x_i}, \frac{x_i}{x_i - 1}, \frac{x_{i+1} - x_i}{1 - x_i}, \ldots, \frac{x_n - x_i}{1 - x_i}\right]. \tag{5.10}$$

(e) If we let $u = \frac{1-v}{1-vx_i}, (i = 1, 2, \ldots, n)$ in (5.6) we find that

$$F_D^{(n)}(A, B_1, B_2, \ldots, B_n\,;\,C\,;\,x_1, x_2, \ldots, x_n) = (1 - x_i)^{C-A-B_i}$$
$$\prod_{\substack{r=1 \\ r\neq i}}^{n}(1 - x_r)^{-B_r} \times F_D^{(n)}\left[C - A, B_1, B_2, \ldots, B_{i-1},\right.$$
$$C - \sum_{k=1}^{n} B_k, B_{i+1}, \ldots, B_n\,;\,C\,;\,\frac{x_i - x_1}{1 - x_1}, \frac{x_i - x_2}{1 - x_2}, \ldots, \tag{5.11}$$
$$\left.\frac{x_i - x_{i-1}}{1 - x_{i-1}}, x_i, \frac{x_i - x_i + 1}{1 - x_{i+1}}, \ldots, \frac{x_i - x_n}{1 - x_n}\right], (i = 1, 2, \ldots, n).$$

6. Solutions of Equations

Solutions of the differential equation (E_5) can be obtained by finding solutions of its equivalent operator equation $(\overline{E}_5)$. By applying property (5.5) with using the equality $D\underset{a}{\overset{x}{I}}{}^{1-w} = \underset{a}{\overset{x}{I}}{}^{-w}$ in $(\overline{E}_5)$ and properties of the inverse operations of the integro-differential operator of generalized order we find that

$$y(x\,;a) = \underset{a}{\overset{x}{I}}{}^{1-w}\prod_{i=1}^{2}(a_i + x)\left[K + \underset{a}{\overset{x}{I}}{}^{w}\prod_{i=1}^{2}(a_i + x)^{-\alpha_i}\underset{a}{\overset{x}{I}}{}^{1-w}C\right] \tag{S}$$

where K and C are arbitrary constants and $\underset{a}{\overset{x}{I}}{}^{1-w}C = C\frac{(x - a)^{w-1}}{\Gamma(w)}$. Thus the fundamental solutions of $(\overline{E}_5)$ or (E_5) may be given by:

$$y_1(x\,;a) = K\,\overset{x}{\underset{a}{I}}{}^{1-w}\,\prod_{i=1}^{2}(a_i + x)^{-\alpha_i-1} \tag{S_1}$$

$$y_2(x\,;a) = \frac{C}{\Gamma(w)}\,\overset{x}{\underset{a}{I}}{}^{1-w}\,\prod_{i=1}^{2}(a_i + x)^{\alpha_i-1}\,\overset{x}{\underset{a}{I}}\,\prod_{i=1}^{2}$$
$$(a_i + x)^{-\alpha_i}(x - a)^{w-1}. \tag{S_2}$$

Solutions $(S_i), (i = 1, 2)$ are determined according to the singular points of the differential equation (E_5). These are $(-a_i)$ and $(-\infty$ or $+\infty), (i = 1, 2)$. Thus the lower limit a of the integrals involved may be replaced by these singular points. Clearly, these solutions are linearly independent. For each lower limit we obtain two solutions (may be called principal solutions) and consequently we have six principal solutions. But according to the transformations (5.8), for $k = 1$, the hypergeometric integrals (5.6) remain unaltered by applying four transformations. Therefore by (5.9), (5.10) and (5.11), for $i = 1$, each principal solution takes four forms of hypergeometric functions F. This would yield a total of twenty-four (may be called) branch solutions. This is the case with solutions of second order differential equations of this type. As to the third order with four singularities there will be six transformations of the type (5.8) and consequently the total of branch solutions is seventy-two ([28], [30], [31]). The fourth order differential equation derived from (E) when $n = 4$ and $m = 4$, the total solutions is one hundred and sixty branch solutions [29]. Also, it has been found that the total number of solutions is three hundred branch solutions for fifth order differential equations with six singularities including $(-\infty)$. It may be concluded, therefore, that in general the number of branch solutions is determined by the order of the equation, the number of singular points and the number of transformations leaving the integrals unaltered. Thuf if in (E) the order is n and the number of singular points is $m + 1$ (including $-\infty$), then the number of transformations of the forms (5.8) is $2m$ and consequently the total number of branch solutions is $2nm(m + 1)$.

Verification of Solutions

To show that $S_i (i = 1, 2)$ are solutions of (E_5) is to verify that they satisfy the equation. It would be sufficient to show that any one of these solutions satisfies the equation. In substitutingg $y_2(x\,;a)$ in $(\overline{E}_5)$ and by using the operational properties of the operator of generalized order, with the fact $\mathop{I}\limits_{a}^{x}{}^{\alpha}\mathop{I}\limits_{a}^{x}{}^{\beta} = \mathop{I}\limits_{a}^{x}{}^{\alpha+\beta}$ as shown in [11], then the left-hand side of $(\overline{E}_5)$ takes the form:

$$\frac{C}{\Gamma(w)}\mathop{I}\limits_{a}^{x}{}^{-w}(x-a)^{w-1} = D_x I^{1-w}C\frac{(x-a)^{w-1}}{\Gamma(w)} = D_x C = 0$$

since $\mathop{I}\limits_{a}^{x}{}^{\alpha}(x-a)^{\beta} = \frac{\Gamma(\beta+1)}{\Gamma(\alpha+\beta+1)}(x-a)^{\alpha+\beta}$, ([11], p. 12).

7. Solutions of Gauss' Equation

In this section we may study Gauss' differential equation (4.5) and its equivalent operator equation (4.5.1) as the operator equation (4.5.2) yields the same results.

As in the previous section the fundamental solutions of (4.5) may be given by (where x_0 may be used to replace (a) as a lower limit):

$$y_1(x\,;x_0) = K\mathop{I}\limits_{x_0}^{x}{}^{1-a}(1-x)^{c-b-1}x^{a-c}, \tag{7.1}$$

and by (5.5)

$$y_2(x\,;x_0) = \frac{C_1}{\Gamma(a)}\mathop{I}\limits_{x_0}^{x}{}^{1-a}(1-x)^{c-b-1}x^{a-c}$$
$$\cdot \mathop{I}\limits_{x_0}^{x}(1-x)^{b-c}x^{c-a-1}(x-x_0)^{a-1}. \tag{7.2}$$

Since $x_0 = 0, 1$ and $-\infty$, then the principal solutions are:

$$\left.\begin{array}{l}y_1(x\,;0) = K\mathop{I}\limits_{0}^{x}{}^{1-a}(1-x)^{c-b-1}x^{a-c}\\[2mm]y_2(x\,;0) = \frac{C_1}{\Gamma(a)}\mathop{I}\limits_{0}^{x}{}^{1-a}(1-x)^{c-b-1}x^{a-c}\mathop{I}\limits_{0}^{x}(1-x)^{b-c}x^{c-2}\end{array}\right\} \tag{7.3}$$

$$\left.\begin{array}{l}y_1(x\,;1) = K\mathop{I}\limits_{1}^{x}{}^{1-a}(1-x)^{c-b-1}x^{a-c}\\[2mm]y_2(x\,;1) = (-1)^{a-1}\frac{C_1}{\Gamma(a)}\mathop{I}\limits_{1}^{x}{}^{1-a}(1-x)^{c-b-1}x^{a-c}\mathop{I}\limits_{1}^{x}(1-x)^{a+b-c-1}x^{c-a-1}\end{array}\right\} \tag{7.4}$$

$$\left.\begin{array}{l}y_1(x\,;-\infty) = K\mathop{I}\limits_{-\infty}^{x}{}^{1-a}(1-x)^{c-b-1}x^{a-c}\\[2mm]y_2(x-\infty) = C_1\mathop{I}\limits_{-\infty}^{x}{}^{1-a}(1-x)^{c-b-1}x^{a-c}\mathop{I}\limits_{-\infty}^{x}(1-x)^{b-c}x^{c-a-1}\end{array}\right\} \tag{7.5}$$

It may be observed here that when $x_0 = 0$, $(x - x_0)^a$ in (7.2) is reduced to x^a, when $x_0 = 1$, then it takes the form $(-1)^{a-1}(1 - x)^{a-1}$ and when $x_0 = -\infty$ then (5.4) may be used where $C_i = 0$ for all $i \neq 1$.

Branch Solutions

To study these solutions we will deal with each one separately. Before we start the discussion it may be useful to point out that the transformations (5.8) for $k = 1$ which will be applied to the integrals in (7.j), (j= 3, 4, 5) are given by the following:

$$\left.\begin{array}{lll}
\text{(i)} & u = v, \\
\text{(ii)} & u = 1 - v, \\
\text{(iii)} & u = \frac{v}{1-(1-v)x}, \\
\text{(iv)} & u = \frac{1-v}{1-vx}
\end{array}\right\} \qquad (7.6)$$

In dealing with branch solutions, the arbitrary constants K and C_1 may be chosen, in such a way, so that it would reduce the coefficients of solutions to unity. These need not be mentioned in detail in the work.

$y(x\,;0)$:

According to (7.3) we have

$$\begin{aligned}
y_1(x\,;0) &= \frac{K}{\Gamma(1 - \alpha)} \int_0^x (x - t)^{-a}(1 - t)^{c-b-1} t^{a-c} dt \\
&= \frac{K\Gamma(a - c + 1)}{\Gamma(2 - c)} x^{1-c} F(a - c + 1, b - c + 1\,;2 - c\,;x) \\
&= x^{1-c} F(a - c + 1, b - c + 1\,;2 - c\,;x)
\end{aligned}$$

where $K = \frac{\Gamma(2-c)}{\Gamma(a-c+1)}$.

Using this result and applying (7.6) as shown in the resulting integrals (5.i), (i= 9, 10, 11) we find that the branch solutions are:

$$y_{11}(x\,;0) = x^{1-c} F(a - c + 1, b - c + 1\,;2 - c\,;x)$$

$$y_{12}(x\,;0) = (1 - x)^{c-b-1} F\left(1 - a, b - c + 1\,;2 - c\,;\frac{x}{x - 1}\right)$$

$$y_{13}(x\,;0) = (1 - x)^{c-a-1} F\left(a - c + 1, 1 - b\,;2 - c\,;\frac{x}{x - 1}\right)$$

$$y_{14}(x\,;0) = (1 - x)^{c-a-b} F(1 - a, 1 - b\,;2 - c\,;x)\,.$$

Also we have

$$y_2(x\,;0) = \frac{C_1}{\Gamma(a)} \int_0^x (x-t)^{-a} t^{a-c} (1-t)^{c-b-1} \left[\int_0^t (1-u)^{b-c} u^{c-2} du \right] dt$$

$$= F(a,b\,;c\,;x)$$

by a suitable choice of the arbitrary constant C_1.

Using this result with the transformations (7.6) and as given by the integrals of the forms (5.i), (i$=9,10,11$), then we have the branch solutions:

$$y_{21}(x\,;0) = F(a,b\,;c\,;x)\,,$$

$$y_{22}(x\,;0) = (1-x)^{c-a-b} F(c-a,c-b\,;c\,;x)$$

$$y_{23}(x\,;0) = (1-x)^{-a} F\left(a,c-b\,;c\,;\frac{x}{x-1}\right)$$

$$y_{24}(x\,;0) = (1-x)^{-b} F\left(c-a,b\,;c\,;\frac{x}{x-1}\right)$$

$y_1(x\,;1)$:

We have

$$y_1(x\,;1) = \frac{K}{\Gamma(1-a)} \int_1^x (x-t)^{-a} t^{a-c} (1-t)^{c-b-1} dt$$

$$= (1-x)^{c-a-b} F(c-a,c-b\,;c-a-b+1\,;1-x)$$

by a suitable choice of K.

Thus, from this and (7.6) together with (5.i), (i$=9,10,11$) we find that

$$y_{11}(x\,;1) = x^{c-a-b} F(c-a,c-b\,;c-a-b+1\,;1-x)$$

$$y_{12}(x\,;1) = x^{1-c}(1-x)^{c-a-b} F(1-b,1-a\,;c-a-b+1\,;1-x)$$

$$y_{13}(x\,;1) = x^{a-c}(1-x)^{c-a-b} F\left(c-a,1-a\,;c-a-b+1\,;\frac{x-1}{x}\right)$$

$$y_{14}(x\,;1) = x^{b-c}(1-x)^{c-a-b} F\left(1-b,c-b\,;c-a-b+1\,;\frac{x-1}{x}\right)\,.$$

Now

$$y_2(x\,;1) = \frac{C_1(-1)^{a-1}}{\Gamma(a)\Gamma(1-a)} \int_1^x (x-t)^{-a} t^{a-c} (1-t)^{c-b-1}$$

$$\left\{ \int_1^t u^{c-a-1}(1-u)^{a+b-c-1} du \right\} dt\,.$$

Since

$$\int_1^t (1-u)^{a+b-c-1} u^{c-a-1} du = \frac{(1-t)^{a+b-c}}{c-a-b} F(a-c+1, a+b-c; a+b-c+1; 1-t)$$

then by a suitable choice of C_1 we find that

$$y_2(x\,;1) = \frac{C_1(-1)^{a-1}}{\Gamma(a)\Gamma(1-a)(c-a-b)} \int_1^x (x-t)^{-a} t^{a-c}(1-t)^{a-1} F(a-c+1, a+b-c; a+b-c+1; 1-t)\,dt$$

$$= F(a,b\,;a+b-c+1\,;1-x).$$

Therefore we have

$$y_{21}(x\,;1) = F(a,b\,;a+b-c+1\,;1-x)$$

$$y_{22}(x\,;1) = x^{1-c} F(a-c+1, b-c+1\,;a+b-c+1\,;1-x)$$

$$y_{23}(x\,;1) = x^{-a} F\left(a, a-c+1\,;a+b-c+1\,;\frac{x-1}{x}\right)$$

$$y_{24}(x\,;1) = x^{-b} F\left(b, b-c+1\,;a+b-c+1\,;\frac{x-1}{x}\right)$$

$y(x\,;-\infty)$:

We have

$$y_1(x\,;-\infty) = \frac{K}{\Gamma(1-a)} \int_{-\infty}^x (x-t)^{-a} t^{a-c}(1-t)^{c-b-1}\,dt$$

$$= x^{-b} F\left(b, b-c+1\,;b-a+1\,;\frac{1}{x}\right)$$

by a suitable choice of K. Thus we have

$$y_{11}(x\,;-\infty) = x^{-b} F\left(b, b-c+1\,;b-a+1\,;\frac{1}{x}\right)$$

$$y_{12}(x\,;-\infty) = x^{-b} F\left(1-\frac{1}{x}\right)^{c-a-b} F\left(1-a, c-a\,;b-a+1\,;\frac{1}{x}\right)$$

$$y_{13}(x\,;-\infty) = x^{-b}\left(1-\frac{1}{x}\right)^{-b} F\left(b, c-a\,;b-a+1\,;\frac{1}{1-x}\right)$$

$$y_{14}(x\,;-\infty) = x^{-b}\left(1-\frac{1}{x}\right)^{c-b-1} F\left(1-a, b-c+1\,;b-a+1\,;\frac{1}{1-x}\right)$$

$$y_2(x\,;-\infty) = C_1 \underset{-\infty}{\overset{x}{\text{I}}}\, ^{1-a} x^{a-c}(1-x)^{c-b-1} \underset{-\infty}{\overset{x}{\text{I}}}\, (1-x)^{b-c} x^{c-a-1}.$$

Now,

$$I = \int\limits_{-\infty}^{x} (1-x)^{b-c} x^{c-a-1} = \int_{-\infty}^{t} (1-u)^{b-c} u^{c-a-1} du \quad \text{and if } u = tz^{-1}$$

we find that

$$I = (-1)^{b-c+1} t^{b-a} \int_0^1 z^{a-b-1} \left(1 - \frac{1}{t}z\right)^{b-c} dz$$

$$= \frac{(-1)^{b-c+1}}{a-b} t^{b-a} F\left(c-b, a-b; a-b+1; \frac{1}{t}\right).$$

Consequently,

$$y_2(x\,;-\infty) = \frac{C_1(-1)^{b-c+1}}{a-b} \int\limits_{-\infty}^{x} {}^{1-a} x^{b-c}(1-x)^{c-b-1}$$

$$F\left(c-b, a-b; a-b+1; \frac{1}{x}\right)$$

$$= \frac{C_1(-1)^{b-c+1}}{(a-b)\Gamma(1-a)} \int_{-\infty}^{x} (x-t)^{-a} t^{b-c}(1-t)^{c-b-1}$$

$$F\left(c-b, a-b; a-b+1; \frac{1}{t}\right) dt.$$

If $z = t^{-1}$ we find that

$$y_2(x\,;-\infty) = \frac{x^{-a} C_1(-1)^{1-a}}{a-b} \int\limits_{0}^{\frac{1}{x}} {}^{1-a} x^{a-1}(1-x)^{c-b-1}$$

$$F(c-b, a-b; a-b+1; x)$$

$$= x^{-a} F\left(a, a-c+1\,a-b+1; \frac{1}{x}\right)$$

with a proper choice of C_1.

Thus we have the second branch solutions when $x_0 = -\infty$:

$$y_{21}(x\,;-\infty) = x^{-a} F\left(a, a-c+1; a-b+1; \frac{1}{x}\right)$$

$$y_{22}(x\,;-\infty) = x^{-a} \left(1-\frac{1}{x}\right)^{c-a-b} F\left(1-b, c-b; a-b+1; \frac{1}{x}\right)$$

$$y_{23}(x\,;-\infty) = x^{-a} \left(1-\frac{1}{x}\right)^{-a} F\left(a, c-b; a-b+1; \frac{1}{1-x}\right)$$

$$y_{24}(x\,;-\infty) = x^{-a} \left(1-\frac{1}{x}\right)^{c-a-1} F\left(1-b, a-c+1; a-b+1; \frac{1}{1-x}\right)$$

Comparing these results with Kummer's branch solutions as shown by the table (p. 88 [5]), we notice that: $p^{(\alpha)}$ corresponds to branch solutions $y_2(x;0), p^{(\alpha')}$ corresponds to $y_1(x;0), p^{(\beta)}$ corresponds to $y_2(x;-\infty), p^{(\beta')}$ corresponds to $y_1(x;-\infty), p^{(\gamma)}$ corresponds to $y_2(x;1)$ and $p^{(\gamma')}$ corresponds to $y_1(x;1)$, where $\{p^{(\alpha)}, p^{(\alpha')}\}, \{p^{(\beta)}, p^{(\beta')}\}$, and $\{p^{(\gamma)}, p^{(\gamma')}\}$, represent the sets of branch solutions of Gauss' equation which correspond to the exponents $\alpha, \alpha', \beta, \beta'$ and γ, γ' of Riemann P-Function, ([5], pp. 83–96 and [10], pp. 206–210).

8. Riemann P-Function and its Extended Form

This branch solutions obtained above are associated with the exponents α, β, γ and α', β', γ' corresponding to the equation's singular points a_2, a_2, a_3. Thus the scheme

$$P \left\{ \begin{array}{l} a_1, a_2, a_3 \\ \alpha, \ \beta, \ \gamma \ x \\ \alpha', \beta', \gamma' \end{array} \right\}$$

is associated with the Gauss' second order differential equation and its solutions and which corresponds to the equation's singular points $0, 1, -\infty$. However, it seems that there exists a scheme associated with equations of higher orders of these types, i.e., Riemann-Gauss-Papperitz equations. These P-Functions are associated with equations of (E) for $m = n = 2$. When $m = n = 3$ we have the extended form of the P-Function. Their solutions are in terms of the hypergeometric functions F_1, [28], [30], [31] which are related to a scheme similar to that given by Riemann's P-Function ([5], pp. 83–92). Such a scheme may be defined by a function of the form

$$M \left\{ \begin{array}{l} a_1 \ a_2 \ a_3 \ a_4 \\ \alpha \ \ \beta \ \ \gamma \ \ \delta \ \ x, y(x) \\ \alpha' \ \beta' \ \gamma' \ \delta' \\ \alpha'' \ \beta'' \ \gamma'' \ \delta'' \end{array} \right\}$$

where a_1, a_2, a_3, a_4 are the singular points of the third order differential equations. In this case at each singular point there are three principal solutions. At $x = a_1$, there are the branches $\{M^{(\alpha)}, M^{(\alpha')}, M^{(\alpha'')}\}$ which are associated with the exponents $(\alpha, \alpha', \alpha'')$. Similarly for other singular points there are the branches: $\{M^{(\beta)}, M^{(\beta')}, M^{(\beta'')}\}, \{M^{(\gamma)}, M^{(\gamma')}, M^{(\gamma'')}\}$ and

$\{M^{(\delta)}, M^{(\delta')}, M^{(\delta'')}\}$. Each branch is represented by six forms of branch solutions. The total is seventy-two compared to 24 Kummer's solutions for the Gauss' second order hypergeometric equation. In a similar manner these functions can be extended and associated with 4th, 5th,..., nth order differential equation. It may be pointed out that the extended form of (3.2) for these third order equations may be given by

$$\begin{bmatrix} w \\ \alpha_1 \\ \alpha_2 \\ \alpha_3 \\ \alpha_4 \end{bmatrix} = U \begin{bmatrix} \alpha + \beta + \gamma + \delta \\ \alpha + \beta + \gamma + \delta \\ \alpha' + \beta' + \gamma' + \delta' \\ \alpha'' + \beta'' + \gamma'' + \delta'' \\ \alpha + \beta + \gamma + \delta \end{bmatrix}$$

where U is 5×5 unit matrix, and $\Sigma\alpha + \Sigma\alpha' + \Sigma\alpha'' = 1$.

Similarly schemes may be extended to include equations of types $m = n$ of (E), i.e., of the nth order differential equation. In this case with $m + 1$ singularities $a_j (i = 1, \ldots, m + 1)$ including the one at $(-\infty)$, the scheme may be given by:

$$M \left\{ \begin{array}{lllll} a_1 & a_2 & a_3 \ldots a_m & a_{m+1} \\ \alpha_1^{(0)} & \alpha_2^{(0)} & \alpha_3^{(0)} \ldots \alpha_m^{(0)} & \alpha_{m+1}^{(0)} \\ \alpha_1^{(1)} & \alpha_2^{(1)} & \alpha_3^{(1)} \ldots \alpha_m^{(1)} & \alpha_{m+1}^{(1)} \quad x, y_1(x), y_2(x), \ldots, y_{n-2}(x) \\ \vdots \\ \alpha_1^{(n-1)} & \alpha_2^{(n-1)} & \alpha_3^{(n-1)} \ldots \alpha_m^{(n-1)} & \alpha_{m+1}^{(n-1)} \end{array} \right\}$$

This M-function, which is the extension of Riemann P-Function, of the generalized Riemann-Papperitz equation has $n(m + 1)$ principal solution sets $M^{(\alpha_j^i)}, (i = 0, 1, 2, \ldots, n - 1), (j = 1, 2, \ldots, m + 1)$ and each set has $2m$ branch solutions which make a total of $2mn(m+1)$ branch solutions for this general case. These solutions are in the forms of the hypergeometric functions $F_D^{(n-1)}$. These M-Functions may be studied in the same lines on which the P-Functions have been studied and analysed ([5], pp. 83–92).

References

1. J. L. Liouville, *Mémoire sur le changement de la variable dans le calcul des differentielle à indices quelconques*, J. Ecole Polytech. **15**, Section 24 (1835).
2. B. Riemann, *Versuch einer allgemeinen Auffassung der Integration und Differentiation*, Gesammelte Mathematische Werke, Leipzig, 1876.

3. H. J. Holmgren, *Om differnntialkalkylen med indices of hvilken natur som helst*, Kingl. Svenska Vetnskaps-Akademins Handlingar, Bd. 5, n. 11, Stockholm (1965–66), pp. 11–83.

4. M. Reisz, *L'integral de Rieman-Liouville et le problem de Cauchy*, Acta Mathematica. **81** (1949) 1–223.

5. E. G. C. Poole, *Introduction to the Theory of Linear Differential Equations*, Dover Publications, New York, 1960.

6. P. E. Appell and J. Kampé de Fériet, *Fonctions Hypergeometrique et Hyperspheriques, Polynomes d'Hermite*, Gauthier-Villars, Paris, 1926.

7. W. N. Bailey, *Generalized Hypergeometric Series*, Cambridge University Press, 1935.

8. E. E. Kummer, Journal für Math., XV (1836), (39–83), (127–172).

9. E. Papperitz, Math. Ann. **XXV** (1885), 212–221.

10. E. T. Whittaker and G. N. Watson, *A Course of Modern Analysis*, Cambridge University Press, 1962.

11. M. A. Al-Bassam, *Some properties of Holmgren-Reisz transform*, Annali della Scoula Normale Superiore di Pisa, Scienze Fisiche e Mathematiche, Serie III, Vol. VX, Fasc. 1–11 (1961) 1–24.

12. M. A. Al-Bassam, *Concerning Holmgren-Reisz transform equations of Gauss-Riemann type*, Rendiconti del Circolo Matematico di Palermo, Serie XI (1962) 1–20.

13. M. A. Al-Bassam, *On certain types of H–R transform equations and their equivalent differential equations*, Journal für die Reine and Angewandte Mathematik, Band **26** (1964) 91–100.

14. M. A. Al-Bassam, *Some existence theorems on differential equations of generalized order*, Journal für die Reine und Angewandte Mathematik, Band **218** (1965) 70–78.

15. M. A. Al-Bassam, *On an integro-differential equation of Legendre-Velterra type*, Portugaliae Math. **25**, Fasc. 1 (1966) 53–61.

16. M. A. Al-Bassam, *H–R transform equations of Laguerre type*, Bulletin, College of Science, University of Baghdad, **9** (1966) 161–184.

17. M. A. Al-Bassam, *On Laplace's second order linear differential equations and their equivalent H–R transform equations*, Journal für die Riene und Angewandte Mathematik, Band **225** (1967) 76–84.

18. M. A. Al-Bassam, *On some differential and integro-differential equations associated with Jacobi's differential equations*, Journal für die Reine und Angewandte Mathematik, Band **288** (1976) 211–217.

19. M. A. Al-Bassam, *H–R transform in two dimensions and some of its applications*, Fractional Calculus and its Applications, Proceedings of the International Conference, Univ. of New Haven, June 1974, ed. B. Ross, Lecture Notes in Mathematics **457**, Springer-Verlag, New York, pp. 91–105.

20. M. A. Al-Bassam, *On fractional calculus and its application to the theory of ordinary differential equations of generalized order*, Nonlinear Analysis and Applications, Vol. 80, Marcel Dekker Inc., New York, 1982, pp. 305–331.

21. K. B. Oldham and J. Spanier, *The Fractional Calculus*, Academic Press, New York, 1974.

22. M. A. Al-Bassam, *On fractional analysis and its applications*, Modern Analysis and its Applications, ed. H. L. Manocha, Prentice Hall of India Ltd., New Delhi, 1986, pp. 269–307.

23. M. A. Al-Bassam, *Some applications of fractional calculus to differential equations*, Research Notes in Mathematics, Pitman Advanced Publishing Program, London, 1985.

24. M. A. Al-Bassam, *On generalized power series and generalized operational calculus and its applications*, Nonlinear Analysis, World Scientific Publishing Co., Pte Ltd., New Jersey, 1987.

25. M. A. Al-Bassam, *Application of fractional calculus to differential equations of Hermite's type*, Indian Journal of Pure and Applied Mathematics (9) **16**, pp. 1009–1016.

26. M. A. Al-Bassam, *Some applications of generalized calculus to differential and integro-differential equations*, Mathematical Analysis and Applications, International Conference, Kuwait University, Pergamon Press, 1988, pp. 61–76.

27. Harold Exton, *Multiple Hypergeometric Functions and Applications*, Ellis Harwood Limited, John Wiley & Sons, Checkester, London & New York, 1976.

28. M. A. Al-Bassam, *On fractional operator equation and solutions of a class of third order differential equation*, Bolletino U. M. I. (7) 3-B (1989).

29. M. A. Al-Bassam, *On Fractional Calculus and Fractional Differential Equations*, Monograph (collection of published papers), Vol. I and II, Kuwait University, May 1990.

30. M. A. Al-Bassam, *Applications of fractional calculus to a class of integro-differential equations of Riemann-Papperitz-Gauss type*, Proceedings of the Third International Conference on "Fractional Calculus", Nihon University, Japan, May 29–June 2, 1989, to appear.

31. M. A. Al-Bassam, *Applications of fractional calculus to a class of third order differential equations*, Proceedings of the Fourth Conference on Differential Equations, Ronsse, Bulgaria, (12–19) August, 1989, to appear.

M. A. Al-Bassam
Department of Mathematics
Kuwait University
P. O. Box 5969
Kuwait

ON THE DIOPHANTINE EQUATION $(X+Y)^A = Z^B$

Leo. J. Alex

In this paper the Diophantine equation $(X+Y)^A = Z^B$ is considered, and all solutions are found for which XYZ has the form $2^a 3^b 5^c 7^d$ where a,b,c,d are non-negative integers and A,B are positive integers. In particular when $(A,B)=1, (X,Y)=1$, it is shown that $B \leq 8$ and the 109 solutions to the equation are determined.

Let A, B be positive integers. Here we consider the Diophantine equation

$$(X + Y)^A = Z^B \tag{1}$$

where the product XYZ has the form $2^a 3^b 5^c 7^d$ where a, b, c, d are non-negative integers without loss of generality we may assume that $(A, B) = 1$.

In [1] all solutions to the related equation,

$$L + M = N, \tag{2}$$

with LMN of the form $2^a 3^b 5^c 7^d$ are determined.

An equation of the form, (2), arises naturally in certain problems involving finite groups. If one assumes that G is a finite group, p is a prime dividing the order of G to the first power only, and the index of the centralizer of a Sylow p-subgroup, S, in the normalizer of S is two, then Brauer's work [2], implies that the degrees of the irreducible characters in any p-block of G of highest defect satisfy Eq. (2), where L, M and N are divisiors of $|G|/p$.

The following Lemma appears as Theorem 3.5 of [1] and is proved in that paper.

Lemma 1. Let $LMN = 2^a 3^b 5^c 7^d$, where a, b, c, d are non-negative integers, $(L, M) = 1$, and $L \leq M$. Then the only positive integral solutions to Eq. (2) are

(L, M, N) = (1, 1, 2), (1, 2, 3), (1, 3, 4), (1, 8, 9), (1, 7, 8), (1, 63, 64),
(1, 27, 28), (1, 6, 7), (1, 48, 49), (1, 4, 5), (1, 15, 16), (1, 9, 10), (1, 80, 81),
(1, 5, 6), (1, 125, 126), (1, 24, 25), (1, 49, 50), (1, 2400, 2401), (1, 4374, 4375),
(1, 20, 21), (1, 14, 15), (1, 224, 225), (1, 35, 36), (3, 4, 7), (2, 7, 9),
(32, 49, 81), (7, 9, 16), (2, 3, 5), (9, 16, 25), (2, 25, 27), (4, 5, 9),
(3, 5, 8), (3, 125, 128), (5, 27, 32), (2, 5, 7), (7, 25, 32), (2, 243, 245), (8, 27, 35),
(3, 32, 35), (5, 16, 21), (5, 1024, 1029), (64, 125, 189), (7, 8, 15), (32, 343, 375),
(7, 128, 135), (5, 9, 14), (3, 25, 28), (3, 7, 10), (7, 243, 250), (5, 7, 12), (5, 49, 54),
(81, 175, 256), (15, 49, 64), (25, 56, 81), (7, 20, 27), (4, 21, 25), (27, 98, 125),
(7, 18, 25), (49, 576, 625), (4, 45, 49), (9, 40, 49), (100, 243, 343), and (24, 25, 49).

We are now in a position to begin the solution of Eq. (1). Let $Z = 2^t 3^u 5^v 7^w$. Then since Z^B is a perfect Ath power, we have $t \equiv u \equiv v \equiv w \equiv 0 \pmod{A}$. If we let $t = At_1, u = Au_1, v = Av_1, w = Aw_1$ we then have

$$X + Y = R \tag{3}$$

where $R = 2^{Bt_1} 3^{Bu_1} 5^{Bv_1} 7^{Bw_1} = Z^{B/A}$. Now we may apply Lemma 1 with $L = X, M = Y, N = R$ to obtain all solutions to Eq. (3) with $(X, Y) = 1$ and $X \leq Y$. Each solution (X, Y, R) of Eq. (3) corresponds to a solution (X, Y, Z) of Eq. (1) with $Z = R^{A/B}$. The following theorem gives the solutions to Eq. (1).

Theorem 2. Let $XYZ = 2^a 3^b 5^c 7^d$, where a, b, c, d are non-negative integers, $(X, Y) = 1$, and $X \leq Y$. Consider the equation $(X + Y)^A = Z^B$, where A, B are positive integers and $(A, B) = 1$.
Then

    (1) $B \leq 8$,
    (2) When $B = 1 (X, Y, Z) = (L, M, N^A)$, where (L, M, N) is one of
        the 63 solutions listed in Lemma 1.
    (3) When $B = 2$, $(X, Y, Z) = (1, 3, 2^A)$, $(1, 8, 3^A)$, $(1, 63, 8^A)$,
        $(1, 48, 7^A)$, $(1, 15, 4^A)$, $(1, 80, 9^A)$, $(1, 24, 5^A)$, $(1, 2400, 49^A)$,
        $(1, 224, 15^A)$, $(1, 35, 6^A)$, $(2, 7, 3^A)$, $(32, 49, 9^A)$, $(7, 9, 4^A)$,

$(9, 16, 5^A)$, $(4, 5, 3^A)$,$(81, 175, 16^A)$, $(15, 49, 8^A)$, $(25, 56, 9^A)$, $(4, 21, 5^A)$, $(7, 18, 5^A)$, $(49, 576, 25^A)$, $(4, 45, 7^A)$, $(9, 40, 7^A)$ or $(24, 25, 7^A)$

(4) When $B = 3$, $(X, Y, Z) = (1, 7, 2^A)$, $(1, 63, 4^A)$, $(2, 25, 3^A)$, $(3, 5, 2^A)$, $(15, 49, 4^A)$, $(7, 20, 3^A)$, $(27, 98, 5^A)$, or $(100, 243, 7^A)$.

(5) When $B = 4$, $(X, Y, Z) = (1, 15, 2^A)$, $(1, 80, 3^A)$, $(1, 2400, 7^A)$, $(32, 49, 3^A)$, $(7, 9, 2^A)$, $(81, 175, 4^A)$, $(25, 56, 3^A)$, $(49, 576, 5^A)$.

(6) When $B = 5$, $(X, Y, Z) = (5, 27, 2^A)$ or $(7, 25, 2^A)$,

(7) When $B = 6$, $(X, Y, Z) = (1, 63, 2^A)$ or $(15, 49, 2^A)$,

(8) When $B = 7$, $(X, Y, Z) = (3, 125, 2^A)$, and

(9) When $B = 8$, $(X, Y, Z) = (81, 175, 2^A)$.

Proof. Consideration of the solutions listed in Lemma 1 shows that B is largest when $Z^B = 256^A = 2^{8A}$. Thus 8 is the largest value possible for B. This proves (1). Then our previous discussion and Lemma 1 yield the solutions as listed for Eq. (1).

References

1. L. J. Alex, *Diophantine equations related to finite groups*, Comm. in Algebra (1) **4** (1976), 77–100.
2. R. Brauer, *On groups whose order contains a prime number to the first power*, I, II, Amer. J. Math. **64** (1942), 401–440.

Leo. J. Alex
Department of Mathematical Sciences
State University College at Oneonta
Oneonta, NY 13820
USA

THE MATH. HERITAGE OF C.F. GAUSS (pp. 35-42)
edited by George M. Rassias
©1991 World Scientific Publ. Co. Singapore

PARTITIONS AND THE GAUSSIAN SUM

George E. Andrews[1]

We examine the partition-theoretic implications of Gauss's identities for Gaussian polynomials.

1. Introduction

Gauss [7], [9; Ch. 9, p. 71] introduced a sum equivalent to

$$G(\zeta) = \sum_{j \bmod p} \zeta^{j^2}, \tag{1.1}$$

where ζ is a primitive p-th root of unity and p is an odd prime.

It is a fairly easy matter to show [9; p. 75, eq. (10.1)] that

$$G(\zeta) = \pm i^{\frac{p-1}{2}} \sqrt{p}; \tag{1.2}$$

however as Rademacher states [9; p. 83]: "The problem of determining the sign of the Gaussian sums for transcendentally specified primitive roots has become famous since the time of Gauss, who devoted a beautiful paper to it ("Smmatio quarumdam serierum singularium"). Since then a number of other quite different methods have been invented to deal with the sign of the Gaussian sums."

Rademacher [9; p. 85], following Gauss [7; p. 18], defines

$$f(q, m) = \sum_{j=0}^{m} (-1)^j \begin{bmatrix} m \\ j \end{bmatrix} \tag{1.3}$$

---

[1]Partially supported by National Science Foundation Grant DMS-870269.

and proves that

$$
f(q,m) = \begin{cases} 0 & \text{if } m \text{ is odd} \\ (1-q)(1-q)^3 \ldots (1-q^{m-1}) & \text{if } m \text{ is even}, \end{cases} \tag{1.4}
$$

where

$$
\begin{bmatrix} m \\ j \end{bmatrix}_q = \begin{bmatrix} m \\ j \end{bmatrix} = \begin{cases} \frac{(1-q^m)(1-q^{m-1})\ldots(1-q^{m-j+1})}{(1-q^j)(1-q^{j-1})\ldots(1-q)}, & \text{if } 0 \leq j \leq m, \\ 0, & \text{otherwise}. \end{cases} \tag{1.5}
$$

From (1.4), the evaluation of the Gaussian sum follows fairly directly.

Actually $f(q,m)$ is a special case of the Rogers-Szegö polynomial [4; p. 49] $H_m(q,t)$ given by

$$
H_m(q,t) = \sum_{j=0}^{m} \begin{bmatrix} m \\ j \end{bmatrix} t^j . \tag{1.6}
$$

Identity (1.3) is then

$$
H_m(q,-1) = \begin{cases} 0 & \text{if } m \text{ odd} \\ (1-q)(1-q^3)\ldots(1-q^{m-1}) & \text{if } m \text{ even}. \end{cases} \tag{1.7}
$$

In addition, there is a second known closed form for $H_m(q,t)$, namely [4; p. 499, Ex. 5]

$$
H_m(q^2,q) = (1+q)(1+q^2)\ldots(1+q^m), \tag{1.8}
$$

Since the Gaussian polynomial $\begin{bmatrix} m \\ j \end{bmatrix}$ is the generating function for partitions with at most j parts each $\leq m - j$, it is natural to ask if (1.6) and (1.7) can be explained in terms of partitions. Such explanations are the object of this paper.

2. Partition-Theoretic Proof of (1.3)

We note that when m is odd the j-th and $(m-j)$-th terms cancel each other. Thus trivially $f(q,m)$ is 0 in this case. Let us now replace m by

$2m, q$ by q^{-1} and multiply both sides of (1.3) by $(-1)^m q^{m^2}$. This yields the equivalent identity

$$\sum_{j=0}^{2m}(-1)^{j+m}q^{(j-m)^2}\begin{bmatrix}2m\\j\end{bmatrix} = (1-q)(1-q^3)\ldots(1-q^{2m-1})\,. \qquad (2.1)$$

Shifting the index of summation, we find

$$\sum_{j=-m}^{m}(-1)^j q^{j^2}\begin{bmatrix}2m\\j+m\end{bmatrix} = (1-q)(1-q^3)\ldots(1-q^{2m-1})\,, \qquad (2.2)$$

or separating even indices and odd indices

$$\sum_{j=-\infty}^{\infty} q^{4j^2}\begin{bmatrix}2m\\m-2j\end{bmatrix} - \sum_{j=-\infty}^{\infty} q^{(2j-1)^2}\begin{bmatrix}2m\\m-2j+1\end{bmatrix} \qquad (2.3)$$
$$= (1-q)(1-q^3)\ldots(1-q^{2m-1})\,.$$

Written in this form we recognize the left-hand side as the polynomial [5; p. 56, eq. (2.10) corrected] $D_{2,1}(0; m, m; q)$. Unfortunately the subscripts 2,1 violate the conditions required in the developments of [5]. Thus in order to understand the left-hand side of (2.3), we must return to the ideas originally expounded in [3] concerning successive ranks.

If π is a partition, we define $r_i(\pi)$ as the number of nodes in the horizontal part of the i-th right angle of the Ferrers graph of π minus the number of nodes in the vertical part of the i-th right angle. Thus if π is $7+6+4+4+3+1+1+1$, then its Ferrers graph is

$$
\begin{array}{c}
\bullet\ \bullet\ \bullet\ \bullet\ \bullet\ \bullet\ \bullet\\
\bullet\ \bullet\ \bullet\ \bullet\ \bullet\ \bullet\\
\bullet\ \bullet\ \bullet\ \bullet\\
\bullet\ \bullet\ \bullet\ \bullet\\
\bullet\ \bullet\ \bullet\\
\bullet\\
\bullet\\
\bullet
\end{array}
$$

$r_1(\pi) = 7-8 = -1, r_2(\pi) = 5-4 = 1, r_3(\pi) = 2-3 = -1, r_4(\pi)1-1 = 0$. We are concerned with the oscillation of the successive ranks between certain positive and negative bounds.

Definition 1. Let h be the largest integer for which there exists a sequence $j_1 < j_2 < \ldots < j_h$ such that $r_{j_1}(\pi) > K - i - 2, r_{j_2}(\pi) \leq$

$-i+1, r_{j_3}(\pi) > K - i - 2, r_{j_4}(\pi) \leq -i + 1$, etc. We define h to be the (K, i)-positive oscillation of π.

Definition 2. Let g be the largest integer for which there exists a sequence $j_1 < j_2 < \ldots < j_g$ such that $r_{j_1}(\pi) \leq -i + 1, r_{j_2}(\pi) > K - 1 - 2, r_{j_3}(\pi) \leq -i + 1, r_{j_4}(\pi) > K - i - 2$, etc. We define the (K, i)-negative oscillation of π to be g.

If, for example, the partition π is $7 + 6 + 4 + 4 + 3 + 1 + 1 + 1$, the (3,1)-positive oscillation is 2 while the (3,1)-negative oscillation is 3.

Definition 3. Let $p_{K,i}(a, b; \mu; N)$ (resp. $m_{K,i}(a, b; \mu; N)$) denote the number of partitions of N with at most b parts, with largest part at most a and with $(K, -i)$-positive (resp. (K, i)-negative) oscillation at least μ.

To these partition funcitons are associated related generating functions:

$$P_{K_i}(a, b; \mu; q) = \sum_{N \geq 0} p_{K,i}(a, b; \mu; N) q^N \tag{2.4}$$

and

$$M_{K,i}(a, b; \mu; q) = \sum_{N \geq 0} m_{K,i}(a, b; \mu; N) q^N . \tag{2.5}$$

Now everything in Sec. 3 of [3] was originally done with $K = 2k+1$ and $1 \leq i \leq k$. However as D. Bressoud [6] has noted all the arguments required to establish (3.15)–(3.21) of [3] remain valid with $k = (K - 1)/2, 1 \leq i \leq K/2$. It should be noted however that Lemma 3.1 of [3] is no longer valid in the case of interest to us, namely $K = 2, i = 1$. It must be replaced by the following:

Lemma. The (2,1)-positive oscillation of any partition π is either 1 larger or 1 smaller than the (2,1)-negative oscillation π unless all the successive ranks of π are zero in which case the two (2,1)-oscillations are equal to the side of Durfee square of π.

Proof. Suppose π has a Durfee square of side S. This means π has S successive ranks. We prove the result by induction on S. If $S = 1$ and the first rank is positive then the (2,1)-positive oscillation is 1 and the (2,1)-negative oscillation is 0; if the first rank is negative then the (2,1)-negative oscillation is 1 and the (2,1)-positive oscillation is 0. If the rank is zero then both oscillations $= 1 = S$. Hence the result is true for $S = 1$.

We now assume the result up to but not including a particular S. Let π' be the partition obtained from π by deleting the outer shell of nodes in the Ferrers graph (i.e., delete the largest part and all 1's from π then subtract 1 from each remaining part).

If all the S ranks of π are equal to zero, then the maximal sequence for both (2,1)-positive and (2,1)-negative oscillation is just $j_i = i$ for $1 \le i \le S$. Consequently, both oscillations $= S$ in this case as desired. Hence we may assume that not all the ranks of π are zero for the remainder of the proof.

Let $a(\pi)$ be the (2,1)-positive oscillation of π and $n(\pi)$ be the (2,1)-negative oscillation of π.

Case 1. $r_1(\pi) > 0$. Clearly then

$$a(\pi) = \max(1 + n(\pi'), a(\pi'))$$
$$n(\pi) = n(\pi').$$

If all the ranks of π' are 0, then $a(\pi) = S$ while $n(\pi) = S - 1$. If not all are 0, then either $n(\pi') = a(\pi') + 1$ in which case $a(\pi) = a(\pi') + 2 = n(\pi) + 1$ or $a(\pi') = n(\pi') + 1$ in which case $a(\pi) = 1 + n(\pi') = 1 + n(\pi)$.

Case 2. $r_1(\pi) < 0$. Clearly then

$$a(\pi) = a(\pi')$$
$$n(\pi) = \max(1 + a(\pi'), n(\pi')).$$

The argument now is exactly like that in Case 1 except the roles of $a(\pi)$ and $n(\pi)$ are exactly reversed.

Case 3. $r_1(\pi) = 0$. Clearly then

$$a(\pi) = \max(1 + n(\pi'), a(\pi'))$$
$$n(\pi) = \max(1 + a(\pi'), n(\pi')).$$

Since $r_1(\pi) = 0$ we must have not all ranks of π' equal zero. Hence $|a(\pi') - n(\pi')| = 1$. If $a(\pi') = n(\pi') + 1$, then $a(\pi) = 1 + n(\pi') = n(\pi) - 1$. If $n(\pi') = a(\pi') + 1$, then $n(\pi) = 1 + a(\pi') = a(\pi) - 1$.

We have now exhausted all possibilities, and we have established the desired result at S. Consequently the lemma holds.

Theorem 1. Equation (2.3) (and consequently (1.3)) holds.

Proof. By (3.28) and (3.29) applied to (3.15)–(3.18) all of [3; p. 1224], we see that

$$M_{2,1}(m, m; 2\mu, q) = q^{4\mu^2} \begin{bmatrix} 2m \\ m - 2\mu \end{bmatrix}, \tag{2.6}$$

$$M_{2,1}(m, m; 2\mu; q) = q^{(2\mu-1)^2} \begin{bmatrix} 2m \\ m - 2\mu + 1 \end{bmatrix}. \tag{2.7}$$

$$P_{2,1}(m, m; 2\mu; q) = q^{4\mu^2} \begin{bmatrix} 2m \\ m - 2\mu \end{bmatrix} \tag{2.8}$$

$$P_{2,1}(m, m; 2\mu - 1; q) = q^{(2\mu-1)^2} \begin{bmatrix} 2m \\ m - 2\mu + 1 \end{bmatrix}. \tag{2.9}$$

Consequently, the left-hand side of (2.3) is just

$$P_{2,1}(m, m; 0; q) + \sum_{\mu \geq 1} (-1)^\mu M_{2,1}(m, m; \mu; q)$$
$$+ \sum_{\mu \geq 1} (-1)^\mu P_{2,1}(m, m; \mu; q)$$
$$= \sum_{n \geq 0} (p_{2,1}(m, m; 0; N) + \sum_{\mu \geq 1} (-1)^\mu m_{2,1}(m, m; \mu; N)$$
$$+ \sum_{\mu \geq 1} (-1)^\mu p_{2,1}(m, m; \mu; N)) q^N.$$

Now any partition π with largest part $\leq m$ and number of parts $\leq m$ makes zero contribution to the coefficient of q^N by precisely the argument given in Lemma 4.1 of [3] unless all successive ranks of π are zero. In this latter case, the contribution is precisely

$$\left(1 + \sum_{\mu=1}^{S} (-1)^\mu + \sum_{\mu=1}^{S} (-1)^\mu \right) = \begin{cases} -1, & S \text{ odd} \\ 1, & S \text{ even} \end{cases}$$
$$= (-1)^S,$$

where S is the side of the Durfee square of π. Therefore the left-hand side of (2.3) with q related by $-q$ is the generating function for partitions with successive ranks $= 0$, largest part $\leq m$ and number of parts $\leq m$, i.e., the generating function for self-conjugate partitions with largest part $\leq m$. But this is clearly by [4; p. 14, Ex. 8]

$$(1+q)(1+q^3)\ldots(1+q^{2m-1}).\tag{2.10}$$

Replacing q by $-q$ we have the desired result.

3. Partition-Theoretic Proof of (1.7)

This identity is immediate if we examine Sylvester's "fish-hook" bijection between partitions with distinct parts and partitions with odd parts as is done in [1; p. 136, Theorem F1, Second proof], see also [5; Sec. 4]. If this proof is restricted to partitions with distinct parts each $\leq m$, then (1.7) is the corresponding generating function assertion.

4. Conclusion

Limiting cases of our polynomial results are also familiar from Gauss's work [8; p. 447, eqs. (14) and (10)]. Letting $m \to \infty$ in (2.3) we find [4; p. 23, eq. (2.2.12)]

$$\sum_{j=-\infty}^{\infty} (-1)^j q^{j^2} = \prod_{j=1}^{\infty} (1-q^j)(1-q^{2j-1}) = \prod_{j=1}^{\infty} \frac{(1-q^j)}{(1+q^j)}.\tag{4.1}$$

If in (1.7) we replace m by $2m$, q by q^{-1}, then multiply by q^{2m^2+m} and replace j by $j+m$, we find that

$$\sum_{j=-m}^{m} q^{j(2j-1)} \begin{bmatrix} 2m \\ j+m \end{bmatrix}_{q^2} = (1+q)(1+q^2)\ldots(1+q^{2m}).\tag{4.2}$$

Now let $m \to \infty$ in (4.2), and we find [4; p. 23, eq. (2.2.13)]

$$\begin{aligned}
\sum_{j=-\infty}^{\infty} q^{j(2j-1)} = \sum_{j=0}^{\infty} q^{\binom{j+1}{2}} &= \prod_{j=1}^{\infty} (1+q^j)(1-q^{2j}) \\
&= \prod_{j=1}^{\infty} \frac{(1-q^{2j})}{(1-q^{2j-1})}.
\end{aligned}\tag{4.3}$$

Identities (4.1) and (4.3) were the subject of a previous partition-theoretic investigation of the implications of Gauss's work [2].

References

1. G. E. Andrews, *On basis hypergeometric series, mock theta functions, and partitions*, II, Quart. J. Math., Oxford Series 2, **17** (1966) 132–143.
2. G. E. Andrews, *Two theorems of Gauss and allied identities proved arithmetically*, Pac. J. Math. **41** (1972) 563–578.
3. G. E. Andrews, *Sieves in the theory of partitions*, Amer. J. Math. **94** (1972) 1214–1230.
4. G. E. Andrews, *The Theory of Partitions*, Vol. 2, The Encyclopedia of Mathematics and Its Applications, G. -C. Rota ed., Addison-Wesley, Reading 1976 (Reissued: Cambridge University Press, London and New York, 1984).
5. G. E. Andrews, *Uses and extensions of Frobenius's representation of partitions*, from Enumeration and Design, D. M. Jackson ans S. Vanstone eds., Academic Press, Toronto and New York, 1984, pp. 51–65.
6. D. M. Bressoud, *Extension of the partition sieve*, J. Number Th. **12** (1980) 87–100.
7. C. F. Gauss, Werke, Vol. 2, Königlichen Gesellschaft der Wissenschaften, Göttingen, 1876.
8. C. F. Gauss, Werke, Vol. 3, Königlichen Gesellschaft der Wissenschaften, Göttingen, 1876.
9. H. Rademacher, *Lectures on Elementary Number Theory*, Blaisdell, New York, 1964.

George E. Andrews
Department of Mathematics
The Pennsylvania State University
University Park, PA 16082
USA

THE MATH. HERITAGE OF C.F. GAUSS (pp. 43-52)
edited by George M. Rassias
©1991 World Scientific Publ. Co. Singapore

THE PERIODICITY OF AN ALGORITHM OVER THE COMPLEX NUMBER FIELD (ACF) AND THE SOLUTION OF THE HERMITE PROBLEM

Malvina Baica

Euler and Lagrange proved that every real quadratic irrational is represented by an infinite periodic fraction or by a periodic Euclidean Algorithm (EA) development and every infinite periodic continued fraction represents a real quadratic irrational. Hermite challenged Jacobi with the question to find a periodic algorithm to develop irrationals of any degree into such periodic sequences. Working on this problem Jacobi and Perron developed an algorithm which produced some periodic developments of high degree irrationals.

In later work by Hasse and Bernstein results of a more general character were obtained. Baica constructing a new algorithm over the complex field (ACF) reproduced all of the previous results as well as many new ones.

In this paper, from the application of the ACF additional higher degree irrationals will be shown to have a periodic sequence development, removing some restrictions that had previously limited the work in this problem.

Introduction

The objective of this paper is to develop an irrational number of any degree into a periodic sequence. This problem of Hermite has been open since the middle of the nineteenth century.

The beginning of this subject is the well-known Euclidean Algorithm (EA). For example, by using ths algorithm it is easy to prove that every rational number, a/b, can be represented as a finite simple continued fraction. In 1737 Euler proved that every real quadratic irrational can be represented by an infinite periodic continued fraction or by a periodic EA development. The converse was proved by Lagrange in 1770.

Of course if the number is not a quadratic irrational, but is a real algebraic number of higher degree or a transcendental real number, then its development by the EA cannot be periodic.

In 1839, Hermite, in one of his letters to Jacobi, challenged Jacobi to find an algorithm to develop irrationals of any degree into periodic sequences. But it was only after thirty years of frustration that Jacobi in 1869 extended EA methods to successfully represent some cubic irrationals by means of simple continued fractions. Then in 1907, Perron generalized the work of Jacobi. This generalization is known as the Jacobi-Perron Algorithm (JPA) and with it, Jacobi exhibited periodic developments for the cube roots of the numbers two, three and five. Advances were slow and difficult, but in 1873 Bachman proved results for other cubic irrationals using the JPA results that were accompanied by many restrictions. It is odd that no more progress occurred on these problems until Hasse and Bernstein turned their attention to them in their paper [6].

In this general form, as defined by Jacobi, an application of the JPA starts with the definition of an initial vector, the components of which are algebraic numbers. By use of the greatest integer function a "companion vector" is defined. A recursive transformation is constructed and applied to these vectors. For good choices of the starting vector and for the transformation, the iteration of the transformation becomes periodic, that is the transformation cycles around a finite set of vectors. In this instance the JPA is said to be periodic, and the results lead to the periodic representation of higher degree irrationals.

The difficulties associated with this work are many. Jacobi's results were confined to a few numerical examples in a cubic field. Perron generalized the method to apply to irrationals of any degree, but since the choices of starting vector and transformation are difficult to make, he was also limited to a few periodic developments of higher degree irrationals. Perron was more successful in showing that if a development is periodic, then the components of the initial vector are algebraic numbers. This latter result was general.

With Perron's work in 1907 on Hermite's Problem progress came to a halt, because of the failure of the JPA to produce new numerical results, that is additional cases in which the transformation becomes periodic were not achieved. Perron and all others recognized that the usual choices for starting vector were too limited.

No results were accomplished between 1907 and 1965, when Hasse and

Bernstein made a broader approach to the peridocity problem associated with the JPA. Hasse and Bernstein were not interested in Hermite's problem and confined their work to searching for units in algebraic number fields.

Hasse and Bernstein started with an algebraic extension of the rational numbers, $Q(w)$, where w takes the form $w = \sqrt[n]{D^n + d}$ with

$$P(x) = \left(\prod_{i=1}^{n} (x^n - D_i^n) \right) - d, d \in Z, D_i \in N \text{ and } d|D_i.$$

$$a^{(0)} = ((w - D_1)(w - D_2)\ldots(w - D_{n-1}), \ldots,$$
$$(w - D_1)(w - D_2), (w - D_2)) \text{ with } b^{(0)} = a^{(0)}(D_1).$$

Hasse and Bernstein showed that certain restrictions on D and d led to a JPA that was purely periodic. For $d > 0$ they proved that the JPA of $a^{(0)}$ is purely periodic when $D \geq (n-2)d$ and $n \geq 3$. For $d < 0$ the sequence is also purely periodic when $D \geq 2(n-1)d$ and $n \geq 3$. The length of the period is $n(n-1)$. For this approach the periodicity remains an open problem since there are bounds on D and the restriction $d|D$ must hold. For example no periodicity for $w = \sqrt[5]{12^5 + 6}$ can be proved under the Hasse-Bernstein restrictions. It will be seen later that some of these conditions can be removed.

When the algorithm becomes periodic an algebraic field unit is generated by the product of the last components of all of the vectors in the cycle. This was a strong result in the theory of units in algebraic number fields. The number of fields in which the Hasse-Bernstein units can be found is tied to the restrictions on their application of the JPA. The Hasse-Bernstein results were limited by their choices of w as real numbers. It should be noted that Hasse and Bernstein were not interested in Hermite's problem even if their results had a strong relation to that problem. Specifically, they did not know that the periodicity of the algorithm leads to a solution of Hermite's Problem for some real algebraic number w.

Baica (1980) defined a modification of the JPA that used the Hasse-Bernstein initial vector, but was not restricted to the real numbers. For the first time the complex numbers were considered. The only differences in the definitions stated above are that the D_i's are now complex. An immediate consequence of this extension is that the bounds on D in the Hasse-Bernstein work are now eliminated and only the divisibility condition, $d|D$, remains. Returning to the example cited above, it can now be seen that $w = \sqrt[5]{12^5 + 6}$ has a periodic development, only $6|12$ is required.

Baica named the extension, the Algorithm for Complex Numbers (ACF). All of the previous results of the JPA and all of the Hasse-Bernstein results are consequences of the Baica modification. All units in algebraic number fields so far discovered are particular cases of the application of Baica's Algorithm for Complex Numbers. The fact that the new algorithm unifies all previous work in the theory of units makes it a partial solution to the problem of Hilbert that asks for a common periodic algorithm from which all units in all algebraic number fields can be derived.

In addition the new algorithm assists in the solution of higher degree Diophantine equations, provides surprising derivations of complicated combinatorial identities, generates n-dimensional Fibonacci numbers, as well as makes possible the algorithmic approximation of higher degree irrationals.

Finally, Baica's Algorithm provides new progress toward the solution of Hermite's Problem. A closer approach to the total solution of Hermite's Problem will be made in which the only remaining barrier is the divisibility condition $d|D$. In addition Hermite's Problem will be extended and restated to include the complex numbers. For the future this new algorithm (ACF) will be named the Generalized Euclidean Algorithm (GEA).

The problem which we investigate here is to obtain a periodic GEA development for more higher degree irrationals.

1. The GEA

JPA is very difficult to prove periodic. Many prominent mathematicians in the modern mathematical world gave various modifications of JPA in order to prove its periodicity, but the problem still remained an open question. The author [1] used a modification which holds for complex fields of any degree (ACF). In the sequel, it will be called GEA. With this GEA for the first time complex numbers not only real numbers were found to have a periodic GEA development. The GEA was described in [1] where the author proved its periodicity and some other important results. For now let us proceed as follows:

Definition 1. Let

$$a^{(0)} = \left(a_1^{(0)}, a_2^{(0)}, \ldots, a_{n-1}^{(0)}\right) \in R^{n-1}, n \geq 1 \tag{1.1}$$

be a fixed, given vector, and

$$\langle a^{(v)} \rangle, \ v = 0, 1, \ldots, \quad a^{(v)} \in R^{n-1}$$

a sequence of vectors in R^{n-1} either given by some rule or calculated from $a^{(0)}$,

$$a^{(v)} = \left(a_1^{(v)}, a_2^{(v)}, \ldots, a_{n-1}^{(v)} \right).$$

Let $\langle b^{(v)} \rangle, v = 0, 1, \ldots; b^{(v)} \in R^{n-1}$ be another sequence of vectors in R^{n-1}, either given by some rule or calculated from $\langle a^{(v)} \rangle$ with the formula

$$\left. a^{(v+1)} = \left(a_1^{(v)} - b_1^{(v)} \right)^{-1} \left(a_2^{(v)} - b_2^{(v)}, \ldots, a_{n-1}^{(v)} - b_{n-1}^{(v)}, 1 \right) \atop v = 0, 1, \ldots . \right\} \qquad (1.2)$$

We say that the GEA holds for $a^{(0)}$. $b^{(v)}$ is called the "companion vector" of $a^{(v)}$.

Definition 2. Let the GEA of an $a^{(0)}$ hold. If there exist two numbers, $s \geq 0, t \geq 1; s, t \in N$ such that $a^{(s+t)} = a^{(t)}$, then the GEA of this $a^{(0)}$ is called "periodic". If contemporarily $\min s = m \geq 0, \min t = \ell \geq 1$, then $\langle a^{(v)} \rangle, v = 0, 1, \ldots, m - 1, \langle a^{(v)} \rangle, v = m, m + 1, \ldots, m + \ell - 1$ are called respectively the "primitive preperiod" and the "primitive period" of the GEA of $a^{(0)}$. If $m = 0$, the GEA of $a^{(0)}$ is called "purely periodic". m and ℓ are called respectively the lengths of the primitive preperiod and primitive period.

The reader should note that if the GEA of some $a^{(0)}$ is periodic then there exists an $a^{(v)}$ in this GEA such that the GEA of $a^{(v)}$ is purely periodic. With this in mind the author [1] has proved

Theorem 1. Let w be an n-th degree irrational $(n \geq 2)$ $\qquad (1.3)$
and $a^{(0)}$ a fixed vector such that

$$\left. a^{(0)} = \left(a_1^{(0)}(w), a_2^{(0)}(w), \ldots, a_{n-1}^{(0)}(w) \right) \atop a_i^{(0)}(w) \text{ algebraic integers } (i = 1, \ldots, n - 1) \right\} \qquad (1.4)$$

Let the GEA of $a^{(0)}$ be purely periodic with length of the primitive period $= \ell$, and let the components of the companion vectors be algebraic integers. Then

$$\left. A_0^{(v\ell)} + a_1^{(0)} A_0^{(v\ell+1)} + a_2^{(0)} A_0^{(v\ell+2)} + \ldots + a_{n-1}^{(0)} A_0^{(v\ell+n-1)} \atop v = 1, 2, \ldots \right\} \qquad (1.5)$$

48

M. Baica

are units, namely

$$A_0^{(\ell)} + a_1^{(0)} A_0^{(\ell+1)} + a_2^{(0)} A_0^{(\ell+2)} + \ldots + a_{n-1}^{(0)} A_0^{(\ell+n-1)} \,.$$

In this context we need the formula used by the author in [1], viz.

$$\prod_{i=1}^{k} a_{n-1}^{(i)} = A_0^{(k)} + a_1^{(k)} A_0^{(k+1)} + \ldots + a_{n-1}^{(k)} A_0^{(k+n-1)} \,. \tag{1.6}$$

If the GEA of $a^{(0)}$ in R_{n-1} is purely periodic with length of the primitive period ℓ, then it follows from (1.6)

$$e = \prod_{i=0}^{\ell-1} a_{n-1}^{(i)} = \sum_{j=0}^{n-1} a_j^{(0)} A_0^{(\ell+j)} \,, \ e \text{ a unit} \tag{1.7}$$

since in this case $a_{n-1}^{(\ell)} = a_{n-1}^{(0)}$ in virtue of (1.5) we have

$$\left. \begin{array}{c} e^f = A_0^{(f\ell)} + a_1^{(0)} A_0^{(f\ell+1)} + \ldots + a_{n-1}^{(0)} A_0^{(f\ell+n-1)} \\ f = 1, 2, \ldots . \end{array} \right\} \tag{1.8}$$

2. GEA – Periodic Development of Irrationals – Case $n = 2$

Though this case is well known from the expansion of real quadratic irrationals as simple continued fractions, we shall include it in our discussion.

Let

$$w = \sqrt{D^2 + 1}, \ D \in N \,, \ w \text{ a quadratic irrational} \,. \tag{2.1}$$

That w is irrational (for $D > 0$) is trivial. We choose the fixed vector.

$$a^{(0)} = w + D \,, \tag{2.2}$$

since here $n - 1 = 1$. Thus $a_1^{(0)} = a_{n-1}^{(0)}$, and we shall generally denote

$$a^{(v)} = a_v \,, \ v = 0, 1, \ldots \,; a_v = a_v(\alpha) \text{ for all GEA of } a^{(0)} \,. \tag{2.3}$$

By (2.3) we denote

$$b^{(v)} = b_v \,, \ v = 0, 1, \ldots .$$

For the calculation of the companion vectors we use the rule

$$b^{(v)} = b_v = a_v(D), \quad v = 0, 1, \ldots \tag{2.4}$$

and have

$$b_0 = (w + D)_{w=D} = 2D. \tag{2.5}$$

hence, by (2.1)

$$a_1 = [(w + D) - 2D]^{-1} \cdot 1 = (w - D)^{-1} = w + D \tag{2.6}$$

since $(w - D)^{-1} = (w + D)$ from $(w^2 - D^2) = 1$. Thus

$$a_0 = a_1 = \ldots = a_v, \ v = 0, 1, \ldots \tag{2.7}$$

and the GEA of $a_0 = w + D$ is purely periodic with length of the primitive period $\ell = 1$. Further,

$$[a_v] = [w + D] = [w] + D = 2D = b_v \tag{2.8}$$

the GEA of $w + D$ coincides with the EA and we have, in the notation of continued fractions

$$a_0 = w + D = \overline{[2D]} \tag{2.9}$$

(2.9) is the periodic GEA development of a quadratic irrational $a_0 = w + D$.

3. The Case $n = 3$

We denote

$$w = \sqrt[3]{D^3 + 1} \tag{3.1}$$

and choose the fixed vector

$$a^{(0)} = (w + 2D, w^2 + Dw + D^2) \tag{3.2}$$

with

$$a^{(0)} = (a_1^{(0)}(w), a_2^{(0)}(w)).$$

We apply the rule for calculating the components of the companion vectors

$$b_i^{(v)} = a_i^{(v)}(D), \ i = 1, 2; \ v = 0, 1, \ldots.$$

We proceed with the GEA of $a^{(0)}$

$$b^{(0)} = (D + 2D, D^2 + D.D + D^2)$$
$$b^{(0)} = (3D, 3D^2).$$
$$a^{(1)} = (w + 2D - 3D)^{-1}(w^2 + Dw + D^2 - 3D^2, 1)$$
$$= (w - D)^{-1}(w^2 + Dw - 2D^2, 1)$$
$$= (w - D)^{-1}((w - D)(w + 2D), 1). \tag{3.3}$$
$$a^{(1)} = (w + 2D, w^2 + Dw + D^2) = a^{(0)}. \tag{3.4}$$

By (3.4) the GEA of

$$a^{(0)} = ((w + 2D), w^2 + wD + D^2), \; w = \sqrt[3]{D^3 + 1}$$

is purely periodic and the length of its primitive period $\ell = 1$. Using the notation in (2.9) we have

$$a_0 = (w + 2D, w^2 + Dw + D^2) = [\overline{3D, 3D^2}]. \tag{3.5}$$

This can be considered the development of the components of a_0 using the peridocity of GEA and those components are algebraic irrationals of third degree. This is not true for all third degree irrationals, but for all those third degree irrationals which are components of a starting vector $a^{(0)}$ that leads to a periodic GEA.

4. The Case $n = 5$

We denote again

$$w = \sqrt[5]{D^5 + 1} \tag{4.1}$$

and choose the fixed vector

$$a^{(0)} = (w + 4D, w^2 + 3wD + 6D^2, w^3 + 2w^2D + 3wD^2 \\ + 4D^3, w^4 + Dw^3 + D^2w^2 + D^3w + D^4). \tag{4.2}$$

Using the previous argument we find

$$b^{(0)} = (5D, 10D^2, 10D^3, 5D^4) \quad \text{or}$$
$$b^{(0)} = \left(\binom{5}{1} D, \binom{5}{2} D^2, \binom{5}{3} D^3, \binom{5}{4} D^4 \right). \tag{4.3}$$

Thus the GEA of $a^{(0)}$ is purely periodic with lengths of primitive period $\ell = 1$ and like in (3.5)

$$a_0 = \left[\overline{5D}, \overline{10D^2}, \overline{10D^3}, \overline{5D^4}\right] . \tag{4.4}$$

This solves the problem for some fifth degree irrationals.

5. The General Case

Let w be the irrational

$$w = \sqrt[n]{D^n + 1}\,;\; n \geq 2,\, D \in N\,; \tag{5.1}$$

and choose the fixed vector

$$\left.\begin{aligned}
a^{(0)} &= \left(a_1^{(0)}, a_2^{(0)}, \ldots, a_s^{(0)}, \ldots, a_{n-1}^{(0)}\right) \\
a_s^{(0)} &= \sum_{i=1}^{s} \binom{n-s-1+i}{i}\, w^{s-i} D^i \\
s &= 1, \ldots, n-1\,.
\end{aligned}\right\} \tag{5.2}$$

The proof that the GEA of the fixed vector $a^{(0)}$ like in (5.24) is purely periodic and the length of its primitive period is $\ell = 1$ was given by the author in [2]. Thuse we find

$$b^{(0)} = \left(\binom{n}{1} D,\, \binom{n}{2} D^2,\, \cdots \binom{n}{n-1} D^{n-1}\right) . \tag{5.3}$$

With (5.3) we denote

$$a_0 = \left[\overline{nD},\, \overline{\frac{n(n-1)}{2} D^2},\, \ldots,\, \overline{nD^{n-1}}\right] . \tag{5.4}$$

(5.4) will solve the problem for some n-th degree irrationals.

Since GEA was not proved periodic for any degree irrational w we cannot get a periodic GEA for any degree irrationals. This GEA does not completely solve Hermite's Problem, which may remain an open question for years to come, but these results are one more step in this direction.

Bibliography

1. M. Baica, *An algorithm in a complex field and its application to the calculation of units*, Pacific J. Math. (1) **110** (1984) 21–40.

2. M. Baica, *n-dimensional Fibonacci numbers and their applications*, The Fibonacci Quaterly (4) **21** (1983) 285–301.

3. M. Baica, *Some new combinatorial identities derived from units in algebraic number fields*, Discrete Mathematics (2) **54** (Amsterdam, 1985) 133–141.

4. M. Baica, *Approximation of irrationals*, Internat. J. Math. Math. Sci. (2) **8** (1985) 303–320.

5. M. Baica, *Diphantine equations and identities*, Internat. J. Math. Math. Sci. (4) **8** (1985) 755–777.

6. L. Bernstein and H. Hasse, *Eineheitenberecnung Mittels des Jacobi-Perronschen algorithmus*, J. Reine Angew. Math. **218** (1965) 51–69.

7. L. Bernstein, *The Jacobi-Perron Algorithm, its Theory and Applications*, Springer, Berlin-Heidelberg-New York, Lect. Notes. Math. Vol. 207, 1971.

8. L. Bernstein, *Representation of $\sqrt[n]{D^n-d}$ as a periodic continued fraction by Jacobi's Algorithm*, Math. Nachr. **29** (1965) 179–200.

9. Ch. Hermite, *Letter to C. G. J. Jacobi*, J.f.d. Reine Angew. Math. **40** (1939) 286.

10. C. G. J. Jacobi, *Allegemine Theorie der kettenbruchaehnlichen algorithmen, In velchen jede Zahl aus drei vorhergehenden gebildet wird*, J.f.d. Reine Angew. Math. **69** (1969), 29–64.

11. O. Perron, *Grundlagen füer eine theorie des Jacobischen kettenbruchalgorithmus*, Math. Ann. **64** (1907) 1–76.

12. O. Perron, *Ein neues Konvergenzkriterium für Jacobi-Ketten 2, Ordnung.* Arch. Math. Phys. (Reine) (3) **17** (1911) 204–211.

Malvina Baica
Department of Mathematics
and Computer Science
University of Wisconsin-Whitewater
Whitewater, WI 53190
USA

THE MATH. HERITAGE OF C.F. GAUSS (pp. 53-65)
edited by George M. Rassias
©1991 World Scientific Publ. Co. Singapore

LORENTZIAN DISTANCE AND CURVATURE

John K. Beem[*]

In this paper, the Lorentzian distance function is used to define the flat bisector condition as well as two point homogeneous. In globally hyperbolic space-times one finds that isotropic implies two point homogeneous. The local version of the flat bisector condition yields constant curvature. A globally hyperbolic space-time with a complete timelike line, flat bisectors and nonpositive curvature is shown to be flat.

1. Introduction

The (positive definite) Riemannian manifold (M, g) is *homogeneous* if for every pair of points p, q there is an isometry of (M, g) taking p to q. For positive definite manifolds homogeneity implies completeness. Here one may use the notion of geodesic completeness or completeness of the induced distance function d on M. These two forms of completeness are equivalent for positive definite spaces by the Hopf-Rinow Theorem [10]. In contrast, the Hopf-Rinow Theorem fails for Lorentzian manifolds and also homogeneity does not imply completeness for these spaces. The Riemannian manifold (M, g) is said to be *two point homogeneous* if given two pairs of poins p_1, p_2 and q_1, q_2 with $d(p_1, p_2) = d(q_1, q_2)$, there is an isometry of the space taking p_1 to q_1 and p_2 to q_2, see [3], [6], [15], [16].

In this paper, we investigate Lorentzian manifolds with particular emphasis on space-times which are the time oriented Lorentzian manifolds. We consider the concepts of isotropic and locally isotropic. Fix a point p in (M, g). If for each pair X, Y of tangent vectors at p with $g(X, X) = g(Y, Y) \neq 0$ there is an isometry of (M, g) taking X to Y, then

---

[*]Supported in part by NSF grant DMS-8803511

we say the space is *isotropic at p*, see Wolf [17, p. 367]. We say (M, g) is *locally isotropic at p* if for each pair X, Y of tangent vectors at p with $g(X, X) = g(Y, Y) \neq 0$ there is some neighborhood U of p and isometry $\phi : (U, g) \to (M, g)$ mapping U to a subset $\phi(U)$ of M and taking X to Y. A space is *isotropic* (or *locally isotropic*) if it is isotropic (locally isotropic) at each point. A Lorentzian manifold is *symmetric* if at each point p there is an isometry which holds the point fixed and maps each tangent vector X at p to its negative $-X$. In other words, there is an isometry corresponding to reflection across the point p. The manifold is *locally symmetric* if each point p has a neighborhood U and an isometry of (U, g) onto itself taking p to itself and each tangent vector X at p to $-X$. It is known [17, p. 377] that locally isotropic implies locally symmetric. For Lorentzian manifolds one may define a local Lorentzian distance function which satisfies a reverse triangle inequality and for space-times a global Lorentzian distance function. Using these functions we may consider the two point homogeneous condition for Lorentzian manifolds. We find that for globally hyperbolic space-times, isotropic implies two point homogeneous. We consider bisectors in Lorentzian spaces and obtain several characterizations of constant curvature. Furthermore, we find that a globally hyperbolic space-time with a complete timelike line, flat bisectors and nonpositive curvature is flat.

2. Preliminaries

Let (M, g) be a connected semi-Riemmanian manifold of dimension n and signature (s, r). At each point p of M, the metric tensor g is a symmetric inner product mapping $T_p M \times T_p M$ to R^1 such that this inner product has r positive eigenvalues and $s = n - r$ negative eigenvalues. The tangent vector X in $T_p M$ is *timelike* if $g(X, X) < 0$, *spacelike* if $g(X, X) > 0$, and *null* if $g(X, X) = 0$. The *nonspacelike* vectors are the timelike and null vectors. These definitions are motivated by General Relativity, but apply to semi-Riemannian manifolds of arbitrary signature.

If $s = 1$, then (M, g) is called *Lorentzian*. A Lorentzian manifold which has a nonvanishing timelike vector field X is said to be a *space-time* and the vector field X is said to *time orient* (M, g). Given X, the timelike or null tangent vector Y at some point p of M is *future directed* if $g(Y, X) < 0$. If (M, g) is a space-time, then there is a chronological relation $<$ and a causal relation $\leq$ induced on M. The point q is in the *chronological future* of p, written $p < q$ if there is some future directed timelike curve $\gamma : [0,1] \to M$ with $\gamma(0) = p$ and $\gamma(1) = q$. If there is a future directed nonspacelike curve γ from p to q or if $p = q$, then q is in the *causal future* of p, written $p \leq q$. The *causal past* of p consists of all the points q such that p is in the causal future of q. Given a point q in the causal future of p, the *Lorentzian distance* $d(p,q)$ from p to q is the supremum of lengths of all future directed nonspacelike curves from p to q. In this paper, we will also set $d(p,q) = d(p,q)$ and thus require that d be symmetric. If q is neither in the causal future or causal past of p, then one sets $d(p,q) = 0$. If $p \leq r \leq q$, one has [3] the *reverse triangle inequality* $d(p,r) + d(r,q) \leq d(p,q)$. Furthermore, the value of $d(p,q)$ may be equal to $+\infty$. In fact, in so called totally vicious sapce-times one has $d(p,p) = +\infty$ for all p in M. A space-time is *strongly causal* if each p in M has arbitrarily small neighborhoods U such that any timelike curve which leaves U fails to ever return to U. Choosing U to be a (sufficiently small) convex normal neighborhood [3] in a strongly causal space-time, yields that d restricted to U is finite valued. Even if (M,g) is not strongly causal, one may obtain a *local distance function* σ on small neighborhoods. Fix p in (M,g) and let U be a convex normal neighborhood of p. Here we assume each q in U is such that the exponential map at q restricted to a neighborhood of the origin of $T_q M$ is a diffeomorphism onto U. For each pair of points q_1, q_2 of U there is a unique geodesic segment $\alpha(q_1, q_2)$ joining q_1 to q_2 and lying in U. Using (U,g) to denote the space U with metric g restricted to U, we let σ be the Lorentzian distance function for (U,g). Then $\sigma(q_1, q_2)$ is finite and is equal to the length of the geodesic segment $\alpha(q_1, q_2)$ when this segment is timelike and zero otherwise. In case (M,g) is strongly causal, one may choose U sufficiently small so that σ agrees with the distance function d of (M,g) restricted to U. A space-time which is strongly causal and has the additional property that given any two points p and q the intersection of causal future of p with the causal past of q is compact is call *globally hyperbolic* [3].

If $s = 0$, then (M,g) is positive definite and there is an induced *distance function* $d : M \times M \to R^1$ which is a metric in the usual sense. The Hopf-Rinow Theorem yields that in the positive definite case (M,g) is geodesically complete iff the induced distance function d is Cauchy complete.

The *semi-Euclidean space* of dimension n and signature $(s, n-s)$ will be denoted by $R^n{}_s$. This is the manifold R^n with the metric tensor given by

$$g = -\left(\sum_{i=1}^{s} dx^i \otimes dx^i\right) + \sum_{i=s+1}^{n} dx^i \otimes dx^i .$$

If $s = 0$, this is the ordinary Euclidean space of dimension n and if $s = 1$, this is the n-dimensional *Minkowski space*. If (M, g) is a semi-Riemannian manifold of dimension n and signature (s, r), then for each fixed p the metric tensor g restricted to $T_p M$ induces a semi-Euclidean metric on $T_p M$ and $(T_p M, g_p)$ is isometric to the standard $R^n{}_s$.

A diffeomorphism $f : M \to M$ is said to be a *conformal transformation* if there is a smooth nonvanishing function $\Omega : M \to R^1$ such that $g(f_* X, f_* Y) = \Omega^2 g(X, Y)$. If Ω is identically one, then f is an *isometry*.

If ∇ represents the usual Levi-Civita connection, then parallel displacement is defined in the usual way. The geodesics are the autoparallel curves and we say (M, g) is *geodesically complete* [3] iff all geodesic segments can be extended to have their domain to be all of R^1. In general semi-Riemannian manifolds one has timelike, null and spacelike geodesics. The manifold (M, g) is *timelike complete* if all timelike geodesics are complete. One defines *null* and *spacelike* (geodesic) *completeness* in similar fashions. Examples by Kundt [12], Geroch [7], and Beem [2] show that, in general, these three types of completenes are logically inequivalent. On the other hand, Lopez [13] has shown that for locally symmetric Lorentzian manifolds, these three types of geodesic completeness are equivalent.

A submanifold N of (M, g) is said to be *totally geodesic* if for each p in N and geodesic γ with $\gamma(0) = p$ and $\gamma'(0)$ tangent to N at p, the geodesic γ is contained in N in some neighborhood of p. A nondegenerate submanifold is totally geodesic iff the second fundamental form is identically zero on N, see [3, p. 55].

A linear subspace of $T_p M$ is nondegenerate if g restricted to this subspace is nondegenerate. Let R denote the curvature of (M, g). A nondegenerate two dimensional linear subspace E of $T_p M$ with basis vectors V and X has *sectional curvature* given by

$$K(p, E) = \frac{g(R(V, X)X, V)}{g(V, V)g(X, X) - [g(V, X)]^2} .$$

3. Isotropic Spaces

Let (M, g) be an isotropic semi-Riemannian manifold. It is well known [17, p. 368] that this implies that (M, g) is goedesically complete. If $\gamma : (-\infty, \infty) \to M$ is a unit speed timelike (respectively, spacelike) geodesic in an isotropic space-time and X is any unit timelike (respectively, spacelike) vector at $\gamma(0)$, then there is an isometry taking $\gamma'(0)$ to X. Thus any timelike (respectively, spacelike) geodesic at a fixed point may be mapped onto any other timelike (respectively, spacelike) geodesic at that point. Furthermore, if p and q are points of the isotropic space-time (M, g) and if there is a timelike (respectively spacelike) geodesic segment $\gamma : [a, b] \to M$ with $\gamma(a) = p$ and $\gamma(b) = q$, then the reflection across the midpoint $\gamma([a + b]/2)$ interchanges p and q. It follows that any timelike (respectively, space-like) geodesic at p may be mapped to any timelike (respectively, spacelike) geodesic through q. Since any two points of (M, g) can be joined by a finite sequence of timelike (or spacelike) geodesic segments we obtain the following result.

Lemma 3.1. If (M, g) is isotropic, then given any two inextendible unit speed timelike geodesics $\gamma_1 : (-\infty, \infty) \to M$ and $\gamma_2 : (-\infty, \infty) \to M$, then there is an isometry ϕ of (M, g) such that $\gamma_2(t) = \phi \circ \gamma_1(t)$ for all $-\infty < t < \infty$.

Let (M, g) be a space of constant sectional curvature such as Minkowski space $R^n{}_1$ and delete one point, the resulting space is an example of a locally isotropic (and locally symmetric) space which fails to be isotropic, fails to be symmetric, and fails to be complete.

Example 3.2. Let M be the upper half plane $\{(x, y) | y > 0\}$ with the metric $g = (dx \otimes dx - dy \otimes dy)y^{-2}$. This is an example of a homogeneous space-time which is locally isotropic and thus also locally symmetric. It is also incomplete, see, [14]. Hence, even the conditions of homogeneity and locally isotropic together do not imply geodesic completeness for Lorentzian manifolds.

Using the fact that for Lorentzian manifolds of dimension three and higher bounded sectional curvature of timelike planes implies constant sectional curvature, we may show that locally isotropic yields constant sectional curvature.

Proposition 3.3. Let (M, g) be a Lorentzian manifold. Assume that for each p in M, the collection of isometries of (M, g) which hold p fixed are transitive on the unit timelike vectors at p. In other words, assume that for any two unit timelike vectors at p there is an isometry taking the first to the second. Then (M, g) has constant sectional curvature.

Proof. If M has dimension two, then it is easy to show that each point p has a neighborhood U such that if q_1 and q_2 are points of U there is a local isometry taking q_1 and q_2. It follows that (M, g) has constant curvature on U. The connectedness of M, then implies (M, g) has constant curvature. Assume that M has dimension at least three. Fix p in M and a unit timelike vector X at p. All two dimensional linear subspaces of $T_p M$ which contain X must be timelike because X is timelike and (M, g) is Lorentzian. It follows that there is some constant C such that the absolute value $|K(p, E)|$ of the sectional curvature $K(p, E)$ is bounded by C for all two planes E containing X. Since sectional curvature is preserved under isometries, it follows that given any other unit timelike vector Y at p the sectional curvatures of all timelike planes containing Y must also be bounded in absolute value by this same constant C. It is known [5], [8] that bounded timelike sectional curvature implies constant sectional curvature for Lorentzian manifolds of dimension at least three. Thus (M, g) has constant sectional curvature at p and Schur's Theorem [17] together with the connectedness of M yields the result.

Corollary 3.4. Let (M, g) be a Lorentzian manifold. The manifold (M, g) is a locally isotropic iff (M, g) has constant sectional curvature.

We now define two point homogeneous.

Definition 3.5. Let (M, g) be a space-time. This space-time is *two-point homogeneous* if, whenever p_1, p_2 and q_1, q_2 are two point pairs with $0 < d(p_q, p_2) = d(q_1, q_2)$, there is an isometry of (M, g) taking p_1 to q_1 and p_2 to q_2.

Note that in Definition 3.5, we only require that there be an isometry when distances are strictly positive (i.e., $d(p_1, p_2) > 0$).

Proposition 3.6. Let (M, g) be a globally hyperbolic space-time. If (M, g) is isotropic, then it is two point homogeneous.

Proof. Assume that (M, g) is isotropic. Let p_1, p_2 and q_1, q_2 be two pairs of points with $0 < d(p_1, p_2) = d(q_1, q_2)$. Since (M, g) is globally hyperbolic, there is at least one timelike geodesic γ_1 from p_1 to p_2 and at least one timelike geodesic γ_2 from q_1 to q_2 with each of these geodesics unit speed of length $d(p_1, p_2)$. Hence, $L(\gamma_1) = L(\gamma_2) = d(p_1, p_2) = d(q_1, q_2)$. Extend γ_1 and γ_2 to complete geodesics γ_3 and γ_4, respectively, and apply Lemma 3.1. One obtains the existence of an isometry taking γ_3 to γ_4 and also the geodesic segment γ_1 to the segment γ_2. This establishes the result.

4. Bisectors

Given points p and q in a positive definite Riemannian manifold, the bisector of the two points is defined to be the set of points equidistant from the two points. In general, the bisector set is not a totally geodesic submanifold of the Riemannian space. Spaces with bisectors which are always totally geodesic have been characterized by Busemann [6] among a large class of (positive definite metric) spaces known as G-spaces.

For semi-Riemannian manifolds, in general, there is no globally defined notion of distance and, consequently, no meaningful global definition of bisector. However, one can use a locally defined notion of separation of points by using convex formal neighborhoods and length of geodesic segments on these neighborhoods, see Beem [1]. On the other hand, for space-times (and locally for Lorentzian manifolds that are not time orientable) one has the Lorentzian distance function. This is a useful tool for causally related points, but is always zero for point pairs which are not causally related. In [1] we established a bisector theorem for semi-Riemannian spaces using the locally defined separation approach mention above. In the present paper we use the Lorentzian distance function to establish a bisector theorem. Our assumption here is less than in the previous paper since we have only zero separation between point pairs which are not causally related. Furthermore, we must use a totally different method of proof. Let (M, g) be a Lorentzian manifold and let U be an open connected subset of M. Assume (U, g) is a space-time and that σ is the Lorentzian distance function for (U, g). Given any two points p, q of u with $\sigma(p, q) > 0$, the bisector $B(U, p, q)$ of the points p and q with respect to U is defined to be the points chronologically between p and q and equidistant from these two points. More precisely,

$$B(U, p, q) = \{r \in U \,|\, \sigma(p, r) > 0 \quad \text{and} \quad \sigma(p, r) = \sigma(r, q)\}.$$

When $M = U$, the bisector $B(M, p, q)$ is defined using the distance function d for all of (M, g). Note that if (M, g) is strongly causal, then each point has a sufficiently small convex normal neighborhood U such that the local distance function σ agrees with the original d of (M, g) on U. For all Lorentzian manifolds, the set U may be chosen as a convex normal neighborhood such that the local distance function σ is the distance function for (U, g). Note that in this case, two points p and q in U joined in U be a (unique) timelike goedesic segment will have the midpoint of this segment as the intersection of this geodesic segment and the bisector of the points. We have defined the Lorentzian distance function to be symmetric and thus it does not matter which time orientation is chosen for (U, g). In general, bisectors are not totally geodesic submanifolds. We now define the global and local flat bisector conditions by requiring the corresponding bisectors be totally geodesic.

Definition 4.1. (a) The space-time (M, g) satisfies the *global flat bisector condition* if, for each pair of points p, q in M with $d(p, q) > 0$, the bisector $B(M, p, q)$ is a totally geodesic submanifold of (M, g).

(b) The Lorentzian manifold (M, g) satisfies the *local flat bisector condition* if for each point of M there is a convex normal neighborhood U about the point such that for all p, q in U with $\sigma(p, q) > 0$, the bisector $B(U, p, q)$ is a totally geodesic submanifold of (U, g).

Consider a Lorentzian manifold (M, g) which satisfies the local flat bisector condition. Let (U, g) be as in Definition 4.1(b). If $p, q \in U$ and $\sigma(p, q) > 0$, then the midpoint r of the timelike geodesic segment $\alpha(p, q)$ lies in the bisector $B(U, p, q)$. Note that $2\sigma(p, r) = \sigma(p, q)$. If r' is any other point of $B(U, p, q)$, then $\sigma(p, r') = \sigma(r', q)$ and the reverse triangle inequality $\sigma(p, r') + \sigma(r', q) < \sigma(p, q)$ yields the inequality $2\sigma(p, r') < \sigma(p, q) = 2\sigma(p, r)$. Hence, r is the unique farthest point from p

on $B(U, p, q)$. It now follows using transversality theory from the calculus of variations applied to the length integral that $B(U, p, q)$ must be orthogonal to the geodesic segment $\sigma(p, q)$ at the point r. In fact, the flat bisector requirement yields that sufficiently near r this bisector must be the collection of (spacelike) geodesics through r which are orthogonal at r to the geodesic segment $\alpha(p, q)$. Note that if r' is any point of $B(U, p, q)$, then one may consider the unit speed geodesic segment $\beta : (-\varepsilon, \varepsilon) \to U$ of (U, g) orthogonal to the bisector at the point r' with $\beta(0) = r'$. For each fixed and sufficiently small t_0 the bisector of the points $\beta(-t_0)$ and $\beta(t_0)$ must be a subset of this same set $B(U, p, q)$ because this is the unique totally geodesic submanifold orthogonal to β at r'. In other words, $B(U, p, q) \supseteq B(U, \beta(-t_0), \beta(t_0))$. Points of β sufficiently near r' may then be reflected across the set $B(U, p, q)$ by $\beta(t) \to \beta(-t)$. Letting r' traverse the set $B(U, p, q)$ one obtains a (locally defined) reflection across this bisector. This is a smooth map near r and its derivative map has full rank at r. The Inverse Function Theorem yields that this reflection is a diffeomorphism sufficiently near r. The next result shows that this reflection is a locally defined isometry.

Lemma 4.2. Let (M, g) be a Lorentzian manifold which satisfies the local flat bisector condition. Let (U, g) be a convex normal neighborhood satisfying the bisector condition and let σ be the Lorentzian distance function for (U, g). If $\sigma(p, q) > 0$, then there is a locally defined isometry f across the bisector $B(U, p, q)$ with the midpoint r of $\alpha(p, q)$ in the domain of f.

Proof. Let f denote the local reflection across $B(U, p, q)$ defined above and assume without loss of generality that f is a diffeomorphism onto its image. Assume further that f is defined on a globally hyperbolic subset W of U with $r \in W$. Choose points p' and q' on the geodesic segment $\alpha(p, q)$ with $p', q' \in W$ and with r the midpoint of the geodesic segment $\alpha(p', q')$. Let W be time oriented with $p' < q'$. Note that the reflection f takes p' to q' and holds the points of the local bisector $B(W, p', q')$ fixed. It follows that f maps the chronological future of p' (with respect to W) onto the chronological past of q'. If $r' \in B(W, p', q')$ let $\beta : (-\varepsilon, \varepsilon) \to W$ be the unit speed future directed timelike geodesic orthogonal to $B(W, p', q')$ with $\beta(0) = r'$. By the same reasoning as for p' and q', we find that for sufficiently small t the map f takes the chronological future of $\beta(-t)$ onto the chronological past of $\beta(t)$ for all sufficiently small positive t. Since f takes chronological futures onto chronological pasts, it takes null vectors onto null vectors. Hence, the pull back metric $f^*(g)$ has the same null directions as the original metric g on W. Using the well known fact that the null directions determine the metric up to a conformal factor [9], it follows that f is conformal. In other words, there is some positive real valued function Ω defined on W which satisfies $g(f_*X, f_*Y) = \Omega^2 g(X, Y)$. To show that Ω is identically equal to one, note that each point is mapped to a point of the orthogonal geodesic at equal distance from the bisector under the reflection f. This implies length is preserved along the geodesics orthogonal to the original spacelike bisector and thus unit tangent vectors to these geodesics are mapped to unit vectors. It follows that f is a local isometry, as desired.

Theorem 4.3. If (M, g) is a Lorentzian manifold, then (M, g) satisfies the local flat bisector condition iff (M, g) has constant sectional curvature.

Proof. If (M, g) has constant sectional curvature, then (M, g) is locally isometric to the model spaces [17] and it is easily seen that locally (M, g) has flat bisectors. Assume that (M, g) satisfies the locally flat bisector condition. Fix p and let U be a convex normal neighborhood such that (U, g) satisfies the global flat bisector condition. Let X be a fixed unit timelike vector at p. The sectional curvature is bounded both below and above on the collection of all (two dimensional) planes which contain X. Let these bounds be denoted by C_1 and C_2. Let Y be a second unit timelike vector at p. Assume for convenience that a time orientation has been chosen for (U, g) and that both X and Y are future directed. The future directed timelike vectors form a convex cone in T_pM and consequently the sum $X + Y$ is also a future directed timelike vector at p. Note that the vector $X + Y$ makes equal semi-Euclidean angles with X and Y. Let H denote the orthogonal complement of $(X + Y)$ in T_pM. Then H is a spacelike hyperplane. Choose an $\varepsilon > 0$ such that the images q_1 and q_2 respectively of $\exp_p(\varepsilon(X + Y))$ and $\exp_p(-\varepsilon(X + Y))$ lie in U. Then the bisector $B(U, q_1, q_2)$ of q_1 and q_2 must be tangent to the hyperplane H at p. The reflection in $B(U, q_1, q_2)$ induces a tangent map on T_pM which is the identity on H and takes the vector $X + Y$ to its negative $-(X + Y)$. It maps the plane of X and Y onto itself and preserves semi-Euclidean angles. Consequently, it maps X to $-Y$ and Y to $-X$. Thus, the sectional curvatures of the collection of all timelike planes through Y (or $-Y$) are bounded by the same bounds C_1 and C_2 used for X. It follows that all timelike planes at p have sectional curvature between C_1 and C_2. However, this implies that the sectional curvature is constant [5], [8] and by Schur's Theorem [17], one finds that (M, g) has constant sectional curvature.

If (M, g) is a globally hyperbolic space-time, then the Lorentzian distance function $d : M \times M \to R^1$ is continuous and $d(p, q)$ is finite valued for each fixed pair of points p, q of M. A complete unit speed timelike geodesic $\gamma : (-\infty, \infty) \to M$ is called a *timelike line* if it is distance maximizing between each pair of points, in other words, if $d(\gamma(t_1), \gamma(t_2)) = |t_1 - t_2|$ for all t_1, t_2 in R^1. It is known [4] that a globally hyperbolic space-time with nonpositive timelike sectional curvature and a complete timelike line splits into a metric product of the form $R^1 \times H$ where R^1 has the negative definite metric $-dt^2$ and where (H, h) is a complete positive definite Riemannian manifold. Using this splitting theorem and Theorem 4.3, one obtains the following corollary.

Corollary 4.4. Let (M, g) be a globally hyperbolic space-time with a complete timelike line and assume that (M, g) satisfies the local flat bisector condition. If each timelike two plane of (M, g) has nonpositive sectional curvature, then (M, g) is flat. Furthermore, there is a complete and flat positive definite Riemannian manifold (H, h) such that $M = R^1 \times H$ and $g = (-dt^2) \otimes h$.

Proof. Theorem 4.3 shows that (M, g) has constant sectional curvature K and Theorem 5.2 of [4, p. 41] shows that M splits into a product $R^1 \times H$ with $g = -(dt^2) \otimes h$ and (H, h) complete. In order to show that $K = 0$, we regard M as a warped product of the form $R^1 \times_f H$ with f identically equal to one. Using coordinates $x_1 = t$ on R^1 and $x_2, \ldots, x_n$ on H, we set $X = \partial/\partial t$ and $V = \partial/\partial x_2$. From Proposition 42 of O'Neill [14, p. 210], we obtain $R(V, X)X = 0$. Thus $K = g(R(V, X)X, V) [g(X, X)g(V, V) - g(X, V))^2]^{-1} = 0$, as desired.

We conclude with a theorem containing several characterizations of constant curvature for Lorentzian manifolds. This theorem combines some of the results of the present paper and theorems from Kulkarni [11] and Harris [8].

Theorem 4.5. Let (M, g) be a Lorentzian manifold. The following three conditions are equivalent.

(a) The manifold (M, g) has constant sectional curvature.

(b) The manifold (M, g) is locally isotropic.

(c) The manifold (M, g) satisfies the local flat bisector condition.

Furthermore, if M has dimension at least three, then the above conditions are also equivalent to each of the following conditions.

(d) The nondegenerate sections have sectional curvatures bounded above at each point.

(e) The nondegenerate sections have sectional curvatures bounded below at each point.

(f) There is both an upper and lower bound on the sectional curvatures of timelike sections at each point.

References

1. J. K. Beem, *Pseudo-Riemannian manifolds with totally geodesic bisectors*, Proc. Amer. Math. Soc. **49** (1975) 212–215.
2. J. K. Beem, *Some examples of incomplete space-times*, Gen. Rel. Grav. **7** (1976) 501–509.
3. J. K. Beem and P. E. Ehrlich, *Global Lorentzian Geometry*, Pure and Applied Math. vol. 67, Marcel Dekker, New York, 1981.
4. J. K. Beem, P. E. Ehrlich, S. Markvorsen and G. J. Galloway, *Decomposition theorems for Lorentzian manifolds with nonpositive curvature*, J. Diff. Geo. **22** (1985) 29–42.
5. J. K. Beem and P. E. Parker, *Values of pseudoriemannian sectional curvature*, Comm. Math. Helv. **59** (1984) 319–331.
6. H. Busemann, *The Geometry of Geodesics*, Academic Press, New York, 1955.
7. R. P. Geroch, *What is a singularity in general relativity*, Ann. Phys. (NY) **48** (1968) 526–540.
8. S. G. Harris, *A triangle comparison theorem for Lorentz manifolds*, Indiana Math. J. **31** (1982), 289–308.
9. S. W. Hawking and G. F. R. Ellis, *The Large Scale Structure of Space-time*, Cambridge University Press, Cambridge, 1973.
10. H. Hopf and W. Rinow, *Über den begriff des vollständigen differentialgeometrischen Fläche*, Comm. Math. Helv. **3** (1931) 209–225.

11. R. S. Kulkarni, *The value of sectional curvature of an indefinite metric*, Comm. Math. Helv. **54** (1979) 173–176.

12. W. Kundt, *Note on completeness of spacetimes*, Zs. für Phys. **172** (1963) 488–489.

13. J. Lafuente Lopez, *A geodesic completeness theorem for locally symmetric Lorentz manifolds*, Revista Mat. Univ. Complut. Madrid 1 (1988) 101–110.

14. B. O'Neill, *Semi-Riemannian Manifolds*, Academic Press, New York, 1983.

15. J. Tits, *Sur certaines classes d'espaces homogenes de groups de Lie*, Memoir Belgian Academy of Sciences, 1955.

16. H. C. Wang, *Two-point homogeneous spaces*, Ann. of Math. **55** (1952) 177–191.

17. J. A. Wolf, *Spaces of Constant Curvature*, 3rd Edition, Publish or Perish Press, Boston, 1974.

John K. Beem
Department of Mathematics
University of Missouri-Columbia
Columbia, MO 65211
USA

THE MATH. HERITAGE OF C.F. GAUSS (pp. 66-99)
edited by George M. Rassias
©1991 World Scientific Publ. Co. Singapore

THE DECOMPOSITION THEORY
AND ITS APPLICATIONS

Neda Bokan

The main purpose of this paper is to give a survey about the decomposition theory of some tensors spaces $\otimes^r V^*$ or their subspaces of a vector space V (V^* is the dual space of a vector space V). More precisely, we show how it is possible to represent some subspaces of $\otimes^r V^*$ as the direct sum of irreducible components under the action of any group G. We choose subspaces of $\otimes^r V^*$ having some geometrical sense and explain why these decompositions do provide insight in some problems of differential geometry, topology etc.

Introduction

Some historians of mathematics (see [STRU]) find that C. F. Gauss is so significant in mathematics as G. W. F. Hegel in philosophy, L. V. Beethoven in music or J. W. Goethe in literature. There is nobody interested in differential geometry who does not know for the notion of the Gauss curvature of a surface and the sectional curvature for a n-dimensional Riemannian manifold, as its generalization. It is also well known that it is possible to determine the Riemannian curvature in terms of a sectional curvature. But, a curvature, as a basic notion in differential geometry, could be possibly considered from different points of view.

The main purpose of this paper is to give a survey about the decomposition theory, or more precisely to explain how it is possible to consider a curvature from the algebraic point of view and why it does provide insight in some problems of differential geometry, topology etc. Let us mention that in this spirit it is possible to study the various curvatures which appear in differential geometry in different contexts (see [KU]).

Moreover, we represent some of the other applications of the decomposition theory to obtain some classification of almost Hermitian manifolds [GR-HE], Riemannian homogeneous structures [TR-VA 3] etc.

Of course all these decompositions are, in principle, consequences of general theorems of group representations (see [WE 2]).

This paper is divided into 3 Chapters:

I Decomposition of a vector space of curvature tensors;

II A decomposition of some space of covariant tensors of order 3;

III Homogeneous structures.

We start in Sec. 1. of Chapter I with a decomposition of the vector space $\mathcal{R}(V)$ consisting of all symmetric linear transformations of the space of 2-vectors of n-dimensional vector space V. This splitting is considered under the action of the orthogonal group $O(V)$. Using this decomposition we characterize, following [SI-THO] an Einstein space. We consider also different applications of this decomposition.

Section 2 is devoted to decompositions of the vector space of Riemannian curvature tensors and the action of the group $U(V)$ [TR-VA 2], including some applications of them.

We consider in the last section of this Chapter decompositions of the vector space of curvature tensors corresponding to a symmetric connection. These curvature tensors are related to projective, affine, holomorphically projective transformation etc. Some connections between these decompositions and groups of transformations are considered too (see [BO 3], [BO 4]).

Chapter II is devoted to the decompositions of a space of covariant tensors of the order 3 which can be geometrically interpreted as the space of tensors which satisfy the same identities as the covariant derivative of 2-form of some manifold with special structure defined by a field of endomorphisms on its tangent bundle. In Sec. 1 we consider in this spirit almost Hermitian manifolds (see [GR-HE]) and in Sec. 2 Riemannian almost-product manifolds [NA]. The most interesting applications of these decompositions are classifications of manifolds in certain sense.

Homogeneous structures are studied in the spirit of the decomposition of a certain vector space in Chapter III. Specially, Riemannian homogeneous structures are treated in Sec. 1 and almost Hermitian homogeneous structures in Sec. 2. We consider the action of the groups $O(V)$ and $U(V)$ respectively (see [TR-VA 3] and [AB-GA]).

I. Decompositions of Vector Spaces of Curvature Tensors

Let V be an n-dimensional vector space and denote by $\mathcal{R}(V)$ the vector space of curvature tensors over V. The development of the theory of the decomposition of $\mathcal{R}(V)$ under the action of some group was initiated by Singer and Thorpe [SI-THO]. Since these results and ideas of Singer and Thorpe are very useful in the studies of some problems in geometry and topology of manifolds, many mathematicians have worked using this algebraic treatment of curvature tensors. Some results are presented in this Chapter.

1. *Curvature Operators and the Action of the Groups $O(V)$ and $SO(V)$*

In the well-known paper [SI-THO] the authors considered the vector space $\mathcal{R}(V)$ consisting all symmetric linear transformations of the space of 2-vectors of n-dimensional vector space V. All tensors having the same symmetries as the Riemannian curvature tensor including the first Bianchi identity also belong to $\mathcal{R}(V)$. They gave a geometrically useful description of the splitting of $\mathcal{R}(V)$ under the action of $O(V)$ into four components. One of the projections gives the Weyl conformal tensor. Namely, their considerations are as follows:

Let V be an n-dimensional vector space over $\mathbb{R}$ with inner product $\langle\,,\,\rangle$. Let Λ^2 denote the space of 2-vectors of V, with inner product given by

$$\langle u_1 \wedge u_2, v_1 \wedge v_2 \rangle = \det[\langle u_i, v_j \rangle], \quad (u_i, v_i \in V).$$

A curvature tensor on V is a symmetric linear transformation $R : \Lambda^2 \to \Lambda^2$. The curvature tensors on V form a vector space $\mathcal{R}(V)$ with inner product given by $\langle R, S \rangle = \operatorname{trace} RS$. Using the usual isomorphisms defined by the inner product, a curvature tensor on V may be regarded as a 2-form on V with values in the vector space of skew symmetric endomorphisms of V. The sectional curvature σ_R of $R \in \mathcal{R}(V)$ is the real valued function defined on Grassman manifold G of oriented 2-dimensional subspaces of V by $\sigma_R(P) = \langle RP, P \rangle$ $(P \in G \subset \Lambda^2)$ where G is identified with the set of decomposable 2-vectors of length 1. The orthogonal group $O(V)$ of V acts isometrically on Λ^2 by $g(u \wedge v) = (gu) \wedge (gv), g \in O(V), u, v \in V. O(V)$ also acts isometrically on $\mathcal{R}(V)$ by $g(R) = g^{-1}Rg : \Lambda^2 \to \Lambda^2$.

The Bianchi map $b : \mathcal{R}(V) \to \mathcal{R}(V)$ is defined by

$$[b(R)](u_1, u_2)(u_3) = \sum_\alpha R(u_{\alpha(1)}, u_{\alpha(2)})(u_{\alpha(3)})$$

where α runs through all cyclic permutations of $(1,2,3)$. The Ricci tensor is the linear map $Q : \mathcal{R}(V) \to \mathcal{J}(V)$, where $\mathcal{J}(V)$ is the space of symmetric linear transformations of V, defined by

$$\rho(V)(v,w) = \langle Q(R)(v), w \rangle = \mathrm{tr} \left\{ u \to R(v,u)(w) \right\}.$$

The trace functional or the scalar curvature τ is the linear map $\tau \colon \mathcal{R}(V) \to$ $\mathbb{R}$. Each of these maps is equivariant with respect to the $O(V)$-action on the spaces involved ($O(V)$ acts trivially on $\mathbb{R}$) and hence the kernel of each of these maps is an $O(V)$-invariant subspace of $\mathcal{R}(V)$. Since $O(V)$ acts isometrically on $\mathcal{R}(V)$, orthogonal complements of invariant subspaces are invariant. If we put

$$\mathcal{R}_1 = (\ker b)^{\perp}, \qquad\qquad \mathcal{R}_2 = (\ker \tau)^{\perp},$$
$$\mathcal{R}_3 = (\ker Q) \cap (\ker b), \qquad \mathcal{R}_4 = (\ker Q)^{\perp} \cap (\ker \tau),$$

we have the decomposition of $\mathcal{R}(V)$ given in the following theorem.

Theorem 1.1. $\mathcal{R}(V) = \mathcal{R}_1 \oplus \mathcal{R}_2 \oplus \mathcal{R}_3 \oplus \mathcal{R}_4$. Moreover,

(i) $R \in \mathcal{R}_1$ if and only if the sectional curvature of $\mathcal{R}$ is identically zero.

(ii) $R \in \mathcal{R}_1 \oplus \mathcal{R}_2$ if and only if the sectional curvature of R is constant.

(iii) $R \in \mathcal{R}_1 \oplus \mathcal{R}_3$ if and only if the Ricci tensor of R is zero.

(iv) $R \in \mathcal{R}_1 \oplus \mathcal{R}_2 \oplus \mathcal{R}_3$ if and only if the Ricci tensor of R is scalar multiple of the identity.

(v) $R \in \mathcal{R}_1 \oplus \mathcal{R}_3 \oplus \mathcal{R}_4$ if and only if the scalar curvature of R is zero.

(vi) $R \in \mathcal{R}_2 \oplus \mathcal{R}_3 \oplus \mathcal{R}_4$ if and only if R satisfies the first Bianchi identity.

Furthermore, the action of $O(V)$ on $\mathcal{R}(V)$ is irreducible on each $\mathcal{R}_i$, $i = 1, 2, 3, 4$.

Let us mention that the $\mathcal{R}_3$-component of a curvature tensor is its Weyl conformal curvature tensor.

Since a curvature tensor R, corresponding to the Levi-Civita connection ∇ of a Riemannian manifold $\mathbb{M}^n$ satisfies the Bianchi identity we have $R \in \mathcal{R}_1^{\perp}$. Statement (iv) of Theorem 1.1 implies a very nice characterization of an Einstein space as follows: a Riemannian manifold $\mathbb{M}$ has curvature tensor in $\mathcal{E} = \mathcal{R}_2 \oplus \mathcal{R}_3$ at each point if and only if $\mathbb{M}$ is an Einstein space.

Now suppose $n = 4$ and let V be given a definite orientation. The star operator $* : \Lambda^2 \to \Lambda^2$ is defined by $\langle *\alpha, \beta \rangle \omega = \alpha \wedge \beta$, where ω is the generator for Λ^4 determined by the inner product and the orientation of V. For $P \in G$, $*P$ is the oriented orthogonal complement $P^\perp$ of P in V. $*$ is symmetric, $*^2 = I$, and $\Lambda^2 = \Lambda^+ \oplus \Lambda^-$ (orthogonal direct sum) where $\Lambda^\pm$ are the (3-dimensional) eigen-spaces of $*$ corresponding to the eigen-values ± 1. Now we have the following results.

Theorem 1.2. Let V be oriented and of dimension 4. Then $\mathcal{R}_1$ consists of all scalar multiples of $*$; $\mathcal{R}_2$ consists of all scalar multiples of I; $R \in \mathcal{R}_3 \iff *R = R*, \tau(R) = 0$, and $\tau(*R) = 0$, and $R \in \mathcal{R}_4 \iff *R = -R*$. In particular, $R \in \mathcal{E} \iff R* = *R$ and $b(R) = 0$.

Moreover, the action of $SO(V)$ on $\mathcal{R}(V) = \mathcal{R}_1 \oplus \mathcal{R}_2 \oplus \mathcal{R}_3^+ \oplus \mathcal{R}_3^- \oplus \mathcal{R}_4$ is irreducible on each of these summands, where the 5-dimensional $SO(V)$-invariant subspaces are defined by $\mathcal{R}_3^\pm = \{R \in \mathcal{R} | R* = \pm R\}$ and hence $\mathcal{R}_3 = \mathcal{R}_3^+ \oplus \mathcal{R}_3^-$.

The decomposition of $\mathcal{R}(V)$ from Theorem 1.1 and specifically from Theorem 1.2 are very useful for the study of Riemannian geometry, topology etc. Let us mention some of these results.

For a compact oriented 4-dimensional Riemannian manifold, the Euler-Poincaré characteristic $\mathcal{N}(\mathbb{M})$ is given by the generalized Gauss-Bonnet formula [CH] which in our terms reads

$$\mathcal{N}(M) = \frac{3}{4\pi^2} \int_M \tau(*R) dV \ .$$

Hence, using Theorem 1.2 we have Berger's result [BER] that the Euler-Poincaré characteristic of a compact 4-dimensional Einstein space is non-negative and is in fact strictly positive unless $\mathbb{M}$ is flat.

It is well known that the diagonalization or normal form of a self adjoint linear transformation T on a real inner product space $\{V, \langle \ , \ \rangle\}$ is equivalent to the analysis of the critical point behavior of the function $v \to \langle Tv, v \rangle$ on the unit sphere (actually projective space) of V. As we have mentioned already the curvature tensor at a point of a Riemannian manifold is completely determined by the sectional curvature σ on the Grassman manifold of 2-planes at the point.

Having in mind these results and using Theorem 1.2, Singer and Thorpe [SI-THO] studied the problem of a "normal form" for the curvature tensor of a 4-dimensional oriented Einstein manifold by analyzing the critical point behavior of the sectional curvature function σ. In this case, the function σ on each 2-plane is equal to its value on the orthogonal complement. Using this characterization, they have shown that the curvature function σ is completely determined by its critical point behavior and they have shown what the locus of critical points looks like.

The relationship between the Euler-Poincaré characteristic $\mathcal{N}(\mathbb{M})$, the Hirzebruch signature $\tau(\mathbb{M})$, and the arithmetic genus $\alpha(\mathbb{M})$, and the decomposition of the space of curvature operators at a point of 4-dimensional compact Riemannian manifold has been studied by Gray [GR-3]. So, he has proved the following theorem.

Theorem 1.3. Let $\mathbb{M}$ be a compact oriented 4-dimensional Einstein manifold. Then

(i)
$$|\tau(\mathbb{M})| \leq \frac{2}{3}\mathcal{N}(\mathbb{M}) - \frac{\tau^2}{288\pi^2}V(\mathbb{M})\,;$$

(ii) if $R \in \mathcal{R}_3^- \oplus \mathcal{R}_2$

$$\tau(\mathbb{M}) = -\frac{2}{3}\mathcal{N}(\mathbb{M}) + \frac{\tau^2}{288\pi^2}V(\mathbb{M})\,;$$

(iii) if $R \in \mathcal{R}_2 \oplus \mathcal{R}_3^+$

$$\tau(\mathbb{M}) = -\frac{2}{3}\mathcal{N}(\mathbb{M}) - \frac{\tau^2}{288\pi^2}V(\mathbb{M})\,,$$

where $\tau = \tau(R)$ is the scalar curvature and $V(\mathbb{M})$ denotes the volume of $\mathbb{M}$.

Hitchin [HI] has also studied inequalities between the signature τ and the Euler-Poincaré characteristic $\mathcal{N}$ of a four-dimensional Einstein manifold.

Using the decomposition of the space of curvature operators Xin [X] has studied characteristic classes in some kinds of Einstein manifolds.

Applications of the decomposition of $\mathcal{R}(V)$, especially of $\mathcal{R}_1^\perp$, involving orthogonal Radon transformations were given by Strichartz [STRI].

Using an algebraic interpretation of the Weyl conformal curvature tensor C due to Singer and Thorpe [SI-THO], Kowalski [KOW] has achieved the generalization of the tensor C for an arbitrary Riemannian vector bundle $E \to \mathbb{M}$ (provided with a metric connection and a solder map $j : T(\mathbb{M}) \to E$) dim $\mathbb{M} \geq 3$. Moreover, if dim $\mathbb{M} \geq 4$, it was shown that all conformally Euclidean vector bundles over $\mathbb{M}$ are characterized by the condition $C = 0$. A simple result also holds for dim $\mathbb{M} = 3$. It is a generalization of the well-known theorem about the characterization of a conformally Euclidean vector bundle by the condition $C = 0$, proved by Schouten [SCHO 1]. This makes possible the development of the theory of submanifolds in conformal differential geometry more up to date (see [KOW]).

Nomizu [NO 1] used the decomposition of Singer and Thorpe [SI-THO] to study generalized curvature tensor fields; in particular he studied proper tensor fields on a Riemannian manifold.

Let V be a finite dimensional vector space with inner product g, and let ∇Curv denote the space of tensors having all the symmetries of the first covariant derivative of a Riemann curvature tensor. Strichartz [STRI] found a decomposition of ∇Curv into irreducible components under the action of $O(V)$ (see also [GR-VA 2]).

2. Riemannian Curvature Tensors and the Action of the Group $U(V)$

When V is a $2n$-dimensional real vector space endowed with a complex structure J compatible with a positive definite inner product $\langle \ , \ \rangle$, let $\mathcal{R}_1(V)$ denote the subspace of the space $\mathcal{R}(V)$ of Riemannian curvature tensors, consisting of tensors satisfying the Kähler identity. The complete decomposition of $\mathcal{R}_1(V)$ under the action of $U(V)$ was treated by Sitaramayya [SIT] (see also [JO], [MO]).

Tricerri and Vanhecke [TR-VA 2] have found the complete decomposition of $\mathcal{R}(V)$ under the action of $U(V)$. Before giving this decomposition we introduce three particular subspaces $\mathcal{R}_i(V), i = 1, 2, 3$, of $\mathcal{R}(V)$, as follows [GR 4]:

$$\mathcal{R}_1(V) = \{R \in \mathcal{R}(V) | R(x, y, z, w) = R(x, y, Jz, Jw)\} ;$$
$$\mathcal{R}_2(V) = \{R \in \mathcal{R}(V) | R(x, y, z, w) = R(Jx, Jy, z, w)$$
$$+ R(Jx, y, Jz, w) + R(Jx, y, z, Jw)\} ;$$
$$\mathcal{R}_3(V) = \{R \in \mathcal{R}(V) | R(x, y, z, w) = R(Jx, Jy, Jz, Jw)\} .$$

Since all of these subspaces are invariant under the action of $U(V)$ and we have $\mathcal{R}_1(V) \subset \mathcal{R}_2(v) \subset \mathcal{R}_3(V)$ (see for example [GR 4], [GR-VA 1]) they are convenient for the consideration of a decomposition. Let us put further

$$\mathcal{R}_1^{\perp}(V) = \text{ orthogonal complement of } \mathcal{R}_1(V) \text{ in } \mathcal{R}_2(V);$$
$$\mathcal{R}_2^{\perp}(V) = \text{ orthogonal complement of } \mathcal{R}_2(V) \text{ in } \mathcal{R}_3(V);$$
$$\mathcal{R}_3^{\perp}(V) = \text{ orthogonal complement of } \mathcal{R}_3(V) \text{ in } \mathcal{R}(V);$$
$$\pi_1(x,y)z = \langle x,z\rangle y - \langle y,z\rangle x;$$
$$\pi_2(x,y)z = 2\langle Jx,y\rangle Jz + \langle Jx,z\rangle Jy - \langle Jy,z\rangle Jx;$$

for all $x,y,z, \in V$.

For a tensor $R \in \mathcal{R}(V)$ we have as usual the Ricci tensor $\rho(R)$ of type $(0,2)$, the Ricci tensor $Q = Q(R)$ and the scalar curvature $\tau = \tau(R)$, and in additiion the Ricci $*$-tensor $\rho^*(R)$ of type $(0,2)$, respectively $Q^* = Q^*(R)$ of type $(1,1)$ and the $*$-scalar curvature $\tau^* = \tau^*(R)$ defined by

$$\rho^*(R)(x,y) = \langle Q^* x, y\rangle = \text{trace } (z \in V \to R(Jz,x)Jy \in V),$$

and

$$\tau^* = \text{trace } Q^*.$$

Now it makes sense to introduce all the components of the decomposition.

$$\mathcal{W}_1 = \mathcal{L}(\pi_1 + \pi_2),$$
$$\mathcal{W}_3 = \{R \in \mathcal{R}_1(V) | \rho(R) = 0\},$$
$$\mathcal{W}_2 = \text{ orthogonal complement of } \mathcal{W}_1 \oplus \mathcal{W}_3 \text{ in } \mathcal{R}_1(V),$$
$$\mathcal{W}_4 = \mathcal{L}(3\pi_1 - \pi_2) \subset \mathcal{R}_1^{\perp}(V),$$
$$\mathcal{W}_6 = \{R \in \mathcal{R}_1^{\perp}(V) | \rho(R) = 0\},$$
$$\mathcal{W}_5 = \text{ orthogonal complement of } \mathcal{W}_4 \oplus \mathcal{W}_6 \text{ in } \mathcal{R}_1^{\perp}(V),$$
$$\mathcal{W}_7 = \mathcal{R}_2^{\perp}(V),$$
$$\mathcal{W}_{10} = \{R \in \mathcal{R}_3^{\perp}(V) | \rho(R) = \rho^*(R) = 0\},$$
$$\mathcal{W}_8 \oplus \mathcal{W}_9 = \text{ orthogonal complement of } \mathcal{W}_{10} \text{ in } \mathcal{R}_3^{\perp}(V),$$
$$\mathcal{W}_8 = \{R \in \mathcal{W}_8 \oplus \mathcal{W}_9 | \rho^*(R) = 0\},$$
$$\mathcal{W}_9 = \text{ orthogonal complement of } \mathcal{W}_8 \text{ in } \mathcal{W}_8 \oplus \mathcal{W}_9.$$

So, we can state the decomposition theorem for $\mathcal{R}(V)$.

Theorem 2.1.
$$\mathcal{R}(V) = \mathcal{W}_1 \oplus \ldots \oplus \mathcal{W}_{10}, \qquad (2.1)$$

where $\mathcal{W}_i$ are orthogonal invariant subspaces. Moreover, we have
  (i)   The decomposition (2.1) is irreducible for $n \geq 4$.
  (ii) For $n = 3, \mathcal{W}_6 = \{0\}$ and the other factors in (2.1) are irreducible.
  (iii) For $n = 2$, $\mathcal{W}_5 = \mathcal{W}_6 = \mathcal{W}_{10} = \{0\}$ and the other factors in (2.1) are irreducible.

Tricerri and Vanhecke [TR-VA 2] introduced the Bochner component $\mathcal{R}_B(V)$ of $\mathcal{R}(V)$ as the subspace formed by the curvature tensors $R \in \mathcal{R}(V)$ such that $\rho(R) = \rho^*(R) = 0$. A tensor belonging to $\mathcal{R}_B(V)$ is called a Bochner tensor and the projection $B(R)$ of $R \in \mathcal{R}(V)$ on $\mathcal{R}_B(V)$ is called the Bochner conformal tensor associated with R. The method to define this tensor is a natural generalization of the Kähler case.

These results are very useful for the study of almost Hermitian manifolds. So, in [TR-VA 2] it was proved that the Bochner tensor associated with the Riemannian curvature tensor is a conformal invariant tensor. Further, the decomposition was illustrated by considering the Calabi-Eckmann manifolds and a Hopf manifold. This generalized Bochner tensor was treated specially on four-dimensional almost Hermitian manifold. It was so proved, for example, that on a Hermitian manifold the Bochner tensor is just the anti-self-dual part of the Weyl conformal tensor.

Applications of the decomposition theorem of Mori [MO] have been studied by Stanilov [STA 1]. Using the decomposition theorem of Tricerri and Vanhecke [TR-VA 2] some results have been obtained by Stanilov [STA 3], [STA 2], Hopteriev [HO].

The decomposition of a vector space of Riemannian curvature tensors of some manifolds under the action of other groups have also been treated by some mathematicians (see for example [CA], [PA-HO], [TR-VA 1], [JA-VA], [G-G-M 2], [GAU]).

3. *Curvature Tensors Corresponding to a Symmetric Connection and the Action of Some Groups*

Let V be an n-dimensional vector space over $\mathbb{R}$ with positive definite inner product $\langle \, , \, \rangle$. A tensor R of type $(1,3)$ over V is a bilinear mapping

$$R : V \times V \to \mathrm{Hom}\,(V, V) : (x, y) \to R(x, y)\,.$$

R is called a curvature tensor over V if it has the following properties for all $x, y, z, w \in V$:

(i) $R(x, y) = -R(y, x)$,

(ii) the first Bianchi identity, i.e., $\sigma R(x, y)z = 0$ where σ denotes the cyclic sum with respect to x, y and z.

We also use the notation

$$R(x, y, z, w) = \langle R(x, y)z, w \rangle\,. \tag{3.1}$$

We denote by $\mathcal{R}(V)$ the vector space of all curvature tensors over V. $\mathcal{R}(V)$ can be geometrically interpreted as the space of tensors which satisfy the same identities as curvature tensors corresponding to an arbitrary symmetric connection ∇ on a Riemannian manifold.

In addition to the Ricci tensor $\rho(R)$ for a curvature tensor $R \in \mathcal{R}(V)$ it makes sense to define the second trace $\hat{\rho}(R)$ by

$$\hat{\rho}(R)(x, y) = \sum_{i=1}^{n} R(e_i, x, e_i, y)\,, \quad x, y \in V\,,$$

where $\{e_i\}$ is an arbitrary orthonormal basis of V. The traces $\rho(R)$ and $\hat{\rho}(R)$ are orthogonal. Moreover, they are neither symmetric nor skew-symmetric in general case. The scalar curvature $\tau = \tau(R)$ of R is defined as the trace of $Q = Q(R)$, given by $\rho(R)(x, y) = \langle Qx, y \rangle$.

Now we can define all the components of the decomposition of $\mathcal{R}(V)$. We put

$$\mathcal{R}^a(V) = \{R \in \mathcal{R}(V) | \rho(R) \text{ and } \hat{\rho}(R) \text{ are skew-symmetric}\}\,,$$

$$\mathcal{R}^s(V) = \{R \in \mathcal{R}(V) | \rho(R) \text{ and } \hat{\rho}(R) \text{ are symmetric}\}\,,$$

$$\mathcal{R}_p(V) = \{R \in \mathcal{R}(V) | \rho(R) \text{ is zero}\}\,.$$

$$\mathcal{R}_0(V) = \mathcal{R}^a(V) \cap \mathcal{R}^s(V) = \{R \in \mathcal{R}(V) | \rho(R) \text{ and } \hat{\rho}(R) \text{ are zero}\}\,,$$

$$\mathcal{W}_4 = \text{ orthogonal complement of } \mathcal{R}_0(V) \text{ in } \mathcal{R}_p(V) \cap \mathcal{R}^a(V)\,,$$

$$\mathcal{W}_5 = \text{ orthogonal complement of } \mathcal{R}_0(V) \text{ in } \mathcal{R}_p(V) \cap \mathcal{R}^s(V)\,,$$

$\mathcal{W}_3 = $ orthogonal complement of $\mathcal{R}_p(V) \cap \mathcal{R}^a(V)$ in $\mathcal{R}^a(V)$,

$\mathcal{W}_1 \oplus \mathcal{W}_2 = $ orthogonal complement of $\mathcal{R}_p(V) \cap \mathcal{R}^s(V)$ in $\mathcal{R}^s(V)$,

$\mathcal{W}_2 = \{R \in \mathcal{W}_1 \oplus \mathcal{W}_2 | \tau(R) \text{ is zero}\}$,

$\mathcal{W}_1 = $ orthogonal complement of $\mathcal{W}_2$ in $\mathcal{W}_1 \oplus \mathcal{W}_2$,

$\mathcal{W}_6 = \{R \in \mathcal{R}_0(V) | R(x,y,z,w) = -R(x,y,w,z); \ x,y,z,w \in V\}$,

$\mathcal{W}_7 = \{R \in \mathcal{R}_0(V) | R(x,y,z,w) = R(x,y,w,z); \ x,y,z,w \in V\}$,

$\mathcal{W}_8 = $ orthogonal complement of $\mathcal{W}_6 \oplus \mathcal{W}_7$ in $\mathcal{R}_0(V)$.

So, we can state the decomposition theorem for $\mathcal{R}(V)$.

Theorem 3.1.
$$\mathcal{R}(V) = \mathcal{W}_1 \oplus \ldots \oplus \mathcal{W}_8, \tag{3.1}$$

where $\mathcal{W}_i$ are orthogonal invariant subspaces under the action of $SO(V)(n \geq 2)$. Moreover,

(i)   The decomposition (3.1) is irreducible for $n \geq 4$.

(ii)   For $n = 4$ we have $\mathcal{W}_6 = \mathcal{W}^+ \oplus \mathcal{W}^-$, where $\mathcal{W}^\pm = \{R \in \mathcal{W}_6 | R_* = \pm R\}$, and the other factors are irreducible.

(iii) For $n = 3$ we have $\mathcal{W}_6 = \mathcal{W}_8 = \{0\}$ and the other factors are irreducible.

(iv) For $n = 2$ we have $\mathcal{W}_4 = \mathcal{W}_5 = \mathcal{W}_6 = \mathcal{W}_7 = \mathcal{W}_8 = \{0\}$ and the other factors are irreducible.

A partial decomposition of $\mathcal{R}(V)$, by the assumption that the Ricci tensor $\rho(R)$, for $R \in \mathcal{R}(V)$ is symmetric in addition, was given by Vanhecke [VA]. We have obtained a partial decomposition of $\mathcal{R}(V)$ in [BO 2], and a complete decomposition of $\mathcal{R}(V)$ in [BO 3] (see also [BO 4]). We have a complete decomposition (3.1) with the algebraic properties of the Weyl projective curvature tensor in mind. Namely, it is well known that the Weyl projective curvature tensor of any n-dimensional manifold $\mathbb{M}^n$ with any symmetric connection ∇ has the form

$$\begin{aligned} P(R)(X,Y)Z =\ & R(X,Y)Z + \frac{1}{n^2-1}[n\rho(X,Z) + \rho(Z,X)]Y \\ & - \frac{1}{n^2-1}[n\rho(Y,Z) + \rho(Z,Y)]X \\ & + \frac{1}{n+1}[\rho(X,Y) - \rho(Y,X)]Z, \end{aligned}$$

for $n > 2$, and for $n = 2$

$$P(R)(X, Y)Z = 0,$$

(see for example [RA], [SCHO 2], [WE 1]). We suppose here $X, Y, Z, \ldots \in \mathfrak{X}(\mathbb{M})$, the algebra of C^∞ vector fields on $\mathbb{M}$ and R, and ρ are the curvature tensor and the Ricci tensor respectively, corresponding to a symmetric connection ∇. $P(R)$ is a tensor that is invariant with respect to each projective transformation of $\mathbb{M}$. $P(R)$ characterizes a space of constant sectional curvature in a very nice way: $P(R) = 0$ if and only if $\mathbb{M}^n (n > 2)$ is a space of constant sectional curvature (in that case R is the Riemannian curvature of $\mathbb{M}^n$). The Weyl projective curvature tensor fulfills the algebraic conditions (i), (ii) and

$$\rho(P(R)) = 0.$$

Hence, the subspace $\mathcal{R}_p(V)$ is called the projective component of $\mathcal{R}(V)$. The projective curvature tensor $P(R)$ associated with R is the orthogonal projection of R on $\mathcal{R}_p(V)$. We recall that the Weyl conformal curvature tensor belongs to the $\mathcal{R}_3$-component from Singer-Thorpe decomposition, which is irreducible under the action of the group $O(V)$. The projective component $\mathcal{R}_p(V)$ is not irreducible under the action of $O(V)$ or $SO(V)$. Namely, using Theorem 3.1 we find

Theorem 3.2. The decomposition of the projective component $\mathcal{R}_p(V)$ of $\mathcal{R}(V)$ into orthogonal irreducible factors under the action of $SO(V)$ is given by

(i) $\mathcal{R}_p(V) = \mathcal{W}_4 \oplus \mathcal{W}_5 \oplus \mathcal{W}_6 \oplus \mathcal{W}_7 \oplus \mathcal{W}_8$, for $n \geq 4$,
(ii) $\mathcal{R}_p(V) = \mathcal{W}_4 \oplus \mathcal{W}_5 \oplus \mathcal{W}^+ \oplus \mathcal{W}^- \oplus \mathcal{W}_7 \oplus \mathcal{W}_8$, for $n = 4$,
(iii) $\mathcal{R}_p(V) = \mathcal{W}_4 \oplus \mathcal{W}_5 \oplus \mathcal{W}_7$, for $n = 3$,
(iv) $\mathcal{R}_p(V) = 0$, for $n = 2$.

Strichartz [STRI] has considered a complete decomposition of $\mathcal{R}(V)$ under the action of the general linear group $GL(V)$ and he has proved the projective component $\mathcal{R}_p(V)$ is irreducible in this case.

The complete decomposition of $\mathcal{R}(V)$, given by Theorem 3.1 is very useful in the study of the group of projective transformations on some manifold $\mathbb{M}^n$ and its subgroups. So, supposing that some of the components vanishes, we have obtained, having in mind [ISH], [ISH-OB], [KOB],

[NO 1], [NOR], [HA], that the corresponding manifold has special groups of transformations (see [BO 3]). Let us mention some of these results.

Theorem 3.3. On Riemannian manifold $\mathbb{M}^n$ there exists the group of equiaffine transformations if and only if $\mathcal{W}_3 = 0$.

Theorem 3.4. If a Riemannian manifold $\mathbb{M}$ is compact and $\mathcal{W}_1 = \mathcal{W}_2 = \mathcal{W}_3 = 0$ then the group $P(\mathbb{M})$ of all projective transformations coincides with its subgroup $A(\mathbb{M})$ of all affine transformations.

Some further applications of the complete decompositions of V can be found in [BO 3].

We have considered also a partial decomposition of the vector space $\mathcal{K}(V)$ of curvature tensors satisfying (i), (ii) and in addition the Kähler identity under the action of the unitary group $U(V)$, where V is a $2n$-dimensional Hermitian vector space (see [BO 1]). A complete decomposition of the vector space $\mathcal{K}(V)$ was given in [MA], [NIK]. Using a complete decomposition of $\mathcal{K}(V)$, the group of projective transformations of Hermitian manifolds have been studied in [B-BO]. The action of the group $U(V) \times 1$ on the tangent bundle of a normal almost contact manifold and the corresponding vector space of curvature tensors for adopted connections was studied by Matzeu [MA].

II. A Decomposition of Some Space of Covariant Tensors of Order 3

In this Chapter we study subspaces of a space $\overset{3}{\otimes} V^*$ and the action of some of the classical groups in order to obtain a classification of some kind of manifolds.

1. *A Classification of Almost Hermitian Manifolds*

In this section we consider a representation of the unitary group $U(V)$ on a certain space $\mathcal{W}$ which can be geometrically interpreted as the space of tensors which satisfy the same identities as the covariant derivative of the Kähler form of an almost Hermitian manifold (see [GR-HE]). The main point of this consideration is to fit all classes of almost Hermitian manifold into a general system, which in a reasonable sense is complete. All of these classes have been obtained by various authors having generalizations of Kähler geometry in mind. In [KOT] Kotō established inclusion relations between various classes. In [GR 1], [GR 2] Gray has shown these inclusion relations are strict by the method of constructing explicit examples.

In order to formulate the main theorem from [GR-HE] we need some notations. Let V be a real vector space of dimension $2n$ with an almost complex structure J and a real positive definite inner product $\langle\ ,\ \rangle$. We assume that J and $\langle\ ,\ \rangle$ are compatible in the sense that $\langle Jx, Jy\rangle = \langle x, y\rangle$ for $x, y \in V$. Let V^* denote the dual space of V, and consider the space $V^* \otimes V^* \otimes V^*$. This space is naturally isomorphic to the space of all trilinear covariant tensors on V. Let $\mathcal{W}$ be the subspace of $V^* \otimes V^* \otimes V^*$ defined by

$$\mathcal{W} = \{\alpha \in V^* \otimes V^* \otimes V^* | \alpha(x, y, z) = -\alpha(x, z, y)$$
$$= -\alpha(x, Jy, Jz) \text{ for all } x, y, z \in V\}\,.$$

There is a natural inner product on $\mathcal{W}$ given by

$$\langle\alpha, \beta\rangle = \sum_{i,j,k=1}^{2n} \alpha(e_i, e_j, e_k)\beta(e_i, e_j, e_k)\,,$$

where $\{e_i, \ldots, e_{2n}\}$ is an arbitrary orthonormal basis of V. Also, for $\alpha \in \mathcal{W}$ let $\overline{\alpha} \in V^*$ be defined by

$$\overline{\alpha}(z) = \sum_{i=1}^{2n} \alpha(e_i, e_i, z)\,,$$

for $z \in V$. We define four subspaces of $\mathcal{W}$ as follows:

$$\mathcal{W}_1 = \{\alpha \in \mathcal{W} | \alpha(x, x, z) = 0 \text{ for all } x, z \in V\}\,,$$
$$\mathcal{W}_2 = \{\alpha \in \mathcal{W} | \alpha(x, y, z) + \alpha(z, x, y) + \alpha(y, z, x) = 0 \text{ for all } x, y, z \in V\}\,,$$
$$\mathcal{W}_3 = \{\alpha \in \mathcal{W} | \alpha(x, y, z) - \alpha(Jx, Jy, z) = \overline{\alpha}(z) = 0 \text{ for all } x, y, z \in V\}\,,$$
$$\mathcal{W}_4 = \{\alpha \in \mathcal{W} | \alpha(x, y, z) = -\frac{1}{2(n-1)}(\langle x, y\rangle\overline{\alpha}(z)$$
$$- \langle x, z\rangle\overline{\alpha}(y) - \langle x, Jy\rangle\overline{\alpha}(Jz) + \langle x, Jz\rangle\overline{\alpha}(Jy)) \text{ for all } x, y, z \in V\}\,.$$

The usual representation of $U(V)$ on V induces a representation of $U(V)$ on $\mathcal{W}$. The next theorem describes the decomposition of this induced representation into irreducible components.

Theorem 1.1. We have $\mathcal{W} = \mathcal{W}_1 \oplus \mathcal{W}_2 \oplus \mathcal{W}_3 \oplus \mathcal{W}_4$. This direct sum is orthogonal, and it is preserved under the induced representation of $U(V)$ on $\mathcal{W}$. The induced representation of $U(V)$ on $\mathcal{W}_i$ is irreducible. For $n = 1, \mathcal{W} = \{0\}$; for $n = 2, \mathcal{W}_1 = \mathcal{W}_3 = \{0\}$, so that $\mathcal{W} = \mathcal{W}_2 \oplus \mathcal{W}_4$. For $n = 2, \mathcal{W}_2$ and $\mathcal{W}_4$ are nontrivial, and for $n \geq 3$ all of the $\mathcal{W}_i$ are nontrivial.

These algebraic results can be applied in differential geometry of almost Hermitian manifolds as follows. Let $\mathbb{M}$ be a C^∞ almost Hermitian manifold with metric $\langle\ ,\ \rangle$, Riemannian connection ∇, and almost complex structure J. Denote by $\mathfrak{X}(\mathbb{M})$ the Lie algebra of C^∞ vector fields on $\mathbb{M}$. Then we have $\langle JX, JY \rangle = \langle X, Y \rangle$ for $X, Y \in \mathfrak{X}(\mathbb{M})$. Also, S will denote the Nijenhuis tensor of $\mathbb{M}$, that is,

$$S(X, Y) = [X, Y] + J[JX, Y] + J[X, JY] - [JX, JY], \quad \text{for } X, Y \in \mathfrak{X}(\mathbb{M}).$$

The Kähler form F is given by $F(X, Y) = \langle JX, Y \rangle$; and the Lee form is the 1-form θ defined by $\theta(X) = ((-1)/(n-1))\delta F(JX)$, where δ denotes the coderivative.

For any almost Hermitian manifold there is a representation of $U(V)$ on each tangent space $\mathbb{M}_m$. Put

$$\mathcal{W}_m = \{\alpha \in \mathbb{M}_m^* \otimes \mathbb{M}_m^* \otimes \mathbb{M}_m^* | \alpha(x, y, z) = -\alpha(x, z, y) = -\alpha(x, Jy, Jz)\}.$$

Then the induced representation of $U(V)$ on $\mathcal{W}_m$ has the four components $\mathcal{W}_{m1}, \mathcal{W}_{m2}, \mathcal{W}_{m3}, \mathcal{W}_{m4}$ as it has been described previously. It is possible to form from these four a total of sixteen invariant subspaces of $\mathcal{W}_m$ (including $\{0\}$ and $\mathcal{W}_m$).

Let $\mathcal{U}$ be one of the sixteen invariant subspaces of $\mathcal{W}$. For an almost Hermitian manifold $\mathbb{M}$ and $m \in \mathbb{M}$, let $\mathcal{U}_m$ denote the corresponding subspace of $\mathcal{W}_m$. Then $\mathcal{U}$ will denote the class of all almost Hermitian manifolds $\mathbb{M}$ such that $(\nabla F)_m \in \mathcal{U}_m$ for all $m \in \mathbb{M}$.

It is easy to show that for any almost Hermitian manifold $\mathbb{M}$, ∇F has all the required symmetries in order that $(\nabla F)_m \in \mathcal{W}_m$ for all $m \in \mathbb{M}$.

The class corresponding to $\mathcal{W}_i$ will be denoted by $\mathfrak{W}_i$, and that corresponding to $\mathcal{W}_i \oplus \mathcal{W}_j$ by $\mathfrak{W}_i \oplus \mathfrak{W}_j$, etc. Also, $\mathfrak{K}$ will corresponds to $\{0\}$ and $\mathfrak{W}$ to $\mathcal{W}$. Some, but not all, of the classes have been studied. We explain how the classes just introduced coincide with classes studied by various authors:

$\mathfrak{K} =$ the class of Kähler manifolds

$\mathfrak{W}_1 = \mathfrak{N}\,\mathfrak{K} =$ the class of nearly Kähler manifolds (also called almost Tachibana spaces) ;

$\mathfrak{W} = \mathfrak{A}\,\mathfrak{K} =$ the class of almost Kähler manifolds ;

$\mathfrak{W}_3 = \mathfrak{H} \cap \mathfrak{S}\,\mathfrak{K} =$ the class of Hermitian semi-Kähler manifolds (also called special Hermitian manifolds) ;

$\mathfrak{W}_4 =$ a class which contains locally conformal Kähler manifolds ;

$\mathfrak{W}_1 \oplus \mathfrak{W}_2 = \mathfrak{Q}\,\mathfrak{K} =$ the class of quasi-Kähler manifolds ;

$\mathfrak{W}_3 \oplus \mathfrak{W}_4 = \mathfrak{H} =$ the class of Hermitian manifolds ;

$\mathfrak{W}_2 \oplus \mathfrak{W}_4 =$ a class which contains locally conformal almost Kähler manifolds ;

$\mathfrak{W}_1 \oplus \mathfrak{W}_2 \oplus \mathfrak{W}_3 = \mathfrak{S}\,\mathfrak{K} =$ the class of semi-Kähler manifolds ;

$\left.\begin{array}{l} \mathfrak{W}_1 \oplus \mathfrak{W}_3 \oplus \mathfrak{W}_4 = \mathfrak{G}_1 \\ \mathfrak{W}_2 \oplus \mathfrak{W}_3 \oplus \mathfrak{W}_4 = \mathfrak{G}_2 \end{array}\right\}$ classes studied by Hervella and Vidal [HE-VI 1], [HE-VI 2] ;

$\mathfrak{W} =$ the class of almost Hermitian manifolds .

Now, using Theorem 1.1 we can describe the sixteen classes, as follows.

N. Bokan

Table I. – Almost Hermitian manifolds of dimension ≥ 6.

Class	Defining conditions
$\mathfrak{K}$	$\nabla F = 0$
$\mathcal{W}_1 = \mathcal{N}\mathfrak{K}$	$\nabla_X(F)(X,Y) = 0$ (or $s\nabla F = dF$)
$\mathcal{W} = \mathfrak{U}\mathfrak{K}$	$dF = 0$
$\mathcal{W}_3 = \mathscr{S}\mathcal{K} \cap \mathfrak{H}$	$\delta F = S = 0$ (or $\nabla_X(F)(Y,Z) - \nabla_{JX}(F)(JY,Z) = \delta F = 0$)
$\mathcal{W}_4$	$\nabla_X(F)(Y,Z) = \frac{-1}{2(n-1)} \{ \langle X,Y \rangle \delta F(Z) - \langle X,Z \rangle \delta F(Y)$ $- \langle X,JY \rangle \delta F(JZ) + \langle X,JZ \rangle \delta F(JY) \}$
$\mathcal{W}_1 \oplus \mathcal{W}_2 = \Omega\mathfrak{K}$	$\nabla_X(F)(Y,Z) + \nabla_{JX}(F)(JY,Z) = 0$
$\mathcal{W}_3 \oplus \mathcal{W}_4 = \mathfrak{H}$	$S = 0$ (or $\nabla_X(F)(Y,Z) - \nabla_{JX}(F)(JY,Z) = 0$)
$\mathcal{W}_1 \oplus \mathcal{W}_3$	$\nabla_X(F)(X,Y) - \nabla_{JX}(F)(JX,Y) = \delta F = 0$
$\mathcal{W}_2 \oplus \mathcal{W}_4$	$dF = F \wedge \theta$ (or $\underset{XYZ}{\sigma} \{ \nabla_X(F)(Y,Z) - \frac{1}{n-1} F(X,Y)\delta F(JZ) \} = 0$)
$\mathcal{W}_1 \oplus \mathcal{W}_4$	$\nabla_X(F)(X,Y) = \frac{-1}{2(n-1)} \{ \|X\|^2 \delta F(Y) - \langle X,Y \rangle \delta F(X)$ $- \langle JX,Z \rangle \delta F(JX) \}$
$\mathcal{W}_2 \oplus \mathcal{W}_3$	$\underset{XYZ}{\sigma} \{ \nabla_X(F)(Y,Z) - \nabla_{JX}(F)(JY,Z) \} = \delta F = 0$
$\mathcal{W}_1 \oplus \mathcal{W}_2 \oplus \mathcal{W}_3 = \mathfrak{S}\mathfrak{K}$	$\delta F = 0$
$\mathcal{W}_1 \oplus \mathcal{W}_2 \oplus \mathcal{W}_4$	$\nabla_X(F)(Y,Z) + \nabla_{JX}(F)(JY,Z) = \frac{-1}{n-1} \{ \langle X,Y \rangle \delta F(Z)$ $- \langle X,Z \rangle \delta F(Y) - \langle X,JY \rangle \delta F(JZ) + \langle X,JZ \rangle \delta F(JY) \}$
$\mathcal{W}_1 \oplus \mathcal{W}_3 \oplus \mathcal{W}_4 = \mathfrak{G}_1$	$\nabla_X(F)(X,Y) - \nabla_{JX}(F)(JX,Y) = 0$ (or $\langle S(X,Y),X \rangle = 0$)
$\mathcal{W}_2 \oplus \mathcal{W}_3 \oplus \mathcal{W}_4 = \mathfrak{G}_2$	$\underset{XYZ}{\sigma} \{ \nabla_X(F)(Y,Z) - \nabla_{JX}(F)(JY,Z) \} = 0$ (or $\underset{XYZ}{\sigma} \langle S(X,Y),JZ \rangle = 0$)
$\mathfrak{W}$	No condition

Theorem 1.2. The defining relations for each of the sixteen classes (in the case that $\dim \mathbb{M} \geq 6$) are given in Table I. The case $\dim \mathbb{M} = 4$ is treated in Table II.

Gray and Hervella [GR-HE] gave also many examples of almost Hermitian manifolds in each of the sixteen classes. A large number of the examples are compact homogeneous spaces. Moreover, they determined all invariants of the representation of $U(V)$ on the space of tensors involving two derivatives of the components of the metric tensor and almost complex structure for each of the 16 classes. This work generalizes that of Gilkey [GI]. See also [DO].

Table II. Almost Hermitian manifolds of dimension $=4$.

Class	Defining conditions
$\mathfrak{K}$	$\nabla F = 0$
$\mathfrak{U}\mathfrak{K} = \mathfrak{W}$	$dF = 0$
$\mathfrak{K} = \mathfrak{W}$	$S = 0$
$\mathfrak{W}$	No condition

2. *A Classification of Riemannian Almost-Product Manifolds*

In order to classify all of the Riemannian almost-product manifolds, Naveira [NA] has studied a representation of the group $O(p) \times O(q)$ on a certain space $\mathcal{W}$ which can be geometrically interpreted as the space of tensors which satisfy the same identities as the covariant derivative of the symmetric 2-covariant tensor Φ corresponding to a Riemannian almost-product structure P on a manifold $\mathbb{M}^n$. Since the metric tensor field g on $\mathbb{M}$ and P are compatible, we have

$$g(PM, PN) = g(M, N),$$

where $M, N \in \mathfrak{X}(\mathbb{M})$.

Let ∇ be the Levi Civita connection and v (respect. h) the vertical (respect. horizontal) projection. The distributions on $\mathbb{M}^n$ defined by v and h will be represented respectively by $\mathbb{V}$ and $\mathbb{H}$.

A symmetric 2-covariant tensor field Φ is given by

$$\Phi(M,N) = g(PM,N),$$

and its covariant derivative $\nabla\Phi$ is defined by

$$\nabla_M(\Phi)(N,O) \equiv (\nabla\Phi)(M,N,O) = g(\nabla_M(P)N,0).$$

The tangent space at each point is a Euclidean vector space, decomposed into the direct sum of two complementary orthogonal subspaces. Since the symmetry properties of $\nabla\Phi$ reduce to those of a 3-covariant tensor on each tangent space, we can model the general situation on an abstract n-dimensional Euclidean space T with two distinguished subspaces V and H, of dimensions p and q respectively so that $T = V \oplus H$.

In the following

$$1 \leq a,b,c,\ldots \leq p; \quad 1 \leq u,v,w,\ldots \leq q; \quad 1 \leq i,j,k,\ldots \leq n,$$

and $A,B,C,\ldots$ will denote vertical vectors; $X,Y,Z,\ldots$ horizontal vectors and $L,M,N,\ldots$ arbitrary vectors on T.

Let T^* be the dual space of T and let us consider the space

$$\oplus^3 T^* = T^* \oplus T^* \oplus T^*,$$

which is naturally isomorphic to the space of all trilinear maps $\alpha : T \times T \times T \to \mathbb{R}$. Then the subspace $\mathcal{W}$ of $\oplus^3 T^*$ is defined by

$$\mathcal{W} = \{\alpha \in \oplus^3 T^* | \alpha(M,N,O) = \alpha(M,N,O) = -\alpha(M,PN,PO)\},$$

for all $M,N,O \in T$.

There is a natural inner product on $\mathcal{W}$ given by

$$\langle \alpha, \beta \rangle = \sum_{i,j,k=1}^{n} \alpha(e_i,e_j,e_k)\beta(e_i,e_j,e_k),$$

where $\{e_i\} = \{e_a, e_u\} = \{e_1, \ldots, e_p, e_{p+1}, \ldots, e_n\}$ is an orthonormal basis of T, where $\{e_a\}$ is a basis of V and $\{e_u\}$ is a basis of H. For any $\alpha \in \mathcal{W}$, let α^v, α^h and $\underline{a}$ be defined by

$$
\alpha^v(M) = \sum_{\alpha=1}^{p} \alpha(e_a, e_a, M) \, ,
$$
$$
\alpha^h(M) = \sum \alpha(e_u, e_u, M) \, ,
$$
$$
\underline{a}(M) = (\alpha^v + \alpha^h)(M) \, .
$$

The next theorem describes the decomposition of the induced representation of $O(\mathrm{p}) \times O(\mathrm{q})$ on $\mathcal{W}$.

Theorem 2.1. The space $\mathcal{W}$ is decomposed as the orthogonal sum $\mathcal{W} = \mathcal{W}^v \oplus \mathcal{W}^h$, where

$$\mathcal{W}^v = \mathcal{W}_1 \oplus \mathcal{W}_2 \oplus \mathcal{W}_3 \, , \quad \mathcal{W}^h = \mathcal{W}_4 \oplus \mathcal{W}_5 \oplus \mathcal{W}_6 \, ,$$

and

$$
\begin{aligned}
\mathcal{W}_1 &= \{\alpha \in \mathcal{W} | \alpha(A, A, X) = 0 \, , \quad \alpha(X, Y, A) = 0\} \, , \\
\mathcal{W}_4 &= \{\alpha \in \mathcal{W} | \alpha(X, X, A) = 0 \, , \quad \alpha(X, Y, A) = 0\} \, , \\
\mathcal{W}_2 &= \{\alpha \in \mathcal{W} | \alpha(A, B, X) = \alpha(B, A, X) \, , \quad \alpha^v = 0 \, , \quad \alpha(X, Y, A) = 0\} \, , \\
\mathcal{W}_5 &= \{\alpha \in \mathcal{W} | \alpha(X, Y, A) = \alpha(Y, X, A) \, , \quad \alpha^h = 0 \, , \quad \alpha(A, B, X) = 0\} \, , \\
\mathcal{W}_3 &= \{\alpha \in \mathcal{W} | \alpha(A, B, X) = \frac{1}{\mathrm{p}} g(A, B)\alpha^v(X) \, , \quad \alpha(X, Y, A) = 0\} \, , \\
\mathcal{W}_6 &= \{\alpha \in \mathcal{W} | \alpha(X, Y, A) = \frac{1}{\mathrm{q}} g(X, Y)\alpha^h(A) \, , \quad \alpha(A, B, X) = 0\} \, .
\end{aligned}
$$

Besides, these six subspaces are invariant and irreducible under the action of $O(\mathrm{p}) \times O(\mathrm{q})$.

Now, it is possible to give a classification of Riemannian almost-product manifolds.

Theorem 2.2. The defining conditions for each of the thirty-six classes in terms of ∇P, and $\mathbb{H}$ are those given in Table III.

N. Bokan

Table III. Riemannian almost-product manifolds.

Class	Representation	Condition of Definition	
		Vertical distribution	Horizontal distribution
$\mathcal{P}$	$(\mathrm{TGF,TGF})$	$\nabla_A(P)=0$	$\nabla_X(P)=0$
$\mathcal{W}_4$	$(\mathrm{TGF,AF})$	$\nabla_A(P)=0$	$\nabla_X(P)X=0$
$\mathcal{W}_5$	(TGF,F_1)	$\nabla_A(P)=0$	$\nabla_X(P)Y=\nabla_Y(P)X,\ \alpha^\lambda=0$
$\mathcal{W}_6$	(TGF,F_2)	$\nabla_A(P)=0$	$\nabla_X(P)Y=\frac{1}{q}g(X,Y)\alpha^\lambda(A)$
$\mathcal{W}_4\oplus\mathcal{W}_5$	(TGF,D_1)	$\nabla_A(P)=0$	$\alpha^\lambda=0$
$\mathcal{W}_4\oplus\mathcal{W}_6$	(TGF,D_2)	$\nabla_A(P)=0$	$g(\nabla_X(P)Y+\nabla_Y(P)X,A)=\frac{2}{q}g(X,Y)\alpha^\lambda(A)$
$\mathcal{W}_1\oplus\mathcal{W}_4$	$(\mathrm{AF,AF})$	$\nabla_A(P)=0$	$\nabla_X(P)X=0$
$\mathcal{W}_2\oplus\mathcal{W}_4$	(F_1,AF)	$\nabla_A(P)B=\nabla_B(P)A,\ \alpha^{*}=0$	$\nabla_X(P)X=0$
$\mathcal{W}_3\oplus\mathcal{W}_4$	(F_2,AF)	$g(\nabla_A(P)B,X)=\frac{1}{p}g(A,B)\alpha^{*}(X)$	$\nabla_X(P)X=0$
$\mathcal{W}_5\oplus\mathcal{W}_6$	(TFG,F)	$\nabla_A(P)=0$	$\nabla_X(P)Y=\nabla_Y(P)X$
$\mathcal{W}_2\oplus\mathcal{W}_5$	(F_1,F_1)	$\nabla_A(P)B=\nabla_B(P)A,\ \alpha^{*}=0$	$\nabla_X(P)Y=\nabla_Y(P)X,\ \alpha^\lambda=0$
$\mathcal{W}_3\oplus\mathcal{W}_5$	(F_2,F_1)	$g(\nabla_A(P)B,X)=\frac{1}{p}g(A,B)\alpha^{*}(X)$	$\nabla_X(P)Y=\nabla_Y(P)X,\ \alpha^\lambda=0$
$\mathcal{W}_3\oplus\mathcal{W}_6$	(F_2,F_2)	$g(\nabla_A(P)B,X)=\frac{1}{p}g(A,B)\alpha^{*}(X)$	$\nabla_X(P)Y=\frac{1}{q}g(X,Y)\alpha^\lambda(A)$
$\mathcal{W}^\perp$	$(\mathrm{TGF},\emptyset)$	$\nabla_A(P)=0$	No condition
$\mathcal{W}_1\oplus\mathcal{W}_4\oplus\mathcal{W}_5$	(AF,D_1)	$\nabla_A(P)A=0$	$\alpha^\lambda=0$
$\mathcal{W}_2\oplus\mathcal{W}_4\oplus\mathcal{W}_5$	(F_1,D_1)	$\nabla_A(P)B=\nabla_B(P)A,\ \alpha^{*}=0$	$\alpha^\lambda=0$
$\mathcal{W}_1\oplus\mathcal{W}_4\oplus\mathcal{W}_6$	(AF,D_2)	$\nabla_A(P)A=0$	$g(\nabla_X(P)Y+\nabla_Y(P)X,A)=\frac{2}{q}g(X,Y)\alpha^\lambda(A)$
$\mathcal{W}_2\oplus\mathcal{W}_4\oplus\mathcal{W}_6$	(F_1,D_2)	$\nabla_A(P)B=\nabla_B(P)A,\ \alpha^{*}=0$	$g(\nabla_X(P)Y+\nabla_Y(P)X,A)=\frac{2}{q}g(X,Y)\alpha^\lambda(A)$

Table III. Cont'd.

$\mathcal{W}_3 \oplus \mathcal{W}_4 \oplus \mathcal{W}_6$	(F_2, D_2)	$g(\nabla_A(P)B, X) = \frac{1}{p} g(A,B)\alpha^*(X)$	$g(\nabla_X(P)Y + \nabla_Y(P)X, A) = \frac{2}{q} g(X,Y)\alpha^\wedge(A)$
$\mathcal{W}_2 \oplus \mathcal{W}_3 \oplus \mathcal{W}_4$	(F, AF)	$\nabla_A(P)B = \nabla_B(P)A$	$\nabla_X(P)X = 0$
$\mathcal{W}_2 \oplus \mathcal{W}_5 \oplus \mathcal{W}_6$	(F_1, F)	$\nabla_A(P)B = \nabla_B(P), A, \alpha^* = 0$	$\nabla_X(P)Y = \nabla_Y(P)X$
$\mathcal{W}_3 \oplus \mathcal{W}_5 \oplus \mathcal{W}_6$	(F_2, F)	$g(\nabla_A(P)B, X) = \frac{1}{p} g(A,B)\alpha^*(x)$	$\nabla_X(P)Y = \nabla_Y(P)X$
$\mathcal{W}_1 \oplus \mathcal{W}^\wedge$	$(AF, \emptyset)$	$\nabla_A(P)A = 0$	No condition
$\mathcal{W}_2 \oplus \mathcal{W}^\wedge$	$(F_1, \emptyset)$	$\nabla_A(P)B = \nabla_B(P)A, \alpha^* = 0$	No condition
$\mathcal{W}_3 \oplus \mathcal{W}^\wedge$	$(F_2, \emptyset)$	$g(\nabla_A(P)B, X) = \frac{1}{p} g(A,B)\alpha^*(X)$	No condition
$\mathcal{W}_1 \oplus \mathcal{W}_2 \oplus \mathcal{W}_4 \oplus \mathcal{W}_5$	(D_1, D_1)	$\alpha^* = 0$	$\alpha^\wedge = 0$
$\mathcal{W}_1 \oplus \mathcal{W}_3 \oplus \mathcal{W}_4 \oplus \mathcal{W}_5$	(D_2, D_1)	$g(\nabla_A(P)B + \nabla_B(P)A, X) = \frac{2}{p} g(A,B)\alpha^*(X)$	$\alpha^\wedge = 0$
$\mathcal{W}_2 \oplus \mathcal{W}_3 \oplus \mathcal{W}_4 \oplus \mathcal{W}_5$	(F, D_1)	$\nabla_A(P)B = \nabla_B(P)A$	$\alpha^\wedge = 0$
$\mathcal{W}_1 \oplus \mathcal{W}_3 \oplus \mathcal{W}_4 \oplus \mathcal{W}_6$	(D_2, D_2)	$g(\nabla_A(P)B + \nabla_B(P)A, X) = \frac{2}{p} g(A,B)\alpha^*(X)$	$g(\nabla_X(P)Y + \nabla_Y(P)X, A) = \frac{2}{q} g(X,Y)\alpha^\wedge(A)$
$\mathcal{W}_2 \oplus \mathcal{W}_3 \oplus \mathcal{W}_4 \oplus \mathcal{W}_6$	(F, D_2)	$\nabla_A(P)B = \nabla_B(P)A$	$g(\nabla_X(P)Y + \nabla_Y(P)X, A) = \frac{2}{q} g(X,Y)\alpha^\wedge(A)$
$\mathcal{W}_2 \oplus \mathcal{W}_3 \oplus \mathcal{W}_5 \oplus \mathcal{W}_6$	(F, F)	$\nabla_A(P)B = \nabla_B(P)A$	$\nabla_X(P)Y = \nabla_Y(P)X$
$\mathcal{W}_1 \oplus \mathcal{W}_2 \oplus \mathcal{W}^\wedge$	$(D_1, \emptyset)$	$\alpha^* = 0$	No condition
$\mathcal{W}_1 \oplus \mathcal{W}_3 \oplus \mathcal{W}^\wedge$	$(D_2, \emptyset)$	$g(\nabla_A(P)B + \nabla_B(P)A, X) = \frac{2}{p} g(A,B)\alpha^*(X)$	No condition
$\mathcal{W}_2 \oplus \mathcal{W}_3 \oplus \mathcal{W}^\wedge$	$(F, \emptyset)$	$\nabla_A(P)B = \nabla_B(P)A$	No condition
$\mathcal{W}$	$(\emptyset, \emptyset)$	No condition	No condition

In order to classify these thirty-six classes in groups with similar geometric properties, we introduce the following order, considering always first the property verified by the vertical distribution

    i)   Totally geodesic foliation, (TGF).

    ii) Foliation, (F).

    iii) Antifoliation, (AF).

    iv) Distribution of type D_1.

    v) Distribution of type D_2.

For more details see [NA] and also [GM], [VI-VIC].

Analogously, a classification of various kind of manifolds were studied by other mathematicians; for example by C. Bejan [BEJ], Bouten [BOU], Chinea and Gonzalez [CHI-GO], Alexiev and Ganchev [AL-GA], Ganchev and Borisov [GA-BO 2] etc. Some applications of Naveira's classification of Riemannian almost-product manifolds were studied by Pavlov [PA].

III. Homogeneous Structures

In this chapter we study homogeneous Riemannian manifolds and some of their generalizations in terms of a tensor T of type (1,2) satisfying certain conditions. These considerations are based on the following facts.

Since these conditions make the algebraic sense, we consider a representation of some of the classical groups on a certain space $\mathcal{W}$ which can be geometrically interpreted as the space of tensors satisfying the same identities as a tensor T. Mainly, we deal with the results of Tricerri and Vanhecke [TR-VA 3] and Abbena and Garbiero [AB-GA].

1. *Homogeneous Structures on Riemannian Manifolds*

It is well known that E. Cartan characterized a symmetric space by the constancy of the curvature tensor under parallel translation. Ambrose and Singer [A-S] extended this theory and gave a characterization of homogeneous Riemannian manifolds by a local condition which has to be satisfied at all points. More precisely, they proved that a connected, complete and simply connected Riemannian manifold $(\mathbb{M}, g)$ is homogeneous, i.e., there exists a transitive and effective group G of isometries of $\mathbb{M}$, if and only if there exists a tensor field T of type (1,2) such that

$$\left.\begin{array}{l} g(T_X Y, Z) + g(Y, T_X Z) = 0\,, \\ (\nabla_X R)_{YZ} = [T_X, R_{YZ}] - R_{T_X YZ} - R_{Y T_X Z}\,, \\ (\nabla_X T)_Y = [T_X, T_Y] - T_{T_X Y}\,, \end{array}\right\} \qquad (1.1)$$

where $X, Y, Z \in \mathfrak{X}(\mathbb{M}), \nabla$ is the Levi Civita connection and R is the Riemannian curvature tensor of $\mathbb{M}$.

These conditions are also used to study weakly locally homogeneous, infinitesimally homogeneous and curvature homogeneous manifolds [A-S], [SI].

In order to explain how it is possible to classify the groups G of which $\mathbb{M}$ is a Riemannian homogeneous space through the T's, let us mention some important facts. Namely, if $(\mathbb{M}, g)$ is a homogeneous Riemannian space with a given group G of isometries, following the method of Ambrose and Singer, this determines a tensor field T. Now, using this T one can determine conversely the group of isometries and this is in general a group G' which is not isomorphic to G. Hence an important problem is to understand for this case the solutions of Eqs. (1.1) give a parameterization of the transitive and effective groups G. More detailed information about this problem have been given in [TR-VA 3].

Following the suggestion of Ambrose and Singer, to classify Riemannian homogeneous manifolds by properties of the T's, Tricerri and Vanhecke [TR-VA 3] studied the decomposition of the space of tensors T which satisfy the condition (1.1.i) into irreducible factors under the action of the orthogonal group. In this way they obtained a set of algebraic conditions for the tensor T. These conditions are invariant under the action of the orthogonal group. Let us consider this decomposition in details.

We say a tensor T of type (1,2) is a homogeneous (Riemannian) structure on $(\mathbb{M}, g)$ if it is a solution of the system (1). Let T be a homogeneous structure on $(\mathbb{M}, g)$ and T' a homogeneous structure on $(\mathbb{M}', g')$. Then T and T' are said to be isomorphic if there exists an isometry $\varphi : (\mathbb{M}, g) \to (\mathbb{M}', g')$ such that $\varphi_*(T_X Y) = T'_{\varphi_* X} \varphi_* Y$, $X, Y \in \mathfrak{X}(\mathbb{M})$.

Let p be a point of $\mathbb{M}$ and $V = T_p \mathbb{M}$. V is a Euclidean vector space over $\mathbb{R}$ with inner product $\langle\ ,\ \rangle$ induced from the metric g on $\mathbb{M}$. We associate the tensor T of type (0,3) to a tensor T of type (1,2) by the isomorphism

$$T_{xyz} = \langle T_x y, z \rangle\,, \quad x, y, z \in V\,.$$

Next we consider the vector space $\mathcal{C}(V)$, subspace of $\otimes^3 V^*$, determined by all the (0,3)-tensors having the same symmetries as a homogeneous structure, i.e.,

$$\mathcal{C}(V) = \{T \in \otimes^3 V^* | T_{xyz} = -T_{xzy}, \ x, y, z \in V\},$$

where V^* denotes the dual vector space of V, as usual. $\mathcal{C}(V)$ is a Euclidean vector space with inner product defined by

$$\langle T, T' \rangle = \sum T_{e_i e_j e_k} T'_{e_i e_j e_k},$$

where $(e_1, \ldots, e_n)$ is an arbitrary orthonormal basis of V.

Further there is a natural action of the orthogonal group $O(V)$ on $\mathcal{C}(V)$ determined by

$$(aT)_{xyz} = T_{a^{-1}x\, a^{-1}y\, a^{-1}z},$$

for $x, y, z \in V$ and $a \in O(V)$.

Let

$$c_{12}(T)(z) = \sum_i T_{e_i e_i z}, \quad z \in V,$$

for an arbitrary orthonormal basis (e_i) of V and put

$$C_1(V) = \{T \in \mathcal{C}(V) | T_{xyz} = \langle x, y \rangle \varphi(z) - \langle x, z \rangle \varphi(y), \ \varphi \in V^*\},$$
$$C_2(V) = \{T \in \mathcal{C}(V) | \underset{x,y,z}{\sigma}\, T_{xyz} = 0, \quad c_{12}(T) = 0\},$$
$$C_3(V) = \{T \in \mathcal{C}(V) | T_{xyz} + T_{yxz} = 0\},$$

where $x, y, z \in V$. Here $\underset{x,y,z}{\sigma}$ denotes the cyclic sum with respect to x, y and z.

We have

Theorem 1.1. Let $\dim V \geq 3$. Then $\mathcal{C}(V)$ is the orthogonal direct sum of the subspaces $C_i(V), i = 1, 2, 3$. Moreover, these spaces are invariant and irreducible under the action of $O(V)$.

Let $\dim V = 2$. Then we have $\mathcal{C}(V) = \mathcal{C}_1(V)$, where $\mathcal{C}(V)$ is irreducible.

It follows from Theorem 1.1 that there are, in general, eight invariant subspaces of $\mathcal{C}(V)$, including the trivial spaces. It implies also the consideration of the same kind of a manifold $(\mathbb{M}, g)$ as follows.

Table IV.

Classes	Defining conditions
Symmetric structures	$T=0$
$\mathcal{C}_1$	$T_{XYZ}=g(X,Y)\varphi(Z)-g(X,Z)\varphi(Y),\varphi\in\Lambda^1\mathbf{M}$
$\mathcal{C}_2$	$\underset{X,Y,Z}{\sigma}\,T_{XYZ}=0,c_{12}(T)=0$
$\mathcal{C}_3$	$T_{XYZ}+T_{YXZ}=0$
$\mathcal{C}_1\oplus\mathcal{C}_2$	$\underset{X,Y,Z}{\sigma}\,T_{XYZ}=0$
$\mathcal{C}_1\oplus\mathcal{C}_3$	$T_{XYZ}+T_{YXZ}=g(X,Y)\varphi(Z)-g(X,Z)\varphi(Y)-g(Y,Z)\varphi(X),\quad\varphi\in\Lambda^1\mathbf{M}$
$\mathcal{C}_2\oplus\mathcal{C}_3$	$c_{12}(T)=0$
$\mathcal{C}=\mathcal{C}_1\oplus\mathcal{C}_2\oplus\mathcal{C}_3$	no conditions

Let $\mathcal{I}(V)$ be an invariant subspace of $\mathcal{C}(V)$. We say that a homogeneous structure T on $(\mathbb{M},g)$ is of type $\mathcal{J}$ when $T_p\in\mathcal{I}(T_p\mathbb{M})$ for all $p\in\mathbb{M}$.

Note that every class $\mathcal{J}$ of the eight classes of homogeneous structures is invariant under isomorphisms of homogeneous structures. These eight classes are given in Table IV. But it also can happen that on the same manifold $(\mathbb{M},g)$ there exist two nonisomorphic homogeneous structures T and T'. The point is that T and T' give rise to two nonisomorphic transitive groups of isometries or to the same group but with different reductive decompositions of the Lie algebra of the group. It is considered in [TR-VA 3 Chap. 7,8] and in addition, all of these classes are considered in details. So, it has shown that the Poincaré half-plane is the only connected, complete and simply connected surface which has a homogeneous Riemannian structure T different from zero. Moreover, up to isomorphism, this nonvanishing structure is unique. Further, the hyperbolic space is the only space (connected, complete, simply connected) having such a structure $T\neq0$. Naturally reductive homogeneous space is characterized by the condition "$T_XX=0$ for al $X\in\mathfrak{X}(\mathbb{M})$".

See also [KOW-TR], [KOW-VA 1], [KOW-VA 2], [NI], [TR-VA 4].

2. *Almost Hermitian Homogeneous Structures*

An almost Hermitian manifold $(\mathrm{I\!M}, g, J)$ is a homogeneous almost Hermitian manifold if there exists a transitive and effective Lie group G of almost complex isometries acting on $\mathrm{I\!M}$.

Sekigawa [SE] proved the analogous theorem to the Theorem of Ambrose and Singer for homogeneous almost Hermitian manifolds. Namely, we have

Theorem 2.1. A connected, simply connected, complete almost Hermitian manifold $(\mathrm{I\!M}, g, J)$ is homogeneous if and only if there is a tensor field T of type $(1,2)$ on $\mathrm{I\!M}$ which satisfies conditions (i), (ii), (iii) of (1.1) and

$$(iv) \qquad\qquad \nabla J = TJ .$$

A tensor T verifying the conditions of the previous theorem is called an almost Hermitian homogeneous structure. Abbena and Garbiero [AB-GA] obtained a classification of almost Hermitian homogeneous structures similar to the classification of Riemannian homogeneous structures given in [TR-VA 3]. In order to obtain this classification they considered the vector space $\mathcal{T}(V)$ of tensors with the same symmetries of the almost Hermitian homogeneous structures. Let us explain it in more detail.

We denote by V a real vector space of dimension $2n$ endowed with a complex structure J and a Hermitian inner product $\langle \, , \, \rangle$, that is,

$$J^2 = I , \quad \langle Jx, Jy \rangle = \langle x, y \rangle ,$$

for any $x, y \in V$ and I the identity isomorphism of V. Then the vector space $\mathcal{T}(V)$ is defined as follows

$$\mathcal{T}(V) = \{T \in \otimes^3 V^* | T_{xyz} = -T_{xzy}\} ,$$

where $T_{xyz} = \langle T_x y, z \rangle$, for any $x, y, z \in V$ (V^* is the dual space of V). $\mathcal{T}(V)$ has a canonical inner product, induced by V, defined in the following way

$$\langle T, \hat{T} \rangle = \sum_{i,j,k=1}^{2n} T_{e_i e_j e_k} \hat{T}_{e_i e_j e_k} ,$$

with $T, \hat{T} \in \mathcal{T}(V)$ and $(e_1, \ldots, e_{2n})$ an arbitrary orthonormal basis of V. The standard representation ρ of the unitary group $U(V)$ on V gives rise to a representation $\tilde{\rho}$ of $U(V)$ on $\mathcal{T}(V)$ defined by

$$(\tilde{\rho}(g)T)_{xyz} = T_{g^{-1}x g^{-1}y g^{-1}z}, \quad x, y, z \in V, \quad g \in U(V), \quad T \in \mathcal{T}(V).$$

Abbena and Garbiero [AB-GA] obtained a classification of almost Hermitian homogeneous structures as follows.

Theorem 2.2. Let $\dim V \geq 6$, then $\mathcal{T}(V)$ is the orthogonal direct sum of the invariant subspaces $\mathcal{H}_i, i = 1, \ldots, 8$ given by

$$\mathcal{H}_1 = \{T \in \mathcal{T}(V) | T_{xyz} = \frac{1}{2}(T_{zxy} + T_{yzx} + T_{JzxJy} + T_{JyJzx}), c_{12}(T) = 0\},$$

$$\mathcal{H}_2 = \{T \in \mathcal{T}(V) | T_{xyz} = \langle x, y \rangle \psi_1(z) - \langle x, z \rangle \psi_1(y) + \langle x, Jy \rangle \psi_1(Jz)$$
$$- \langle x, Jz \rangle \psi_1(Jy) - 2 \langle Jy, z \rangle \psi_1(Jx), \psi_1 \in V^*\},$$

$$\mathcal{H}_3 = \{T \in \mathcal{T}(V) | T_{xyz} = -\frac{1}{2}(T_{zxy} + T_{yzx} + T_{JzxJy}$$
$$+ T_{JyJzx}), c_{12}(T) = 0\},$$

$$\mathcal{H}_4 = \{T \in \mathcal{T}(V) | T_{xyz} = \langle x, y \rangle \psi_2(z) - \langle x, z \rangle \psi_2(y) + \langle x, Jy \rangle \psi_2(Jz)$$
$$- \langle x, Jz \rangle \psi_2(Jy) + 2 \langle Jy, z \rangle \psi_2(Jx), \psi_2 \in V^*\},$$

$$\mathcal{H}_5 = \{T \in \mathcal{T}(V) | T_{xyz} = -T_{yxz} = -T_{xJyJz}\},$$

$$\mathcal{H}_6 = \{T \in \mathcal{T}(V) | T_{xyz} = \underset{xyz}{\sigma} T_{xyz} = 0, T_{xyz} = -T_{xJyJz}\},$$

$$\mathcal{H}_7 = \{T \in \mathcal{T}(V) | T_{xyz} - T_{xJyJz} = T_{JxJyz}, c_{12}(T) = 0\},$$

$$\mathcal{H}_8 = \{T \in \mathcal{T}(V) | T_{xyz} = \langle x, y \rangle \psi_3(z) - \langle x, z \rangle \psi_3(y) + \langle x, Jz \rangle \psi_3(Jy)$$
$$- \langle x, Jy \rangle \psi_3(Jz), \psi_3 \in V^*\},$$

where $c_{12}(T)(x) = \sum_{i=1}^{2n} T_{e_i e_i x}$ for any $x \in V$.

If $\dim V = 2$, then $\mathcal{T}(V) = \mathcal{H}_4$.

If $\dim V = 4$, then $\mathcal{T}(V) = \mathcal{H}_2 \oplus \mathcal{H}_3 \oplus \mathcal{H}_4 \oplus \mathcal{H}_6 \oplus \mathcal{H}_8$.

Moreover, each $\mathcal{H}_i$ is irreducible under the action of $U(V)$.

Let $(\mathbb{M}, g, J)$ be a homogeneous almost Hermitian manifold of dimension $2n$. For every $p \in \mathbb{M}$, $(T_p\mathbb{M}, g_p, J_p)$ is a Hermitian vector space. Given an orthonormal basis $(e_i, \ldots, e_n, Je_1, \ldots, Je_n)$ of $T_p\mathbb{M}$, there is a standard representation of $U(V)$ on $T_p\mathbb{M}$. Hence it is possible to decompose the vector space $T(T_p\mathbb{M})$ as in Theorem 2.2. Let $\mathcal{H}$ denote a subspace of $T(T_p\mathbb{M})$ invariant with respect to the representation of $U(T_p\mathbb{M})$.

We say that $\mathbb{M}$ is of type $\mathcal{H}$ if $T_p \in \mathcal{H}$, for all $p \in \mathbb{M}$, where T is the corresponding homogeneous almost Hermitian structure (see Theorem 2.1). Then we simply write $\mathbb{M} \in \mathcal{H}$.

From the decomposition given in Theorem 2.2 it is possible to deduce the characterization of a Kähler manifold as a homogeneous almost Hermitian manifold of type $\mathcal{H}_1 \oplus \mathcal{H}_2 \oplus \mathcal{H}_3 \oplus \mathcal{H}_4$.

Since some subspaces of the decomposition are equivalent in the sense of the representation theory, it follows that there is an infinite number of invariant subspaces of $T(V)$. As a consequence there exist several decompositions. The choice among the different possibilities is motivated by geometrical considerations. In order to obtain a classification of naturally reductive homogeneous almost Hermitian manifolds in eight classes, Abbena and Garbiero [AB-GA] studied a second decomposition considering a connection with a decomposition of $T(V)$ under the action of the orthogonal group $O(V)$. Some other geometrical results and examples in details can be seen in [AB-GA].

References

[AB-GA]   E. Abbena and S. Garbiero, *Almost Hermitian homogeneous structures*, Proc. Edinburgh Math. Soc. **31** (1988) 375–395.

[AL-GA]   V. Alexiev and G. Ganchev, *A classification of almost contact metric manifolds*, Int. Conf. Geom. Applic., Smolyan (Bulgaria), July 6–12, 1986, Summaries, pp. 43–44.

[A-S]   W. Ambrose and I. M. Singer, *On homogeneous Riemannian manifolds*, Duke Math. J. **25** (1958) 647–669.

[BEJ]   C. Bejan, *A classification of the almost parahermitian manifolds*, Proc. Int. Conf. Diff. Geom. Appl., Dubrovnik (Yugoslavia), June 26–July 3, 1988, pp. 23–27.

[BER]   M. Berger, *Sur les variétés d'Einstein compactes*, C. R. IIIe Réunion du Groupment des Mathematicien d'Expression Latine, Louvain-Belgique (1966), pp. 35–55.

[B-BO]   N. Blažić and N. Bokan, *The group of projective transformations of Hermitian manifolds*, preprint 1989.

[BO 1]   N. Bokan, *Curvature tensors on Hermitian manifolds*, Colloquia Mathematica Societatis János Bolyai, 46. Topics in Diff. Geom., Debrecen (Hungary), (1984), pp. 213–239.

[BO 2]   N. Bokan, *Curvature tensors of Riemannian manifold with connection without torsion*, Matematički vesnik **37** (1985) 356–364.

[BO 3]   N. Bokan, *On the complete decomposition of curvature tensors of Riemannian manifolds with symmetric connection*, Rendiconti de Circolo Matematico di Palermo (3) **39** (1990) (to appear).

[BO 4]   N. Bokan, *Quadratic invariants and locally affine symmetric manifolds*, Proc. Int. Conf. Diff. Geom. Appl., Dubrovnik (Yugoslavia), June 26–July 3, 1988, pp. 69–82.

[BOU]   S. Bouten, *IX österreichische mathematiker Kongress*, Salzburg, 1977, Abstracts, p. 83.

[CA]   F. J. Carreras, *On the Riemannian curvature tensor of an almost-product manifold*, Czech. J. Math. **36** (1986) 72–86.

[CH]   S. S. Chern, *On the curvature and characteristic classes of a Riemannian manifold*, Abh. Math. Sem. Univ. Hamburg **20** (1956) 117–126.

[CHI-GO] D. Chinea and C. Gonzalez, *A classification of almost contact metric manifolds*, preprint 1985.

[DO]   H. Donnelly, *Invariance theory of Hermitian manifolds*, Proc. Amer. Math. Soc. **58** (1976) 229–233.

[GA-BO 1] G. Ganchev and A. Borisov, *Isotropic sections and curvature properties of hyperbolic Kaehlerian manifolds*, Publ. de l' Inst. math. (52) **38** (1985) 183–192.

[GA-BO 2] G. Ganchev and A. Borisov, *Note on the almost complex manifolds with a Norden metric*, Comptes rendus de l' Academie bulgarie de Sciences (5) **39** (1985) 31–34.

[G-G-M 1] G. Ganchev, K. Gribachev and V. Mihova, *Holomorphically conformal invariants on conformally Kaehler manifolds with Norden metric*, Int. Conf. Geom. Applic., Smolyan (Bulgaria), July 6–12, 1986, Summaries, pp. 60–61.

[G-G-M 2] G. Ganchev, K. Gribachev and V. Mihova, *The connections and their conformal invariants on conformally Kaehlerian manifolds with B-metric*, Publ. de l' Inst. Math. **42** (1987), 107–121.

[GAU]   P. Gauduchon, *La 1-forme de torsion d'une variété hermitienne compacte*, Math. Ann. **267** (1984) 495–518.

[GM]   O. Gil-Medrano, *On the geometric properties of some classes of almost-product structures*, Rend. Circ. Mat. di Palermo, to appear.

[GI]   P. Gilkey, *Spectral geometry and the Kähler condition for complex manifolds*, Invent. Math. **26** (1974) 231–258.

[GR 1]   A. Gray, *Minimal varieties and almost Hermitian submanifolds*, Michigan Math. J. **12** (1965) 273–279.

[GR 2]   A. Gray, *Some examples of almost Hermitian manifolds*, Illinois J. Math. **10** (1969) 353–366.

[GR 3]   A. Gray, *Invariants of curvature operators of four-dimensional Riemannian manifolds*, Proc. Thirteenth Biennial Semin. Canad. Math. Cong., Halifax, vol. 2 (1971), pp. 42–65.

[GR 4]   A. Gray, *Curvature identities for Hermitian and almost Hermitian manifolds*, Tôhoku Math. J. **28** (1976) 601–612.

[GR-HE]   A. Gray and L. M. Hervella, *The sixteen classes of almost Hermitian manifolds*, Ann. Mat. Pura Appl. **123** (1980) 35–58.

[GR-VA 1]   A. Gray and L. Vanhecke, *Almost Hermitian manifolds with constant holomorphic sectional curvature*, Czechoslovak Math. J. **104** (1979) 170–179.

[GR-VA 2]   A. Gray and L. Vanhecke, *Decomposition of the space of covariant derivatives of curvature operators*, preprint.

[HA]   J. Hano, *On affine transformations of a Riemannian manifold*, Nagoya Math. Jour. **9** (1955) 99–109.

[HE-VI 1]   L. M. Hervella and E. Vidal, *Nouvelles géométries pseudo-kählériennes G_1 et G_2,* C. R. Acad. Sci. Paris **283** (1976) 115–118.

[HE-VI 2]   L. M. Hervella and E. Vidal, *New pseudo-Kähler geometries*, to appear.

[HI]   N. Hitchin, *Compact four-dimensional Einstein manifolds*, J. Diff. Geom. **9** (1974) 435–441.

[HO]   H. Hopteriev, *A vector space of the curvature tensors and its decomposition*, Inter. Conf. Geom. Applic., Smolyan (Bulgaria), July 6–12, 1986, Abstracts, p. 66.

[ISH]   S. Ishihara, *Groups of projective transformations and groups of conformal transformations*, Jour. Math. Soc. of Japan, (2) **9** (1957) 195–227.

[ISH-OB]   S. Ishihara and M. Obata, *Affine transformations in a Riemanian manifold*, Tôhoku Math. Jour. **7** (1975) 146–150.

[JA-VA]   D. Janssens and L. Vanhecke, *Almost contact structures and curvature tensors*, Kodai Math. J. (1) **4** (1981) 1–27.

[JO]   D. L. Jonson, *Sectional curvature and curvature normal forms*, Michigan Math. J. **27** (1980) 275–294.

[KOB]   S. Kobayashi, *A theorem on the affine transformation group of a Riemannian manifold*, Nagoya Math. Jour. **9** (1955) 39–41.

[KOT]   S. Koto, *Some theorems on almost Kählerian space*, J. Math. Soc. Japan **12** (1960) 422–433.

[KOW]   O. Kowalski, *Partial curvature structures and conformal geometry of submanifolds*, J. Diff. Geom. (1) **8** (1973) 53–70.

[KOW-TR]   O. Kowalski and F. Tricerri, *Riemannian manifolds of dimension $n \leq 4$ admitting a homogeneous structure of class T_2,* Conferenze del Seminario di Matematica Univ. di Bari, 222, Laterza, Bari, 1987.

[KOW-VA1]   O. Kowalski and L. Vanhecke, *Classification of five dimensional naturally reductive spaces*, Math. Proc. Cambridge Phil. Soc. **37** (1985) 445–463.

[KOW-VA2]   O. Kowalski and L. Vanhecke, *Four-dimensional naturally reductive homogeneous spaces*, Rend. Sem. Mat. Torino, Fasc. Speciale (1985), 223–232.

[KU]   R. S. Kulkarni, *On the Bianchi identities*, Math. Ann. **199** (1972) 175–204.

[MA]   P. Matzeu, *La decomposizione completa di $\mathcal{R}(V)$*, preprint 1989.

[MO]   H. Mori, *On the decomposition of generalized K-curvature tensor fields*, Tôhoku Math. J. **25** (1973) 225–235.

[NA]   A. M. Naveira, *A classification of Riemannian almost-product manifolds*, Rendiconti di Matematica, Vol. 3 Serie VII (1983) 577–592.

[NI]   L. Nicoldi, *Strutture riemanniane omogenee*, Tesi, Univ. of Trento, 1985, and preprint 1988.

[NIK]   S. Nikčević, *On the decomposition of curvature tensor fields on Hermitian manifolds*, Coll. Diff. Geom., August 20–26, 1989, Eger (Hungary), Abstracts, p. 39, preprint 1989.

[NO 1]   K. Nomizu, *Studies on Riemannian homogeneous spaces*, Nagoya Math. Jour. **9** (1955) 43–56.

[NO 2]   K. Nomizu, *On the decomposition of generalized curvature tensor fields*, Differential geometry (in honour of K. Yano), Kinokuniya, Tokyo (1972), pp. 335–345.

[NOR]   A. P. Norden, *Prostranstwa Affinnoj Swyaznosti*, Nauka, Moskwa, 1976.

[PA]   E. Pavlov, *On the geometry of almost B-manifolds*, Int. Conf. Diff. Geom. Appl. Dubrovnik (Yugoslavia), June 26–July 3, 1988, Abstracts.

[PA-HO]   E. Pavlov, and H. Hopteriev, *Curvature tensors on generalized B-manifolds*, preprint 1986.

[RA]   P. K. Rašewskij, *Rimanowa Geometriya i Tenzorniy Analiz*, Gostehizdat, Moskwa, 1953.

[SCHO 1]   J. A. Schouten, *Über die konforme Abbildung n-dimensionaler Mannigfaltigkeiten mit quadratischer Massbestimmung auf eine Mannigfaltigkeit mit euclidischer Massbestimmung*, Math. Z. **11** (1921) 58–88.

[SCHO 2]   J. A. Schouten, *Ricci-Calculus*, 2nd. ed., Springer Verlag, Berlin, 1954.

[SE]   K. Sekigawa, *Notes on homogeneous almost Hermitian manifolds*, Hokkaido Math. J. **7** (1978) 203–213.

[SI]   I. M. Singer, *Infinitesimally homogeneous spaces*, Comm. Pura Appl. Math. **13** (1960) 685–697.

[SI-THO]   I. M. Singer and J. A. Thorpe, *The curvature of 4-dimensional Einstein space*, Global Analysis (paper in honour of K. Kodaira), Univ. of Tokyo Press, Tokyo (1969), pp. 355–365.

[SIT]   M. Sitaramayya, *Curvature tensors in Kähler manifolds*, Trans. Amer. Math. Soc. **183** (1973) 341–351.

[STA 1]   G. Stanilov, *Generalized characteristics of some almost Hermitian manifolds*, Comptes rendus de l' Académie bulgare des Sciences, (1) **31** (1978) 11-14.

[STA 2]   G. Stanilov, *Geometric interpretation of some transformations of curvature-like tensors*, Tensor N. S. **37** (1982) 155–161.

[STA 3]   G. Stanilov, *Almost Hermitian manifolds of constant type*, Proc. Conf. Diff. Geom. Applic., Dubrovnik (Yugoslavia), June 26–July 3, 1988, pp. 349–351.

[STRI]   R. S. Strichartz, *Linear algebra of curvature tensors and their covariant derivatives*, Can. J. Math. (5) **15** (1988) 1105–1143.

[STRU]   D. J. Struik, *Abriss der Geschichte der Mathematik*, VEB deutscher Verlag der Wissenschaften, Berlin, 1963.

[TR-VA 1]   F. Tricerri and L. Vanhecke, *Decomposition of a space of curvature tensors on a quaternionic Kähler manifold*, Simon Stevin, A Quaterly J. of Pure and Appl. Math. (1-2) **53** (1979) 163–173.

[TR-VA 2]   F. Tricerri and L. Vanecke, *Curvature tensors on almost Hermitian manifolds*, Trans. of the Amer. Math. Soc. (2) **267** (1981) 365–398.

[TR-VA 3]   F. Tricerri and L. Vanhecke, *Homogeneous structures on Riemannian manifolds*, London Math. Soc., Lect. Notes 83, Cambridge Univ. Press., 1983.

[TR-VA 4]   F. Tricerri and L. Vanhecke, *Special homogeneous structures on Riemannian manifolds*, Colloquia Mathematica Soc. J. Bolyai, 46. Topics in Diff. Geom. Debrecen (Hungary), 1984.

[VA]   L. Vanhecke, *Curvature tensors*, J. Korean Math. Soc. (1) **14** (1977) 143–151.

[VI-VIC] E. Vidal and E. Vidal-Costa, *Special connections and almost foliated metrics*, J. Diff. Geom. **8** (1973) 297–304.

[WE 1]   H. Weyl, *Zur Infinitesimalgeometrie, Einordnung der projektiven und der konformen Auffassung*, Gottinger Nachrichten (1921) 99–112.

[WE 2]   H. Weyl, *Classical Groups, Their Invariants and Representations*, Princeton Univ. Press, Princeton, N. J., 1946.

[X]   Y. L. Xin, *Remarks on characteristic classes of four-dimensional Einstein manifolds*, J. Math. Phys. (2) **21** (1980) 343–346.

Neda Bokan
Faculty of Mathematics
University of Belgrade
Studentski trg 16, P.B. 550
11000 Belgrade, Yugoslavia

THE MATH. HERITAGE OF C.F. GAUSS (pp. 100-118)
edited by George M. Rassias
©1991 World Scientific Publ. Co. Singapore

SPACES OF MORPHISMS BETWEEN
ALGEBRAIC SCHEMES

Paul Cherenack and Lucio Guerra

We consider the problem whether a universal structure of a locally ringed space H exists on the set of morphisms between two quasi-projective schemes X, Y of finite type over a ground field K. We give some additional information in the classical case where X, Y are projective over K. We prove that, for $X=Y=\mathbb{A}^1, H$ does not exist. In this particular case, however, we point out that there is a locally ringed space structure L which satisfies some weaker representation property.

1. Introduction

We denote with S the category of locally noetherian schemes over a ground field K, with R the category of locally ringed spaces over K. In this paper, we are concerned with the following:

Problem I. If X, Y are the quasi-projective schemes of finite type over K, to determine whether a locally ringed space H over K exists such that the (Set-valued) functor on S defined by $T \mapsto \mathrm{Hom}_S(T \times X, Y)$ is R-represented by H, in the sense that there is a natural isomorphism of functors on S:

$$\mathrm{Hom}_S(T \times X, Y) \xrightarrow{\sim} \mathrm{Hom}_R(T, H).$$

When this happens, we simply say that $H = \mathrm{Hom}_S(X, Y)$ does exist in R.

In addition, one might ask whether there exists some "evaluation" morphism $\eta : H \times X \to Y$ having the following universal property:

(U) for every morphsim $F : T \times X \to Y$, T any locally noetherian scheme over K, there is a unique morphism $u : T \to H$ such that $F = \eta \circ (u \times 1)$.

When η exists, the association $F \mapsto u$ defines the natural bijection above.

This is here considered as a first step in the direction of solving:

Problem II. To determine some suitable category C which contains S as a full subcategory and is cartesian closed (i.e., has products, and every functor $T \mapsto \mathrm{Hom}_C(T \times X, Y)$ is represented by some object H of C).

As concerns Problem I, it is known that H exists when both X, Y are projective over K, and H is the disjoint sum of a countable family of quasi-projective schemes of finite type over K. In Sec. 1 we shortly review that theory and show how some further information on H can be deduced from the theory of Hilbert and Picard schemes (Theorem 1.4). For X, Y affine over K, instead, the space H does not exist, in general. This is proved in Sec. 2 for $X = Y = \mathbb{A}^1$ (Theorem 2.1). In this case, however, there is a ringed space L which R-represents the functor defined in Problem I, when restricted to reduced schemes T (Proposition 2.8). Finally, a short Sec. 3 contains a few informal considerations concerning possibilities for a category C in which the analogue of Problem I might be positively answered (a candidate thus for a solution to Problem II); and also connects our work to some existing work [1], [3] on the subject.

The present subject has been taken up here mainly in view of its usefulness in the construction of a purely algebraic homotopy theory for algebraic schemes over K. In fact, when one has these hom-spaces H, one may then use the approach of Huber [7] to homotopy theory, which requires a cartesian closed ambient category (where, in addition, some kinds of quotients must exist). Some recent works in this field is [2], [8].

Finally, such interests occur in other parts of mathematics. Frölicher [5] has identified a cartesian closed category, the category of convenient vector spaces, in which a meaningful study of the differential calculus can be pursued. Nel [9] has treated functional analysis in a similar way.

1. Projective Schemes

The leading fact is the following easily shown assertion:

Lemma 1.1. For every family of morphisms $F : T \times X \to Y$ the collection of all $\Gamma_t = $ graph of $F_t = F(t, -) : X \to Y$ is a flat family of subschemes of $X \times Y$.

Proof. If Γ is the graph of F and thus a closed subscheme of $T \times X \times Y$, for the projection $p : \Gamma \to T$ and $t \in T$, one has $p^{-1}(t) = \Gamma_t$. As p is the morphism composed of the projection $\Gamma \to T \times X$ (an isomorphism) followed by projection $T \times X \to T$, p is flat.

Remark 1.2. One cannot conclude from Lemma 1.1 that $Y_t = F_t(X)$ is a flat family of subschemes of Y (this can be explicitly seen in Example 1.5). Also note that the assumption of flatness for F itself is not necessary.

Therefore, assuming that the Hilbert scheme $\mathrm{Hilb}(X \times Y)$ does exist, one is lead to expect that the set theoretic map $f \mapsto \Gamma = $ graph of f, from the set $\mathrm{Hom}_S(X, Y)$ into $\mathrm{Hilb}(X \times Y)$, will identify the former with the set of closed points of some subscheme H of the latter, which will be the required solution to Problem I. In fact, this direction can be taken up when both X and Y are projective schemes (of finite type over K) since Hilbert schemes of projective schemes are known to exist (see [6], IV).

Theorem 1.3. Assume that X and Y are projective schemes of finite type over K. Then the solution H to Problem I does exist. Moreover, $H = \mathrm{Hom}_S(X, Y)$ is an open subscheme of $\mathrm{Hilb}(X \times Y)$.

Proof. See [4], Sec. 10.

In particular, let $\mathcal{O}_X(1), \mathcal{O}_Y(1)$ be very ample sheaves on X, Y, respectively. Then the Hilbert polynomial P of the graph Γ of a morphism $F : X \to Y$ is of the form

$$(*) \qquad P(n) = \text{Euler characteristic of} \quad \mathcal{O}_X(n) \otimes F^*(\mathcal{O}_Y(n)).$$

Note again that P being constant in a family of morphisms $\{F_t\}$ does not imply that the Hilbert polynomial of the $F_t(X)$ is constant too. We obtain in this way a decomposition of H into the disjoint sum of H_P, where P ranges over the Hilbert polynomials of the type appearing in $(*)$ above, each H_P being a quasi-projective scheme of finite type over K.

We now consider the scheme $\text{Hom}_S(X, Y)$ in the important case when Y is a projective space $\mathbb{P}^n$. For an arbitrary Y any closed embedding $Y \to \mathbb{P}^n$ gives rise to a closed embedding $\text{Hom}_S(X, Y) \to \text{Hom}_S(X, \mathbb{P}^n)$.

The following description of the scheme $H = \text{Hom}_S(X, \mathbb{P}^n)$ is suggested by the well-known description of any morphism $f : X \to \mathbb{P}^n$ as determined by giving some invertible sheaf $\mathcal{L}$ on X and sections $s_0, \ldots, s_n \in H^0(X, \mathcal{L})$ having no common zero, two systems $(\mathcal{L}, (s_i)), (\mathcal{L}', (s_i'))$ giving rise to the same f if and only if $\mathcal{L}$ and $\mathcal{L}'$ are isomorphic sheaves and, up to some isomorphism $\mathcal{L} \approx \mathcal{L}'$, the ordered system (s_i) becomes proportional to the ordered system (s_i'). When f is given, the sheaf $f^*(\mathcal{O}(1))$ may be taken as $\mathcal{L}$, and the pullbacks of the cannonical sections $x_0, \ldots, x_n$ of $\mathcal{O}(1)$ may be taken as $s_0, \ldots, s_n$.

Theorem 1.4. If X is projective over K, there is a natural morphism $H = \text{Hom}_S(X, \mathbb{P}^n) \to \text{Pic}(X)$ sending a closed point $\xi \in H$, representing a morphism $f : X \to \mathbb{P}^n$, into the isomorphism class of the sheaf $f^*\mathcal{O}(1)$, whose image is a locally closed subscheme $\text{Pic}^\sigma(X)$ of $\text{Pic}(X)$, having as closed points the classes of sheaves possessing $n + 1$ global sections having no common zero. In this way, H becomes a quasi-projective scheme over $\text{Pic}^\sigma(X)$.

Proof. The set $\text{Div}(X)$ of points of $\text{Hilb}(X)$ which correspond to (effective) Cartier divisors is a union of irreducible components of $\text{Hilb}(X)$. Furthermore, the natural map $\text{Div}(X) \to \text{Pic}(X)$ is a projective morphsim (cf. [6], V, p. 11, 4.4 plus p. 17, 6.6). This means that its image is a closed subscheme $\text{Pic}^\tau(X)$ of $\text{Pic}(X)$ and that there is a coherent algebra E over $\text{Pic}^\tau(X)$ such that $\text{Div}(X) = \text{Proj}(E)$. Thus a closed point $\lambda \in \text{Pic}^\tau(X)$ is the class of some invertible sheaf $\mathcal{L}$ on X possessing non-zero global sections and the fibre of $\text{Proj}(E)$ over λ is isomorphic to $\mathbb{P}(H^0(X, \mathcal{L}))$. Let us consider the affine morphisms $\pi : \text{Spec}(E) \to \text{Pic}^\tau(X)$, whose fibre over λ can therefore be identified with the vector space $H^0(X, \mathcal{L})$ itself, and write the relative product $\text{Spec}(E) \times_\pi \ldots \times_\pi \text{Spec}(E)$ $(n + 1$ times) as $\text{Spec}(F) \to \text{Pic}^\tau(X)$, where $F = E \otimes \ldots \otimes E$ $(n+1$ times$)$, the tensor product being taken over $\text{Pic}^\tau(X)$. The fibre over λ of the projective morphism

$\mathrm{Proj}(F) \to \mathrm{Pic}^\tau(X)$ can therefore be identified with the space of ordered systems of sections $s_0, \ldots, s_n \in H^0(X, \mathcal{L})$, up to proportionality.

Consider now the universal family of divisors $I \to \mathrm{Div}(X) = \mathrm{Proj}(E)$. There is a family I' over $\mathrm{Spec}(E)$ such that, over the open subscheme $\mathrm{Spec}(E) - 0$, where 0 denotes the zero section, I' agrees with the pullback of I under $\mathrm{Spec}(E) - 0 \to \mathrm{Proj}(E)$. Indeed, I being projective over $\mathrm{Div}(X)$, we may write $I = \mathrm{Proj}(G)$, for some coherent algebra G over $\mathrm{Pic}^\tau(X)$ (G may also be assumed to contain E as a subalgebra), and take $I' = \mathrm{Spec}(G)$. As the fibre of I over any point $\delta \in \mathrm{Div}(X)$ is the corresponding divisor I_δ, viewed as a subscheme of X, the fibre of I' over $s \in \mathrm{Spec}(E) - 0$ is the divisor I_δ, if s lies over $\delta \in \mathrm{Div}(X)$, s being then identified with a section of the invertible sheaf $\mathcal{O}(I_\delta)$ such that $\mathrm{div}(s) = I_\delta$. Furthermore, since $I \subset \mathrm{Div}(X) \times X$, we also have $I' \subset \mathrm{Spec}(E) \times X$. Therefore, together with the map $p : I \to \mathrm{Pic}^\tau(X) \times X$, we have a similar projection $p' : I' \to \mathrm{Pic}^\tau(X) \times X$, which is a restriction of $\pi \times 1$.

Consider then the relative product $J' = I' \times_{p'} \ldots \times_{p'} I' \to \mathrm{Pic}^\tau(X) \times X$. The scheme J' is clearly a homogeneous closed subscheme of the product $\pi \overset{\times}{\times} 1 (\mathrm{Spec}(E) \times X) = (\overset{\times}{\pi} \mathrm{Spec}(E)) \times X = \mathrm{Spec}(F) \times X$. Thus J' maps onto a homogeneous closed subscheme J of $\mathrm{Spec}(F)$. As the fibre of J' over $(\lambda, x) \in \mathrm{Pic}^\tau(X) \times X$ consists of all systems $(x, s_0, \ldots, s_n)$ where each $s_i \in H^0(X, \mathcal{L})$ and $s_i(x) = 0$, the fibre of J over $\lambda \in \mathrm{Pic}^\tau(X)$ consists of all systems $s_0, \ldots, s_n \in H^0(X, \mathcal{L})$ having some common zero $x \in X$.

Write $J = \mathrm{Spec}(F')$ and $H = \mathrm{Proj}(F) - \mathrm{Proc}(F')$. Let $\mathrm{Pic}^\sigma(X)$ denote the image of H under the map $\mathrm{Proj}(F) \to \mathrm{Pic}^\tau(X)$. The fibre of H over $\lambda \in \mathrm{Pic}^\tau(X)$ is clearly bijective to the set of morphisms $f : X \to \mathbb{P}^n$ such that $f^*\mathcal{O}(1) \approx \mathcal{L}$. The scheme H is $\mathrm{Hom}_S(X, \mathbb{P}^n)$. Let us prove this by checking that H together with an appropriate η satisfies property (U) in the introduction.

As the scheme I is a Cartier divisor on $\mathrm{Div}(X) \times X$, correspondingly I' is a Cartier divisor on $\mathrm{Spec}(E) \times X$. We may consider the invertible sheaf $\mathcal{M} = \mathcal{O}(I')$ on $\mathrm{Spec}(E) \times X$ and choose a global section σ of $\mathcal{M}$ such that $\mathrm{div}(\sigma) = I'$. It is possible then to find an invertible sheaf $\overline{\mathcal{F}}$ on $\mathrm{Spec}(F) \times X$ and global sections $\overline{\sigma}_0, \ldots, \overline{\sigma}_n$ of $\overline{\mathcal{F}}$ such that $J' = \mathrm{div}(\overline{\sigma}_0) \cap \ldots \cap \mathrm{div}(\overline{\sigma}_n)$. Indeed, using the ith projection $p_i : \mathrm{Spec}(F) \times X \to \mathrm{Spec}(E) \times X$, it suffices to take $\overline{\mathcal{F}} = p_i^*(\mathcal{M})$, which is independent of i, and $\overline{\sigma}_i = p_i^*(\sigma)$ for $i = 0, \ldots, n$. Note that the ordered system $(\overline{\sigma}_0, \ldots, \overline{\sigma}_n)$ is determined up to proportionality. Finally, one obtains an invertible sheaf $\mathcal{F}$ on $\mathrm{Proj}(F) \times X$

and global sections $\sigma_0, \ldots, \sigma_n$ of $\mathcal{F}$ having no common zero on $H \times X$, since $\operatorname{div}(\sigma_0) \cap \ldots \cap \operatorname{div}(\sigma_n)$ is the image of J' in $\operatorname{Proj}(F) \times X$. In this way, using $\sigma_0, \ldots, \sigma_n$, the morphism $\eta : H \times X \to \mathbb{P}^n$ is defined.

Let $F : T \times X \to \mathbb{P}^n$ be any family of morphisms parametrized by a locally noetherian scheme T over K. The morphism F is defined by some invertible sheaf $\mathcal{N} = F^* \mathcal{O}(1)$ on $T \times X$, plus global sections $s_0, \ldots, s_n$ of $\mathcal{N}$ having no common zero. Because of the universal properties defining $\operatorname{Pic}(X), \operatorname{Div}(X), \operatorname{Spec}(E)$, there are commutative diagrams, one for each $i = 0, \ldots, n$:

$$
\begin{array}{ccc}
& & \operatorname{Spec}(E) \\
& \overset{\alpha_i}{\nearrow} & \downarrow \\
T & \longrightarrow & \operatorname{Div}(X) \\
& \searrow & \downarrow \\
& & \operatorname{Pic}(X)
\end{array}
$$

where the bottom arrow sends $t \in T$ to the class of the invertible sheaf $\mathcal{N}(t)$ which is the restriction of $\mathcal{N}$ to $\{t\} \times X$, where the top arrow α_i sends t to the section $s_i(t)$ of $\mathcal{N}(t)$ which is the restriction of s_i to $\{t\} \times X$ and where the middle arrow sends t to $\operatorname{div}(s_i(t))$. In particular we obtain a morphism $\alpha = (\alpha_i) : T \to \operatorname{Spec}(F)$ sending t to $(s_0(t), \ldots, s_n(t))$ which clearly factors through a morphism $u : T \to H$.

From the universal property of $\operatorname{Spec}(E)$, it follows that under each morphism $\alpha_i \times 1 : T \times X \to \operatorname{Spec}(E) \times X$ one has $\mathcal{N} = (\alpha_i \times 1)^*(\mathcal{M})$ and $s_i = (\alpha_i \times 1)^*(\sigma)$. Therefore under $\alpha \times 1$ one finds that $\mathcal{N} = (\alpha \times 1)^*(\overline{\mathcal{F}})$ and $s_i = (\alpha \times 1)^*(\overline{\sigma}_i)$. Finally under $u \times 1 : T \times X \to H \times X$ one has $\mathcal{N} = (u \times 1)^*(\mathcal{F}), s_i = (u \times 1)^*(\sigma_i)$. This means precisely that the diagram

$$
\begin{array}{ccc}
& u \times 1 & \\
T \times X & \dashrightarrow & H \times X \\
{\scriptstyle F} \searrow & & \nearrow {\scriptstyle \eta} \\
& \mathbb{P}^n &
\end{array}
$$

commutes.

In order to visualize the preceding description we present a special case.

Example 1.5. We consider $H = \operatorname{Hom}_S(\mathbb{P}^n, \mathbb{P}^m), n \leq m$. According to Theorem 1.4, H is quasi-projective over $\operatorname{Pic}^\sigma(\mathbb{P}^n) = \mathbb{Z}_{\geq 0}$, by means of a map sending f to $f^*\mathcal{O}(1) = \mathcal{O}(d)$ for some integer $d \geq 0$. Consider, for fixed $d \geq 0$, the fibre H_d. Since sections of the invertible sheaf $\mathcal{O}(d)$ are identified with homogeneous polynomials of degree d in the variables $x_0, \ldots, x_n$, then, using the same notation as in the proof of Theorem 1.4, the space $\mathrm{Proj}(F)$ is identified with the projective space $\mathbb{P}^N$ consisting of all ordered systems $(f_0, \ldots, f_m)$ of homogeneous polynomials of the same degree d in $x_0, \ldots, x_n$, where the dimension $N = (m+1)\binom{n+d}{d} - 1$, while the subscheme $\mathrm{Proj}(F')$ is identified with the "resultant" subscheme R of $\mathbb{P}^n$ (an irreducible hypersurface if $n = m$). Thus $H_d = \mathbb{P}^N - R$ is an irreducible open subscheme of $\mathbb{P}^N$, and H has precisely as many irreducible components H_d as $\mathrm{Pic}^\sigma(\mathbb{P}^N)$.

Remark 1.6. As a corollary of Example 1.5 we find that, when $n = m$, since H_d is then an affine scheme for $d > 0$, the only families $T \times \mathbb{P}^n \to \mathbb{P}^n$, with T *complete*, are families of constant maps.

Finally, let us shortly discuss what happens if one tries to apply analogous arguments to the case where X and Y are arbitrary quasi-projective schemes (of finite type over K) using some projective completions $\overline{X}$ and $\overline{Y}$, respectively.

Any family $T \times X \to Y$ gives rise to a rational map $T \times \overline{X} \to \overline{Y}$, but the family of all Γ_t, where Γ_t is the closure in $\overline{X} \times \overline{Y}$ of the graph of $F_t : \overline{X} \to \overline{Y}$, is *not necessarily flat* anymore. However, there is a *flattening stratification* $U_i = T_i - T_{i+1}$ of T, induced by a filtration $T = T_0 \supset T_1 \supset \ldots \supset T_i \supset \ldots$ such that every family $\{\Gamma_t\}_{t \in U_i}$ is a flat one. Also, there is a subscheme $\overline{R}$ of $\mathrm{Hilb}(\overline{X} \times \overline{Y})$ parametrizing rational maps $\overline{X} \to \overline{Y}$ which are regular on X and one obtains a collection of morphism $U_i \to \overline{R}$, but different U_i are sent into different connected components of $\overline{R}$, so that one does not even obtain a continuous map $T \to \overline{R}$. In fact a ringed space $H = \mathrm{Hom}_S(X, Y)$, if it exists, must have the image of U_{i+1} in the boundary of the image of U_i. In the next section we will show how this fact actually makes Problem I unsolvable in general.

2. Affine Schemes

When X, Y are affine schemes of finite type over K, a ringed space $H = \mathrm{Hom}_S(X, Y)$ as in Problem I does not exist, in general. More precisely, for $X = Y = \mathbb{A}^1$, we prove:

Theorem 2.1. The functor $T \mapsto \mathrm{Hom}_S(T \times \mathbb{A}^1, \mathbb{A}^1)$, on the category S of locally noetherian schemes over K, is not represented by any object in R.

Consider, for each $d \geq 0$, the affine space $L_d = \mathrm{Spec}(R_d)$, where $R_d = K[\eta_0, \ldots, \eta_d]$. Then the set of its K-valued points $(b_0, \ldots, b_n)$ is canonically bijective to the set of those morphism $\mathbb{A}^1 \to \mathbb{A}^1$ which are represented by polynomials $f(X) = b_0 + b_1 X \times + \ldots + b_d X^d \in K[X]$ of degree $\leq d$, and there are natural embeddings $L_d \hookrightarrow L_{d+1}$. Furthermore, each L_d parametrizes a family of morphisms $\lambda_d : L_d \times \mathbb{A}^1 \to \mathbb{A}^1$ which, up to the previous identification of K-valued points of L_d with morphisms $f : \mathbb{A}^1 \to \mathbb{A}^1$, acts via evaluation $(f, x) \mapsto f(x)$. Namely, the dual K-morphism $K[t] \to K[\eta_0, \ldots, \eta_d] \otimes K[X] = K[\eta_0, \ldots, \eta_d, X]$ sends $t \mapsto \omega_d = \eta_0 + \eta_1 X + \ldots + \eta_d X^d$.

The important fact here is that:

2.2. for any family of morphisms $F : T \times \mathbb{A}^1 \to \mathbb{A}^1$, T any connected locally noetherian scheme over K, there exist some $d \geq 0$ and a unique morphism $u : T \to L_d$ such that $F = \lambda_d \circ (u \times 1)$.

Proof. For every affine open subset $U = \mathrm{Spec}(D)$ of T, let the restriction of F to $U \times \mathbb{A}^1$ be dual to the K-morphism $K[t] \to D \otimes K[X] = D[X]$, where $t \mapsto$ some θ_U, and let d_U be the degree of θ_U as a polynomial in X. For every $d \geq 0$, the union T_d of those U such that $d_U = d$ is clearly open and closed in T, and T is the disjoint union of the T_d. Since T is connected, then $T = T_d$ for some d. For every U, write then $\theta_U = \sum_{i=0}^{d} \theta_i X^i$, all $\theta_i \in D$. The K-morphism $K[\eta_i] \to D$ sending $\eta_i \mapsto \theta_i$ gives rise to a morphism $U \to L_d$. Clearly two such morphisms agree where both are defined, so that we have a well-defined morphism $u : T \to L_d$ which is easily seen to satisfy $F = \lambda_d \circ (u \times 1)$.

If we consider the ringed space $L = \varinjlim L_d$ (let us recall that the underlying topological space of L is the inductive limit of the underlying topological spaces of L_d; calling $i_d : L_d \to L$ the canonical maps, a subset $F \subseteq L$ is closed iff every $i_d^{-1}(F)$ is closed in L_d; the structure sheaf is then

$\mathcal{O}_L = \varprojlim i_{d*}(\mathcal{O}_{L_d})$) then it is possible to define a natural transformation of functors:

$$\mathrm{Hom}_S(T \times \mathbb{A}^1,\ \mathbb{A}^1) \xrightarrow{\phi} \mathrm{Hom}_R(T, L)\,. \tag{2.3}$$

For every $F : T \times \mathbb{A}^1 \to \mathbb{A}^1$, with T connected, let $\phi(F) = i_d \circ u$ where u is the map defined in 2.2 (note that the composition $i_d \circ u$ is clearly independent of d). For arbitrary T, let $\phi(F) = v$ be the unique morphism such that, for every connected component T' of T, if F' denotes the restriction of F to $T \times \mathbb{A}^1$, then $v|_{T'} = \phi(F')$ as defined before.

This suggests that the ringed space L will play a privileged role among all possible solutions to Problem I in the category R. In fact, one has:

2.4. If the functor in 2.1 were represented by some object in R, then it were represented by L.

Suppose that a ringed space H represents our functor. By definition, we then have a family of compatible morphisms $j_d : L_d \to H$, hence a morphism $j : L \to H$ such that every $j_d = j \circ i_d$. For every morphism $F : T \to \mathbb{A}^1 \to \mathbb{A}^1$, with T connected, if $u : T \to L_d$ is as in 2.2, then by functoriality the assciated morphism $T \to H$ is the composed map $T \xrightarrow{u} L_d \xrightarrow{j_d} H$, where $j_d \circ u = j \circ i_d \circ u = j \circ \phi(F)$. When T is not connected and $v = \phi(F)$ then, by a similar componentwise argument, one sees that the associated morphism $T \to H$ is the composed map $T \xrightarrow{v} L \xrightarrow{j} H$.

This means that the representation map $\mathrm{Hom}_S(T \times \mathbb{A}^1,\ \mathbb{A}^1) \to \mathrm{Hom}_S(T, H)$ factors as:

$$\mathrm{Hom}_S(T \times \mathbb{A}^1,\ \mathbb{A}^1) \xrightarrow{\phi} \mathrm{Hom}_R(T, L) \xrightarrow{\psi} \mathrm{Hom}_R(T, H) \tag{2.5}$$

where $\psi(v) = j \circ v$.

To prove 2.4, we have therefore to show that:

2.6. $\psi \circ \phi$ being bijective implies both ϕ, ψ bijective.

Furthermore, because of 2.4, Theorem 2.1 is equivalent to a more explicit statement, namely that the ringed space L does not represent the functor in question. Even more precisely, because of 2.6, it is equivalent to the following:

2.7. The natural transformation of functors ϕ in 2.3, which is clearly injective, is not an isomorphism.

It is important to point out that, however:

Proposition 2.8. The functor $T \mapsto \mathrm{Hom}_S(T \times \mathbb{A}^1, \mathbb{A}^1)$, when restricted to the category S_{red} of reduced locally noetherian schemes T over K, is R-represented by L.

Equivalently, the map ϕ is bijective for T reduced.

The proof of 2.6 relies upon the proof of 2.8. Before entering into the proofs of statements 2.6, 2.7, 2.8, which will require some more information concerning the structure of the ringed space L, we are in a position to make a few comments which further elucidate the present situation.

The collection of all evaluation morphisms $\lambda_d : L_d \times \mathbb{A}^1 \to \mathbb{A}^1$ only determines a morphism $\lambda = \varinjlim \lambda_d : \varinjlim(L_d \times \mathbb{A}^1) \to \mathbb{A}^1$, while there is no morphism $\eta : L \times \mathbb{A}^1 \to \mathbb{A}^1$ satisfying $\lambda_d = \eta \circ (i_d \times 1)$.

Indeed, if η would exist, then the map ϕ defined in 2.3 would clearly be surjective, any $v : T \to L$ coming from $T \times \mathbb{A}^1 \overset{v \times 1}{\to} L \times \mathbb{A}^1 \overset{\eta}{\to} \mathbb{A}^1$.

This agrees with the fact that the natural morphism $\varinjlim(L_d \times \mathbb{A}^1) \to (\varinjlim L_d) \times \mathbb{A}^1$, which is an homeomorphism of the underlying topological spaces, is not an isomorphism of ringed spaces.

The induced map of global sections of their structure sheaves is indeed:

$$\varprojlim(K[\eta_0, \dots, \eta_d] \otimes K[X]) \leftarrow (\varprojlim K[\eta_0, \dots, \eta_d]) \otimes K[x] \qquad (2.9)$$

(using the basic fact that the ring of global sections of the structure sheaf of $X \times \mathbb{A}^1$ is $\Gamma(X, \mathcal{O}_x) \otimes \Gamma(\mathbb{A}^1, \mathcal{O}\mathbb{A}^1)$), which is easily seen to be injective but not bijective using, for instance, the following easy and useful description of any ring of the kind $\varinjlim A[\eta_0, \dots, \eta_d]$, for any ring of coefficients A. (Recall that the projective limit $\varprojlim R_d$ of a projective system of rings R_d is the subring of their product ΠR_d consisting of all ordered systems (a_d) such that $g_{d\delta}(a_d) = a_\delta$ whenever $d \geq \delta$ and $g_{d\delta}$ is the corresponding homomorphism $R_d \to R_\delta$.)

Consider the ring $A[[\eta]]$ of formal power series in the infinite set of variables $\eta_0, \eta_1, \dots, \eta_d, \dots$, and denote by M_d the ideal consisting of those power series in which a monomial occurs iff it contains some of the variables $\eta_{d+1}, \dots$; in particular $A[[\eta]]/M_d = A[[\eta_0, \dots, \eta_d]]$. Any power series $P(\eta)$ may be written in a unique way as $P = P' + P''$, where $P' \in A[[\eta_0, \dots, \eta_d]]$ and $P'' \in M_d$.

Theorem 2.10. The ring $R = \varprojlim A[[\eta_0, \dots, \eta_d]]$ is cannonically isomorphic to the subring R' of $A[[\eta]]$ consisting of all power series $P(\eta)$ such that, for every d, P is congruent to a polynomial in $A[\eta_0, \ldots, \eta_d]$ modulo the ideal M_d or, in other words, the power series $P' \in A[[\eta_0, \ldots, \eta_d]]$ actually belongs to the polynomial ring $A[\eta_0, \ldots, \eta_d]$.

Proof. There are obvious projections $R' \to A[\eta_0, \ldots, \eta_d], P \mapsto P_d$, hence there is an induced morphism $\sigma : R' \to R, P \mapsto (P_d)$. Let (F_d) be any element of R (see the second remark in brackets after 2.9). Define a power series P in the following way: for every monomial m in the variables $\eta_0, \ldots, \eta_d, \ldots$, let d be any integer $\geq$ degree of m; take then the coefficient $a(m)$ of m in F_d; it is independent of d since, for any other $d' \geq$ degree of m, supposing for instance $d \leq d', F_{d'}$ is congruent to F_d modulo M_d, hence m occurs in $F_{d'}$ with the same coefficient $a(m)$; let the power series P be defined as $\sum_m a(m)m$. Then $P \in R'$ since $P \equiv F_d \bmod M_d$ and, therefore, $\sigma(P) = (F_d)$. Thus σ is surjective and obviously injective, hence an isomorphism.

Applying 2.10 to 2.9 we see that elements of the left side are identified with power series in the variables $\eta_0, \ldots, \eta_d, \ldots$ having arbitrary coefficients in $K[X]$ and satisfying the above defined property; an element of the right side, instead, has coefficients in $K[X]$ whose degrees only take a finite number of values, since it is obtained as a finite linear combination of power series with coefficients in K – satisfying the same property – multiplied by polynomials in $K[X]$. Thus 2.9 is a proper embedding.

Note again that, after 2.10, the morphism $\lambda = \varinjlim \lambda_d$ defined above induces the map of global sections of structure sheaves $R' \cong \varprojlim(K[\eta_0, \ldots, \eta_d] \otimes [X]) \leftarrow K[t]$ where t is sent into the element $\omega = \sum_{d=0}^{\infty} \eta_d X^d$, which is not an element of $(\varprojlim K[\eta_0, \ldots, \eta_d]) \otimes K[X]$, in accordance to what has already been said.

Let us now come to a more thorough description of subspaces of the ringed space L and ideals of its structure sheaf $\mathcal{O} = \mathcal{O}_L$.

If $\mathcal{J}$ is an ideal of $\mathcal{O}_L$, consider the restriction $\mathcal{O} \to i_* \mathcal{O}_d$, where $\mathcal{O}_d = \mathcal{O}_{L_d}$ and for simplicity of notation we use the same symbol i for all inclusions $L_d \hookrightarrow L$, as no confusion can occur. Let $\mathcal{F}_d$ be the image of $\mathcal{J}$ in $i_* \mathcal{O}_d$, a submodule of $i_* \mathcal{O}_d$. Putting $\mathcal{J}_d = i^* \mathcal{F}_d$, an ideal of $\mathcal{O}_d$, we may also write $\mathcal{F}_d = i_* \mathcal{J}_d$, because $i : L_d \hookrightarrow L$ is a closed embedding (in general, there only is a morphism $\mathcal{G} \to i_* i^* \mathcal{G}$). The projective limit $\overline{\mathcal{J}} = \varprojlim i_* \mathcal{J}_d$ is then an ideal of $\mathcal{O}$ containing $\mathcal{J}$, which we call the *saturation* of $\mathcal{J}$; the ideal $\mathcal{J}$ will be said to be *saturated* iff $\mathcal{J} = \overline{\mathcal{J}}$.

Note that $\overline{\mathcal{J}} = \mathcal{O}$ iff every $\mathcal{J}_d = \mathcal{O}_d$. Furthermore, for a saturated $\mathcal{J}$, one has $\mathcal{O}/\mathcal{J} = \varprojlim i_* \mathcal{O}_d / \mathcal{J}_d$.

In the same way, for the ring $R = \varprojlim K[\eta_0, \dots , \eta_d]$ of global sections of $\mathcal{O}$, we may define the saturation of any ideal I of R as $\bar{I} = \varprojlim I_d, I_d =$ image of I in $R_d = K[\eta_0, \dots , \eta_d]$, and the concept of a saturated ideal $I = \bar{I}$.

The global sections of a saturated ideal $\mathcal{J} = \varprojlim i_* \mathcal{J}_d$ of $\mathcal{O}$ constitute a saturated ideal $I = \varprojlim I_d$ of R, where $I_d = \Gamma(L_d, \mathcal{J}_d)$. Conversely, a saturated ideal $\mathcal{J}$ such that every $\mathcal{J}_d$ is quasi-coherent is in turn determined by its module of global sections I, since $\mathcal{J}_d = \tilde{I}_d, \tilde{I}_d$ denoting as usual the quasi-coherent sheaf on the affine space L_d whose module of global sections is I_d.

The following lemma is easily checked:

Lemma 2.11. If X is a closed ringed subspace of L, with inclusion mapping $i : X \to L$, then there is a unique inductive system of closed ringed subspaces X_d of L_d such that $X = \varinjlim X_d$. The structure sheaf is then $\mathcal{O}_X = \varinjlim i_* \mathcal{O}_{X_d}$, the associated ideal is $\mathcal{J} = \varprojlim i_* \mathcal{J}_d$, where $\mathcal{J}_d =$ ideal of X_d in $\mathcal{O}_d$. This establishes a bijection between closed ringed subspaces of L and saturated ideals of $\mathcal{O}_L$. Furthermore, every X_d is a closed subscheme of L_d iff every $\mathcal{J}_d$ is a quasi-coherent ideal of $\mathcal{O}_d$.

Proof of 2.7. We will show that, for $T = \operatorname{Spec} K[\varepsilon]/\varepsilon^2$, there is some morphism $f : T \to L$ (a tangent vector at a K-valued point of L) which does not come from any morphism $T \times \mathbb{A}^1 \to \mathbb{A}^1$, since it does not factor as $T \xrightarrow{u} L_d \xrightarrow{i_d} L$.

First of all, let us describe how any $f : T \to L$ is made up. It gives rise to a morphism $f' ; T' = \operatorname{Spec} K \to L$. The image of the single point of T' will be some point x of L belonging to some L_d and there is a K-morphism of sheaves $\mathcal{O}_L \to f'_*(K)$, necessarily surjective, whose kernel $\mathcal{J}$ is therefore the ideal of the closure of x. Hence $\mathcal{J}$ is a saturated maximal ideal wth $\mathcal{O}_L/\mathcal{J} = f_*(K)$ and such that every $\mathcal{J}_d$ is quasi-coherent (2.11).

It is determined by its module of global sections $I = \Gamma(L, \mathcal{J})$, which is a saturated maximal ideal for $R = \Gamma(L, \mathcal{O}_L)$ with $R/I = K$.

Consider now all the (surjective) projections $p_d : R/I \to K[\eta_0, \dots, \eta_d]/I_d$. They cannot all be zero-maps, otherwise all $I_d = 1$, and this implies $I = 1$ (I saturated). Since $R/I = K$ is a field, all non-zero projections are isomorphisms. Furthermore, if p_d is an isomorphism then p_δ is so too, for every $\delta \geq d$. Let therefore d be such that $p_\delta = 0$ for $\delta \leq d - 1$, and $p_\delta =$ isomorphism for $\delta \geq d$. This implies $I_\delta = 1$ for $\delta \leq d - 1$. Also, I_d must be a maximal ideal M of $K[\eta_0, \dots, \eta_d]$ with $K[\eta_0, \dots, \eta_d]/M = K$. Furthermore, since $K[\eta_0, \dots, \eta_{d+1}]/I_{d+1} \to K[\eta_0, \dots, \eta_d]/I_d$ is an isomorphism (it is defined by $\eta_{d+1} \mapsto 0$), then $I_{d+1} =$ ideal generated by M and η_{d+1}. Similarly, $I_\delta =$ ideal generated by $M, \eta_{d+1}, \dots, \eta_\delta$ for every $\delta \geq d+1$. Hence, I is the ideal consisting of those power series P such that, writing $P = P' + P''$ with $P' \in K[\eta_0, \dots, \eta_d]$ and $P'' \in M_d$ (as in 2.10), then $P' \in M$. Thus, x is a closed point of L_d and the map $T' \to L$ actually factors as $T' \to L_d \to L$.

The original morphism $f : T \to L$ also determines a morphism $\mathcal{O}_L \to f_*(K[\varepsilon]/\varepsilon^2)$ sending $\mathcal{J}$ into (ε), whose kernel $\mathcal{N}$ contains $\mathcal{J}^2$, hence a morphism $\mathcal{J}/\mathcal{N} \to K$, and f is determined by $\mathcal{J}$ and $\mathcal{N}$. When $\mathcal{N}$ is a saturated ideal such that every $\mathcal{N}_d$ is quasi-coherent, then f is determined by the two modules of global sections $I = \Gamma(L, \mathcal{J}), N = \Gamma(L, \mathcal{N})(N \supseteq I^2)$. It is immediate to see that, if f factors as $T \to L_\delta \to L(L_\delta$ containing $x)$ then the induced map $\mathcal{J}/\mathcal{N} \to K$ factors as $\mathcal{J}/\mathcal{N} \to i_*(\mathcal{J}_\delta/\mathcal{N}_\delta) \to K$ and conversely. Equivalently, by taking global sections, we obtain a K-linear map $I/N \to K$ which factors as $I/N \to I_\delta/N_\delta \to K$ iff f factors through $L_\delta, I_\delta/N_\delta$ being a finite-dimensional vector space over K.

The sequence of projections $\dots \to I_{\delta+2}/N_{\delta+2} \to I_{\delta+1}/N_{\delta+1} \to I_\delta/N_\delta \to \dots$ may happen to be "stationary", i.e., all projections are isomorphisms for $\delta \geq$ some k which is necessarily $\geq d$. In this case $I/N = \varprojlim I_\delta/N_\delta \to I_{k-1}/N_{k-1}$ is an isomorphism, and linear maps $I/N \to K$ are the same as tangent vectors to L_k at x $(k \geq d)$.

Suppose, instead, that the above sequence is non-stationary. We may identify at the same time each vector space I_δ/N_δ with some K^n (n depending on δ) and each projection $I_{\delta'}/N_{\delta'} \to I_\delta/N_\delta(\delta' \geq \delta)$ with some coordinate projection $K^{n'} \to K^n(n' \geq n)$. Since the sequence of dimensions n is by hypothesis increasing non-stationary, the space $I/N = \varprojlim I_\delta/N_\delta$ turns out to be identified with $K^{\mathbb{N}} = \varprojlim K^n$. By elementary infinite-dimensional linear algebra, there are K-linear maps $K^{\mathbb{N}} \to K$ which do not factor through any K^n.

In conclusion: take, for instance, as $\mathcal{J}$ the ideal of the K-valued point x of L representing the zero-map $\mathbb{A}^1 \to \mathbb{A}^1$, whose module I of global sections consists of those power series $P(\eta) \in R$ having no constant term; as $\mathcal{N}$ the saturated ideal which corresponds as in 2.11 to the submodule N of I consisting of those $P(\eta)$ having neither constant nor linear terms. The space I_δ/N_δ is identified to the space of homogeneous linear polynomials in the variables $\eta_0, \ldots, \eta_\delta$, hence the sequence of dimensions $n(\delta) = \delta + 1$ is strictly increasing. Take then any K-linear map φ from $I/N = K^{\mathbb{N}}$ to K which does not factor through any I_δ/N_δ. The induced K-morphism Φ from $R/N = K \oplus I/N$ into $K[\varepsilon]/\varepsilon^2$ (which is defined by $\Phi(c + P(\eta)) = c + \varepsilon\varphi(P)$) determines a morphism $f : T \to L$ (sending the single point of T into x) which, according to what has just been said, does not factor through any L_δ, hence does not come from any $T \times \mathbb{A}^1 \to \mathbb{A}^1$.

Proof of 2.8. The statement means that the map ϕ in 2.3 is surjective for T reduced. Clearly, we may assume T irreducible. In this case, it is equivalent to show that every morphism $f : T \to L$ factors as $T \to L_d \to L$, for some d.

Let x be the image of the generic point of T and let $x \in L_d$ but $x \notin L_{d-1}$. Then, as concerns the underlying topological spaces, we have $f(T) \subseteq$ closure of $x \subseteq L_d$. Let F be the field of rational functions on T (local ring of the generic point of T). Then a morphism $T' = \operatorname{Spec} F \to L$ is induced by f, sending the single point of T' into x. The kernel $\mathcal{J}$ of the dual morphism of sheaves $\mathcal{O}_L \to f_*F$ (the last being the constant sheaf with support equal to the closure of x and stalk equal to F) is the ideal of the closure of x, hence $\mathcal{J}$ is saturated and every $\mathcal{J}_d$ is quasi-coherent by 2.11. On the other hand, $\mathcal{J}$ is easily seen to agree with the kernel of $\mathcal{O}_L \to f_*\mathcal{O}_T$, the morphism of sheaves associated with f.

If $\mathcal{P}_d$ is the ideal of the closure of x in $\mathcal{O}_d$, one immediately sees that $\mathcal{J}_\delta = \mathcal{O}_\delta$ for $\delta \le d-1, \mathcal{J}_d = \mathcal{P}_d$, and $\mathcal{J}_\delta =$ ideal generated by $\mathcal{P}_d, \eta_{d+1}, \ldots, \eta_\delta$ in $\mathcal{O}_\delta$ for $\delta \ge d+1$. It follows that $\mathcal{O}_\delta/\mathcal{J}_\delta = 0$ for $\delta \le d-1$ and $\mathcal{O}_\delta/\mathcal{J}_\delta \xrightarrow{\sim} \mathcal{O}_d/\mathcal{J}_d$ for $\delta \ge d$. Hence, $\mathcal{J}$ being saturated, one has $\mathcal{O}_L/\mathcal{J} = \varprojlim i_*\mathcal{O}_\delta/\mathcal{J}_\delta \xrightarrow{\sim} i_*\mathcal{O}_d/\mathcal{P}_d$. This implies that $\mathcal{O}_L/\mathcal{J} \to f_*\mathcal{O}_T$ factors as $\mathcal{O}_L/\mathcal{J} \to i_*\mathcal{O}_d/\mathcal{P}_d \to f_*\mathcal{O}_T$, whence $\mathcal{O}_L \to f_*\mathcal{O}_T$ factors as

$\mathcal{O}_L \to i_* \mathcal{O}_d \to f_* \mathcal{O}_T$, and $f : T \to L$ factors as $T \to L_d \to L$.

We need the following information concerning the residue fields of the structure sheaf $\mathcal{O}_L$ at any point of L.

Lemma 2.12. Let x be any point of L. The closure of x is an algebraic subvariety of some L_d, whose field of rational functions we denote by F. Then the residue field of L at x is $\mathcal{O}_{L,x}/M_{L,x} = F$.

Proof. By definition, $\mathcal{O}_{L,x} = \varinjlim_{x \in U} \mathcal{O}_L(U) = \varinjlim_{x \in U} \varprojlim_{\delta} \mathcal{O}_\delta(U_\delta)$, where $U_\delta = i_\delta^{-1}(U)$. Let $\mathcal{J}$ be the saturated ideal of the closure of x in L. Assume that $x \in L_d$ but $x \notin L_{d-1}$. If $\mathcal{P}_d$ is the ideal of the closure of x in L_d, we have seen in the proof of 2.8 that $\mathcal{O}_L/\mathcal{J} = \varprojlim \mathcal{O}_\delta/\mathcal{J}_\delta \cong \mathcal{O}_d/\mathcal{P}_d$. The maximal ideal $M_{L,x}$ is the stalk of $\mathcal{J}$ at x, hence $M_{L,x} = \varinjlim_{x \in U} \mathcal{J}(U) = \varinjlim_{x \in U} \varprojlim_{\delta} \mathcal{J}_\delta(U_\delta)$, and $\mathcal{O}_{L,x}/M_{L,x} = \varinjlim_{x \in U} \varprojlim_{\delta} \mathcal{O}_\delta(U_\delta)/\mathcal{J}_\delta(U_\delta)$.

For any doubly indexed system of rings A_{ij}, which is direct in i and inverse in j, there is a canonical map $\varinjlim_i \varprojlim_j A_{ij} \to \varprojlim_j \varinjlim_i A_{ij}$, which is easily seen to be an isomorphism provided that $A_{ij} \to A_{ij'}$ is surjective for $j \geq j'$ and all i, and there is some j_0 such that $A_{ij} \to A_{ij_0}$ is an isomorphism for $j \geq j_0$ and all i. Both double limits are then isomorphic to $\varinjlim_i A_{ij_0}$.

In the present case, we have a map:

$$\mathcal{O}_{L,x}/M_{L,x} = \varinjlim_{x \in U} \varprojlim_{\delta} \mathcal{O}_\delta(U_\delta)/\mathcal{J}_\delta(U_\delta) \to \varprojlim_{\delta} \varinjlim_{x \in U} \mathcal{O}_\delta(U_\delta)/\mathcal{J}_\delta(U_\delta)\,,$$

where $\varinjlim_{x \in U} \mathcal{O}_\delta(U_\delta)/\mathcal{J}_\delta(U_\delta)$ is 0 for $\delta \leq d-1$, and is F for $\delta \geq d$, so that the ring on the right side is F.

The statement will follow from the remark above as soon as we show that there is a basis of neighborhoods U of x (a cofinal system of U containing x) such that every $\mathcal{O}_{\delta+1}(U_{\delta+1}) \to \mathcal{O}_\delta(U_\delta)$ is surjective. Then clearly every $\mathcal{O}_{\delta+1}/\mathcal{J}_{\delta+1}(U_{\delta+1}) \to \mathcal{O}_\delta/\mathcal{J}_\delta(U_\delta)$ is surjective too and is an isomorphism for $\delta \geq d$.

Let U be any neighborhood of x, and let $X = \varinjlim X_\delta$ be its closed complement. Let $\mathcal{F}$ be the saturated ideal of X. Since X_δ is simply the complement of some open U_δ in L_δ, it is a reduced closed subscheme of L_δ, hence $\mathcal{F}_\delta$ is quasi-coherent. The ideal $\mathcal{F}$ is therefore determined by its saturated module of global sections $\Gamma(L, \mathcal{F}) = \varprojlim F_\delta$. The same clearly happens for the ideal $\mathcal{J}$ of the closure of x in L, which we have already described before, and let $\Gamma(L, \mathcal{J}) = \varprojlim I_\delta$. Then $x \notin X$ means $\mathcal{F} \subset \mathcal{J}$, hence there is some $f \in \Gamma(L, \mathcal{F}) - \Gamma(L, \mathcal{J})$. This means that some $f_\delta \in F_\delta - I_\delta$. But note that $f_\delta \in I_\delta$ implies $f_{\delta'} \in I_{\delta'}$, for $\delta' \geq \delta \geq d$ (cfr. proof of 2.8). Hence we have $f_\delta \in F_\delta - I_\delta$ for $\delta \geq d$. Let V_δ be the open subset of L_δ defined by $f_\delta \neq 0$. Then V_δ is an open neighborhood of x contained in U_δ for every δ, so that $V = \varinjlim V_\delta$ is an open neighborhood of x contained in U, and $\mathcal{O}_{\delta+1}(V_{\delta+1}) = (R_{\delta+1})_{f_{\delta+1}} \to \mathcal{O}_\delta(V_\delta) = (R_\delta)_{f_\delta}$ is clearly surjective for every δ.

Proof of 2.6. After 2.8, we know that, when G is reduced, the map ϕ is bijective. Hence, $\psi \circ \phi$ bijective implies ψ bijective too. In general, $\psi \circ \phi$ bijective only implies ψ surjective. We will prove ψ injective for arbitrary T by showing that $j : L \to H$ is necessarily a monomorphism in the category R, i.e., it is set-theoretically injective with epimorphic map of sheaves $\mathcal{O}_H \to j_* \mathcal{O}_L$.

Let $x, x' \in L$ and let d be such that $x, x' \in L_d$. Then x, x' are the generic points of subvarieties of L_d, whose fields of rational functions we denote by F, F'. We have morphisms $T = \operatorname{Spec} F \xrightarrow{f} L, T' = \operatorname{Spec} F' \xrightarrow{f'} L$, and the residue fields of x, x' in L are $\mathcal{O}_{L,x}/M_{L,x} = F, \mathcal{O}_{L,x'}/M_{L,x'} = F'$, by Lemma 2.12. Assume now that $j(x) = j(x') = y \in H$, and put $g = j \circ f, g' = j \circ f'$. The composed morphism g induces an injective map of residue fields $\overline{F} = \mathcal{O}_{H,y}/M_{H,y} \to \mathcal{O}_{L,x}/M_{L,x} = F$ which, for simpler notation, we write as an inclusion $\overline{F} \subseteq F$. Putting $\overline{T} = \operatorname{Spec} \overline{F}$, we thus have a morphism $\alpha : T \to \overline{T}$ such that $g = \overline{g} \circ \alpha$, where $\overline{g}$ is the map $\overline{T} \to H$ which is associated to the $\overline{F}$-valued point y of H. Since $\overline{T}$ is reduced and ψ is bijective, there is $\overline{f} : \overline{T} \to L$ such that $\overline{g} = j \circ \overline{f}$. Both maps f and $\overline{f} \circ \alpha$ are such that $\psi(f) = \psi(\overline{f} \circ \alpha) = g$. Again, since T is reduced and ψ is bijective, we have $f = \overline{f} \circ \alpha$. This means that the map $F = \mathcal{O}_{L,x}/M_{L,x} \to \overline{F}$ induced by $\overline{f}$ is a right inverse of the inclusion $\overline{F} \hookrightarrow F$, whence $\overline{F} = F$. Similarly, we also have $\overline{F} = F'$. Hence, there is an isomorphism $F \xrightarrow{\sim} F'$ which makes the identity $\overline{F} = \mathcal{O}_{H,y}/M_{H,y} = \mathcal{O}_{L,x}/M_{L,x} = F$ isomorphic to the identity $\overline{F} = \mathcal{O}_{H,y}/M_{H,y} = \mathcal{O}_{L,x'}/M_{L,x'} = F'$. Thus, the dual isomorphism $\theta : \operatorname{Spec} F' \to \operatorname{Spec} F$ is such that $g' = g \circ \theta$. Both maps f'

and $f \circ \theta$ are such that $\psi(f') = \psi(f \circ \theta) =$ the associated map $\bar{g}$ to the $\overline{F}$-valued point y of H, defined before. Again, it follows that $f' = f \circ \theta$ and therefore $x' = x$.

We have thus proved that $j : L \to H$ is injective. It is, furthermore, surjective. Let $y \in H$ and let $\overline{F} = \mathcal{O}_{H,y}/M_{H,y}$ be its residue field. The morphism $\bar{g} : \mathrm{Spec}\,\overline{F} \to H$ factors through L, i.e., there is $\bar{f} : \mathrm{Spec}\,\overline{F} \to L$ such that $\bar{g} = j \circ \bar{f}$. Then $y = j(x)$, for $x =$ image under $\bar{f}$ of the single point of $\mathrm{Spec}\,\overline{F}$.

Finally, we are now prepared to show that the map of sheaves $\mathcal{O}_H \to j_* \mathcal{O}_L$ is necessarily an epimorphism. Let $y = j(x)$ be an arbitrary point of H. We have a commutative diagram of local homomorphisms of local rings:

$$
\begin{array}{ccc}
\mathcal{O}_{H,y} & \longrightarrow & (j_* \mathcal{O}_L)_y \\
& \searrow & \downarrow \\
& & \mathcal{O}_{L,x}
\end{array}
$$

where both downward arrows are clearly surjective. Taking quotients by the maximal ideals and using earlier notation, we obtain a commutative diagram of injective maps:

$$
\begin{array}{ccc}
\overline{F} = \mathcal{O}_{H,y}/M_{H,y} & \longrightarrow & (j_* \mathcal{O}_L)_y/M \\
& \searrow & \downarrow \\
& & \mathcal{O}_{L,x}/M_{L,x} = F
\end{array}
$$

where the arrow $\overline{F} \to F$ is an identity map, and $(j_* \mathcal{O}_L)_y/M \to F$ is an isomorphism, hence $\overline{F} \to (j_* \mathcal{O}_L)_y/M$ is an isomorphism too, and $\mathcal{O}_{H,y} \to (j_* \mathcal{O}_L)_y$ is therefore an epimorphism.

3. Considerations for Further Work

A number of consequences can be drawn from the negative result of Sec. 2, concerning possible approaches to Problem II. A preliminary step should be the following generalization of Problem I.

Problem I$'$. To determine some category C, containing S as a full subcategory, such that every functor $T \mapsto \mathrm{Hom}_S(T \times X, Y)$, for objects X, Y of S, is C-represented by some object H of C, i.e., there is a natural isomorphism of functors $\mathrm{Hom}_S(T \times X, Y) \xrightarrow{\sim} \mathrm{Hom}_C(T, H)$.

At this point, two different directions may be followed:

(A) The category C may still be a subcategory of R, having the same objects as R, but then, for the space L of Sec. 2, $\mathrm{Hom}_C(T, L)$ must be equal to the image in $\mathrm{Hom}_R(T, L)$ of the natural transformation occurring in 2.3, for any object T of S. In particular, C must be a non-full subcategory of R. The author are not yet in a position to provide a useful description of morphisms in C.

(B) The category C may not be defined as a subcategory of R, for instance by taking as objects ringed spaces plus some additional data.

It may be shown, for instance, that the category $\mathrm{Ind}(S)$ whose objects are inductive systems $(X_i)_{i \in I}$ of locally noetherian schemes over K (I an up-directed set), and where $\mathrm{Hom}_{\mathrm{Ind}(S)}((X_i), (Y_j)) = \varprojlim_i \varinjlim_j \mathrm{Hom}_S(X_i, Y_j)$,

is a solution to Problem I'. It is not expected, however, to be a solution to the more general Problem II, since the category $\mathrm{Ind}\,(\mathrm{Ind}(S))$ is probably non-isomorphic to $\mathrm{Ind}(S)$ itself. One may perhaps successfully look at suitable subcategories C of $\mathrm{Ind}(S)$.

Form this point of view, a partial answer to Problem II is given in [1], where a cartesian closed extension of the category A of affine schemes of finite type over K is constructed, which is called *Ind-Aff* (not to be confused with $\mathrm{Ind}\,(A)$).

Something more can be said concerning possible features of categories C which satisfy Problem I'. One may set a more special requirement on C, analogous to (U) in the introduction, namely that:

(U*) For objects X, Y of S, there is some morphism $\eta : H \times X \to Y$ in C which is universal in the sense that for every $F : T \times X \to Y$ in S there is a unique $u : T \to H$ in C such that $F = \eta \circ (u \times 1)$.

Then, in particular, the assignment $F \mapsto u$ defines a natural isomorphism of functors on S, $\mathrm{Hom}_S(T \times X, Y) \xrightarrow{\sim} \mathrm{Hom}_C(T, H)$, hence C satisfies Problem I'.

Accordingly, a *strong* concept of cartesian closedness may be defined and then required for C in Problem II.

Because of what we have seen in the first part of Sec. 2, one does not expect such a universal family to exist if the category C is constructed as in case (A) above. This shows that the two points of view (A), (B) above described present some qualitative difference.

Acknowledgement

The first author acknowledges financial support from the South African Council for Scientific and Industrial Research via the Topology Research Group at the University of Cape Town.

References

1. P. Cherenack, *A cartesian closed extension of a category of affine schemes*, Cahiers Top. Geom. Diff. (3) **23** (1982) 291–318.
2. P. Cherenack, *Cones and comparisons in ind-affine homotopy theory*, Questiones Math. **6** (1983) 49–66.
3. P. Cherenack, *Convenient affine algebraic varieties*, preprint, 1987.
4. A. Douady, *Le probleme des modules pour les sous-espaces analytiques compacts d'un espace analytique donne*, Ann. Inst. Fourier, Grenoble **16** (1966) 1–95.
5. A. Frölicher, *Smooth structures*, in Category Theory, Proc. Int. Conf. Cummersbach 1981, Lecture Notes in Math. 962, Springer, Berlin, 1982.
6. A. Grothendieck, *Techniques de construction et theoremes d'existence en geometrie algebrique*, I to V, Sem. Bourbaki, vol. 1959 to 1962.
7. P. J. Huber, *Homotopy theory in general categories*, Math. Ann. **144** (1961) 361–385.
8. J. F. Jardine, *Algebraic homotopy theory, groups and K-theory*, thesis, The University of British Columbia, 1981.
9. L. D. Nel, *Topological universes and smooth Gelfand-Naimark duality*, Contemporary Mathematics **30** (1984) 244–276.

Paul Cherenack
Department of Mathematics
University of Cape Town
7700 Rondebosch
South Africa

Lucio Guerra
Dipartimento di Matematica
Via Vanvitelli, 1
06100 Peguria
Italy

THE MATH. HERITAGE OF C.F. GAUSS (pp. 119-132)
edited by George M. Rassias
©1991 World Scientific Publ. Co. Singapore

UNIQUENESS OF SOLUTIONS OF ELECTROMAGNETIC INTERACTION PROBLEMS ASSOCIATED WITH SCATTERING BY BIANISOTROPIC BODIES COVERED WITH IMPEDANCE SHEETS

David K. Cohoon

Bianisotropic materials are more general than either anisotropic or chiral materials. We write down the frequency domain Maxwell equations for a bianisotropic material and develop conditions on tensors appearing in these equations which guarantee uniqueness of the solution of the electromagnetic interaction problem. The primary tools here are the use of Silver Mueller radiation conditions identities involving integrals of field quantities over the interior and surface of the scattering body derived from impedance sheet boundary conditions, and some theorems of Gauss. The integral equation formulation of the electromagnetic interaction problem is provided in three and seven dimensional space.

1. Introduction

Bianisotropic materials (Kong [7]), because of their greater complexity, have greater potential for reliably modeling physical materials which respond linearly to stimulating electromagnetic radiation. Chiral properties are a special case of bianisotropic materials. With chiral materials there is a special scalar ξ_c (Jaggard and Engetta [6]) such that

$$\vec{D} = \epsilon\vec{E} + i\xi_c\vec{B} \tag{1.1}$$

and

$$\vec{B} = \mu\vec{H} - i\xi_c\mu\vec{E}. \tag{1.2}$$

With the more general bianisotropic materials there are tensors $\boldsymbol{\alpha}$ and $\boldsymbol{\beta}$

120 D. K. Cohoon

with the property that

$$\vec{D} = \epsilon \vec{E} + \boldsymbol{\alpha} \vec{H}/(i\omega) \tag{1.3}$$

and

$$\vec{B} = \boldsymbol{\mu} \vec{H} + \boldsymbol{\beta} \vec{E}/(i\omega) \tag{1.4}$$

where ϵ and $\boldsymbol{\mu}$ are tensors. Here Maxwell's equations have the form

$$\mathrm{curl}(\vec{E}) = -i\omega \vec{B} \tag{1.5}$$

and

$$\mathrm{curl}(\vec{H}) = i\omega \vec{D} + \boldsymbol{\sigma}\vec{E}. \tag{1.6}$$

Using these notions we make Maxwell's equations look like the standard Maxwell equations with complex sources by introducing the generalized electric and magnetic current densities by the relations,

$$\mathrm{curl}(\vec{E}) = i\omega\mu_0 \vec{H} - \vec{J}_m \tag{1.7}$$

and

$$\mathrm{curl}(\vec{H}) = i\omega\epsilon_0 \vec{E} + \vec{J}_e \tag{1.8}$$

where

$$\vec{J}_e = i\omega\epsilon\vec{E} + \boldsymbol{\alpha}\vec{H} - i\omega\epsilon_0\vec{E} \tag{1.9}$$

and

$$\vec{J}_m = i\omega\boldsymbol{\mu}\vec{H} + \boldsymbol{\beta}\vec{E} - i\omega\mu_0\vec{H}. \tag{1.10}$$

We also assume that there is an impedance sheet current density given by

$$\vec{J}_s = \sigma_s(\vec{E} - (\vec{n} \cdot \vec{E})\vec{n}). \tag{1.11}$$

The formulation of integral equations for bianisotropic materials, therefore, is carried out by the analysis of the following coupled system of integral equations based on the notion of electric and magnetic charges defined by the two continuity equations

$$\mathrm{div}(\vec{J}_e) + \frac{\partial \rho_e}{\partial t} \tag{1.12}$$

and

$$\operatorname{div}(\vec{J}_m) + \frac{\partial \rho_m}{\partial t} \, . \tag{1.13}$$

Having developed this the coupled system of integral equations describing the interaction of electromagnetic radiation with a bounded *bianisotropic* body Ω is given by the following relations. The electric field integral equation is given by

$$
\begin{aligned}
\vec{E} - \vec{E}^i = {}& - \operatorname{grad}\left(\int_\Omega \frac{\operatorname{div}(\vec{J}_e)}{\omega \epsilon_0} G(r,s) dv(s) \right) \\
& + \frac{i}{\omega \epsilon_0} \operatorname{grad}\left(\int_{\partial\Omega} (\vec{J}_e \cdot \mathbf{n}) G(r,s) da(s) \right) \\
& + \left(\frac{-i}{\omega \epsilon_0} \right)\left(\int_{\partial\Omega} \{ k_0^2 (\vec{J}_s) G(r,s) + \operatorname{div}(\vec{J}_s)\operatorname{grad}(G(r,s)) \} da(s) \right) \\
& - i\omega\mu_0 \int_\Omega \vec{J}_e G(r,s) dv(s) - \operatorname{curl}\left(\int_\Omega \vec{J}_m G(r,s) dv(s) \right)
\end{aligned}
\tag{1.14}
$$

and the magnetic field integral equation may be expressed as

$$
\begin{aligned}
\vec{H} - \vec{H}^i = {}& - \operatorname{grad}\left(\int_\Omega \frac{\operatorname{div}(\vec{J}_m)}{\omega \mu_0} G(r,s) dv(s) \right) \\
& - \frac{i}{\omega \mu_0} \operatorname{grad}\left(\int_{\partial\Omega} (\vec{J}_m \cdot \mathbf{n}) G(r,s) da(s) \right) \\
& - \left(\int_{\partial\Omega} (\vec{J}_s \times (\operatorname{grad}(G(r,s)))) da(s) \right) \\
& - i\omega\epsilon_0 \int_\Omega \vec{J}_m G(r,s) dv(s) \\
& + \operatorname{curl}\left(\int_\Omega \vec{J}_e G(r,s) dv(s) \right)
\end{aligned}
\tag{1.15}
$$

where $G(r,s)$ is the rotation invariant, temperate fundamental solution of the Helmholtz equation,

$$(\Delta + k_0^2)G = \delta \tag{1.16}$$

given by

$$G(r,s) = \frac{\exp(-ik_0|r-s|)}{4\pi|r-s|} \, . \tag{1.17}$$

122   *D. K. Cohoon*

Substituting (1.9) through (1.11) into equations (1.14) and (1.15), we obtain the coupled integral equations for bianisotropic materials. The electric field integral equation for a bianisotropic material is given by,

$$
\vec{E} - \vec{E}^i = -\,\mathrm{grad}\left(\int_\Omega \frac{\mathrm{div}(i\omega\epsilon\vec{E} + \boldsymbol{\alpha}\vec{H} - i\omega\epsilon_0\vec{E})}{\omega\epsilon_0} G(r,s)dv(s) \right)
$$

$$
+ \frac{i}{\omega\epsilon_0}\mathrm{grad}\left(\int_{\partial\Omega} (i\omega\epsilon\vec{E} + \boldsymbol{\alpha}\vec{H} - i\omega\epsilon_0\vec{E} \cdot \mathbf{n})G(r,s)da(s) \right)
$$

$$
+ \left(\frac{-i}{\omega\epsilon_0}\right)\left(\int_{\partial\Omega} \left\{ k_0^2(\sigma_s(\vec{E} - (\vec{n} \cdot \vec{E})\vec{n}))G(r,s) + \right. \right.
$$

$$
\mathrm{div}(\sigma_s(\vec{E} - (\vec{n} \cdot \vec{E})\vec{n}))\mathrm{grad}(G(r,s))\Big\}da(s) \Bigg)
$$

$$
- i\omega\mu_0 \int_\Omega (i\omega\epsilon\vec{E} + \boldsymbol{\alpha}\vec{H} - i\omega\epsilon_0\vec{E})G(r,s)dv(s) +
$$

$$
- \mathrm{curl}\left(\int_\Omega (i\omega\boldsymbol{\mu}\vec{H} + \boldsymbol{\beta}\vec{E} - i\omega\mu_0\vec{H})G(r,s)dv(s) \right). \tag{1.18}
$$

The magnetic field integral equation for a bianisotropic material covered by an impedance sheet is given by

$$
\vec{H} - \vec{H}^i = -\,\mathrm{grad}\left(\int_\Omega \frac{\mathrm{div}(i\omega\boldsymbol{\mu}\vec{H} + \boldsymbol{\beta}\vec{E} - i\omega\mu_0\vec{H})}{\omega\mu_0} G(r,s)dv(s) \right)
$$

$$
- \frac{i}{\omega\mu_0}\mathrm{grad}\int_{\partial\omega} (i\omega\boldsymbol{\mu}\vec{H} + \boldsymbol{\beta}\vec{E} - i\omega\mu_0\vec{H} \cdot \mathbf{n})G(r,s)da(s)
$$

$$
- \left(\int_{\partial\Omega} (\sigma_s(\vec{E} - (\vec{n} \cdot \vec{E})\vec{n}) \times (\mathrm{grad}(G(r,s)))da(s) \right)
$$

$$
- i\omega\epsilon_0 \int_\Omega (i\omega\boldsymbol{\mu}\vec{H} + \boldsymbol{\beta}\vec{E} - i\omega\mu_0\vec{H})G(r,s)dv(s)
$$

$$
+ \mathrm{curl}\left(\int_\Omega (i\omega\epsilon\vec{E} + \boldsymbol{\alpha}\vec{H} - i\omega\epsilon_0\vec{E})G(r,s)dv(s) \right). \tag{1.19}
$$

While we have obtained exact solutions for layered materials, most of the problems are so complex that one must formulate the interaction problems using integral equations. The primary focus of this paper is to demonstrate the equivalence of integral equation and Maxwell equation formulations of the problem for suitable function spaces by demonstrating uniqueness. Then we can carry out the design of complex materials using an improvement of classical spline methods (Tsai, Massoudi, Durney, and Iskander [9], pp 1131–1139) and Li [8]. The Tsai, Massoudi, Durney, and Iskander paper is unusual in that comparisons are made between internal fields predicted from moment method computations and Mie solution computations. In Li [8] this verification was carried out analytically. There are several methods of solving coupled integral equations of the form (1.18) and (1.19) by approximate methods. However, as the scattering bodies become more complex the computational requirements become larger and larger. With exact finite rank integer equation theory ([3]), if one has a discretization that enables one to closely approximate the solution, then refinements can be made by a process based on the concept that the norm of the difference between an approximate integral operator and the actual integral operator is simply smaller than one, not necessarily close enough to give answers of acceptable accuracy. Then the answer is improved by an iterative process to any desired precision without the use of additional computer memory.

When solving electromagnetic scattering problems for isotropic bodies, one can proceed in a fairly direct way with the discretization of specializations of the integer equations (1.18) and (1.19), but in modeling classes of the more complex materials such as liquid crystals for electron camera shutters or piezoelectric materials for micromotors, greater care has to be taken that the model of the interaction has only one solution.

2. Uniqueness

If $\vec{E}$ and $\vec{H}$ are electric and magnetic fields in a bianisotropic material, then there exist tensors μ, ϵ, α and β such that

$$\operatorname{curl}(\vec{E}) = -i\omega\mu\vec{H} - \beta\vec{E} \tag{2.1}$$

and

$$\operatorname{curl}(\vec{H}) = (i\omega\epsilon + \sigma)\vec{E} + \alpha\vec{H}. \tag{2.2}$$

We assume that if α is a complex tensor that α^* denotes its complex conjugate. We assume that a bounded bianisotropic body Ω with a smooth normal is embedded in three dimensional free space and is subjected to a remote source of radiation whose electric field is $\vec{E^i}$ and whose magnetic field is $\vec{H^i}$. If $\vec{E}$ denotes the difference between two solutions of the form $\vec{E^i} + \vec{E^s}$, where $\vec{E^s}$ denotes the scattered radiation, in the exterior of Ω, or simply the difference of two solutions in the interior of Ω, then the solution is unique if we can prove that $\vec{E}$ is identically zero in the exterior of Ω or everywhere inside Ω.

The starting point for proofs of uniqueness of solutions of electromagnetic scattering problems in the Silver Mueller radiation condition which demand that if C_R is a sphere of radius R centered at a point in the scattering body, that then

$$\lim_{R \to \infty} \int_{C_R} (\vec{n} \times \mathrm{curl}(\vec{E}) - ik_0\vec{E}) \cdot (\vec{n} \times \mathrm{curl}(\vec{E^*}) + ik_0\vec{E^*})da = 0. \quad (2.3)$$

Thus, we note that

$$\int_{C_R} |\vec{n} \times \mathrm{curl}(\vec{E}) - ik_0\vec{E}|^2 da = \int_{C_R} (|\vec{n} \times \mathrm{curl}(\vec{E})|^2 + k_0^2|E|^2)da$$
$$+ ik_0 \int_{C_R} ((\vec{n} \times \mathrm{curl}(\vec{E})) \cdot \vec{E^*})da \quad (2.4)$$
$$- ik_0 \int_{C_R} \vec{E} \cdot (\vec{n} \times \mathrm{curl}(\vec{E^*}))da.$$

Focusing our attention on the last two terms in this equation, we see that

$$ik_0 \int_{C_R} (\vec{n} \times \mathrm{curl}(\vec{E})) \cdot \vec{E^*} da = ik_0 \int_{C_R} \vec{n} \cdot (\mathrm{curl}(\vec{E}) \times \vec{E^*})dv$$
$$= ik_0 \int_{V_1} \mathrm{div}(\mathrm{curl}(\vec{E}) \times \vec{E^*})dv \quad (2.5)$$
$$+ (ik_0) \int_{S_2} \vec{n} \cdot (\mathrm{curl}(\vec{E}) \times \vec{E^*})da.$$

In the previous equations S_2 is the surface bounding the bianisotropic body and V_1 is the region between the bounded bianisotropic body and the sphere C_R centered at a point in the bianisotropic material. We will assume that V_2 represents the bounded bianisotropic body covered by an impedance sheet. Continuing our analysis, and replacing $\mathrm{curl}(\vec{E})$ by $-i\omega\mu_0\vec{H}$ we find that

$$ik_0 \int_{C_R} (\vec{n} \times \text{curl}(\vec{E})) \cdot \vec{E}^* da = ik_0 \int_{C_R} \text{div}(\text{curl}(\vec{E}) \times \vec{E}^*) dv$$
$$+ k_0 \omega \mu_0 \int_{S_2} \vec{n} \cdot (\vec{H} \times \vec{E}^*) da \,. \tag{2.6}$$

Thus, making use of the impedance sheet boundary condition which states that

$$\vec{n} \times \vec{H} = \vec{n} \times \vec{H}_2 + \sigma_s(\vec{E}_2 - (\vec{n} \cdot \vec{E}_2)\vec{n}) \tag{2.7}$$

we find that

$$ik_0 \int_{C_R} (\vec{n} \times \text{curl}(\vec{E})) \cdot \vec{E}^* da = ik_0 \int_{C_R} \text{div}(\text{curl}(\vec{E}) \times \vec{E}^*) dv$$
$$+ k_0 \omega \mu_0 \int_{S_2} \vec{n} \cdot (\vec{H}_2 \times \vec{E}_2^*) da$$
$$+ k_0 \omega \mu_0 \int_{S_2} \{\sigma_s(\vec{E}_2 - (\vec{n} \cdot \vec{E}_2)\vec{n}) \cdot \vec{E}_2^*\} da \tag{2.8}$$

where $\vec{H}_2$ and $\vec{E}_2$ are the electric and magnetic fields just inside the impedance sheet on the surface S_2. First using the Gauss divergence theorem we find that

$$ik_0 \int_{C_R} (\vec{n} \times \text{curl}(\vec{E})) \cdot \vec{E}^* da = ik_0 \int_{C_R} \text{div}(\text{curl}(\vec{E} \times \vec{E}^*)) dv$$
$$+ k_0 \omega \mu_0 \int_{V_2} \text{div}(\vec{H}_2 \times \vec{E}_2^*) dv$$
$$+ k_0 \omega \mu_0 \int_{S_2} \{\sigma_s(\vec{E}_2 - (\vec{n} \cdot \vec{E}_2)\vec{n}) \cdot \vec{E}_2^*\} da \,. \tag{2.9}$$

We now make use of the vector calculus identity,

$$\text{div}(\vec{A} \times \vec{B}) = \vec{B} \cdot (\text{curl}(\vec{A})) - \vec{A} \cdot (\text{curl}(\vec{B})) \,. \tag{2.10}$$

We find that

$$ik_0 \int_{C_R} (\vec{n} \times \text{curl}(\vec{E})) \cdot \vec{E}^* da =$$
$$ik_0 \int_{V_1} (\vec{E}^* \cdot (\text{curl}(\text{curl}(\vec{E}))) - \text{curl}(\vec{E}) \cdot \text{curl}(\vec{E}^*)) dv$$
$$+ k_0 \omega \mu_0 \int_{V_2} (\vec{E}_2^* \cdot \text{curl}(\vec{H}_2) - \vec{H}_2 \cdot \text{curl}(\vec{E}_2^*)) dv$$

$$+k_0\omega\mu_0 \int_{S_2} \sigma_s \{(\vec{E}_2 \cdot \vec{E}_2^*) - (\vec{E}_2 \cdot \vec{n})(\vec{E}_2^* \cdot \vec{n})\}da. \qquad (2.11)$$

Substituting in the constitutive relations we find that

$$ik_0 \int_{C_R} (\vec{n} \times \mathrm{curl}(\vec{E})) \cdot \vec{E}^* da = -ik_0 \int_{V_1} \{\vec{E}^* \cdot \Delta\vec{E} + |\mathrm{curl}(\vec{E})|^2\}dv$$

$$+ k_0\omega\mu_0 \int_{V_2} \vec{E}_2^* \cdot \{(i\omega\epsilon + \sigma)\vec{E}_2 + \alpha\vec{H}_2\}dv$$

$$- k_0\omega\mu_0 \int_{V_2} \vec{H}_2 \cdot (i\omega\mu^* \vec{H}_2^* - \beta^* \vec{E}_2^*)dv$$

$$+ k_0\omega\mu_0 \int_{S_2} \sigma_s \{|(\vec{E}_2 - (\vec{E}_2 \cdot \vec{n}) \cdot \vec{n})|^2\}da. \qquad (2.12)$$

Considering the conjugate term of this form we observe that

$$-ik_0 \int_{C_R} (\vec{n} \times \mathrm{curl}(\vec{E}^*)) \cdot \vec{E}\, da = ik_0 \int_{V_1} \{\vec{E} \cdot \Delta\vec{E}^* + |\mathrm{curl}(\vec{E})|^2\}dv$$

$$+ k_0\omega\mu_0 \int_{V_2} \vec{E}_2 \cdot \{(-i\omega\epsilon + \sigma^*)\vec{E}_2^* + \alpha^* \vec{H}_2^*\}dv$$

$$- k_0\omega\mu_0 \int_{V_2} \vec{H}_2^* \cdot (-i\omega\mu\vec{H}_2 - \beta\vec{E}_2)dv$$

$$+ k_0\omega\mu_0 \int_{S_2} \sigma_s^* \{|(\vec{E}_2 - (\vec{E}_2 \cdot \vec{n}) \cdot \vec{n})|^2\}da. \qquad (2.13)$$

Adding these equations we find that the solution is unique provided that either a quadratic form is positive definite or another form is either negative or positive definite. Indeed it may be easier to prove uniqueness for the more complex material than when the scatterer has a simpler form. We find that

$$2\mathrm{Re}\left(ik_0 \int_{C_R} (\vec{n} \times \mathrm{curl}(\vec{E})) \cdot \vec{E}^* da\right) =$$

$$k_0\omega\mu_0 \int_{V_2} \left\{(\vec{E}_2^* \cdot (i\omega\epsilon + \sigma)\vec{E}_2) + (\vec{E}_2 \cdot (-i\omega\epsilon^* + \sigma^*)\vec{E}_2^*)\right\}dv$$

$$+ k_0\omega\mu_0 \int_{V_2} \left\{(\vec{E}_2^* \cdot (\alpha)\vec{H}_2) + (\vec{E}_2 \cdot (\alpha^*)\vec{H}_2^*)\right\}dv$$

$$- k_0\omega\mu_0 \int_{V_2} \left\{(\vec{H}_2 \cdot (i\omega\mu^* \vec{H}_2^*)) + (\vec{H}_2^* \cdot (-i\omega\mu\vec{H}_2))\right\}dv$$

$$+ k_0\omega\mu_0 \int_{V_2} \left\{ (\vec{H}_2 \cdot \boldsymbol{\beta}\vec{E}_2) + (\vec{H}_2^* \cdot \boldsymbol{\beta}\vec{E}_2) \right\} dv$$
$$+ k_0\omega\mu_0 \int_{S_2} (\sigma_s^* + \sigma_s) \left\{ (\vec{E}_2 \cdot \vec{E}_2^*) - (\vec{E}_2 \cdot \vec{n})(\vec{E}_2^* \cdot \vec{n}) \right\} da\,. \tag{2.14}$$

Note that if the permeability, μ, the permittivity, ϵ, and the tensors $\boldsymbol{\alpha}$ and $\boldsymbol{\beta}$ are scalars times the identity matrix then sufficient conditions for uniqueness are that the real part of σ is positive and the imaginary parts of ϵ and μ are negative and that the real part of σ_s is positive and that the quadratic form associated with the matrix $\mathbf{Q}$ defined by

$$\mathbf{Q} = \begin{pmatrix} A_e & 0 & 0 & \mathrm{Re}(\alpha) & 0 & 0 \\ 0 & A_e & 0 & 0 & \mathrm{Re}(\alpha) & 0 \\ 0 & 0 & A_e & 0 & 0 & \mathrm{Re}(\alpha) \\ \mathrm{Re}(\beta) & 0 & 0 & \omega\mathrm{Im}(\mu) & 0 & 0 \\ 0 & \mathrm{Re}(\beta) & 0 & 0 & \omega\mathrm{Im}(\mu) & 0 \\ 0 & 0 & \mathrm{Re}(\beta) & 0 & 0 & \omega\mathrm{Im}(\mu) \end{pmatrix} \tag{2.15}$$

where

$$A_e = \omega\mathrm{Im}(\epsilon) + \mathrm{Re}(\sigma)\,, \tag{2.16}$$

is positive definite. Thus, in particular, if there is enough domination of the α and β terms by the positive diagonal terms, then this form is positive definite and we do indeed have a unique solution of the electromagnetic interaction problem in a variety of naturally arising function spaces if the impedance sheet conductivity σ_s has a positive real part.

The general uniqueness result is therefore derived by observing that

$$\lim_{R\to\infty} \int_{C_R} (\vec{n} \times \mathrm{curl}(\vec{E}) - ik_0\vec{E}) \cdot (\vec{n} \times \mathrm{curl}(\vec{E}^*) + ik_0\vec{E}^*) da =$$
$$\lim_{R\to\infty} \int_{C_R} (|\vec{n} \times \mathrm{curl}(\vec{E})|^2 + k_0^2|E|^2) da$$
$$+ k_0\omega\mu_0 \int_{V_2} \left\{ (\vec{E}_2^* \cdot (i\omega\epsilon + \sigma)\vec{E}_2) + (\vec{E}_2 \cdot (-i\omega\epsilon^* + \sigma^*)\vec{E}_2^*) \right\} dv$$
$$+ k_0\omega\mu_0 \int_{V_2} \left\{ (\vec{E}_2^* \cdot (\boldsymbol{\alpha})\vec{H}_2) + (\vec{E}_2 \cdot (\boldsymbol{\alpha}^*)\vec{H}_2^*) \right\} dv$$
$$- k_0\omega\mu_0 \int_{V_2} \left\{ (\vec{H}_2 \cdot (i\omega\mu^*\vec{H}_2^*)) + (\vec{H}_2^* \cdot (-i\omega\mu\vec{H}_2)) \right\} dv$$

$$+ k_0\omega\mu_0 \int_{V_2} \left\{ (\vec{H}_2 \cdot \boldsymbol{\beta}^* \vec{E}_2^*) + (\vec{H}_2^* \cdot \boldsymbol{\beta} \vec{E}_2) \right\} dv$$

$$+ k_0\omega\mu_0 \int_{S_2} (\sigma_s^* + \sigma_s)\left\{ (\vec{E}_2 \cdot \vec{E}_2^*) - (\vec{E}_2 \cdot \vec{n})(\vec{E}_2^* \cdot \vec{n}) \right\} da. \tag{2.17}$$

The uniqueness is established by observing that upon taking the limit of all terms as the radius R of C_R becomes infinite that if the difference $\vec{E}$ between two solutions were a nonzero function, then we would get effectively two equations by taking the real and imaginary parts of both sides of the relationship

$$0 = C^2 + k_0\omega\mu_0 \int_{V_2} \left\{ (\vec{E}_2^* \cdot (i\omega\epsilon + \sigma)\vec{E}_2) + (\vec{E}_2 \cdot (-i\omega\epsilon^* + \sigma^*)\vec{E}_2^*) \right\} dv$$

$$+ k_0\omega\mu_0 \int_{V_2} \left\{ (\vec{E}_2^* \cdot (\boldsymbol{\alpha})\vec{H}_2) + (\vec{E}_2 \cdot (\boldsymbol{\alpha})\vec{H}_2^*) \right\} dv$$

$$- k_0\omega\mu_0 \int_{V_2} \left\{ (\vec{H}_2 \cdot (i\omega\boldsymbol{\mu}^*\vec{H}_2^*) + (\vec{H}_2^* \cdot (-i\omega\boldsymbol{\mu}\vec{H}_2) \right\} dv$$

$$+ k_0\omega\mu_0 \int_{V_2} \left\{ (\vec{H}_2 \cdot \boldsymbol{\beta}^* \vec{E}_2^*) + (\vec{H}_2^* \cdot \boldsymbol{\beta}\vec{E}_2) \right\} dv$$

$$+ k_0\omega\mu_0 \int_{S_2} (\sigma_s^* + \sigma_s)\left\{ (\vec{E}_2 \cdot \vec{E}_2^*) - (\vec{E}_2 \cdot \vec{n})(\vec{E}_2^* \cdot \vec{n}) \right\} da \tag{2.18}$$

where C^2 is the real number given by

$$C^2 = \lim_{R\to\infty} \int_{C_R} (|\vec{n} \times \mathrm{curl}(\vec{E})|^2 + k_0^2|E|^2)da. \tag{2.19}$$

Since this is not possible if the electromagnetic parameters are such that the body is dissipative in the sense that the bilinear form acting on the function $(\vec{E}, \vec{H})$ that is defined by

$$b(\vec{E}, \vec{H}) = k_0\omega\mu_0 \int_{V_2} \left\{ (\vec{E}_2^* \cdot (i\omega\epsilon + \sigma)\vec{E}_2) + (\vec{E}_2 \cdot (-i\omega\epsilon^* + \sigma^*)\vec{E}_2^*) \right\} dv$$

$$+ k_0\omega\mu_0 \int_{V_2} \left\{ (\vec{E}_2^* \cdot (\boldsymbol{\alpha})\vec{H}_2) + (\vec{E}_2 \cdot (\boldsymbol{\alpha}^*)\vec{H}_2^*) \right\} dv$$

$$- k_0\omega\mu_0 \int_{V_2} \left\{ \vec{H}_2 \cdot (i\omega\boldsymbol{\mu}^*\vec{H}_2) + \vec{H}_2^* \cdot (i\omega\boldsymbol{\mu}\vec{H}_2) \right\} dv$$

$$+ k_0\omega\mu_0 \int_{V_2} \left\{ (\vec{H}_2 \cdot \boldsymbol{\beta}^* \vec{E}_2^*) + (\vec{H}_2^* \cdot \boldsymbol{\beta}\vec{E}_2) \right\} dv$$

$$+ k_0\omega\mu_0 \int_{S_2} (\sigma_s^* + \sigma_s)\left\{ (\vec{E}_2 \cdot \vec{E}_2^*) - (\vec{E}_2 \cdot \vec{n})(\vec{E}_2^* \cdot \vec{n}) \right\} da \tag{2.20}$$

is positive definite.

The proof of uniqueness can now be completed in a variety of naturally arising function spaces where the integrals are defined. We see that for an isotropic material that the Silver Mueller radiation conditions, Eq. (2.3), and our final relation, which is embodied in Eqs. (2.18) and (2.19), imply that if $\vec{E}$ and $\vec{H}$ denote the difference between two solutions of the electromagnetic interaction problem then we have the relationship

$$0 = C^2 + 2k_0\omega\mu_0 \int_\Omega \{(\omega\epsilon'') + \mathrm{Re}(\sigma)\}|\vec{E}|^2 dv + 2k_0\omega\mu_0 \int_\Omega (\omega\mu'')|\vec{H}|^2 dv$$

$$+ 2k_0\omega\mu_0 \int_\Omega \mathrm{Re}(\sigma_s)|(\vec{E} - (\vec{E} \cdot \vec{n})\vec{n})|^2 da .$$

$$\tag{2.21}$$

Since we usually write the permeability in the form

$$\mu = \mu' - i\mu'' \tag{2.22}$$

where μ'' is positive and write the permittivity in the form

$$\epsilon = \epsilon' - i\epsilon'' \tag{2.23}$$

where ϵ'' is positive, and assume that the real parts of σ and σ_s are not negative, we see that for normal isotropic physical materials, there is only one solution of the electromagnetic interaction problem, but that, while there are many interesting situations in which uniqueness can be established for bianisotropic materials, there is no such simple separate condition on each individual tensor by itself which would guarantee uniqueness of the interaction problem for bianisotropic materials, in the sense that, for isotropic materials, (2.21) tells us immediately that the difference between the electric vectors of the two solutions is identically zero.

130   *D. K. Cohoon*

3. Seven Dimensions

We suggest, here, 7 dimensional theory as a method of solving electromagnetic interaction problems. It is clear that any electromagnetic interaction problem in 7 dimensional space translates into a scattering problem in three dimensional space. Also, problems which are simple in 7 dimensional space often translate into electromagnetic scattering problems in three dimensional space which are seemingly very complex. What remains open is a systematic method of going from problems in 3 dimensional space that we really want to solve into solvable problems in 7 dimensional space. If we consider a seven dimensional vector field

$$\vec{E} = \sum_{j=1}^{7}\{E_j\,\vec{e}_j\} \tag{3.1}$$

where the components are smooth functions of the spatial variables

$$\mathbf{x} = (x_1, x_2, \cdot, \cdot, \cdot, x_7) \tag{3.2}$$

than we have

$$\mathrm{curl}(\vec{E}) = \sum_{i=1}^{7}\left[\left(\frac{\partial E_{i+3}}{\partial x_{i+1}} - \frac{\partial E_{i+1}}{\partial x_{i+3}}\right) + \left(\frac{\partial E_{i+6}}{\partial x_{i+2}} - \frac{\partial E_{i+2}}{\partial x_{i+6}}\right) + \left(\frac{\partial E_{i+5}}{\partial x_{i+4}} - \frac{\partial E_{i+4}}{\partial x_{i+5}}\right)\right]\vec{e}_i \tag{3.3}$$

where $\vec{e}_i$ is the unit vector in the direction of the ith coordinate axis in 7 dimensional space and

$$E_{i+7} = E_i \tag{3.4}$$

and

$$x_{i+7} = x_i \tag{3.5}$$

for all i in $\{1, 2, 3, 4, 5, 6, 7\}$. The main body of the theory which shows that every vector field in 7 dimensional space is a *curl* plus a *gradient* and is the relation

$$\mathrm{curl}(\mathrm{curl}(\vec{E})) = \mathrm{grad}(\mathrm{div}(\vec{E})) - \Delta\vec{E}. \tag{3.6}$$

This follows from the fact that all nonempty open subsets Ω of $\mathbb{R}^7$ are Δ-*convex* in the sense of Hormander ([5], Corollary 3.5.2, page 82) and that consequently if $\vec{F}$ is a vector field in $C^\infty(\Omega, \mathbb{C}^7)$, it follows that there is another vector field $\vec{G}$ such that

$$\Delta \vec{G} = \vec{F}\,. \tag{3.7}$$

Using the previous identity we see that

$$\vec{F} = \text{grad}(\text{div}(\vec{G})) + \text{curl}(\text{curl}(-\vec{G}))\,. \tag{3.8}$$

Some work will show that the uniqueness theory also carries over in a natural way to 7 dimensional space without the use of exterior differentials to represent curl operations. The self adjointness of the 3 and 7 dimensional curl on $C_c^\infty(\Omega, \mathbb{C}^n)$ for $n = 3$ and $n = 7$, respectively, permits variational formulations of the problems and the study of weak solutions of interaction problems.

4. Summary

At the present time computational electromagnetism is advancing rapidly as complex analysis and embedding of problems in algebras, such as the Cartan algebra mentioned above, has the potential of helping us understand the interaction of electromagnetic waves with new man made materials such as liquid crystals, and the birefringent and piezoelectric materials that have been a source of fascination for centuries.

References

1. John G. Burr, David K. Cohoon, Earl L. Bell, and John W. Penn, *Thermal response model of a simulated cranial structure exposed to radiofrequency radiation*, IEEE Transactions of Biomedical Engineering, vol. BME-27, No. 8 (August, 1980) 452–460.
2. D. K. Cohoon, J. W. Penn, E. L. Bell, D. R. Lyons, and A. G. Cryer, *A Computer Model Predicting the Thermal Response to Microwave Radiation, SAM-TR-82-22*, Books AFB, Tx 78235: USAF School of Aerospace Medicine. (RZ) Aerospace Medical Division (AFSC) (December, 1982).
3. D. K. Cohoon, *An exact formula for the accuracy of a class of computer solutions of integral equation formulations of electromagnetic scattering problems*, Electromagnetics, (2) **7** (1987) 153–165.

4. Harry Hochstadt, *The Functions of Mathematical Physics*, New York: Dover, 1986.

5. Lars Hormander, *Linear Partial Differential Operators*, New York: Academic Press, 1963.

6. D. L. Jaggard and N. Engheta, *Chirosorb^{TM} as an invisible medium*, Electronic Letters, (3) **25** (February 2, 1989) 173–174.

7. J. A. Kong, *Electromagnetic Wave Theory*, New York: John Wiley, 1986.

8. Shu Chen Li, *Interaction of electromagnetic fields with simulated biological structures*, Ph. D. Thesis (Temple University, Department of Mathematics 038–16, Philadelphia, Pa 19122) (1986) 454 pages.

9. Chi-Taou Tsai, Habib Massoudi, Carl H. Durney, and Magdy F. Iskander, *A procedure for calculating fields inside arbitrarily shaped, inhomogeneous dielectric bodies using linear basis functions with the moment method*, IEEE Transactions on Microwave Theory and Techniques, Vol. MTT-34, No. 11 (November, 1986) 1131–1139.

10. E. T. Whittaker, and G. N. Watson, *A Course of Modern Analysis*, London: Cambridge University Press, 1986.

David K Cohoon
Mathematics Dept.
West Chester University
West Chester PA 19383
USA

THE MATH. HERITAGE OF C.F. GAUSS (pp. 133-136)
edited by George M. Rassias
©1991 World Scientific Publ. Co. Singapore

ON THE TOPOLOGICAL STRUCTURE OF CURVES

P. J. Collins

Non-compact curves are discussed, in particular, as perfect images of standard spaces and as unions of increasing sequences of locally connected continua.

1. Introduction

We discuss here, in the spirit of the investigations of C. F. Gauss, some aspects of the topological theory of curves, when compactness is no longer an absolute requirement. The article is largely expository. Some of the work, important to the discussion, appeared first in [1]. The interesting results on tree-coverings come, with his permission, from D. R. Hartley's thesis [5].

A (*generalised*) *Peano curve* is a (locally) compact, connected and locally connected, metrisable space. For convenience, we shall use the term *sigma curve* to describe a locally and σ-compact*, connected and locally connected space. A *continuum* will be a compact, connected set and *a perfect* (*or proper*) mapping will be closed and have compact fibres.

2. Growth Sequences of Locally Connected Continua

Various types of expanding sequences of locally connected continua are central to results throughout this article. The following elementary

---

*A space is $\sigma-compact$ if it can be expressed as a countable union of compact subsets (the space $\mathbb{R}$ of real numbers being an obvious non-compact example).

134   *P. J. Collins*

observation sets the scene.

Proposition 1. X is a locally compact, σ-compact (and connected) space if and only if there is a sequence (C_i) of compact (and connected) subsets of X such that

$$X = \bigcup_i C_i \tag{1}$$

$$C_i \subseteq \operatorname{int} C_{i+1}{}^* \qquad (i = 1, 2, \ldots) \tag{2}$$

Let us agree to call a sequence (C_i) of locally connected continua a *growth sequence* if it satisfies (1), (2) of Proposition 1. Unfortunately, the existence of a locally connected continuum with no proper non-degenerate locally connected sub-continuum, guaranteed to us *under CH* by an example of S. Mardesic [6], and recently without that condition by G. Gruenhage [3], does not permit every locally and σ-compact, connected *and locally connected* space to have a growth sequence. The distinction to be drawn is underlined by the following straightforward result.

Proposition 2. A space X has a growth sequence of locally connected continua if and only if X is σ-compact, connected and each point in X has a locally connected, compact neighbourhood.

3. Mapping Theorems and Tree-Coverings

The starting point here is the following re-statement of the celebrated characterisation of Peano curves by H. Hahn [4] and S. Mazurkiewicz [7].

Hahn-Mazurkiewicz Theorem

A topological space is the image of the closed unit interval under a *perfect* mapping if and only if it is a non-empty Peano curve.

An appropriate extension of this theorem to the case where the domain is the whole of $\mathbb{R}$ may be stated as follows:

Theorem 1. A necessary and sufficient condition for a space X to be a

---

*int A denotes the topological interior of A.

perfect image of $\mathbb{R}$ is that X is non-compact, locally connected, metrisable and has a growth sequence (C_i) of Peano curves C_i such that $C_{i+1} - \operatorname{int} C_i$ has components which are locally connected and at most two in number for $i = 1, 2, \ldots$.

Thus, Euclidean n-space $\mathbb{R}^n$ is the perfect image of $\mathbb{R}$. D. R. Hartley's very satisfactory general approach to the study of non-compact curves uses trees. Suppose that $<$ partially orders the set T, that t^+ denotes the set of immediate successors of t in T, and that

$$\hat{t} = \{s \in T \,:\, s < t\}\,.$$

For our purposes, T is a *tree* if
    (a) each $\hat{t}$ is finite and linearly ordered,
    (b) each t^+ is finite and non-empty,
    (c) T has an infinium.
Note that we do more than demand that each $\hat{t}$ be well-ordered. Of particular interest is the 'universal tree'

$$E = \cup\{2^n \,:\, n \in \omega\}\,.$$

Joining each element of a tree T to each of its successors by a copy of the unit interval, denote the resulting space, with the (weak) topology induced by the usual topologies on the inserted arcs, by $\mathbb{R}_T$.

A *tree-covering* of a space X is a locally finite, compact covering $\{C_t \,: t \in T\}$ of X, where T is a tree and $C_s \cap C_t \neq \emptyset$ if and only if s, t are adjacent elements in the ordering on T.

Theorem 2. For a space X, the following are equivalent:
(i) X is a non-compact generalised Peano space,
(ii) X has a tree-covering by Peano spaces,
(iii) X is a perfect image of $\mathbb{R}_E$.

The non-metrisable case is of special interest and again involves growth sequences.

Theorem 3. For a space X, the following are equivalent:

(i) X has a tree-covering by locally connected continua,

(ii) X is non-compact, connected and has a locally finite covering by locally connected continua,

(iii) X has a growth sequence (C_i) of locally connected continua and a sequence (V_i) of open sets such that, for each i, $V_i \subseteq C_i \subseteq V_{i+1}$ and $C_{i+1} - V_i$ is locally connected and non-empty.

References

1. P. J. Collins, *Long Peano curves*, Math. Colloq. Univ. Cape Town **XI** (1977) 1–6.
2. H. Freudenthal, *Neuaufbau der Endentheorie*, Ann. Math. **43** (1942) 261–279.
3. G. Gruenhage, *Two non-metric locally connected continua*, pre-print.
4. H. Hahn, *Mengentheoretische Characterisierung der stetigen Kurven*, Sitzungsberichte der Akad. der Wissenschaften **123** (1914) 2433.
5. D. R. Hartley, *Problems in analytic topology*, D. Phil thesis, Oxford (1982).
6. S. Mardesic, *A locally connected continuum which contains no proper locally connected subcontinuum*, Glasnik Mat. Ser. III, (2) **22** (1967) 167–178.
7. S. Mazurkiewicz, *Sur les lignes de Jordan*, Fund. Math. 1 (1920) 166–209.

P. J. Collins
St Edmund Hall
Oxford
OX1 4AR
U.K.

THE MATH. HERITAGE OF C.F. GAUSS (pp. 137-156)
edited by George M. Rassias
©1991 World Scientific Publ. Co. Singapore

GAUSS, BAYES, KALMAN:
STATE-SPACE MODELS

Pete Daffer

We trace briefly the evolution of the classical linear model into one of the many forms that have evolved from it: the dynamic linear model, or state-space model. We survey some problems arising in connection with Bayesian recursive estimation and examine some solutions which have been proposed. We describe several proposed filtering techniques for rendering linear state-space models more robust with respect to basic model assumptions. Among these are dynamic generalized linear models, multi-state models, the use of contaminated normals or other heavy-tailed distributions such as Student-t, and nonparametric methods using splines.

1. Introduction

The extent of the contributions of C. F. Gauss to the linear model has been documented by historians of science [49, 50, 52] and been found to be so great that it has been suggested that the linear model be named after him (Seal [49]). The story of how the youthful Gauss established his reputation as one of the leading mathematicians of Europe by using the methods of least squares to determine the location where astronomers would find the planetoid Ceres belongs to the folklore [7, 12]. His investigations on the subject were published in austere Latin over a period of decades [14, 15, 16] and were so exhaustive that certain of his results were rediscovered and thought to be new in the 20th century; see Seal [49] for an account of these. For a charming first-hand account of a course given by Gauss on the method of least-squares, see Dedekind [8].

We describe briefly the relationship of the classical linear model to one of the many forms of the linear model which have appeared fairly recently:

the dynamic linear model, or state-space model introduced in the engineering literature in the 1950's and 1960's [24, 25]. We then examine certain generalizations to the classical linear recursive filter known as the Kalman filter. In particular, a pressing issue has been to come up with techniques to render the filter more robust by getting away from the traditional assumptions of Gaussian noise distributions, while maintaining some degree of tractability. As the Kalman filter was discovered by statisticians and its uses greatly proliferated in the 1970's, more of such refinements and generalizations came from the statistics community.

In our discussion we limit ourselves to the case of Bayesian recursive filtering, mentioning smoothing and prediction only in references. One quick disclaimer: No attempt has been made to be representative or even moderately exhaustive in the references. Extensive bibliographies already exist; for example Kailath [23] lists 390 references. We list only a few which may serve as an entrance to the interested reader.

2. Estimation in the Classical Linear Model

The classical linear model of Gauss and his contemporaries is $y = X\theta + \epsilon$, where y and ϵ are $(k \times 1)$-vectors, θ is a $(p \times 1)$ vector and X a $(k \times p)$-matrix. Here ϵ is a vector of uncorrelated random variables, centered, with covariance matrix $\sigma^2 I$, where I is the identity matrix; this is denoted by $\epsilon \sim \mathrm{WS}[0, \sigma^2 I]$ where "WS" stands for "wide sense", and indicates that only the first two moments are given. The least-squares estimate $\hat{\theta}$ of θ is given by $\hat{\theta} = (X'X)^{-1}X'y$ (X' denotes the transpose of X) and by the Gauss-Markov theorem is the unique minimum variance linear unbiased estimator of θ. The generalization to $\epsilon \sim \mathrm{WS}[0, \Sigma]$, where Σ is positive definite, was treated by Aitken [1]. Gauss-Markov theory has been developed extensively, to cover the cases where X is not of full rank (as in analysis of variance) or A is singular, or both (see Rao [45], Chapter 4; see also [44, 41]), where generalized inverses or equivalent constraint conditions [30] are utilized. Mixed models where some regression parameters are random and others are constant, are treated by Sallas and Harville [47].

The state-space model and Kalman filter made its appearance in the late 1950's in the engineering literature as an alternative to the frequency-domain filtering techniques developed during World War II by Kolmogorov [28] and Wiener [59], and which depended upon stationarity. See Kailath

[23] for an exhaustive account of its development. Least-squares techniques were used to derive the filter, still the preferred method in the engineering literature [2, 3, 22, 46, 17]. One of the first demonstrations that the Kalman filter could be dervied using least-squares regression techniques for the classical linear model was made by Duncan and Horn [11] (see also Plackett [42]). This involved as a first step considering the case of a random parameter θ. This had been done before in other contexts, e.g., "random effects" in analysis of variance (Scheffe [48]). A Gauss-Markov theory paralleling the classical one was developed in [11] by expanding the model. If $\theta = \theta_0 + u$, with u a random vector with $Eu = 0$, then consider

$$\tilde{y} = \tilde{X}\theta + \tilde{\epsilon}, \tag{2.1}$$

where $\tilde{y} = \binom{\theta_0}{y}, \tilde{X} = \binom{I}{X}, \tilde{\epsilon} = \binom{-u}{\epsilon}$.

A dynamic linear model is introduced by allowing (θ_n) to vary with time according to the first-order autoregressive process $\theta_n = G_n\theta_{n-1} + w_n$, with G_n a matrix and (w_n) white noise. Gauss-Markov theory developed for (2.1) is then used to demonstrate that, given vectors $y_1, \ldots, y_{n-1}$ of observations up to $n-1$, the least-squares estimate $\hat{\theta}_n$ of θ_n is given by the recursive Kalman filter expressions. See also Diderrich [10], who attributes this technique to Goldberger and Theil.

In fact, the notion of a recursive algorithm for obtaining updated estimates $\hat{\theta}$ of θ as observations y_i arrived sequentially goes back to Gauss [15], who derived such algorithms for the classical linear model; this is noted in Peña and Guttman [40]. It is also interesting to note that Gauss, in his treatment of the linear model, made use of a prior probability distribution on the parameter θ, rendering it a random variable and foreshadowing the dynamic linear model. Diaconis and Ylvisaker [9] refer to Gauss as a "classical Bayesian", as opposed to more modern types.

3. Some Modern Approaches to Bayesian Recursive Estimation

3.1. *Notation*

A symbol like $\theta|y$ will denote the conditional probability distribution of the random variable θ, given the random variable y. The symbol p will denote a probability density, with respect to some σ-finite measure on an appropriate Euclidean space. Use will be made of the following abuses of notation: $p(\theta|y)$ or $p_\theta(\theta|y)$ or $p_{\theta|y}(\theta|y)$, for example, will denote a conditional distribution of θ, where the same symbol is used to indicate, as a subscript, the random variable in question; or, the subscript is left off and the symbol θ in the argument indicates the random variable of which p is the probability distribution. This device is very widely employed (see [58, 26] for example). Far from breeding confusion, this explicit use in the argument of the symbol for the random variable appears to increase the readability of complicated expressions.

The Gaussian distribution with mean m and covariance matrix C is denoted by $N(m; C)$.

3.2. *Observation of a Random Process*

Consider a Markov process (θ_n) taking values in $\mathbf{R}^p$ together with a related random process (y_n) which depends on it and takes values in $\mathbf{R}^k$: $y_n = f_n(\theta_n, v_n)$ where v_n is a random variable independent of θ_n. The process (y_n) represents what is available in order to make inferences about the underlying process (θ_n). One natural recursive estimation procedure is provided by a sequential Bayesian formulation. Writing $y^n = \{y_n, y_{n-1}, \ldots\}$ for the data up to time n and taking as point of departure the conditional probability distribution of θ_{n-1} given y^{n-1}, denoted by $\theta_{n-1}|y^{n-1}$, or $p(\theta_{n-1}|y^{n-1})$, we have

$$p(\theta_n|y^{n-1}) = \int \pi(\theta_n|\theta_{n-1})p(\theta_{n-1}|y^{n-1})d\theta_{n-1} \qquad (3.1)$$

where $\pi(\theta_n|\theta_{n-1})$ is the transition probability of (θ_n) from $n-1$ to n. Viewing this as a prior distribution of θ at time n, Bayes' formula provides the update:

$$p(\theta_n|y^n) = \frac{p(y_n|\theta_n, y^{n-1})p(\theta_n|y^{n-1})}{p(y_n|y^{n-1})} \qquad (3.2)$$

where the prediction distribution for y_n is obtained by:

$$p(y_n|y^{n-1}) = \int p(y_n, |\theta_n, y^{n-1})p(\theta_n|y^{n-1})d\theta_n . \qquad (3.3)$$

In practice a prior distribution would be required at some point, say $n = 0$; thus $p(\theta_0)$ is specified and we have the random processes $(\theta_n)_{n \geq 0}$ and $(y_n)_{n \geq 1} \cdot p(\theta_0|y^0)$ is identified with $p(\theta_0)$, and $p(\theta_1|y^0) = p(\theta_1) = \int \pi(\theta_1|\theta_0)p(\theta_0)d\theta_0$. With $p(y_1|y^0) = p(y_1) = \int p(y_1|\theta_1)p(\theta_1)d\theta_1$, we have $p(\theta_1|y^1) = p(y_1|\theta_1)p(\theta_1)/p(y_1)$ and the recursion is set into motion. This formulation assumes that the effect of $y_{n-1}, y_{n-2}, \ldots,$ on the probability distribution of y_n does not reach back beyond y_1, and considers the evolution of $(\theta_n)_{n \geq 0}$, as seen by (y_n) via $y_n = f_n(\theta_n, v_n)$, to be determined by the prior $p(\theta_0)$ on θ_0. This simplification, together with the accuracy with which the prior $p(\theta_0)$ estimates the true probability distribution of θ_n at $n = 0$ will influence the quality of $p(\theta_n|y^n)$ as an estimate of the true probability distribution of θ_n.

3.3. *State Space Model*

A crucial simplification results when the processes take the standard state space form:

$$\begin{aligned}
\theta_n &= g_n(\theta_{n-1}) + w_n \\
y_n &= f_n(\theta_n) + v_n
\end{aligned} \tag{3.4}$$

where f_n and g_n are deterministic functions and $(w_n), (v_n)$ are zero-mean, independent random processes in $\mathbf{R}^p$ and $\mathbf{R}^k$ respectively, which are independent of each other. Here we can assume a starting point $n = 0$ since θ_n depends on $\theta_0, w_n, w_{n-1}, \ldots, w_1$ and y_n depends only on the current θ_n and v_n. We assume that w_1 is independent of θ_0.

We assume that the observation process (y_n) depends upon the true evolution of the state process (θ_n). Let $p(\theta_0)$ denote the true probability distribution of θ_0. Then based on $p(\theta_0)$, the evolution of the estimates of (θ_n) proceeds according to (3.1) and (3.2). The forms of the observation equation and system equation (3.4) now yield $p(y_n|\theta_n, y^{n-1}) = p(y_n|\theta_n) = p_{v_n}(y_n - f_n(\theta_n))$ and $\pi(\theta_n|\theta_{n-1}) = p_{w_n}(\theta_n - g_n(\theta_{n-1}))$, where p_{v_n} and p_{w_n} are the probability distributions of v_n and w_n, respectively. We thus have

$$p(\theta_1) = \int p_{w_1}(\theta_1 - g_1(\theta_0))p(\theta_0)d\theta_0$$

$$p(y_1) = \int p_{v_1}(y_1 - f_1(\theta_1))p(\theta_1)d\theta_1$$

$$p(\theta_1|y^1) = \frac{p_{v_1}(y_1 - f_1(\theta_1))p(\theta_1)}{p(y_1)}$$

and for $n \geq 2$,

$$p(\theta_n|y^{n-1}) = \int p_{w_n}(\theta_n - g_n(\theta_{n-1}))p(\theta_{n-1}|y^{n-1})d\theta_{n-1} \tag{3.5}$$

$$p(\theta_n|y^n) = \frac{p_{u_n}(y_n - f_n(\theta_n))p(\theta_n|y^{n-1})}{p(y_n|y^{n-1})} \tag{3.6}$$

where

$$p(y_n|y^{n-1}) = \int p_{v_n}(y_n - f_n(\theta_n))p(\theta_n|y^{n-1})d\theta_n \, .$$

3.4. *The Kalman Filter*

We now specialize to the dynamic linear model so useful in practice, the state-space model:

$$\theta_n = G_n\theta_{n-1} + w_n$$
$$y_n = F_n\theta_n + v_n \, , \tag{3.7}$$

$n \geq 1$, where G_n is a $p \times p$ matrix, F_n a $k \times p$ matrix, and where $(w_n), (u_n)$ still satisfy the independence hypotheses of (3.4). If now $w_n \sim N(0, W_n)$, $v_n \sim N(0, V_n)$, for covariance matrices W_n, V_n and if the prior on θ_0 is Gaussian, then all distributions will be Gaussian and, starting with the form of the Gaussian distribution of θ_{n-1}, given $y^{n-1} = \{y_{n-1}, \ldots, y_1\}$, the Bayesian arguments of Meinhold and Singhpurwalla [35] (see also [34]) yield the recursive algorithm leading to $\theta_n|y^{n-1}$ known as the Kalman filter. Given

$$\theta_{n-1}|y^{n-1} \sim N(\hat{\theta}_{n-1}, P_{n-1}) \, ,$$

we get

$$\theta_n|y^{n-1} \sim N(G_n\hat{\theta}_{n-1}, P_{n|n-1})$$

where

$$P_{n|n-1} = G_nP_{n-1}G_n' + W_n \, , \tag{3.8}$$

and from this,

$$\theta_n|y^n \sim N(G_n\hat{\theta}_{n-1} + K_n(y_n - \hat{y}_n), (I - K_nF_n)P_{n|n-1}) \, ,$$

with $\hat{y}_n = F_nG_n\hat{\theta}_{n-1}$, the Kalman gain

$$K_n = p_{n/n-1}F_n'(F_nP_{n/n-1}F_n' + V_n)^{-1} \, ,$$

and where finally the updated covariance matrix P_n of $\hat{\theta}_n$ is given by

$$P_n = (I - K_nF_n)P_{n|n-1} \, . \tag{3.9}$$

Thus, the state estimate $\hat{\theta}_{n-1}$ at time n is predicted forward by $G_n\hat{\theta}_{n-1}$ and then updated (filtered) at time n as

$$\hat{\theta}_n = G_n\hat{\theta}_{n-1} + K_n(y_n - F_nG_n\hat{\theta}_{n-1}) \, , \tag{3.10}$$

and the covariance matrix prediction and update are given by (3.8) and (3.9).

This same algorithm was first obtained without the assumption of Gaussian distribution, using linear least-squares techniques [24, 25]; see also [11, 46, 2, 17]. Here it is assumed that $w_n \sim \text{WS}[0, W_n]$ and $v_n \sim \text{WS}[0, V_n]$. In all cases the validity of the algorithm depends strongly upon the assumptions made on the "noise" sequences $(w_n), (v_n)$ (see [39] for what happens when the independence or uncorrelation assumptions are relaxed).

The Kalman algorithm involves a one-step prediction followed by a filtering step. We restrict ourselves here to considering this situation, but note that similar algorithms are available for smoothing and forecasting. See [36] for smoothing algorithms with applications to setting tolerance limits for toxic substances and to life testing. Asymptotic distribution theory for the Kalman filter is treated by Spall and Wall [51].

3.5. *Generalized Dynamic Linear Models*

Consider again the problem of recursive estimation of θ_n in the general state space model where the state process (θ_n) is Markov and the observation process (y_n) has the form $y_n = f_n(\theta_n, v_n)$, where f_n is a given (deterministic) function and v_n is independent of θ_n. Assume further that the process (v_n) consists of independent random variables.

West, Harrison and Migon [58] considered a dynamic generalization of the Generalized Linear Model (GLM) of Nelder and Wedderburn [38, 33] which, when applicable, would permit the use of the classical Kalman recursion formulas to obtain prior distributions $\theta_n | y^{n-1}$. Let $p_n(y_n | \theta_n)$ denote the conditional distribution of y_n, given θ_n, and assume (as in the GLM) the existence of a linear relationship between some given function g of θ_n and a further random process (β_n) which is assumed to evolve linearly according to $\beta_n = G_n \beta_{n-1} + w_n$, where (w_n) are independent and $w_n \sim \text{WS}[0, W_n]; \beta_n$ has q components, G_n is a $q \times q$ matrix. Thus,

$$g(\theta_n) = \lambda_n = F_n \beta_n \,,$$

where F_n is a $p \times q$ matrix, p being the dimension of θ_n. Thus, (β_n) evolves like (θ_n) did in the dynamic linear model (3.7), and the mean and covariance matrix of a specified prior probability distribution of β_n can be predicted forward linearly, using the matrix F_n.

If the prior distribution $p(\theta_n | y^{n-1})$ is given then (3.2) yields the posterior given the new observation y_n:

$$p(\theta_n | y^n) = \frac{p(y_n | \theta_n) p(\theta_n | y^{n-1})}{p(y_n | y^{n-1})} \,. \tag{3.11}$$

Since no explicit expression is assumed for the evolution of the state process (θ_n), (3.1) cannot be used to calculate the prediction, or prior distribution $p(\theta_{n+1}|\theta_n)$ since the transition probabilities $\pi(\theta_{n+1}|\theta_n)$ are not known. The predicted mean and covariance matrix of the process (β_n) are used for this purpose.

Thus, it is assumed that $\beta_n \sim \text{WS}[\hat{\beta}_n, B_n]$, and β_{n-1} is predicted forward linearly in the usual fashion: denoting by β^* this prediction, we have $\beta_n^* \sim \text{WS}[G_n\hat{\beta}_n, G_n B_{n-1} G_n' + W_n]$. Now, the mean and covariance matrix of $F_n\beta_n^*$ are calculated and related to θ_n via the function g; thus $g(\theta_n) = F_n\beta_n$. This yields two moments which can be used toward determining a prior distribution $p(\theta_{n+1}|y^n)$ for θ_{n+1}. If these first two moments should suffice to determine the probability distribution $p(\theta_{n+1}|y^n)$, then the iteration can proceed forward.

The iteration cannot be completed, however, until a new (wide-sense) posterior distribution for β_{n+1} is obtained. This is achieved by first updating $p(\theta_{n+1}|y^n)$ by (3.11) to obtain the posterior distribution $p(\theta_{n+1}|y^{n+1})$. Then, the first two moments of $p(\theta_{n+1}|y^{n+1})$ are mapped by the function g, via $\lambda = g(\theta)$, where they are used in the update equations (3.8), (3.9), (3.10) of the classical Kalman filter to obtain the desired posterior $\beta_{n+1} \sim \text{WS}[\hat{\beta}_{n+1}, B_{n+1}]$, finally closing the iteration. The update equations of the Kalman filter are expressed in terms of the moments of $F_n\beta_n$, rather than those of β_n, for this purpose.

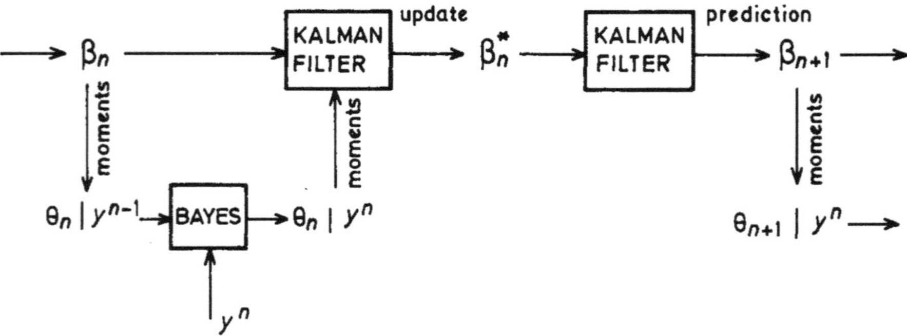

Fig. 1.

The algorithm thus consists of two interlocking iterative procedures, which complement each other by providing each other quantities required to calculate prediction and update expressions. (β_n) can be viewed as an auxiliary process which, in the absence of an explicit relationship of the observations (y_n) to (θ_n), allows one to get at the estimation of the process of interest, (θ_n). An attempt to illustrate this schematically is shown in Fig. 1.

Of course a great deal is begged by this scheme.

(1) (θ_n) must be transformable into a linear relationship $g(\theta_n) = \lambda_n = F_n \beta_n$ with a first-order autoregressive process (β_n). Conditions, such as one-to-oneness, are needed in order to map mean vectors and covariance matrices back and forth between the processes (θ_n) and $(F_n \beta_n)$.

(2) Restrictions of the prior probability distributions to a parametric family must proceed *a priori*: $p(\theta_n | y^{n-1})$ must belong to a family in which the probability distributions are determined by the first two moments.

(3) The update procedure (3.11) via Bayes formula must be tractable.

In West, Harrison and Migon [58] the observation process (y_n) is assumed to consist of scalar random variables. Though no conditions appear to be placed on g in [58], in the GLM it is usually assumed to be monotonic and differentiable ([33], p. 20). The question of tractability of the Bayes procedure (3.11) is nicely handled by restricting the probability distribution of the observation y_n to belong to the same subset of the exponential family as was used in the GLM:

$$p(y_n | \theta_n, \phi) = b(y_n; \phi) \exp[\phi(\theta_n y_n - a(\theta_n))] \qquad (3.12)$$

where ϕ is a scale parameter. (3.10) meshes with a conjugate prior of the following form:

$$p(\theta_n | y^{n-1}) = C(\alpha_n, \beta_n) \exp[\alpha_n \theta_n - \beta_n a(\theta_n)] \qquad (3.13)$$

to yield a posterior $p(\theta_n | y^n)$ of the form (3.13).

The form (3.12) includes most of the standard parametric families used in applied statistics. Details of the algorithm for the choices (3.12) and (3.13) are in [58] (see also [42, 54, 55, 56]).

3.6. *Nonparametric Approach*

In the general setup of Sec. 3.3, the recursive algorithm for estimation of the state process (θ_n) given by expression (3.1) for prediction and (3.2) for update can be approached non-parametrically using numerical methods, if the probability distributions involved have density functions which are continuous or differentiable to some degree, and the transition probabilities $\pi(\theta_n|\theta_{n-1})$ for (θ_n) are known. All the probability densities $p(\theta_n|y^{n-1}), \pi(\theta_n|\theta_{n-1}), p(\theta_n|y^n), p(y_n|\theta_n)$ are approximated with splines. Kitagawa [26] obtains explicit expressions for the case of a linear state space model (3.7) and first-order (i.e., continuous, piecewise linear) splines. Kohn and Ansley [27] demonstrate how to extend his analysis to cases where the densities can be degenerate.

3.7. *Robustness*

The reasons why Gauss declined to employ in his linear model the probability distribution that so often bears his name are outlined in [49], pp. 3, 4. In the development of the modern dynamic successors to his linear model there is a similar trend.

It had long been recognized that the classical Kalman filter estimation algorithm based on Gaussian distributions is not robust with respect to basic model assumptions, and many modifications and improvements have been devised to enhance the robustness of the filter [30, 31, 37, 40, 19, 53, 54, 18].

There are many different types of robustness or non-robustness. Robustness can be construed as stability of the process evolution with respect to initial conditions, i.e., to a perturbation of the probability distribution of (θ_n) at a given time point, say $n = 0$. It can be construed as the behaviour of the model with respect to changes with time of the basic probability structure of the process (θ_n); such changes could be sudden changes in the mean of θ_n, or in its variance, or both, and the changes could be lasting or transient. If the true random behaviour of the parameter, or state process (θ_n) deviates substantially from that given by the model, it can result in the occurrence of outliers in the observation process (y_n), i.e., values for the observations which are very improbable within the assumed model. Robustness can be thought of as the capability of the model to recognize, disregard or otherwise deal with spurious observations, whose origin can be attributed to fluke values of the realization of the state process (θ_n) or of the observations (y_n).

3.8. *Stability*

For a general formulation of the problem of robustness from the point of view of stability with respect to a prior, assuming that the state space model represents correctly the evolution of the process, consider the situation resulting when the true probability distribution $p(\theta_0)$ is not known, and is replaced by an assigned prior distribution, say $p'(\theta'_0)$. Writing $\delta_0 = p - p'$ we have

$$p(\theta_1) = \int p_{w_1}(\theta_1 - g_1(\xi))p'(\xi)d\xi + \delta_1(\theta_1)$$
$$= p'(\theta_1) + \delta_1(\theta_1),$$

with

$$\delta_1(\theta_1) = \int p_{w_1}(\theta_1 - g_1(\xi))\delta_0(\xi)d\xi.$$

Then,

$$p(\theta_1|y^1) = \frac{p_{v_1}(y_1 - f_1(\theta_1))[p'(\theta_1) + \delta_1(\theta_1)]}{p(y_1)}$$
$$= p'(\theta_1|y^1) + \Delta_1(\theta_1, y_1),$$

where

$$p'(\theta_1|y^1) = \frac{p_{v_1}(y_1 - f_1(\theta_1))p'(\theta_1)}{p'(y_1)}$$

and

$$p'(y_1) = \int p_{v_1}(y_1 - f_1(\theta_1))p'(\theta_1)d\theta_1$$

and

$$\Delta_1(\theta_1, y_1) = \frac{p_{v_1}(y_1 - f_1(\theta_1))}{p(y_1)}\left[\delta_1(\theta_1) - \frac{p'(\theta_1)[p(y_1) - p'(y_1)]}{p'(y_1)}\right].$$

In general, for $n \geq 2$:

$$p(\theta_n|y^{n-1}) = \int p_{w_n}(\theta_n - g_n(\theta_{n-1}))p'(\theta_{n-1}|y^{n-1})d\theta_{n-1} + \delta_n(\theta_n, y_{n-1})$$

with

$$\delta_n(\theta_n, y_{n-1}) = \int p_{w_n}(\theta_n - g_n(\theta_{n-1}))\Delta_{n-1}(\theta_{n-1}, y^{n-1})d\theta_{n-1}$$

and

$$p(\theta_n|y^n) = p'(\theta_n|y^n) + \Delta_n(\theta_n, y^n),$$

where

$$p'(\theta_n|y^n) = \frac{p_{v_n}(y_n - f_n(\theta_n))p'(\theta_n|y^{n-1})}{p'(y_n|y^{n-1})}$$

and

$$\Delta_n(\theta_n, y_n) = \frac{p_{v_n}(y_n - f_n(\theta_n))}{p(y_n|y^{n-1})}\left[\delta_n(\theta_n, y^{n-1}) - \frac{p'(\theta_n|y^{n-1})[p(y_n|y^{n-1}) - p'(y_n|y^{n-1})]}{p'(y_n|y^{n-1})}\right]$$

with

$$p(y_n|y^{n-1}) = \int p_{v_n}(y_n - f_n(\theta_n))p(\theta_n|y^{n-1})d\theta_n$$

and

$$p'(y_n|y^{n-1}) = \int p_{v_n}(y_n - f_n(\theta_n))p'(\theta_n|y^{n-1})d\theta_n \ .$$

We thus have

$$p(\theta_y|y^{n-1}) = p'(\theta_n|y^{n-1}) + \delta_n(\theta_n, y^{n-1})$$
$$p(\theta_n|y^n) = p'(\theta_n|y^n) + \Delta_n(\theta_n, y^n)$$

where the primed expressions represent the estimates of the probability distributions of the states resulting from the prior $p'(\theta_0')$ and the unprimed expressions the estimates of the state distributions that would result from the true state probability distribution $p(\theta_0)$ at $n = 0$. The accuracy of the estimates will depend on the behaviour of (δ_n) and (Δ_n).

3.9. *Robustness against Departures from Gaussian Distributions*

It is remarkable that closed-form expressions are available for the extension of the Kalman filter for the linear model (3.7) to the case where either the state noise or the observation noise (but not both) is non-Gaussian. The filter expressions for these cases are derived by Masreliez [30], where it is shown that the role of the Kalman gain and innovations in the update expressions (3.10) and (3.9) is assumed respectively by the vector of scores of the observation prediction density $p(y_n|y^{n-1})$, $g_n(y_n) := -\frac{\partial}{\partial y_n}\log p(y_n|y^{n-1})$, and its derivative, the information, $G_n(y_n) := \frac{\partial}{\partial y_n}g_n(y_n)$. Thus, for example, for the case in which the state θ_n is Gaussian and the observation noise v_n is not, (3.10) takes on the form:

$$\hat{\theta}_n = \bar{\theta}_n + P_{n|n-1}F_n'g_n(y_n) \tag{3.14}$$

where $\bar{\theta}_n$ is the mean of $p(\theta_n|y^{n-1})$.

Masreliez and Martin [31] treat the problem of constructing filters which are robust against departures of either the state or the observation noise from Gaussian distributions. The first step is to investigate robustness of estimates of θ in the random parameter observation equation $y = X\theta + v$, with respect to certain families of probability distributions which contain the Gaussian distributions, such as symmetric contaminated Gaussian densities. In the case where θ is Gaussian, the estimates considered are of the form (3.14):

$$\hat{\theta} = \bar{\theta} + PF'\Psi \tag{3.15}$$

where Ψ is the (bounded) influence function of a min-max stochastic approximation estimator with respect to the given family $\mathcal{F}$, which contains the Gaussian densities. Here $\bar{\theta}$ is the mean of a Gaussian prior. For this setup, where departures from Gaussian densities have the form of contaminated Gaussian densities, bounds are derived for the covariance matrix P of the estimator $\hat{\theta}$ of θ.

To construct a robust filter, an iterative procedure is used employing updating expressions of the form (3.15), robustness resulting from the bounds on the covariance P_n. The update equation (3.15) does not yield a Gaussian distribution for $\hat{\theta}$, but when deviations from Gaussian distributions are small the robust filter functions well as is.

3.10. *The Multi-Process Model*

A Bayesian update-and-filtering algorithm can be exploited to introduce refinements into the model which allow the model to produce estimates which react reasonably and efficiently to stark or sudden changes in the observation sequences, in particular to outliers. Such a refined model was developed under the rubrik of Bayesian forecasting by Harrison and Stevens ([20, 21]; see also [4, 5, 6, 13]). The idea is to allow the model to change, to accommodate a detected change in the observation sequence. This is modelled by allowing the covariance matrix W_n of the state θ_n to be randomly chosen during each iteration of the algorithm, combining new information so as to optimize the choice of W_n at each time n.

The multiprocess model known as Bayesian forecasting can be developed in the general setting (3.4) without linearity or Gaussian distributions. Added to the model is an independent random process (S_n), consisting of independent and identically distributed random variables taking values in the set $1, \ldots, N$. (As in [21] and [4], we restrict ourselves to the case where (S_n) takes finitely many values to avoid measurability questions.) The value of S_n determines which of N models holds sway (or which "state" the model is in) at time n: it determines the value of the covariance matrix W_n of θ_n. The useful case is where the distribution of w_n is one of a family indexed by a scale parameter. When $S_n = i$, the conditional distribution of θ_n given θ_{n-1} has mean zero and covariance matrix $W_n^{(i)}$; the covariance matrix depends on i and can change over time.

The recursion begins at $n-1$ with N conditional probability distributions $p_{\theta_{n-1}}(.|i, y^{n-1})$ given for the parameter θ_{n-1}, given the state of the model (value of S_{n-1}) and the set of observations up to and including that at time $n-1$. Also given are N conditional probabilities $q_{n-1}^{(i)} = P[S_{n-1} = i|y^{n-1}], i = 1, \ldots, N$. The recursion will be complete with the update of these quantities to time n.

First we obtain N^2 prediction distributions as conditional probabilities for θ_n given $S_{n-1} = i$ and $S_n = j$ as follows (we write simply $p_{\theta_n}(\theta_n|i, j, y^{n-1})$ for $p_{\theta_n}(\theta_n|S_{n-1} = i, S_n = j, y^{n-1})$ to unburden the notation: i goes with S_{n-1}, j with S_n):

$$\begin{aligned}
p_{\theta_n}(\theta_n|i, j, y^{n-1}) &= \int p_{\theta_n}(\theta_n|\theta_{n-1}, i, j, y^{n-1}) p_{\theta_{n-1}}(\theta_{n-1}|i, j, y^{n-1}) d\theta_{n-1} \\
&= \int p_{\theta_n}(\theta_n|\theta_{n-1} i, j) p_{\theta_{n-1}}(\theta_{n-1}|i, y^{n-1}) d\theta_{n-1} \\
&= \int p_{w_n^{(j)}}(\theta_n - g_n(\theta_{n-1})) p_{\theta_{n-1}}(\theta_{n-1}|i, y^{n-1}) d\theta_{n-1}
\end{aligned}$$

$$(3.16)$$

where we have used the independence of the sequence (S_n) to conclude that $p_{\theta_{n-1}}(.|i, j, y^{n-1})$ does not depend on j, the state of the model at time n, and is the given conditional distribution $p_{\theta_{n-1}}(.|i, y^{n-1})$, and where $p_{w_n^{(j)}}$ denotes the probability distribution of the random variable w_n under the model given by $S_n = j$, which has covariance matrix $W_n^{(j)}$. Once the observation y_n comes in, these yield the N^2 updates by an application of Bayes' formula:

$$\begin{aligned}
p_{\theta_n}(\theta_n|i, j, y^n) &\propto p(y_n|\theta_n, i, j, y^{n-1}) p(\theta_n|i, j, y^{n-1}) \\
&= p(y_n|\theta_n) p(\theta_n|i, j, y^{n-1}) \\
&= p_{v_n}(y_n - f_n(\theta_n)) p(\theta_n|i, j, y^{n-1})
\end{aligned}$$

noting that y_n, given θ_n, depends only on the probability distribution of its "error" v_n, and not on previous observations or values of S_n. Thus,

$$p_{\theta_n}(\theta_n|i, j, y^n) = \frac{p_{v_n}(y_n - f_n(\theta_n)) p(\theta_n|i, j, y^{n-1})}{p(y_n|i, j, y^{n-1})} \qquad (3.17)$$

with

$$p(y_n|i,j,y^{n-1}) = \int p_{v_n}(y_n - f_n(\theta_n))p(\theta_n|i,j,y^{n-1})d\theta_n \, ,$$

the prediction distribution of the observation given the two model states $S_{n-1} = i, S_n = j$. We now arrive at the desired update by summing over i:

$$p(\theta_n|S_n = j, y^n) = \sum_{i=1}^{N} p(\theta_n, i|j, y^n)$$
$$= \sum_{i=1}^{N} p(\theta_n|i, j, y^n)p(S_{n-1} = i|S_n = j, y^n) . \tag{3.18}$$

The last loose end is to figure out what $P(S_{n-1} = i|S_n = j, y^n)$ is and to obtain the updates $q_n^{(j)} = P[S_n = j|y^n]$. Note that if $P[S_{n-1} = i, S_n = j|y^n]$ can be determined both these quantities will be known, $q_n^{(j)}$ from $P[S_n = j|y^n] = \Sigma_{i=1}^{N} P[S_{n-1} = i, S_n = j|y^n]$ and then

$$P[S_{n-1} = i|, S_n = j, y^n] = \frac{P[S_{n-1} = i, S_n = j|y^n]}{P[S_n = j|y^n]} \, .$$

Using Bayes' formula again, we have

$$P[S_{n-1} = i, S_n = j|y^n] \propto$$
$$P[y_n|S_{n-1} = i, S_n = j, y^{n-1}]P[S_{n-1} = i, S_n = j|y^{n-1}]$$

and the first factor on the right is just $p(y_n|i, j, y^{n-1})$, the denominator of (3.17), and for the second factor, we have

$$P[S_{n-1} = i, S_n = j|y^{n-1}] = P[S_n = j]P[S_{n-1} = i|y^{n-1}] \, ,$$

again by the independence of the random variables (S_n), since the observations $y_k, 1 \leq k \leq n-1$, involve only $\theta_0, \theta_1, \ldots, \theta_{n-1}$ and (v_n) and are independent of S_n. Now the unconditional probability distributions of $(S_n)_{n\geq 1}$ are given with the model, and $q_{n-1}^{(i)} = P[S_{n-1} = i|y^{n-1}], i = 1, \ldots, N$, began the recursion. the algorithm is now complete.

These ideas are developed very nicely for the case of a linear state space model (3.7) and Gaussian distributions in [4]. Here a basic problem which is not apparent in the general formulation comes to light: that of actually determining the probability distributions in (3.18). In the Gaussian case, the N^2 predictions of (3.16) go forward in the classical way, without the need of actually executing any convolutions, by applying the transition matrix G_n. The classical Kalman filter is likewise available for the Bayesian updates of (3.17). But the expression (3.18) for N updates at time n is a sum of N Gaussian densities. At this point Gaussian distributions are needed to complete the recursion. In [21] and [4] the problem is resolved by replacing the density $p(y_n|S_n = j, y^{n-1})$ with that of the Gaussian density having the same mean and covariance matrix as the density represented by the sum in (3.18).

3.11. *Weighted Sums of Densities*

In some situations it may be expedient to express probability densities as convex combinations of densities from a given family $\mathcal{F}$. Such techniques are useful in rendering tractable the recursive Bayesian estimation algorithm (3.5), (3.6) for the non-linear state space model (3.4) (the "extended Kalman filter"). One such scheme is to use a Gaussian sum approximation (see Anderson and Moore [2]) for the posterior density, and to apply the classical Kalman filter for the linear model (3.7) to each term of the sum to arrive at a similar representation for the updated posterior. Thus,

$$p(\theta_{n-1}|y^{n-1}) = \sum_{i=1}^{m} \alpha_i p^{(i)}(\theta_{n-1}|y^{n-1}) \tag{3.19}$$

where each $p^{(i)}$ is a multivariate Gaussian density and $\alpha_i \geq 0, \Sigma_{i=1}^{m}\alpha_i = 1$. The Kalman filter applied componentwise to each $p^{(i)}$ again yields Gaussian densities and one constructs for the updated posterior $p(\theta_n|y^n)$ the same convex combination, of these updated densities.

Representations of the form (3.19) have been used to arrive at robust filters for linear models (3.7), where the densities are members of some parametric family $\mathcal{F}$. Typically, heavy-tailed densities, with influence functions which are bounded or asymptotically zero are used to achieve robustness. If the densities are not Gaussian, and if the same termwise updating scheme is to be used as in the Gaussian sum case, then the prediction $p^{(i)}(\theta_n|y^{n-1})$ and update $p^{(i)}(\theta_n|y^n)$ densities will need to be approximated from within $\mathcal{F}$. Suppose that the densities in $\mathcal{F}$ are determined by their first two moments. Then, if $\hat{\theta}_{n-1}^{(i)}, P_{n-1}^{(i)}$ are the estimates from $p^{(i)}(\theta_{n-1}|y^{n-1})$, then $p^{(i)}(\theta_n|y^{n-1})$ is chosen as that density from $\mathcal{F}$ with mean $G_n\hat{\theta}_{n-1}^{(i)}$ and covariance $G_nP_{n-1}G_n' + W_n$.

For the application of the Bayes' formula update (3.6), the $p^{(i)}$ will need to be members of some conjugate prior family with respect to the distributions of the observations, or else some tractable approximation method for replacing each updated posterior with a member of $\mathcal{F}$ must be available. With such a procedure in place, each $p^{(i)}(\theta_n|y^{n-1})$ is replaced by a member of $p^{(i)}(\theta_n|y^n)$ of $\mathcal{F}$, the recursion is complete.

Meinhold and Singhpurwalla [37] employed sums (3.19) consisting of Student-t densities in a carefully balanced scheme designed to produce a filter that is outlier robust by using the appropriate relative heaviness of the tails of the densities of the posteriors and likelihoods. In this scheme the observations (y_n) are also Student-t distributed. The filtered densities given by (3.6) will not be Student-t, but are known to be unimodal or bimodal. An approximation method is employed which replaces each filtered density with either one or two Student-t densities. Here the number of terms in the sum (3.19) can grow, and an ad hoc method is needed to combine densities and keep the number of terms to a manageable level. This is reminiscent of the situation in the multi-process model of Harrison and Stevens (cf. [4]) except that there the rate of proliferation of densities was determined and constant.

Meinhold and Singhpurwalla [37] contains an analysis of robustness, philosophies for dealing with outliers, and an explanation of the non-robustness of the Gaussian model.

References

1. A. C. Aitken, *On least squares and linear combination of observations*, Proceedings of the Royal Society of Edinburgh, Sec. A, **55** (1934) 42–47.
2. B. D. O. Anderson and J. B. Moore, *Optimal Filtering*, Prentice-Hall, 1979.
3. M. Aoki, *Optimization of Stochastic Systems; Topics in Discrete-Time Systems*, Academic Press, New York, 1967.
4. W. M. Bolstad, *Harrison-Stevens forecasting and the multiprocess dynamic linear model*, American Statistician (2) **40** (1986) 129–135.
5. W. M. Bolstad, *Estimation in the multiprocess dynamic generalized linear model*, Communications in Statistics, Part A, **17** (1988) 4179–4184.
6. W. M. Bolstad, *The multiprocess dynamic linear model with biased perturbations*, Biometrika **75** (1988) 685–692.
7. W. K. Bühler, *Gauss, A Biographical Study*, Springer-Verlag, 1981.
8. R. Dedekind, *Gauss in seiner Vorlesung über die Methode der kleinsten Quadrate*, Gesammelte Mathematische Werke II, Chelsea, 1969, pp. 293–306.

9. P. Diaconis and D. Ylvisaker, *Quantifying prior opinion*, in Bayesian Statistics 2, Eds. Bernardo, DeGroot, Lindley, Smith, Elsevier, North-Holland, 1985, pp. 133–156.

10. G. T. Diderrich, *The Kalman filter from the perspective of Goldberger-Theil estimators*, American Statistician (3) **39** (1985) 193–198.

11. D. B. Duncan and S. B. Horn, *Linear dynamic recursive estimation from the viewpoint of regression analysis*, Journal of the American Statistical Association **67** (1972) 815–821.

12. G. W. Dunnington, *Carl Friedrich Gauss: Titan of Science*, Hafner, 1955.

13. R. Fildes, *An evaluation of Bayesian forecasting*, Journal of Forecasting **2** (1983) 137–150.

14. C. F. Gauss, *Theoria Motus Corporum Coelestium in Sectionibus Conicis Solem Ambientum*, Werke, Volume VII, Herausgegeben von der Königlichen Gesellschaft der Wissenschaften, B. G. Teubner, Leipzig, 1906.

15. C. F. Gauss, *Disquisitio Palladis*, Werke, Volume VI, Herausgegeben von der Königlichen Gesellschaft der Wissenschaften, B. G. Teubner, Leipzig, 1906, pp. 1–24.

16. C. F. Gauss, *Theoria Combinationis*, Werke, Volume IV, Herausgegeben von der Königlichen Gesellschaft der Wissenschaften, B. G. Teubner, Leipzig, 1906, pp. 1–93.

17. A. Gelb, ed. *Applied Optimal Estimation*, M.I.T. Press, 1984.

18. K. Gordan and A. F. M. Smith, *Modelling and monitoring discontinuous changes in time series*, Bayesian Analysis of Time Series and Dynamic Models, ed. James C. Spall, Marcel Dekker, New York and Basel, pp. 359–391.

19. I. Guttman and D. Pena, *Robust filtering*, Comment on "Dynamic generalized linear models and Bayesian forecasting", by West, Harrison and Migon, Journal of the American Statistical Association **80** (1985) 91–92.

20. P. J. Harrison and C. F. Stevens, *A Bayesian approach to short-term forecasting*, Operational Research Quarterly **22** (1971) 341–362.

21. P. J. Harrison and C. F. Stevens, *Bayesian forecasting*, Journal of the Royal Statistical Society (B), **38** (1976) 205–247.

22. A. H. Jazwinski, *Stochastic Processes and Filtering Theory*, Academic Press, New York, 1970.

23. T. Kailath, *A view of three decades of linear filtering theory*, IEEE Transactions on Information Theory, Vol. IT-20, No. 2 (1974) 146–181.

24. R. E. Kalman, *A new approach to linear filtering and prediction problems*, Journal of Basic Engineering, Trans. ASME, Series D (1) **82** (1960) 35–45.

25. R. E. Kalman and R. Bucy, *New results in linear filtering and prediction theory*, Journal of Basic Engineering, Trans. ASME, Series D (3) **83**(1961) 95–108.

26. G. Kitagawa, *Non-Gaussian state-space modelling of nonstationary time series* (with discussion), Journal of the American Statistical Association (400) **82** (1987) 1032–1041.

27. R. Kohn and C. F. Ansley, Comment on: "Non-Gaussian state-space modelling of nonstationary time series", Journal of the American Statistical Association (400) **82** (1987) 1041–1044.

28. A. N. Kolmogorov, *Interpolation and extrapolation of stationary random sequences*, Izv. Akad. Nauk SSSR, Ser. Mat. 5 (1941) 3–14.

29. S. Kourouklis and C. C. Paige, *A constrained least squares approach to the general Gauss-Markov linear model*, Journal of the American Statistical Association (375) **76** (1981) 620–635.

30. C. J. Masreliez, *Approximate non-Gaussian filtering with linear state and observation relations*, IEEE Transactions on Automatic Control AC-20 (1975) 107–110.

31. C. J. Masreliez and R. D. Martin, *Robust Bayesian estimation for the linear model and robustifying the Kalman filter*, IEEE Transactions on Automatic Control AC-22(3) (1977) 361–371.

32. K. O. May, Gauss, Carl Friedrich, In *Dictionary of Scientific Biography*, Vol. V, Scribner, New York, 1972, pp. 298–315.

33. P. McCullaugh and J. A. Nelder, *Generalized Linear Models*, Chapman and Hall, 1983.

34. R. K. Mehra, *Kalman filters and their applications to forecasting*, Forecasting, eds. Makridakis & Wheelwright, Studies in the Management Sciences, North-Holland, 1979, 75–94.

35. R. J. Meinhold and N. D. Singhpurwalla, *Understanding the Kalman filter*, American Statistician (2) **37** (1983) 123–127.

36. R. J. Meinhold and N. D. Singhpurwalla, *A Kalman filter smoothing approach for extrapolations in certain dose-response, damage-assessment and accelerated-life-testing studies*, American Statistician (2) **41** (1987) 101–106.

37. R. J. Meinhold and N. D. Singhpurwalla, *Robustification of Kalman filter models*, Journal of the American Statistical Association (406) **84** (1989) 479–486.

38. J. A. Nelder and R. W. M. Wedderburn, *Generalized linear models*, Journal of the Royal Statistical Society (A) **135** (1972) 370–384.

39. P. L. Odell and T. O. Lewis, *Best linear recursive estimation*, Journal of the American Statistical Association (336) **66** (1971) 893–896.

40. D. Peña and I. Guttman, *Bayesian approach to robustifying the Kalman filter*, Bayesian Analysis of Time Series and Dynamic Models, ed. James C. Spall, Marcel Dekker, New York and Basel, 1988, pp. 227–253.

41. D. Pfefferman, *On extensions of the Gauss- Markov theorem in the case of stochastic regression coefficients*, Journal of the Royal Statistical Society (B) **46** (1984) 239–148.

42. R. L. Plackett, *Some theorems in least squares*, Biometrika **37** (1950) 149–157.

43. A. Pole, M. West and P. J. Harrison, *Nonnormal and nonlinear dynamic Bayesian modeling*, Bayesian Analysis of Time Series and Dynamic Models, ed. James C. Spall, Marcel Dekker, New York and Basel, 1988, pp. 167–198.

44. S. Puntanen and G. P. H. Styan, *The equality of the ordinary least squares estimator and the best linear unbiased estimator*, American Statistician **43** (1989) 153–161.

45. C. R. Rao, *Linear Statistical Inference and its Applications*, Wiley & Sons, 1973.

46. A. P. Sage and J. L. Melsa, *Estimation Theory with Applications to Communication and Control*, Robert E. Krieger, Florida, 1971.

47. W. M. Sallas and D. A. Harville, *Best linear recursive estimation for mixed linear models*, Journal of the American Statistical Association **76** (1981) 860–869.

48. H. Scheffe, *The Analysis of Variance*, Wiley, New York, 1959.

49. H. L. Seal, *The historical development of the Gauss linear model*, Biometrika **54** (1967) 1–24.

50. H. W. Sorenson, *Least-squares estimation: from Gauss to Kalman*, IEEE Spectrum, July, 1970, pp. 63–68.

51. J. C. Spall and K. D. Wall, *Asymptotic distribution theory for the Kalman filter state estimator*, Communications in Statistics, Theory & Methods **13** (1981) 1981–2003.

52. S. M. Stigler, *Gauss and the invention of least squares*, Annals of Statistics (3) **9** (1981) 465–474.

53. M. West, *Robust sequential approximate Bayesian estimation*, Journal of the Royal Statistical Society (B) **43** (1981) 157–166.

54. M. West, *Outlier models and prior distributions in Bayesian linear regression*, Journal of the Royal Statistical Society (B) **46** (1984) 431–439.

55. M. West, *Bayesian model monitoring*, Journal of the Royal Statistical Society (B) (1) **48** (1986) 70–78, 157–166.

56. M. West and P. J. Harrison, *Monitoring and adaption in Bayesian forecasting models*, Journal of the American Statistical Association **81** (1986) 741–750.

57. M. West and P. J. Harrison, *Subjective intervention in formal models*, Journal of Forecasting (1) **8** (1989) 33–53; **80**, 91–92.

58. M. West, P. J. Harrison and H. Migon, *Dynamic generalized linear models and Bayesian forecasting* (with discussion), Journal of the American Statistical Association **80** (1985) 73–97.

59. N. Wiener, *Interpolation, Extrapolation and Smoothing of Stationary Time Series*, M.I.T. Press, Cambridge, Mass., 1949.

Pete Daffer
Department of Mathematics
University of South Alabama
Mobile, Alabama 36688
USA

THE MATH. HERITAGE OF C.F. GAUSS (pp. 157-171)
edited by George M. Rassias
©1991 World Scientific Publ. Co. Singapore

WAVES, QUANTA, $E = mc^2$, AND PERIHELION SHIFTS: A NEW SCIENCE-HISTORICAL PERSPECTIVE AND MATHEMATICAL UNIFICATION

Kyril Demys

1. Some History of Science

The work and genius of Carl Friedrich Gauss immediately leads one into the history of science as well as to the applications of mathematics to physical theory. Gauss was the first, as his journals show, to discover self-consistent alternatives to the Euclidean geometry of the plane, and also the first to conceive of the idea of a hypercomplex number as his profound discussions of biquadratic residues show — thoughts he did not see fit to publish, although they certainly belonged there, in his now classic work *Disquisitiones Arithmeticae.*

The reason for Gauss' reluctance was one all discoverers ahead of their times must disregard at their peril: the fear of reprisals, including the deadly conspiracy of silence ploy, against them by envious or simply fearful peer groups or claques seated in places of sociological power in the given field. It is easy for us today to reprimand Gauss' reluctance at our safe distance, conveniently forgetting the tragedies in their own lifetimes of Evariste Galois and Oliver Heaviside to name but two innovators whose insights and genius were spurned by jealous and mediocre comtemporaries. It turns out from the Royal Society records that the well-known William Burnside was the rather unscrupulous referee who, in somewhat cowardly anonymity, stopped the Royal Society's publication of the third part of Heaviside's brilliant discussion of the factorial function of unrestricted argument, Gauss'

$\prod(x)$. Heaviside, in justified contempt, resigned his membership. Burnside also has the dubious distinction of excluding from publication the details of Thorold Gossett's pioneering discoveries in higher dimensional geometry. Burnside was reputed in the establishment of his day and evidently resented anyone who did not enter mathematics through the path of embedded conditioning in the curricula of conventional schooling. Note well that both Heaviside and Gossett were self-taught geniuses: a breed for whom Burnside by his hostile and unscientific behaviour showed an unadmirable envy. Unfortunately, unlike Śrinivasa Ramanujan, who otherwise would have received the same treatment, neither Heaviside nor Gossett as young men ever found their Hardy.

Still, as Shakespeare says, "Murder, though it hath no tongue, yet speaks with most miraculous organ." And it does also with the murder of people's potential and societal effectiveness. Thus Léon Brillouin, renowned 20th century physicist and the discoverer of Brillouin zones, concludes that "Heaviside was the forgotten genius of physics" [1]. Donald Coxeter has rediscovered Gossett and given him his justly deserved honour [2].

Another such puzzle falls into place when we consider that a delusory situation arises when people really believe that lengths when measured from one moving object to another actually change simply because the measuring signals of finite speed are of course then distorted. This was shown as early as 1937–38 in two impeccably clear articles on "Relativity, Clocks, and Cameras" by E. Colthurst in *The Mathematical Gazette*. The use of retarded potentials in Maxwellian theory at once yields the Lorentz transformation, just as their use in Newtonian gravitational perturbation theory (treating such perturbations as propagated at finite speed c) yields the perihelion shifts of general relativity (announced in 1915).

The almost forgotten genius, the physicist Paul Gerber,* using Maxwellian-Newtonian perturbation theory with retarded potentials, had arrived at the identical perihelion-shift equations in 1898 in a stunning historical priority [3], and had given further details in 1902, three years before even the first publication on special relativity, let alone general. The renowned German physicist Ernst Mach, one of Einstein's intellectual heroes, whom the latter carefully studied, had specifically cited Gerber's work, for example on p. 201 of the 5th edition of Mach's greatest work, *Die Mechanik in Ihrer Entwicklung*. Gerber was also the first who proved

---

*of Stargard-in-Pommern, now part of Poland.

that gravitational perturbations must travel at a finite speed equal to that of electromagnetic impulses. This was all before the perihelion shift equation appeared (with merely a change in the letters used for parameters) seventeen years later than Gerber's published result.

Paul Gerber showed that the equation is much more simply and directly derived from considerations of retarded potential, i.e., a consequence of the finite velocity of a gravitational impulse. There was no attempt to dress up the phenomena of force and acceleration as "curvature" in a pseudo-space of Minkowskian type. By using the much more physically direct idea of retarded potential, Gerber could retain the real and undeniable phenomenological difference between space and time. So his approach was actually more physically realistic in addition to being preferable on grounds of Occam's razor: logical parsimony. These facts do Einstein's work no discredit, but simply place it in fuller historical perspective.* Yet to this date, Gerber's priority is buried in an era supposedly welcoming scientific objectivity but in practice very dominated and censored by controlling cliques.

It must also be noted that the recent attempt by N. Y. Roseveare [*Mercury's Perihelion*, Clarendon Press, Oxford, 1982, p. 142] to denigrate Gerber's perihelion-shift equation of 1898 (exactly the same as Einstein's of 1915), is naively fallacious. Roseveare's other demurral to Gerber's discovery on the basis of light deflection is equally ill-advised because Gerber's work fully allowed for the finite speed of gravitational perturbations: in fact, he was the first to conceive of that seminal idea. Moreover, by a genial application of retarded potentials, Gerber fully took into account the effects on measurements between moving systems of the finite velocity of light. Thus Gerber's equation for the perihelion shift of Mercury is fully validated physically, quite apart from the fact that it is the identical equation Einstein published 17 years later.

The historical research tells much the same with $E = mc^2$, which can be shown to a consequence of the brilliant nineteenth century theory of James Clerk Maxwell, based on the laboratory discoveries of Michael Fara-

---

*It is now also known (as reported at the 1990 annual meeting of the American Association for the Advancement of Science by historian Senta Troemel-Ploetz and physicist Evan Harris Walker) that Albert Einstein's first wife Mileva Maric (or Marity) collaborated substantively and creatively with him in his work on special relativity and Brownian motion, published in 1905. He left her in 1914, but felt conscience-bound to give her all of his 1921 Nobel Prize money which, however, she had to spend on their mentally disturbed son and actually died in poverty.

day. The latter is, incidentally, a good antidote to the unscientific hogwash one reads too often, to the effect that physics "is equations, not words." Yet Faraday's words explained and represented new facts and breakthrough concepts which Maxwell then clarified and elegantly rendered in mathematical form. Hendrik Antoon Lorentz, still following Maxwell's lead, derived his famous transformations (read "special relativity theory") by seeking the most parsimonious expression of the fact that the electromagnetic phenomenology should leave Maxwell's equations invariant. This was in 1895.

The other postulate of special relativity (that the speed of light cannot be exceeded) though it follows from the Lorentz transformations, is much more questionable. For we know that c depends on the dielectric in which the radiation is. Also, it has long been known, via consequences from Newton and Maxwell, that a gravitational field could affect the mass and momentum of an electromagnetic impulse by effectively changing the dielectric constants of space. Already by Maxwell's analysis, so-called "empty" space has a substantial impedance of about 377 ohms. On other counts also, c is not a limit. For even in rarefied gases, it has been shown that particles may be accelerated to speeds exceeding that of light.

Oliver Heaviside first predicted this possibility and before 1900 worked out the same equations for this predicted effect, published in 1934 by I. Frank and I. Tamm with no mention of Heaviside then or since. The superluminal speed of an electron was first demonstrated in water in 1933 by Piotr Čerenkov. It was later confirmed and extended to even rarefied gases by other experimental physicists.

All these experimental and observational facts make one realize that the Lorentz transformations are necessary but not sufficient to the problem, and that they shift into another form as the particle reaches and then exceeds the speed of light in a given dielectric with a given impedance. The so-called "vacuum" (which seems more like a plenum, as Paul Dirac began to note) with its substantial impedance of 377 Ω would appear as no ultimate exception here. We must go back to the laboratory and observe more exactly just how the virtual mass of a moving particle changes as the particle passes the light barrier. We predict that those mass curves, plotted against velocity, will be roughly S-shaped or else show a rise to a peak and then a subsidence with leveling off.

1.1. *Planck, de Broglie, Compton, and Maxwell*

The following sections deal much with quantum theory, the essence of which is the concern for whole numbers. In this way this paper honours Gauss because of his deep interest in the science of integers, which he calls "the queen of mathematics." Although the uncertainty principle and the noncommutative equation $pq - qp = ih/2\pi$ lie close to the heart of quantum theory, it would be a mistake to believe they are its essence. The essence of quantum theory is its domination by the mathematics of integers and half-integers which, when physically analyzed, reflect resonances and antiresonances. What quantum physics has ushered in is what may be called with minute exactitude *the resonant universe*. This principle extends deeply into quantum biology as well where, for example, molecular field resonances lie at the heart of enzymatic catalysis.

Outstanding among the great contributions of Carl Friedrich Gauss to mathematics were his accomplishments in developing the applications of mathematics to physics. In that spirit, this paper develops the most fundamental treatment of the mass-energy relation in mathematical-physical terms. To do this needs citing the findings of C. Musès, first discovered in 1963, then published in Europe in 1965 and made public in the United States in 1970 during an invited lecture at the NASA Ames Research Center in Moffett Field, California. It was published, under the editorship of Norbert Wiener, rather obscurely in 1965 [4]; however, not so obscurely as to prevent those findings from appearing later with no original source reference. This paper is thus inevitably concerned in some measure with the history of science. Parallels are the ascription of the discovery of antibiotics to Ian Fleming, instead of to the much earlier work of René Dubos [5].

Ever since P. Lebedev's [6] experimental discovery of radiation pressure in 1900 (anticipating E. Nichols and G. Hull [7] in 1903) — it was realized that mass was to be associated with radiation. M. Planck's quantum theory, announced in 1900, led to the same conclusion since the size of radiation quanta was proportional to the energy of the radiation, in turn long known to be proportional to its frequency. Planck showed the constant of proportionality to be h, amounting to some 6.6×10^{-27} erg-sec, far too small to be apparent in the phenomenology of macrophysics.

Hence, if ν is the frequency and E the energy, we have the Planck-Einstein relation, $h\nu = E$, which dimensionally is mv^2 as C. Musès first noted, where v is the lineal velocity associated with the energy and m, the associated mass. Following Musès, in the case of radiation $v = c$, the propagational velocity of a light signal. Hence $h\nu = mc^2$. But $\nu = c/\lambda$ where λ is the wavelength of the given frequency. Thus $hc/\lambda = mc^2$ or

$$\lambda = h/mc \qquad (1)$$

which is the shortest and most perspicuous derivation of L. de Broglie's relation*, which was actually obvious since Lebedev's and Planck's work, though Louis de Broglie did not publish it until 1924 — a lapse of time that measures the extent to which the comprehension of a new paradigm is blocked by the mental obstruction of a prior and narrower conditioning.

What has not been clearly realized is that the Compton wavelength is also the de Broglie wavelength for an electron travelling through a given medium at the speed of light in that medium. That particles can not only do this, but even exceed the speed of light in any physical medium, was first predicted by Oliver Heaviside [8] in 1898, but not observed until Piotr Čerenkov did so in 1933. The mathematical theory by I. Frank and I. Tamm in 1934, giving the semivertex angle of the Čerenkov cone, had already been found also by Heaviside whose priority, however, was lost in the shuffle of human history, even though he was credited for his equally brilliant prediction of an ionospheric layer that reflects radio waves of sufficiently short wavelength so that they may travel around the earth.

The semivertex angle of the Čerenkov cone, as first calculated (before Čerenkov's observation) by O. Heaviside and a generation later by Frank and Tamm, is $\cos \theta = c/v$ where c is the speed of light in the medium through which the given mass moves with velocity $v > c$. This could be also derived from the formula for the semivertex angle of the Mach cone $\sin \theta = c/v$, where a mass is moving through air at velocity $v > c$, the latter being the speed of sound in air. Since sound is a longitudinal and light a transverse wave, the sine and cosine would interchange, and the Čerenkov cone relation emerges at once from the Mach cone.

Returning to the de Broglie wavelength, given by $\lambda = h/mv$ for a mass m travelling at velocity v, we can see that a Čerenkov electron at speed c, there is a de Broglie wave $\lambda = h/mc$. But that is also the Compton wavelength for the electron-scattering of radiation.

---

*The above is the Compton wavelength. The de Broglie relation is obtained by substituting v for c in Eq. (1).

By the same token, there is also, for any radiation at velocity c and of wavelength λ, an associated electromagnetic mass m, given by $m = h/\lambda c$ or $m = h\nu/c^2$ where ν is the frequency of the radiation and $h\nu$ its energy. Thus both the Compton and de Broglie wavelengths have a common physical root in the mass-energy relation.

It is to be noted that the Čerenkov effect has been observed even when the medium is a very rarified gas. The observations do not show the virtual mass of the particle approaching infinity as the Čerenkov or luminal barrier is reached and then surpassed. In fact, there is no physically demonstrable principle making the physical vacuum "sacred" in this sense; and already in 1898 Heaviside had observed that with $v > c$ the expression $m/\sqrt{1 - (v^2/c^2)}$ is no longer valid. In modern terms, the usual Pomeranchuk trajectory for scattering cross sections does not hold at Čerenkov speeds as observations of the mass of moving bodies with $v > c$ show.

In other words, given sufficiently powerful experimental apparatus, there is no demonstrable reason why protons, say, should not travel faster than radiation in any medium including the physical vacuum which, in P. Dirac's conclusion, does possess a physical structure and properties. That implication was also present, however, in Maxwell's theory, where the impedance of the physical vacuum was long known to be about 377 ohms, and neither infinity nor zero. The observations of the zero-point energy of the vacuum which are associated with the anomalous Lamb effect also bear this out. Again quantum physics and Maxwell's theory point the same way.

2. Lorentz's Dependence on Maxwell

Before presenting our main results, another introductory comment is indicated; namely, that Hendrik Antoon Lorentz [9] precisely *designed* his transformation (the heart of special relativity theory) in 1895 so that Maxwell's equations for the electrodynamics of the electromagnetic field should remain invariant under such transformation. Hence, the whole of Maxwell's theory and findings is implied in the Lorentz transformations which were taken over unchanged as the basic assumption of special relativity theory a decade later. In fact, in 1967 [10] N. Seshagiri of the Tata Institute of Fundamental Research (Bombay) had already shown that Maxwell's equations and the Lorentz transformations are equivalent. We found Seshagiri's work after the first draft of this paper had been written. It is important and too little known; hence its citation here.

Now it was also known, ever since the electromagnetic energy density and momentum vectors were specified by Poynting and Poincaré respectively, that a deep and general relation between mass and energy was implied directly in Maxwell, though unrealized by him as a conscious insight.

Historically, let us turn to a current university physics text of the 1930's, the excellent though little known work [11] of Alexander Marcus, then professor of physics at the present City University of New York, pages 92 and 94:

> If the energy density is E and we suppose that it is moving along with velocity v, we have R [the Poynting vector] $= Ev$ and G [the Poincaré vector or momentum density] $=(E/C^2)v$. From this we infer that the energy (per unit volume) of the field has the mass $E/c^2=M$, since this mass mulitplied by the velocity v is the momentum G. If we attribute the property of mass to this type of energy we may endow all forms of energy with mass, for, any kind of energy is convertible into any other kind ... In Einstein's reasoning these results are necessary conclusions from the Special Relativity Theory of 1905. Although as we have seen it is not difficult to arrive at these conclusions without the use of the Relativity Theory.

Thus Marcus astutely realized* that $E = mc^2$ was derivable from Maxwell's theory, though he did not yet see that its logical necessity in special relativity theory arose by way of the heart of that theory, Lorentz's transformations, which in turn were devised to keep the Maxwellian equations invariant.

It is clear, then, that $E = mc^2$ was a direct implication of classical electromagnetic wave theory; and that the reason it arises in special relativity theory is that the mathematical substance of the latter is the Lorentz transformations, which are the direct consequence of assuming that Maxwell's theory applies to radiation signals between moving bodies.

---

*As Max Born noted [p. 283, *Theory of Relativity*, revised and enlarged edition, Dover, New York 1962], Einstein himself realized this, although $E=mc^2$ continued to be touted as the most important result of the relativistic assumptions, from which, however, it is logically independent, as we see. It is interesting, that though Heaviside had predicted particle travel faster than light in the 1890s, and Čerenkov had proved it in 1933, yet in Born's book, even in the 1962 revised edition, there is not a single mention of Heaviside's or Čerenkov's work. It probably would have been embarrassing and caused too much uncomfortable revision. But history cannot forever be ignored.

Mathematically, then, that $E = mc^2$ emerges from the Lorentz transformations is no more surprising than the fact that those transformations themselves are a mathematical consequence of Maxwell's theory if it is to remain valid for radiation signals between bodies moving past each other.

3.1. *Quantum Theory and the Mass-Energy Relation*

We should hence also be entitled to seek the mass energy relation in quantum theory as a legitimate implication embedded in it (just as the theory of lasers and masers) since there is in fact no contradiction but rather a deep overlapping between Maxwellian electromagnetic theory and quantum electrodynamics. Let us start with an analysis, in terms of mass, space and time, of two basic constants in quantum theory, heretofore considered "fundamental" in the sense of unanalyzable into more basic constituents. We speak of Planck's constant h and the Fine Structure Constant α, which is "dimensionless" and about $1/137$ in magnitude.

By analysing various expressions involving these constants, we have found it possible to demonstrate them in more basic physical terms, involving only mass, length, and duration — the three ultimate dimensions of physics. This has not heretofore been done. Even a dimensionless constant can be made more physically meaningful by considering it as the ratio of two physical quantities of like dimension. Thus the "dimensionless" transcendental number π can be considered as the ratio of the length of the circumference of any circle to the length of its diameter.

Likewise, suitable analysis shows that the ubiquitous Fine Structure Constant α is the relation of the square root of the length of the classical electron radius (r) to the square root of the length of the radius of the first Bohr orbit (a) — a remarkable simple physical structure; which was first announced by C. Musès in 1965 [4]:

$$\alpha^2 = r/a \ . \tag{2}$$

Similarly, it may be shown that the fundamental constant h may itself be even more fundamentally analyzed (published by Musès also in 1965 [4]) into constitutents of length, duration and mass; namely,

$$h = 2\pi mc\sqrt{ar} \tag{3}$$

where a and r are as before and m is the rest mass of the electron and c here is the speed of light in the dielectric or medium through which the electron is moving. From (2) and (3) it follows that

$$R_\infty = \alpha/4\pi a \ . \tag{4}$$

The background research that led to Eqs. (2) and (3) was given in the course of an invited lecture on the Dirac matrices at the NASA Ames Research facility at Moffett Field, CA, in 1970 and in the references of [12]. Neither Eqs. (2) or (3) have appeared before Musès' publication of 1965 in the literature. The Fine Structure Constant is usually rather inexactly defined by $\alpha = 2\pi e^2/hc$ whereas the complete statement, though still more physically opaque and less mathematically elegant than Eq. (2) is $\alpha = 2\pi e^2/hc\varepsilon$ where ε is the permittivity of the medium and c the radiation velocity in that medium. Note that if we then substitute from Eq. (3) our basic expression for h in the foregoing equation we obtain $\alpha = (e^2/mc^2\varepsilon) \cdot (1/\sqrt{ar})$. Observing that r may be expressed as $e^2/mc^2\varepsilon$, we obtain at once Eq. (2) for α as its most basic expression, exhibiting its fundamental physcial dimensionality immediately as the ratio of two lengths.*

Now since the speed of light depends on the index of refraction of the dielectric and since h remains observably constant in all media, we see that there must be a compensating change in a and r if the dielectric, and hence c, should change. Before proceeding further, let us clarify Eq. (3) more. Using Eq. (2), Musès writes (3) in the form:

$$h = 2\pi rmc/\alpha = 2\pi amc\alpha \tag{5}$$

where c/α and $c\alpha$ can be considered as a phase (v_p) and a group (v_g) velocity respectively since

$$v_p v_g = (c/\alpha)(c\alpha) = c^2 \ . \tag{6}$$

which expresses the fundamental relation between phase and group velocity. Actually, v_g here is the lineal velocity in the 1st Bohr orbit, now seen simply as $c\alpha$.

---

*The great advantage of Eqs. (2) and (3) appear when confronted, for example by the usual cumbersome equation for the Rydberg Number $R_\infty = me^4/ch^3\varepsilon^2$, which as in (4) we can at once now write as $R_\infty = \alpha/4\pi a$. Note that even though the basic quantum of action h and the basic coupling factor α are universal constants, the simplest expressions for them are based on the hydrogen atom, the proton/electron system that is the basis of ordinary matter.

3.2. *Some Testable Predictions*

It should be no surprise that $v_p > c$, since in all cases of wave phenomena, when the group velocity is less than the signal propagation speed, the phase velocity exceeds that speed. Returning to the remarks following Eq. (3), we can now further elucidate them in the light of Eq. (4) which, together with Eq. (3) shows that if c decreases both r and a must increase together in the same proportion, since $h, 2\pi, m$, and α remain constant during a change of dielectric. That means that for a neutral hydrogen atom moving from air into water, for example, both the size of its electron and the distance of the first stable orbit from the nucleus increase in the same ratio that the speed of light decreases from air to water. This prediction tests Musès' equations.

Indeed, since radiation has associated mass in Maxwell's theory, as elucidated by Poynting, Poincaré and Heaviside*, it should, by Newton's laws, be subject to gravitational perturbations (in turn subject to the Lorentz transformations) and hence slow down when passing from a less into a more intense gravitational field. This amounts to a gravitationally induced change of dielectric, and hence the interactions between matter and light in the physical vacuum must change if the matter in question experiences a change of gravitational field intensity. Such changes are in principle experimentally observable, and the herein predicted size-change of a neutral hydrogen atom as it moves from air into water should also be observable through radiation-scattering experiments.

3.3. *The Key Derivation*

Proceeding to our chief enquiry into the mass-energy relation as analyzed by C. Musès in both published and unpublished material to which we had access, we have from Eq. (5) that the frequencies corresponding to c/α and $c\alpha$ are respectively given by the basic relation, too often overlooked, that the physical dimensions of frequency can be expressed as distance divided by lineal velocity. Let ν_r and ν_a be respectively the frequencies associated with c/α and $c\alpha$. Then, also using Eq. (3), we have

$$\nu_r = (c/\alpha)/2\pi r = c/2\pi r\alpha \tag{7}$$

---

*Oliver Heaviside (1850–1925) among other notable achievements, first expounded the vector form of Maxwell's theory, and was the first to demonstrate the mathematical theory of electrons exceeding the speed of light in any medium (in his *Electromangetic Theory*, 3 volumes) in 1898, anticipating the 1933 discovery of P. Čerenkov and its subsequent mathematical exposition by Frank and Tamm in 1934.

and the associated much lower frequency

$$\nu_a = c\alpha/2\pi a \tag{8}$$

or, using Eq. (2)

$$\nu_a/\nu_r = \alpha^4 \ . \tag{9}$$

Both these frequencies are thus shown to decrease as the electron or atom moves from a dielectric of greater to one of lesser transparency to radiation; i.e., ν_r and ν_a decrease if c decreases through such a change of medium. This effect of Eqs. (7) and (8) is related to the concomitant increases in a and r with such change, as Eq. (5) shows.

From Eq. (1) and the preceding discussion, we have $mc^2 = h\nu$ and, since $h\nu$ is the energy, we finally obtain

$$E = mc^2 \ . \tag{10}$$

Since any form of energy is interconvertible to any other by a suitable transducer, Eq. (10) is general. It shows how the fundamental mass-energy relation (carried over into general relativity theory when Maxwell's theory was embedded in the former by its being founded in the Maxwellian invariance of the Lorentz transformations) can also be derived out of fundamental quantum physics, without recourse to special relativity theory, from which $E = mc^2$ was logically independent in the first place, as has been shown.

4.1. *Waves, Elasticity Theory, and the Mass-Energy Relation*

Pursuing Muses' logical analysis of fundamental physical theory, we now address the wider problem of the physical origin of the mass-energy relation.

In every wave motion, transverse or longitudinal in any medium, there is a fundamental relation between the energy density and the mass density per unit volume. Note that "wave" implies some elastic medium, for without elasticity there can be no wave motion. In this connection, it is necessary to observe that the dimensions of pressure, force per unit area, are $(ms/t^2)/s^2$, where m, s, t are mass, length, and duration respectively.

Thus

$$p = m/st^2 \tag{11}$$

where p is pressure.

Also, we observe that the dimensions of elasticity are the same as pressure, namely m/st^2. Finally, energy density has the dimensions $(ms^2/t^2)/s^3$ or, again, m/st^2. Thus, by dimensional analysis, of which Lord Rayleigh made such good use and rightly found so valuable [13], we see that — in terms of experimental phenomenology — pressure, elasticity and energy density are physical expressions of the same phenomenon. There is also no wave of any kind without some inherent viscosity, internal friction, or elastic hysteresis — all terms for the same effect. That fact means that all radiation suffers with time a decrease in frequency and a concomitant increase in wavelength. Such red shifts would increase with distance of travel and be essentially nonlinear. Thus to accurately assess red shifts due to gravitational fields or Doppler effects, especially over great distances, the by no means negligible *hysteresis* red shift would first have to be allowed for. This is another testable prediction following from this analysis, and it is very relevant to the phenomena of quasars which would thus not have to be mythologized with such ridiculously high velocities. It is as naive to say light is frictionless as it was once said to be instaneously (before Olaf Roemer's discovery in 1675 of light's finite velocity, the ultimate basis of the Lorentz transformations).

It is now necessary to recall that the fundamental relation governing all wave motion, to which we have already alluded, though not yet specified, is given by the relation: square root of the quotient of energy density divided by mass density. That quantity has the dimensions of a velocity, in fact the wave propagation velocity in the dielectric in the case of radiation. Hence we can write:

$$E/M = v^2 \tag{12}$$

where E and M are energy density and mass density respectively. The dimensional proof of this comes from the fact that, dimensionally

$$E = (ms^2/t^2)/s^3 = m/st^2 \tag{13}$$

and

$$M = m/s^3 \ . \tag{14}$$

Hence

$$\sqrt{E/M} = s/t \tag{15}$$

that is, a velocity.

Often, we see Eq. (15) in a form such as that the square root of the quotient of the specific elasticity divided by the specific inertia is the velocity of the propagation, and the like. But the physics of all these situations is the same and the basic equation is (12) which, in the case of an electromagnetic wave, with propagation velocity c, transforms into Eq. (10). And since all energy is convertible we see that (13) is universally applicable. Its quantum physical form in Eq. (10) is more fundamental than its artificial appearance in special relativity via Maxwell's theory inherent in the Lorentz transformations (see prior discussion and [10]).

As we now see, the quantum derivation of Eq. (10) was not its physically most fundamental derivation. The deepest physical source of the mass-energy relation, as shown in Eqs. (12) and (15), springs from the nature of the elasticity of wave motion itself, thus shown to be involved in the bases of all physical phenomena. Those bases are charge or matter, together with their alternate form, radiation; and the physical vacuum with which they both interact.

4.2. *Waves and the Mathematics of Integers*

The outstanding characteristic of quantum theory, Musès notes, has always been its dependence on whole numbers or integers (or half integers). That was Niels Bohr's epoch-making discovery, and the addition of Max Born's noncommutative relation and Werner Heisenberg's uncertainty relation changed that fundamental characteristic not a whit. This fact gives us the clue that wave-motion — the analysis of which depends on resonances (reinforcing or interfering, i.e., on a whole or half number of completed cycles of some kind) — is fundamental to quantum theory and hence to the physically observable universe.

5. Conclusion

The founder of quantum theory, Max Planck himself, concluded that for a new basic concept to be accepted, a whole generation of more restricted minds must pass away. Let us hope, now that three generations have elapsed, that this draconic sociological law of science history is enough fulfilled, and the unclouded physical origins of the mass-energy relation are now clearly seen. Gauss would have been glad of this.

References

1. Léon Brillouin, *Relativity Reexamined*, Acad. Press, New York, 1970, p. 103.
2. H. S. M. Coxeter, *Regular Polytopes*, 2nd ed., Macmillan, New York, 1963.
3. Paul Gerber, Zeitscheift f. Math. u. Physik **43**, p. 93.
4. C. Musès, *Aspects of some crucial problems in medical and biological cybernetics*, in Progress in Biocybernetics, vol. 2, Norbert Wiener and J. P. Schadé, eds., Elsevier, Amsterdam, 1965, pp. 241–242.
5. René Dubos, Science **246** (1989) 883.
6. P. Lebedev, *Lichtdruck*, Lexikon der Physik, Stuttgart, 1959 p. 836.
7. E. E. Nichols, and G. F. Hull, article *Lichtdruck*, Lexikon der Physik, Stuttgart, 1959, p. 836.
8. O. Heaviside, *Electromagnetic Theory*, vol. 2., London, 1899.
9. H. A. Lorentz, *Versuch einer Theorie der elektrischen und optischen Erscheinungen in bewegten Körpern*, Leiden, 1895.
10. N. Seshagiri, J. Inst. Telecom Engrs. (India), **13** (1967) 337–344.
11. A. Marcus, *The Electromagnetic Field and Electric Circuits*, New York, 1930, pp. 92, 94.
12. C. Musès, J. Appl. Math. Comput. **6** (1980) 63.
13. J. W. Strutt, (Lord Rayleigh), *The Theory of Sound*, vol. 2, New York, 1945, p. 429.

Kyril Demys
Centre de Mathématique et Morphogenèse
Case Postale 370
1211 Genève, Switzerland

THE MATH. HERITAGE OF C.F. GAUSS (pp. 172-178)
edited by George M. Rassias
©1991 World Scientific Publ. Co. Singapore

UNIFORMLY CONTINUOUS MULTI-VALUED MAPPINGS

Doitchin Doitchinov

In the first section of the paper the notion of a uniformly continuous multi-valued mapping is introduced and two theorems on extending of such a mapping are proved. The second section contains some considerations concerning several topological notions and their mutual relations.

The results in the first section have, of course, a sense independently of the reflexions in the second one.

1. A Definition and Two Theorems

An appropriate notion of a uniformly continuous multi-valued mapping must surely generalize the standard notion of a uniformly continuous single-valued mapping. On the other hand it should enable us to obtain some interesting results and preferably such ones which are analogous to certain known theorems. Finally, its relations to the well-known notions of the semicontinuity of a multi-valued mapping have to be investigated.

We will use the following, very naturally sounding, definition (here and everywhere in the sequel, all multi-valued mappings are supposed to have non-empty images):

Definition 1. Let $(X, \mathcal{U}_X)$ and $(Y, \mathcal{U}_Y)$ be two uniform spaces. A multi-valued mapping $f : X \to Y$ is called *uniformly continuous* if for any $V \in \mathcal{U}_Y$ there exists a $U \in \mathcal{U}_X$ such that $(x', x'') \in U$ implies $f(x)' \subset V(f(x''))$.

Here, as usual, we denote

$$V(A) = \cup\{V(y) \mid y \in A\}$$

for $A \subset Y$, where

$$V(y) = \{z \in Y \mid (y, z) \in V\}$$

for $V \in \mathcal{U}_Y$.

It is clear, by the way, that, after Definition 1, a multi-valued mapping f is or is not uniformly continuous simultaneously with the mapping $\bar{f}$, defined by setting $\bar{f}(x) = \overline{f(x)}$. In other words, a uniformly continuous mapping can always be supposed to have (non-empty) closed sets for images of the points of its domain.

Definition 1 is obviously a generalization of the usual definition of a uniformly continuous single-valued mapping. Moreover, in the metric case, i.e., in the case when both X and Y are metric spaces, these two definitions simply coincide if one considers f as a single-valued mapping from X into the power space 2^Y endowed by the usual Hausdorff metric. This observation enables us to obtain immediately the following theorem.

Theorem 1. Let (X, d_X) and (Y, d_Y) be metric spaces, and $A \subset X, \bar{A} = X$. If the space (Y, d_Y) is complete, then any uniformly continuous multi-valued mapping $f : A \to Y$ admits a uniformly continuous multi-valued extension $f^* : X \to Y$ (in the sense that $f^*(x) = f(x)$ for $x \in A$).

This theorem is, in fact, a special case of the well-known theorem on extension of uniformly continuous single-valued mappings for metric spaces. Indeed, as explained above, we can consider f as a single-valued mapping of the form $f : A \to 2^Y$ (replacing, if it is needed, f by $\bar{f}$, where $\bar{f}(x) = \overline{f(x)}$). Then it suffices to recall Hahn theorem [3] stating that the completeness of (Y, d_Y) implies the completeness of the space 2^Y (with respect to the Hausdorff metric).

Another extending theorem can also be proved in a rather standard manner.

Theorem 2. Let $(X, \mathcal{U}_X)$ and $(Y, \mathcal{U}_Y)$ be uniform spaces, and $A \subset X, \bar{A} = X$. If the space $(Y, \mathcal{U}_Y)$ is uniformly locally compact, then every uniformly continuous multi-valued mapping $f : A \to Y$ possesses a uniformly continuous multi-valued extension $f^* : X \to Y$.

(Here the condition of the uniform local compactness of the space $(Y, \mathcal{U}_Y)$ means that there is a $V \in \mathcal{U}_Y$ such that $\overline{V(y)}$ is compact for every $y \in Y$. The mapping f^* is understood to be an extension of f again in the sense that $f^*(x) = f(x)$ for $x \in A$.)

Proof. We admit that, for every $x \in A$, the (non-empty) set $f(x)$ is closed. (The general case can be reduced to this in an obvious manner.) We define the mapping

$$f^* : X \to Y$$

by setting, for every $x \in X$,

$$f^*(x) = \cap\{\overline{f(U(x) \cap A)} \mid U \in \mathcal{U}_X\} .$$

(Here and in the sequel we use the standard notation $f(B) = \cup\{f(x) | x \in B\}$ for $B \subset A$.) It is easy to check, using the uniform continuity of f, that $f^*(x) = f(x)$ for $x \in A$. We have to prove the uniform continuity of f^*.

First of all let us fix a $V^* \in \mathcal{U}_Y$ such that $\overline{V^*(y)}$ is compact for each $y \in Y$. Let now $V_0 \in \mathcal{U}_Y$. Choose a $V' \in \mathcal{U}_Y$ such that $V' \circ V' \circ V' \subset V_0 \cap V^*$ and a $U' \in \mathcal{U}_X$ for which $(x', x'') \in U'$, where $x', x'' \in A$, implies $f(x') \subset V'(f(x''))$. Finally, let $U_0 \in \mathcal{U}_X$ be such that $U_0 \circ U_0 \circ U_0 \subset U'$. We will show that, for $x_1, x_2 \in X, (x_1, x_2) \in U_0$ implies $f^*(x_1) \subset V_0(f^*(x_2))$.

Suppose that $(x_1, x_2) \in U_0$. Pick out an arbitrary point y' of the set $f(U_0(x_1) \cap A)$. Then $y' \in f(x')$ for some $x' \in U_0(x_1) \cap A$. As a first step of the proof we will show that

$$V'(y') \cap f(U(x_2) \cap A) \neq \emptyset \tag{1}$$

for each $U \in \mathcal{U}_X$. Indeed, if $x'' \in U(x_2) \cap U_0(x_2) \cap A$, then we have $(x'', x_2) \in U_0, (x_2, x_1) \in U_0, (x_1, x') \in U_0$, and therefore $(x'', x') \in U'$. Hence $f(x') \subset V'(f(x''))$. Then there is a $y'' \in f(x'')$ such that $y' \in V'(y'')$, and so $y'' \in V'(y')$. On the other hand $y'' \in f(x'') \subset f(U(x_2) \cap A)$ and thus relation (1) is verified.

Our second step is to see that there exists a point $y^* \in \overline{V'(y')}$ such that

$$V(y^*) \cap f(U(x_2) \cap A) \neq \emptyset \tag{2}$$

whenever $U \in \mathcal{U}_X, V \in \mathcal{U}_Y$. Suppose the opposite. Then for every $y \in \overline{V'(y')}$ there are a $U_y \in \mathcal{U}_X$ and a $V_y \in \mathcal{U}_y$ such that $V_y(y) \cap f(U_y(x_2) \cap A) = \emptyset$. Choosing a finite collection of points $y_1, y_2, \ldots, y_k$ of the compact set $\overline{V'(y')}$ such that $\overline{V'(y')} \subset \bigcup_{i=1}^{k} V_{y_i}(y_i)$, we have $\overline{V'(y')} \cap f(U^*(x_2) \cap A) = \emptyset$ for $U^* = \bigcap_{i=1}^{k} U_{y_i}$, which contradicts the relation (1).

In order to accomplish the third — the achieving — step of our proof, let us pick a point $y^* \in \overline{V'(y')}$ satisfying (2). It is clear that

$$y^* \in \cap\left\{ \overline{f(U(x_2) \cap A)} \mid U \in \mathcal{U}_X \right\} = f^*(x_2) \ . \tag{3}$$

(By the way, from (3) one concludes that $f^*(x) \neq \emptyset$ for every $x \in X$.) Clearly we have $y^* \in V' \circ V'(y')$ and hence $y' \in V' \circ V'(y^*) \subset V' \circ V'(f^*(x_2))$. Because of the choice of the point y' we obtain $f(U_0(x_1) \cap A) \subset V' \circ V'(f^*(x_2))$, and therefore

$$f^*(x_1) \subset \overline{f(U_0(x_1) \cap A)} \subset V' \circ V' \circ V'(f^*(x_2)) \subset V_0(f^*(x_2)) \ .$$

This completes the proof.

Let us note, in addition, that in both theorems above the extension f^* of the mapping f is uniquely determined if one requires that the set $f^*(x)$ is closed for each $x \in X \backslash A$. The easy proof of this fact can be left to the reader.

It is clear that the theorems stated above are far from giving a more or less complete treatment of the problem of extending of uniformly continuous muilti-valued mappings. They rather represent only first steps in this direction. It would be, of course, very desirable to find a general result comprising both these theorems as special cases. In particular, it is interesting to investigate whether the completeness of the space $(Y, \mathcal{U}_Y)$ is sufficient in the general case. (Let us note, by the way, that this condition, sufficient in Theorem 1, follows also from the condition of the uniform local compactness in Theorem 2.)

2. Continuity of Multi-Valued Mappings — Some Reflections

If $(X, \mathcal{U}_X)$ and $(Y, \mathcal{U}_Y)$ are two uniform spaces, then every uniformly continuous single-valued mapping $f : X \to Y$ is also continuous in respect of the topologies induced on X and Y by the uniformities $\mathcal{U}_X$ and $\mathcal{U}_Y$, respectively. It would be, of course, desirable that the situation would be similar with the multi-valued mappings. In order to treat this question, however, we need a revision of some basic notions.

There exist two well-known notions of continuity of a multi-valued mapping, namely those of upper semicontinuity and lower semicontinuity. Both of them are basically connected with the concept of a topological space and use essentially the notion of an open set. Here another approach is followed in which the notion of a neighbourhood plays a basic role.

A single-valued mapping $f : X \to Y$, where $(X, \mathcal{T}_X)$ and $(Y, \mathcal{T}_Y)$ are topological spaces, is continuous at a point $x \in X$ if for any $\mathcal{T}_Y$-neighbourhood V of the point $f(x)$ there exists a $\mathcal{T}_X$-neighbourhood U of x such that $f(x') \in V$ for every $x' \in U$.

If f above is a multi-valued mapping, then $f(x)$ is a set and this definition can only be used provided that it is known what a "neighbourhood" of $f(x)$ is. A possible (and very common) way is, when $(X, \mathcal{T})$ is a topological space and $A \subset X$, to consider any $\mathcal{T}$-open set containing A (or — more generally — any set containing some $\mathcal{T}$-open set containing A) as a neighbourhood of A. Then the "neighbourhood approach" expressed by the definition mentioned above yields in the multi-valued case the standard definition of the upper semicontinuity of f. The concept of a supertopological space (cf. [1] where the notion of a supertopology is introduced under the name of a generalized topology; cf. also [2]) proposes another way.

A *supertopology* $(\mathcal{M}, \mathcal{V})$ on a set X is defined by means of a family $\mathcal{M}$ of subsets of X, which always includes all single-point subsets of X, and an operator $\mathcal{V}$ assigning to each $A \in \mathcal{M}$ a filter $\mathcal{V}(A)$ on X — the filter of the "$\mathcal{V}$-neighbourhoods" of A. This filter satisfies some conditions (in particular, it is required that A is contained in each $U \in \mathcal{V}(A)$) such that $\mathcal{V}(\{x\})$, for every $x \in X$, is nothing but the neighbourhood filter for the point x in a fixed topology on X. So the notion of a supertopology coincides with the usual notion of a topology in the special case when $\mathcal{M}$ is the family of all single-point subsets of X. In the general case a supertopological space, i.e., a set X equipped with a supertopology $(\mathcal{M}, \mathcal{V})$, can be regarded as a topological space together with an additional structure on it. The filter $\mathcal{V}(A)$, for any $A \in \mathcal{M}$, possesses a base consisting of open sets with respect to the topology induced on X by the supertopology $(\mathcal{M}, \mathcal{V})$.

For our purpose it is important to point out two special kinds of supertopologies.

i) When $(X, \mathcal{T})$ is a topological space, we can consider the supertopology $(\mathcal{P}_X, \mathcal{V}_{\mathcal{T}})$ on X, where $\mathcal{P}_X$ is the family of all subsets of X and $\mathcal{V}_{\mathcal{T}}(A)$, for any $A \in \mathcal{P}_X$, is the filter having for a base the collection of all $\mathcal{T}$-open sets containing A, provided that $A \neq \emptyset$. (It is admitted that $\mathcal{V}_{\mathcal{T}}(\emptyset) = \mathcal{P}_X$.) This supertopology will be called *the $\mathcal{T}$-supertopology* on X.

ii) When $(X, \mathcal{U})$ is a uniform space, one can consider the supertopology $(\mathcal{P}_X, \mathcal{V}_{\mathcal{U}})$ on X, where $\mathcal{P}_X$ is again the family of all subsets of X and, for $A \in \mathcal{P}_X, A \neq \emptyset, \mathcal{V}_{\mathcal{U}}(A)$ is the filter having for a base the collection $\{U(A) | U \in \mathcal{U}\}$. (Here also it is admitted that $\mathcal{V}_{\mathcal{U}}(\emptyset) = \mathcal{P}_X$.) This supertopology will be called *the $\mathcal{U}$-supertopology* on X. It obviously induces on X the topology induced by the uniformity $\mathcal{U}$.

Now, when we wish to define the notion of continuity of a multi-valued mapping $f : X \to Y$, where $(X, \mathcal{T}_X)$ and $(Y, \mathcal{T}_Y)$ are topological spaces, it seems very convenient to consider the space Y together with a super-topology $(\mathcal{M}_Y, \mathcal{V}_Y)$ on it, inducing the given topology $\mathcal{T}_Y$ and such that $f(x) \in \mathcal{M}_Y$ for every $x \in X$. Then f will be called *continuous in respect of the supertopology* $(\mathcal{M}_Y, \mathcal{V}_Y)$ at the point $x \in X$ — or simply *$\mathcal{V}_Y$-continuous at x* — if for any $\mathcal{V}_Y$-neighbourhood V of $f(x)$, i.e., for any $V \in \mathcal{V}_Y(f(x))$, there exists a $\mathcal{T}_X$-neighbourhood U of x such that $f(x) \subset V$ for every $x \in U$.

Evidently this definition coincides with the definition of the upper semicontinuity of f in the special case when $(\mathcal{M}_Y, \mathcal{V}_Y)$ is the $\mathcal{T}_Y$-supertopology on Y.

On the other hand, if $(X, \mathcal{U}_X)$ and $(Y, \mathcal{U}_Y)$ are two uniform spaces and the multi-valued mapping $f : X \to Y$ is uniformly continuous (in the sense of Definition 1 from Sec. 1), then obviously f is continuous at any point $x \in X$ in respect of the $\mathcal{U}_Y$-supertopology on Y.

According to what is said above, it is clear that there is only one special case in which the uniform continuity of f implies its upper semicontinuity. This is the case when the $\mathcal{U}_Y$-supertopology coincides with the $\mathcal{T}(\mathcal{U}_Y)$-supertopology on Y, where $\mathcal{T}(\mathcal{U}_Y)$ is the topology induced on Y by the uniformity $\mathcal{U}_Y$. And this happens when the topology $\mathcal{T}(\mathcal{U}_Y)$ is paracompact and $\mathcal{U}_Y$ is the finest uniformity on the topological space $(Y, \mathcal{T}(\mathcal{U}_Y))$.

Somewhat surprisingly, it turns out that the situation is simpler with respect to the lower semicontinuity. Namely, it is easy to verify that a uniformly continuous multi-valued mapping is always lower semicontinuous.

Indeed, let $(X, \mathcal{U}_X)$ and $(Y, \mathcal{U}_Y)$ be two uniform spaces and let $f : X \to Y$ be a uniformly continuous multi-valued mapping. Fix a $x_0 \in X$

                                    D. Doitchinov

and suppose that for the open set G in the topological space $(Y, \mathcal{T}(\mathcal{U}_Y))$ we have $G \cap f(x_0) \neq \emptyset$. Pick out a point $y_0 \in f(x_0) \cap G$. We have $V(y_0) \subset G$ for some $V \in \mathcal{U}_Y$. Choose a $U \in \mathcal{U}_X$ such that from $(x', x'') \in U$ it follows that $f(x') \subset V(f(x''))$. If now we suppose that for some $x_1 \in X$ we have $(x_0, x_1) \in U$ and $f(x_1) \cap G = \emptyset$, then we would have $f(x_1) \cap V(y_0) = \emptyset$, hence $y_0 \notin V(f(x_1))$, and therefore $f(x_0) \not\subset V(f(x_1))$, a contradiction. So for every point $x \in U(x_0)$ we have $f(x) \cap G \neq \emptyset$. This proves the lower semicontinuity of f.

References

1. D. Doitchinov, *A unified theory of topological, proximity and uniform spaces*, Dokl. Acad. Nauk SSSR **156** (1964) 21–24 (in Russian).
2. D. Doitchinov, *Supertopologies and some classes of extensions of topological spaces*, General Topology and its Relations to Modern Analysis and Algebra V (Proc Fifth Prague Topol. Symp. 1981), ed. J. Novak, Berlin, 1982, pp. 151–155.
3. H. Hahn, *Reelle Funktionen I*, Leipzig 1932.

Doitchin Doitchinov
Department of Mathematics
University of Sofia
boul. A. Ivanov 5
Sofia 1126
Bulgaria

THE MATH. HERITAGE OF C.F. GAUSS (pp. 179-194)
edited by George M. Rassias
©1991 World Scientific Publ. Co. Singapore

COMPUTERS IN ALGEBRAIC TOPOLOGY

E. Domínguez and J. Rubio

Introduction

Nowadays there exists a strong interrelation between Mathematics and Computer Science. Although Topology is not one of the fields in which this influence has been more profound, the contributions, in one way or another, are beginning to be noteworthy. There exist developments in topological techniques which can be applied to Computer Science (see [BEY]), interpretations of structures of the latter which solve eminently topological problems (see [BRIN]) and, in both situtations, there are reciprocal influences.

In this paper, we wish to present a panorama of how the solution to some problems in Algebraic Topology is approached, and how these problems are influenced by the development of Computer Science.

In an informal way, we could define Computer Science as a science which deals with problems that admit a solution by means of a finite number of steps, and develops concepts and techniques to find an efficient solution by means of a computer.

The concept of an efficient solution is ambiguous and will require a more detailed explanation. The first approach we can make is that it is not possible to consider as efficient the solution given in some cases by means of the classical theorems on existence. One cannot consider the proof of the existence of something as a solution, if that something is not explicitly presented. For example, Brouwer's well-known Fixed Point Theorem which states that every continuous function from Euclidean interval to itself has a fixed point. One of the usual proofs presented in Algebraic Topology consists of the following: if there is no fixed point, one proves easily that there exists a continuous retraction over the interval boundary, the impossibility of which is an immediate consequence of the number of connected components.

This proof, from the Computer Science point of view, cannot be considered as a solution to the problem. Another type of solution which is closer to our objective is the one which is deduced from the fact that the graph of the aforesaid function f must be cut by the principal diagonal of the Euclidean plane. The problem is then reduced to finding a solution to the functional equation $f(t) = t$. Given such a function, does there exist a method such that, after a finite number of elementary operations, we can obtain a point of intersection? The authors of this paper do not know what degree of complexity the aforesaid problem has in the general case which has been established. However, if we restrict ourselves to a more particular case of piecewise linear functions, the problem is equivalent to calculating on the plane the points of intersection of a family of segments. It is well known that, if the number of segments is n, the time it would take to find the solution is in the order of n^2. This is considered an efficient solution.

This is an example of how a solution coming from a field of Computer Science (specifically, from Computational Geometry) and impelled by other necessities different from those of the establishment of Brouwer's Fixed Point Theorem, provides us with a satisfactory solution in a traditional field of Mathematics.

1. Decidable and Undecidable Problems

It appears that until 1935–1936 the problem of non-algorithmic solubility was not established. It came about through A. Turing [TUR] and A. Church [CHU], who independently gave precise definitions of the intuitive notion of algorithm. With this, the concept of a "more difficult" problem, from the computational point of view, was formalized. To be specific, a problem is said to be insoluble (undecidable, in the case of decision problems) if there does not exist an algorithm which solves it.

The most significant undecidable problem, from the topological point of view, is the word problem [NOV] (see [STI 3]), which consists of deciding if two words in a finitely presented group represent in the same element or, in a particular case, the triviality problem [MAR 1], which tries to decide if one word in a group represent the identity element.

From the two previous problems, one can deduct the undecidability of the problem of homeomorphism between two compact combinatorial n-manifolds with $n \geq 4$ (proved by Markov [MAR 2]) and the undecidability of knowing if a 2-complex is simply connected. In particular, given a simplicial closed curve in a simplicial 2-complex, it is undecidable to know if it is null-homotopic.

In general, it is not easy to prove that a problem is undecidable unless, with more or less difficulty, a logical equivalence can be established with an undecidable problem which is already known. This is the case in the latter example in which, by means of a well-known Dehn's technique [DEH] by which every group which is finitely presented can be represented by the fundamental group of a 2-complex, the equivalence can easily be proved with the triviality problem.

Another non-evident result has been obtained by J. C. Stillwell [STI 2] who proved that it is undecidable to determine if a closed simplicial curve in a 2-complex is isotopical to one point. This fact stands out in comparison with the affirmation of W. Haken [HAK] that isotopical problems are easier than homotopical ones.

Another problem of undecidability in Algebraic Topology, which is interesting to point out, is the one obtained by D. J. Anick [ANI 1] which shows that, in the class of CW-complexes which only two and four dimension cells have, it is undecidable to find out if the loop space of two such CW-complexes has the same rational homotopy type.

Finally, let us show a famous open problem:

Poincaré's Conjecture. The only 3-manifold which is compact, simply connected and without a boundary is the sphere S^3.

Although some specialists consider that probably this is an undecidable problem, up to now no one has been able to prove it (see [HAK]).

It is less complex, comparatively speaking, to prove the existence of an algorithm when this is the case. This type of problems, meaning those that admit an algorithmic solution, receive the name "soluble" or "computable". In this case the difficulty is in proving the existence of an efficient solution.

2. Tractable and Intractable Problems

The problem which we are now interested in is to define some sort of hierarchy among the soluble problems to somehow find from among these problems, those which admit an efficient solution. Such a hierarchy can be formally established by the Complexity Theory which develops techniques in which one can correctly express the concepts of problem, solution, algorithm and input data, using one of the models of theoretic computation. Since formal definitions would excessively prolong this paper, we will limit ourselves to giving ideas, which, although they are informal, are close to practical reality. So let us suppose that the algorithm is written in the form of a programme in some language of structured programming, for example, in Pascal.

One of the basic concepts which must be considered is the size of the data in an algorithmic problem. In general terms, the size is a natural number which attempts to reflect the quantity of data necessary to describe the input of the algorithm which solves the problem. To run the programme, the data have to be codified in the programming language and the computer, to process them, will codify them in binary. We can consider as size the sum of the lengths of the words in binary which are the result of the codification of the data of the problem.

Another necessary basic concept is that of the running time of an algorithm. Every programming language admits a series of elementary operations (of calculation such as addition, subtraction, ...; of decision such as equality of characters, ...) the knowledge of which is considered as inherent for the construction of programmes. The running time of a programme is the name given to the time the programme takes until it finds the solutions, giving to each elementary operation a time cost equal to one unit. If the programme is general (if it solves a problem for all the data which belong to a set of more than one element), as occurs in the cases which we are interested in, the running time will be expressed as a function of the data of the problem, and, accordingly, of the size. On many occasions, it will not be easy to find such an explicit function, but, in the case which we are interested in, it will be sufficient to find one which is a superior bound to the running time. In practical cases the procedure is, once the running time is calculated, to extract from the information of the input data of the programme one or more natural variables which, reflecting the size of the data, allow the discovery of a function of aforesaid variables which will be superior, as close as possible, to the running time.

It is said that the algorithm is polynomial if there exists a superior bound which is of the form n^k, k being a constant and n the size of the problem. This situation indicates that, given such a problem, for any data the programme or algorithm takes to run a time which at most is a polynomial function of the size of the input data. It is said that a problem is polynomial if it admits, as a solution, a polynomial algorithm. An algorithm is exponential if its running time cannot be bounded by a polynomial function.

The class of polynomial problems is represented by **P** and it is said that a problem Q is a **P**-problem if Q admits a polynomial algorithm. These are the problems which are considered as tractable problems. Accordingly, what is considered as an intractable problem is one for which no polynomial algorithm exists.

If a computable problem does not naturally admit a polynomoial algorithm, generally it is difficult to determine if it is a case of a tractable or intractable problem. In the majority of cases an exponential algorithm is known but the existence of a polynomial algorithm is not known. In this class of problems one would consider a sub-class formed by the problems for which it is efficient to verify if a possible solution is a real solution to the problem. That is to say, there exists a polynomial algorithm which has, as input data, those of the problem together with a possible solution and this algorithm verifies if it is the case of a correct solution. In such a situation, it is said to be a case of an **NP**-problem and **NP** represents the class formed by such problems.

In other words, **P** is the class of problems which admits a tractable solution while **NP** is the class of problems for which it is tractable to verify the solution. Evidently $\mathbf{P} \subset \mathbf{NP}$, but finding out if every **NP**-problem is a **P**-problem is one of the main open problems in the field of Complexity Theory. This leads us to consider a class of more difficult problems with respect to an efficient solution; problems called **NP**-completes.

A problem is said to be a decision problem if its answer is a Boolean variable: meaning, its answer can only be YES or NO. If we restrict ourselves to this class of problems we can say that a problem **P** is polynomically transformable over another **P'** if there exists a polynomial algorithm which changes the input data **I** of **P** into input data **I'** of **P'**, in such a way that the answer of **P** to **I** is YES if and only if the answer of **P'** to **I'** is the same. So it is said that a decision problem is **NP**-complete if it is **NP** and every other **NP**-problem is polynomically transformed over it. Such problems are the most difficult in class **NP**, and, evidently, if it is proved that any **NP**-complete problem is polynomial, then automatically every **NP**-problem would be polynomial. Meaning, we would have to show that **P=NP**.

In this sense it is important to note J. C. Stillwell's result [STI 1] which states that the problem to decide if a simplicial closed simple curve in a 2-complex is the boundary of a disk, is an **NP**-complete problem. Given that the 2-complex is finite, evidently there exists a non-polynomial algorithm which solves the problem. From the characterization of the polyhedral disks as surfaces with a boundary which have Euler number equal to 1, it is easy to see the existence of a polynomial algorithm which decides if a given polyhedral disk is the boundary of a given curve. Stillwell manages to solve the problem, proving that the problem of the Hamiltonian path in a graph, can be polynomically reduced to the given problem. This is another example of how a problem in Computer Science provides us with a solution in the field of Topology.

The theory of **NP**-completeness has been designed to be applied to decision problems. However, although in Algebraic Topology this kind of problems are numerous and varied, it is also true that the principal objective is the computation of invariants. Specifically it is necessary to evaluate the complexity of computational functions. In many cases, the invariant is a finite set, a finitely presented group or a finitely generated Abelian group. In the two latter cases it is sufficient to find a finite set of data so that the structure is totally determined. An approach to these is through the so-called enumeration problems which consist of computing the cardinality of the images of any multi-valued function. For this type of problems there exist notions, called $\sharp$**P** and $\sharp$**P**-completeness similar to those of **NP** and **NP**-completeness. It is difficult to give an approximate idea of these concepts without using a model of theoretic computation. However we consider it interesting to point out that D. J. Anick [ANI2] has proved that the problem of computing the dimension of the rational homotopy groups of finite simply connected CW-complexes belongs to the class of $\sharp$**P**-hard problems, a class which contains the $\sharp$**P**-completes.

3. The Calculation of Invariants in Algebraic Topology Using a Computer

The simplest invariant in Algebraic Topology, from the computational point of view, is the simplicial homology. From the beginning of Algebraic Topology (see [LEF], [VEB]) an algorithm to compute the simplicial homology in the case of finite simplicial complexes, has been known. This method consists basically of the diagonalization of a matrix of integers. Given that this algorithm was already known, it is not strange that with the development of computers, the topologists tried to programme it. This was done by T. Pinkerton [PIN] in 1966. More recently, J. Saludes [SAL] introduced a new programme using the same algorithm and adding the case in which a cell is joined to the simplicial complex. Different algorithms to compute the homology in special cases have been studied by I. K. Svethichnova [SVE] (in the case of complexes in Euclidean 3-dimensional space) and by G. Tourlakis [TOU] (for the $n-1$ homology group of certain complexes in Euclidean n-dimensional space).

One type of invariants which have shown themselves to be very difficult to compute, is the homotopy groups. E. H. Brown [BRO] proved in 1957 that the homotopy groups of finite, simply connected simplicial sets are computable (this result has recently been generalized by Weld [WEL] to the case of nilpotent simplicial sets). Despite Brown's theorem, practical computation, even for the simplest space-spheres, is still unsolved. Precisely the homotopy groups of spheres have been and are still widely studied and partial results obtained (see some of them for S^2 in the adjoining table) show that there does not exist any simple relation of "arithmetic" type between the degrees and the order of the homotopy groups.

n	0	1	2	3	4	5	6	7	8	9	10	11	12	13	14	15	16	17	18	19	20
$|\pi_n S^2|$	1	1	∞	∞	2	2	12	2	2	3	15	2	4	24	316	4	6	30	30	12	48

Given the difficulty of computing the above groups, it is not strange that since the development of computers, they have been used as a basic tool. In this field there exist early publications by M. D. MacLaren and M. E. Mahowald [MAC], and by A. Liulevicius [LIU 2]. However, computers were used more frequently in specific cases, to obtain tables, but these works were poorly described and they appeared in publications which were not related to Computer Science. See, for example, [LIU 1], [MAH], [CUR 1]. A recent example of how these more or less "hidden" applications of Computer Science to Topology are still valid, is the important book by

D. C. Ravenel [RAV 2] on the stable homotopy groups of spheres: it would have been impossible to construct the difficult tables and diagrams which appear there without using computational methods.

However, recently an important group of American topologists including M. C. Tangora, M. E. Mahowald, D. Ravenel and E. Curtis, who are dedicated to the computation of the homotopy of spheres, decided to publish papers related to their work in programming. To carry out their programmes they apply all the mechanisms used in Algebraic Topology: EHP spectral sequence, May spectral sequence, Brown-Peterson methods, etc. However the most used tool is the Adams spectral sequence [ADA]. To be able to extract information about homotopy groups of spheres from this spectral sequence three steps are required:

    (a) compute the cohomology of the Steenrod algebra,

    (b) find the differentials in the spectral sequence,

    (c) solve the extension problems.

Computational techniques were directed in the first place to solving problem (a), but they ran into enormous size of data to be processed in Steenrod algebra. Different approaches were looked for and in this way the lambda-algebra appeared [BOU]. This retains all the necessary information but is substantially smaller. M. C. Tangora's publications [TAN] is dedicated to calculating with a computer of the lambda-algebra homology and may be considered as the "official" recognition of the use of computers in the calculation of homotopy groups of spheres.

The conclusions which can be arrived at after analysing these applications of computers to the calculation of homotopy groups of spheres is ambivalent: on the one hand, we can only be surprised and congratulate ourselves on the great quantity of information obtained, taking into account the difficulty of the proposed problem; on the other hand, the partial results obtained seem to indicate that the question will never be solved by Computer Science methods (because of the complexity of the algorithms, meaning, because of the necessary running time). This latter point of view, has made D. Ravenel [RAV 1] state:

Mahowald's Uncertainty Principle

    (i) Any spectral sequence which converges to the homotopy groups of spheres and whose second term is computable, will necessarily have differentials which can only be computed by ad-hoc methods.

    (ii) No reasonable length programme to find the homotopy groups of spheres can be run with a reasonable speed without human intervention.

Again, this principle can be interpreted in two ways: in the obvious negative way or accepting that this abstract problem (computing homotopy groups) presents a challenge in the manner of applying Computer Science methods. In fact, the same D. Ravenel in [RAV 3], putting into practice point (ii) of the Uncertainty Principle, describes a novel programme which, when it finds difficulties, stops and requests the necessary data from a human "expert" (a topologist, in this case). This approach does not appear very far from Artificial Intelligence techniques (expert systems, heuristics, etc.) and could be a surprising field of application from which both disciplines would emerge strengthened.

There also exists a group of Japanese mathematicians who have made calculations of algebraic invariants using a computer both in Topology (O. Nakamura [NAK1], [NAK 2], dealing with the cohomology of May complex) and in Differential Geometry (K. Shibata [SHI 1], [SHI 2], dealing with Gelfand-Fuks cohomology).

4. Two General Theories on Computability in Algebraic Topology

To make a systematic study of the problem of computability in Algebraic Topology it is necessary to determine a class of topological spaces (to be exact, of simplicial sets which are more suitable for working with algorithms) for which the invariants of homotopy type are computable and which have a good behavior with respect to the usual constructions in Algebraic Topology. Schön and Sergeraert independently established two general theories on computability in Algebraic Topology, the former using the notion of simplicial set with extended computable homology and the latter using the notion of simplicial set with effective homology. Among the results found by both authors we point out the results which appear sufficient to prove that a certain class C of simplicial sets is suitable for the study of computability in which we are interested (see the papers by Schön and Sergeraert; in the case of effective homology, see also the papers by Rubio and Rubio-Sergeraert). The results are the following:

(1) Evidently the first condition required is that the homology of the elements in C can be computed by means of an algorithm. C should also contain all the finite simplicial sets (meaning, with a finite number of simplexes in each dimension).

(2) If in a fiber space $F \to E \to B$, two of the components are in C, the

third component must also be in $\mathcal{C}$ (it is also necessary that the fibration is simplicial, that its base is simply connected, that eventually the fiber is connected, etc.; in the following points these technical questions will not be mentioned).

(3) If a simplicial set is in $\mathcal{C}$, its iterated loop spaces are also in $\mathcal{C}$.

(4) If a simplicial group is in $\mathcal{C}$, its classifying space is also in $\mathcal{C}$. This case includes the Eilenberg-MacLane spaces $K(\pi, n)$ (once it has been proved that $K(\pi, 1)$ is in $\mathcal{C}$).

Finally,

(5) The homotopy groups of the elements of $\mathcal{C}$ (at least in the simply connected case) are computable.

These two theories present a review of Homological Algebra from the algorithmic point of view. In particular, in both theories algorithms have been developed which have the role of the main mechanism to compute in Homological Algebra and Algebraic Topology the spectral sequences.

The main reason why Brown's algorithm to compute homotopy groups is not only inapplicable in practice, but also difficult to understand, is that the tools used (Postnikov systems or Whitehead towers, ...) employ simplicial sets which are not finite. So it is necessary to substitute the aforesaid sets and use other finite simplicial sets which contain sufficient information. This same difficulty, i.e., the manipulation of infinite sets, also appears in the case of loop spaces.

Given that one of the main objectives, if not the main one, of any mathematician interested in computability in Algebraic Topology, is to find simpler or more suitable algorithms to compute homotopy groups, then this is logical that in the basis of any theory of computability a treatment of the infinite sets is found. These sets are those which appear in Topology or, from the algebraic point of view, are the chain complexes of free Abelian groups which are not necessarily finitely generated. The solutions found by Schön and Sergeraert are different, although they pursue the same objective: to store together with a chain complex C, certain information of a finite type which allows the recuperation of the homotopy type of C, and to work with C in an algorithmic context.

Let us comment briefly on the ideas of each of these two authors without giving technical details. A complex C with extended computable homology is defined (in an algorithmic way) as the colimit of a sequence of finitely generated complexes in such a way that there exists an algorithm to find an index from which the homology of C is determined by that of the complexes in the sequence (note that the latter homology is always computable).

On the other hand, a chain complex C with effective homology is a complex in which a homotopy equivalence between C and a finite type complex is given. In addition, all the data of the equivalence (the equivalence, its homotopic inverse and the homotopies between its composites and the respective identities) are defined by algorithms.

Although these two concepts might appear separated from each other, the fact that both give the same type of results, shows us the opposite (in fact, if a complex is defined as a colimit of a sequence of finitely generated complexes and it has effective homology, it is simple to verify that its homology is extended computable).

Each of the two theories is simpler or more elegant in some of the aforesaid results (in particular, in Schön's theory, the treatment of the complexes with trivial homology is very simple, while proving the acyclicity of a complex in effective homology theory is a long way from being easy: see the example developed in the third chapter of [RUB 2]). However, an essential difference between both theories is that while Schön's theory seems rather difficult in practical implementations, Sergeraert's effective homology is already being programmed in LISP. This is possible using a new programming technique, called functional coding, also introduced by Sergeraert ([SER 2] [SER 4]). This new technique permits the solutions of problems related to the manipulation of infinite sets to working with computers.

Other topics studied by these two aforementioned authors are the computability of k-invariants [SCH5], the homotopy problem for maps [SCH 6], the computability of the homology of free loop spaces [SER 3] and the computability of certain K-theory groups [SER 5].

5. Perspectives in the Application of Computers to Topology

Although the main objective in Algebraic Topology is to solve specific decision problems and, above all, to develop tools for the computation of invariant which would help in the classification problems of topological structures, the use of computers in this branch of Mathematics does not only produce results of this type, but also on occasions other unexpected results. Below, we indicate some of these applications, which we believed are general for any field of pure Mathematics in which problems are studied from the algorithmic or computational point of view.

1. *The use of results produced by computer to suggest the principal steps in a proof.* An application of this type is explained by Davis in [DAV]. Another example is given by Shibata in [SHI 2].

2. *Proofs of theorems.* We do not refer here to the automatic proof of theorems (a branch which has already proved itself fruitful in, for example, Geometry: see [CHO]), but rather to proofs which depend on the exhaustive and mechanical study of a finite number of cases, but this number is so great that it is outside the sphere of human capacity. The best known example is the proof given by Appel and Hankel [APP] of the famous 4-color theorem.

3. *Exhaustive search using a computer in discrete structures which represent manifolds.* This type of applications has been made especially in low dimensional topology. Some papers of this type are [GIL], in which it is verified where counter-examples to Poincaré's Conjecture cannot be found; or [LIN], which studied the problem of the classification of 3-manifolds coded by certain graphs with less than 30 vertices.

4. *Generalization of theoretical results.* The work of programming requires a very close observation of the results with which mathematicians work. This allows us to find similarities between questions which at first appear to be unrelated. This situation presented itself, for example, when the second of the two authors of this paper and F. Sergeraert, were programming the acyclic models technique (introduced by Eilenberg and MacLane [EIL]) and were led to find a generalized statement of this method. This generalization was developed in [RUB 4].

5. *Use of the computer as a laboratory to make conjectures.* By studying results obtained with the help of the computer in some concrete cases, one can discover formulae or relations between the objects which are manipulated. On some occasions, these formulae or relations are attributed only with difficulty to chance, and they allow, therefore, the establishment of conjectures. Some examples of this type can also be found in [RUB 4].

References

[ADA]   J. F. Adams, *On the structure and applications of the Steenrod algebra.* Com. Math. Helv. (1958) 180–214.

[ANI 1] D. J. Anick, *Diophantine equations, Hilbert series and undecidable spaces*, Ann. of Math. **122** (1985) 87–112.

[ANI 2] D. J. Anick, *The computation of rational homotopy groups is ♯P-hard*, Computers in Geometry and Topology, ed. M. Tangora, pp. 1–57, Marcel Dekker, 1989.

[APP] K. Appel, and W. Haken, *Every planar map is 4-colorable*, III. J. Math. **21** (1977) 429–467.

[BEY] W. T. Beyer, *Recognition of topological invariants by iterative arrays*, Massch. Inst. of Technology, 1969.

[BOU] A. K. Bousfield, E. B. Curtis, D. M. Kan, D. G. Quillen, D. L. Rector, and J. W. Schlesinger, *The mod-p lower central series and the Adams spectral sequence*, Topology **5** (1966) 331–342.

[BRIN] L. W. Brinn, *Computing topologies*, Math. Magaz. **58** (1985) 67–77.

[BRO] E. H. Brown, *Finite computability of Postnikov complexes*, Ann. of Math. (1) **65** (1957) 1–20.

[BRU 1] R. R. Bruner, *Calculation of Large Ext modules*, Preprint.

[BRU 2] R. R. Bruner, *Machine calculation of the cohomology of the Steenrod algebra*, preprint.

[CHO] S. C. Chou, *Mechanical Geometry Theorem Proving*, Reidel, 1988.

[CHU] A. Church, *An unsolvable problem of elementary number theory*, Amer. J. of Math. **56** (1936) 345–363.

[CUR 1] E. Curtis, *Simplicial Homotopy Theory*, Aarhus Lect. Notes Series, no. 10, 1968.

[CUR 2] E. B. Curtis, M. E. Mahowald, and R. J. Milgram, *Some calculations of unstable Adams E_2-terms for spheres*, Proceedings of the Seattle Emphasis Year in Homotopy Theory, Springer-Verlag.

[CUR 3] E. B. Curtis, and M. E. Mahowald, *EHP Computations of $E_2(s^m)$*, Computers in Geometry and Topology ed. M. Tangora, pp. 105–120, Marcel Dekker, 1989.

[DAV] D. M. Davis, *The use of a computer to suggest key steps in the proof of a theorem in Topology*, Computers in Geometry and Topology, ed. M. Tangora, pp. 121–130, Marcel Dekker, 1989.

[DEH] M. Dehn, *Uber die Topologie des dreidemensionalen Raumes*, Math. Ann. **69** (1910) 137–168.

[EIL] S. Eilenberg, and S. MacLane, *Acyclic models*, Amer. J. Math. **75** (1953) 189–199.

[GIL] D. Gillman, and P. Laszlo, *A computer search for contractible 3-manifolds*, Top. and its App. **16** (1983) 33–41.

[HAK]   W. Haken, *Connections between topological and group theoretical decision problems* in Word Problems, eds. W. W. Boone, F. B. Cannonito, R. C. Lyndon, North-Holland, 1973, pp. 427–441.

[HOL]   D. F. Holt, *A computer program for the calculation of a covering group of a finite group*, J. Pure Appl. Algebra **35** (1985) 287–295.

[HOL]   D. F. Holt, *The mechanical computation of first and second cohomology groups*, J. Symbolic Comput. **1** (1985) 351–361.

[HOM]   T. Homma, M. Ochiai, and M. Takahashi, *An algorithm for recognizing S^3 in 3-manifolds with Heegaard splittings of genus two*, Osaka J. Math. **17** (1980) 625–298.

[LEF]   S. Lefschet, *Topology*, A. M. S. Coll. Publ. **12**, 1930.

[LIN]   S. Lins, and C. Durand, *A complete catalogue of rigid graph-encoded 3-manifolds up to 28 vertices*, Notas e Com. Mat., 168, Univ. Pernambuco, 1989.

[LIU 1]   A. Liulevicius, *The factorization of cyclic reduced powers by secondary cohomology operations*, Memo A. M. S., no. 42, 1962.

[LIU 2]   A. Liulevicius, *Coalgebras, resolutions and the computer*, Math. Algorithms **1** (1966) 4–12.

[MAC]   M. D. MacLaren, *A machine calculation of a spectral sequence*, Proc. Symp. Applied Math. **19** (1967) 117–124.

[MAH]   M. Mahowald, *The metastable homotopy of S^n*, Memo A. M. S., no. 72, 1967.

[MAR 1]   A. A. Markov, *Impossibility of algorithms for recognizing some properties of associative systems*, Dokl. Akad. Nauk SSSR **77** (1951) 953–956.

[MAR 2]   A. A. Markov, *Insolubility of the problem of homeomorphy*, Proc. Inter. Congr. of Math. 1958, Cambridge Univers. Press, pp. 300–306.

[NAK 1]   O. Nakamura, *On the cohomology of May complex III*, Res. Rep. Kochi Univ. **35** (1986) 129–139.

[NAK 2]   O. Nakamura, *On the cohomology of the May complex*, Congress on Computational Geometry and Topology and Computation in Teaching Mathematics, Sevilla, 1987.

[NOV]   P. S. Novikov, *On the algorithmic unsolvability of the word problem in group theory*, Trudy Mat. Inst. Steklov, **44** (1955).

[PIN]   T. Pinkerton, *An algorithm for the automatic computational of integral homology groups*, Math. Algorithms **1** (1966) 27–44.

[RAV 1]   D. C. Ravenel, *Localization and periodicity in Homotopy theory*, Proceedings of the Dunham Symposium on Homotopy Theory, 1985.

[RAV 2]   D. C. Ravenel, *Complex cobordism and stable homotopy of spheres*, Academic Press, 1986.

[RAV 3] D. C. Ravenel, *Homotopy groups of spheres on a small computer*, Computers in Geometry and Topology, ed. M. Tangora, pp. 259–285, Marcel Dekker, 1989.

[RUB 1] J. Rubio, *Codage fonctionnel des ensembles simpliciaux infinis: les espaces de lacets*, Actas III Seminario de Topología, 1987, pp. 79–88.

[RUB 2] J. Rubio, *Homología efectiva y sucesión espectral de Eilenberg-Moore*, Dissertation. Publ. Sem. Mat. García Galdeano, sección 2, no. 20, 1988.

[RUB 3] J. Rubio, and F. Sergeraert, *Homologie effective et suites spectrales d'Eilenberg-Moore*, C. R. Acad. Sc. Paris **306** (1988) 723–726.

[RUB 4] J. Rubio, and F. Sergeraert, *Supports acycliques et algorithmique*, to appear in Astérisque.

[RUB 5] J. Rubio, and F. Sergeraert, *Introduction à l'homologie effective*, Inst. Fourier, 1988.

[SAL] J. Saludes, *Un programa per a calculer l'homologie simplicial*, Bu. Soc. Cat. Cien. III (1984) 127–146.

[SCH 1] R. Schön, *The homology of iterated loop spaces*, to appear in Mem. A.M.S.

[SCH 2] R. Schön, *Effective algebraic topology*, to appear in Mem. A.M.S.

[SCH 3] R. Schön, *Fibrations with calculable homology*, to appear in Mem. A.M.S.

[SCH 4] R. Schön, *An algorithm for calculating homotopy groups*, to appear in Mem. A.M.S.

[SCH 5] R. Schön, *The effective computability of k-invariants*, to appear in Mem. A.M.S.

[SCH 6] R. Schön, *The homotopy problem for maps*, to appear in Mem. A.M.S.

[SER 1] F. Sergeraert, *Homologie effective I, II*, C. R. Acad. Sc. Paris **304** (1987) 279–282, 319–321.

[SER 2] F. Sergeraert, *Functional coding and effective homology*, Congress on Computational Geometry and Topology and Computation in Teaching Mathematics, Sevilla, 1987, to appear in Astérisque.

[SER 3] F. Sergeraert, *Free loop spaces and computability in Algebraic Topology*, VI International Colloquium and Differential Geometry, Santiago de Compostela, 1988, pp. 239–245.

[SER 4] F. Sergeraert, *The computability problem in Algebraic Topology*, Prep. Institut Fourier, no. 119, 1988, to appear in Advamies in Math.

[SER 5] F. Sergeraert, *From a noncomputability result to new interesting definitions and computability results*, First Int. Conf. of ISAAC-88 and AAECC-6, July 1988, Roma, Lect. Notes in Computer Science, no. 358, 26–32.

[SHI 1]   K. Shibata, *Applications of the programming language Prolog to linear and homological algebra (III)*, J. Saitama Univ. **4** (1986) 43–73.

[SHI 2]   K. Shibata, *Micro-computer Prolog as a handy tool for formal algebraic computations*, Congress on Computational Geometry and Topology and Computation in Teaching Mathematics, Sevilla, 1987.

[STA]   S. Stahl, *Homology as a tool in integer programming*, Siam J. Alg. Dis. Meth. **6** (1985) 641–642.

[STI 1]   J. C. Stillwell, *Isotopy in surface complexes from the computational viewpoint*, Bull. Austral. Math. Soc **20** (1979) 1–6.

[STI 2]   J. C. Stillwell, *Unsolvability of the knot problem for surface complexes*, Bull. Austral. Mat. Soc. **20** (1979) 131–137.

[STI 3]   J. C. Stillwell, *The word problem and the isomorphism problem for groups*, Bull. AMS. **6** (1982) 33–56.

[SVE]   I. K. Svetlichnova, *Calculation of homology groups for systems with large numbers of elements*, Mosc. Univ. Math. Bull. (5) **39** (1984) 100–103.

[TAN]   M. C. Tangora *Computing the homology of the Lambda algebra*, Mem. A. M. S., no. 337, 1985.

[TOU]   G. Tourlakis, *Homological methods for the classification of discrete euclidean structure*, Siam J. Appl. Math. (1) **33** (1977) 51–54.

[TUR]   A. Turing, *On computable numbers with an application to the Entschidungsproblem*, Proc. of London Math. Soc. **42** (1936) 230–265 **43** (1937) 544–546.

[VEB]   O. Veblen, *Analysis Situs*, A. M. S. Coll Publ. 5, 1931.

[WAL]   F. Waldhausen, *The word problem in fundamental groups of sufficiently large irreducible 3-manifolds*, Ann. of Math. (2) **88** (1968) 272–280.

[WEL]   K. Weld, *Computability of homotopy groups of nilpotent complexes*, J. Pure Appl. Algebra **48** (1987) 39–53.

E. Domínguez and J. Rubio
Dept. de Matemáticas
Facultad de Ciencias
Universidad de Zaragoza
50.009 Zaragoza
Spain

THE MATH. HERITAGE OF C.F. GAUSS (pp. 195-203)
edited by George M. Rassias
©1991 World Scientific Publ. Co. Singapore

FACTORIZATION IN QUATERNION ORDERS OVER NUMBER FIELDS

Dennis R. Estes

The relationships between element factorization, ideal theory, and quadratic lattice theory are derived for a quaternion order over the ring of integers in a number field F. In particular, those orders for which factorization of the norm of an element induces a corresponding factorization of the element (FNF) are classified by properties of the genus of the order.

The quaternion orders over $\mathbb{Z}$ which satisfy FNF (see Sec. 1 for definition) and other forms of element factorization were shown in [2] to have specific local structures, and are therefore classified by properties of their genus and spinor genus. The classication provided the authors with an algorithm which produced the list of all positive definite quaternion orders over $\mathbb{Z}$ having various forms of factorization. Moreover, factorization properties of indefinite orders are easily detected in their local Jordon splittings. Over a general number field, the units, the class number, and the ramification at 2 are impediments to the general theory. Nevertheless, the theory transfers with suitable modifications, as we see in this note. The investigation is motivated by Hurwitz's proof of Lagrange's 4-square theorem, G. Pall's investigations factorization of generalized quaternions in [7], and the determination of all quaternion norms whose class and spinor genus coincide.

1. Preliminaries

Let A denote a quaternion algebra over a number field F and let $*$

denote the involution on A for which $\alpha + \alpha^* = \operatorname{tr}\alpha, \alpha\alpha^* = Q(\alpha)$ are the reduced trace and norm on A. We view A as a regular quadratic space over F under the bilinear form $(\alpha, \beta) = (\operatorname{tr}\alpha\beta^*)/2$. $O(A)$ denotes the corresponding orthogonal group and $SO(A)$ the subgroup of isometries of A having determinant 1. Thus $\vartheta \in SO(A)$ if and only if there exist $\eta, \tau \in A^\times$, the group of units of A, such that $\vartheta(x) = \eta x \tau$ and $Q(\eta) = 1/Q(\tau)$ [5]. The spinor norm of such an isometry is $Q(\eta)^2(F^\times)^2$ and $O'(A)$ denotes the subgroup of $O(A)$ of those isometries having spinor norm $1 = (F^\times)^2$.

Let D be ring of all integers in F and let L denote a D-lattice on A. We assume throughout that $LF = A$. For each prime P of F, archimedean and nonarchimedean (finite), L_P is the completion of L with respect to the topology defined by P. The class, $\operatorname{Cls}(L)$ and proper class, $\operatorname{Cls}^+(L)$, are the orbits of L under $O(A)$ and $SO(A)$ respectively. The genus, $\operatorname{Gen}(L)$ consists of all D-lattices K on A such that $K_P = \sigma_P(L_P)$ with $\sigma_P \in SO(A_P)$ and the spinor genus, $\operatorname{Spn}(L)$ $(\operatorname{Spn}^+(L))$ consists of those K such that there exist $\sigma \in O(A)$ $(SO(A))$ and for each prime $P, \sigma_P \in O'(A)$ such that $\sigma(K_P) = \sigma_P(L_P)$. Since $* \in O(\Lambda)$ with Λ a D-order on A and $*$ has determinant -1, class and proper class coincide for Λ as do spinor genus and proper spinor genus.

A D-order Λ on A is said to have *factorization induced b norm factorization* (FNF) if for $\gamma \in \Lambda$ with $Q(\gamma) = ab \neq 0, a, b \in D, \ni \alpha, \beta \in \Lambda, u \in D^\times$ such that $\gamma = \alpha\beta$ and $Q(\alpha) = au$. Λ has *factorization induced by local factorization* (FLF) if there is such a global factorization whenever $\ni v \in D^\times$ and for each prime $P, \alpha_P, \beta_P \in \Lambda_P$ such that $\gamma = \alpha_P\beta_P$ and $Q(\alpha_P) = av$. The *principal genus*, $\operatorname{PGen}(\Lambda)$, consists of all fractional right Λ ideals L such that $\lambda L \in \operatorname{Gen}(\Lambda)$ for some $\lambda \in A^\times$.

2. Theorems

Theorem 1. The following are equivalent:
  a) Λ has FLF,
  b) $L \in \operatorname{PGen}(\Lambda)$ implies L is principal, and
  c) Every $L \in \operatorname{Spn}(\Lambda)$ represents a unit.

Proof. a $\implies$ b: Let $L \in \operatorname{PGen}(\Lambda)$ ana multiply by $t \in F^\times$ to insure $\Lambda \supset L$. Let $\lambda \in A^\times$ with $\lambda^{-1}L \in \operatorname{Gen}(\Lambda)$, set $a = Q(\lambda)$, and let S denote the set of primes P in D dividing a. For $P \in S$, let $\eta_P, \tau_P \in A_P^\times$

satisfy $\lambda^{-1}L_P = \eta_P \Lambda_P \tau_P$ with $Q(\eta_P) = Q(\tau_P)^{-1}$, since L is a right Λ ideal, $\Lambda_P \tau_P = \tau_P \Lambda_P$ and we can assume $\tau_P = 1, Q(\eta_P) = 1$. By the strong approximation theorem, $\exists \delta \in \lambda^{-1}L$ such that $\delta \Lambda_P = \eta_P \Lambda_P$, for $P \in S$. Set $\gamma = \lambda \delta$. Then $\gamma \in L$ and thus in Λ. Since $L_P = \lambda \Lambda_P = \Lambda_P S$, for $P \notin S$, and $L_P = \lambda \eta_P \Lambda_P = \gamma \Lambda_P, P \in S, N(L) = aD, a \in D$, and γ has a local left factor having norm a at all P. The hypothesis implies that $\gamma = \alpha \beta, \alpha, \beta \in \Lambda$ with $Q(\alpha) = au$ for some $u \in D^\times$. A determinant argument now shows that $\gamma \Lambda_P = \alpha \Lambda_P$ at all primes P, hence $L = \alpha \Lambda$.

b $\implies$ c: Let $L \in \mathrm{Spn}(\Lambda)$ and adjust L by an isometry of A to insure that $L_P = \eta_P \Lambda \tau_P$ with $\eta_P, \tau_P \in A_P^\times, Q(\eta_P) = A(\tau_P)^{-1} = r_P^2$ with $r_P \in F_P^\times$ on a finite set S of primes and $L_P = \Lambda_P$ elsewhere. We replace η_P with $\eta_P/r_P, \tau_P$ with $r_P \tau_P$ so that the two elements have norm 1. Now let K denote the D-lattice on A which satisfies $K_P = \eta_P \Lambda_P$ for $P \in S$ and $K_P = \Lambda_P$ elsewhere. Since $K \in \mathrm{PGen}\,(\Lambda), K = \eta \Lambda$ for some $\eta \in A^\times$. Since $\eta = \eta_P \xi_P$ with $\xi_P \in \Lambda_P^\times, \eta L_P \eta^{-1} = \Lambda_P \xi_P^{-1} \tau_P \eta_P = \Lambda_P \delta_P$ with $Q(\delta_P) = 1$. By replacing L with $\eta^{-1}L\eta$, we can assume $\eta_P = 1$ for all P. Then $L_P^* = \tau_P^* \Lambda_P$ and, as previously, $L^* = \tau^* \Lambda$ for some $\tau \in A^\times$. Since $\tau \in L$ and $D = N(L) = Q(\tau)D$, c) follows.

c) $\implies$ a): Let $a, u, \gamma, \alpha_P, \beta_P$ be as in the definition of FLF and set $L = a\Lambda + \gamma \Lambda$. By the Hasse principle, $\exists \lambda \in A$ such that $Q(\lambda) = av$. Note that $\lambda^{-1} \alpha_P \Lambda_P \supset \lambda^{-1}L_P$ for all primes P. Let K denotes the D-lattice on A satisfying $K_P = \lambda^{-1} \alpha_P \Lambda_P$ for all P. Since $Q(\lambda^{-1}\alpha_P) = 1, K \in \mathrm{Spn}(\Lambda)$. Let $\mu \in K$ with $\mathrm{Q}(\mu) = u \in D^\times$. Since $K_P = \mu \Lambda_P$ for all P, $\lambda \mu \Lambda = \lambda K \supset L$, and $\alpha = \lambda \mu$ is the required left divisor.

Theorem 1 appears in parts in [6]. Since the spinor genus and class coincide for indefinite lattices [4, 104:5], we have

Corollary 1. If A is indefinite then each D-order on A has FLF.

Let $Q_{\mathrm{gen}}(\Lambda)$ denote the set of integers in D represented by the genus of Λ; i.e., $Q_{\mathrm{gen}}(\Lambda) = \{b \in D : b \in Q(\Lambda_P)$ for all primes $P\}$. Set $D(\Lambda) = \{b \in D : b$ is prime to $16\nu(\Lambda)\}, \nu(\Lambda)$ the volume of Λ, and note that if P does not divide $16\nu(\Lambda)$ then $Q(\Lambda_P) \supset D_P$ [4,91.3 and 63.18]. Let $h(F)$ denote the class number of F. We note for the proof below that a (right) Λ ideal L is invertible if and only if L_P is principal [3].

Theorem 2. The following are equivalent for a D-order Λ on A:

  a) invertible fractional right Λ ideals are principal;

  b) each $L \in \mathrm{Gen}(\Lambda)$ represents a unit, $h(F) = 1$ and $D^{\times}Q_{\mathrm{gen}}(\Lambda) \supset D(\Lambda)$;

  c) for $\gamma \in \Lambda$ with $Q(\gamma) = ab, a, b(\neq 0) \in D$ and $aD + bD = D, \exists \alpha, \beta \in \Lambda, u \in D^{\times}$ with $\gamma = \alpha\beta, Q(\alpha) = au$.

Proof. a $\Longrightarrow$ b: The proof that each $L \in \mathrm{Gen}(\Lambda)$ represents a unit parallels the proof of b $\Longrightarrow$ c of Theorem 1. Let I denote an ideal in D, hence $I_P = i_P D_P$ at each finite prime P. Let $\alpha_P \in A_P$ with $Q(\alpha_P) = i_P(A_P$ is universal). Since we can take $\alpha_P = 1$ a.e., there is a lattice L on A such that $L_P = \alpha_P \Lambda_P$ for all P. Since L is an invertible Λ ideal, $L = \alpha\lambda$ for some $\alpha \in A^{\times}$. Thus, $N(L) = Q(\alpha)D$. Since $N(L_P) = Q(\alpha)D = I_P, I = N(\alpha)D$ is principal.

Now let $b \in D(\Lambda)$. For each prime $P|b$, let $\alpha_P \in A_P$ have norm b. Let L be the D-lattice on A such that $L_P = \alpha_P \Lambda_P$ for $P|b$ and $L_P = \Lambda_P$ otherwise. As above, $L = \alpha\Lambda$ for some $\alpha \in A^{\times}$. Since $N(L) = bD, Q(\alpha) = bu$ for some $u \in D^{\times}$. Since $\alpha \in \Lambda_P$ for $P \nmid b$ and since $Q(\Lambda_P) = D_P$ for $P|b, bu \in Q_{\mathrm{gen}}(\Lambda)$.

b $\Longrightarrow$ c: Let $L = a\Lambda + \gamma\Lambda$. Clearly, $L_P = \Lambda_P$ if P does not divide a and $L_P = \gamma\Lambda_P$ for P for dividing a. Thus $N(L) = aD$ and it suffices to show that L is principal. Set $\lambda_P = 1$ or γ according as to which generates L_P and use the strong approximation theorem to exhibit a $\delta \in L$ such that $\delta L_P = \alpha_P \Lambda_P$ for all $P|16a\nu(\Lambda)$. Note that $\delta^* L = aK$ with $\Lambda \supset K$ and $N(K) = (N(\delta)/a)D = rD$ is prime to $16\nu(\Lambda)$. Select $\rho \in K$ such that $K_P = \rho\Lambda_P$ for $P|r$. Therefore $K = r\Lambda + \rho\Lambda$ and it suffices to see that K is principal. Let $u \in D^{\times}$ with $ru \in Q_{\mathrm{gen}}(\Lambda)$. The Hasse principal implies the existence of $\xi \in A$ such that $Q(\xi) = ru$. Now $\xi^{-1}K_P = \xi^{-1}\Lambda_P$ for $P \nmid r$ and $\xi^{-1}\rho\Lambda_P$ for $P|r$. Since there is a $\xi_P \in \Lambda_P^{\times}$ such that $Q(\xi_P) = ru$ for $P|r$ and $Q(\xi_P) = u^{-1}$ for $p|r, \xi^{-1}K \in \mathrm{Gen}(\Lambda)$. If $\mu \in \xi^{-1}K$ has norm a unit, then $\mu\Lambda = \xi^{-1}K$ and K is principal.

c $\Longrightarrow$ a: $h(F) = 1$ implies that $N(L)$ is principal for an invertible right Λ ideal L. We can assume $\Lambda \supset L$. Set $N(L) = aD$ and use strong approximation to get $\gamma \in L$ with $\gamma\Lambda_P = L_P$ for all $P|a$. Then $L = a\Lambda + \gamma\Lambda$ with $Q(\gamma) = ab$ and b prime to a. It follows from c) that L is principal.

Remark 1. The condition $D^{\times}Q_{\mathrm{gen}}(\Lambda) \supset D(\Lambda)$ and the approximation theorem implies that F contains units having independent signs at the real primes of F such that A is definite; i.e., for each such prime $\nu, \exists u \in D^\times$ such that u is negative at ν while positive at all other real primes of F.

Remark 2. The condition that $h(F) = 1$ in the above theorem can be deleted if we replace a) with the condition that invertible right ideals having principal norm are principal.

For a finite prime P and D_P lattice L with $D_P \supset N(L)$, we set $J(L) = \{x \in L : 2(x, L) \text{ and } Q(x) \equiv 0 \bmod P\}$. Note that $J(L) = PL$ if $2^4\nu(L) = D$. It is shown in [2] that if $\gamma \in \Lambda_P - J(\Lambda_p)$ has norm divisible by $p, pD_P = P$, then γ has a right factor having norm pu_p, u_p a p-adic unit.

We use the following property of quaternion algebras in the proofs below: if K is a two dimensional regular subspace of A containing 1, then K is a subalgebra, $A = K \perp L$, and if $\beta \in L$ with $Q(\beta) \neq 0$ then $A = K + \beta K$ and conjugation by β induces $*$ on K. Also, a maximal order on K is the (unique) lattice L on K maximal with respect to $D_P \supset N(L)$ and $1 \in L$. We say that Λ_P has FNF if for each $\gamma \in \Lambda_P$, with $\mathrm{ord}_P\gamma \geq 2, \exists \alpha, \beta \in \Lambda_P$ such that $\gamma = \alpha\beta$ and $\mathrm{ord}_P\alpha = 1$.

Lemma 1. Let P be a dyadic prime of $F, p \in D_P$ with $\mathrm{ord}_P p = 1, K$ a 2 dimensional semisimple subalgebra of A_P, and O_K the maximal order of K. If $\beta \in K^\perp$ and $\mathrm{ord}_P Q(\beta) = 0$ or 1, then $\Lambda_P = O_K + \beta O_K$ has FNF if and only if K is unramified (i.e., $p \nmid 4\nu(O_K)$), or K is a totally ramified field extension of F_P, $\mathrm{ord}_P Q(\beta) = 0, -Q(\beta) \notin Q(O_K) \bmod P^2$, and O_K is $p/2$ modular.

Proof. In the unramified case $J(\Lambda_P) = p\Lambda_P$ when $Q(\beta) \in D_P^\times$. In this event, it suffices to show that if $p^2|Q(\gamma), \gamma \in \Lambda_P$, then γ has a right factor having norm $pu, u \in D_P^\times$. Since this is the case for $\gamma \notin J(\Lambda_P)$, we have only to exhibit an element of norm p. However, $Q(\Lambda_P) = D_P[4, 91.3]$. If $p|Q(\beta)$ then $\sigma + \beta\sigma' \in J(\Lambda_P)$ if and only if $\sigma \in J(O_K) = pO_K$, in which case β is the desired left divisor.

Assume now that K is ramified. Then K is a totally ramified field extension of F_P (see [4, 82:21] which shows that a maximal integral lattice on an isotropic space is unramified). Thus, $O_K = D_P[\pi]$ with the minimal polynomial $x^2 + pax + pb$ for π Eisenstein irreducible [1, p. 23, Theorem 1]. Thus, $\pi^2 O_K = pO_K = \pi^* O_K, p \nmid b$, and O_K is $p^j/2$ modular when

$j = \mathrm{ord}_P(pa) \leq \mathrm{ord}_P 2$. Let $\gamma = \sigma + \beta\sigma' \in \Lambda_P$ have norm divisible by P^2. If $-Q(\beta) \notin Q(O_K) \bmod p^2$ then both $Q(\sigma)$ and $Q(\sigma')$ are divisible by p, and it follows that π is the desired divisor. On the other hand, if $-Q(\beta) \equiv Q(\sigma)$ $\bmod p^2$ then $\gamma = \sigma + \beta$ has p^2 dividing its norm but does not have left factor having norm exactly divisible by p, for if $(\sigma_1 + \beta\sigma_2)(\sigma + \beta) \equiv 0 \bmod p$ and the first factor has norm divisible by p but not p^2 then σ_2 is a unit; otherwise π divides both σ_1, σ_2 and we would have the contradiction $\pi(\sigma + \beta) \equiv 0$ $\bmod p^2$. We can assume therefore that $\sigma_2 = 1$. Since the coefficient $\sigma_1^* + \sigma$ of β in the product is divisible by $p, Q(\sigma_1) = Q(-\sigma + p\delta) \equiv Q(\sigma) \bmod p^2$. But then $p^2 | Q(\sigma_1 + \beta)$, contradicting our assumption.

Now assume that $p | Q(\beta)$. Since $Q(\beta)/Q(\pi) \equiv t^2 \bmod p$ is solvable (D_P/pD_P is perfect), $-Q(\beta) \equiv Q(\pi t) \bmod p$. Finally, assume that $Q(\beta)$ is a unit and that $p | a$. We claim that $-Q(\beta) \equiv Q(x + \pi y) \bmod p^2$ with $x + \pi y \in O_K$. Since $Q(x + \pi y) = x^2 - paxy + pby^2 \equiv x^2 + pby^2 \bmod p^2$, select x so that $x^2 \equiv -Q(\beta) \bmod p$ and solve $by^2 \equiv (-Q(\beta) - x^2)/p \bmod p$.

In the ramified case above the condition $-Q(\beta) \notin Q(O_K) \bmod p^2$ forces the space A_P to be anisotropic. To see this, note that A_P is isotropic if and only if $-Q(\beta) \in Q(K)$ and since K is anisotropic and O_K is a maximal integral lattice on $K, Q(O_K) = Q(K) \cap D_P$ [4, 91:1].

The quaternion order having basis $1, i, j, k$ over $\mathbb{Z}_2[\sqrt{2}]$ where $i^2 = -(\sqrt{2})i - \sqrt{2}, j^2 = 1 - \sqrt{2}, ij = -ji = k$ provides an example which is $1/\sqrt{2}$ modular and has FNF, and the ordinary quaternions $i^2 = j^2 = -1, ij = -ji = k$ provide a unimodular example over $\mathbb{Z}_2$.

Lemma 2. Let K, P, p be as in Lemma 1 and let R denote a rank 2, D_P order on K. If $\Lambda_P = R \perp B$ is a D_P order on A_P and $N(B) \supset pD_P$, then $B = \beta R$ or $\beta R'$ with R' a D_P order on K, $\mathrm{ord}_P Q(\beta) = 1, R' \supset R \supset pR$. Moreover, if $\Lambda_P = R + \beta R$ has FNF then R is maximal.

Proof. Let $\beta \in B$ with $\mathrm{ord}_P Q(\beta)$ minimal. Since βK is the orthogonal complement of K in $A_P, B = \beta C$ with C a fractional R ideal in K. Moreover, $C \supset R \supset Q(\beta)CC^* \supset Q(\beta)C \supset pR$. If $Q(\beta)$ is a unit then $C = R$. Otherwise, since R/pR is 2 dimensional over $D_P/pD_P, CC^* = (1/p)R$ or C. In the second event, $C = C^*$ and C is therefore an order on K. In the first event, C is invertible and C is principal since R is semilocal. Let $C = \rho R$.

Then $\mathrm{ord}_P Q(\rho) = -1$. Since $B = \beta\rho R, \beta\rho \in B$. But $\mathrm{ord}_P Q(\beta\rho) = 0$, contradicting the minimal choice for β.

Assume now that $\Lambda_P = R + \beta R$ has FNF and let $O_K = D_P + D_P\omega$ be the maximal order on K. Then $R = D_P + D_P p^s \omega$ for some $s \geq 0$. If $s > 0$ then FNF implies that $\lambda = p^s \omega$ has a left factor $\rho + \beta\rho'$ having norm exactly divisible by $p, \rho, \rho \in R$. If $p|Q(\rho')$ then $p|Q(\rho)$, in which case, $\rho, \rho' \in pO_K$, contradicting the fact that one must not have norm divisible by p^2. Therefore $\rho' \in R^\times$. Since $(\rho^* - \beta\rho')\lambda \equiv 0 \bmod pR, \rho'\lambda \equiv 0 \bmod pR$ and hence $p|\lambda$, a contradiction.

Theorem 3. Let Λ_P be a D_P-order on A_P, P a finite prime, $p \in D_P$ with $\mathrm{ord}_P p = 1$. Then Λ_P has FNF if and only if:

a) P is nondyadic and (i) Λ_P is unimodular;

   (ii) $\Lambda_P = R \perp \beta R$ with R a maximal D_P order in a regular quadratic subalgrbra of A_P, R is unimodular and $\mathrm{ord}_P Q(\beta) = 1$; or

   (iii) $\Lambda_P = R \perp \beta R, R = D_P[\sqrt{p}], \mathrm{ord}_P Q(\beta) = 1$, and $-Q(\beta) \notin Q(R)$ $\bmod\ p^2$.

b) P is dyadic and (i) Λ_P is 1/2-modular; or there is a D_P order R in a degree 2 field extension K of F_P in A_P such that

   (ii) $\Lambda_P = R \perp \beta R, R$ is maximal and K totally ramified, R is $p/2$ modular, $p{\nmid}Q(\beta)$, and $-Q(\beta) \notin Q(R) \bmod p^2$; or

   (iii) $\Lambda_P = R \perp \beta R', R'$ maximal and K unramified, $\mathrm{ord}_P Q(\beta) = 1, R' \supset R \supset pR$.

Proof. The result for P nondyadic appears in [2]. Assume therefore that P is dyadic.

Case 1. $s(\Lambda_P) = (\Lambda_P, \Lambda_P) \not\subset D_P$: Since $(\alpha, \beta) = (\alpha\beta^*, 1), \exists \gamma \in \Lambda_P$ such that $(\gamma, 1)$ generates $s(\Lambda_P)$. Let R denote the subring $D_P + \gamma D_P$. Then $\Lambda_P = R \perp B$ [4, 82:15]. Let $\beta \in B$ with $\mathrm{ord}_P Q(\beta)$ minimal. If $\mathrm{ord}_P Q(\beta) \geq 2$ we claim that β does not have a left divisor having norm exactly divisible by p. For if $\rho + \tau, \rho \in R, \tau \in B$, were such a divisor then $(\rho^* - \tau)\beta \equiv 0 \bmod p$. Consequently, $\rho^*\beta \equiv 0 \bmod p$. since $Q(\rho+\tau) \equiv Q(\rho)$ $\bmod\ p^2, \mathrm{ord}_P Q(\rho^*\beta/p) < \mathrm{ord}_P(\beta)$, a contradiction. Thus, $\mathrm{ord}_P(\beta) \leq 1$.

Thus R is maximal and Λ_P is one of the 1/2 modular (the unramified case) or $p/2$ modular orders described in Lemma 1, or $\Lambda_P = R + \beta R'$ with $R' \supset R \supset pR'p|Q(\beta)$. In this last event Λ_P has FNF if and only if Q is anisotropic on R'/pR'. This is a consequence of te observations: $Q(\rho + \beta\rho') \equiv 0 \bmod p^2$ implies that $p|Q(\rho), R = D_P + pR'$ hence $p|Q(\rho)$ implies that $p \in pR'$, and $\beta\Lambda_P \supset pR' + p\beta R' \supset (pR' + \beta R)^2$. Finally, Q is anisotropic on R'/pR' if and only if R' is maximal and its ring of quotients K is a unramified field extension of F_P.

Case 2. $D_P \supset s(\Lambda_P)$: Then $\Lambda_P = D_P \perp L$. Select $\alpha \in L$ with $\mathrm{ord}_P Q(\alpha)$ minimal and note that p fails to have a factor having norm exactly divisible by p if $\mathrm{ord}_P Q(\alpha) \geq 2$. We claim that αD_P is an orthogonal summand of L. If this is not the case then $\exists \alpha' \in L$ such that $(\alpha, \alpha') \not\equiv 0 \bmod N(\alpha)$. Thus, $p|N(\alpha)$ and $p\nmid(\alpha, \alpha')$. Set $\alpha\alpha' = r + \lambda, r \in D_P, \lambda \in L$. Then $0 \equiv Q(\alpha\alpha') \equiv r^2 \bmod p$, but $r = (\alpha\alpha', 1) = -(\alpha, \alpha')$ which is not divisible by p.

Thus $R = D_P + \alpha D_P$ splits Λ_P and an element β in its complement with $\mathrm{ord}_P Q(\beta)$ minimal must have $\mathrm{ord}_P Q(\beta) = 0,1$ as the arguments in the previous case show. By Lemma 2, $\Lambda_P = R + \beta R$ or $R + \beta R'$ with R' an order containing $R, R \supset pR'$, and $\mathrm{ord}_P Q(\beta)=1$. In the first event, R is maximal and, by Lemma 1, $p\nmid Q(\beta), -Q(\beta) \notin Q(R) \bmod p^2$, and R is unimodular. The second event is treated as in case 1, the main distinction is that the condition $D_P \supset s(\Lambda_P)$ forces 2 to be unramified in F_P.

It is possible that different choices in the above theorem give isomorphic orders. Since an order is determined up to isomorphism by its isometry class, one can distinguish the isomorphism classes by the dyadic classification theorem for lattices [4, 93: 28].

Theorem 4. A D order Λ on A has FNF if and only if it has done one of the local structures prescribed in Theorem 3, $L \in \mathrm{Gen}(\Lambda)$ implies L represents a unit, and $D^\times Q_{\mathrm{gen}}(\Lambda) \supset D(\Lambda)$.

Proof. Let $\gamma \in \Lambda, Q(\gamma) = ab \neq 0, a, b \in D$. We can assume that a is not a unit. For $P|a$, let $\alpha_P \in \Lambda_P$ denote a left divisor of γ having norm $au_P, u_P \in D_P^\times$. Pick $\delta \in \Lambda$ with $\delta\Lambda_P = \alpha_P\Lambda_P$ at all $P|a$. Note that γ is in the invertible right ideal $L = a\Lambda + \delta\Lambda, N(L) = aD$, and $Q(\delta) = ab$ with $aD + bD = D$. Remark 1 implies the existence of a factorization $\delta = \alpha\beta$ in Λ with $Q(\alpha) = au$ for some $u \in D^\times$. Since $\alpha\Lambda \supset L, \alpha$ is the desired left divisor of γ.

The condition $D^\times Q_{\mathrm{gen}}(\Lambda) \supset D(\Lambda)$ excludes, for example, the nondyadic configuration (ii) of Theorem 3 since only squares prime to p are locally represented.

References

1. J. W. S. Cassels and A. Fröhlich, *Algebraic Number Theory*, Academic Press, New York, 1967.
2. D. R. Estes and G. Nipp, *Factorization in quaternion orders*, J. Number Th., to appear.
3. I. Kaplansky, *Submodules of quaternion algebras*, Proc. London Math. Soc. **19** (1969) 219–232.
4. O. T. O'Meara, *Introduction to Quadratic Forms*, Springer-Verlag, New York, 1963.
5. G. Nipp, *Quaternion orders associated with ternary lattices*, Pacific J. Math. **53** (1974) 525–537.
6. G. Nipp, *The spinor genus of quaternion orders*, Trans. A. M. S. **215** (1975) 299–309.
7. G. Pall, *On generalized quaternions*, Trans. A. M. S. **59** (1946) 503–513.

Dennis R. Estes
Department of Mathematics
University of Southern California
Los Angeles, CA 90089-1113
U. S. A.

THE MATH. HERITAGE OF C.F. GAUSS (pp. 204-224)
edited by George M. Rassias
©1991 World Scientific Publ. Co. Singapore

THUE INEQUALITIES WITH A SMALL NUMBER OF SOLUTIONS

J.-H. Evertse[a] *and K. Győry*[b]

Let $F(X,Y) \in \mathbb{Z}[X,Y]$ be a binary form with exactly $n \geq 3$ pairwise linearly independent linear factors in its factorization over $\mathbb{C}$, and let $M \geq 1$. We shall prove (cf. Theorem 1) that if the inequality (*) $0 < |F(x,y)| \leq M$ in $x,y \in \mathbb{Z}$ with $\gcd(x,y)=1$ and $y>0$ or $(x,y)=(1,0)$ has more than $N := \max(6 \times 7^{2\binom{400}{3}}, 6n)$ solutions, then there is a matrix $\begin{pmatrix} a & b \\ c & d \end{pmatrix}$ in $\mathrm{SL}(2,\mathbb{Z})$ such that the maximum of the absolute values of the coefficients of $F(aX+bY, cX+dY)$ is bounded above by an effectively computable number depending only on M and $\deg F$. We first show (cf. Theorem 3) that if (*) has more than N solutions and F has no multiple factors, then the absolute value of the discriminant $D(F)$ of F is bounded, by using the arguments of Bombieri and Schmidt in [3]. Then we prove by means of a result of Győry [8] on equivalence of polynomials that for some $\begin{pmatrix} a & b \\ c & d \end{pmatrix} \in \mathrm{SL}(2,\mathbb{Z})$, the coefficients of $F(aX+bY, cX+dY)$ can be bounded above in terms of $|D(F)|$ and M. Finally, on combining Theorems 1 and 3 with a result of Győry and Papp [9] on the Thue equation we show (cf. Theorem 2) that if (*) has more than N solutions, one of which is $(1,0)$, then for each solution (x,y), $|y|$ is bounded above by an effective constant which depends only on M and n, but not on the coefficients of F.

---

[a]This research has been made possible by a fellowship of the Royal Netherlands Academy of Arts and Sciences (K.N.A.W.).

[b]This research was supported in part by Grant 273 from the Hungarian National Foundation for Scientific Research.

1. Introduction

Let $F(X,Y) \in \mathbb{Z}\,[X,Y]$ be a binary form. Denote by $\omega(F)$ the maximal number of pairwise non-proportional linear forms in $\mathbb{C}\,[X,Y]$ dividing F. Put $r = \deg F$ and assume that $\omega(F) \geq 3$. Let $M \geq 1$ be an integer and consider the *Thue inequality*

$$0 < |F(x,y)| \leq M \quad \text{in} \quad x,y \in \mathbb{Z} \tag{1}$$

with $\gcd(x,y) = 1$ and $y > 0$ or $(x,y) = (1,0)$. This inequality can be reduced to a finite number of equations of the form

$$F(x,y) = a \quad \text{in} \quad x,y \in \mathbb{Z}\,, \tag{2}$$

where a is a given integer with $0 < |a| \leq M$. In 1909, Thue [19] proved that (2) has only finitely many solutions if F is irreducible and $r \geq 3$. From this one can easily deduce that (2) has only finitely many solutions if F is any binary form in $\mathbb{Z}\,[X,Y]$ with $\omega(F) \geq 3$. Hence inequality (1) can have only finitely many solutions. Since Thue's result became known, much work has been done to estimate the numbers of solutions of (1) and (2) from above. In [7], a more detailed history about the numbers of solutions of (1), (2) (and more generally, the Thue-Mahler equation) has been given, and we only give here a brief survey. From results of Mahler [12] in 1933 and Lewis and Mahler [11] in 1961 it follows that the number of solutions of (2) has an upper bound depending only on r, the number t of distinct primes dividing a, and the coefficients of F. In 1983, Evertse ([6], Theorem 6.4) proved that this upper bound can be replaced by a bound which depends only on r and t and hence is independent of the coefficients of F. Later, similar bounds but with a much better dependence on r, were obtained by Bombieri and Schmidt [3] and Bombieri [2]. From a result of Mahler [13] in 1933 it follows that for irreducible F, (1) has at most $c(F)M^{2/r}$ solutions, where $c(F)$ depends on F only. Mahler mentioned that this was in fact an unpublished result of Siegel. If $F(X,Y) = a_0 X^r + a_1 X^{r-1} Y + \ldots + a_r Y^r$ then each pair $(x,y) \in \mathbb{Z}^2$ with $x > 0, y > 0, \gcd(x,y) = 1$ and $\max(x,y) \leq \left\{ M/(|a_0| + \ldots + |a_r|) \right\}^{1/r}$ is a solution of (1) (if M is sufficiently large), so the dependence on M of Mahler's bound is best possible. The results of Evertse, Bombieri and Schmidt and Bombieri mentioned above suggested that $c(F)$ has an upper bound depending only on r. Recently, Mueller and Schmidt [16] proved that (1) has at most $c \cdot s^2 M^{2/r} (1 + \log M^{1/r})$ solutions, where c is an absolute constant and s is the number of non-zero coefficients of F.

If $F(X,Y)$ is a binary form and $A = \begin{pmatrix} a & b \\ c & d \end{pmatrix}$ is a matrix, then we put $F_A(X,Y) := F(aX + bY, cX + dY)$. By $\mathrm{SL}(2, \mathbb{Z})$ we denote the multiplicative group of 2×2-matrices with entries in $\mathbb{Z}$ and determinant 1. Two binary forms $F, G \in \mathbb{Z}[X,Y]$ are called *equivalent* if $G = F_U$ for some U in $\mathrm{SL}(2, \mathbb{Z})$. It is obvious that the number of solutions of (1) does not change when F is replaced by an equivalent form.

The purpose of this paper is not to give a uniform upper bound for the number of solutions of (1) for each F, but to give an as good as possible upper bound for the number of solutions of (1) if we are allowed to exclude a finite number of equivalence classes of binary forms F. In [7] (Theorem 1) we showed that for every finite normal extension $L/\mathbb{Q}$ and for every $r \geq 3$ and $M \geq 1$, there are at most finitely many equivalence classes of binary forms F of degree r with splitting field L and with $\omega(F) \geq 3$, such that (1) has more than *two* solutions. The proof of this result does not allow us to compute representatives for these equivalence classes effectively. The bound 2 is best possible. However, if we vary the splitting field L of F, this result does not remain valid. For if $a_1, \ldots, a_r \in \mathbb{Z}$ are distinct and $F(X,Y) = (X - a_1 Y) \ldots (X - a_r Y) + Y^r$ then $F(X,Y) = 1$ has the r solutions $(a_1, 1), \ldots, (a_r, 1)$. In [7] we already mentioned that, by using the arguments of Bombieri and Schmidt in [3] and applying an effective result of Győry [8] on equivalence of polynomials, one could prove that (1) has at most $c \cdot r$ solutions for all but finitely many equivalence classes of binary forms F, and that a representative for each exceptional class can be determined effectively. In this paper we make this result explicit. The *height* $H(P)$ of a polynomial P with integral coefficients is the maximum of the absolute values of its coefficients.

Theorem 1. Let $r \geq n \geq 3$ be integers, let $F(X,Y) \in \mathbb{Z}[X,Y]$ be a binary form of degree r with $\omega(F) = n$, and put $N(n) = 6n \times 7^{2\binom{n}{3}}$ if $n < 400$ and $N(n) = 6n$ for $n \geq 400$. If (1) has more than $N(n)$ solutions then there is a matrix U in $\mathrm{SL}(2, \mathbb{Z})$ such that

$$H(F_U) \leq \exp\left\{ c_1(r) M^{10(n-1)^4} \right\}, \tag{3}$$

where $c_1(r)$ is an effectively computable number depending only on r. Further, if $0 < |F(1,0)| \leq M$, then (3) holds for some U of the form $\begin{pmatrix} 1 & b \\ 0 & 1 \end{pmatrix}$ with $b \in \mathbb{Z}$.

For the equation

$$|F(x,y)| = 1 \quad \text{in} \quad x,y \in \mathbb{Z} \quad \text{with} \quad y > 0 \quad \text{or} \quad (x,y) = (1,0) , \qquad (4)$$

better results than that mentioned in Theorem 1 have been recently established. Bombieri and Schmidt [3] proved that for *every* irreducible binary form $F \in \mathbb{Z}[X,Y]$ of degree $r \geq 3$ (so without the exception of finitely many equivalence classes) (4) has at most $c \cdot r$ solutions, where c is an absolute constant which can be taken as 215 if r is large. Quite recently, Brindza, van der Poorten and Waldschmidt [4] proved that for all but finitely many equivalence classes of irreducible binary forms $F \in \mathbb{Z}[X,Y]$ of given degree $r \geq 3$, representatives of which can be effectively determined, (4) has at most $2r^2$ solutions.

By combining Theorem 1 with an effective result of Győry and Papp [9] on the Thue equation one can show that if (1) has more than $N(n)$ solutions then for some $U \in \mathrm{SL}(2, \mathbb{Z})$, the solutions of $0 < |F_U(x,y)| \leq M$ in $x,y \in \mathbb{Z}$ with $\gcd(x,y) = 1$ and $y > 0$ or $(x,y) = (1,0)$ satisfy $\max(|x|,|y|) \leq c(M,n)$, where $c(M,n)$ is effectively computable and depends on M,n only. If $(1, 0)$ is a solution of (1), then U can be chosen to be of the form $\begin{pmatrix} 1 & b \\ 0 & 1 \end{pmatrix}$ with $b \in \mathbb{Z}$, according to Theorem 1, and such a matrix U does not affect y. Thus we have $|y| \leq c(M,n)$ for each solution (x,y) of (1) itself. More precisely, we shall prove the following.

Theorem 2. Let r, n and $F(X,Y)$ be as in Theorem 1, and suppose that (1) has more than $N(n)$ solutions, one of which is $(1, 0)$. Then for each solution (x,y) we have

$$|y| < \exp \left\{ c_2(n) M^{10n^4} \right\} , \qquad (5)$$

where $c_2(n)$ is some effectively computable number depending only on n.

In other words, if (1) has solution $(1, 0)$ and a solution for which $|y|$ exceeds the right-hand side of (5), then (1) has at most $N(n)$ solutions. It is important to observe that the bound in (5) is *independent* of the coefficients of F.

Theorem 1 will be deduced from Theorem 3 below and an effective result of Győry [8] (cf. Lemma 9 in this paper) on equivalence of polynomials. The *discriminant* of $F(X,Y) = \prod_{i=1}^{r} (\alpha_i X - \beta_i Y)$ is defined by $D(F) = \prod_{i>j} (\alpha_i \beta_j - \alpha_j \beta_i)^2$. As before, F is a binary form of degree r in $\mathbb{Z}[X,Y]$. Then $D(F) \in \mathbb{Z}$.

Theorem 3. Let $r \geq 3$, put $\delta(r) = \frac{5}{6}r(r-1)$ if $r < 400$ and $\delta(r) = 120(r-1)$ if $r \geq 400$ and assume that

$$|D(F)| \geq \exp\{80r(r-1)\} \cdot M^{\delta(r)}.$$

Then (1) has at most $N(r)$ solutions.

In 1983, Evertse ([6], Theorem 6.4) proved that if $|D(F)| \geq (13M)^{5r(r-1)/6}$ then (1) has at most $6r \times 7^{2\binom{r}{3}}$ solutions. This implies Theorem 3 for $r < 400$, so we have to prove it only for $r \geq 400$. Theorem 3 for $r \geq 400$ can be proved in essentially the same way as the above-mentioned result of Bombieri and Schmidt on Eq. (4). We shall not repeat all arguments of Bombieri and Schmidt in this paper, but only indicate the necessary modifications in their original arguments.

In certain special cases, results much better than Theorem 3 have been obtained. Siegel [17] proved that if F is an irreducible cubic form in $[X,Y]$ of positive discriminant exceeding some bound depending only on M, then (1) has at most *eighteen* solutions. In a student's paper of 1949, Gel'man improved this to 10 (cf. [5], Chap. 5). In 1937, Siegel [18] showed that if a,b,r,m are integers with $r \geq 3, (ab)^{1/2} \geq 188rM^4$ and $\sqrt[r]{a/b} \notin \mathbb{Q}$ then the inequality $0 < |ax^r - by^r| \leq M$ in $x, y \in \mathbb{Z}$ with $x > 0, y > 0$ has at most *one* solution. This result, together with that of Mueller and Schmidt [16] mentioned before, leads us to the *conjecture* that is possible to prove a result similar to Theorem 3, with $N(r)$ replaced by a quantity depending only on the number of non-zero coefficients of F.

2. Proof of Theorem 3

Let $F(X,Y) \in \mathbb{Z}[X,Y]$ be a binary form of degree $r \geq 3$ with non-zero discriminant $D(F)$, and let $M \geq 1$ be an integer. We consider the inequality

$$0 < |F(x,y)| \le M \quad \text{in} \quad x,y \in \mathbb{Z} \text{ with } \gcd(x,y) = 1 \text{ and } y > 0$$
$$\text{or } (x,y) = (1,0) \ . \tag{1}$$

As was mentioned before, if $r < 400$ then Theorem 3 is a direct consequence of Theorem 6.4 of [6]. Therefore, it suffices to prove the following

Proposition. If $r \ge 400$ and $|D(F)| \ge e^{80r(r-1)}M^{120(r-1)}$, then (1) has at most $6r$ solutions.

Here, $e = 2.7182\ldots$. We mention that the exponents 80 and 120 in the lower bound for $|D(F)|$ can be reduced if r is much larger than 400. Further, one can show that there exist numbers $c_3(r), c_4(r)$ depending on r, and an absolute constant c_5, such that (1) has at most $5r + c_5\sqrt{r}$ solutions if $|D(F)| \ge c_3(r)M^{c_4(r)}$. We shall not work this out.

The Proposition can be proved in essentially the same way as the result of Bombieri and Schmidt [3], mentioned in Sec. 1, that the equation

$$|F(x,y)| = 1 \quad \text{in} \quad x,y \in \mathbb{Z} \tag{4}$$

with $\gcd(x,y) = 1$ and $y > 0$ or $(x,y) = (1,0)$ has at most $c \cdot r$ solutions. Bombieri and Schmidt proved this in the following way. First they showed that (4) has only few "large solutions" by applying a "Thue-Siegel principle"; this argument was more or less standard. Then, by developing some new ideas, they also succeeded in giving a good upper bound for the number of "small" solutions. Bombieri and Schmidt assumed that F is irreducible. This assumption is not necessary when dealing with the small solutions. When dealing with the large solutions following the Thue-Siegel principle mentioned above, it is in principle not necessary to assume that F is irreducible, but then one might have to consider some subcases. Instead of this, we derive, by elementary means, an explicit upper bound for the solutions of (1) for reducible F, and prove directly that in this case, (1) has only few large solutions.

If $F(X,Y)$ is replaced by an equivalent form, then neither the discriminant of F, nor the number of solutions of (1) changes. Each binary form is equivalent to one with $F(1,0) \ne 0$. Therefore it is no restriction to assume that

$$F(X,Y) = a_0(X - \alpha_1 Y)\ldots(X - \alpha_r Y)$$

with $a_0 \in \mathbb{Z} \setminus \{0\}$ and distinct algebraic numbers $\alpha_1, \ldots, \alpha_r$, and we shall do so in the sequel. The *Mahler measure* of F is defined by

$$M(F) = |a_0| \prod_{i=1}^{r} \max\left(1, |\alpha_i|\right) .$$

Then

$$|D(F)| \leq r^r M(F)^{2r-2} \tag{6}$$

(cf. [15]). Further, if $F, G \in \mathbb{Z}[X, Y]$ are binary forms then

$$M(FG) = M(F) \cdot M(G) . \tag{7}$$

Finally, if $F(X, Y) = a_0 X^r + a_1 X^{r-1} Y + \ldots + a_r Y^r$, then

$$|a_i| \leq \binom{r}{i} M(F) \quad \text{for} \quad i = 0, \ldots, r . \tag{8}$$

Lemma 1. Let $F(X, Y) \in \mathbb{Z}[X, Y]$ be a reducible binary form of non-zero discriminant with $F(1, 0) \neq 0$. Then every solution (x, y) of (1) satisfies

$$\max(|x|, |y|) \leq 2^{r^2/2}(M(F) + M)^r . \tag{9}$$

Proof. Suppose that $F(X, Y) = D(X, Y)E(X, Y)$, where

$$D(X, Y) = d_0 X^m + \ldots + d_m Y^m , \quad E(X, Y) = e_0 X^n + \ldots + e_n Y^n$$

are non-constant binary forms in $\mathbb{Z}[X, Y]$. Let (x, y) be a solution of (1). Then there are integers m_1, m_2 with $0 < |m_1| \leq M, 0 < |m_2| \leq M$ such that

$$D(x, y) = m_1 , \quad E(x, y) = m_2 .$$

We claim that the two polynomials $d(X, Y) := D(X, Y) - m_1$ and $e(X, Y) := E(X, Y) - m_2$ have no non-constant common factor. For let $A(X, Y) \in \mathbb{Q}[X, Y]$ be a polynomial dividing both $d(X, Y)$ and $e(X, Y)$. Then $A(X, Y)$ is also a common divisor of $\left\{m_1^{-1} D(X, Y)\right\}^n - 1$ and $\left\{m_2^{-1} E(X, Y)\right\}^m - 1$. Hence

$$A(X, Y) | B(X, Y) := \left(m_1^{-1} D(X, Y)\right)^n - \left(m_2^{-1} E(X, Y)\right)^m .$$

$B(X, Y)$ is non-zero, since $D(X, Y)$ and $E(X, Y)$ have no common factor. Further, B is homogeneous of degree mn. Hence $A(X, Y)$ must be homogeneous. But since $d(0,0) = -m_1 \neq 0, e(0,0) \neq 0$, this is possible only if A is a constant. This proves our claim.

We consider the solutions of

$$d(x, y) = 0 \quad \text{and} \quad e(x, y) = 0 \quad \text{in} \quad x, y \in \mathbb{Z} \tag{10}$$

and eliminate y from these equations. Consider $d(X, Y)$ and $e(X, Y)$ as polynomials in Y with coefficients in $\mathbb{Q}[X]$, and let $R(X)$ be the resultant of these polynomials in Y. Then $R(X)$ is non-zero and

$$R(X) = \begin{vmatrix} d_m & d_{m-1}X & \cdots & d_1 X^{m-1} & d_0 X^m - m_1 & & & \\ & d_m & d_{m-1}X & & d_1 X^{m-1} & d_0 X^m - m_1 & & \\ & & \ddots & \ddots & & \ddots & & \ddots \\ & & d_m & d_{m-1}X & & & d_1 X^{m-1} & d_0 X^m - m_1 \\ e_n & e_{n-1}X & \cdots & e_1 X^{n-1} & e_0 X^n - m_2 & & & \\ & \ddots & \ddots & & \ddots & & \ddots & \\ & & e_n & e_{n-1}X & & e_1 X^{n-1} & e_0 X^n - m_2 & \end{vmatrix}$$

$$\tag{11}$$

Each solution of (10) satisfies $R(x) = 0$. We shall determine an upper bound for the height $H(R)$ of $R(X)$. Then each solution (x, y) of (10) satisfies $|x| \leq H(R)$. Suppose that the coefficient of X^j in $R(X)$ has the largest absolute value. Then Cauchy's integral formula implies that

$$H(R) = \left| \frac{1}{2\pi i} \oint_{|z|=1} \frac{R(z)}{z^{j+1}} dz \right| \leq \sup_{|z|=1} |R(z)| .$$

Hence there is a $z_0 \in \mathbb{C}$ with $|z_0| = 1$ such that $H(R) \leq |R(z_0)|$. By substituting z_0 in (11) and using Hadamard's inequality we get

$$|R(z_0)| \leq \left(\sum_{i=1}^{m} |d_i|^2 + |d_0 z_0^m - m_1|^2 \right)^{n/2}$$

$$\times \left(\sum_{i=1}^{n} |e_i|^2 + |e_0 z_0^n - m_2|^2 \right)^{m/2} .$$

Using that $\left|d_0 z_0^m - m_1\right|^2 \leq (|d_0| + |m_1|)^2$ and $\left|e z_0^n - m_2\right|^2 \leq (|e_0| + |m_2|)^2$ we infer

$$H(R) \leq |R(z_0)|$$
$$\leq \left\{ \left(\sum_{i=0}^m |d_i|^2 \right)^{1/2} + |m_1| \right\}^n \left\{ \left(\sum_{i=0}^n |e_i|^2 \right)^{1/2} + |m_2| \right\}^m .$$

From (8) we deduce that

$$\left(\sum_{i=0}^m |d_i|^2 \right)^{1/2} \leq \left\{ \sum_{i=0}^m \binom{m}{i}^2 \right\}^{1/2} M(D) \leq 2^m M(D) ,$$
$$\left(\sum_{i=0}^n |e_i|^2 \right)^{1/2} \leq 2^n M(E) .$$

Together with $M(D) \leq M(F), M(E) \leq M(F), |m_1|, |m_2| \leq M, m + n = r$, this yields

$$H(R) \leq \left\{ 2^m M(F) + M \right\}^n \left\{ 2^n M(F) + M \right\}^m$$
$$\leq 4^{mn} \{ M(F) + M \}^{m+n} \leq 2^{r^2/2} \{ M(F) + M \}^r .$$

Hence $|x| \leq 2^{r^2/2}(M(F) + M)^r$. It follows similarly that $|y| \leq 2^{r^2/2}(M(F) + M)^r$. $\blacksquare$

In what follows we assume that $F(X,Y) \in \mathbb{Z}[X,Y]$ is a binary form of degree $r \geq 400$ with $F(1,0) \neq 0$ and discriminant $D(F)$ satisfying

$$|D(F)| \geq e^{80r(r-1)} M^{120(r-1)} . \tag{12}$$

Together with (6) this implies that

$$M(F) \geq r^{-r/(2r-2)} e^{40r} M^{60} . \tag{13}$$

The height of a pair $(x, y) \in \mathbb{Z}^2$ with $\gcd(x, y) = 1$ is defined by $H(x,y) = \max(|x|, |y|)$. We first deal with the large solutions of (1).

Large solutions

We shall use

Lemma 2. If $(x, y) \in \mathbb{Z}^2$ and $y > 0$, then

$$
\min_{\alpha} \min \left(1, \left| \alpha - \frac{x}{y} \right| \right) \leq \frac{\left\{ 2r^{1/2} M(F) \right\}^r |F(x, y)|}{|D(F)|^{1/2} H(x, y)^r} ,
$$

where the first minimum on the left is taken over the roots α of $F(X, 1)$.

This is exactly Lemma 1 of [3], apart from the factor $|D(F)|^{1/2}$ in the denominator. Bombieri and Schmidt obtained their Lemma 1 by slightly modifying the proof of a result of Lewis and Mahler ([11], Theorem 1), specialized to the case that no non-Archimedean valuations are involved. It is easy to check that in this special case, the arguments of Lewis and Mahler yield an extra factor $|D(F)|^{1/2}$. ∎

From 12 it follows that $|D(F)| \geq M^2$. Hence for every solution (x, y) of (1) there is, by Lemma 2, a root α_0 of $F(X, 1)$ such that

$$
\left| \alpha_0 - \frac{x}{y} \right| \leq \frac{C}{H(x, y)^r} \quad \text{with} \quad C = \left\{ 2r^{1/2} M(F) \right\}^r . \tag{14}
$$

By exactly the same argument as on p. 73 of [3], which remains valid for reducible forms F, we obtain

Lemma 3. (Strong gap principle). If $(x_1, y_1), (x_2, y_2), \ldots$ are solutions of (1) satisfying (14) for the same α_0 and

$$
C^{1/r} \leq H(x_1, y_1) \leq H(x_2, y_2) \leq \ldots ,
$$

then

$$
H(x_n, y_n) \geq \left\{ (2C)^{-\frac{1}{r-2}} H(x_1, y_1) \right\}^{(r-1)^{n-1}} \quad \text{for} \quad n = 1, 2, \ldots .
$$

This can be used to estimate the number of large solutions.

Lemma 4. (1) has at most $3r$ solutions for which

$$H(x,y) \geq \left(\frac{7}{2}\right)^{0.3r} M(F)^{1.92} \ .$$

Proof. We apply the Thue-Siegel principle as formulated in [3]. Let t, τ be positive integers such that

$$t < \sqrt{2/r} \ , \quad \sqrt{2 - rt^2} < \tau < t \ , \quad \lambda := \frac{2}{t - \tau} < r \ , \tag{15}$$

and put

$$A_1 := \frac{t^2}{2 - rt^2}\left(\log M(F) + \frac{r}{2}\right) \ , \quad Y_0 = (2C)^{\frac{1}{r-\lambda}}\left(4e^{A_1}\right)^{\lambda/(r-\lambda)} \ ,$$

$$\delta = \frac{rt^2 + \tau^2 - 2}{r - 1} \ .$$

Then

$$Y_0 = \left\{2 \cdot (2r^{1/2})^r\right\}^{1/(r-\lambda)} \cdot \left\{4e^{(r/2)[t^2/(2-rt^2)]}\right\}^{\lambda/(r-\lambda)}$$
$$\cdot M(F)^{\left\{\frac{r}{r-1}\right\}+\left\{\frac{r^2}{2-rt^2}\right\}\left\{\frac{\lambda}{r-\lambda}\right\}} \ . \tag{16}$$

First suppose that F is irreducible. Then the arguments on pp. 74/75 of [3] can be used without any modifications, and we infer that the number of solutions of (1) with $H(x,y) \geq Y_0$ is at most

$$\left(2 + \left[\frac{\log \delta^{-1}(\lambda - 2)^{-1}}{\log(r - 1)}\right]\right) r \ , \tag{17}$$

where $[\alpha]$ denotes the largest integer $\leq \alpha$. Now suppose that F is reducible. If $(x_1, y_1), (x_2, y_2), \ldots$ are the solutions of (1) with $H(x,y) \geq Y_0$ satisfying (14) for some fixed root α_0 of $F(X, 1)$, and if $H(x_1, y_1) \leq H(x_2, y_2) \leq \ldots$ then, by Lemmas 1 and 3,

$$2^{r^2/2}(M(F) + M)^r \geq H(x_n, y_n) \geq \left\{(2C)^{-1/(r-2)}H(x_1, y_1)\right\}^{(r-1)^{n-1}}$$
$$\geq \left\{(2C)^{-1/(r-2)}Y_0\right\}^{(r-1)^{n-1}} \ .$$

By (13), we have $2^{r^2/2}(M(F)+M)^r \le M(F)^{2r}$. Further, it is easy to see, in view of (14), that

$$(2C)^{-1/(r-2)}Y_0 \ge M(F)^{\{r/(r-\lambda)\}-\{r/(r-2)\}} = M(F)^{\frac{r(\lambda-2)}{(r-\lambda)(r-2)}} .$$

Hence

$$(r-1)^{n-1} \le \frac{2(r-\lambda)(r-2)}{\lambda-2} . \tag{18}$$

By assumption, $r \ge 400$. Thus we have

$$\delta^{-1} = \frac{r-1}{rt^2+\tau^2-2} \ge \frac{r-1}{\tau^2} > \frac{r-1}{t^2} > \frac{r(r-1)}{2} > 2(r-\lambda) .$$

Together with (18) this implies that

$$(r-1)^{n-2} \le \delta^{-1}(\lambda-2)^{-1} .$$

Since we have r possibilities for the root α_0 of $F(X,1)$, this implies that the set of solutions of (1) with $H(x,y) \ge Y_0$ is also bounded above by the number given in (17).

We now choose t and τ. Put $a = 0.55$ and choose

$$t = \sqrt{\frac{2}{r+a^2}} , \quad \tau = \left(1 - \sqrt{\frac{4}{5}\,(1-a)}\right)t .$$

Then

$$\lambda = \frac{1}{1-a}\sqrt{\frac{5}{2}\,(r+a^2)} .$$

Using $r \ge 400$, it can be verified by straightforward computation that

$$\lambda < r , \quad \delta^{-1}(\lambda-2)^{-1} < (r-1)^2 , \quad Y_0 \le \left(\frac{7}{2}\right)^{0.3r} M(F)^{1.92}$$

Together with (17) this proves Lemma 4. ∎

Small solutions

We change Bombieri's and Schmidt's notion of normalised form as follows. A binary form $F(X,Y) \in \mathbb{Z}(X,Y]$ is called *normalised* if $1 \le$

$|F(1,0)| \leq M$. If (1) has a solution (x_0, y_0) then there are $a, b \in \mathbb{Z}$ such that $ax_0 - by_0 = 1$, the form $G(X, Y) = F(x_0 X + bY, y_0 X + aY)$ is equivalent to F and $1 \leq |G(1,0)| = |F(x_0, y_0)| \leq M$. Hence every binary form for which (1) is solvable is equivalent to a normalised form. We call a binary form $F(X, Y) \in \mathbb{Z}[X, Y]$ *reduced* if F is normalised and if $M(G) \geq M(F)$ for every normalised binary form G that is equivalent to F. In what follows we assume that F is a reduced form of degree $r \geq 400$ satisfying $|D(F)| \geq e^{80r(r-1)} M^{120(r-1)}$. It suffices to prove the Proposition for such forms, by what was mentioned above.

We can express F as

$$F(X, Y) = a_0 L_1(X, Y) \ldots L_r(X, Y) \, ,$$

where a_0 is an integer with $1 \leq |a_0| \leq M$ and $L_i(X, Y) = X - \alpha_i Y$ for $i = 1, \ldots, r$. We shall denote pairs (x, y) by $\underline{x}$. For $\underline{x} = (x, y), \underline{x}_0 = (x_0, y_0)$ we put $D(\underline{x}, \underline{x}_0) = xy_0 - x_0 y$.

Lemma 5.   Suppose that $\underline{x}, \underline{x}_0$ are solutions of (1). Then for $1 \leq i, j \leq r$

$$\frac{L_i(\underline{x}_0)}{L_i(\underline{x})} - \frac{L_j(\underline{x}_0)}{L_j(\underline{x})} = (\beta_i - \beta_j) D(\underline{x}, \underline{x}_0) \, , \tag{19}$$

where $\beta_1, \ldots, \beta_r$ depend on $\underline{x}$, and have the property that the form $G(V, W) = F(\underline{x})(V - \beta_1 W) \ldots (V - \beta_r W)$ is equivalent to F.

Remark. Note that $G(V, W)$ is normalised.

Proof. Exactly the same as that of Lemma 3 of [3].   ■

By taking $\underline{x}_0 = (1, 0)$, formula (19) gives

$$\frac{1}{L_i(\underline{x})} - \frac{1}{L_j(\underline{x})} = (\beta_i - \beta_j) y \, . \tag{20}$$

For every solution $\underline{x}$ of (1) we have

$$|L_1(\underline{x})| \ldots |L_r \underline{x})| = \frac{|F(\underline{x})|}{|a_0|} \geq \frac{1}{M} \, .$$

Hence there is an integer $j = j(\underline{x})$ such that $\left|L_j(\underline{x})\right| \geq M^{-1/r}$. Together with (20) this implies that

$$\frac{1}{\left|L_i(\underline{x})\right|} \geq \left|\beta_j - \beta_i\right|\left|y\right| - M^{1/r} .$$

We also have $\left|\overline{L_j(\underline{x})}\right| \geq M^{-1/r}$. Hence $\left|L_i(\underline{x})\right|^{-1} \geq \left|\bar{\beta}_j - \beta_i\right| \cdot \left|y\right| - M^{1/r}$. Therefore,

$$\frac{1}{\left|L_i(\underline{x})\right|} \geq \left|\mathrm{Re}(\beta_j) - \beta_i\right| \cdot \left|y\right| - M^{1/r} .$$

Choose an integer $m = m(\underline{x})$ such that $\left|m - \mathrm{Re}(\beta_j)\right| \leq \frac{1}{2}$. Then

$$\frac{1}{\left|L_i(\underline{x})\right|} \geq \left(\left|m - \beta_i\right| - \frac{1}{2}\right)\left|y\right| - M^{1/r} \quad \text{for} \quad i = 1, \ldots, r . \qquad (21)$$

Put $Y_0 = (7/2)^{0.3r} M(F)^{1.92}$. For $1 \leq i \leq r$, let $\mathcal{X}_i$ be the set of solutions of (1) with $1 \leq y \leq Y_0$ and $\left|L_i(\underline{x})\right| \leq 1/(2y)$. Then we have

Lemma 6. Suppose that $\underline{x} = (x, y), \underline{\tilde{x}} = (\tilde{x}, \tilde{y})$ are pairs in $\mathcal{X}_i$ with $y \leq \tilde{y}$. Then

$$\tilde{y}/y \geq \frac{1}{\frac{5}{2} + M^{1/r}} \cdot \max\left(1, \left|\beta_i - m\right|\right) ,$$

where $\beta_i = \beta_i(\underline{x}), m = m(\underline{x})$.

Proof. By following the proof of Lemma 4 of [3] one obtains

$$\tilde{y}/y \geq \max\left(1, \frac{1}{2}\left|m - \beta_i\right| - \frac{1}{4} - \frac{1}{2}M^{1/r}\right) .$$

Now Lemma 6 follows by using that for each pair of reals $\xi \geq 0$, $a \geq 0$, $\max\left(1, \frac{1}{2}\xi - a\right) \geq \{1/(2a + 2)\}\max(1, \xi)$ holds. $\blacksquare$

Lemma 7. Let $\underline{x} = (x, y)$ be a solution of (1) with $1 \leq y \leq Y_0$, $\left|L_i(\underline{x})\right| > 1/2y$. Then

$$\left|m - \beta_i\right| \leq \frac{5}{2} + M^{1/r} ,$$

where again $\beta_i = \beta_i(\underline{x})$ and $m = m(\underline{x})$.

Proof. Exactly the same as that of Lemma 5 of [3].   ■

We obtain the Proposition by combining Lemma 4 with Lemma 8 below.

Lemma 8. (1) has at most $3r$ solutions (x, y) with $H(x, y) \leq Y_0$.

Proof. For each set $\mathcal{X}_i$ defined above that is non-empty, choose a solution $\underline{x}(i)$ with maximal value of y. Let $\mathcal{X}_0$ be the set of these solutions $\underline{x}(i)$, and let $\mathcal{X}$ be the set of solutions of (1) with $1 \leq y \leq Y_0$, with the solutions from $\mathcal{X}_0$ excluded. By following the arguments of [3], p. 77 one obtains

$$\prod_{\underline{x} \in \mathcal{X}} \left(\frac{1}{\frac{5}{2} + M^{1/r}} \cdot \max\left(1, |\beta_i(\underline{x}) - m(\underline{x})|\right) \right) \leq Y_0 \quad \text{for} \quad i = 1, \ldots, r \, .$$

$$(22)$$

The form $G(V, W) = F(\underline{x}) \prod_{i=1}^r \left(V - \beta_i W\right)$ obtained in Lemma 5 is equivalent to F, hence so is the form $\widehat{G}(V, W) = F(x) \prod_{i=1}^r \left(V - (\beta_i - m)W\right)$. Further, this form $\widehat{G}$ is normalised. Since F was reduced we have $M(\widehat{G}) \geq M(F)$. Hence

$$\prod_{i=1}^r \max\left(1, |\beta_i(x) - m(x)|\right) = \frac{M(\widehat{G})}{|F(\underline{x})|} \geq \frac{M(F)}{M} \, .$$

By taking the product of the terms in (22) for $i = 1, \ldots, r$ and substituting this inequality we obtain

$$\left(\frac{1}{\left(\frac{5}{2} + M^{1/r}\right)^r} \frac{M(F)}{F} \right)^{|\mathcal{X}|} \leq Y_0^r \, .$$

Hence

$$|\mathcal{X}| \leq \frac{r \log Y_0}{\log M(F) - \log M - r \log\left(\frac{5}{2} + M^{1/r}\right)}$$

$$\leq r \cdot \frac{1.92 \log M(F) + 0.3r \log\left(\frac{7}{2}\right)}{\log M(F) - 2 \log M - r \log\left(\frac{7}{2}\right)} \, .$$

Putting $V := \log\{(\frac{7}{2})^r M^2\}/\log M(F)$ this gives

$$|\mathcal{X}| \le r \cdot \frac{1.92 + 0.3V}{1 - V} \,. \tag{23}$$

From (13) and $r \ge 400$ it follows that $V \le 1/30$. By substituting this into (23) and using $r \ge 400$, we get $|\mathcal{X}| \le 2r - 1$. The set $\mathcal{X}$ contains all solutions of (1) with $H(x, y) \le Y_0$, except $(1, 0)$ and those belonging to $\mathcal{X}_0$. The set $\mathcal{X}_0$ contains one solution from each non-empty set $\mathcal{X}_i$, and there are at most r such non-empty sets. Hence (1) has at most $2r - 1 + r + 1 = 3r$ solutions with $H(x, y) \le Y_0$. This proves Lemma 8. ∎

3. Proof of Theorem 1

In the proof of Theorem 1 we shall need two lemmas. Two binary forms $F, G \in \mathbb{Z}[X, Y]$ are called *strongly equivalent* if $G(X, Y) = F(X + bY, Y)$ for some $b \in \mathbb{Z}$. Obviously, if F, G are strongly equivalent then $D(F) = D(G)$ and $F(1, 0) = G(1, 0)$.

Lemma 9. Let $F(X, Y) \in \mathbb{Z}[X, Y]$ be a binary form of degree $r \ge 3$ with $F(1, 0) \ne 0$ and discriminant $D \ne 0$. Then F is strongly equivalent to a binary form G for which

$$H(G) < \exp\{c_6(r)(|F(1, 0)|^{(r-1)(r-2)}|D|)^{5(r-1)(r-2)}\}\,,$$

where $c_6(r)$ is an effectively computable number depending only on r.

Proof. This is a special case of Corollary 1.5 of Győry [8]. It was proved by means of Baker's method on linear forms in logarithms. We mention that Corollary 1.5 was formulated in [8] in terms of polynomials $f(X) = F(X, 1)$ instead of binary forms $F(X, Y)$. ∎

A *primitive* binary form is a binary form $F \in \mathbb{Z}[X, Y]$ such that $F(1, 0) > 0$ and the coefficients of F are relatively prime. Every primitive binary form F has a unique factorization

$$F = P_1^{k_1} \dots P_u^{k_u} \,.$$

where $P_1, \ldots, P_u$ are distinct, primitive irreducible binary forms and $k_1, \ldots, k_u$ are positive integers. The *square-free* part of F is defined by

$$F' = P_1 \ldots P_u \, ,$$

by Gauss' lemma, F' is primitive, and is uniquely determined by F.

Lemma 10. Let (X, Y) be a primitive binary form of degree r with square-free part F'. Then

$$H(F) \leq (3r)^r H(F')^r \, .$$

Proof. We use Mahler measures. Let $F = P_1^{k_1} \ldots P_u^{k_u}$ be the unique factorization of F into irreducible primitive binary forms. Then, by (7),

$$M(F) = M(P_1)^{k_1} \ldots M(P_u)^{k_u} \leq M(P_1 \ldots P_u)^{\max(k_1, \ldots, k_u)} \leq M(F')^r \, .$$

By (8), we have $H(F) \leq 2^r M(F)$. Further, by a result of Mahler [14], we have $M(F') \leq \sqrt{\deg F' + 1} \cdot H(F') \leq \sqrt{r + 1} \cdot H(F')$. Hence

$$H(F) \leq 2^r M(F')r \leq 2^r (r + 1)^{r/2} H(F')^r \leq (3r)^r H(F')^r \, . \qquad \blacksquare$$

Proof of Theorem 1. Let $F(X, Y)$ be a binary form with $\deg F = r$ and $\omega(F) = n$, such that (1) has more than $N(n)$ solutions. As we observed in Sec. 3 (subsection on small solutions), F is equivalent to a form G for which $0 < |G(1, 0)| \leq M$. Hence it suffices to prove the part of Theorem 1 in which $0 < |F(1, 0)| \leq M$ and we assume from now on that F satisfies this condition. Further, we assume that F is primitive, which is obviously no restriction either. Let F' be the square-free part of F. Then F' has degree n, non-zero discriminant, $0 < |F'(1, 0)| \leq M$ holds and the inequality $0 < |F'(x, y)| \leq M$ in $x, y \in \mathbb{Z}$ with gcd $(x, y) = 1$ and $y > 0$ or $(x, y) = (1, 0)$ has more than $N(n)$ solutions. Now Theorem 3 implies that the discriminant D of F' satisfies

$$|D| \leq e^{80n(n-1)} M^{\delta(n)} \, .$$

By Lemma 9, F' is strongly equivalent to a binary form G' for which

$$H(G') \leq \exp\{c_6(n)(|F'(1,0)|^{(n-1)(n-2)}|D|)^{5(n-1)(n-2)}\}$$
$$\leq \exp\{c_7(n)(M^{(n-1)(n-2)+\delta(n)})^{5(n-1)(n-2)}\}$$
$$\leq \exp\{c_7(n)M^{10(n-1)^4}\}\,,$$

where $c_7(n)$ is an effectively computable number, depending only on n. Let $U = \begin{pmatrix} 1 & b \\ 0 & 1 \end{pmatrix}$ $(b \in \mathbb{Z})$ be the matrix such that $G' = F'_U$. Then G' is the square-free part of F_U. Now Lemma 10 yields

$$H(F_U) \leq (3r)^r H(G')^r \leq \exp\{c_1(r)M^{10(n-1)^4}\}\,.$$

This proves Theorem 1. ∎

4. Proof of Theorem 2

We need the following lemma.

Lemma 11. Let $F(X,Y) \in \mathbb{Z}[X,Y]$ be a binary form of degree $r \geq 3$ with non-zero discriminant D and with $F(1,0) \neq 0$. Further, let $M \geq 1$. Then all solutions of the inequality

$$0 < |F(x,y)| \leq M \quad \text{in} \quad x,y \in \mathbb{Z} \tag{24}$$

satisfy

$$\max(|x|,|y|) < \exp\{c_8(r)|D|(\log|2D|)^{2r}(1 + \log(H(F)M))\}\,, \tag{25}$$

where $c_8(r)$ is an effectively computable positive number depending only on r.

We note that the proof of this lemma will be based on an effective result of Győry and Papp [9] on Thue equations which was proved by means of Baker's method concerning linear forms in logarithms.

Proof. First suppose that F is reducible, and that (x,y) is a solution of (24) with $\gcd(x,y) = d$. Then $x = dx_1, y = dy_1$, where (x_1,y_1) is a solution of (1) and $d^r \leq M$, and Lemma 11 follows from Lemma 1 and Mahler's inequality $M(F) \leq \sqrt{r+1} \cdot H(F)$. Now assume that F is irreducible, and let α be one of the roots of $F(X, 1)$ in C. Put $K = \mathbf{Q}(\alpha)$, and denote by D_K the discriminant of K. Then it follows from the Corollary of Győry and Papp [9] that all solutions of (24) satisfy

$$\max(|x|, |y|) < \exp\{c_9(r)|D_K|(\log|2D_K|)^{2r}(1 + \log(H(F) \cdot M))\} , \quad (26)$$

where $c_9(r)$ is an effectively computable number depending only on r. Further, it is well-known (cf. [10]) that D_K divides D. This together with (26) implies (25).   ∎

Proof of Theorem 2. Let $F(X, Y) \in \mathbf{Z}[X, Y]$ be a binary form with $\omega(F) = n \geq 3$ such that (1) has more than $N(n)$ solutions, one of which is (1, 0). In the proof of Theorem 2 we may assume that F is primitive. Further, if F' is the square-free part of F, then each solution (x, y) of (1) satisfies $0 < |F'(x, y)| \leq M$. Hence we may assume that $F = F'$ has non-zero discriminant and that $\deg F = n$. By Theorem 1, there is a matrix $U = \begin{pmatrix} 1 & b \\ 0 & 1 \end{pmatrix}$ $(b \in \mathbf{Z})$ such that $H(F_U) \leq C_1 := \exp\{c_1(n)M^{10(n-1)^4}\}$. The matrix U does not affect y, hence it suffices to prove Theorem 2 for F_U instead of F. To summarize, we may assume that $\deg F = n, F$ has non-zero discriminant D, and $H(F) \leq C_1$. By Theorem 3 we have $|D| \leq C_2 := e^{80n(n-1)}M^{\delta(n)}$. By substituting C_1 for $H(F)$ and C_2 for $|D|$ in (25) and using that $n \geq 3$, we obtain from Lemma 11 that

$$|y| < \exp\{c_{10}(n)M^{10n^4}\}$$

with some effectively computable number $c_{10}(n)$ which depends only on n.   ∎

Acknowledgement

The results in this paper were obtained during the first author's stay at the University of Debrecen, in the fall of 1988. The first author wants to thank the University of Debrecen for its hospitality.

References

1. E. Bombieri, *On the Thue-Siegel-Dyson theorem*, Acta Math. **148** (1982) 255–296.

2. E. Bombieri, *On the Thue-Mahler equation*, in "Diophantine Approximation and Transcendence Theory", Seminar, Bonn, 1985 (G. Wüstholz, ed.), Lecture Notes in Math. 1290, Springer Verlag (1987) 213–243.

3. E. Bombieri and W. M. Schmidt, *On Thue's equation*, Invent. Math. **88** (1987) 69–81.

4. B. Brindza, A. J. van der Poorten and M. Waldschmidt, in preparation.

5. B. N. Delone and D. K. Faddeev, *The theory of irrationalities of the third degree*, Amer. Math. Soc. Transl. of Math. Monographs 10, Providence, USA (1964).

6. J. H. Evertse, *Upper bounds for the numbers of solutions of diophantine equations*, MC-tract 168, Centre of Mathematics and Computer Science, Amsterdam, 1983.

7. J. H. Evertse and K. Győry, *Thue-Mahler equations with a small number of solutions*, J. Reine Angew. Math. **399** (1989) 60–80.

8. K. Győry , *Sur les polynômes à coefficients entiers et de discriminant donné III*, Publ. Math. Debrecen **23** (1976) 141–165.

9. K. Győry and Z. Z. Papp, *Norm form equations and explicit lower bounds for linear forms with algebraic coefficients*, Studies in Pure Mathematics to the Memory of Paul Turán, Birkhäuser, Basel (1983) pp. 245–257.

10. D. Hilbert, *Über diophantische Gleichungen*, Göttinger Nachr. (1987) 48–54.

11. D. J. Lewis and K. Mahler, *Representation of integers by binary forms*, Acta Arith. **6** (1961) 333–363.

12. K. Mahler, *Zur Approximation algebraischer Zahlen II. Über die Anzahl der Darstellungen ganzer Zahlen durch Binärformen*, Math. Ann.**108** (1933) 37–55.

13. K. Mahler, *Zur Approximation algebraischer Zahlen III. Über die mittlere Anzahl der Darstellungen grosser Zahlen durch binäre Formen*, Acta Math. **62** (1933) 91–166.

14. K. Mahler, *On some inequalities for polynomials in several variables*, J. London Math. Soc. **37** (1962) 341–344.

15. K. Mahler, *An inequality for the discriminant of a polynomial*, Michigan Math. J. **11** (1964) 257–262.

16. J. Mueller and W. M. Schmidt, *Thue's equation and a conjecture of Siegel*, Acta Math. **160** (1988) 207–247.

17. C. L. Siegel, *Über einige Anwendungen diophantischer Approximationen*, Abh. Preuss. Akad. Wiss. Phys.-Math. Kl., No. 1 (1929).

18. C. L. Siegel, *Die Gleichung $ax^n - by^n = c$*, Math. Ann.114 (1937) 57–68.
19. A. Thue, *Über Annäherungswerte algebraischer Zahlen*, J. Reine Angew. Math. **135** (1909) 284–305.

J. H. Evertse
University of Leiden
Department of Mathematics and
Computer Science
P O Box 9512
2300 RA Leiden
The Netherlands

K. Győry
Kossuth Lajos University
Mathematical Institute
4010 Debrecen
Hungary

THE MATH. HERITAGE OF C.F. GAUSS (pp. 225-266)
edited by George M. Rassias
©1991 World Scientific Publ. Co. Singapore

A LOAD BALANCED ALGORITHM FOR THE CALCULATION OF THE POLYNOMIAL KNOT AND LINK INVARIANTS

Bruce Ewing and Kenneth C. Millett*

In this paper we develop effective computational algorithms which allow the determination of the oriented and semi-oriented polynomial invariants associated to classical knots and links. This is especially important since these calculations are known to be NP-hard. A brief introduction to the fundamental concepts of classical knot theory and, especially, to issues concerning the representation of knots and links (the "Gauss" code and alternatives) and the recursive approach to the polynomial invariants which are associated to them is provided. The basic structures of several recursive algorithms will be described and discussed. We shall give a precise description of the load balanced algorithm, its implementation and, an evaluation of the theoretical and observed growth in complexity associated to the algorithms and their implementations.

Introduction

In this paper we describe the fundamental combinatorial nature of the knot and link invariants and the development of effective computational algorithms which allow their determination for a significant range of cases. A brief introduction to the fundamental concepts of classical knot theory and, especially, to issues concerning the representation of knots and links (the "Gauss" code and alternatives) and to the recursive approach to the polynomial invariants which are associated to these knots and links, is provided in the first section. The basic structures of several recursive algorithms will be described and discussed in the second section of the paper. In the third

---

*Partially supported by grants from UCSB Office of Research Development Administration and Academic Senate

section of the paper we shall give a precise description of the load balanced algorithm and its implementation. The fourth section is devoted to an evaluation of the theoretical and observed growth in complexity associated to the algorithms and their implementation. This evaluation is of fundamental importance since it is known that the calculation of these polynomial invariants in NP-hard. [Ja1, 3] and [Th2]. The fifth, and final section, gives a summary of our conclusions and some remarks on remaining questions. The computer codes used in the implementations of the load balanced algorithm for the calculation of the oriented and semi-oriented polynomials are given in appendices A and B, respectively. The appendices are available, separately, from the authors.

The family of invariants discussed in this paper are generalizations of finite Laurent polynomials associated to oriented knots and links in Euclidean space. The first was developed by J. W. Alexander [Al] in about 1926, while the second was discovered in the spring of 1984 by V. F. R. Jones [Jon1]. The first generalization, the oriented polynomial, was discovered almost immediately following the announcement of Jones' discovery by four sets of authors [FYHLMO]. Subsequently, another distinctly different generalization was discovered by Brandt, Lickorish, and Millett [BLM] and Ho [Ho]. This, in turn, was immediately extended by Kauffman [Kau1] to give the definition of the semi-oriented polynomial. The effective calculation of the oriented and semi-oriented polynomials is the goal of this work.

One of the vehicles to define these invariants is a recursive formula relating the polynomial associated to one knot or link to those associated to potentially simpler knots. An alternative approach is to alter the presentation of the knot or link so that it is in the "braid" position. One can then exploit algebraic methods to complete the calculation. There are, however, difficulties with each of these methods. The recursive approach has exponential growth in the number of cases that have to be calculated. In the special case of "braid" presentations one may exploit the algebraic structure of the presentation to gain calculational simplicity. In the general case such an approach is limited by the intrinsic complexity of translating a general knot or link into a braid presentation and the limit on the braid index of such presentations required in current algorithms, cf [MoS1].

The recursive approach to the calculation of the polynomials, especially in the topological case, is the focus of an ongoing search for effective algorithms. In addition significant effort is devoted to the search for alternative mathematical theories which would provide a comprehensive conceptual approach to their study. Despite the recent theoretical advances via Yang-Baxter statistical mechanical states models derived from quantum mechanics, [Jon3], or the functional integral formulation recently described by Ed Witten [W1,2], these recursive methods remain the most effective calculational method for general presentations of knots and links.

In this paper we describe and compare the theoretical "standard ascender", its implementation (one implementation by Hoste/Jenkins, commonly known as "Knotter"), the implementation of a modification of the standard ascender algorithm which includes some reduction in complexity of representation, and the implementation of the load balancing algorithm which involves additional reduction in complexity. The load balancing algorithm and its underlying philosophy will be presented in greater detail. The usefulness of this approach is demonstrated through the presentation of supporting data from the implementations of these algorithms and their application to large classes of knots and links as well as a collection of more challenging examples.

Other examples of algorithms enjoying the same general properties arise in the study of the various chromatic polynomials in graph theory, their applications to scheduling problems, and their generalizations arising out of the study of the knot and link polynomials, [Bi], [Bo], [Ja1,2], [Kau3,4], [Th2–6], [Tra], (including the generalizations by Tutte and Whitney, [Tut1–4]), matching polynomials, and network reliability problems, [Co1,2], [Po], [PoB], [SaC]. Similar considerations for implementation of recursion algorithms apply in these contexts as well.

The development of the algorithms discussed in this paper began in collaboration with Robert Brandt, who, with W. B. R. Lickorish and the second author, was a codiscoverer of one of the polynomials whose computation is the goal of this effort, continued in collaboration with Peter Pacheco, who wrote a Pascal implementation of an early algorithm utilizing the Reidemeister moves and circuit reductions, and reached the present development of the load balanced algorithm in the collaboration of the authors, who developed the load balancing approach and the associated study of "triple", "gamma", and "skinny circuit" configurations in the reduction of knot or link presentations.

Finally, we wish to acknowledge our great gratitude to Morwen Thistlethwaite for the use of his table enumerating all knots through 13 crossings. This provided an important resource in the evaluation of the effectiveness of our algorithms.

1. A Brief Introduction to Combinatorial Knot Theory

The purpose of this section is to provide a brief survey of the fundamental aspects of the study of knots and links in 3 space, i.e., the spacial analysis of disjoint simple closed curves, whereby one allows these "strings" to move about from one position to another so long as no portion passes through another. Knots or links which can be moved, one to the other, are said to be equivalent or "to be the same knot or link". Fuller accounts of the theory of knots and links than we shall attempt to provide here include recent books of Burde and Zieschang, [BZ], and Kauffman [Kau5] as well as those of Crowell and Fox, [CrF], and Rolfsen, [Rol], and the expository articles of Lickorish [L], Lickorish and Millett [LM2], Fox, [F], or Thistlethwaite [Th1].

One fundamental goal of this theory is to develop mathematical methods which allow one to distinguish between different knots or links, as opposed to different presentations of the same knot or link. Thus one is lead to try to associate numbers or equations to presentations of knots or links which are to be unchanged when a spatial movement occurs taking one configuration to another equivalent one. A traditional way in which to begin is to consider regular projections of rather nice models of the knots or links onto a fixed plane and to study the allowable changes of the presentation which are sufficient to be able to change from one presentation of a knot or link to another presentation of the same knot or link. The projections are broken so as to indicate places where one strand crosses over another strand, examples are shown in Fig. 1.1, and non-generic behavior, such as shown in Fig. 1.2, is not permitted.

Although C. F. Gauss, [G], left little written record of his interest in knots he did describe a notation for the representation of a regular projection of an alternating knot. This was rediscovered, in a different form, by Tait and was embellished by Dowker and Thistlethwaite in their enumeration efforts. The Gauss method is simply to select a point on the image of the regular projection of the knot and to proceed along the strand in some chosen direction labelling the crossing points in order, A, B, C until all are uniquely labelled without repetitions and recording the sequence in which the crossings are encountered until one returns to the starting point.

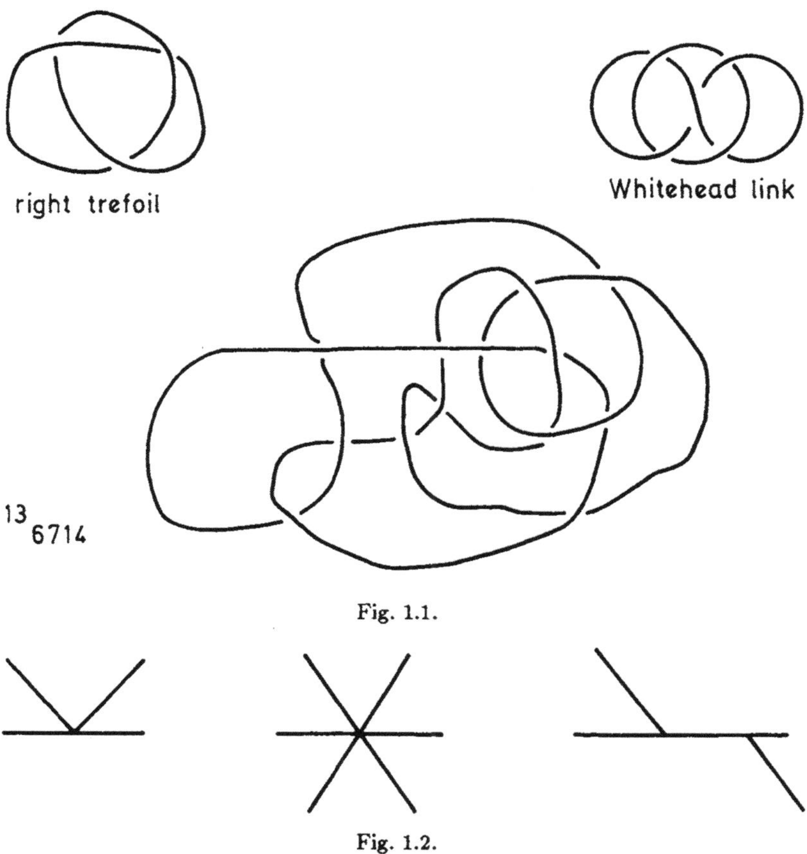

Fig. 1.1.

Fig. 1.2.

An example is shown in Fig. 1.3. This knot, the figure 8 knot, would have
the Gauss code ABCDBADC. Gauss posed the problem of determining necessary and sufficient conditions such that a sequence of n letters in which
each letter occurs exactly twice represent a regular projection of a knot.
An efficient algorithm for doing this was developed by Dowker and Thisthlethwaite and plays an important role in their enumeration efforts, [DTh].
Other algorithms have been described by Dehn [D], a topologist, and by
graph theorists, [RR].

Gauss' student, Listing, went further in the development of a notation
for alternating regular projections creating a rather different notation from
that suggested by Gauss. Listing used an alternating coloring of the complementary regions of the knot diagram, say black and white, and listed the
numbers of regions of each color bounded by each number of arcs. Although

                          B. Ewing & K. C. Millett

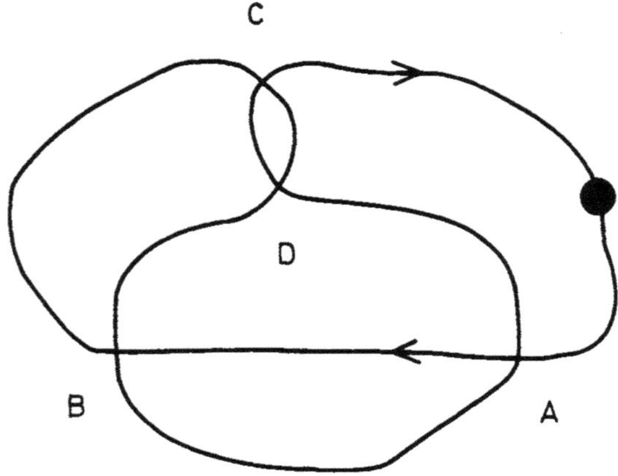

Fig. 1.3. The Gauss code.

there are fundamental problems with this approach it did have significant impact on the efforts of later knot enumerators.

In the Dowker-Thistlethwaite approach one labels crossing points, there are $2n$ of them (two for each crossing), starting from any over crossing and proceeding in a chosen direction. There is an involution of $\{1, 2, 2n\}$, denoted by a, taking "i" to $a(i)$, the "other" crossing number. One sees easily that a reverses parity so that the sequence of even numbers, $a(1), a(3), \ldots$ $a(2n-1)$ determines the entire sequence and the involution. The Dowker-Thistlethwiate standard representation of a given knot is the sequence associated to presentation for which one has the minimum in the lexicographic ordering. Thus, the Dowker-Thistlethwaite code for the figure 8 knot, shown in Fig. 1.4, is 4682. This is completely satisfactory for alternating knots but if one has a non-alternating projection, an additional element of information is required to indicate that the crossing or crossings in question differ from that of the alternating knot, e.g. one may append a "−" to the even integer to indicate this fact.

Although the Dowker-Thistlethwaite notation is rather effective for the enumeration problem it does not appear to be well adapted to the problem of the calculation of the oriented and semi-oriented knot and link invariants that we shall describe below. As a consequence we have developed another notation for the presentation designed for this purpose. Recall that all components are oriented and all crossings are generic, i.e., there are no

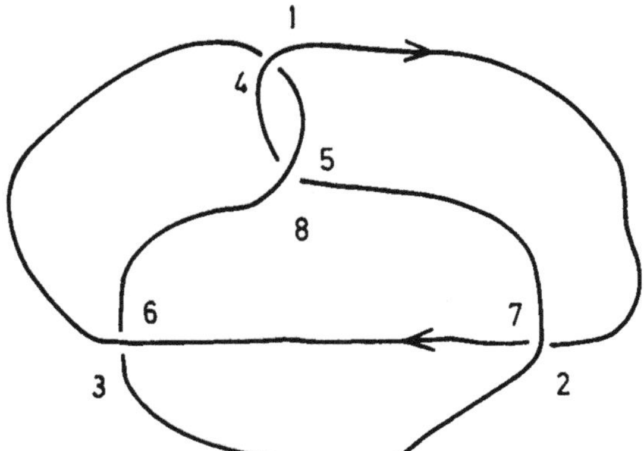

Fig. 1.4. The Dowker-Thistlethwaite code.

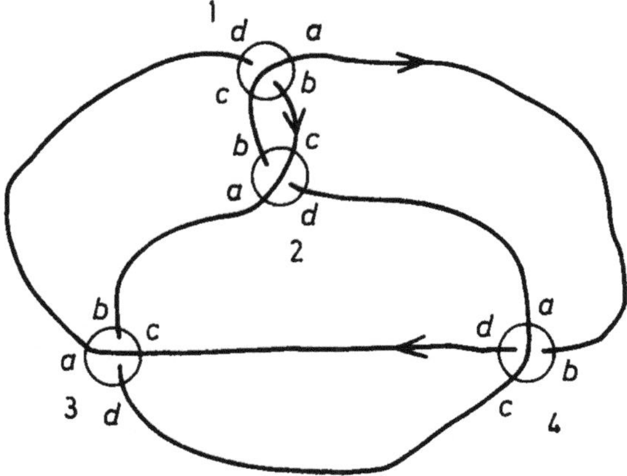

Fig. 1.5.

degenerate crossings as shown in Fig. 1.2, occurring in the given presentation. An example of our coding scheme is shown in Fig. 1.5. Here one sees that each of the crossings is assigned a specific number and at each of the crossings letters are assigned to the directions occurring at the crossing according to the rule: the outward pointing direction on the over crossing segment is assigned the letter "*a*" and the remaining directions are assigned the letters "*b*", "*c*", and "*d*", proceeding in a clockwise direction from "*a*". Furthermore, each crossing is assigned a sign, "+", or "−", following the convention shown in Fig. 1.10. The data associated to a given presentation

                    B. Ewing & K. C. Millett

consists of the ordered list, for each crossing, of the sign of the crossing followed by the connection points of the "a", "b", "c", and "d" directions of the crossing. Thus, data for the example shown in Fig. 1.5, are given in Table 1.6.

$$- \, 4b2c2b3a$$
$$- \, 3b1c1b4a$$
$$+ \, 1d2a4d4c$$
$$+ \, 2d1a3d3c$$

Table 1.6. Figure 8 knot data.

Other types of data structures have been employed to describe various aspects of the knot and link enumeration problem. Notable among these is the method of Conway, [Con]. Although it is rather more compact than ours, ours enjoys the benefit over the Dowker-Thistlethwaite and Conway presentations of supporting very rapid recognition of the Reidemeister moves of types I and II, supporting rapid deletion of crossing, and supporting the rapid evaluation of the potential benefits of the various possible crossing changes without becoming too large during the course of a typical calculation so that copying of tables, etc, becomes too time consuming. These moves allow one to simplify the presentations of arbitrary knots and links as described below. This ease of recognition explains the use of Reidemeister moves in our modified version of the standard ascender algorithm.

Planar motions or modifications of the pictures which do not change the crossing and connection relationships of the strands will take a presentation of a knot or link to a presentation of an equivalent knot or link because these can be understood as shadows of permissible spatial movements. It can be shown that sequences of only three additional elementary local alterations, called Reidemeister moves of types I, II and III, of the presentations are sufficient to determine the equivalence of any two presentations of the same knot or link. These Reidemeister moves are shown in Fig. 1.7.

A knot is unknotted if it is equivalent to the standard round circle in the plane. An unknotted knot is often called an unknot. The unlink is one equivalent to the disjoint union of a family of round circles in the plane. By choosing a direction on each strand, or component, of a knot or link, indicated by an arrow on each component, one defines an oriented knot or

link. These oriented links are the natural domain of the invariants that we
shall want to calculate. A given presentation can be changed to one of a
possibly distinct knot by simply reversing all the crossings in the picture,
i.e., reflecting in a mirror determined by the plane of the presentation.
This knot or link is called the mirror reflection of the first. Two knots
can be "summed" together by the processes illustrated in Fig. 1.8. With
respect to this summing operation, a knot is said to be prime if it cannot
be represented as the sum of two knots, neither of which is unknotted.

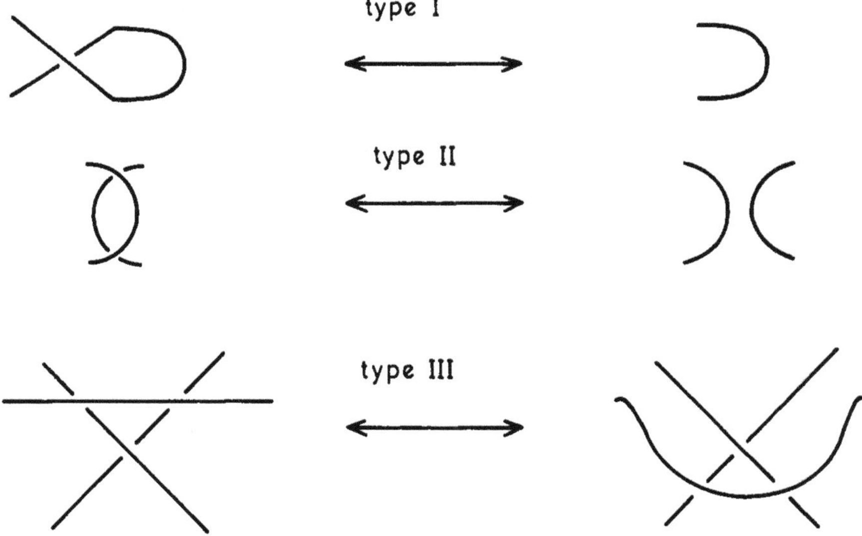

Fig. 1.7. Reidemeister Moves.

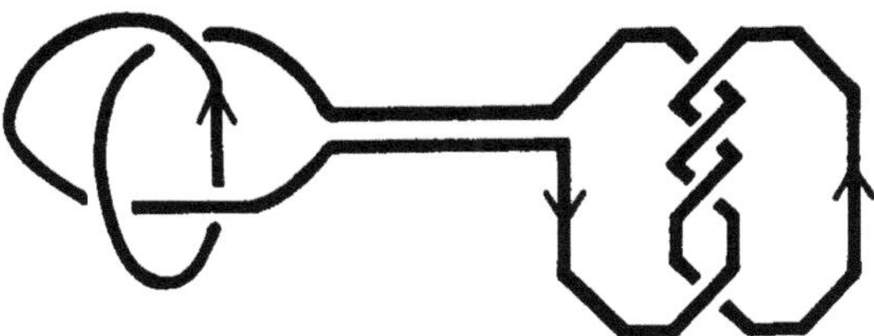

Fig. 1.8. Connected sum construction.

Knots are usually tabulated by listing the prime knots and by listing
once those which only differ by string orientation or by mirror reflection.

234   *B. Ewing & K. C. Millett*

Thus, a single entry in a knot may stand for as many as four distinct knots and even more, if one is considering the enumeration of links. Tabulation of the prime knots of few crossings goes back to the 19th century. With the advent of computers, C. H. Dowker and M. B. Thistlethwaite [DTh] and, subsequently, Thisthlethwaite have extended the tabulations to include all 12 and 13 crossing prime knots. From the results of H. Doll and J. Hoste, we also include the number of prime links through 9 crossings in which we do not list those having distinct string orientations. The number of prime knots, up to string orientation and mirror reflection, is given in Table 1.9. The numbers in parentheses represent the tabulation where one differentiates between distinct string orientations and mirror images. In addition, we estimate the number of distinct prime knots above 13 crossings as follows: (14) 63959, (15) 179333, (17) 3236500. A comparison of these figures demonstrates the desirability of an effective tool to determine the identity of an arbitrary knot presentation.

♯crossings	1	2	3	4	5	6	7	8	9	10	11	12	13	14
♯knots*	0	0	1(2)	1	2(4)	3(5)	7(14)	21(38)	49(102)	165(3795)	552*	2176*	9988*	63960*est
♯links**	0	1	0	1(2)	1	6(10)	9(14)	29(63)	82(103+?)					
♯total	0	1	1(2)	2(3)	3(5)	9(15)	16(28)	50(101)	131(205+?)					

* from 11 crossings, knots are not distinguished by virtue of orientation or mirror reflection.
** Links are distinguished by virtue of orientation but not mirror reflection.
? Non-alternating links are not included in this number

Table 1.9. Oriented prime knots and links.

As mentioned in the introduction, these polynomials can be calculated by means of a simple recursive process. Indeed, this process can be employed to define the polynomials. The fundamental fact that makes this works is that one may change a knot or link to the unknot or unlink by simply reversing some of the crossings in a given presentation. The polynomials to be calculated are finite Laurent polynomials with two variables having integer coefficients. Laurent polynomials allow both negative and positive powers of the variables. The following theorem describes the fundamental properties of the oriented polynomial discovered almost simultaneously by four groups of researchers following the first announcement of the Jones polynomial which it generalizes.

Theorem. The *oriented polynomial.* $P_L(l,m)$, of an oriented knot or link, L, is the unique Laurent polynomial in the variables, "l" and "m", which satisfies the following fundamental formulae:

(i) if U denotes the standard unknotted circle in the plane, then $P_U(l, m) = 1$, and

(ii) if L_+, L_-, and L_0 are planar pictures of oriented links in each of which we have identified a small circular region of the picture containing either a single crossing or, in the last case, no crossing at all, and such that outside these small circular regions shown in Fig. 1.10, the planar pictures are *exactly* the same, then the oriented polynomial satisfies the formula

$$lP_{L+}(l, m) + l^{-1}P_{L-}(l, m) + mP_{L_0}(l, m) = 0.$$

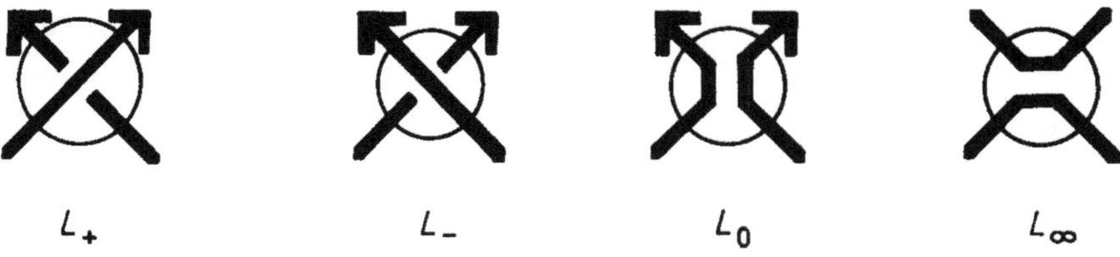

$$L_+ \qquad L_- \qquad L_0 \qquad L_\infty$$

Fig. 1.10.

Subsequently, another polynomial was discovered by Brandt, Lickorish, and Millett, and Ho which required no orientations of any sort (hence the name 'absolute polynomial') but involved the consideration of the other possible way of removing the crossing, denoted by L_∞ in Fig. 1.10.

Theorem. The absolute polynomial, $Q_L(x)$, is the unique Laurent polynomial in the variable "x", which satisfies the following fundamental formulae:

(i) if U denotes the standard unknotted circle in the plane, then $Q_U(x) = 1$, and

(ii) if L_+, L_-, L_0, and L_∞ are planar pictures of oriented links in each of which we have identified a small circular region of the picture in Fig. 6 containing either a single crossing or, in the last cases, no crossing at all, according to the convention shown in Fig. 1.10 (with orientations ignored), and such that outside these small circular regions the planar pictures are *exactly* the same, then the absolute polynomial satisfies the formula

$$Q_{L+}(x) + Q_{L-}(x) = x[Q_{L_0}(x) + Q_{L\infty}(x)].$$

The absolute polynomial was extended by L. Kauffman who explained how to introduce a second variable, "a", into the absolute polynomial. As a result, one has another polynomial, distinct from the previous ones and which satisfies a somewhat more complicated recursive formulation involving two versions of the crossing relation depending upon the number of distinct strands involved in the crossings. The number of strands, or components, in a link L will be denoted by $c(L)$ and let $\langle X, Y \rangle$, X and Y mutually disjoint oriented links, denote the algebraic linking number of X and Y. This number can be calculated from a presentation of the links as one-half the sum of $+1$'s and -1's for each of the crossings between X and Y, according to whether the crossing is positive or negative as indicated in Fig. 1.11. The orientation conventions that are to be employed in the recursive calculation are indicated in Fig. 1.11 according to whether the crossing in L_+ involves the same or distinct strands of the link.

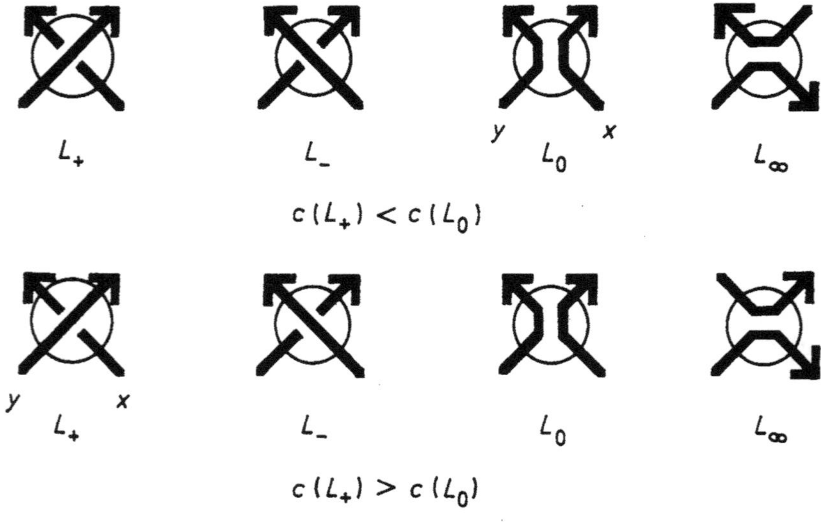

Fig. 1.11.

Theorem. The semi-oriented polynomial, $F_L(a, x)$, of an oriented knot or link, L, is the unique Laurent polynomial in the variables, "a" and "x", which satisfies the following fundamental formulae:

   (i)   if U denotes the standard unknotted circle in the plane, then $F_U(a, x) = 1$, and

   (ii)   if L_+, L_-, and L_0 are planar pictures of oriented links in each of which we have identified a small circular region of the picture containing either a single crossing or, in the last cases, no crossing at all, and such that outside these small circular regions shown in Fig. 1.11, the planar pictures are *exactly* the same, then the semi-oriented polynomial satisfies one of the formulae, according as the crossing in L_+ involves the same or distinct components:

(I) If $c(L_+) < c(L_0)$, let $\lambda = \langle x, L_0 - x \rangle$ and
$$aF_{L+}(x) + a^{-1}F_{L-}(x) = x[F_{L_0}(x) + a^{-4\lambda}F_{L\infty}(x)].$$

(II) If $c(L_+) > c(L_0)$, let $\mu = \langle x, L_+ - x \rangle$ and
$$aF_{L+}(x) + a^{-1}F_{L-}(x) = x[F_{L_0}(x) + a^{-4\mu+2}F_{L\infty}(x)].$$

So as to illustrate the complexity of the calculations that one encounters in the use of these polynomial invariants and to make more concrete the general calculational method that we shall describe in the next section, we shall show how one may calculate the semi-oriented polynomial associated to the right-handed trefoil. Consider the configurations, shown in Fig. 1.12. Here we find that the trefoil is the case L_+. In the case of a general presentation of a knot or link, the selection of a crossing to be changed in order to significantly simplify the resulting presentation is an important and difficult question, especially algorithmically. In the case of the trefoil knot, any crossing change will result in an unknot. Indeed, both L_- and L_∞ are trivial knots. This is seen by use of Reidemeister move I, twice, in the case of L_∞ and by use of Reidemeister moves I and II, once each, in the case of L_-.

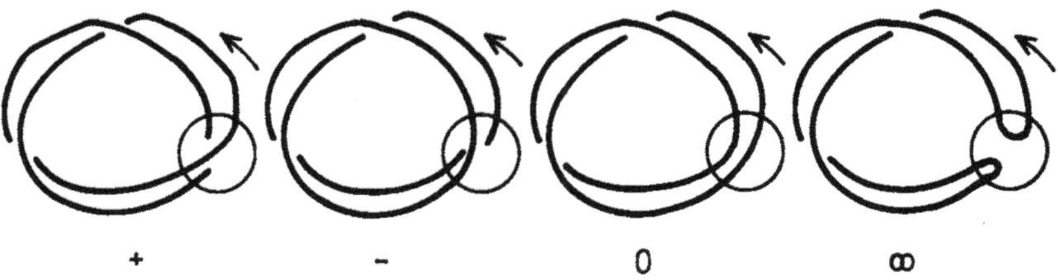

Fig. 1.12.

These moves may be exploited, in more general cases, to indicate the optimal selection of the crossing to be changed so as to insure that the number of crossings in the presentations of the subsidiary knots or links can be significantly reduced by their application. It is not possible, in general, to select crossing changes which provide such immediate and dramatic reductions. The overall goal of the recursive method is to select crossing changes which result in the greatest possible overall reduction of complexity.

By definition, the polynomials associated to L_- and L_∞ are identically equal to 1. Next we must compute the polynomial associated to L_0. For simple presentations it is possible to develop a table which associates the desired polynomial with the presentation. Indeed, this is exactly the method that we employ for presentation having fewer than six crossings and which cannot be further reduced by means of the Reidemeister moves. In the present situation we show that the recursion procedure can be used to reduce all configurations to those involving a trivial knot or link. Here, this is accomplished by using the set of configurations shown in Fig. 1.13.

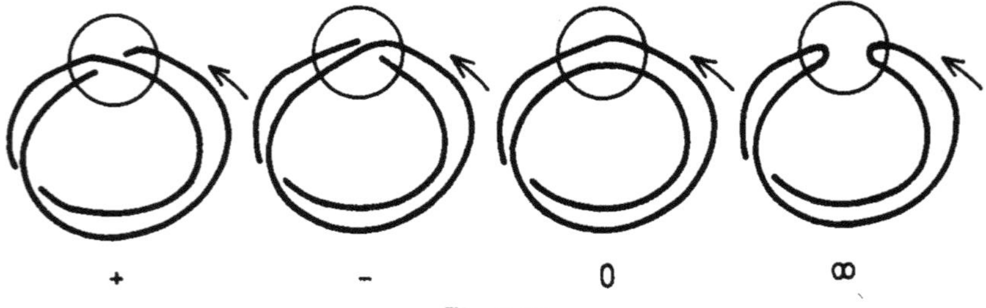

Fig. 1.13.

Here we discover that L_0 and L_∞ are both trivial knots, again by using Reidemeister move I, once in each case. L_- is the trivial link. This is shown by using Reidemeister move II once. Thus, in order to complete the calculation, we need only determine the polynomial associated to the trivial link of two separated components, which is convenient to denote by U^2. For this we employ the set of configurations shown in Fig. 1.14.

Here we find that $L_0 = U^2$ is the configuration whose associated polynomial we wish to calculate and that all the other configurations are equivalent to the trivial knot, using Reidemeister move I for each of L_+ and L_-. As a consequence of the fundamental formula we have:

$$aF_U(a, x) + a^{-1}F_U(a, x) = x[F_{U^2}(a, x) + a^0 F_U(a, x)]$$

so that

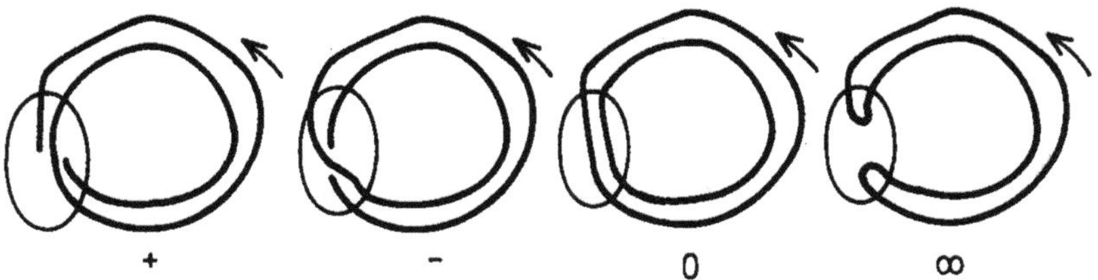

Fig. 1.14.

$$a + a^{-1} = x[F_{U^2}(a, x) + 1]$$

and, therefore,

$$F_{U^2}(a, x) = (a + a^{-1})x^{-1} - 1.$$

To compute the polynomial associated to the L_+ depicted in Fig. 1.13 we have:

$$aF_{L_+}(a, x) + a^{-1}F_{U^2}(a, x) = x[F_U(a, x) + a^{-4+2}F_U(a, x)]$$

so that

$$F_{L_+}(x) = -a^{-2}(a + a^{-1})x^{-1} + a^{-2} + a^{-1}(1 + a^{-2})x \, .$$

Finally to complete the calculation of the polynomial associated to the right-handed trefoil we return to the set of configurations in Fig. 1.12, from which we compute, using the fact that the $F_{L_+}(x)$ just computed is the $F_{L_0}(x)$ in that diagram:

$$aF_T(x) + a^{-1}F_U(x) = x[F_{L_0}(x) + a^{-4}F_U(x)]$$

so that

$$F_T(x) = -a^{-2} + a^{-1}x[-a^{-2}(a + a^{-1})x^{-1} + a^{-2} + a^{-1}(1 + a^{-2})x + a^{-4}]$$
$$= -2a^{-2} - a^{-4} + (a^{-3} + a^{-5})x + (a^{-2} + a^{-4})x^2 \, .$$

This sample calculation contains many of the elementary aspects that are employed to achieve a rescursive calculation in the general case. First, note that the appropriate linking index was computed in each of the cases so as to be able to apply the recursion formulae. The first principle is that one can change a crossing in a diagram in order to relate a given knot or link to a simpler one at the cost of having to consider one or two additional knots or links. The altered knot or link and the auxiliary ones all are, in some measurable way, simpler than the initial case and the calculations of their polynomials can be made completely independently so long as one remembers how to reassemble the result to calculate the polynomial of the desired knot or link. The essential calculational differences between the oriented and semi-oriented polynomial calculations are that in the first case one has only one auxiliary calculation to make while, in the second case, there are two cases to be considered and one must determine linking numbers in order to reassemble the information. Finally, one can see that there may be many alternative computational strategies that may be employed to calculate the same invariant. Specifically, the choice of which crossing is to be changed initially could have significant impact upon the complexity of the resulting calculation. These are the fundamental concerns to be addressed in the next section where we describe the various calculational algorithms which we have implemented and their underlying philosophical approaches.

2. The Algorithms

The work leading to the development of the load balanced algorithm which we shall describe later in this section is based upon several fundamental premises: First, within the range of problems we proposed to study, i.e., knots and links whose presentations contained no more than 150 crossings and which did not possess any special structure such as being given by braid presentation, the space required to store their descriptions would not provide a barrier to the calculation. Second, the data structure should be designed so as to optimize the speed of the fundamental elements of the algorithm such as the recognition of those elementary configurations appearing within the presentation which we wish to exploit. Third, the data structure should provide for fast changes such as crossing switches and removals of the sort appearing in the fundamental recursion relation. Fourth, the data structure should provide for fast evaluation of the benefit associated with a variety of crossing change strategies associated to the elementary Reidemeister moves and their generalizations which are developed and evaluated in our effort to reduce the complexity of the calculation.

Briefly, at each stage of the process one generates a small collection (often only two or three) of associated subsidiary problems of the same type. The complexity of these subsidiary problems is measurably simpler and their individual solutions can therefore be calculated more readily than for the original problem, [LM 1,4]. There are two basic ways in which we measure this reduction of complexity. The first is simply the total number of crossings in the presentation while the second involves the identification of a sequence of crossing changes which would result in a significant change in the knot or link. A fundamental quality typical of these subsidiary problems is that, in addition to being structurally simpler than the original problem, they can be solved completely independently of one another. The solutions found for these subproblems can then be combined in known manner to yield the desired result. In most cases this re-assembly is easy, with a few simple algebraic operations yielding the correct answer, the actual implementation of the process requires a prior estimate of the size of the final answer and traditional methods to ensure that the calculation is accomplished efficiently.

The critical issue is that the recursive application of this reduction process results in a regular, exponential growth, computational tree with nodes representing individual instances of the calculations, edges representing the reduction process, and the root node representing the solution to the original problem. In the case of the polynomial invariants associated to the classical knots and links there are several useful recursive algorithms for their calculation which have been developed. We will describe the general approaches and the specific implementations which we have developed in this section of the paper. In the subsequent section we shall describe the special attributes of the load balanced algorithm.

First it is useful to consider one of the algorithms derived from the article of J. H. Conway and which played a fundamental role in one of the theoretical approaches to the definition of the polynomial invariants. This method is described implicity in Conway [Con] and is exploited in the papers of Hoste [Hos] and Lickorish and Millett [LM1]. The basic concept is that of the "standard ascending position". Actually Hoste utilizes a standard descending position, but this is an equivalent method. Consider a specific projection of a given knot into a plane, to be thought of as the floor of the room, and imagine taking a length of rope (or several lengths, if one is looking at the projection of a link of several components) which is exactly long enough to cover the projection. Beginning at any point in the projection, which is either an over or under crossing, and proceeding in the direction given by the orientation, cover the projection with the rope.

When returning to the starting point, join the two ends together. Proceed with any other components in exactly the same fashion, if there are any. This physical process introduces under and over crossings which may differ from those of the given presentation and thereby defines and distinguishes the standard ascending position from that of the given knot. One can easily observe that the standard ascending position always represents the trivial knot or link. The specific representation is, however, dependent upon the choice of starting point or points and their order, if there are several components.

The *standard ascender algorithm* is described as follows: Proceed in the direction determined by the orientation from any starting point, chosen as in the definition of the standard ascending position, change crossings as necessary to achieve the standard ascending position. With each crossing change, the new knots or links defined by removing crossings are added to a list of cases to be resolved (attached to each of which is the information needed to calculate their specific contribution to the final polynomial) and one proceeds to change crossings until the standard ascending position is reached. At this point one updates a polynomial table and continues with the last unresolved case by means of the same procedure until there are no further cases. The recursion formulae have thereby provided the means to calculate the polynomial by reducing a knot or link to the associated standard ascender and computing the other terms in the relation, each of which involves knots or links having fewer crossings in their presentations. The number of examples on the list need never be more than half the total number of crossings. The polynomials of the standard ascenders are determined by a table rather than calculated for each of these simple cases. The size of the table is bounded as a function of the number of components and the number of crossings of the original example. Showing that this process leads to a single polynomial, independent of the choices made in its recursive definition is one of the fundamental tasks of the theoretical development of the polynomial invariant theory.

We have implemented the standard ascender algorithm and will report the results of test calculations later. The current version of "Knotter", a commonly used program for the calculation of the Conway (and, thereby, the Alexander polynomial), the Jones polynomial, and the oriented polynomial employs a similar standard ascender algorithm. It was written by John H. Jenkins of UC Berkeley following the algorithm of a BASIC program, Q version 6.2, written by Jim Hoste in 1985. This program does not systematically use any simplification techniques associated to possible Reidemeister moves which arise in the presentations. There is one case where a simplification is made in the sense that it does not change a crossing if that crossing is associated with a Reidemeister type I configuration.

The *modified standard ascender* algorithm employs the Reidemeister moves of types I and II and, in addition, their generalizations developed as part of the load balanced algorithm to be described later. They are employed to reduce the complexity of the knot or link presentation under consideration and, thereby, reduce the overall complexity of the calculation. This algorithm exploits two of the fundamental aspects of the combinatorics. First, one searches the given presentation of a knot for instances of Reidemeister moves of types I or II or their generalizations and immediately reduces the presentation according to these moves as soon as they are discovered. With our database, instances of these moves can be identified and the presentation modified to represent the result of the move extremely rapidly. Since these moves reduce the number of crossings in the presentations of all subsequent knots appearing in the applications of the recursion, there is a significant savings in the complexity of the calculation which is reflected in the running time and the total number of cases to be considered. When there are no further moves of types I and II or their generalizations in a given case, the standard ascender algorithm is applied to select a crossing to be changed. The ancillary cases are added to the list and process of seeking type I or II moves is begun again.

The fundamental issue with respect to standard ascender algorithms is the fact that one subsidiary example, in the oriented case, and two subsidiary examples, in the semi-oriented case, are created with each crossing change invoked in the algorithm. As a consequence one expects roughly expotential growth in complexity proportional to 2^n, in the case of the oriented polynomial, and proportional to 3^n, in the case of the semi-oriented polynomial. Thus the effect of the reduction in the number of crossings has significant potential for reducing the running time of the implementation if this can be accomplished rapidly. In some versions of Knotter, an attempt was made to include the Reidemeister moves but, according to some remarks of Jenkins, removing this decreased the running time of the program. This, however, was not our experience as our database is well adapted to the identification of these moves. In addition, it may be that the Jenkins' test involved only smaller knots and links where the exponential growth of the calculation is acceptable and the computation required to discover the Reidemeister moves which may be possible is greater than the associated savings in the complexity that they might achieve.

Because the growth inherent in the complexity of the calculation of the semi-oriented polynomial is much greater than that of the oriented polynomial we shall focus the remainder of our discussion of the algorithms on that case. The same principles, however, hold for our implementation of the analogous algorithm for the oriented polynomial calculation.

Before giving a technical description of the *load balanced algorithm* we shall first describe the underlying reasons that this approach is expected to yield important reductions in complexity and associated decreases in running time. As in the modified standard ascender algorithm, the first step is to search for applications of the first two Reidemeister moves. The next, and crucial step, is the evaluation of each crossing change for potential simplification and pruning of the computational tree through the reduction of the number of crossings. Load balancing across the three ancillary cases is used to minimize the exponential growth of the calculation by selecting crossing changes which would yield the greatest *weighted* reduction in the total complexity. A simple example illustrates the nature of this evaluation and the potential for reduction of complexity: Suppose that the complexity of a calculation is proportional to α^n, where n is the size of the data and that a reduction is possible which reduces the resulting complexity to 0 while giving 2 auxiliary calcuations of complexity $\rho\alpha^{n-1}$ each. We shall say that this is a reduction of type $(n, 1, 1)$. Suppose that another choice of reduction provides $\rho\alpha^{n-2}$ and gives 2 auxiliary cases of the same complexity, thus having complexity type $(2, 2, 2)$. Since

$$\rho\alpha^0 + \rho\alpha^{n-1} + \rho\alpha^{n-1} = \rho(2\alpha + \alpha^{2-n}) \cdot \alpha^{n-2} > 3 \cdot \rho\alpha^{n-2}$$
$$= \rho\alpha^{n-2} + \rho\alpha^{n-2} + \rho\alpha^{n-2}$$

the second choice would be preferable to the first if, for example, $\alpha > 1.5$. The systematic evaluation and choice of crossing changes based upon this strategy is what we shall mean by *load balancing* in this context.

We next outline the structure of the load balancing algorithm for the case of the semi-oriented polynomial, assuming that the necessary initialization and loading of the oriented knot or link has been completed:

> First, search for and immediately remove any Reidemeister type I and type II configurations until no further reductions of these types remain.

Next, search for and evaluate potential complexity type of triples:
A triple is a configuration whose projection has the motif indicated in Fig. 2.1. Null triples of types 1, 2, and 3 are immediately reduced by means of a generalized Reidemeister II move and the reduction process is immediately restarted by seeking instances of Reidemeister type I and II moves, above. None of the remaining possibilities can be reduced in this elementary fashion. The complexity types of the remaining possibilities are evaluated for their potential benefit.

Next search for and evaluate potential complexity type of *gammas*: A gamma is a configuration whose basic motif is shown in Fig. 2.2. Note that in cases 1 and 3, a crossing can be removed by a simple twist. This gives a bigon configuration shown in the same figure. These twists are performed immediately and the algorithm is restarted with the search for Reidemeister moves of types I and II above.

Next search for and identity the potential complexity type of *circuits*: Here one is searching for large portions of the knot or link which can be simplified by changing relatively few crossings. Furthermore, circuits will always occur and thereby provide crossing changes of "last resort". A circuit is either an entire strand of the link which never crosses itself, as indicated in Fig. 2.3, or a generalization of the gamma configuration described above.

This later is defined as a segment of the knot or link presentation in which a strand crosses the segment precisely once but may cross the other strands or the exterior of the segment in any arbitrary fashion. An example of this is also shown in Fig. 2.4.

The process begins by following a strand as it leaves a crossing and continues, listing the crossings encountered and their type, i.e., over or under, either it returns to the initial crossing or encounters an earlier crossing on the list. In the later case one does not have a cricuit beginning at the initial crossing in that direction (a smaller segment having been identified as a circuit) so that the analysis goes back to the initial crossing and considers another direction, proceeding through the set of crossings and directions in this manner. Each of the circuits is evaluated to determine their potential contribution to the reduction processes. Of special importance are cases

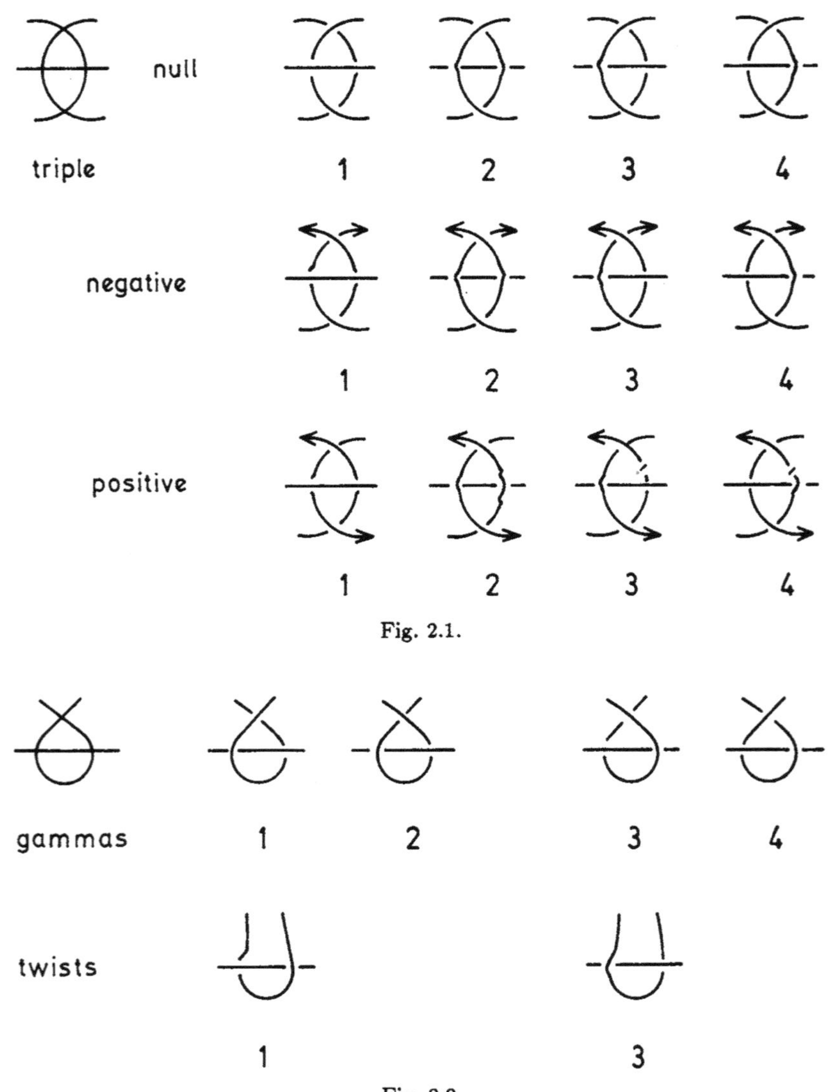

Fig. 2.1.

Fig. 2.2.

where half or more of crossings of the circuit occur with the same type, i.e., under or over, in a sequence as one travels around the circuit. Such occurrences allow us to do a global movement of the presentation and create a *skinny circuit* in whose interior there are no further crossings. A potential skinny circuit is assigned a complexity based upon the number of potential

type I circuit

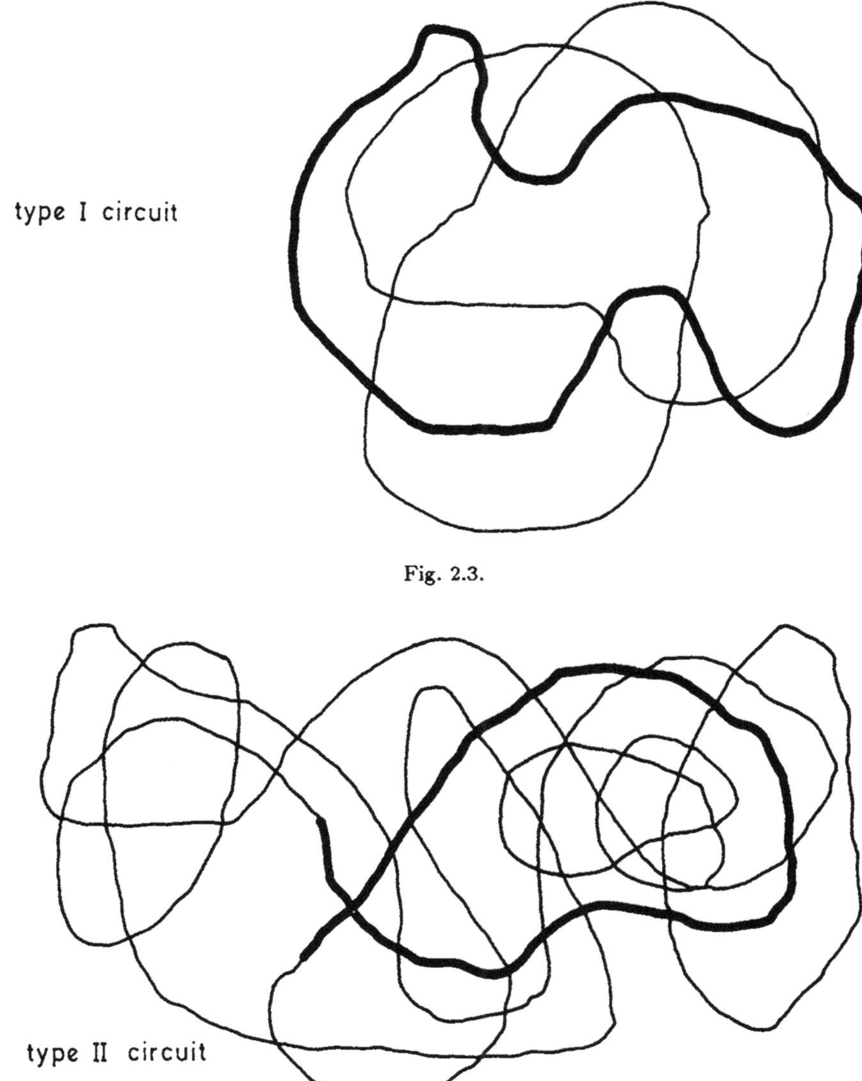

Fig. 2.3.

type II circuit

Fig. 2.4.

immediate type II reductions following its creation. This situation provides for particularly effective crossing changes and associated reductions. An example of this type of modification is shown in Fig. 2.5.

Fig. 2.5.

In the first situation one has a sequence of under crossings numbering half or more of the total while in the second a global transformation, another "generalized Reidemeister type III move", has been performed to achieve a skinny circuit configuration. These transformations have several essential properties. First, they never transform a skinny circuit that may already exist in the presentation into one which is no longer skinny and, second, although these transformations do not reduce the number of crossings they do create opportunities for advantageous reduction. If all the crossings were of the same type, i.e., under or over, this allows for the removal of the entire circuit via a "generalized Reidemeister type II move". Any time a skinny circuit is created we immediately restart the analysis by searching again for instance of Reidemeister type I and II moves.

Next   search for and evaluate the potential complexity type of *bigons*: A bigon is a configuration whose projection has the motif indicated in Fig. 2.6. Bigons occur frequently in knot and link presentations are created as a byproduct of the creation of skinny circuits.

Next,   if the knot or link presentation has five or fewer crossings the associated polynomial is determined in a table and stored in a table which accumulates to the final polynomial. If the knot or link has six or more crossings, the crossing change having the greatest potential advantage for reducing the level of complexity is performed thereby giving a modified presentation and two auxiliary presentations. These are added to the list according to the number of crossings in the presentation. The whole process is then restarted with the last presentation and continued until there are no knots or links remaining in the list.

At this point the calculation of the polynomial is completed and the result is reported to the user.

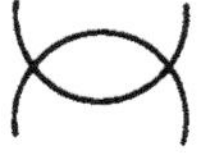

positive   negative

bigons

Fig. 2.6.

3. The Load Balancing Algorithm, A Technical Description

In this section we shall briefly describe some of the technical considerations which determine the load balancing aspects of our load balanced algorithm. We shall focus on the case of the semi-oriented polynomial because the oriented polynomial case is similar but less complicated. Recall that we are considering a computation, in the case of the modified ascender, whose complexity or running time appears to be given by a function of the form $\rho\alpha^n$, where n is the number of crossings in the presentation of the knot or link and ρ is some constant of proportionality. This growth assumption is supported by the data developed in a variety of calculations which we shall report in the next section. Given an n crossing presentation and a possible crossing change, three auxiliary cases are created. Suppose that these have presentations $(\pm, 0, \infty)$ having $(n-k, n-l, n-m)$ crossings, respectively. One calculates the total complexity of this crossing change as

$$\rho\alpha^{n-k} + \rho\alpha^{n-l} + \rho\alpha^{n-m} = \rho\alpha^n(\alpha^{-k} + \alpha^{-l} + \alpha^{-m})$$

$$k, l, m \geq 0 .$$

Since the last factor is independent of the number of crossings, it is a useful normalized index of the relative benefit in comparisons between the competing crossing changes under consideration. We say that the crossing change has complexity type (k, l, m) and define/evaluate its potential for reduction as

$$\beta(k, l, m) = \alpha^{-k} + \alpha^{-l} + \alpha^{-m} .$$

Our experimental calculations indicate that the value of the α which would appear to be associated to the standard ascender algorithm, in the range which we have studied, is approximately 3.411 whereas the α associated to the modified standard ascender, in which we have included the Reidemeister moves of types I and II, is approximately 1.670, for the average 13 crossing knot. For some larger knots and links, in the 25 to 35 crossing range, we have observed α's of 1.791 to 1.924 for the modified ascender algorithm. Because our implementation of the load balanced algorithm makes initial use of these Reidemeister moves, we use $\alpha = 1.9$ for our estimates of the reduction potential of the crossing change. In Table 3.1 we show three examples of values of α. One can compare these and observe the relative sensitivity of the load balancing strategy to this value. The load balanced strategy is based upon the selection of crossings changes which, among all the potential changes for a given presentation, rank highest in the scale. The first, or least desirable rank, is given to a crossing change which does not reduce the number of crossings.

rank	$\alpha = 1.9$	$\alpha = 1.59$	$\alpha = 3$
1	001	001	001
2	002	002	002
3	003	011	003
4	011	003	011
5	012	012	012
6	013	111	013
7	111	013	022
8	022	022	023
9	023	112	033
10	112	023	111
11	033	113	112
12	113	033	113
13	122	122	122
14	123	123	123
15	125	222	125
16	222	133	133
17	133	223	222
18	223	233	223
19	233	333	233

Table 3.1. Relative ranking of load balanced complexities: α=1.59, 1.9, 3.

Part of the theoretical justification for the load balanced algorithm is based upon the identification of each state in the calculation with a position in a high dimensional space. Each potential crossing change is associated to a movement in that space from one position to an adjacent one. Each of the positions have an associated complexity and each movement, or crossing change, yields a corresponding change in complexity. Each movement is a unit step so that the change in complexity corresponds to a derivative and the choice of direction yielding the greatest decrease in complexity, locally, is therefore a "gradient descent" choice. In short, the load balanced algorithm can be interpreted as a gradient descent algorithm in space of all states. Thus, what one can assert is that the load balanced algorithm yields the greatest local reduction in complexity.

In addition, however, there is a global aspect of the implementation of the algorithm involving large scale changes in the presentation. Specifically, the identification of cirucits and their evaluations with respect to the potential gain to be achieved by the creation of a "skinny circuit" represents global changes going beyond the one step change intrinsic to the gradient descent aspect of the algorithm. Thus, the current implementation of the load balanced algorithm represents the intergration of local reduction strategies with key global reduction strategies.

We shall now review the specific elements of the load balanced algorithm, as presented in the previous section, in order to describe those aspects related to the load balancing effort. It is important to remark at this point that the data structure which we utilize in our implementation was designed to allow the rapid identification and evaluation of the specific elements of the presentation that play an important role in the load balancing effort. In other data structures, this type of algorithm may not lead to the increased speed that we have experienced due to the increased effort required to identify the targeted crossing changes. Indeed, this appears to have been the case with the Hoste/Jenkins implementations of the (standard ascender/modified standard ascender) algorithms in which the Dowker-Thistlethwaite data structure, [Dth], was utilized. Recall that the addition of Reidemeister moves of types I and II distinguishes the standard ascender algorithm from the modified standard ascender and load balanced algorithms.

Continuing the technical description of the load balanced algorithm, we assume, therefore, that the presentation contains no Reidemeister moves of types I and II. The next step is the identification and evaluation of "triples", shown in Fig. 2.1. The crossing reductions that are implied by the proposed crossing change create a 3-tuple of natural numbers, in the order "crossing change" followed by the "0" and "∞" crossing eliminations, in either order. We list these possibilities for the case of the semi-oriented polynomial in Table 3.2. For example, null 4 gives the triple $(2, 1, 1)$ since the crossing change in the middle segment provides a reduction of 2 via the generalized Reidemeister move of type II whereas the 0 and ∞ crossing eliminations insure only a reduction of 1 in the crossing number. It may be that, in specific cases, more crossings are eliminated but these are the minimal reductions assured in the general case. Recall that null triples of types 1, 2, and 3 are immediately reduced and the process is then returned to the search for Reidemeister moves of types I and II. Thus all triples are identified and the most promising triple is retained as the potential target.

The next step is the identification, evaluation, and reduction of those "gammas", shown in Fig. 2.2, which do not require the creation of the ancillary cases. This is followed by the identification and evaluation of

	1	2	3	4
null	reduce	reduce	reduce	(2,1,1)
negative	(2,1,3)	(2,1,3)	(2,1,2)	(2,1,2)
positive	(2,3,1)	(2,3,1)	(2,2,1)	(2,2,1)

Table 3.2.

circuits, shown in Figs. 2.3 and 2.4, in certain cases we can immediately reduce the presentation via generalized Reidemeister type I move or can significantly simplify the presentation through the creation of skinny circuits. This element of the algorithm is invoked when the presentation has 12 or more crossings. The skinny circuits insure the existence of bigons, shown in Fig. 2.6. These bigons are, in addition, quite common in "generic" presentations of knots and links. In the next section we shall present data which indicates the frequency with which they, as well as the other configurations, arise in the course of the calculations. The bigon changes provide reductions of (2,1,2) and (2,2,1), respectively.

The crossing change which would result in the greatest load balanced reduction of complexity is performed if the presentation has 6 or more crossings, otherwise the presentation corresponds to a knot or link in our table of simple knots and links, and the process continues with the last knot or link on the list of cases to be calculated.

4. Evaluation and Comparison of the Algorithms

In this section we shall present data for the standard ascender, modified ascender, and the load balanced algorithms applied to a variety of test cases: all 12 and 13 crossing knots in the Thistlethwaite enumeration and a 25 crossing test knot, respectively. The goal is to demostrate amount and origin of the reduction achieved by the modified ascender and load balanced algorithms for both the oriented and semi-oriented polynomials. We compare the number of knots or links that are produced by the three algorithms, as a function of the number of crossings n, and those given by theoretical estimates for the standard ascender algorithm.

An extremely rough theoretical estimate of the number of k crossing presentations that are considered in the calculation of the oriented polynomial of an n crossing presentation of a knot or link is given by recursion on the number of crossings and the observation that one need change no more than half the crossings in a presentation to achieve either an ascending or descending presentation. Thus, for n crossing knot or link presentations, the standard ascender algorithm has a worst case theoretical prediction of $(n!)/(2^{n-k+1}(k-1)!)\,k$ crossing knot or link presentations. The total number of cases that are considered, in this worst case analysis, is given by

$$(n!/2^n)\Sigma^{n-1}(2^j)/j! \approx e^2(n!/2^n) \approx e^2(2\pi)^{-1/2}(n+1)^{n+1/2}e^{-n-1}2^{-n}$$
$$\approx e^{3/2}(\pi)^{-1/2}[(n+1)/(2e)]^{n+1/2}.$$

where we have invoked Stirling's formula for the estimate of the factorial.

For the semi-oriented polynomial, the standard ascender gives the worst case theoretical prediction of $(n!)/(2(n-k-1)!)\,k$ crossing knot or link presentations. The total number of cases that are considered, in this worst case analysis of the semi-oriented polynomial, is given by

$$(n!/2)\Sigma^{n-1}1/j! \approx (e/2)n! \approx e^{1/2}2^{-3/2}\pi^{-1/2}[(n+1)/(e)]^{n+1/2}.$$

where we have again invoked Stirling's formula.

We note that these asymptotic formulae for the standard ascender algorithms do not satisfy our hypothesis of exponential growth with constant "α" as the "α" in these formulae are functions of the number of crossings in presentation, i.e., $(n+1)/(2e)$ and $(n+1)/3$ for the oriented and semi-oriented polynomials, respectively. Since, however, we apply the load balancing method to the modified standard ascender algorithm for which the experimental data strongly support the exponential growth assumptions, these estimates do not impeach the appropriateness of the load balancing methodology. We have not been able to develop theoretical asymptotic formulae of the same sort for the modified standard ascender algorithm.

The following tables give the average data for the 9988 13 crossing knots in the Thistlethwaite enumeration for the oriented and semi-oriented polynomials. The entries in the table are the numbers of times the appropriate recursion formula is invoked to calculate the polynomial for a knot or link having the indicated number of crossings, in its given presentation, during the course of the calculation of the polynomial of the average 13 crossing knot. From the table of semi-oriented polynomial calculations, one observes that the run times of the modified ascender and the standard ascender implementations are 1.483 and 245.352 times as long as the load balanced program, respectively. Furthermore, the modified ascender and the standard ascender require 1.336 and 449.967 times as many recursions, respectively. From the table of oriented polynomial calculations, one observes that the run times of the modified ascender and the standard ascender implementations are 1.293 and 3.865 times as long as the load balanced program, respectively. Furthermore, the modified ascender and the standard ascender require 1.493 and 10.986 times as many recursions, respectively.

crossings	load balncd	mod ascdr	std ascdr	th std ascdr
13	1	1.3	6.324	6.5
12	.967	1.963	35.226	78
11	2.25	3.758	159.130	858
10	3.808	6.212	600.337	8580
9	6.808	10.458	1900.113	77220
8	12.009	17.637	5097.424	617760
7	21.301	27.35	10679.355	4324320
6	36.672	44.628	19686.077	25945920
TOTAL	84.815	113.306	38163.986	30974742.5
TIME	.210 s	.311 s	51.524 s	

Table 4.1. Semioriented polynomial: 13 crossing knots.

crossings	load balncd	mod ascdr	std ascdr	th std ascdr
13	1	1.3	6.3	6.5
12	.154	1.117	13.628	39
11	1.297	1.767	20.957	214
10	.910	2.147	27.469	1072
9	1.865	2.519	27.734	4826
8	1.961	3.529	25.727	19305
7	2.914	3.852	20.202	67567
6	4.180	5.098	14.857	202702
TOTAL	14.282	21.329	156.898	295731.5
TIME	.041	.053	.160	

Table 4.2. Oriented polynomial: 13 crossing knots.

The following tables give the average data for the 2176 12 crossing knots in the Thistlethwaite enumeration for the oriented and semi-oriented polynomials. From the table of semi-oriented polynomial calculations, one observes that the run times of the modified ascender and the standard ascender implementations are 1.185 and 91.345 times as long as the load balanced program, respectively. Furthermore, the modified ascender and the standard ascender require 1.269 and 178.066 times as many recursions, respectively. From the table of oriented polynomial calculations, one observes that the run times of the modified ascender and the standard ascender implementations are 1.125 and 3.013 times as long as the load balanced program, respectively. Furthermore, the modified ascender and the standard ascender require 1.365 and 9.182 times as many recursions, respectively.

crossings	load balncd	mod ascdr	std ascdr	th std ascdr
12	.994	1.242	5.874	6
11	.980	1.828	29.530	66
10	2.229	3.373	118.037	660
9	3.782	5.477	391.443	5940
8	6.726	9.139	1090.613	47520
7	12.004	14.842	2356.381	332640
6	20.839	24.455	4475.882	1995840
TOTAL	47.554	60.357	8467.760	2382672
TIME	.119 s	.141 s	10.870 s	

Table 4.3. Semioriented polynomial: 12 crossing knots.

crossings	load balncd	mod ascdr	std ascdr	th std ascdr
12	.994	1.243	5.874	6
11	.111	.934	10.928	33
10	1.300	1.577	15.054	165
9	.869	1.772	18.714	742.5
8	1.813	2.235	17.082	2970
7	2.076	2.818	15.040	10395
6	3.253	3.634	10.954	31185
TOTAL	10.416	14.213	95.636	45496.5
TIME	.032	.036	.097	

Table 4.4. Oriented polynomial: 12 crossing knots.

Although the growth rate in the 12 and 13 crossing cases illustrates the increases in speed associated to the load balancing approach to these calculations, the real purpose of this effort is the application to significantly larger topologically interesting examples. Indeed, one notes that the ratio of increases in speed and reduction in the cases considered increases in passing from the average 12 crossing case to the average 13 crossing case. These increases represent the increasing impact of the load balancing methodology on the calculations. We have completed calculations of the oriented and semi-oriented polynomials for knots and links whose minimal presentations contain as many as 90 crossings. The following data reports on the calculation the semi-oriented and oriented polynomials, respectively, of a certain 25 crossing knot. The calculation of the semi-oriented polynomial via the standard ascender algorithm is not included because we estimate that it would require 300 years of cpu time on our SUN 3/50 to complete the calculation.

crossings	load balncd	mod ascdr	load balncd	mod ascdr	std ascdr
25	2	2	2	5	12
24	5	19	2	9	80
23	10	52	3	18	312
22	17	129	3	20	988
21	29	261	3	20	1924
20	62	460	9	30	3686
19	110	915	7	44	5350
18	213	1772	10	60	7443
17	392	3039	7	68	8122
16	776	5348	27	112	8471
15	1379	8928	20	120	7620
14	2624	15016	34	129	6347
13	4820	24563	38	133	4605
12	10016	39433	62	156	3206
11	17950	71642	72	196	2131
10	37583	116959	104	323	1174
9	59272	186233	89	383	788
8	120373	285851	130	301	495
7	223086	421430	150	426	247
6	363754	655769	179	645	178
TOTALS	842473	1837824	951	3198	63179
TIME	2331.65 s	5424.76 s	4.76 s	13.93 s	104.08 s

Table 4.5. A 25 crossing example.

The calculation of the semi-oriented polynomial via the modified ascender algorithm takes 2.327 times as long as via the load balanced algorithm and requires 2.181 as many recursions. The calculation of the oriented polynomial via the modified ascender algorithm and the standard ascender take 2.926 and 21.866 times as long as via the load balanced algorithm, respectively, and requires 3.363 and 66.434, respectively, as many recrusions.

In general, the experimental data, shown in Graphs 4.6 and 4.7, predicts that the calculation of the semi-oriented polynomial of an n crossing knot requires $[0.034](1.827)^n$, $[0.027](1.904)^n$ and, $[0.00059718](3.954)^n$ recursions via the load balanced, modified standard ascender, and standard ascender algorithms, respectively. For the oriented polynomials, these growths are $[0.174](1.404)^n$, $[0.122](1.489)^n$ and, $[0.215](1.662)^n$, respectively. We note that the data for the standard ascender algorithm implementation appears to satisfy the exponential growth assumptions for the implementation of the load balancing method, i.e., is significantly less than the growth predicted by the theoretical estimate.

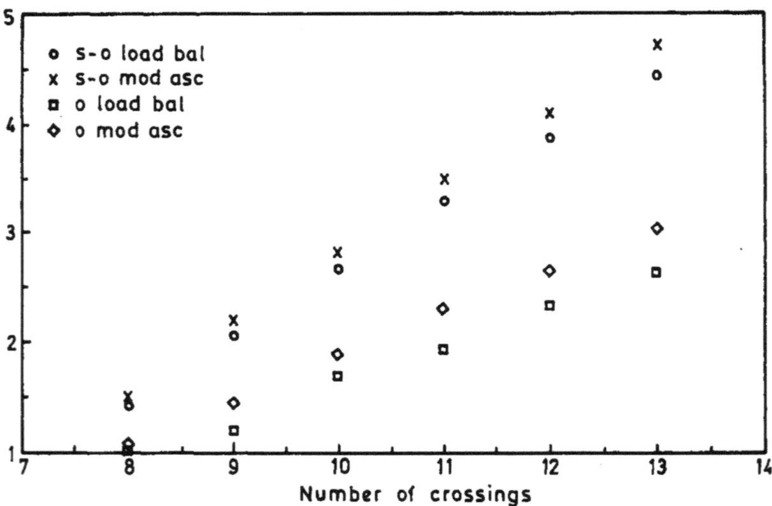

Graph 4.6. Ln graph for load balanced and modified ascender algorithm recursions.

Similarly, the experimental data, shown in Graphs 4.8 and 4.9, predicts that the calculation of the semi-oriented polynomial of an n crossing knot requires $[0.00012415](1.772)^n$, $[0.000043186](1.975)^n$ and, $[0.000001271](3.800)^n$ seconds via the load balanced, modified standard ascender, and standard ascender algorithms, respectively. For the oriented polynomials, these growths are $[0.00042676](1.425)^n$, $[0.00048746](1.434)^n$ and, $[0.0011516](1.449)^n$, respectively. One can use these formulae to compare the running time of the standard ascender to the load balanced algorithm, i.e., the standard ascender is $[.010238](2.144)^n$ times as long as the load balanced algorithm. Thus, for example, the semi-oriented polynomial of the 25 crossing example required 2331.65 cpu seconds to complete. This

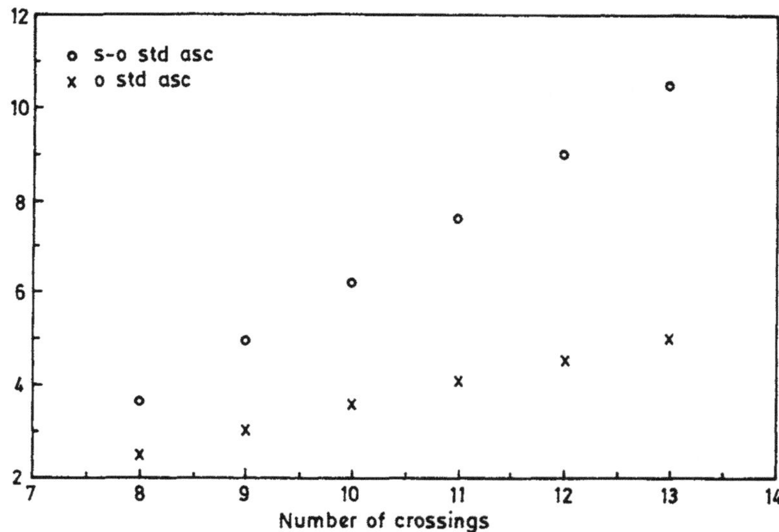

Graph 4.7. Ln graph for standard ascender algorithm recursions.

method would predict that the standard ascender algorithm would require 2331.65 [.010238]$(2.144)^{25}$ seconds, i.e., 312.285 cpu years as opposed to the 38.85 cpu minites that the load balanced algorithm required to complete the calculation!

From the perspective supplied by this experimental data, one sees that the load balanced algorithm provides an effective computational tool for a significant range of interesting examples well beyond that possible with the standard ascender algorithm.

To understand the influence of the various analyses employed in the implementation of the load balancing algorithm, we present and discuss experimental data exhibiting the frequency with which the fundamental cases evaluated in selecting crossing change in the algorithm are invoked in the course of the computations for the 15 crossing and a 25 crossing example. We shall see that these become more influential as the number of crossings in the presentation of the knot or link increases and that this influence cannot be attributed alone to the impact of the skinny circuit analysis.

Although the Reidemeister moves and their generalizations invoked in the modified standard ascender and load balanced algorithms make a significant contribution to the improvement over the classical standard ascender algorithm, the load balancing selection procedure and the circuit

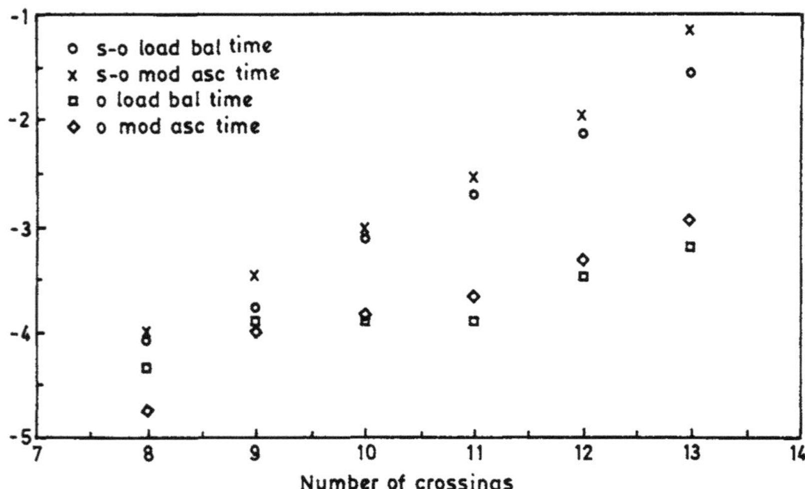

Graph 4.8. Ln graph for load balanced and modified ascender algorithm time.

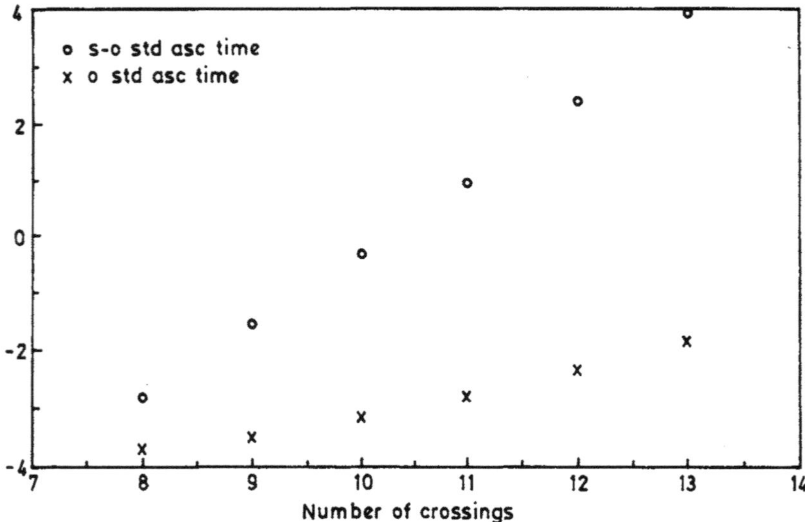

Graph 4.9. Ln graph for standard ascender algorithm time.

procedures provide additional significant improvement in the load balanced algorithm. Furthermore, the latter becomes even more important for larger knots and links. This accounts for the 57% reduction of running time for the calculation of the semi-oriented polynomial from that required by the modified ascender for the 25 crossing example, whose data we presented in

	semi-oriented	mod ascdr	oriented
Total recursions	763	1190	40
monogons	980	2841	34
bigons	852	1929	52
triples	228	557	20
gammas	63	223	7
skinny circuits			
	15 1		1
	14 2		2
Total	3		3

Table 4.10. A 15 crossing example.

	semi-oriented	mod ascdr	oriented	mod ascdr
Total recursions	842473	1837824	951	3198
monogons	1096253	4560361	1098	3934
bigons	918843	2840833	1205	4344
triples	258133	765718	525	2275
gammas	43798	424456	68	685
skinny circuits				
	25 1		1	
	24 2		1	
	23 3		1	
	22 6		1	
	21 13		2	
	20 23		5	
	19 42		3	
	18 51		3	
	17 88		2	
	16 191		15	
	15 351		13	
	14 563		3	
	13 910		19	
	12 2064		28	
Total	4308		99	

Table 4.11. A 25 crossing example.

Table 4.5. The data in Tables 4.10 and 4.11 for the 15 crossing and the 25 crossing examples illustrate these phenomena. Recall that the utilization of reductions via monogons, bigons, triples, and gammas provide simplifications which do not create ancillary cases but, instead, simply reduce the number of crossings in the given presentation.

The utilization of monogon, bigon, triple, and gamma reductions is proportionally fewer, when normalized for the total number of recursions, in the load balanced algorithm data compared to that of the modified standard ascender algorithm, see Table 4.12 for both semi-oriented and oriented polynomial cases. It is important to normalize for the total number of recursions because we are concerned with the impact of this analysis on each recursion step; the same analysis being repeated for each such recursion application. Therefore, since the creation of skinny circuits creates many bigons and there are proportionally fewer, one may conclude that the principal contributor to the increased speed is the load balancing aspect of the algorithm as opposed to the modification of the presentations via the creation of skinny circuits.

	semi-oriented	mod ascdr	oriented	mod ascdr
monogons	1.30	2.48	1.16	1.23
bigons	1.09	1.55	1.27	1.36
triples	.306	.417	.552	.711
gammas	.052	.231	.072	.214

Table 4.12. Normalized comparisons.

5. Conclusions

In this paper three algorithms, the standard ascender, the modified ascender, and the load balanced algorithms, to provide recursive calculations of the oriented and semi-oriented polynomials associated polynomials associated to classical knots and links were described. A new data structure, distinct from the Dowker-Thistlethwaite extension of the Gauss code, is developed and utilized in the implementation of these algorithms enabling the effective application of Reidemeister moves of types I and II, as well as their generalizations, and thereby providing significant increases in speed in our modification of the standard ascender algorithm as compared to the traditional standard ascender algorithm. Furthermore, the load balancing capability of the third algorithm provides an additional increase in speed beyond that achieved by the modification of the standard ascender. The load balancing algorithm, thereby, provides a substantial extension of the effective range of knot and link presentation complexity.

The effect of the implementation of the Reidemeister moves and their generalizations and the load balancing capability is demonstrated by the data presented in Sec. 3. Although the calculation of these knot and link polynomials is of exponential growth in the number of crossings in a presentation, it may be possible, with the development of an appropriate data structure well adapted to the implementation of this algorithm, to further reduce the growth rate of the calculation by taking advantage of duplicate cases that arise in the generation of the ancillary cases.

The load balancing approach to recursive algorithms of the same fundamental structure can be expected to lead to similar increases in the effective range and speed of recursive calculations of other like mathematical quantities.

References

[A]   J. W. Alexander, *Topological invariants of knots and links*, Trans. Amer. Math. Soc. **20** (1928) 275–306.

[AB]   J. W. Alexander and G. B. Briggs, *On types of knotted curves*, Ann. of Math. **2** (1927) 562–586.

[Bi]   N. L. Biggs, *Algebraic Graph Theory*, Cambridge University Press, 1974.

[Bo]   B. Bollabas, *Combinatorics*, Cambridge University Press, 1986.

[BLM]   R. D. Brandt, W. B. R. Lickorish, and K. C. Millett, *A polynomial invariant for nonoriented knots and links*, Invent. Math. **84** (1986) 563–573.

[BZ]   G. Burde, H. Zieschang, *Knots*, de Gruyter (1986).

[Col1]   C. J. Colbourn, *The reliability polynomial*, Ars Combinatoria **21A** (1986) 31–58.

[Col2]   C. J. Colbourn, *The Combinatorics of Network Reliability*, Oxford University Press, New York, 1987.

[Con]   J. H. Conway, *An enumeration of knots and links and some of their algebraic properties*, Computational Problems in Abstract Algebra, (John Leech, ed.), Pergamon Press, Oxford and New York, 1969, pp. 329–358.

[CrF] R. H. Crowell and R. H. Fox, *Introduction to knot theory*, Graduate Texts in Mathematics 57, Springer, 1977.

[D] M. Dehn, *Uber kombinatorische topologie*, Acta Math. **67** (1936) 123–168.

[Dth] C. H. Dowker and M. B. Thistlethwaite, *Classification of knot projections*, Topology and its Applications **16** (1893) 19–31.

[ES] C. Ernst, D. W. Summers, *The growth number of prime knots*, Math. Proc. Camb. Phil. Soc. **102** (1987) 303–315.

[F] R. H. Fox, *A quick trip through knot theory*, Topology of 3-manifolds (ed. M. K. Fort Jr.) Prentice Hall, 1962, pp. 120–167.

[FYHLMO] P. Freyd, D. Yetter; J. Hoste; W. B. R. Lickorish, K. Millett; A. Ocneanu, *A new polynomial invariant for knots and links*, Bulletin (New Series) of the American Mathematical Society, Vol 12, No. 2, April 1985, pp 239–246.

[G] C. F. Gauss, *Zur mathematischen Theorie der electrodynamischen Wirkungen*, Werke, Koniglichen Gesellschaft der Wissinchaften zu Gottingen, **5** (1877) 602–629.

[Ho] C. F. Ho, *A new polynomial for knots and links – preliminary report*, Abstracts AMS **6**, 4(1985), abstract 821–57–16.

[Hos] J. Hoste, *A polynomial invariant of knots and links*, Pacific J. Math. **124** (1986) 295–320.

[Ja1] F. Jaeger, *On Tutte polynomials and link polynomials*, Proc. Amer. Math. Soc. **103** (1988) 647–654.

[Ja2] F. Jaeger, *Composition products and models for the HOMFLY polynomial*, preprint, 1988.

[Ja3] F. Jaeger, D. L. Vertigan and D. J. A. Welsh, *On the computational complexity of the Jones and Tutte polynomials*, preprint, 1988.

[Jon1] V. F. R. Jones, *A polynomial invariant for knots via von Neumann algebras*, Bull. Am. Math Soc. **12** (1985) 103–111.

[Jon2] V. F. R. Jones, *Hecke algebra representations of braid groups and link polynomials*, Ann. of Math. **126** (1987) 335–388.

[Jon3] V. F. R. Jones, *On knot invariants related to some statistical mechanical models*, preprint, 1988.

[Kau1] L. Kauffman, *A invariant of regular isotopy*, preprint.

[Kau2] L. Kauffman, *Invariants of regular isotopy*, preprint.

[Kau3] L. Kauffman, *State models and the Jones polynomial*, Topology **26** (1987) 395–407.

[Kau4] L. Kauffman, *A Tutte polynomial for signed graphs*, preprint, 1987.

[Kau5]   L. Kauffman, *On knots*, Ann. of Math. Studies **115**, Princeton University Press (1987).

[Li]   J. B. Listing, *Vorstudien zur Topologie*, Gottinger Studien, 1847.

[LM1]   W. B. R. Lickorish and K. C. Millett, *A polynomial invariant for oriented links*, Topology **26** (1987) 107–141.

[LM4]   W. B. R. Lickorish and K. C. Millett, *The new polynomials for knots and links*, Mathematics Magazine **61** (1988) 3–23.

[Mos2]   H. Morton and H. Short, *Calculating the 2-variable polynomial for knots presented as closed braids*, preprint.

[O]   A. Ocneanu, *A polynomial invariant for knots: a combinatorial and an algebraic approach*, preprint.

[Po1]   J. S. Provan, *Polyhedral combinatorics and network reliability*, Mathematics of Operations Research **11** (1986) 36–61.

[PoB]   J. S. Provan, M. O. Ball, *Computing network reliablity in time polynomial in the number of cuts*, Operations Research **32** (1984) 516–526.

[PT2]   J. Przytycki and P. Traczyk, *Invariants of the Conway type*, Kobe J Math. **4** (1987) 115–139.

[ReaR]   R. C. Read and P. Rosenstiehl, *Principal edge tripartition of a graph*, Annals of Discrete Math. **3** (1978) 195–226.

[Re]   N. Y. Reshetikhin, *Quantized universal enveloping algebras, the Yang-Baxter equation and invariants of links, I and II*, LOMI reprints E–4–87 and E–17–87, Steklov Institite, Leningrad.

[Rol]   D. Rolfsen, *Knots and Links*, Publish or Perish Press, 1976.

[SaC]   A. Satyanarayana, M. K. Chang, *Network reliability and the factoring theorem*, Networks **13** (1983) 107–120.

[Su]   D. W. Sumners, *The knot enumeration problem*, Proceedings of an international course and conference on the interfaces between mathematics, chemistry, and computer science, Dubrovnik, Yugoslavia, Studies in Physical and Theoretical Chemistry, 1988, Vol 54, pp. 67–82.

[Th1]   M. Thistlethwaite, *Knot tabulations and related topics*, Aspects of Topology, Ed. I. M. James and E. H. Kronheimer, L. M. S. Lecture Notes 93 (1985) 1–76.

[Th2]   M. Thistlethwaite, *A spanning tree expansion of the Jones polynomial*, Topology **26** (1987) 297–309.

[Th3]   M. Thistlethwaite, *Kauffman's polynomial and prime knots*, preprint.

[Th4]   M. Thistlethwaite, *Kauffman's polynomial and alternating links*, Topology, to appear.

[Th5]   M. Thistlethwaite, *An upper bound for the breadth of the Jones polynomial*, preprint.

[Th6]   M. Thistlethwaite, *On the Kauffman polynomial of an adequate link*, Inventiones Math, to appear.

[Tra]   L. Traldi, *A dichromatic polynomial for weighted graphs, and link polynomials*, preprint, 1988.

[Tu2]   V. G. Turaev, *The Yang-Baxter equations and invariants of links*, LOMI preprint E-3-87, Steklov Institute, Leningrad.

[Tu3]   V. G. Turaev, *The Conway and Kauffman modules of the solid torus with a appendix on the operator invariants of tangles*, LOMI preprint E-6-88, Steklov Institute, Leningrad.

[Tut1]   W. T. Tutte, *A contribution to the theory on chromatic polynomials*, Canadian J. Math **6** (1954) 80–91.

[Tut2]   W. T. Tutte, *Codichromatic graphs*, Jour. Combinatorial Theory **B16** (1974) 168–174.

[Tut3]   W. T. Tutte, *The dichromatic polynomial*, Proc. 5th Brit. Comb. Conf. 1975, pp. 605–635.

[Tut4]   W. T. Tutte, *Graph Theory*, Encyclopedia of Mathematics, 1985.

[Wen1]   H. Wenzl, *Derived link invariants and subfactors*, preprint, 1987.

[Wen2]   H. Wenzl, *Representations of braid groups and the quantum Yang-Baxter equation*, preprint 1988.

[W1]   E. Witten, *Address to the Centennial Meeting of the American Mathematical Society*, Prividence, R. I., August 1988.

[W2]   E. Witten, *Quantum field theory and the Jones polynomial*, preprint, 1988.

Bruce Ewing and Kenneth C. Millett
Department of Mathematics
University of California
Santa Barbara
CA 93106
USA

CONVEX POLYHEDRAL MODELS
FOR THE FINITE THREE-DIMENSIONAL
ISOMETRY GROUPS

Lorraine L. Foster

The finite three-dimensional isometry groups are classified geometrically as belonging to fourteen general categories. In this paper twenty-seven convex polyhedral models are described which exemplify all possible cases associated with these categories.

1. Introduction

The isometries, or distance preserving bijections, of Euclidean three-space form a group $\mathcal{G}$ with respect to composition of mappings. A finite subgroup G of $\mathcal{G}$ must fix at least one point O. Also, it can contain only rotations σ about lines k containing O and rotatory reflections, i.e., mappings of the form $\sigma\rho$, where ρ is a reflection in a plane containing O and perpendicular to k. The rotations, or performable isometries (including the identity, or trivial, rotation) in G, form a subgroup H of G of index one or two. If G contains a central inversion λ in O (reflection in the point O) then it is isomorphic to the direct product $H \times \langle\lambda\rangle$. If G does not contain a central inversion but does contain rotatory reflections, it is isomorphic to a rotation group $H \cup \lambda H$ and designated in Weyl's notation [4] by GH. We let A_n, (resp. S_n, C_n, D_n) represent the alternating group, (resp. symmetric, cyclic, dihedral group) of degree n. (Since $D_1 \cong C_2$ we will suppose that D_n has order greater than two.) The following remarkable theorem ([2], [4]) identifies all possible three-dimensional isometry groups.

Theorem A. In the above notation, all possible finite three-dimensional isometry groups are represented abstractly by the symbols in the following three lists:

A_4, S_4, A_5, C_n, D_n (groups containing rotations only)

$A_4 \times \langle \lambda \rangle, S_4 \times \langle \lambda \rangle, A_5 \times \langle \lambda \rangle, C_n \times \langle \lambda \rangle, D_n \times \langle \lambda \rangle$ (groups with a central inversion)

$S_4 A_4, C_{2n} C_n, D_n C_n, D_{2n} D_n$ ("mixed" groups).

The elements of $\mathcal{G}$, which when restricted to the points of a three-dimensional figure F are bijections, form a subgroup of $\mathcal{G}$ known as the symmetry group of F. An actual three-dimensional object F (e.g., a cone or sphere) can have infinitely many symmetries. But if F has only finitely many symmetries, F must be a model for one of the groups listed in Theorem A.

The polyhedra (solid figures with surfaces composed entirely of polygons) are among the most fascinating and beautiful three-dimensional solid objects. The problem of finding and constructing a polyhedral model for each of the cases listed in Theorem A is an intriguing one. If it is required additionally that the models constructed be convex and that the number of faces be kept to a reasonable minimum, some calculus is introduced into the problem. To illustrate the theory of finite three-dimensional symmetry groups effectively in the classroom and at colloquia, the author has constructed many such models. The best of these (which were chosen for their relative ease of construction and a certain amount of artistic merit) are the subject of this paper. The twenty seven types of polyhedra shown in Figs. 1, 6 and 9 represent all possible cases. These and some alternate models are described in Secs. 2–4. The validity of each model is explained at a level appropriate for a student who would be handling the model and who has knowledge of elementary group theory. The Appendix contains instructions for constructing patterns for the non-standard polyhedra in the list.

2. Models with Rotation Group A_4, S_4 or A_5

There are seven cases to consider in this category, five of which are associated with familiar polyhedra. In general, symmetries permute the faces, as well as the edges, vertices and diagonals of a polyhedron. Also, any axis of rotation of a polyhedron F involved in a symmetry must intersect the surface of F in a vertex, in the midpoint of an edge, or in the center of a face.

We begin with the (regular) tetrahedron T (Fig. 1(d)). It is easy to see that T has eight rotations about four *three-fold* axes (i.e., axes belonging to rotations of maximal order 3), each through a vertex and the center of the opposite face. It also has three *two-fold* axes through midpoints of opposite edges. Each rotation of T in the rotation subgroup H clearly corresponds to a unique even permutation of the four faces of T. The complete symmetry group G of T is thus $H \cup \rho H$, where ρ is a reflection in a plane through an edge of T corresponding to a transposition (which is an odd permutation) of its faces. Thus H and G represent the abstract groups A_4 and S_4. Since T does not have a central inversion, Weyl's symbol for the group of T is $S_4 A_4$.

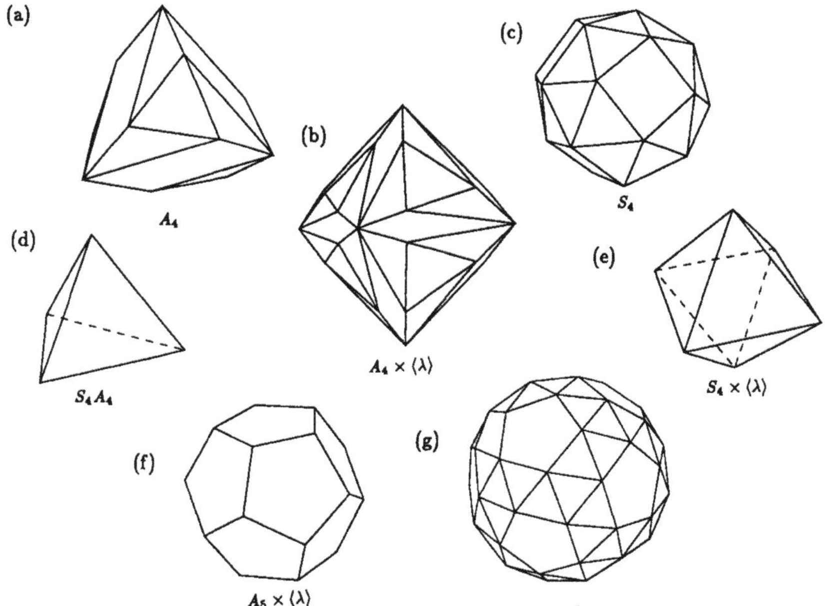

Fig. 1. Models with symmetries or alternating rotation groups: (a) a "mock tetrahedron"; (b) a "mock icosahedron"; (c) a snub cube; (d) the (regular) tetrahedron; (e) the octahedron; (f) the dodecahedron and (g) a snub dodecahedron.

One model with group G isomorphic to A_4 can be visualized as the union of T with four copies of the rather strange "pyramid" $\mathcal{M}$ (in actuality a variant of an antiprism), shown in Fig. 2 (view from above), joined to the four faces of T. (Here $\triangle PQR$ and $\triangle STU$ are concentric and equilateral, so that $\mathcal{M}$ has a *three-fold* axis of rotation.) We assume that the edges of T have unit length and that $P = (0,0,0)$, $Q = (1,0,0)$ and $S = (x,y,z)$ in Fig. 2, where $x > 1/2$. If the ratio $s = z/y$ is suitably chosen, adjacent copies of $\triangle PQS$ on the surface of T will be coplanar and a rather attractive "mock tetrahedron" T' will be obtained, as shown in Fig. 3. Since the dihedral angle of $\mathcal{M}$ at an outer edge and the dihedral angle of T are $\alpha = \arccos(s^2+1)^{-1/2}$ and $\beta = 2\arcsin 3^{-1/2}$, respectively, and we wish to have $\beta + 2\alpha = \pi$, we choose $s = 2^{1/2}$. Since $x > 1/2$, $\mathcal{M}$ has no reflections and thus T' has no rotatory reflections. (A one-piece pattern for T' is shown in Fig. 10(a).)

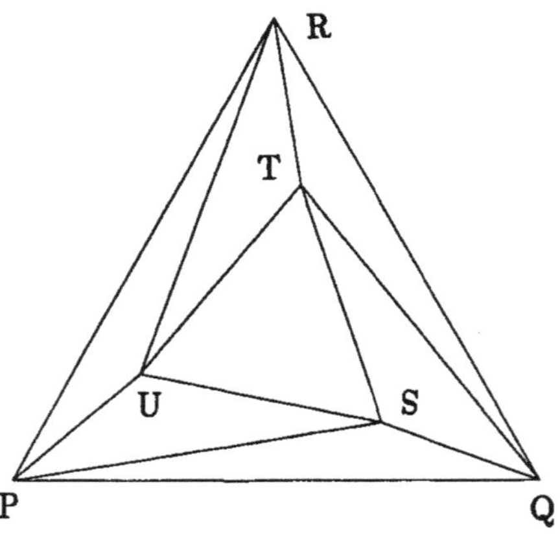

Fig. 2.

A hands-on examination of the octahedron $\mathcal{O}$ (Fig. 1(e)) reveals four *three-fold* axes through centers of opposite faces, three *four-fold* axes through pairs of opposite vertices, and six *two-fold* axes through midpoints of opposite edges, corresponding to twenty-four rotations. Furthermore, each of these rotations permutes the four *three-fold* axes among themselves. The rotation about a *two-fold* axis through edges E and F interchanges the two *three-fold* axes passing through the ends of E and F and takes the other two axes to themselves. Since any two *three-fold* axes are endpoints of a pair of edges and since in general the symmetric group on n letters is generated by its transpositions, H is isomorphic to S_4. Since $\mathcal{O}$ is centrally symmetric, its complete group is reprsented by $S_4 \times \langle \lambda \rangle$.

The usual polyhedral model with group $G \cong S_4$ (rotations only) is the Archimedean snub cube (Fig. 1(c)) with four *three-fold* axes through

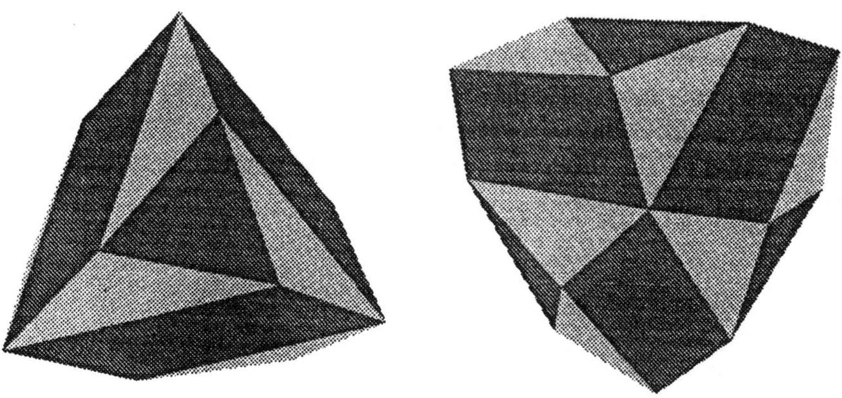

Fig. 3. A "mock tetrahedron" (two views) with symmetry group A_4.

the centers of triangles which have no edge in common with a square. It occcurs in two ("right- and left-handed") forms which are mirror images of each other and can therefore admit no non-performable symmetries. An alternate and perhaps more obvious example is constructed by attaching eight copies of the "pyramid" $\mathcal{M}$ of Fig. 2 to the faces of an octahedron $\mathcal{O}$. Since the dihedral angle of $\mathcal{O}$ is $\alpha = 2\arcsin(2/3)^{1/2}$, if we choose the ratio s above to be $2^{-1/2}$, a "snub mock octahedron" with twelve faces which are parallelograms and thirty-two triangular faces is formed (Fig. 4).

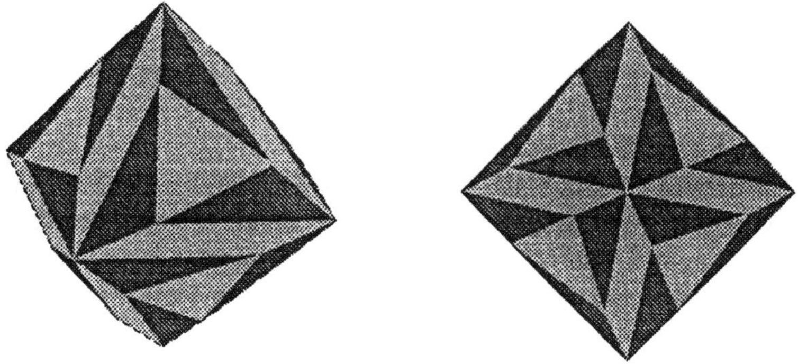

Fig. 4. A "snub mock octahedron" with symmetry group A_4.

Another unusual model is produced by attaching copies of $\mathcal{M}$ and its mirror image $\mathcal{M}'$ alternately to the faces of the octahedron $\mathcal{O}$, again with $s = 2^{-1/2}$, to obtain the "mock octahedron" of Fig. 5. The resulting figure retains the *two-* and *three-fold* axes of $\mathcal{O}$ only and remains centrally symmetric. Its group is therefore represented by $A_4 \times \langle \lambda \rangle$. (The author has also constructed one-piece patterns for striking alternate models for this group, based on the cuboctahedron and the icosahedron, each using four copies of $\mathcal{M}$ and $\mathcal{M}'$.)

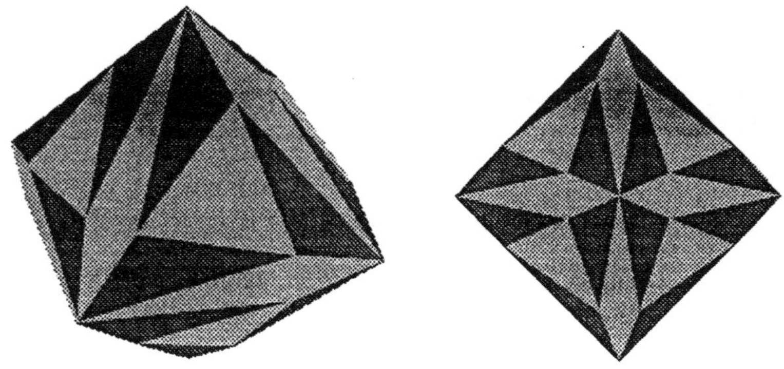

Fig. 5. A "mock octahedron" with symmetry group $A_4 \times \langle \lambda \rangle$.

The dodecahedron $\mathcal{D}$ (Fig. 1(f)) has exactly sixty rotations, namely about six *five-fold* axes, ten *three-fold* axes and fifteen *two-fold* axes $(24 + 20 + 15 + 1 = 60)$. From Sylow theory, the rotation group H of $\mathcal{D}$ is simple. For, a subgroup K of H of order 5 is a Sylow subgroup. The set of its conjugates must therefore contain twenty four elements. Similarly, the set of conjugates of a subgroup of order three contains twenty elements. It follows that a non-trivial normal subgroup K of H must be of order two and must be the intersection of all n Sylow two-subgroups of order four, where n divides 60. Since there are fifteen elements of order two, no such K can exist. It is well known (cf., [1]) that a simple group of order sixty is isomorphic to A_5. Thus $H \cong A_5$. (An alternate argument involving more elementary group theory is given in [3]). Since $\mathcal{D}$ has central symmetry, its group is designated by $A_5 \times \langle \lambda \rangle$.

The usual polyhedron exemplifying the group A_5 (rotations only) is the snub dodecahedron (Fig. 1(g)), which can be seen without much difficulty to have the same number of *n-fold* axes for $n = 5$ (resp. 3,2) as does $\mathcal{D}$ (so that the argument of the preceding paragraph immediately applies) and occurs in right- and left-handed forms. (Again it is possible — albeit rather tedious — to construct an alternate model by adjoining copies of $\mathcal{M}$ to all faces of an icosahedron. If the ratio s is taken to be $((7-3 \cdot 5^{1/2})/2)^{1/2}$, a figure with 110 faces will be obtained!)

3. Models with Rotation Group C_n or D_n for $n \geq 3$

Models for this section fall into eleven categories. Five of these are related to right regular pyramids $\mathcal{P}_n$ (i.e., pyramids with n sloping congruent isosceles triangular "sides" and a regular n-gonal base, such as $\mathcal{P}_8$ of Fig. 6(a)). (We require that $\mathcal{P}_3 \neq \mathcal{T}$.) It is easy to see that $\mathcal{P}_n$ has a single n-*fold* axis and n planes of reflection through this axis. Since, in general, a group of order $2n$ containing a cyclic subgroup of order n and n (additional) elements of order two is isomorphic to the dihedral group D_n, it follows that the group of $\mathcal{P}_n$ is denoted by $D_n C_n$.

If congruent pyramids with non-isosceles faces are removed (sliced off) at each vertex of $\mathcal{P}_n$ to form $\mathcal{P}'_n$, shown in Fig. 6(b) (with $n = 8$), a model with group C_n (rotations only) is obtained. Let us require in addition that the faces of $\mathcal{P}'_n$ be parallel to its axis. One such $\mathcal{P}'_n$, for example, has vertices given in cylindrical coordinates by $(0,0,h), (1, 2k\pi/n, h-s)$ and $(r, 2k\pi/n + \pi/(2n), 0)$, where $k = 0, 1, \ldots, n-1$ and where $s > 0$,

$$h = s \cdot \cos \frac{\pi}{2n} \left(\sin \frac{\pi}{2n} - \sin \left(\frac{\pi}{2n} - \frac{2\pi}{n} \right) \right) \Big/ \left(\sin \frac{2\pi}{n} \cos \frac{\pi}{n} \right),$$

$$r = h \cdot \cos \frac{\pi}{n} \Big/ \left(s \cdot \cos \frac{\pi}{2n} \right).$$

If two copies of $\mathcal{P}'_n$ are joined base-to-base in such a fashion that the triangular faces join to form parallelograms, a model is obtained with n additional axes which are *two-fold*, passing through vertices on the base of $\mathcal{P}'_n$ and/or centers of parallelograms. This polyhedron, shown in Fig. 6(c) (for the case $n = 8$), exists in right- and left-handed forms and therefore has group D_n.

A different group is illustrated if we join $\mathcal{P}'_n$ and its mirror image through the base plane II (Fig. 6(d) and 6(e)). The "mock dipyramid" $\mathcal{P}''_n$ so obtained has a single n-*fold* axis (and rotation group $H \cong C_n$). Further, the reflection ρ in II commutes with the rotations of $\mathcal{P}''_n$, so that G is the inner direct product $H \times \langle \rho \rangle$. Since $\lambda \in G$ if and only if n is even and since

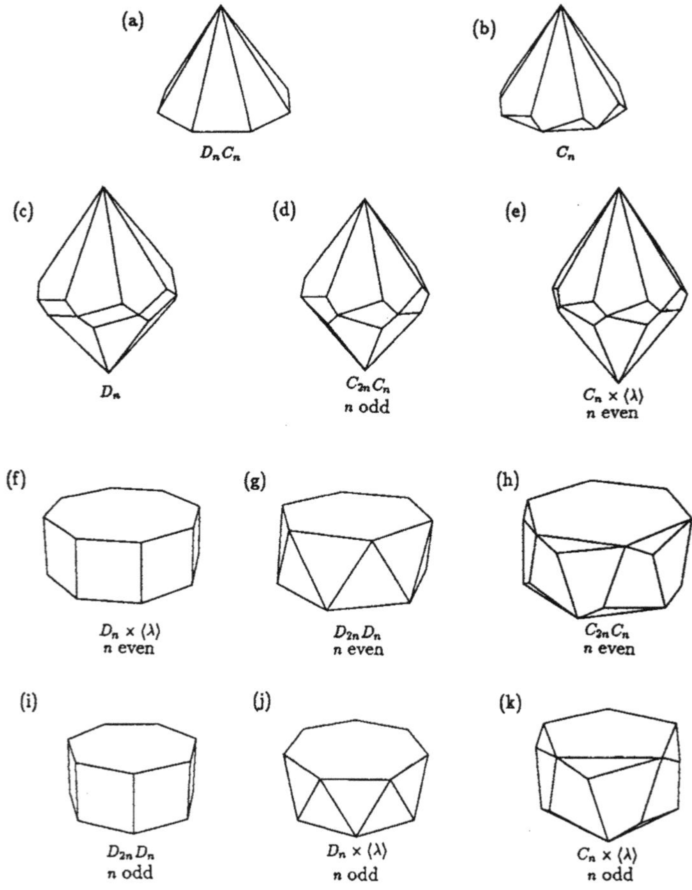

Fig. 6. Representative models with rotation group C_n or D_n, $n \geq 3$.

$C_{2n} \cong C_n \times C_2$ if n is odd, it follows that the group of $\mathcal{P}_n''$ is represented by $C_{2n} C_n$ if n is odd and $C_n \times \langle \lambda \rangle$ if n is even.

The remaining models of this section are prisms, antiprisms and modified antiprisms. A right regular prism $\mathcal{R}_n$ (i.e., a prism with n congruent rectangular "sides" and two regular n-gonal bases) which (in the case $n = 4$) is not a cube has by inspection a single *n-fold* axis of rotation perpendicular to the central plane Π and n *two-fold* axes through midpoints of edges and/or centers of opposite sides (cf., Fig. 6(f) and 6(i)). It therefore has rotation group $H \cong D_n$. Here again the complete symmetry group G is $H \times \langle \rho \rangle$. Also, $\lambda \in G$ if and only if n is even. Since $D_{2n} \cong D_n \times C_2$ if n is odd, it follows that $\mathcal{R}_n$ has group $D_{2n}D_n$ if n is odd and $D_n \times \langle \lambda \rangle$ if n is even.

A right regular antiprism $\mathcal{W}_n$ (cf., Fig. 6(g) and 6(j)) which has $2n$ isosceles triangular sides and two regular n-gonal bases has a single n-*fold* axis k, as well as n *two-fold* axes through pairs of opposite edges. (If the sides of $\mathcal{W}_n$ are regular then it is uniform. We will suppose for the moment that $\mathcal{W}_3$ is not uniform, so that $\mathcal{W}_3 \neq \mathcal{O}$.) Therefore $\mathcal{W}_n$ also has rotation group D_n. A rotatory reflection which is the product of a reflection in the central plane and a rotation through the angle π/n generates a cyclic subgroup of order $2n$ of the group G of $\mathcal{W}_n$. The remaining $2n$ symmetries of $\mathcal{W}_n$ are the n rotations about *two-fold* axes and the n reflections in planes through k (all of order two). It follows in both cases that G is isomorphic to the abstract group D_{2n} so that, since $\lambda \in G$ if and only if n is odd, G is represented by $D_{2n}D_n$ if n is even and $D_n \times \langle \lambda \rangle$ if n is odd.

We will realize two more cases of this section with polyhedral $\mathcal{U}_n$ derived by attaching pyramids $\mathcal{N}$ with equilateral bases shown in Fig. 7 (view from above) to the faces of uniform antiprisms $\mathcal{W}_n$ in such a fashion that copies of the edge PQ lie along the bases of $\mathcal{W}_n$. We will suppose that $\mathcal{W}_n$ has edges of unit length, and that the vertices of the prototypic $\mathcal{N}$ shown are as follows: P $= (0,0,0)$, Q $= (0,1,0)$, R $= (1/2, 3^{1/2}/2, 0)$ and S $= (x, x/3^{1/2}, z)$, where $x < 1/2$. An unusual "mock antiprism" $\mathcal{U}_n$ is created if x and z are chosen so that adjacent copies of $\triangle$PSR and $\triangle$RSQ along a lateral edge of $\mathcal{W}_n$ are coplanar (cf., Fig. 8). Such a polyhedron has $4n+2$ faces, $2n$ of which are irregular quadrilaterals. With a little calculation, it can be verified that such a figure is produced if the ratio $t = z/x$ is chosen with $\cos\beta = (3t^2 + 1)^{1/2}$ and $(\pi - \alpha_n)/2 < \beta < \pi - \alpha_n$, where α_n, the dihedral angle of $\mathcal{W}_n$, is given by $\alpha_n = \arccos((1 - 4\cos(\pi/n))/3)$. The value of z is then determined by the expression:

$$ z = \frac{3}{4t^{-1} + 2\sqrt{3}\cot(\pi - \alpha_n - \beta)} . $$

It is easy to see that $\mathcal{U}_n$ has rotation group C_n (as does $\mathcal{W}_n$) and a rotatory inversion of order $2n$. The group of $\mathcal{U}_n$ is thus given by $C_{2n}C_n$ if n is even and $C_n \times \langle \lambda \rangle$ if n is odd.

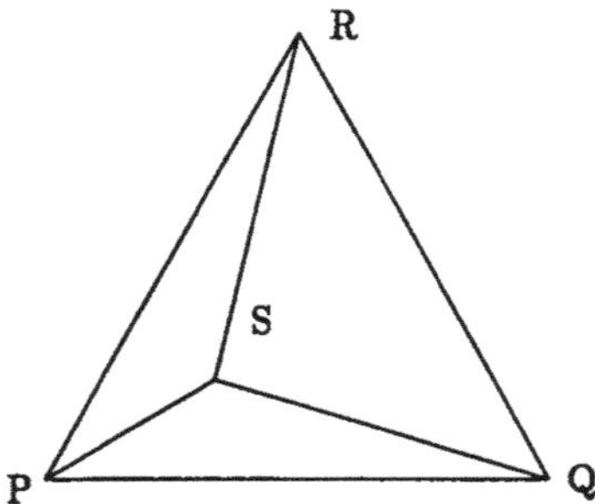

Fig. 7.

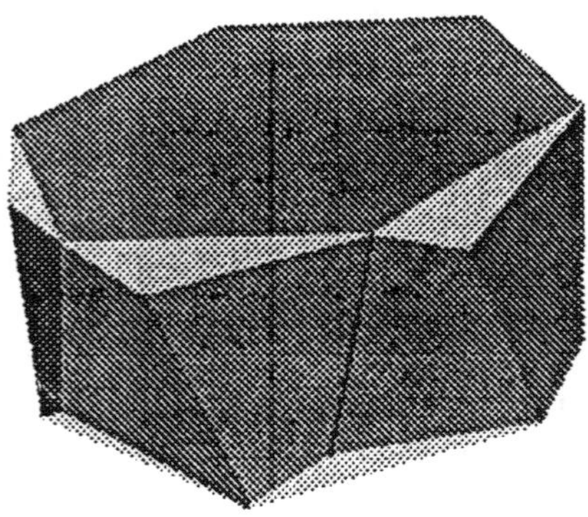

Fig. 8. A "mock antiprism" with symmetry group $C_{12}C_6$.

4. Models with Cyclic or Dihedral Rotation Group — The Special Cases

If n is 1 or 2 in Theorem A, there are nine more groups demanding models. Seven of these models are generated by a single "primitive element", which we will take to be the pyramid $\mathcal{X}$ of Fig. 9(a), with non-square rectangular base and four right triangular sides (and with trivial symmetry group C_1). If $\mathcal{X}$ and a mirror image $\mathcal{X}'$ of $\mathcal{X}$ are adjoined base-to-base to form the polyhedron of Fig. 9(b), a model with group $C_1 \times \langle \lambda \rangle$ is obtained. If on the other hand a copy of $\mathcal{X}$ is rotated 180° and joined base-to-base to $\mathcal{X}$ as in Fig. 9(c), a model with group C_2 is produced. A pyramid $\mathcal{X}''$ with one reflection and hence with group C_2C_1 is obtained by joining the vertical triangular faces of $\mathcal{X}$ and $\mathcal{X}'$ as shown in Fig. 9(d). Two copies of $\mathcal{X}''$ joined base-to-base as in Fig. 9(e) have group $C_2 \times \langle \lambda \rangle$. If instead their vertical faces are joined, a pyramid $\mathcal{Y}$ (Fig. 9(f)) with two reflections and two rotations and hence with group $D_2 C_2$ is created. To complete this "family" we join the bases of two copies of $\mathcal{Y}$ to obtain a model with group $D_2 \times \langle \lambda \rangle$ (Fig. 9(g)).

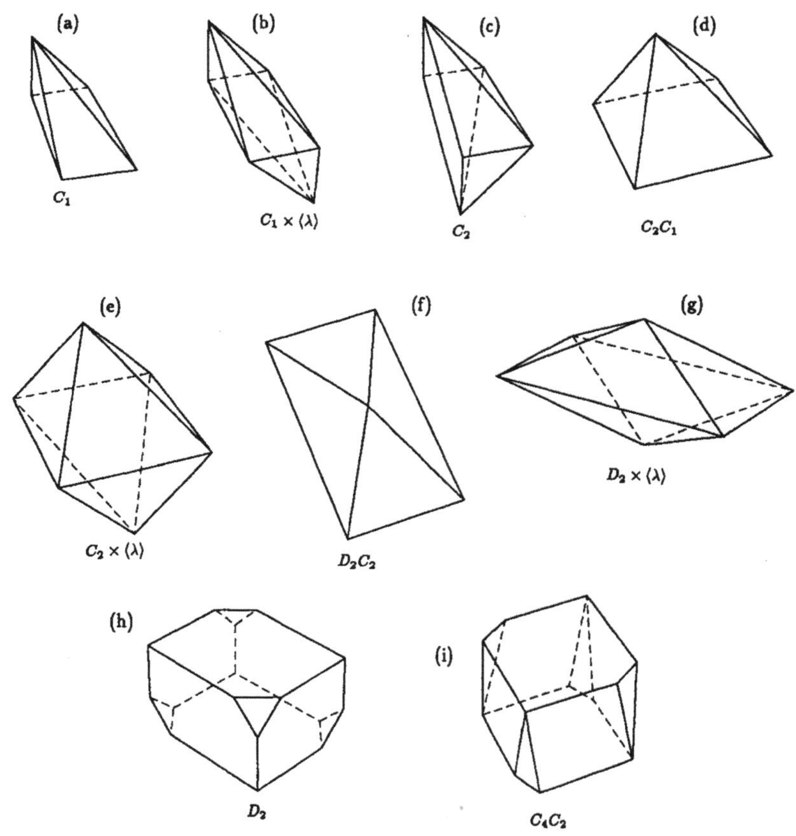

Fig. 9. Models with rotation group C_1, C_2 or D_2.

Figure 9(h) depicts a rectangular prism with no square faces which has been truncated by the removal of congruent right regular pyramids at alternating vertices. This figure has three *two-fold* axes, rotation group D_2, and no other symmetries.

The most exotic model in this category is obtained from the cube by removing four congruent pyramids with non-isosceles faces as shown in

Fig. 9(i). A rotatory reflection of order four generates the group of this figure (which admits no proper reflection) which is thus symbolized by $C_4 C_2$.

Appendix

The non-standard models shown in Secs. 2–4 can be constructed using patterns suggested by Figs. 10–13. All patterns can be generated by rotating and reflecting the basic pattern elements shown (and parts thereof) a sufficient number of times. Superb patterns can be computer drawn, using software that permits rotations in $0.01°$ increments, and plotted or printed directly on heavy colored stock. All models mentioned in this paper have been constructed by the author from such patterns. (Especially striking models — which are particularly effective in the classroom — were created by overlaying various types of faces with different colored copies.) Illustrations shown in Secs. 2 and 3 were created by computer-tracing scanned images of photographs of these models. Illustrations for the models of Sec. 4, on the other hand, were created by tracing computer generated three-dimensional images. (The author is happy to respond to all inquiries regarding software and hardware used.) Additional information for specific patterns is given below. (In some cases more information is given than is necessary from a strictly mathematical viewpoint.)

Three copies of Fig. 10(d) meet to form a central equilateral triangle. The approximate values of $(\phi, \theta, \psi, c, d)$ which were used to construct patterns 10(a) and 10(b) were $(15°, 30°, 33.23°, 0.3660, 0.7071)$ and $(7°, 16°, 41.12°, 0.3119, 0.7054)$, respectively. Patterns 10(b) and 10(c) have the same generating element.

All of the patterns in the pyramidal family shown in Fig. 11 were constructed using the generating element (Fig. 11(d)) with $(\phi, \theta, \psi, c, d) = (97.22°, 127.93°, 25.53°, 0.3412, 0.7724)$. The corresponding values for the model with *seven-fold* axis shown in Fig. 6(d) are $(98.51°, 131.52°, 29.01°, 0.3680, 0.7806)$.

The "mock antiprism" with pentagonal base shown in Fig. 6(k) was constructed by taking $(\phi, \theta, d) = (15.28°, 47.34°, 0.7378)$ in the generating element of Fig. 12. For the corresponding hexagonal model (Fig. 6(h)) these values are $(13.73°, 47.97°, 0.7468)$.

The models shown in Fig. 13 are all very easy to construct. Patterns

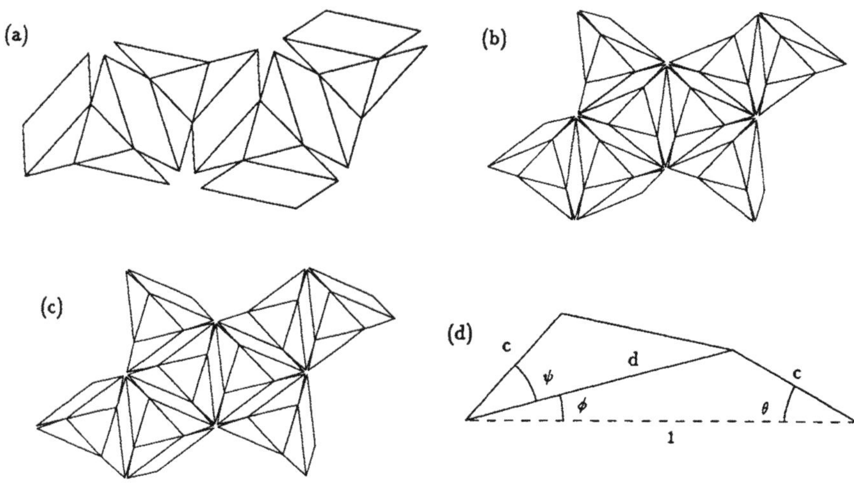

Fig. 10. Patterns for: (a) a "mock tetrahedron" (Fig. 2); (b) a "mock octahedron" (Fig. 4); (c) a "snub mock octahedron" (Fig. 3) and (d) a pattern-generating element (see text).

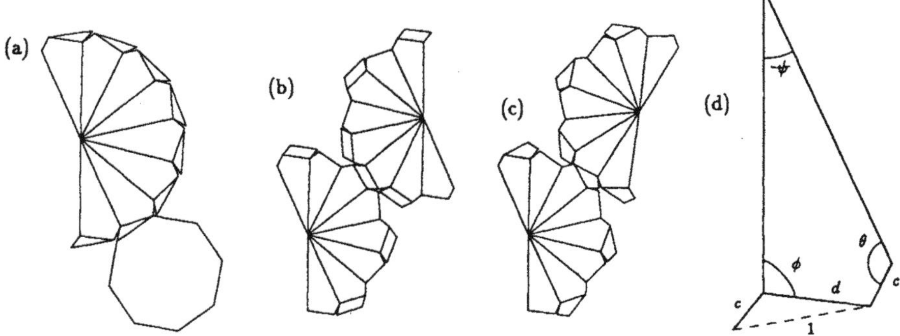

Fig. 11. Patterns for the models of: (a) Fig. 6(b); (b) Fig. 6(c); (c) Fig. 6(e) and (d) the basic element of all of these patterns (see text).

13(a)–13(g) involve essentially only right and isosceles triangles and rectangles. Patterns 13(b) and 13(c) are derived from 13(a) in an obvious fashion, as are 13(e) from 13(d) and 13(g) from 13(f). Pattern 13(h) produces a $3 \times 4 \times 5$ prism which has been truncated one unit from alternating corners. The bases of the cubical prismoid in pattern 13(i) are obtained by

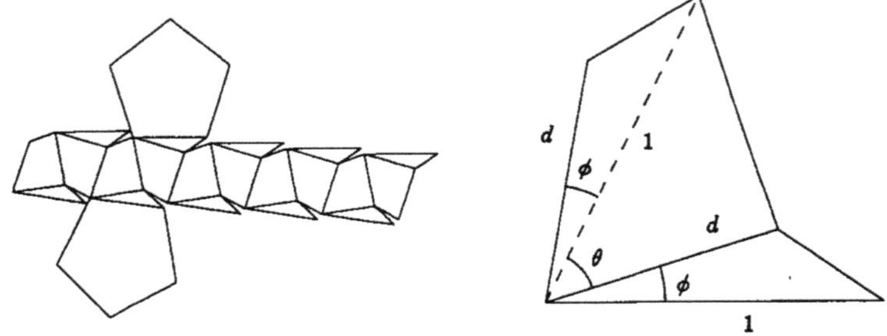

Fig. 12. A pattern for the model of Fig. 6(h) and the basic element of the pattern (see text).

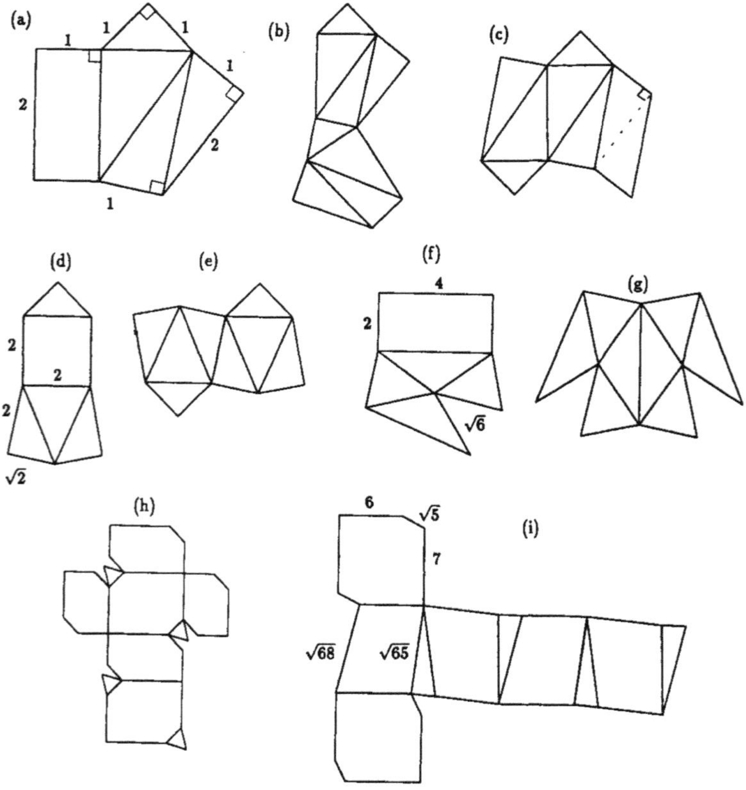

Fig. 13. Patterns for the models of Figs. 9(a)–9(i).

removing right triangles with legs of lengths 1 and 2 units from a square of length 8.'

References

1. R. B. J. T. Allenby, *Rings, Fields and Groups*, Edward Arnold, London, 1983, pp. 270–271.
2. H. S. M. Coxeter, *Introduction to Geometry* (2nd ed.), Wiley, New York, 1969.
3. Lorraine L. Foster, *On the symmetry group of the dodecahedron*, Math. Magazine **63** (1990), pp. 106–107.
4. H. Weyl, *Symmetry*, Princeton University Press, New Jersey, 1952.

Lorraine L. Foster
Department of Mathematics
California State University
Northridge, California 91330
USA

THE LORENTZIAN MODULAR GROUP AND NONLINEAR LATTICES

Glenn J. Fox and Phillip E. Parker

The discrete group $SO_1^{2+}(\mathbb{Z})$ is called the Lorentzian modular group by analogy with the usual elliptic modular group $SL_2(\mathbb{Z})$. We begin a detailed study of its arithmetic and analytic properties. In particular, we characterize the orbit structure of nonlinear integral lattices on the hyperboloids $z^2 - x^2 - y^2 = k > 0$.

Introduction

The discrete group $SL_2(\mathbb{Z})$ is called the modular group. It has an incredibly rich history reaching back at least to Gauss [2, p. 339]. In this paper we shall consider the Lorentzian analogue, a discrete isometry group of Minkowski 3-space $\mathfrak{M}$. The classical modular group is a discrete isometry group of the Poincaré upper half-plane $\mathfrak{H}$.

Consider $\mathbb{R}^3$ with the nondegenerate indefinite quadratic form $\eta = z^2 - x^2 - y^2$. If we regard η as a metric tensor of differential geometry, then $(\mathbb{R}^3, \eta)$ is called Minkowski 3-space $\mathfrak{M}$. It has constant (sectional) curvature zero; one says that it is a flat Lorentzian space form [11]. By comparison, $\mathfrak{H}$ is a Riemannian space form of constant negative curvature -1. The isometry group of $\mathfrak{M}$ is the 3-dimensional *Lorentz group* $O_1^2(\mathbb{R}) = \{A \in GL_3(\mathbb{R}); A\eta A^{\mathrm{T}} = \eta\}$. As usual, $SO_1^2(\mathbb{R})$ is the subgroup consisting of those matrices of determinant $+1$. Unlike the familiar rotation group $SO_3(\mathbb{R})$, however, $SO_1^2(\mathbb{R})$ has two connected components: the subgroup $SO_1^{2+}(\mathbb{R})$ preserves the z-orientation and the other component reverses

MSC (1985) *primary* 11E36; *secondary* 11F06, 11Y70, 20H10, 30F35 it. The 3-dimensional Lorentz group occurs in many applications; two of current interest are 3-dimensional gravity [e.g., 3; 4; 10] and Lie optics [7].

The level surfaces $\{\eta = k > 0\}$ are model space forms for Riemannian surfaces of constant negative curvature $-\frac{1}{k}$. They are also the *mass shells* of theoretical physics (in dimension 3). We shall be interested in those where k is an integer, in line with our attention to nonlinear analogues of arithmetic lattices in vector spaces, or briefly, *nonlinear lattices*. These are of course related to the so-called lattices in Lie groups [6].

We turn now to the customary brief description of contents. Sec. 1 contains various results describing some general properties of the group $SO_1^{2+}(\mathbb{Z})$. The main result characterizes the inverse image Γ of $SO_1^{2+}(\mathbb{Z})$ in $SL_2(\mathbb{R})$ under the usual double covering map $SL_2(\mathbb{R}) \twoheadrightarrow SO_1^{2+}(\mathbb{R})$. As a consequence, we find that $SO_1^{2+}(\mathbb{Z})$ is isomorphic to the Hecke group $G(\sqrt{2})$. For future reference, we include an explicit calculation of the Schwarz mapping function s of Γ.

Section 2 begins with a characterization of all integral solutions of $\eta = k$ for positive integral k. Next, we find essentially unique orbit representatives for these integral solutions. Then we give a computationally effective proof that the number of orbits on each sheet $\{\eta = k\}$ is finite.

Finally, Sec. 3 includes the output from several machine calculations, together with ancillary results needed to validate the procedures used.

Fox thanks Paul Sinclair for conversations and suggestions. Parker thanks Harvey Cohn for helpful conversations, the Wichita State University for partial support *via* a Faculty Summer Research Award in 1988, and C. T. J. Dodson and the Department of Mathematics of the University of Lancaster for hospitality and partial support during June of 1988.

1. General Properties

We begin by observing that the usual double covering map $SL_2(\mathbb{R}) \twoheadrightarrow SO_1^{2+}(\mathbb{R})$ intertwines the actions of $SL_2(\mathbb{R})$ on $\mathfrak{H}$ by linear fractional transformations and of $SO_1^{2+}(\mathbb{R})$ on surfaces $\{\eta = k\}$ by Lorentz transformations in $\mathfrak{M}$.

Now, this double covering extends to a map $GL_2 \to GL_3$ in which the image of $\left[\begin{smallmatrix} a & b \\ c & d \end{smallmatrix}\right]$ is

$$\begin{bmatrix} \frac{a^2+b^2+c^2+d^2}{2} & \frac{a^2-b^2+c^2-d^2}{2} & -ab-cd \\ \frac{a^2+b^2-c^2-d^2}{2} & \frac{a^2-b^2-c^2+d^2}{2} & -ab+cd \\ -ac-bd & -ac+bd & ad+bc \end{bmatrix}.$$

It is easy to see that this map preserves transpose and inverse. Straightforward but tedious computations show that the determinants are $D = ad-bc$ and D^3, respectively, and the traces are $a+d$ and $(a+d)^2-D$, respectively.

The first objective is a description of the inverse image of $SO_1^{2+}(\mathbb{Z})$. In order that the image of an element from GL_2 be integral in GL_3, these must be even integers:

$$\begin{aligned} & a^2 + b^2 \pm (c^2 + d^2) \\ & a^2 - b^2 \pm (c^2 - d^2). \end{aligned} \tag{1}$$

Adding and subtracting once, $a^2 \pm b^2$ and $c^2 \pm d^2$ must be integers. Repeating the process, it follows that $2a^2, 2b^2, 2c^2$, and $2d^2$ must be integers whence

$$a = \pm\sqrt{\frac{k}{2}}, \; b = \pm\sqrt{\frac{l}{2}}, \; c = \pm\sqrt{\frac{m}{2}}, \; d = \pm\sqrt{\frac{n}{2}} \tag{2}$$

for some nonnegative integers k, l, m, n. If $(ad - bc)^3$ is an integer, then it now follows that $ad - bc$ must be an integer. Thus there are the following additional conditions on a, b, c, d:

$$\begin{aligned} & ab \pm cd \\ & ac \pm bd \\ & ad \pm bc \end{aligned} \tag{3}$$

must be integers.

Using the representations (2) and adding and subtracting conditions (3), all sums and differences of the types $\sqrt{kl} \pm \sqrt{mn}, \sqrt{kn} \pm \sqrt{lm}$, etc., must be even integers. Adding and subtracting yet again, it follows that all products $\sqrt{kl}, \sqrt{ln}$, etc., must be integers.

We find now several cases.

If any one of k, l, m, n is a square, then all are; e.g., if $k = h^2$ with $h \geq 0$, then $\sqrt{kl} = h\sqrt{l}$ is an integer whence l is also a square, and similarly for the other possibilities.

If any one of k, l, m, n is twice a square, then all are by a similar argument.

Otherwise, we may assume that none of k, l, m, n is either a square or twice a square. Now, if any one is even, then all are. Let $D = ad - bc$ and consider the two subcases.

If all of k, l, m, n are even, factor each of $\frac{k}{2}, \frac{l}{2}, \frac{m}{2}, \frac{n}{2}$ into square and square-free parts according to this pattern: $\frac{k}{2} = k_1^2 k_2$. Since all products such as $\sqrt{kl}$ are integers, all products such as $\sqrt{k_2 l_2}$ must be integers. Thus if p is a prime and $p|k_2$, then $p|l_2, m_2, n_2$ whence $p|D$. Therefore $k_2 = l_2 = m_2 = n_2|D$ and

$$a = \pm k_1 \sqrt{s}, \ b = \pm l_1 \sqrt{s}, \ c = \pm m_1 \sqrt{s}, \ d = \pm n_1 \sqrt{s} \tag{4}$$

where s is a square-free divisor of D.

If all of k, l, m, n are odd, factor each into square and square-free parts as before and proceed to obtain

$$a = \pm k_1 \sqrt{\frac{s}{2}}, \ b = \pm l_1 \sqrt{\frac{s}{2}}, \ c = \pm m_1 \sqrt{\frac{s}{2}}, \ d = \pm n_1 \sqrt{\frac{s}{2}} \tag{5}$$

where s is now an odd square-free divisor of D.

Combining all these cases, we have proved

Theorem 1.1. If the matrix $\begin{bmatrix} a & b \\ c & d \end{bmatrix}$ has an integral image, then the entries are either of form (4) where $s = 1$ or s is a square-free divisor of $D = ad - bc$, or of form (5) where $s = 1$ or s is an odd square-free divisor of D.

Corollary 1.2. The inverse image of $SO_1^{2+}(\mathbb{Z})$ in $SL_2(\mathbb{R})$ is the subgroup Γ generated by $\begin{bmatrix} 1 & 2 \\ 0 & 1 \end{bmatrix}$ and $\frac{1}{\sqrt{2}} \begin{bmatrix} 1 & -1 \\ 1 & 1 \end{bmatrix}$.

Proof. Here $D = 1$, so one has $s = 1$ in forms (4) and (5). From conditions (1) above, there must be an even number of odd integers occurring as coefficients in (4) or (5). If all four are odd, then it must be form (5), and conversely. Since $D = 1$, in form (4) there must be exactly two odd integers. The result now follows.

Theorem 1.3. Γ is conjugate to the Hecke group $G(\sqrt{2})$ generated by $\begin{bmatrix} 1 & \sqrt{2} \\ 0 & 1 \end{bmatrix}$ and $\begin{bmatrix} 0 & -1 \\ 1 & 0 \end{bmatrix}$ *via* the matrix

$$\alpha = \begin{bmatrix} 2^{1/4} & -2^{-1/4} \\ 0 & 2^{-1/4} \end{bmatrix},$$

as $\Gamma = \alpha G(\sqrt{2})\alpha^{-1}$.

The proof is an easy calculation which we leave to the reader.

Observe that the binary quadratic form Q given by $aX^2 + bXY + cY^2$ corresponds to a point (z, x, y) on $\{\eta = -\text{dis}\,Q\}$ *via* $z = a+c, x = a-c, y = -b$. Thus integral quadratic forms correspond to integral lattice points. It is easy to see that equivalence under the dual (transposed) action of $SO_1^{2+}(\mathbb{Z})$ corresponds to ordinary equivalence under $\Gamma \cong G(\sqrt{2}) \leq SL_2(\mathbb{Z}[\sqrt{2}])$. Consequently, equivalence under $SO_1^{2+}(\mathbb{Z})$ is related to ordinary (integral) equivalence over $\mathbb{Q}(\sqrt{2})$. This relation to the arithmetic of $\mathbb{Q}(\sqrt{2})$ is also seen in the classification of elliptic curves E_τ corresponding to $\tau \in \{\eta = k\}$: equivalence under $SO_1^{2+}(\mathbb{Z})$ corresponds to isomorphism over $\mathbb{Q}(\sqrt{2})$. These relations will be discussed more in a subsequent paper.

We recall that a Schwarz mapping function maps the upper half-plane onto a hyperbolic triangle in the unit disc. When extended to the entire plane *via* his reflection principle, it provides a uniformization for a certain Riemann surface.

Theorem 1.4. Γ is a triangle group $(\frac{1}{4}, \frac{1}{2}, 0)$. Thus its Schwarz mapping function $s(\frac{1}{4}, \frac{1}{2}, 0; w)$ is transcendental.

Proof. The first part follows from [1, Sec. 11.3]. We note that such a group is also said to have signature $(0 : 4, 2, \infty)$ or to be of type $(\frac{\pi}{4}, \frac{\pi}{2}, 0)$. A fundamental domain for Γ in $\bar{\mathfrak{H}}$ has sides $x = \pm 1$ and bottom $y = \sqrt{2 - (x \pm 1)^2}$, the two circular arcs meeting at i and the corners where the sides meet the bottom being at $\pm 1 + i\sqrt{2}$.

The second part follows from the famous result of Schwarz [8]. What is relevant here is that the mapping function for the hyperbolic triangle with vertex angles $\lambda\pi, \mu\pi, \nu\pi$ is transcendental whenever $\lambda + \mu + \nu \leq 1$, and we have $\frac{1}{4} + \frac{1}{2} + 0 < 1$.

We proceed to compute s. From [2, pp. 101, 95], it follows that $\{s, w\} = 2p$ is the Schwarzian equation for s where

$$p = \frac{1}{64}\frac{12w^2 - 11w + 15}{w^2(w-1)^2}.$$

Thus, Γ is the monodromy group of $y'' + py = 0$ [2, Sec. 1.4]. Consulting

Secs. V.7 and VI.5 of [5], it follows from the formulas on p. 315 there that

$$z = \frac{\Gamma(\frac{1}{4})^2 w^{1/4} F(\frac{3}{8}, \frac{7}{8}, \frac{5}{4}; w)}{2\sqrt{2}\Gamma(\frac{3}{4})^2 F(\frac{1}{8}, \frac{5}{8}, \frac{3}{4}; w)}$$

maps $\bar{\mathfrak{H}}$ to the appropriate hyperbolic triangle in the unit disc *via* $0 \mapsto 0, 1 \mapsto 1$, and $i\infty \mapsto (\sqrt{2} - 1)e^{i\pi/4}$. Now, s should map $\bar{\mathfrak{H}}$ to the same hyperbolic triangle, but *via* $1 \mapsto 0, i\infty \mapsto 1$, and $0 \mapsto (\sqrt{2}-1)e^{i\pi/4}$. Therefore, using a standard identity for the hypergeometric function [e.g., 9, 60:5:3], we obtain

Theorem 1.5.

$$s(w) = z\left(\frac{w-1}{w}\right) = \frac{\Gamma(\frac{1}{4})^2 (w-1)^{1/4} F(\frac{3}{8}, \frac{3}{8}, \frac{5}{4}; 1-w)}{2\sqrt{2}\Gamma(\frac{3}{4})^2 F(\frac{1}{8}, \frac{1}{8}, \frac{3}{4}; 1-w)}.$$

Now form the composition

$$\tau = i\frac{1+s}{1-s}$$

so that the image hyperbolic triangle has vertices $i, -1 + i\sqrt{2}, i\infty$ in $\bar{\mathfrak{H}}$. The inverse map $J(\tau) = w$ is the absolute or Hecke *modular invariant* for Γ. Observe that $J^{1/2}$ and $(J-1)^{1/4}$ are both holomorphic on $\mathfrak{H}$.

2. Nonlinear Lattices

For each $k > 0, \{\eta = k\}$ is a hyperboloid of two sheets, each characterized by the sign of z. Note that $z \neq 0$ and $|z| > |x|, |y|$. We shall be concentrating on the sheets with $z > 0$.

Our first main result in this section is a characterization of all integral solutions of $\eta = k$ for $k > 0$.

Theorem 2.1. For $k > 0$, every solution $(z, x, y) \in \mathbb{R}^3$ of $\eta = z^2 - x^2 - y^2 = k$ can be written in the form

$$z = \frac{k + a^2}{2n} + n - a$$

$$x = \frac{k + a^2}{2n} - a$$

$$y = n - a$$

where $n, a \in \mathbb{R}$ with $n \neq 0$. Moreover, $(z, x, y) \in \mathbb{Z}^3$ if and only if $k, a, n \in \mathbb{Z}$, $n \neq 0$, and $k \equiv -a^2 \pmod{2n}$.

Proof. Let $z, x, y \in \mathbb{R}$ such that $\eta = k$ and define

$$n = z - x$$
$$a = z - x - y.$$

Then $n, a \in \mathbb{R}$, and since $k > 0$, we have $z \neq x$ whence $n \neq 0$. It follows that

$$z = \frac{k + a^2}{2n} + n - a$$
$$x = \frac{k + a^2}{2n} - a$$
$$y = n - a.$$

If $z, x, y \in \mathbb{Z}$, then from the definition of n and a we see that $n, a \in \mathbb{Z}$. Note that we must also have $\frac{k+a^2}{2n} \in \mathbb{Z}$ since $\frac{k+a^2}{2n} = z - y$. Therefore $k \equiv -a^2 \pmod{2n}$ and $k \in \mathbb{Z}$.

The remainder of the proof follows directly.

Note that the proof of the theorem implies that for $k > 0$ there is a one-to-one correspondence between the integral solutions (z, x, y) of $\eta = k$ and the integer pairs (n, a) satisfying $k \equiv -a^2 \pmod{2n}$, $n \neq 0$.

The hyperboloid of two sheets can now be characterized by the sign of n. Since $|z| > |x|$, $z > 0$ implies $z - x > 0$, thus $n > 0$. Similarly, $z < 0$ implies $n < 0$.

We next wish to find $SO_1^{2+}(\mathbb{Z})$- and $O_1^{2+}(\mathbb{Z})$-orbit representatives for the integral solutions (z, x, y) of $\eta = k$ where k is a positive integer. Note that we can have integral solutions if and only if $k \in \mathbb{Z}$. Now, the matrix $\begin{bmatrix} 1 & 0 & 0 \\ 0 & 0 & 1 \\ 0 & -1 & 0 \end{bmatrix} \in SO_1^{2+}(\mathbb{Z})$ relates solutions $(z, x, y), (z, -y, x), (z, -x, -y)$, and $(z, y, -x)$ as being in the same $SO_1^{2+}(\mathbb{Z})$-orbit. Similarly, (z, y, x), $(z, -x, y), (z, -y, -x)$, and $(z, x, -y)$ are in the same $SO_1^{2+}(\mathbb{Z})$-orbit. Therefore, given solutions $(z, \pm x, \pm y)$ and $(z, \pm y, \pm x)$, we have at most two distinct $SO_1^{2+}(\mathbb{Z})$-orbit representatives contained within these; for example, (z, x, y) and (z, y, x). However, since $\begin{bmatrix} 1 & 0 & 0 \\ 0 & 0 & 1 \\ 0 & 1 & 0 \end{bmatrix} \in O_1^{2+}(\mathbb{Z})$, we see that

(z, x, y) and (z, y, x) are in the same $O_1^{2+}(\mathbb{Z})$-orbit. Thus, these solutions contain exactly one $O_1^{2+}(\mathbb{Z})$-orbit representative.

Note that (z, x, y) and (z, y, x) are in the same $SO_1^{2+}(\mathbb{Z})$-orbit if either $x = 0$ or $y = 0$. Also, since $\begin{bmatrix} 3 & -2 & -2 \\ 2 & -2 & -1 \\ 2 & -1 & -2 \end{bmatrix} \in SO_1^{2+}(\mathbb{Z})$, they are in the same $SO_1^{2+}(\mathbb{Z})$-orbit if $z = x + y$.

Therefore, we need consider only the solutions (z, x, y) of $\eta = k$ such that $k \in \mathbb{Z}, k > 0, z, x, y \in \mathbb{Z}$, and $z > x, y \geq 0$.

Lemma 2.2. Let $n, a, k \in \mathbb{Z}$ with $n, k > 0$ such that $k \equiv -a^2 \pmod{2n}$. Then

$$\sqrt{2}\sqrt{k + a^2} \leq \frac{k + a^2}{2n} + n \leq \frac{1}{2}(k + a^2) + 1 .$$

Proof. Since $n \in \mathbb{Z}$ and $n > 0$, then $n \geq 1$. Also, since $k \equiv -a^2 \pmod{2n}$, then $\frac{1}{2}(k + a^2) \geq n \geq 1$. Therefore, we wish to find the minimum and maximum of $f(x) = \frac{k + a^2}{2x} + x$ on the interval $[1, \frac{1}{2}(k + a^2)]$. From the calculus we obtain an absolute minimum at $x = \sqrt{\frac{1}{2}(k + a^2)}$ and the absolute maximum at the endpoints of the interval. The result follows.

Theorem 2.3. If $k \in \mathbb{Z}$ and $k > 0$, then each $SO_1^{2+}(\mathbb{Z})$-orbit has a representative $(z, x, y) \in \mathbb{Z}^3$ with $z, x, y \geq 0$ and $z \geq x + y$.

Proof. Suppose that there is an $SO_1^{2+}(\mathbb{Z})$-orbit ξ which does not have a representative $(z, x, y) \in \mathbb{Z}^3$, with $z, x, y \geq 0$, of the form $z \geq x + y$. Let $\varphi = \xi \cap \{(z, x, y); z, x, y \geq 0, z < x + y\}$. Note that φ is non-empty. Since $k > 0$, for each $(z, x, y) \in \varphi$ we must have $z \geq 1$. Thus there would exist some $(z_0, x_0, y_0) \in \varphi$ such that $z_0 \leq z$ for every $(z, x, y) \in \varphi$.

Now, there exist $a_0, n_0 \in \mathbb{Z}, n_0 > 0, k \equiv -a_0^2 \pmod{2n_0}$, such that

$$z_0 = \frac{k + a_0^2}{2n_0} + n_0 - a_0$$

$$x_0 = \frac{k + a_0^2}{2n_0} - a_0$$

$$y_0 = n_0 - a_0 .$$

Note that $\begin{bmatrix} 3 & -2 & -2 \\ 2 & -2 & -1 \\ 2 & -1 & -2 \end{bmatrix} \in SO_1^{2+}(\mathbb{Z})$ and let

$$\begin{bmatrix} z_1 \\ x_1 \\ y_1 \end{bmatrix} = \begin{bmatrix} 3 & -2 & -2 \\ 2 & -2 & -1 \\ 2 & -1 & -2 \end{bmatrix} \begin{bmatrix} z_0 \\ x_0 \\ y_0 \end{bmatrix},$$

from which we obtain

$$z_1 = \frac{k + a_0^2}{2n_0} + n_0 + a_0$$

$$x_1 = n_0 + a_0$$

$$y_1 = \frac{k + a_0^2}{2n_0} + a_0.$$

From Lemma 2.2, we have

$$z_1 \geq \sqrt{2}\sqrt{k + a_0^2} + a_0,$$

so $z_1 > 0$. However, we also have

$$z_1 = z_0 + 2a_0.$$

Since $a_0 = z_0 - x_0 - y_0$, it follows that $a_0 < 0$ whence $z_1 < z_0$.

Now, one of the matrices $\begin{bmatrix} 1 & 0 & 0 \\ 0 & 0 & 1 \\ 0 & -1 & 0 \end{bmatrix}^m \in SO_1^{2+}(\mathbb{Z}), m = 1, 2, 3, 4$, will produce a representative $(z_1, x_2, y_2) \in \varphi$ from the representative (z_1, x_1, y_1). Thus we would have a representative $(z_1, x_2, y_2) \in \varphi$ such that $z_1 < z_0$, which is a contradiction. Therefore, every orbit must contain a representative $(z, x, y) \in \mathbb{Z}^3$ with $z, x, y \geq 0$ such that $z \geq x + y$.

Following the discussion preceding Lemma 2.2, we obtain

Corollary 2.4. Every $O_1^{2+}(\mathbb{Z})$-orbit contains a unique representative satisfying the additional requirement that $z > x \geq y$.

Proof. We know that each $O_1^{2+}(\mathbb{Z})$-orbit has a representative $(z, x, y) \in \mathbb{Z}^3$ with $z > x \geq y \geq 0$ and $z \geq x + y$. Suppose there is an $O_1^{2+}(\mathbb{Z})$-orbit that has two such representatives $\xi_1 = (z_1, x_1, y_1)$ and $\xi_2 = (z_2, x_2, y_2)$. Then there would exist an $A = [a_{ij}] \in O_1^{2+}(\mathbb{Z})$, such that

$$\xi_2 = A\xi_1 .$$

Without loss of generality, we may assume that $z_1 \geq z_2$. Note that for each such matrix A, we have $a_{11} \equiv 1 \pmod 2$ and $a_{11} > 0$.

If $a_{11} = 1$ or $a_{11} = 3$, then by inspection we see that $\xi_1 = \xi_2$.

Now assume $a_{11} \geq 5$. Then

$$z_2 = a_{11}z_1 + a_{12}x_1 + a_{13}y_1 ,$$

and since $x_1 \geq y_1 \geq 0$, it follows that

$$z_2 \geq a_{11}z_1 - \max\{|a_{12}|, |a_{13}|\}x_1 - \min\{|a_{12}|, |a_{13}|\}y_1 .$$

Without loss of generality, $|a_{12}| \geq |a_{13}|$. Hence

$$z_2 \geq a_{11}z_1 - |a_{12}|x_1 - |a_{13}|y_1 .$$

Note that $|a_{12}| < a_{11}$.

Case 1. $|a_{12}| = a_{11} - 1$. Then

$$a_{11}^2 - a_{12}^2 - a_{13}^2 = 1$$
$$a_{11}^2 - (a_{11} - 1)^2 - a_{13}^2 = 1$$
$$a_{13}^2 = 2(a_{11} - 1)$$
$$|a_{13}| = \sqrt{2}\sqrt{a_{11} - 1}$$
$$\leq \frac{1}{2}(a_{11} + 1)$$
$$\leq a_{11} - \frac{1}{2}a_{11} + \frac{1}{2}$$
$$\leq a_{11} - 2 .$$

Therefore,

$$a_{11}z_1 + a_{12}x_1 + a_{13}y_1 \geq a_{11}z_1 - (a_{11} - 1)x_1 - (a_{11} - 2)y_1$$
$$\geq z_1 + (a_{11} - 1)(z_1 - x_1 - y_1) + y_1 .$$

If $z_1 = x_1 + y_1$, then $y_1 > 0$. Hence $z_2 > z_1$.

Case 2. $|a_{12}| \leq a_{11} - 2$. Then also $|a_{13}| \leq a_{11} - 2$. Thus

$$a_{11}z_1 + a_{12}x_1 + a_{13}y_1 \geq a_{11}z_1 - (a_{11} - 2)x_1 - (a_{11} - 2)y_1$$
$$\geq 2z_1 + (a_{11} - 2)(z_1 - x_1 - y_1)$$
$$> z_1.$$

In both cases we see that $z_2 > z_1$, a contradiction. Thus if $a_{11} \geq 5$, there exists no such A.

Therefore we have $\xi_1 = \xi_2$, and uniqueness is established.

Our last goal in this section is finiteness of the number of orbits on each sheet $\{\eta = k\}$.

Lemma 2.5. Let $k > 0$. If $\eta = k$ for $x, y, z \in \mathbb{Z}$ with $x, y, z \geq 0$ and $x + y \leq z$, then $z \leq \frac{1}{2}k + 1$.

Proof. First consider the case where $x + y < z$. Then

$$(x + y)^2 \leq (z - 1)^2.$$

Now, we also have

$$(x + y)^2 = z^2 - k + 2xy.$$

Therefore

$$(z - 1)^2 \geq z^2 - k + 2xy$$

which simplifies to $z \leq \frac{1}{2}(k + 1)$.

Now consider $x + y = z$. We know that there exist $n, a \in \mathbb{Z}$ with $n \neq 0$ and $k \equiv -a^2 \pmod{2n}$ such that

$$z = \frac{k + a^2}{2n} + n - a.$$

From Lemma 2.2, we see that

$$z \leq \frac{1}{2}(k + a^2) + 1 - a.$$

Since $a = z - x - y, a = 0$. Thus $z \leq \frac{1}{2}k + 1$ again, as desired.

At this point, we actually have established the desired finiteness. However, we choose to incorporate into the proof computationally more effective bounds for later use.

Lemma 2.6. Let $k > 0$. If $\eta = k$ for $x, y, z \in \mathbb{Z}$ with $x, y, z \geq 0$ and $x + y \leq z$, then

$$0 \leq a \leq \sqrt{k}$$
$$a \leq n \leq \frac{1}{2}(k+1),$$

where $a = z - x - y$ and $n = z - x$. If $x \geq y$, then $n \leq \sqrt{\frac{1}{2}(k + a^2)}$.

Proof. As a result of Theorem 2.1, we have

$$x = \frac{k + a^2}{2n} - a$$
$$y = n - a.$$

Since $x, y \geq 0$,

$$n \geq a$$
$$k + a^2 \geq 2an.$$

Now, $a \geq 0$ since $z \geq x + y$. Therefore,

$$k + a^2 \geq 2a^2$$
$$a \leq \sqrt{k}.$$

The general upper bound on n can be derived from three cases. First, if $a = 0$, then $2n|k$ so $n \leq \frac{1}{2}k$. If $a = 1$, then $2n|k+1$ and $n \leq \frac{1}{2}(k+1)$. Finally, if $a \geq 2$, then $\frac{k+a^2}{2n} \geq a$ and $n \leq \frac{k+a^2}{2a}$ whence $n \leq \frac{1}{2}k$. The bound on n for $x \geq y$ follows easily.

Theorem 2.7. For each integer $k > 0$, there are a finite number of $SO_1^{2+}(\mathbb{Z})$-orbits on $\{\eta = k\}$.

Proof. From Theorem 2.3, we know that any $SO_1^{2+}(\mathbb{Z})$-orbit must have a representative $(z, x, y) \in \mathbb{Z}^3$ such that $z, x, y \geq 0$ and $z \geq x + y$.

Also, from Theorem 2.1 and Lemma 2.6, we know that any representative of this form can equivalently be expressed in terms of integers n and a with $n = z - x$ and $a = z - x - y$, where n and a must satisfy the bounds $0 \leq a \leq \sqrt{k}$ and $a \leq n \leq \frac{1}{2}(k + 1)$. Therefore there are a finite number of different representatives of this form, hence a finite number of distinct orbits.

3. Numerical Results

We begin this section with an improved bound.

Lemma 3.1. Let $k > 0$. If $\eta = k$ for $x, y, z \in \mathbb{Z}$ with $x, y, z \geq 0$ and $z \geq x + y$, then

$$\min\{x, y\} \leq \sqrt{\frac{1}{2}k} \, .$$

Proof. Without loss of generality, we may assume $y \leq x$. From Lemma 2.6, we know that $n \leq \sqrt{\frac{1}{2}(k + a^2)}$ where $n = z - x$ and $a = z - x - y$. Therefore, since $y = n - a$, we have

$$y \leq \sqrt{\frac{1}{2}(k + a^2)} - a \, .$$

Note that the bounds on a are $0 \leq a \leq \sqrt{k}$. Thus we want to maximize $f(x) = \sqrt{\frac{1}{2}(x^2 + k)} - x$ on the interval $[0, \sqrt{k}]$. From the calculus, we see that we have a maximum at $x = 0$. Thus, $y \leq \sqrt{\frac{1}{2}k}$.

From Theorem 2.7 we know that there is a finite number m of $O_1^{2+}(\mathbb{Z})$-orbits on $\{\eta = k\}$ for each integer $k > 0$. The following lemma provides an upper bound for this m, given k.

Lemma 3.2. Let $k > 0$ and let m be the number of $O_1^{2+}(\mathbb{Z})$-orbits for k. Then

$$m \leq \frac{7\sqrt{2}}{24}k + \frac{4 + 3\sqrt{2}}{8}\sqrt{k} + \frac{6 + \sqrt{2}}{12} \, .$$

Proof. Consider the function

$$f(x) = \frac{\sqrt{2}}{4\sqrt{k}}x^2 + \frac{\sqrt{2}}{2}\sqrt{k} - \sqrt{\frac{1}{2}(x^2+k)}$$

on the interval $[0, \sqrt{k}]$. From the calculus we see that the minimum value is 0 at $x = 0$. Therefore $f(x) \geq 0$ on $[0, \sqrt{k}]$. Thus we have

$$\sqrt{\frac{1}{2}(k+a^2)} \leq \frac{\sqrt{2}}{4\sqrt{k}}a^2 + \frac{\sqrt{2}}{2}\sqrt{k}$$

for $0 \leq a \leq \sqrt{k}$.

An upper bound on m can then be derived by estimating the number of lattice points contained within the region

$$\{(n,a) | 0 \leq a \leq \sqrt{k},\, a \equiv k \,(\mathrm{mod}\ 2),\, a \leq n \leq \frac{\sqrt{2}}{4\sqrt{k}}a^2 + \frac{\sqrt{2}}{2}\sqrt{k},\, n \neq 0\}.$$

More precise estimates on the upper bound of m should be possible.

The following FORTRAN subroutine can be used to obtain those $O_1^{2+}(\mathbb{Z})$-orbit representatives specified in Theorem 2.3 and Corollary 2.4. It generates representatives by searching for divisors n of $\frac{1}{2}(k+a^2)$, incorporating the bounds found in Lemma 2.6 with the additional requirements that $n \neq 0$ and $k \equiv a \,(\mathrm{mod}\ 2)$. To use the subroutine, the value of k, represented by the parameter K, and the maximum value of m, represented by the parameter MMAX, must be specified beforehand. The appropriate value of MMAX can be approximated *via* Lemma 3.2. The subroutine will then generate and return the corresponding x-,y-, and z-coordinates of the orbit representatives (the parameters X, Y, and Z), and the number of orbits m (the parameter M).

```
      SUBROUTINE ORBIT (K, MMAX, M, X, Y, Z)
      INTEGER MMAX, Z(MMAX), X(MMAX), Y(MMAX)
      INTEGER K, M, N, NL, NU, A, AL, AU, Q
      M = 0
      AL = K-2*(K/2)
      AU = IFIX (SQRT (FLOAT (K)) +. 5)
      IF(AU*AU. GT. K) AU = AU - 1
      NL = 1
      DO 200 A = AL, AU, 2
          Q = (K + A*A)/2
          NU = IFIX (SQRT (FLOAT (Q)) +. 5)
          IF (NU*NU. GT. Q) NU = NU - 1
          DO 100 N = NL, NU
             L = Q/N
             IF (N*L. EQ. Q) THEN
                M = M + 1
                Z(M) = L + N - A
                X(M) = L - A
                Y(M) = N - A
             END IF
100       CONTINUE
          NL = A + 2
200   CONTINUE
      RETURN
      END
```

We conclude with three tables giving sample outputs from the subroutine.

Table 1 lists one representative for each $O_1^{2+}(\mathbb{Z})$-orbit on each sheet $\{\eta = k\}$ for $1 \leq k \leq 100$. $SO_1^{2+}(\mathbb{Z})$-orbit representatives are mostly the same, the exceptions being (z,x,y) with both $x, y \neq 0, x \neq y$, and $z \neq x + y$. In this case, there are two $SO_1^{2+}(\mathbb{Z})$-orbit representatives: (z,x,y) and (z,y,x).

Table 2 lists the number of $O_1^{2+}(\mathbb{Z})$-orbits m for each $k \leq 1000$.

Table 3 lists the values of $k \leq 10000$ for which there are $m, O_1^{2+}(\mathbb{Z})$-orbits for $1 \leq m \leq 6$. It provides support for the conjecture that there are at most finitely many values of $k > 0$ for which are number of $O_1^{2+}(\mathbb{Z})$-orbits is m.

Table 1. $O_1^{2+}(\mathbb{Z})$-orbit representatives

k	orbit representatives (z,x,y)
1	$(1,0,0)$
2	$(2,1,1)$
3	$(2,1,0)$
4	$(2,0,0), (3,2,1)$
5	$(3,2,0)$
6	$(4,3,1)$
7	$(3,1,1), (4,3,0)$
8	$(3,1,0), (4,2,2), (5,4,1)$
9	$(3,0,0), (5,4,0)$
10	$(6,5,1)$
11	$(4,2,1), (6,5,0)$
12	$(4,2,0), (5,3,2), (7,6,1)$
13	$(7,6,0)$
14	$(4,1,1), (8,7,1)$
15	$(4,1,0), (5,3,1), (8,7,0)$
16	$(4,0,0), (5,3,0), (6,4,2), (9,8,1)$
17	$(5,2,2), (9,8,0)$
18	$(6,3,3), (10,9,1)$
19	$(6,4,1), (10,9,0)$
20	$(5,2,1), (6,4,0), (7,5,2), (11,10,1)$
21	$(5,2,0), (11,10,0)$
22	$(12,11,1)$
23	$(5,1,1), (6,3,2), (7,5,1), (12,11,0)$
24	$(5,1,0), (7,4,3), (7,5,0), (8,6,2), (13,12,1)$
25	$(5,0,0), (13,12,0)$
26	$(6,3,1), (14,13,1)$
27	$(6,3,0), (8,6,1), (14,13,0)$
28	$(6,2,2), (8,6,0), (9,7,2), (15,14,1)$
29	$(7,4,2), (15,14,0)$
30	$(8,5,3), (16,15,1)$
31	$(6,2,1), (7,3,3), (9,7,1), (16,15,0)$
32	$(6,2,0), (7,4,1), (8,4,4), (9,7,0), (10,8,2), (17,16,1)$
33	$(7,4,0), (17,16,0)$
34	$(6,1,1), (18,17,1)$
35	$(6,1,0), (8,5,2), (10,8,1), (18,17,0)$
36	$(6,0,0), (7,3,2), (9,6,3), (10,8,0), (11,9,2), (19,18,1)$
37	$(19,18,0)$
38	$(8,5,1), (20,19,1)$
39	$(7,3,1), (8,4,3), (8,5,0), (11,9,1), (20,19,0)$
40	$(7,3,0), (9,5,4), (11,9,0), (12,10,2), (21,20,1)$
41	$(7,2,2), (9,6,2), (21,20,0)$
42	$(10,7,3), (22,21,1)$
43	$(12,10,1), (22,21,0)$
44	$(7,2,1), (8,4,2), (9,6,1), (12,10,0), (13,11,2), (23,22,1)$
45	$(7,2,0), (9,6,0), (23,22,0)$
46	$(8,3,3), (24,23,1)$
47	$(7,1,1), (8,4,1), (9,5,3), (10,7,2), (13,11,1), (24,23,0)$
48	$(7,1,0), (8,4,0), (10,6,4), (11,8,3), (13,11,0), (14,12,2), (25,24,1)$
49	$(7,0,0), (9,4,4), (25,24,0)$
50	$(10,5,5), (10,7,1), (26,25,1)$

G. J. Fox & P. E. Parker

Table 1. Cont'd.

k	orbit representatives (z,x,y)
51	$(8,3,2)$, $(10,7,0)$, $(14,12,1)$, $(26,25,0)$
52	$(9,5,2)$, $(14,12,0)$, $(15,13,2)$, $(27,26,1)$
53	$(11,8,2)$, $(27,26,0)$
54	$(8,3,1)$, $(12,9,3)$, $(28,27,1)$
55	$(8,3,0)$, $(9,5,1)$, $(10,6,3)$, $(15,13,1)$, $(28,27,0)$
56	$(8,2,2)$, $(9,4,3)$, $(9,5,0)$, $(11,7,4)$, $(11,8,1)$, $(15,13,0)$, $(16,14,2)$, $(29,28,1)$
57	$(11,8,0)$, $(29,28,0)$
58	$(30,29,1)$
59	$(8,2,1)$, $(10,5,4)$, $(12,9,2)$, $(16,14,1)$, $(30,29,0)$
60	$(8,2,0)$, $(10,6,2)$, $(11,6,5)$, $(13,10,3)$, $(16,14,0)$, $(17,15,2)$, $(31,30,1)$
61	$(9,4,2)$, $(31,30,0)$
62	$(8,1,1)$, $(12,9,1)$, $(32,31,1)$
63	$(8,1,0)$, $(9,3,3)$, $(10,6,1)$, $(11,7,3)$, $(12,9,0)$, $(17,15,1)$, $(32,31,0)$
64	$(8,0,0)$, $(9,4,1)$, $(10,6,0)$, $(12,8,4)$, $(17,15,0)$, $(18,16,2)$, $(33,32,1)$
65	$(9,4,0)$, $(13,10,2)$, $(33,32,0)$
66	$(10,5,3)$, $(14,11,3)$, $(34,33,1)$
67	$(18,16,1)$, $(34,33,0)$
68	$(9,3,2)$, $(10,4,4)$, $(11,7,2)$, $(13,10,1)$, $(18,16,0)$, $(19,17,2)$, $(35,34,1)$
69	$(11,6,4)$, $(13,10,0)$, $(35,34,0)$
70	$(12,7,5)$, $(36,35,1)$
71	$(9,3,1)$, $(10,5,2)$, $(11,5,5)$, $(11,7,1)$, $(12,8,3)$, $(14,11,2)$, $(19,17,1)$, $(36,35,0)$
72	$(9,3,0)$, $(11,7,0)$, $(12,6,6)$, $(13,9,4)$, $(15,12,3)$, $(19,17,0)$, $(20,18,2)$, $(37,36,1)$
73	$(9,2,2)$, $(37,36,0)$
74	$(10,5,1)$, $(14,11,1)$, $(38,37,1)$
75	$(10,4,3)$, $(10,5,0)$, $(14,11,0)$, $(20,18,1)$, $(38,37,0)$
76	$(9,2,1)$, $(11,6,3)$, $(12,8,2)$, $(20,18,0)$, $(21,19,2)$, $(39,38,1)$
77	$(9,2,0)$, $(15,12,2)$, $(39,38,0)$
78	$(16,13,3)$, $(40,39,1)$
79	$(9,1,1)$, $(12,7,4)$, $(12,8,1)$, $(13,9,3)$, $(21,19,1)$, $(40,39,0)$
80	$(9,1,0)$, $(10,4,2)$, $(11,5,4)$, $(12,8,0)$, $(13,8,5)$, $(14,10,4)$, $(15,12,1)$, $(21,19,0)$, $(22,20,2)$, $(41,40,1)$
81	$(9,0,0)$, $(11,6,2)$, $(15,12,0)$, $(41,40,0)$
82	$(10,3,3)$, $(42,41,1)$
83	$(10,4,1)$, $(12,6,5)$, $(16,13,2)$, $(22,20,1)$, $(42,41,0)$
84	$(10,4,0)$, $(11,6,1)$, $(13,7,6)$, $(13,9,2)$, $(17,14,3)$, $(22,20,0)$, $(23,21,2)$, $(43,42,1)$
85	$(11,6,0)$, $(43,42,0)$
86	$(12,7,3)$, $(16,13,1)$, $(44,43,1)$
87	$(10,3,2)$, $(11,5,3)$, $(13,9,1)$, $(14,10,3)$, $(16,13,0)$, $(23,21,1)$, $(44,43,0)$
88	$(13,9,0)$, $(15,11,4)$, $(23,21,0)$, $(24,22,2)$, $(45,44,1)$
89	$(11,4,4)$, $(13,8,4)$, $(17,14,2)$, $(45,44,0)$
90	$(10,3,1)$, $(14,9,5)$, $(18,15,3)$, $(46,45,1)$
91	$(10,3,0)$, $(12,7,2)$, $(24,22,1)$, $(46,45,0)$
92	$(10,2,2)$, $(11,5,2)$, $(12,6,4)$, $(14,10,2)$, $(17,14,1)$, $(24,22,0)$, $(25,23,2)$, $(47,46,1)$
93	$(17,14,0)$, $(47,46,0)$
94	$(12,5,5)$, $(12,7,1)$, $(48,47,1)$
95	$(10,2,1)$, $(11,5,1)$, $(12,7,0)$, $(13,7,5)$, $(14,10,1)$, $(15,11,3)$, $(18,15,2)$, $(25,23,1)$, $(48,47,0)$
96	$(10,2,0)$, $(11,4,3)$, $(11,5,0)$, $(13,8,3)$, $(14,8,6)$, $(14,10,0)$, $(16,12,4)$, $(19,16,3)$, $(25,23,0)$, $(26,24,2)$, $(49,48,1)$
97	$(13,6,6)$, $(49,48,0)$
98	$(10,1,1)$, $(14,7,7)$, $(18,15,1)$, $(50,49,1)$
99	$(10,1,0)$, $(12,6,3)$, $(14,9,4)$, $(18,15,0)$, $(26,24,1)$, $(50,49,0)$
100	$(10,0,0)$, $(15,10,5)$, $(15,11,2)$, $(26,24,0)$, $(27,25,2)$, $(51,50,1)$

Table 2. Number of $O_1^{2+}(\mathbb{Z})$-orbits m for $k \leq 1000$

k	m	k	m	k	m	k	m	k	m	k	m	k	m
1	1	51	4	101	4	151	8	201	4	251	11	301	3
2	1	52	4	102	2	152	10	202	2	252	15	302	4
3	1	53	2	103	6	153	5	203	7	253	2	303	11
4	2	54	3	104	10	154	3	204	12	254	5	304	14
5	1	55	5	105	4	155	7	205	3	255	14	305	5
6	1	56	8	106	2	156	11	206	6	256	12	306	7
7	2	57	2	107	5	157	2	207	11	257	5	307	5
8	3	58	1	108	9	158	3	208	10	258	3	308	13
9	2	59	5	109	2	159	11	209	6	259	7	309	4
10	1	60	7	110	4	160	11	210	4	260	13	310	3
11	2	61	2	111	9	161	6	211	5	261	6	311	20
12	3	62	3	112	8	162	4	212	9	262	2	312	10
13	1	63	7	113	3	163	2	213	3	263	14	313	3
14	2	64	7	114	3	164	12	214	2	264	15	314	7
15	3	65	3	115	4	165	4	215	15	265	3	315	12
16	4	66	3	116	9	166	3	216	15	266	6	316	12
17	2	67	2	117	4	167	12	217	4	267	4	317	3
18	2	68	7	118	2	168	10	218	3	268	6	318	4
19	2	69	3	119	12	169	3	219	7	269	6	319	11
20	4	70	2	120	10	170	4	220	11	270	6	320	20
21	2	71	8	121	3	171	9	221	5	271	12	321	6
22	1	72	8	122	3	172	6	222	4	272	17	322	4
23	4	73	2	123	4	173	4	223	8	273	4	323	7
24	5	74	3	124	8	174	4	224	18	274	4	324	15
25	2	75	5	125	4	175	9	225	6	275	9	325	5
26	2	76	6	126	5	176	14	226	3	276	13	326	6
27	3	77	3	127	6	177	2	227	8	277	2	327	13
28	4	78	2	128	11	178	3	228	8	278	4	328	8
29	2	79	6	129	4	179	8	229	3	279	17	329	8
30	2	80	10	130	2	180	12	230	6	280	10	330	4
31	4	81	4	131	8	181	3	231	14	281	6	331	5
32	6	82	2	132	8	182	4	232	5	282	3	332	15
33	2	83	5	133	2	183	9	233	4	283	5	333	4
34	2	84	8	134	4	184	8	234	6	284	16	334	4
35	4	85	2	135	10	185	5	235	4	285	6	335	19
36	6	86	3	136	8	186	4	236	15	286	4	336	20
37	1	87	7	137	3	187	4	237	4	287	16	337	3
38	2	88	5	138	3	188	12	238	4	288	17	338	5
39	5	89	4	139	5	189	6	239	16	289	4	339	10
40	5	90	4	140	12	190	2	240	15	290	6	340	8
41	3	91	4	141	3	191	14	241	4	291	7	341	8
42	2	92	8	142	2	192	13	242	4	292	7	342	6
43	2	93	2	143	11	193	2	243	8	293	5	343	10
44	6	94	3	144	14	194	6	244	9	294	5	344	15
45	3	95	9	145	3	195	8	245	5	295	9	345	4
46	2	96	11	146	5	196	9	246	4	296	15	346	3
47	6	97	2	147	5	197	3	247	7	297	6	347	8
48	7	98	4	148	4	198	4	248	13	298	2	348	15
49	3	99	6	149	4	199	10	249	4	299	13	349	4
50	3	100	6	150	4	200	13	250	4	300	15	350	7

G. J. Fox & P. E. Parker

Table 2. Cont'd.

k	m	k	m	k	m	k	m	k	m	k	m	k	m
351	18	401	6	451	10	501	5	551	27	601	6	651	14
352	11	402	5	452	12	502	4	552	15	602	7	652	6
353	5	403	4	453	4	503	22	553	4	603	9	653	4
354	5	404	19	454	4	504	23	554	6	604	16	654	8
355	7	405	7	455	22	505	3	555	8	605	8	655	13
356	17	406	5	456	15	506	8	556	15	606	4	656	30
357	4	407	17	457	3	507	8	557	5	607	14	657	7
358	2	408	10	458	7	508	12	558	8	608	24	658	4
359	20	409	5	459	14	509	8	559	17	609	6	659	17
360	20	410	5	460	12	510	6	560	28	610	4	660	16
361	4	411	10	461	8	511	16	561	6	611	16	661	5
362	5	412	12	462	4	512	20	562	3	612	20	662	6
363	8	413	6	463	8	513	6	563	14	613	3	663	18
364	12	414	7	464	23	514	5	564	13	614	9	664	15
365	6	415	11	465	6	515	10	565	4	615	22	665	8
366	4	416	24	466	3	516	18	566	8	616	15	666	9
367	10	417	4	467	11	517	4	567	20	617	4	667	7
368	16	418	3	468	17	518	5	568	8	618	4	668	24
369	8	419	14	469	5	519	19	569	9	619	8	669	4
370	4	420	16	470	6	520	10	570	6	620	21	670	4
371	13	421	3	471	17	521	9	571	8	621	10	671	31
372	8	422	3	472	10	522	4	572	23	622	4	672	22
373	3	423	17	473	4	523	8	573	5	623	24	673	4
374	8	424	10	474	6	524	24	574	6	624	23	674	7
375	14	425	9	475	9	525	8	575	23	625	5	675	17
376	13	426	7	476	24	526	4	576	27	626	10	676	11
377	5	427	4	477	6	527	20	577	3	627	8	677	8
378	6	428	15	478	3	528	20	578	6	628	9	678	6
379	5	429	6	479	26	529	5	579	13	629	10	679	20
380	19	430	4	480	22	530	8	580	13	630	8	680	20
381	6	431	22	481	5	531	15	581	8	631	14	681	6
382	3	432	21	482	6	532	8	582	5	632	13	682	4
383	18	433	4	483	8	533	4	583	9	633	6	683	8
384	21	434	8	484	11	534	6	584	23	634	4	684	27
385	4	435	8	485	6	535	15	585	9	635	16	685	4
386	6	436	9	486	8	536	20	586	5	636	23	686	10
387	9	437	6	487	8	537	4	587	11	637	5	687	13
388	7	438	3	488	15	538	3	588	15	638	6	688	14
389	6	439	16	489	6	539	15	589	5	639	23	689	11
390	6	440	20	490	5	540	22	590	6	640	21	690	6
391	16	441	9	491	14	541	3	591	23	641	8	691	8
392	16	442	3	492	12	542	7	592	10	642	5	692	19
393	4	443	8	493	4	543	13	593	7	643	5	693	9
394	3	444	19	494	8	544	18	594	10	644	24	694	3
395	13	445	3	495	23	545	9	595	8	645	6	695	25
396	18	446	9	496	16	546	8	596	19	646	5	696	20
397	2	447	15	497	8	547	5	597	4	647	24	697	4
398	6	448	14	498	3	548	12	598	3	648	18	698	7
399	18	449	6	499	5	549	9	599	26	649	6	699	16
400	14	450	8	500	18	550	5	600	20	650	9	700	19

Table 2. Cont'd.

k	m	k	m	k	m	k	m	k	m	k	m
701	9	751	16	801	11	851	16	901	7	951	27
702	6	752	24	802	4	852	13	902	8	952	16
703	15	753	4	803	16	853	3	903	18	953	9
704	28	754	6	804	18	854	12	904	13	954	9
705	8	755	19	805	6	855	27	905	7	955	7
706	7	756	26	806	8	856	10	906	8	956	32
707	10	757	3	807	15	857	9	907	5	957	6
708	8	758	6	808	10	858	6	908	24	958	5
709	3	759	26	809	9	859	11	909	12	959	38
710	9	760	10	810	11	860	31	910	6	960	27
711	27	761	11	811	11	861	8	911	32	961	6
712	13	762	4	812	21	862	3	912	20	962	8
713	8	763	7	813	4	863	22	913	4	963	15
714	8	764	28	814	4	864	35	914	10	964	17
715	8	765	8	815	31	865	5	915	14	965	12
716	24	766	7	816	28	866	12	916	14	966	8
717	5	767	23	817	4	867	11	917	6	967	12
718	4	768	23	818	8	868	14	918	6	968	18
719	32	769	6	819	18	869	9	919	20	969	8
720	30	770	10	820	13	870	6	920	30	970	4
721	6	771	10	821	8	871	23	921	6	971	23
722	6	772	7	822	6	872	15	922	5	972	24
723	7	773	7	823	10	873	7	923	16	973	4
724	14	774	9	824	28	874	6	924	30	974	10
725	9	775	17	825	10	875	20	925	5	975	23
726	7	776	28	826	4	876	21	926	11	976	23
727	14	777	6	827	11	877	3	927	27	977	6
728	20	778	4	828	23	878	6	928	11	978	7
729	9	779	16	829	6	879	23	929	10	979	13
730	4	780	24	830	6	880	23	930	8	980	22
731	19	781	6	831	29	881	11	931	12	981	9
732	19	782	8	832	20	882	10	932	17	982	3
733	4	783	26	833	10	883	5	933	5	983	28
734	11	784	21	834	5	884	23	934	7	984	20
735	21	785	5	835	10	885	8	935	30	985	7
736	18	786	5	836	28	886	5	936	30	986	12
737	6	787	8	837	6	887	30	937	6	987	14
738	7	788	14	838	4	888	20	938	5	988	15
739	8	789	9	839	34	889	6	939	13	989	10
740	23	790	5	840	20	890	7	940	12	990	12
741	8	791	34	841	5	891	16	941	12	991	18
742	3	792	20	842	7	892	16	942	4	992	31
743	22	793	3	843	10	893	8	943	18	993	4
744	20	794	11	844	15	894	8	944	35	994	6
745	5	795	8	845	9	895	17	945	12	995	13
746	7	796	20	846	12	896	36	946	5	996	18
747	15	797	8	847	13	897	6	947	8	997	4
748	12	798	6	848	23	898	4	948	18	998	7
749	9	799	18	849	8	899	22	949	4	999	34
750	8	800	30	850	7	900	23	950	12	1000	20

G. J. Fox & P. E. Parker

Table 3. Values of $k \leq 10000$ with m, $O_1^{2+}(\mathbb{Z})$-orbits

$m = 1$ (10 *values of* k)

1 2 3 5 6 10 13 22 37 58

$m = 2$ (50 *values of* k)

4 7 9 11 14 17 18 19 21 25 26 29 30 33 34 38 42 43 46 53 57 61 67 70 73
78 82 85 93 97 102 106 109 118 130 133 142 157 163 177 190 193 202 214 253
262 277 298 358 397

$m = 3$ (78 *values of* k)

8 12 15 27 41 45 49 50 54 62 65 66 69 74 77 86 94 113 114 121 122 137 138
141 145 154 158 166 169 178 181 197 205 213 218 226 229 258 265 282 301 310
313 317 337 346 373 382 394 418 421 422 438 442 445 457 466 478 498 505 538
541 562 577 598 613 694 709 742 757 793 853 862 877 982 1093 1213 1318

$m = 4$ (148 *values of* k)

16 20 23 28 31 35 51 52 81 89 90 91 98 101 105 110 115 117 123 125 129 134
148 149 150 162 165 170 173 174 182 186 187 198 201 210 217 222 233 235 237
238 241 242 246 249 250 267 273 274 278 286 289 302 309 318 322 330 333 334
345 349 357 361 366 370 385 393 403 417 427 430 433 453 454 462 473 493 502
517 522 526 533 537 553 565 597 606 610 617 618 622 634 653 658 669 670 673
682 685 697 718 730 733 753 762 778 802 813 814 817 826 838 898 913 942 949
970 973 993 997 1030 1033 1038 1042 1090 1117 1138 1162 1177 1198 1222 1237 1258
1282 1285 1297 1402 1453 1558 1597 1618 1642 1873 1978 2017 2062 2293

$m = 5$ (145 *values of* k)

24 39 40 55 59 75 83 88 107 126 139 146 147 153 185 211 221 232 245 254 257
283 293 294 305 307 325 331 338 353 354 362 377 379 402 406 409 410 469 481
490 499 501 514 518 529 547 550 557 573 582 586 589 625 637 642 643 646 661
717 745 785 786 790 834 841 865 883 886 907 922 925 933 938 946 958 1002 1018
1129 1142 1149 1153 1178 1201 1306 1345 1357 1366 1373 1417 1438 1450 1462 1465
1474 1477 1498 1510 1537 1549 1578 1621 1657 1698 1717 1738 1822 1842 1882 1933
1945 1957 2038 2053 2077 2113 2122 2137 2221 2242 2302 2353 2377 2410 2437 2458
2533 2557 2566 2578 2605 2773 2797 2878 2893 2902 2962 2998 3037 3217 3238 3322
3733 4678 5413

$m = 6$ (256 *values of* k)

32 36 44 47 76 79 99 100 103 127 161 172 189 194 206 209 225 230 234 261 266
268 269 270 281 285 290 297 321 326 342 365 378 381 386 389 390 398 401 413
429 437 449 465 470 474 477 482 485 489 510 513 534 554 561 570 574 578 590
601 609 633 638 645 649 652 662 678 681 690 702 721 722 737 754 758 769 777
781 798 805 822 829 830 837 858 870 874 878 889 897 910 917 918 921 937 957
961 977 994 1005 1006 1009 1021 1037 1045 1054 1057 1065 1066 1078 1081 1082 1102
1105 1110 1113 1114 1122 1126 1137 1165 1182 1185 1189 1246 1257 1261 1270 1273
1290 1302 1317 1333 1338 1342 1353 1354 1369 1378 1390 1393 1398 1429 1486 1489
1493 1497 1513 1522 1534 1542 1570 1582 1593 1605 1626 1633 1645 1653 1654 1662
1677 1678 1693 1702 1705 1714 1753 1758 1761 1765 1774 1786 1798 1810 1813 1857
1858 1870 1885 1893 1897 1906 1918 1930 1942 1969 1981 2002 2013 2073 2074 2101
2118 2158 2170 2173 2182 2202 2230 2233 2281 2290 2326 2374 2398 2422 2433 2473
2494 2542 2577 2638 2641 2677 2722 2818 2830 2842 2857 2917 2965 3070 3082 3117
3118 3133 3178 3193 3298 3382 3397 3418 3490 3502 3637 3697 3802 3817 3922 3958
3973 4057 4102 4153 4162 4342 4558 4873 4918 5077 5098 5317 5377 6637

References

1. A. F. Beardon, *The Geometry of Discrete Groups*, Springer, 1983.
2. J. Gray, *Linear Differential Equations and Group Theory from Riemann to Poincaré*, Birkhäuser, 1986.
3. G. S. Hall, T. Morgan, and Z. Perés, *Three dimensional space-times*, Gen. Rel. Grav. **19** (1987) 1137–1147.
4. A. Holz, *Geometry and action of arrays of disclinations in crystals and relation to (2 + 1)-dimensional gravity*, Class. Quant. Grav. **5** (1988) 1259–1282.
5. Z. Nehari, *Conformal Mapping*, Dover, 1975.
6. M. S. Raghunathan, *Discrete Subgroups of Lie Groups*, Springer, 1972.
7. J. Sanchez Mondragon and K. B. Wolf, *Lie Methods in Optics* (LNP 250), Springer, 1986.
8. H. A. Schwarz, *Über diejenigen Fälle, in welchen die Gaußische hypergeometrische Reihe eine algebraischen Function ihres vierten Elementes darstellt*, J. reine angew. Math. **75** (1872) 292–335.
9. J. Spanier and K. B. Oldham, *An Atlas of Functions*, Hemisphere, 1987.
10. E. Witten, *2 + 1 dimensional gravity as an exactly soluble system*, Nuclear Physics B **311** (1988/89) 46–78.
11. J. A. Wolf, *Spaces of Constant Curvature*, 5th ed., Publish or Perish, 1984.

Glenn J. Fox and Phillip E. Parker
Mathematics Department
Wichita State University
Wichita, KS 67208
USA
pparker@twsuvm.bitnet

Current address of G. J. Fox:
Mathematics Department
University of Georgia
Athens, GA 30602
USA
fox@joe.mathiuga.edu

THE MATH. HERITAGE OF C.F. GAUSS (pp. 304-320)
edited by George M. Rassias
©1991 World Scientific Publ. Co. Singapore

INTEGRAL OPERATORS FOR HARMONIC FUNCTIONS

Allan Fryant

A generating function for the spherical harmonics in $\mathbb{R}^n$ is obtained, and used to develop an integral operator which transforms functions $h(z,t), z \in \mathbb{C}$, $t \in \mathbb{R}^{n-1}$, to harmonic functions in $\mathbb{R}^n$. The associate functions $h(z,t)$ are analytic in the complex variable z, and continuous in t. The Bergman-Whittaker B_3 integral operator appears as a special (three dimensional) case of this result.

1. Introduction

In [12] E. T. Whittaker showed that if H is a function which is harmonic in a neighborhood of the origin in $\mathbb{R}^3$, then H has the representation

$$H(x,y,z) = \frac{1}{2\pi} \int_0^{2\pi} h(x + iy\cos t + iz\sin t, t)dt, \qquad (1)$$

where $h(u,t)$ is an analytic function of the complex variable u, and is continuous in t on $[0, 2\pi]$.

Whittaker's result provides an integral which transforms to analytic functions to harmonic functions in $\mathbb{R}^3$. This transform is both bounded and linear. Thus it is analogous to the operation of "taking the real part", $u(x,y) = \operatorname{Re}\{f(x + iy)\}$, which transforms analytic functions of a single complex variable to harmonic functions in $\mathbb{R}^3$.

Pursuing this analogy, Bergman used Whittaker's integral transform to obtain results for harmonic functions in $\mathbb{R}^3$ by drawing extensively on classical complex analysis. Expressing the transform (1) as a contour integral, Bergman called the result the B_3 integral operator, and wrote

1980 Mathematics Subject Classification. Primary 33A35, 31B10.

$H = B_3(h)$. This was the first of Bergman's integral operators for solutions of elliptic partial differential equations [1]. Arguing on the transformation $H = B_3(h)$, Bergman was particularly successful in studying multiply valued harmonic functions [2], characterizing their singularities [3], and in obtaining coefficient properties of harmonic functions [4]. The B_3 integral operator has also been used successfully in the study of entire harmonic functions and the interpolation and approximation of harmonic functions by harmonic polynomials [6, 7, 8].

In this paper we obtain a new generating function for the spherical harmonics, and use it to develop an integral operator for harmonic functions of n real variables. The integral operator developed reduces to the B_3 operator in the case $n = 3$.

2. Generating Function for the Spherical Harmonics in $\mathbb{R}^n$

A homogeneous polynomial of degree k in the real variables $x_1, x_2, \ldots ,$ x_n which satisfies Laplace's equation,

$$\Delta H = \frac{\partial^2 H}{\partial x_1^2} + \frac{\partial^2 H}{\partial x_2^2} + \ldots + \frac{\partial^2 H}{\partial x_n^2} = 0 , \tag{2}$$

is called a spherical harmonic of degree k in $\mathbb{R}^n$. The set of all spherical harmonics of degree k in $\mathbb{R}^n$ forms a vector space over the field of complex numbers $\mathbb{C}$. We denote this vector space by $\mathcal{H}_n^k$.

The dimension of $\mathcal{H}_n^k$ is

$$d_n^k = \dim \mathcal{H}_n^k = (n + 2k - 2)\frac{(n + k - 3)!}{k!(n - 2)!} ,$$

[11, p. 140]. An elementary argument by induction on k shows that for any $n = 3, 4, 5, \ldots ,$

$$d_n^k = \sum_{j=0}^{k} d_{n-1}^j . \tag{3}$$

This dimensional equality suggests a deep relationship between the spherical harmonics of degree less than or equal to k in $\mathbb{R}^{n-1}$, and the spherical harmonics of degree k in $\mathbb{R}^n$. We investigate this relationship here.

Let $\sum_{n-1}$ denote the unit sphere in $\mathbb{R}^n$, and $\langle f, g \rangle$ be the usual inner product on $\sum_{n-1}$. That is,

$$\sum_{n-1} = \{(x_1, x_2, \ldots, x_n) : x_1^2 + x_2^2 + \ldots + x_n^2 = 1\} \subset \mathbb{R}_n \,,$$

and

$$\langle f, g \rangle = \int_{\sum_{n-1}} f(x)\overline{g(x)}dx \,, \tag{4}$$

where $f, g \in L^2(\sum_{n-1})$. It is well known that spherical harmonics of different degrees are orthogonal with respect to the inner product (4). That is, if $S \in \mathcal{H}_n^j$ and $T \in \mathcal{H}_n^k$, then $\langle S, T \rangle = 0$ when $j \neq k$, [11, p. 144].

An orthonormal set of d_{n-1}^k spherical harmonics can be chosen as a basis for each of the vector spaces $\mathcal{H}_{n-1}^k$. Thus, considering the dimensional equality (3), any set $\{P_1, P_2, \ldots, P_{d_n^k}\}$ of d_n^k orthonormal spherical harmonics in $\mathbb{R}^{n-1}$ of degree less than or equal to k forms a basis for the vector space of all harmonic polynomials in $\mathbb{R}^{n-1}$ of degree not exceeding k. Our first theorem shows how such spherical harmonics in $\mathbb{R}^{n-1}$ can be used to constructively generate the spherical harmonics of degree k in $\mathbb{R}^n$.

Theorem 1. (Generating function). Let $P_1, P_2, \ldots, P_{d_n^k}$ be orthonormal spherical harmonics in $\mathbb{R}^{n-1}$ of degree less than or equal to k. If $x = (x_1, x_2, \ldots, x_n) \in \mathbb{R}^n$ and $t = (t_1, t_2, \ldots, t_{n-1}) \in \sum_{n-2}$, then

$$(x_1 + ix_2t_1 + ix_3t_2 + \ldots + ix_nt_{n-1})^k = \sum_{j=1}^{d_n^k} Y_k^j(x)P_j(t) \,, \tag{5}$$

where $Y_k^j, j = 1, 2, \ldots, d_n^k$ are orthogonal spherical harmonics in $\mathbb{R}^n$ of degree k.

Proof. For $t \in \mathbb{R}^{n-1}$ consider the multinomial expansion

$$(x_1 + ix_2t_1 + ix_3t_2 + \ldots + ix_nt_{n-1})^k = \sum_{\nu=1}^{l} c_\nu x^{\alpha_\nu} t^{\beta_\nu} \,, \tag{6}$$

where

$$\alpha_\nu = (\alpha_1^\nu, \alpha_2^\nu, \dots, \alpha_n^\nu), x^{\alpha_\nu} = x_1^{\alpha_1^\nu} x_2^{\alpha_2^\nu} \dots x_n^{\alpha_n^\nu},$$

$$\beta_\nu = (\alpha_2^\nu, \alpha_3^\nu, \dots, \alpha_n^\nu), t^{\beta_\nu} = t_1^{\alpha_2^\nu} t_2^{\alpha_3^\nu} \dots t_{n-1}^{\alpha_n^\nu},$$

$$\alpha_1^\nu + \alpha_2^\nu + \dots + \alpha_n^\nu = k, \text{ and } l = \frac{(n+k-1)!}{(n-1)!k!}.$$

Any polynomial of degree j in $t_1, t_2, \dots, t_{n-1}$ is, when restricted to the unit sphere $\sum_{n-2}$, a linear combination of spherical harmonics of degree less than or equal to j, [11, p. 140]. Thus for t restricted to the unit sphere $\sum_{n-2}$, there exist constants $k_{\nu_j} \in \mathbb{C}$ such that

$$c_\nu t^{\beta_\nu} = \sum_{j=1}^{d_n^k} k_{\nu_j} P_j(t), \ \nu = 1, 2, \dots, l.$$

On substituting these into the multinomial expansion (6), a rearrangement yields

$$(x_1 + ix_2 t_1 + \dots + ix_n t_{n-1})^k = \sum_{j=1}^{d_n^k} Y_k^j(x) P_j(t).$$

The Y_k^i are clearly homogeneous polynomials of degree k in $x_1, x_2, \dots, x_n$. Further, since $t_1^2 + t_2^2 + \dots + t_{n-1}^2 = 1$, these polynomials are harmonic. For,

$$0 = \Delta(x_1 + ix_2 t_1 + \dots + ix_n t_{n-1})^k$$

$$= \Delta \left[\sum_{j=1}^{d_n^k} Y_k^j(x) P_j(t) \right]$$

$$= \sum_{j=1}^{d_n^k} [\Delta Y_k^j(x)] P_j(t),$$

and it follows that $\Delta Y_k^j(x) = 0, j = 1, 2, \dots, d_n^k$, as a consequence of the linear independence of the spherical harmonics $P_j(t)$ on $\sum_{n-2}$.

It remains to show that the homogeneous harmonic polynomials Y_k^j, $j = 1, 2, \dots, d_n^k$, are orthogonal on $\sum_{n-1}$, and thus from a basis for $\mathcal{H}_n^k$.

Demonstration of this orthogonality will depend on the following two lemmas:

Lemma 1. If $(x_2, x_3, \ldots, x_n)$ is restricted to the unit sphere $\sum_{n-2}$, then

$$Y_k^j(x_1, x_2, \ldots, x_n) = \lambda_\nu(x_1)\overline{P_j(x_2, x_3, \ldots, x_n)},$$

where λ_ν is a polynomial depending on $\nu = \deg P_j$.

Proof.

$$(x_1 + ix_2 t_1 + \ldots + ix_n t_{n-1})^k = \sum_{l=0}^{k} i^l \binom{k}{l} x_1^{k-l}(x_2 t_1 + \ldots + x_n t_{n-1})^l$$

$$= \sum_{l=0}^{k} i^l \binom{k}{l} x_1^{k-l}(\tilde{x}, t)^l$$

$$= F[(\tilde{x}, t)],$$

where $\tilde{x} = (x_2, x_3, \ldots, x_n), (\tilde{x}, t) = x_2 t_1 + x_3 t_2 + \ldots + x_n t_{n-1}$, and F is a polynomial of degree k. In particular, F is a continuous function of its real variable. Thus, by the Funk-Hecke theorem [5, p. 247],

$$\int_{\sum_{n-2}} F[(\tilde{x}, t)]\overline{P_j(t)}dt = \lambda_\nu \overline{P_j(\tilde{x})},$$

where

$$\lambda_\nu = \frac{\omega_{n-3}}{C_\nu^{n/2-3/2}(1)} \int_{-1}^{1} F(u)C_\nu^{n/2-3/2}(u)(1-u^2)^{n/2-2}du,$$

$\nu = \deg P_j, \omega_{n-3}$ is the surface area of the unit sphere $\sum_{n-3}$, and C_ν^μ are the Gegenbauer (ultraspherical) polynomials of degree ν and order μ defined by

$$(1 - 2ut + t^2)^{-\mu} = \sum_{\nu=0}^{\infty} C_\nu^\mu(u)t^\nu.$$

Further, since $F(u) = (x_1 + iu)^k, \lambda_\nu$ is a polynomial in x_1. That is,

$$\lambda_\nu(x_1) = \frac{\omega_{n-3}}{C_\nu^{n/2-3/2}(1)} \int_{-1}^{1} (x_1 + iu)^k C_\nu^{n/2-3/2}(u)(1-u^2)^{n/2-2}du.$$

Also, since the spherical harmonics P_j are by hypothesis orthonormal on $\sum_{n-2}$, we have

$$\int_{\sum_{n-2}} F[(\tilde{x},t)]\overline{P_j(t)}dt = \int_{\sum_{n-2}} (x_1 + ix_2t_1 + \ldots + ix_nt_{n-1})^k \overline{P_j(t)}dt$$

$$= \int_{\sum_{n-2}} \left[\sum_{l=1}^{d_n^k} Y_k^l(x)P_l(t)\right]\overline{P_j(t)}dt$$

$$= Y_k^j(x),$$

which completes the proof.

Lemma 2. If $(t_1,t_2,\ldots,t_{n-1})$ is restricted to the unit sphere $\sum_{n-2}$, then

$$Y_k^j(1,-it_1,-it_2,\ldots,-it_{n-1}) = \gamma_j\overline{P_j(t_1,t_2,\ldots,t_{n-1})},$$

where γ_j is a constant.

Proof. Appealing to the result of Lemma 1, we have

$$Y_k^j(1,-it_1,-it_2,\ldots,-it_{n-1}) = \lambda_\nu(1)\overline{P_j(-it_1,-it_2,\ldots,-it_{n-1})}.$$

Since P_j is a homogeneous polynomial of degree ν, this yields

$$Y_k^j(1,-it_1,-it_2,\ldots,-it_{n-1}) = \lambda_\nu(1)i^\nu\overline{P_j(t_1,t_2,\ldots,t_{n-1})},$$

and the proof is complete.

We are now in a position to finish the proof of Theorem 1; that is, to establish the orthogonality of the homogeneous harmonic polynomials Y_k^j given by (5).

We begin by considering the function

$$G(u,b) = \int_{\sum_{n-1}} (\xi_1 u_1 + \xi_2 u_2 + \ldots + \xi_n u_n)^k(\xi_1 b_1 + \xi_2 b_2 + \ldots + \xi_n b_n)^k \, d\xi$$

$$= \int_{\sum_{n-1}} (u,\xi)^k(b,\xi)^k d\xi.$$

$$(7)$$

310   *A. Fryant*

Since the integration is carried out over the whole unit sphere $\sum_{n-1}$, G is an orthogonal invariant. That is, $G(Ou, Ob) = G(u, b)$ for any orthogonal transformation O of $\mathbb{R}^n$. Thus there exists a polynomial $\phi(\lambda, \mu, \nu)$ such that

$$G(u, b) = \phi[(u, u), (u, b), (b, b)] \,,$$

[5, pp. 244–245]. Further, since $G(u, b)$ is homogeneous and of degree k in the components of u and b, it follows that

$$\phi[(u, u), (u, b), (b, b)] = \sum c_{\alpha\beta\gamma}(u, u)^\alpha (u, b)^\beta (b, b)^\gamma$$

where the sum is over all non-negative integers α, β, γ such that $2\alpha + \beta = k$ and $2\gamma + \beta = k$. Thus we have the polynomial equation

$$\sum c_{\alpha\beta\gamma}(u, u)^\alpha (u, b)^\beta (b, b)^\gamma = \int_{\sum_{n-1}} (u, \xi)^k (b, \xi)^k d\xi \,. \tag{8}$$

Now let

$$u = (1, it_1, it_2, \ldots, it_{n-1}), \, t \in \sum_{n-2}$$

and

$$b = (1, -is_1, -is_2, \ldots, -is_{n-1}), \, s \in \sum_{n-2} \,.$$

Then $(u, u) = (b, b) = 0$, and Eq. (8) becomes

$$c_k(u, b)^k = \int_{\sum_{n-1}} (\xi_1 + i\xi_2 t_1 + \ldots + i\xi_n t_{n-1})^k \,.$$

$$(\xi_1 - i\xi_2 s_1 - \ldots - i\xi_n s_{n-1})^k d\xi$$

$$= \int_{\sum_{n-1}} \left[\sum_{j=1}^{d_n^k} Y_k^j(\xi) P_j(t) \right] \overline{\left[\sum_{l=1}^{d_n^k} Y_k^l(\xi) P_l(s) \right]} d\xi$$

$$= \sum_{j=1}^{d_n^k} \sum_{l=1}^{d_n^k} \left[\int_{\sum_{n-1}} Y_k^j(\xi) \overline{Y_k^l(\xi)} d\xi \right] P_j(t) \overline{P_l(s)} \,.$$

But

$$(u, b)^k = (1 + s_1 t_1 + s_2 t_2 + \ldots + s_{n-1} t_{n-1})^k \,,$$

and since

$$(x_1 + ix_2t_1 + \ldots + ix_nt_{n-1})^k = \sum_{j=1}^{d_n^k} Y_k^j(x)P_j(t),$$

it follows that

$$(u,b)^k = \sum_{j=1}^{d_n^k} Y_k^j(1, -is_1, -is_2, \ldots, -is_{n-1})P_j(t)$$

$$= \sum_{j=1}^{d_n^k} \gamma_j P_j(t)\overline{P_j(s)},$$

where the latter equality is obtained by appealing to the result of Lemma 2. Thus,

$$\sum_{j=1}^{d_n^k} \sum_{l=1}^{d_n^k} \left[\int_{\sum_{n-1}} Y_k^j(\xi)\overline{Y_k^l(\xi)}d\xi \right] P_j(t)\overline{P_l(s)}$$

$$= C_k \sum_{j=1}^{d_n^k} \gamma_j P_j(t)\overline{P_j(s)}, \tag{9}$$

from which it follows that

$$\int_{\sum_{n-1}} Y_k^j(\xi)\overline{Y_k^l(\xi)}d\xi = 0, \text{ if } j \neq l. \tag{10}$$

This completes the proof of Theorem 1.

Note that the result of Theorem 1 provides a means of inductively generating the spherical harmonics. That is, we use the spherical harmonics in $\mathbb{R}^{n-1}$ to constructively generate the spherical harmonics in $\mathbb{R}^n$. Thus, for example, starting with the spherical harmonics in $\mathbb{R}^2$, $\text{Re}\{z^j\}, \text{Im}\{z^j\}, j = 0, 1, 2, \ldots, k$, successive use of the generating function (5) allows the explicit formal computation of the spherical harmonics through degree k in $\mathbb{R}^n$ for any $n \geq 3$.

While the generating function given in Theorem 1 provides a convenient computational device for obtaining the spherical harmonics in $\mathbb{R}^n$, its greatest importance here lies with its theoretical applications. For example, our proof of Theorem 1 yields an addition formula for the spherical harmonics:

312   *A. Fryant*

Corollary 1. (Addition formula). If $P_1, P_2, \ldots, P_{d_n^k}$ are orthonormal spherical harmonics in $\mathbb{R}^{n-1}$ of degree less than or equal to k, and $Y_k^j, j = 1, 2, \ldots, d_n^k$ are spherical harmonics of degree k in $\mathbb{R}^n$ as given by the generating function (5), then

$$c_k(1 + s_1 t_1 + s_2 t_2 + \ldots + s_{n-1} t_{n-1})^k = \sum_{j=1}^{d_n^k} \|Y_k^j\|_2^2 P_j(t) \overline{P_j(s)} \tag{11}$$

for all $s, t \in \sum_{n-2}$, where c_k is a real constant.

Proof. As shown in the proof of Theorem 1,

$$c_k(1 + s_1 t_1 + s_2 t_2 \ldots + s_{n-1} t_{n-1})^k = \sum_{j=1}^{d_n^k} \sum_{l=1}^{d_n^k} \Big[\int_{\sum_{n-1}} Y_k^j(\xi) \overline{Y_k^l(\xi)} d\xi \Big] P_j(t) \overline{P_l(s)}.$$

Thus, appeal to the orthogonality (10) of the spherical harmonics Y_k^j yields the addition formula (11), where

$$\|Y_k^j\|_2 = \left[\int_{\sum_{n-1}} Y_k^j(\xi) \overline{Y_k^j(\xi)} d\xi \right]^{1/2}.$$

To see that the constant c_k in the addition formula (11) is real, consider the definition of the function $G(u, b)$ in the proof of Theorem 1, or merely let $s = t$ in (11).

The preceding addition formula yields an immediate Pythagorean identity for the spherical harmonics:

Corollary 2. (Pythagorean identity). If $P_1, P_2, \ldots, P_{d_n^k}$ are orthonormal spherical harmonics in $\mathbb{R}^{n-1}$ of degree less than or equal to k, and $Y_k^j, j = 1, 2, \ldots, d_n^k$ are spherical harmonics of degree k in $\mathbb{R}^n$ as given by the generating function (5), then

$$\sum_{j=1}^{d_n^k} \|Y_k^j\|_2^2 |P_j(t)|^2 = c_k 2^k \tag{13}$$

for all $t \in \sum_{n-2}$.

Proof. Merely let $s = t$ in the addition formula (11).

In the following, we let ω_{n-1} denote the surface area of the unit sphere $\sum_{n-1} \subset \mathbb{R}^n$.

Corollary 4. If $Y_k^j, j = 1, 2, \ldots, d_n^k$ are spherical harmonics given by the generating function (5), then

$$\sum_{j=1}^{d_n^k} \|Y_k^j\|_2^2 \leq \omega_{n-1}\omega_{n-2}. \tag{14}$$

Proof. Appealing to the result of Theorem 1, we have

$$\Big[\sum_{j=1}^{d_n^k} Y_k^j(x)P_j(t)\Big]\Big[\sum_{j=1}^{d_n^k} \overline{Y_k^j(x)P_j(t)}\Big] = (x_1^2 + (x_2t_1 + x_3t_2 + \ldots + x_nt_{n-1})^2]^k,$$

which by the Cauchy-Schwarz inequality is

$$\leq [x_1^2 + (x_2^2 + x_3^2 + \ldots + x_n^2)(t_1^2 + t_2^2 + \ldots + t_{n-1}^2)]^k$$
$$= (x_1^2 + x_2^2 + \ldots + x_n^2)^k = |x|^{2k}.$$

Thus,

$$\int_{\sum_{n-1}} \Big[\sum_{j=1}^{d_n^k} Y_k^j(x)P_j(t)\Big]\Big[\sum_{j=1}^{d_n^k} \overline{Y_k^j(x)}\overline{P_j(t)}\Big] dx \leq \int_{\sum_{n-1}} |x|^{2k} dx.$$

Further, since $|x| = 1$ on $\sum_{n-1}$, and the Y_k^j are orthogonal on $\sum_{n-1}$, carrying out the integration yields

$$\sum_{j=1}^{d_n^k} \|Y_k^j\|_2^2 |P_j(t)|^2 \leq \omega_{n-1}.$$

Hence

$$\int_{\sum_{n-2}} \sum_{j=1}^{d_n^k} \|Y_k^j\|_2^2 |P_j(t)|^2 dt \leq \int_{\sum_{n-2}} \omega_{n-1} dt$$

from which the result (14) is immediate as a consequence of the orthonormality of the spherical harmonics $P_j, j = 1, 2, \ldots, d_n^k$.

Integrating each side of the Pythagorean identity (13) over the unit sphere $\sum_{n-2}$ yields

$$\sum_{j=1}^{d_n^k} \|Y_k^j\|_2^2 = c_k 2^k \omega_{n-2}\,.$$

This result together with the result of Corollary 4 provides an upper bound on the constant c_k appearing in the addition formula (11). That is, $c_k \leq \omega_{n-1}/2^k$.

3. Integral Operators for Harmonic Functions

As an application of the generating function given in Theorem 1, we develop an integral operator for harmonic functions in $\mathbb{R}^n, n \geq 3$.

Let H be harmonic in a neighborhood Ω of the origin in $\mathbb{R}^n$. That is, suppose H satisfies Laplace's equation (2) throughout Ω. If $\rho > 0$ is chosen so that the sphere $|x| = \rho$ is contained in Ω, then H has the spherical harmonic (Fourier) expansion

$$H(x) = \sum_{k=0}^{\infty} \sum_{j=1}^{d_n^k} a_{kj} r^k Y_k^j(x/r)\,, \ r = |x|\,, \tag{15}$$

where

$$a_{kj} = \frac{1}{\rho^{2k+n-1}} \int_{|x|=\rho} H(x) Y_k^j(x) dx\,, \tag{16}$$

[10, p. 45]. The series converges in the L^2 sense on the sphere $|x| = \rho$, and thus converges uniformly to H on compact subsets of the open ball $|x| < \rho$.

It has recently been shown [9] that a spherical harmonic expansion (15) in fact converges uniformly to H on compact subsets of the open ball $B_R = \{x \in \mathbb{R}^n : |x| < R\}$, where

$$R^{-1} = \varlimsup_{k \to \infty} \left(\sum_{j=1}^{d_n^k} |a_{kj}|^2 \|Y_k^j\|_2^2 \right)^{1/2k}\,, \tag{17}$$

and that such convergence fails to obtain within any open ball of greater radius centered at the origin. That is, R as given by (17) is the distance from the origin to the nearest singularity of H.

Appealing to this result, and using the generating function (5), we obtain an operator that transforms functions $h(z, t)$ which are analytic in the complex variable z and continuous in $t \in \sum_{n-2}$ to functions which are harmonic in $\mathbb{R}^n$.

Theorem 2. (Integral operator). Let H be harmonic in a neighborhood of the origin in $\mathbb{R}^n$, and suppose R is the distance from the origin to the nearest singularity of H. Then there exists a function $h(z, t)$, analytic in the complex variable z in the disk $|z| < R$ and continuous for $t \in \sum_{n-2}$, such that

$$H(x) = \int_{\sum_{n-2}} h(x_1 + ix_2 t_1 + ix_3 t_2 + \ldots + ix_n t_{n-1}, t) dt \qquad (18)$$

for all $x \in B_R = \{x : |x| < R\}$.

In fact, if

$$H(x) = \sum_{k=0}^{\infty} \sum_{j=1}^{d_n^k} a_{kj} Y_k^j(x),$$

where Y_k^j are spherical harmonics as given by the generating function (5), then the associate function $h(z, t)$ is

$$h(z, t) = \sum_{k=0}^{\infty} \sum_{j=1}^{d_n^k} a_{kj} \overline{P_j(t)} z^k,$$

and this series converges absolutely and uniformly for z on compact subsets of the disk $|z| < R$ and for all $t \in \sum_{n-2}$.

Proof. Since R is the distance from the origin to the nearest singularity of the harmonic function H, H has the spherical harmonic expansion

$$H(x) = \sum_{k=0}^{\infty} \sum_{j=1}^{d_n^k} a_{kj} Y_k^j(x), \qquad (19)$$

where, as shown in [9],

$$R^{-1} = \overline{\lim_{k \to \infty}} \left(\sum_{j=1}^{d_n^k} |a_{kj}|^2 \|Y_k^j\|_2^2 \right)^{1/2k},$$

and the series (19) converges uniformly to H on compact subsets of the open ball $B_R = \{x : |x| < R\}$. Thus, appealing to the result of Theorem 1, we have for all $x \in B_R$

$$H(x) = \sum_{k=0}^{\infty} \sum_{j=1}^{d_n^k} a_{kj} \int_{\sum_{n-2}} (x_1 + ix_2 t_1 + \ldots + ix_n t_{n-1})^k \overline{P_j(t)} dt$$

$$= \sum_{k=0}^{\infty} \int_{\sum_{n-2}} \left[(x_1 + ix_2 t_1 + \ldots + ix_n t_{n-1}) \sum_{j=1}^{d_n^k} a_{kj} \overline{P_j(t)} \right] dt . \tag{20}$$

We next show that the series

$$\sum_{k=0}^{\infty} \sum_{j=1}^{d_n^k} a_{kj} \overline{P_j(t)} (x_1 + ix_2 t_1 + \ldots + ix_n t_{n-1})^k \tag{21}$$

converges absolutely and uniformly for x on compact subsets of B_R and for all $t \in \sum_{n-2}$, thus permitting the interchange of the order of summation and integration in (20). Now,

$$\left| \sum_{k=0}^{\infty} \sum_{j=1}^{d_n^k} a_{kj} \overline{P_j(t)} (x_1 + ix_2 t_1 + \ldots + ix_n t_{n-1})^k \right|$$

$$\leq \sum_{k=0}^{\infty} \left| \sum_{j=1}^{d_n^k} a_{kj} \overline{P_j(t)} \right| [x_1^2 + (x_2 t_1 + \ldots + x_n t_{n-1})^2]^{k/2}$$

$$\leq \sum_{k=0}^{\infty} \left| \sum_{j=1}^{d_n^k} a_{kj} \overline{P_j(t)} \right| |x|^k .$$

Further, using the Cauchy-Schwarz inequality,

$$\left| \sum_{j=1}^{d_n^k} a_{kj} \overline{P_j(t)} \right| = \left| \sum_{j=1}^{d_n^k} a_{kj} \|Y_k^j\|_2 \frac{P_j(t)}{\|Y_k^j\|_2} \right|$$

$$\leq \left(\sum_{j=1}^{d_n^k} |a_{kj}|^2 \|Y_k^j\|_2^2 \right)^{1/2} \left(\sum_{j=1}^{d_n^k} \frac{|P_j(t)|^2}{\|Y_k^j\|_2^2} \right)^{1/2} .$$

Thus the series

$$\sum_{k=0}^{\infty} \left(\sum_{j=1}^{d_n^k} |a_{kj}|^2 \|Y_k^j\|_2^2 \right)^{1/2} \left(\sum_{j=1}^{d_n^k} \frac{|P_j(t)|^2}{\|Y_k^j\|_2^2} \right)^{1/2} z^k \, , \; z \in \mathbb{C} \qquad (22)$$

dominates the series (21) for all $t \in \sum\limits_{n-2}$. The radius of convergence ρ of the power series (22) is given by

$$\rho^{-1} = \varlimsup_{k \to \infty} \left[\left(\sum_{j=1}^{d_n^k} |a_{kj}|^2 \|Y_k^j\|_2^2 \right)^{1/2} \left(\sum_{j=1}^{d_n^k} \frac{|P_j(t)|^2}{\|Y_k^j\|_2^2} \right)^{1/2} \right]^{1/k}$$

$$= R^{-1} \varlimsup_{k \to \infty} \left(\sum_{j=1}^{d_n^k} \frac{|P_j(t)|^2}{\|Y_k^j\|_2^2} \right)^{1/2k} . \qquad (23)$$

Again using the generating function (5), we have

$$P_j(t) = \frac{1}{\|Y_k^j\|_2^2} \int_{\sum_{n-1}} (x_1 + ix_2 t_1 + \ldots + ix_n t_{n-1})^k \overline{Y_k^j(x)} dx \, ,$$

and thus

$$\frac{|P_j(t)|^2}{\|Y_k^j\|_2^2} = \frac{\overline{P_j(t)}}{\|Y_k^j\|_2^4} \int_{\sum_{n-1}} (x_1 + ix_2 t_1 + \ldots + ix_n t_{n-1})^k \overline{Y_k^j(x)} dx \, .$$

We next sum over j, appeal to the Cauchy-Schwarz inequality, and to the orthogonality of the spherical harmonics Y_k^j:

$$\sum_{j=1}^{d_n^k} \frac{|P_j(t)|^2}{\|Y_k^j\|_2^2} = \left| \int_{\sum_{n-1}} (x_1 + ix_2 t_1 + \ldots + ix_n t_{n-1})^k \sum_{j=1}^{d_n^k} \frac{\overline{P_j(t)} Y_k^j(x)}{\|Y_k^j\|_2^4} dx \right|$$

$$\leq \left(\int_{\sum_{n-1}} |x_1 + ix_2 t_1 + \ldots + ix_n t_{n-1}|^{2k} dx \right)^{1/2}$$

$$\times \left(\int_{\sum_{n-1}} \sum_{j=1}^{d_n^k} \frac{P_j(t) Y_k^j(x)}{\|Y_k^j\|_2^4} \sum_{j=1}^{d_n^k} \frac{\overline{P_j(t) Y_k^j(x)}}{\|Y_k^j\|_2^4} dx \right)^{1/2}$$

318 A. Fryant

$$\leq \Big(\int_{\sum_{n-1}} |x|^{2k} dx \Big)^{1/2} \Big(\sum_{j=1}^{d_n^k} \frac{|P_j(t)|^2 \|Y_k^j\|_2^2}{\|Y_k^j\|_2^4} \Big)^{1/2}$$

$$= \sqrt{\omega_{n-1}} \Big(\sum_{j=1}^{d_n^k} \frac{|P_j(t)|^2}{\|Y_k^j\|_2^2} \Big)^{1/2}.$$

That is,

$$\sum_{j=1}^{d_n^k} \frac{|P_j(t)|^2}{\|Y_k^j\|_2^2} \leq \omega_{n-1}.$$

Using this result in (23), we have

$$\rho^{-1} \leq R^{-1} \, \overline{\lim_{k \to \infty}} \, \omega_{n-1}^{1/2k} = R^{-1}.$$

Thus the power series (22) has radius of convergence $\rho \geq R$, and it follows that the series (21) converges absolutely and uniformly for x on compact subsets of B_R and for all $t \in \sum_{n-2}$. Interchanging the order of summation and integration in (20) then yields

$$H(x) = \int_{\sum_{n-2}} h(x_1 + ix_2 t_1 \ldots + ix_n t_{n-1}, t) dt, \tag{24}$$

where

$$h(z,t) = \sum_{k=1}^{\infty} \sum_{j=1}^{d_n^k} a_{kj} \overline{P_j(t)} z^k$$

is an analytic function of the complex variable z in the disk $|z| < R$, and is continuous in t on $\sum_{n-2}$. Thus the integral representation (24) of the harmonic function H obtains for all $x \in B_R$. This completes the proof of Theorem 2.

We abbreviate the integral (18) for harmonic functions in $\mathbb{R}^n, n = 3, 4, 5, \ldots$, by

$$H = \mathcal{F}_n(h).$$

In the special case $n = 3$, the $\mathcal{F}_N$ integral operator becomes the Bergman-Whittaker B_3 integral operator. Just as Bergman used the B_3 integral operator to obtain results for harmonic functions in $\mathbb{R}^3$ by direct appeal to classical complex analysis (e.g. [2, 3, 4]), we are now in a position to do likewise for harmonic functions in $\mathbb{R}^n, n > 3$.

Moreover, we note that harmonic functions do not form an algebra with respect to ordinary multiplication. That is, if $H_1(x)$ and $H_2(x)$ are harmonic, the product $H_1(x)H_2(x)$ need not be harmonic. The integral representation $H = \mathcal{F}_n(h)$ obtains for all functions which are harmonic within a ball B_R centered at the origin in $\mathbb{R}^n$. Further the corresponding associate functions $h(z,t)$ are analytic on the disk $|z| < R$. Thus the greatest potential value of the $\mathcal{F}_n$ integral operator may lie with the fact that it provides a tranformation from an *algebra* of associate functions $h(z,t)$ which are analytic in the disk $|z| < R$ onto the space of functions $H(x)$ which are harmonic in the ball $|x| < R$. The possibility of obtaining a Banach algebra of harmonic functions in $\mathbb{R}^n$ thus presents itself.

References

1. S. Bergman, *Integral Operators in the Theory of Linear Partial Differential Equations*, Springer-Verlag, New York, 1969.

2. S. Bergman, *Zur Theorie der ein – und mehrwertigen harmonischen Funktionen des dreidimensionalen Raumes*, Math. Zeit. **24** (1926), 641–669.

3. S. Bergman, *On solutions with algebraic character of linear partial differential equations*, Trans. Amer. Math. Soc. **68** (1950), 461–507.

4. S. Bergman, *Some properties of a harmonic function of three variables given by its series development*, Arch. Rat. Mech. Anal. **8** (1961), 207–222.

5. A. Erdelyi, *Higher Transcendental Functions*, Vol. 2, McGraw-Hill, New York, 1953.

6. A. Fryant, *Growth of entire harmonic functions in* $\mathbb{R}^3$, J. Math. Anal. Appl. **66** (1978), 599–605.

7. A. Fryant, *Ultraspherical expansions and pseudo analytic functions*, Pacific J. Math. **94** (1981), 83–105.

8. A. Fryant and M. Marden, *Green's function with ring pole and applications to interpolation*, Pure and Appl. Math. Sci. **12** (1980), 101–112.

9. A. Fryant and H. Shankar, *Fourier coefficients and growth of harmonic functions*, International J. Math. and Math. Sci. **10** (1987), 433–452.

     A. Fryant

10. N. du Plessis, *An Introduction to Potential Theory*, Oliver and Boyd, Edinburg, 1970.
11. E. M. Stein and G. Weiss, *Fourier Analysis on Euclidean Spaces*, Princeton University Press, Princeton, New Jersey, 1971.
12. E. T. Whittaker, *On the partial differential equations of mathematical physics*, Math. Ann. **57** (1903), 333–355.

Allan Fryant
Jamestown College
Jamestown, North Dakota 58401
U.S.A

THE MATH. HERITAGE OF C.F. GAUSS (pp. 321-325)
edited by George M. Rassias
©1991 World Scientific Publ. Co. Singapore

QUADRATURE AND HARMONIC L^1-APPROXIMATION

M. Goldstein

1. Introduction

Let $B(a, R)$ denote an open ball centered at a and radius R in N-dimensional Euclidean space R^N and let $\overline{B}(a, R)$ denote its closure. Let $S(a, R) = \partial B(a, R)$. Gauss' famous theorem of the arithmetic mean (c.f. page 223 in [8] where the three dimensional version of Gauss' theorem is given) says that if a function u is harmonic on $\overline{B}(a, R)$, then $u(a)$ is the arithmetic mean or surface mean value of its values on $S(a, R)$. This implies that $u(a)$ is also equal to its volume mean value over $B(a, R)$. It is not difficult to prove that in the volume mean value theorem, it suffices to assume that u is harmonic and integrable in $B(a, R)$.

2. Analytic Characterization by Means of Quadrature Formulae

The following question naturally arises. Suppose D is an open subset of $R^N, N \geq 2$, such that $0 \in D$ and $\lambda(D) < \infty$ where λ denotes N-dimensional Lebesgue measure. If the mean value equality

$$(\lambda(D))^{-1} \int_D h d\lambda = h(0) \tag{1}$$

holds for every integrable harmonic function h in D, is D a ball centered at 0? This question was first posed by Epstein [4] who showed the answer is yes if $N = 2$ and D is simply connected. In a later paper, Epstein and Schiffer [5] extended the above result to open sets D with the property that $R^N \backslash D$ has non-empty interior and to all dimensions N. Next Ow and the author [7] answered the above question in the affirmative in dimension two if D has at least one boundary component which is a continuum. In 1972, Kuran [10] answered the above question in the affirmative in all dimensions and in complete generality. Nevertheless, further questions still arise. Thus an obvious question is this. Does one need to assume that (1) holds for all integrable functions in D? Interestingly enough, the answer turns out to be no. In fact, D. H. Armitage and the author [2] have shown that Kuran's result holds if we assume the (1) holds only for all positive integrable harmonic functions in D. We further showed that if we assume that (1) holds only for all positive bounded harmonic functions in D, then $D = B\backslash E$ where B is an open ball centered at 0 and E is a polar set. Actually, we showed more. Thus we proved that if p is a fixed real number satisfying $0 < p < N/(N-2)$, where N denotes the dimension, and if (1) holds for all positive harmonic functions h belonging to the Lebesgue class $L^p(D)$, then D is a ball centered at 0. The conclusion fails when $p = N/(N-2)(= +\infty$ when $N = 2$). Let $B(x,r)$ denote the open ball in R^N with center x and radius r. For each number p satisfying $0 \leq p \leq N - 2$, let $M_p(D)$ denote the set of all positive integrable harmonic functions in D satisfying

$$\int_{B(y,r)\cap D} hd\lambda \leq Cr^{N-p} \tag{2}$$

for all $y \in \partial D$ and all $r > 0$, where C depends only on h. Then $M_0(D)$ equals the set of positive bounded harmonic functions in D. If $0 \leq p < q \leq N-2$, then $M_p(D) \subseteq M_q(D)$. Let m_α denote α-dimenisonal Hausdorff measure on R^N. Suppose that $0 \leq p \leq N - 2$, and that (1) holds for each $h \in M_p(D)$. Then if $N \geq 3$ and $p = N - 2$, D is a ball centered at 0. If $N \geq 3$ and $0 < p < N - 2$, then $D = B/E$ where B is an open ball centered at 0 and E is a relatively closed subset of B with $m_{N-2-p}(E) = 0$. If $N \geq 2$ and $p = 0$, then $D = B\backslash E$ where B is an open ball centered at 0 and E is a relatively closed polar subset of B.

Now suppose we drop the assumption that $0 \in D$. What can we conclude? The author, Haussman, and Rogge [6] showed that if D is a non-empty bounded open set in $R^N, N \geq 2$, such that $R^N\backslash\overline{D}$ is connected, and if there exists a point $x_0 \in R^N$ such that for every function h harmonic in R^N, either (1) holds or (1) holds with D replaced by $\overline{D}$, then $\overline{D}$ is a closed ball centered at x_0. Trivial examples show that D itself need not be an open ball and that the assumption that $R^N\backslash\overline{D}$ is connected cannot be replaced by the assumption that $R^N\backslash D$ is connected.

If further, we drop the assumption that $R^N \backslash \overline{D}$ is connected, now with $N \geq 3$, what can we say? Armitage and the author [1] have shown that if D is a non-empty open set in R^N such that $\lambda(\overline{D}) < \infty$, if $c > 0$, and if for all $x \in R^N \backslash \overline{D}$, we have

$$(\lambda(\overline{D})^{-1}) \int_D h_x(y) d\lambda(y) \in \{h_x(0), c\} \tag{3}$$

where $h_x(y)$ denotes the fundamental superharmonic function $\|x - y\|^{2-N}$ with pole at x, then $\overline{D}$ is a closed r-annulus with $r = c^{-1/(N-2)}$ or $\overline{D}$ is a closed ball with center at 0. A closed r-annulus is the closure of an annulus $\{x \in R^N : r_1 < \|x\| < r_2\}$ with

$$r = \left(\frac{2}{N} \frac{r_2^N - r_1^N}{r_2^2 - r_1^2} \right)^{1/(N-2)}. \tag{4}$$

Similar results can be found in the same paper [1].

Thus balls and annuli can be characterized analytically by quadrature formulae. Kondraskov [9] has characterized an ellipsoid analytically by a quadrature formula. A natural question to ask is are there any domains of infinite N-dimensional Lebesgue measure which can be characterized analytically by quadrature formulae? Armitage and the author have proved the following [3]. Denote a point $x \in R^N$ where $N \geq 2$ by $x = (x_1, \ldots, x_N)$. If $-\infty < a < b < +\infty$, let $\Omega(a, b) = \{x \in R^N : a < x_1 < b\}$. Let D be an open subset of R^N such that $\overline{D} \subset \Omega(a, b)$ and such that each connected component of $R^N \backslash \overline{D}$ contains either $\Omega(-\infty, a)$ or $\Omega(b, +\infty)$. If a function f is integrable with respect to $(N - 1)$-dimensional Lebesgue measure λ^* on the hyperplane $\{x \in R^N : x_1 = t\}$, then we write

$$M(f, t) = \int_{R^{N-1}} f(t, x_2, \ldots, x_N) d\lambda^*(x_2, \ldots, x_N). \tag{5}$$

Now suppose that

$$\int_D h d\lambda = M(h, c) \tag{6}$$

for every function h harmonic in R^N and integrable in $\Omega(a, b)$. Then $\overline{D} = \Omega(c - \frac{1}{2}, c + \frac{1}{2})$.

324   *M. Goldstein*

3. Best Harmonic L^1 Approximation

We shall now turn our attention to best L^1 approximation of a subharmonic function by a harmonic function. The connection of this topic with quadrature formulae is this. For every domain which can be characterized analytically by a quadrature formula, one can prove a best harmonic L^1 approximation theorem. Thus let $s \in C^2(B(0,1)) \cap C(\overline{B(0,)})$ with $\Delta s > 0$ a.e. in B. A function h^* harmonic in $B(0,1)$ and continuous on $\overline{B(0,1)}$ is a best harmonic L^1-approximant to s, i.e., $\|s - h^*\|_1 \leq \|s - h\|_1$ for all functions h harmonic in $B(0,1)$ and continuous on $\overline{B}(0,1)$ where

$$\|v\|_1 = \frac{1}{\lambda(\overline{B(0,1)})} \int_{\overline{B(0,1)}} |v| d\lambda \quad (v \in L^1(\overline{B(0,1)}) \tag{7}$$

if and only if $h^*|_{\partial B(0,1/\sqrt[n]{2})} = s|_{\partial B(0,1/n\sqrt{2})}$ and $s - h^* > 0$ a.e. on $\overline{B(0,1)} \backslash \overline{B(0,1/\sqrt[n]{2})}$. If both of these conditions are satisfied then h^* is the unique best L^1-approximant to s. This theorem is due to the author, Haussman and Rogge [6]. A similar theorem [1] holds for an annulus. Another similar theorem [3] of this type holds for a strip. The proofs of all these theorems depend intimately on the analytic characterizations of these domains by quadrature formulae.

References

1. D. H. Armitage and M. Goldstein, *Quadrature and harmonic L^1-approximation in annuli*, Trans. Amer. Math. Soc. **312** (1989) 141–154.

2. D. H. Armitage and M. Goldstein, *The volume mean-value property of harmonic functions*, Complex Variables (to appear).

3. D. H. Armitage and M. Goldstein, *Quadrature and harmonic approximation of subharmonic functions in strips* (to appear).

4. B. Epstein, *On the mean value property of harmonic functions*, Proc. Amer. Math. Soc. **13** (1962) 830.

5. B. Epstein and M. Schiffer, *On the mean value property of harmonic functions*, J. Analyse Math. **14** (1965) 109–111.

6. M. Goldstein, W. Haussmann, and L. Rogge, *On the mean value proerty of harmonic functions and best harmonic L^1-approximation*, Trans. Amer. Math. Soc. **305** (1988) 505–515.

7. M. Goldstein and W. H. Ow, *On the mean value property of harmonic functions*, **29** (1971) 341–344.

8. O. D. Kellog, *Foundations of Potential Theory*, Frederick Unger Publishing Company, New York, 1929.

9. A. V. Kondraskov, *On the Uniqueness of the Reconstruction of Certain Regions From Their Exterior Gravitational Potentials*, Ill-Posed Mathematical Problems and Problems of Geophysics, Novosibirsk, 1976, pp. 122-129 (Russian).

10. Ü. Kuran, *On the Mean Value Property of Harmonic Functions*, Bull. London Math. Soc. 4 (1972) 311–312.

M. Goldstein
Department of Mathematics
Arizona State University
Tempe, Arizona 89287
USA

THE MATH. HERITAGE OF C.F. GAUSS (pp. 326-333)
edited by George M. Rassias
©1991 World Scientific Publ. Co. Singapore

THE THEOREMA EGREGIUM OF GAUSS FROM A VIEWPOINT OF PARTIAL DIFFERENTIAL EQUATIONS

*Chong-Kyu Han**

The Gaussian curvature is invariant under the change of isometric imbeddings. As a generalization, the notion of proper invariant of the imbedding equation is defined. The author's results on the regularity of isometric imbeddings are surveyed.

Introduction

Let M be a surface in R^3 and $\eta : M \to S^2$ be the Gauss map. The Theorema Egregium of Gauss states that the Gaussian curvature of M, originally defined as the ratio $d\sigma/dA$ where $d\sigma$ is the (signed) area of the spherical image of area element dA of the Gauss map η, is independent of imbedding, and depends only on the metric of M. In this paper we will define a notion of proper invariant of isometric imbeddings, of which the Gaussian curvature is an example, and study the regularity of imbeddings as an application.

We fix notations and definitions first: Let Ω be an open neighborhood of the origin of R^n and let $g : [g_{ij}]$ be a C^∞ *Riemannian metric* on Ω. That is, for each integer $i, j = 1, \dots, n, g_{ij}$ is a real valued C^∞ function on Ω, and the matrix $g = [g_{ij}]_{n \times n}$ is symmetric and positive definite. A *local isometric imbedding* of (Ω, g) into R^{n+1} is a system of C^1 functions $u = (u^1, \dots, u^{n+1})$ which satisfies the following system of partial differential

---

*Research supported by grant H91175 of Korea Research Foundations

equations:

$$\sum_{\lambda=1}^{n+1} \frac{\partial u^\lambda}{\partial x_i} \frac{\partial u^\lambda}{\partial x_j} = g_{ij}, \quad i,j = 1, \dots, n. \tag{1}$$

Now we fix a C^2 solution u of (1) and express the Gaussian curvature K_u of $u(\Omega)$ in terms of partial derivatives of u. Let $(e^1, \dots, e^n)$ be a C^∞ orthogonal frame over Ω. Set

$$A_{ij}(x) = \langle \nabla'_{u_* e_j} \eta, u_* e_j \rangle \quad \text{at} \quad u(x), \tag{2}$$

where $\eta = (\eta^1, \dots, \eta^{n+1})$ is the normal vector field over $u(\Omega)$ and ∇' is the covariant differentiation in R^{n+1}.

Since

$$M = \begin{bmatrix} e_1 u^1 & \dots & e_1 u^{n+1} \\ \vdots & & \vdots \\ e_n u^1 & & e_n u^{n+1} \\ \eta^1 \circ u & & \eta^{n+1} \circ u \end{bmatrix} \tag{3}$$

is an orthogonal matrix, we see that for each $\lambda = 1, \dots, n+1$, the function $\eta^\lambda \circ u$ is equal to its cofactor in M and therefore

$$\eta^\lambda \circ u = \eta^\lambda(x, u, D^\alpha u : |\alpha| \leq 1), \tag{4}$$

a C^∞ function in the arguments x, u and the first order derivatives of u. Thus we see that

$$\begin{aligned} A_{ij} &\equiv \langle \nabla'_{u_* e_i} \eta, u_* e_j \rangle \circ u \\ &= \sum_{\lambda=1}^{n+1} [(u_* e_i)\eta^\lambda \cdot (u_* e_j)\eta^\lambda] \circ u \\ &= \sum_{\lambda=1}^{n+1} e_i(\eta^\lambda \circ u)e_j(\eta^\lambda \circ u). \end{aligned} \tag{5}$$

Substituing (4) for $\eta^\lambda \circ u$ we see that each A_{ij} is a C^∞ function in the arguments x, u and the partial derivatives of order up to 2. Then we have

$$K_u(u(x)) = \det[A_{ij}(x)]_{i,j=1,\dots,n}.$$

Now we state a version of the Theorema Egregium as follows:

Theorem 1. [5, p. 268] Let $\Omega \subseteq R^n$ be an open set and g a C^∞ Riemannian metric on Ω. Suppose that u and v are any C^∞ isometric imbeddings of (Ω, g) into R^{n+1}. If n is even we have

$$K_u(u(x)) = K_v(v(x)), \quad \forall x \in \Omega.$$

It is shown in [1] that if $n \geq 4$ an imbedding of codimension 1 is under a generic assumption unique up to rigid motion, and the above theorem is rather obvious. Thus Theorem 1 is most interesting when $n = 2$, which is the original Theorema Egregium of Gauss.

The above theorem motivates the intrinsic definition of curvature. The relation between the intrinsic and extrinsic notions of curvature is given by the Gauss equation

$$A_{ii}A_{jj} - A_{ij}^2 = K_{ij} \tag{6}$$

where K_{ij} is the sectional curvature of the plane section $e_i \wedge e_j$.

Here, we observe that the left side of (6) is a C^∞ function in the arguments $(x, u, D^\alpha u : |\alpha| \leq 2)$, and the right side a C^∞ function of the derivatives of g up to order 2. However, when we differentiate g_{ij} in (1) twice, third order derivatives of u appear on the left side of (1) and all cancel out. Thus we are led to define a notion of proper invariant of (1).

1. Invariants of the Imbedding Equation

(1) is a system of PDE's. The nunber of equation is $\frac{n(n+1)}{2}$. Consider the following operations on the right side of (1):

1) multiplication by a C^∞ function
2) sum
3) product
4) inversion for nonzero expressions
5) $\sqrt[n]{}$
6) partial differentiation

By an invariant we shall mean the expression obtained by a composite of operations 1)–6) on the g_{ij}'s. Such an expression is certainly invariant under changes of isometric imbedding u.

The set of all invariants cna be graded as follows: let I_k be the set of invariants involving the partial derivatives of g_{ij} of order k and not those of order $\geq k + 1$. The set I_k will be called the set of invariants of order k. For each invariant a of order k we get a partial differential equation

$$L_a(x, u, D^\alpha u : |\alpha| \leq k + 1) = a(x), \tag{7}$$

where L_a is the expression obtained from the left sides of (1) by the same process that gave $a(x)$ from the right sides of (1).

Definition 2. The function $a \in I_k$ is said to be a proper invariant of order k if L_a in (7) involves only the partial derivatives up to order k.

As an example we will show that the Gaussian curvature is a proper invariant of (1) in the case $n = 2$.

Proposition 3. Let Ω be an open set in R^2 and $g = (g_{ij}), i, j = 1, 2$ be a C^∞ Riemannian metric on Ω, we write E, F and G for g_{11}, g_{12} and g_{22}, respectively. Then

$$\frac{1}{H}\left\{\left(\frac{FE_2 - EG_1}{2HE}\right)_1 + \left(\frac{2EF_1 - FE_1 - EE_2}{2HE}\right)_2\right\}, \qquad (8)$$

where $H = (EG - F^2)^{1/2}$ and the subscripts 1 and 2 denote the partial derivatives, is a proper invariant of order 2.

Proof. (8) is the Gaussian curvature of the metric, which involves the second derivatives of g_{ij}'s. We will show that all the third order derivatives of u cancel out as the left side of (1) passes through the same operations as in (8). All the third order derivatives appear in the following:

$$\frac{1}{H}\frac{1}{2HE}(FE_{21} - EG_{11} + 2EF_{12} - FE_{12} - EE_{22})$$

$$= \frac{1}{2H^2}(-G_{11} + 2F_{12} - E_{22})$$

$$= \frac{1}{2H^2}\{-(2u_1^1 u_{122}^1 + 2u_1^2 u_{122}^2 + 2u_1^3 u_{122}^3)$$

$$+ 2(u_{112}^1 u_2^1 + u_1^1 u_{212}^1 + u_{112}^2 u_2^2 + u_1^2 u_{122}^2 + u_{112}^3 u_2^3 + u_1^3 u_{122}^3)$$

$$- (2u_2^1 u_{211}^1 + 2u_2^2 u_{211}^2 + 2u_2^3 u_{211}^3)\}$$

$$= 0,$$

which proves the proposition.

2. Ellipticity of Imbedding Equations

By a prolongation of a given system of PDE's we mean a new system obtained by differentiating the original equations and combining them by the operations 1)–6). A merit of finding proper invariants is that the properties of solutions that are due to the lower order terms of (1) may be uncovered by eliminating the principal part in the prolongations. For instance, ellipticity of a system of PDE's is determined by the principal part. Although (1) is not elliptic, under certain conditions on some proper invariants one can prolong it to an elliptic system of order ≥ 2. In particular, we have the following:

Theorem 4. [3] Let Ω be an open set in $R^n, n \geq 2$, and let $g = (g_{ij})$ be a C^∞ Riemannian metric on Ω. Suppose that all the sectional curvatures of (Ω, g) are positive. Then (1) can be prolonged to an elliptic system of second order.

Theorem 5. [4] Let Ω be an open set in $R^n, n \geq 3$, and let $g = (g_{ij})$ be a C^∞ Riemannian metric. Let $u = (u^1, \ldots, u^{n+1})$ be an isometric imbedding of class C^2. Suppose that $u(\Omega)$ has at least three nonzero principal curvatures. Then (1) can be prolonged to a system of second order equations which is elliptic at u.

By the theory of regularity of elliptic equations, this implies the following:

Corollary 6. Under teh hypertheses of Theorem 4 or Theorem 5, an isometric imbedding u is C^∞ if u is C^2.

The idea of the proof is that a new system of PDE's consisting of (1) differentiated once and the new equations given by proper invariants of order 2 (curvature, in these cases) forms an elliptic system of order 2. We will sketch the proof of Theorem 4 for the case $n = 2$. Namely, we prove

Proposition 7. Let Ω be an open neighborhood of the origin of R^2 and let $g = (g_{ij}), i, j = 1, 2$, be a C^∞ Riemann metric of Ω. Suppose that the curvature of g is positive. Then the imbedding equation (1) can be prolonged to a second order elliptic system.

Proof. The summation convention will be used for repeated Greek indices. Without loss of generality, we assume that

$$\frac{\partial u^i}{\partial x_j}(0) = \delta_{ij} \qquad \text{(Kronecker delta), and}$$

$$\frac{\partial u^3}{\partial x_i}(0) = 0 \qquad i,j = 1,2 .$$

Let (e_1, e_2) be an orthonormal frame such that $e_j(0) = \frac{\partial}{\partial x_j}, j = 1,2$. Then in (3) we see that

$$\eta^j \circ u = (e_j u^3)B^j + (e_\lambda u^3)\zeta^\lambda , \quad j = 1,2$$

and

$$\eta^3 \circ u = (e_1 u^1)(e_2 u^2)B^j + \zeta ,$$

where B^j is a C^∞ function in $(x, D^\alpha u^k : |\alpha| \leq 1, k \neq 3), B^j = 1$ at $(0, D^\alpha u^k(0))$, and each ζ^λ, ζ are C^∞ functions in $(x, D^\alpha u^k : |\alpha| \leq 1, k \neq 3)$ which vanish at $(0, D^\alpha u(0))$. Therefore, in (5)

$$A_{ij}(x) = (e_i e_j u^3)B_j(e_j u^j) + (e_\lambda e_\mu u^\nu)\zeta_\nu^{\lambda\mu}$$

where each $\zeta_\nu^{\lambda\mu}$ is a C^∞ function in $(x, D^\alpha u^k : |\alpha| \leq 1)$ and $\zeta_\nu^{\lambda\mu} = 0$ at $(0, D^\alpha u(0))$. Then the Gaussian curvature is $K(x) = A_{11}(x)A_{22}(x) - A_{12}(x)^2$. In terms of coordinates (x_1, x_2) the above equation becomes

$$\begin{aligned}
G(x, D^\alpha u) =&(u_{11}^3 B_1 u_1^1 + u_{\lambda\mu}^\nu \zeta_\nu^{\lambda\mu} + \zeta_1)(u_{22}^3 B_2 u_2^2 + u_{\lambda\mu}^\nu \zeta_\nu^{\prime\lambda\mu} + \zeta_2) \\
&- (u_{12}^3 B_2 u_2^2 + u_{\lambda\mu}^\nu \zeta_\nu^{\prime\prime\lambda\mu} + \zeta_{12})^2 - K(x) \\
=&0
\end{aligned}$$

where B's are $1, \zeta$'s are 0 at $(0, D^\alpha u(0))$.

On the other hand, differentiating (1), we get

$$\begin{aligned}
H_{ij}(x, D^\alpha u) \equiv & u_{ii}^1 u_j^1 + u_i^1 u_{ij}^1 + u_{ii}^2 u_j^2 + u_i^2 u_{ij}^2 + u_{ii}^3 u_j^3 + u_i^3 u_{ij}^3 \\
= & 0
\end{aligned}$$

Let $\tilde{H}_{ij}$ and $\tilde{G}$ be the linearizations at u of H_{ij} and G, respectively. Namely,

$$\tilde{H}_{ij}w = \sum_{|\alpha|\leq 2} \frac{\partial H_{ij}}{\partial(D^\alpha u^k)} D^\alpha w^k \qquad \text{and}$$

$$\tilde{G}w = \sum_{|\alpha|\leq 2} \frac{\partial G}{\partial(D^\alpha u^k)} D^\alpha w^k \qquad \text{where } w = (w^1, w^2, w^3)$$

Then the principal symbol $\sigma(x,\xi), x \in \Omega, \xi \in R^3$, of the system of linear partial differential operators $\tilde{H}_{11}, \tilde{H}_{12}, \tilde{H}_{21}, \tilde{H}_{22}, \tilde{G}$ is a 5×3 matrix. An easy computation shows that

$$\sigma(0,\xi) = \begin{bmatrix} 2(\xi_1)^2 & 0 & 0 \\ \xi_1\xi_2 & (\xi_1)^2 & 0 \\ (\xi_2)^2 & \xi_1\xi_2 & 0 \\ 0 & 2(\xi_2)^2 & 0 \\ 0 & 0 & * \end{bmatrix},$$

where $*$ is equal to $u_{22}^3(0)(\xi_1)^2 + u_{11}^3(0)(\xi_2)^2 - 2u_{12}^3(0)\xi_1\xi_2$. Since $K(0) = u_{11}^3(0)u_{22}^3(0) - (u_{12}^3(0))^2 > 0, *$ never vanishes unless $\xi = 0$. Thus we see that $\sigma(0,\xi)$ is of rank 3, if $\xi \neq 0$, which completes the proof.

References

1. E. Berger, R. Bryant and P. Griffiths, *The Gauss equations and rigidity of isometric imbeddings*, Duke Math. J. **50** (1983) 803–892.
2. R. Bryant, P. Griffiths and D. Yang, *Characteristics and existence of isometric imbeddings*, Duke Math. J. **50** (1983) 893–994.
3. C. K. Han, *Regularity of isometric immersions of positively curved Riemannian manifolds and its analogy with CR geometry*, J. Diff. Geom. **28** (1988) 477–484.

4. C. K. Han, *Regularity of certain rigid isometric immersion of n-dimensional Riemannian manifolds into R^{n+1}*, Michigan Math. J. **36** (1989) 245–250.
5. S. Sternberg, *Lecture on Differential Geometry*, Chelsea, New York, 1983.
6. N. Tanaka, *Rigidity for isometric imbeddings*, J. Math. Kyoto Univ. **18** (1978) 1–70.
7. N. Tanaka, *Rigidity for elliptic isometric imbeddings*, Nagoya Math. J. **51** (1973) 137–160.

Chong-Kyu Han
Pohang Institute of Science and Technology
Department of Mathematics
Pohang, 790-330
South Korea

THE MATH. HERITAGE OF C.F. GAUSS (pp. 334-341)
edited by George M. Rassias
©1991 World Scientific Publ. Co. Singapore

A NEW COSINE FUNCTIONAL EQUATION

Hiroshi Haruki

The purpose of this paper is to solve a new cosine functional equation on the Gaussian plane.

1. Introduction

We consider the following equation (see [1, pp. 117–128], [2, pp. 103–113], [3]–[13])

$$2f(x)f(y) = f(x+y) + f(x-y),\qquad(1)$$

where f is an entire function of a complex variable and x, y are complex variables. There are several cosine functional equations (see [1, pp. 24–25], [1, pp. 177–179], [2, p. 112], [5, pp. 25–26], [6]). Now we give a new cosine functional equation in the following way. Setting $x = \frac{s+it}{2}$ and $y = \frac{s-it}{2}$ in (1) where s, t are real variables yields

$$2f\left(\frac{s+it}{2}\right) f\left(\frac{s-it}{2}\right) = f(s) + f(it).\qquad(2)$$

If $f(z) = \cos z$ in (2), by $f\left(\frac{s-it}{2}\right) = \cos\left(\frac{s-it}{2}\right) = \overline{\cos(\frac{s+it}{2})} = \overline{f(\frac{s+it}{2})}$ we obtain $2f\left(\frac{s+it}{2}\right)\overline{f(\frac{s+it}{2})} = f(s) + f(it)$, and so

$$\left| f\left(\frac{s+it}{2}\right) \right|^2 = \frac{f(s) + f(it)}{2}.\qquad(3)$$

Taking the absolute values of both sides of (3) yields

$$\left| f\left(\frac{s+it}{2}\right) \right|^2 = \left| \frac{f(s) + f(it)}{2} \right|.\qquad(4)$$

1980 Mathematics Subject Classification: Primary 39B40, Secondary 30D05.

Thus we have obtained a new cosine functional equation (4).

The purpose of this paper is to solve (4) under the hypotheses that f is an entire function of a complex variable and s, t are real variables, i.e., to prove the following theorem.

Theorem. Suppose that f is an entire function of a complex variable z. Then, the only entire solutions of (4) are

$$f(z) \equiv 0 \,, f(z) = e^{i\theta} \cos(\alpha z) \text{ and } f(z) = e^{i\theta} \cosh(\alpha z) \,,$$

where θ, α are arbitrary real constants.

2. Lemmas

To prove Theorem in Sec. 1 we shall apply the following two Lemmas.

Lemma 1. Suppose that f is an entire function of a complex variable z. Then,

(i) $\overline{f(\bar{z})}$ is also entire function of z,

(ii) if $\overline{f(\bar{z})} = f(z)$, then, $f(z) \in R$ when $z \in R$.

Proof. Since each of the proofs of (i), (ii) is easy, we omit them.

Lemma 2. Suppose that f is a real-valued continuous function of a real variable x in R. Then, the only continuous solutions of (1) in R are

$$f(x) \equiv 0 \,, \ f(x) = \cos(\alpha x) \ \text{ and } \ f(x) = \cosh(\alpha x) \,,$$

where α is an arbitrary real constant.

Proof. See [1, pp. 117–120], [2, pp. 103–105] and [12].

3. Proof of the Theorem

Throughout the proof we may assume that

$$f(z) \not\equiv 0 \,. \tag{5}$$

We show first that

$$f(0) \neq 0. \tag{6}$$

The proof is by contradiction. Assume the contrary. Then

$$f(0) = 0.$$

Setting $t = 0$ in (4) and using the above equality yields

$$|f(s)| = 2|f(\tfrac{s}{2})|^2 \tag{7}$$

for all real s. We shall prove that

$$|f(s)| = 2^{2^n - 1}|f(\tfrac{s}{2^n})|^{2^n} \quad (n = 1, 2, 3, \ldots) \tag{8}$$

for all real s. By (7) the result (8) is true for $n = 1$. Let r be an arbitrary positive integer. Now we prove that the result (8) is true for $n = r + 1$ under the assumption that the result (8) is true for $n = r$. By the induction hypothesis we obtain

$$|f(s)| = 2^{2^r - 1}|f(\tfrac{s}{2^r})|^{2^r} \tag{9}$$

for all real s. Substituting the left side of (9) into the right side of (7) yields

$$|f(s)| = 2 \left(2^{2^r - 1}|f\left(\frac{\frac{s}{2}}{2^r} \right)|^{2^r} \right)^2$$
$$= 2^{2^{r+1} - 1}|f\left(\frac{s}{2^{r+1}} \right)|^{2^{r+1}}$$

for all real s. So the result (8) is true for $n = r + 1$. Thus (8) is true.

There exists a real number s_0 such that

$$f(s_0) \neq 0, \text{ i.e., } |f(s_0)| > 0. \tag{10}$$

The proof is by contradiction. Assume the contrary. Then $f(s) = 0$ for all real s. Hence, by the Identity Theorem we obtain $f(z) \equiv 0$ for all complex z which contradicts (5).

By (8) we obtain

$$|f(s_0)| = 2^{2^n - 1}|f\left(\frac{s_0}{2^n} \right)|^{2^n} \quad (n = 1, 2, 3, \ldots). \tag{11}$$

Taking the 2^nth root of both sides of (11) yields

$$|f(s_0)|^{1/2^n} = 2^{2^n - 1/2^n}|f\left(\frac{s_0}{2^n} \right)| \quad (n = 1, 2, 3, \ldots). \tag{12}$$

We let $n \to +\infty$ in (12). Then, taking (10) into account and recalling $f(0) = 0$ yields

$$1 = 0,$$

which is a contradiction. Consequently (6) is true.

We show next that

$$|f(0)| = 1. \tag{13}$$

Setting $s = 0, t = 0$ in (4) yields

$$|f(0)| = 1 \quad \text{or} \quad f(0) = 0.$$

However, by (6) the second case does not occur. Hence we have (13). We normalize f.

If we define θ to be

$$\theta \stackrel{\text{def}}{=} \arg\left(f(0)\right), \tag{14}$$

then, by (6) this definition makes sense. If we set

$$F(z) \stackrel{\text{def}}{=} e^{-i\theta} f(z) \tag{15}$$

for all complex z, by (13), (14) we obtain

$$F(0) = 1. \tag{16}$$

Furthermore, by (4), (15) F satisfies the functional equation

$$\left| F\left(\frac{s+it}{2}\right) \right|^2 = \left| \frac{F(s) + F(it)}{2} \right| \tag{17}$$

for all real s, t. Squaring both sides of (17) yields

$$\left| F\left(\frac{s+it}{2}\right) \right|^4 = \left| \frac{F(s) + F(it)}{2} \right|^2. \tag{18}$$

By the formula $|\gamma|^2 = \gamma \bar{\gamma}$ for all complex γ and by (18) we obtain

$$\left(F\left(\frac{s+it}{2}\right) \overline{F\left(\frac{s+it}{2}\right)} \right)^2 = \frac{F(s) + F(it)}{2} \frac{\overline{F(s)} + \overline{F(it)}}{2}. \tag{19}$$

338   H. Haruki

Here we introduce the function G defined by

$$G(z) \stackrel{\text{def}}{=} \overline{F(\bar{z})} \quad \text{(and so } F(z) = \overline{G(\bar{z})} \text{ holds)} \tag{20}$$

for all complex z.

Since, by (15), F is an entire function of a complex variable z, by Lemma 1 (i) G is also an entire function of z. By (20), (19) and by observing that $\bar{i} = -i$ we obtain

$$\left(F\left(\frac{s+it}{2} \right) G\left(\frac{s-it}{2} \right) \right)^2 = \frac{1}{4}(F(s) + F(it))(G(s) + G(-it)) \tag{21}$$

for all real s, t. By (21) and by the Identity Theorem we have

$$4\left(F\left(\frac{x+y}{2} \right) G\left(\frac{x-y}{2} \right) \right)^2 = (F(x) + F(y))(G(x) + G(-y)) \tag{22}$$

for all complex x, y. Replacing x and y by $x + y$ and $x - y$ in (22), respectively, yields

$$4(F(x)G(y))^2 = (F(x+y) + F(x-y))(G(x+y) + G(-x+y)) \tag{23}$$

for all complex x, y. If we set $y = 0$ in (23), then we obtain

$$2(F(x)G(0))^2 = F(x)(G(x) + G(-x)) \tag{24}$$

for all complex x. By (16), (20), (24) we have

$$2F(x)^2 = F(x)(G(x) + G(-x)) \tag{25}$$

for all complex x. By (5), (15) we obtain $F(z) \not\equiv 0$. Since the ring of all entire functions has no divisors of zero, we can divide both sides of (25) by $2F(x) \not\equiv 0$. Therefore, we have

$$F(x) = \frac{1}{2}(G(x) + G(-x)) \tag{26}$$

for all complex x. By (26), we obtain

$$F(-x) = F(x) \tag{27}$$

for all complex x. By (27) $F(x)$ is an even function of a complex variable x. Hence, by (20) $G(x)$ is also an even function of x, i.e.,

$$G(-x) = G(x) \tag{28}$$

for all complex x. By (26), (28) we have

$$F(x) = G(x) \tag{29}$$

for all complex x. Since $G(z)$ is an even function of a complex variable z, by (23) we otbain

$$4(F(x)G(y))^2 = (F(x+y) + F(x-y))(G(x+y) + G(x-y)) \tag{30}$$

for all complex x, y. By (29), (30) we have

$$4(F(x)F(y))^2 = (F(x+y) + F(x-y))^2 \tag{31}$$

for all complex x, y. Since the ring of all entire functions has no divisors of zero, by (31) we obtain

$$2F(x)F(y) = F(x+y) + F(x-y) \tag{32}$$

for all complex x, y, or

$$2F(x)F(y) = -(F(x+y) + F(x-y)) \tag{33}$$

for all complex x, y.

However, the second case (33) does not occur. Because, setting $y = 0$ and using (16) yields $F(x) \equiv 0$, and so, by (15), $f(z) \equiv 0$ which contradicts (5). Thus (32) holds. Furthermore, by (20), (29) we obtain

$$\overline{F(\bar{z})} = F(z)$$

for all complex z.

Hence, by Lemma 1 (ii) $F(z)$ is a real-valued function when z is a real variable. Since analyticity implies continuity, F is continuous on the whole real line. F satisfies the functional equation (32) for all real x, y. Hence, by Lemma 2 we have, when z is a real variable,

$$F(z) \equiv 0 \text{ or } F(z) = \cos(\alpha z) \text{ or } F(z) = \cosh(\alpha z), \tag{34}$$

where α is a real constant. By (15), (34) we obtain, when z is a real variable,

$$f(z) \equiv 0 \text{ or } f(z) = e^{i\theta}\cos(\alpha z) \text{ or } f(z) = e^{i\theta}\cosh(\alpha z). \tag{35}$$

By (35) and by the Identity Theorem we obtain

$$f(z) \equiv 0 \text{ or } f(z) = e^{i\theta}\cos(\alpha z) \text{ or } f(z) = e^{i\theta}\cosh(\alpha z) \tag{36}$$

for all complex z. Direct substitution shows that each of (36) satisfies our original functional equation (4).

Acknowledgement

The author wishes to thank Professor J. Aczél for his helpful suggestions. This work was supported by the Natural Sciences and Engineering Research Council of Canada Nr. A-4012.

References

1. J. Aczél, *Lectures on Functional Equations and Their Applications*, Academic Press, New York-London, 1966.

2. J. Aczél, and J. Dhombres, *Functional Equations in Several Variables*, Cambridge University Press, Cambridge, New York, New Rochelle, Melbourne, Sydney, 1989.

3. T. M. Flett, *Continuous solutions of the functional equation $f(x+y)+f(x-y)=2f(x)f(y)$*, Amer. Math. Monthly **70** (1963) 392–397.

4. H. Haruki, *Another proof of Kaczmarz's theorem*, Sci. Rep. Osaka Univ. **11** (1962) 1–2.

5. H. Haruki, *Studies on certain functional equations from the standpoint of analytic function theory*, Sci. Rep. Osaka Univ. **14** (1965) 1–40.

6. H. Haruki, *An application of Nevanlinna-Pólya theorem to a cosine functional equation*, J. Austral. Math. Soc. **11** (1970) 325–328.

7. H. Haruki, *On a functional equation for Jacobi's elliptic function $cn(z;k)$*, Aequationes Math. **27** (1984) 317–321.

8. S. Kaczmarz, *Sur l'équation functionnelle $f(x)+f(x+y)=\varphi(y)f(x+\frac{y}{2})$*, Fund. Math. **6** (1924) 122–129.

9. Pl. Kannappan, *On the functional equation $f(x+y)+f(x-y)=2f(x)f(y)$*, Amer. Math. Monthly **72** (1965) 374–377.

10. Pl. Kannappan, *The functional equation $f(xy)+f(xy^{-1})=2f(x)f(y)$ for groups*, Proc. Amer. Math. Soc. **19** (1968) 69–74.

11. S. Kurepa, *A cosine functional equation in Banach algebras*, Acta Sci. Math. (Szeged) **23** (1962) 255-267.

12. F. J. Papp, *The D'Alembert functional equation*, Amer. Math. Monthly **92** (1985) 273–275.

13. W. H. Wilson, *On certain related functional equations*, Bull. Amer. Math. Soc. **26** (1919–1920) 300–312.

Hiroshi Haruki
Department of Pure Mathematics
University of Waterloo
Waterloo, Ontario
N2L 3G1
Canada

THE MATH. HERITAGE OF C.F. GAUSS (pp. 342-348)
edited by George M. Rassias
©1991 World Scientific Publ. Co. Singapore

ON PARALINDELÖF AND METALINDELÖF SPACES*

Hasan Z. Hdeib

The concepts of locally countable and point countable families outside closed discrete sets were introduced. Then we use these concepts to obtain new characterizations of paralindelöf and metalindelöf spaces.

1. Introduction

In this note we introduce the concepts of locally countable and point countable families outside closed discrete sets. Then we prove that a space X is paralindelöf (metalindelöf) if and only if every open cover of X has an open refinement which is locally (point) countable outside closed discrete sets. We also obtain a generalization of a theorem of Tall [6].

Throughout this paper X denotes a topological space. The word space corresponds to a topological space. For the concepts not defined here we refer the reader to Willard [7].

1.1. Definition. Let X be a space and let $\underset{\sim}{U}$ be a family of subsets of X. Then

(a) $\underset{\sim}{U}$ is a locally countable family if for each $x \in X$ there is an open neighborhood which intersects at most countably many members of $\underset{\sim}{U}$.

(b) $\underset{\sim}{U}$ is a point countable family if each $x \in X$ belongs to at most countably many members of $\underset{\sim}{U}$.

---

*Dedicated to the memory of C. F. Gauss (1777–1855).

(c) U is a σ-locally countable family if $\underset{\sim}{U} = \bigcup\limits_{n=1}^{\infty} \underset{\sim}{U}_n$ where each $\underset{\sim}{U}_n$ is a locally countable family in X.

1.2. Definition.

(a) A space X is paralindelöf if each open cover has an open locally countable refinement.

(b) A space X is metalindelöf if each open cover has an open point countable refinement.

(c) A space X is σ-paralindelöf if each open cover has an open σ-locally countable refinement.

2. Paralindelöf and Metalindelöf Spaces

In [1], Andenaes used the concepts of locally finite and point finite families outside closed sets to characterize paracompact and metacompact spaces. Here we define analogue concepts for locally countable and point countable families. Therefore it is natural to ask whether similar characterizations to those given by Andenaes can be obtained for paralindelöf and metalindelöf spaces. We provide such characterizations using the concept of maximal distinguished sets, due to Aull [2].

2.1. Definition. A subset M of a space X is called a distinguished set with respect to a cover $\underset{\sim}{U}$ of X if for each pair $x, y \in M$ with $x \neq y$, $x \in U \in \underset{\sim}{U}$ implies $y \notin U$. We say M is a maximal distinguished set with respect to a cover $\underset{\sim}{U}$ of X if it is not a subset of any distinguished subset of X with respect to $\underset{\sim}{U}$.

2.2. Theorem [2]. Let X be a space and let $\underset{\sim}{U}$ be an open cover of X. Then the following holds:

(a) If M is a distinguished subset of X with respect to $\underset{\sim}{U}$ then M is discrete in X.

(b) There exists a maximal distinguished subset of X with respect to $\underset{\sim}{U}$.

(c) If M is a maximial distinguished subset of X with respect to $\underset{\sim}{U}$, and $\underset{\sim}{V} = \{U | U \in \underset{\sim}{U}, U \cap M \neq \phi\}$ then $\underset{\sim}{V}$ covers X.

2.3. Definition. Let X be a space and let $\underset{\sim}{U}$ be a family of subsets of X. Then

(a) For each subset A of X, we put

$$\underset{\sim}{U}(A) = \{U \in \underset{\sim}{U} \, | U \cap A \neq \phi\}.$$

(b) $\underset{\sim}{U}$ is said to be locally countable outside closed (closed discrete) sets if for each closed (closed discrete) set F in X and for each $x \in X - F$ there exists an open neighborhood U of x intersecting at most countably many members of $\underset{\sim}{U}(F)$.

(c) $\underset{\sim}{U}$ is said to be point countable outside closed (closed discrete) sets if for each closed (closed discrete) set F in X, a point $x \in X - F$ is contained in at most countably many members of $\underset{\sim}{U}(F)$.

2.4. Proposition. Let $\underset{\sim}{U}$ be an open cover of a T_1-space X which is point countable outside closed discrete sets. Then $\underset{\sim}{U}$ has an open point countable refinement.

Proof. Let $\underset{\sim}{U}$ be an open cover of X such that $\underset{\sim}{U}$ is point countable outside closed discrete sets. Let M be the maximal distinguished subset of X with respect to $\underset{\sim}{U}$. Then M is a closed discrete subset of X. By Theorem 2.2, $\underset{\sim}{U}(M)$ covers X.

Let $\underset{\sim}{V} = \{U \backslash M | U \in \underset{\sim}{U}(M)\}$. Since $\underset{\sim}{U}$ is point countable outside closed discrete sets, it follows that $\underset{\sim}{V}$ is point countable. For each $x \in M$, choose $U_x \in \underset{\sim}{U}(M)$, such that $x \in U_x$, then $\underset{\sim}{H} = \{U_x | x \in M\}$ covers M. Again since $\underset{\sim}{U}$ is point countable outside closed discrete sets, it follows that $\underset{\sim}{H}$ is point countable. Hence $\underset{\sim}{V} \cup \underset{\sim}{H}$ is an open point countable refinement of $\underset{\sim}{U}$.

2.5. Proposition. Let U be an open cover of a T_1-space X which is point countable outside closed discrete sets. Then U is either point countable at $x \in X$ or $\cap\{U | x \in U, U \in U\} = \{x\}$.

The proof is obvious.

2.6. Proposition. Let U be an open cover of a space X such that U is locally countable outside closed discrete sets. Then U has a locally countable open refinement.

Proof. Let $U = \{U_\alpha | \alpha \in \Lambda\}$ be an open cover of X such that U is locally countable outside closed discrete sets. Well order the set Λ; for convenience, then suppose $\Lambda = \{1, 2, \ldots, \alpha, \ldots\}$. By Proposition 2.4, U has an open point countable refinement K. Construct $M = \{M_\alpha | \alpha \in \Lambda\}$ by transfinite induction as follows: let $M_1 = \cup\{K \in K | K \subseteq U_1\}$, then $M_1 \subseteq U_1$. Let $M_2 = \cup\{K \in K | K \subseteq U_2, K \nsubseteq U_1\}$, then $M_2 \subseteq U_2$. Suppose that M_β has been defined for each $\beta < \alpha$, let $M_\alpha = \cup\{K \in K | K \subseteq U_\alpha, K \nsubseteq U_\beta$ for all $\beta < \alpha\}$. Then M is point countable since K is point countable. Moreover for each $\alpha \in \Lambda, M_\alpha \subseteq U_\alpha$, therefore $M = \{M_\alpha | \alpha \in \Lambda\}$ is locally countable outside closed discrete sets. Let D be the maximal distinguished subset of X with respect to M. Then D is closed and discrete. By Theorem 2.2, $M(D)$ covers X. Since M is locally countable outside closed discrete sets, it follows that for each $x \in X - D$ there exists an open neighborhood which intersects countably many members of $M(D)$. Let $y \in D$, then $D \backslash \{y\}$ is a closed subset of D, hence it is a closed subset of X. Again since M is locally countable outside closed discrete sets, there exists an open neighborhood G of y such that G intersects countably many members of $M(D \backslash \{y\})$. It is obvious that $M(D) = M(D \backslash \{y\}) \cup M(\{y\})$. Also $M(\{y\})$ is countable since M is point countable. Therefore G intersects countably many members of $M(D)$. Hence we have proved that $M(D)$ is locally countable. But $M(D)$ covers X. Therefore $M(D)$ is a locally countable open refinement of U.

The following two theorems provide characterizations of metalindelöf and paralindelöf spaces:

2.7. Theorem. For a space X, the following are equivalent:

(a) X is metalindelöf.

(b) Every open cover of X has an open refinement which is point countable outside closed sets.

(c) Every open cover of X has an open refinement which is point countable outside closed discrete sets.

Proof.

(a) $\Rightarrow$ (b) and (b) $\Rightarrow$ (c) are obvious.

(c) $\Rightarrow$ (a) follows from Proposition 2.4.

2.8. Theorem. For a space X, the following are equivalent:

(a) X is paralindelöf.

(b) Every open cover of X has an open refinement which is locally countable outside closed sets.

(c) Every open cover of X has an open refinement which is locally countable outside closed discrete sets.

Proof.

(a) $\Rightarrow$ (b) and (b) $\Rightarrow$ (c) are obvious.

(c) $\Rightarrow$ (a) follows from Proposition 2.6.

Following similar techniques one can obtain the following theorem which generalizes Andenaes results [1].

2.9. Theorem.

(a) A space X is metacompact if and only if every open cover of X has an open refinement which is point finite outside closed discrete sets.

(b) A space X is paracompact if and only if every open cover of X has an open refinement which is locally finite outside closed discrete sets.

2.10. Corollary [1].

(a) A space X is metacompact if and only if every open cover of X has an open refinement which is point finite outside closed sets.

(b) A space X is paracompact if and only if every open cover of X has an open refinement which is locally finite outside closed sets.

In [6], Tall showed that a countably paracompact, σ-paralindelöf space X is paralindelöf. Here we generalize this theorem using spaces weaker than countably paracompact. For this purpose we need the following definition:

2.11. Definition.

(a) A subset A of a space X is said to be S-open if it can be written in the form $U \backslash S$, where U is open and S is separable.

(b) A collection $\underset{\sim}{U}$ of subsets of a space X is said to be S-locally finite if for each point x of X there exists an S-open set V containing x such that V intersects finitely many members of $\underset{\sim}{U}$.

(c) A space X is said to be S-paracompact (countably S-paracompact) if every open cover (countable open cover) of X has an S-locally open refinement.

The concept of S-paracompact spaces was suggested first by the author (see [4]). In [4], Nyikos obtained some results concerning S-paracompact spaces.

Now we have the following theorem:

2.12. Theorem. A countably S-paracompact, σ-paralindelöf space X is paralindelöf.

The proof follows by a slight modification of the argument used in the proof of Theorem 3.1 of [3].

Observe that every countably paracompact space is countably S-paracompact. However the converse need not be true. Indeed the Niemytzki's space (see [5], example 82) is countably S-paracompact, but not countably paracompact. Now it is natural to ask the following question.

Question. Is every normal space countably S-paracompact?

Acknowledgement

The author would like to thank Prof. C. M. Pareek for helpful discussions.

References

1. P. R. Andenaes, *Note on metrization*, Fund. Math. **62** (1968) 287–292.
2. C. E. Aull, *A generalization of a theorem of Aquaro*, Bull. Austral. Math. Soc. **9** (1973) 105–108.
3. W. G. Fleissner and G. M. Reed, *Paralindelöf spaces and spaces with a countable σ-locally countable base*, Topology Proceedings, Vol. 2, No. 1, (1977) 89–110.
4. P. Nyikos, *A survey of two problems*, Topology Proceedings, Vol. 3, No. 2, (1978) 461–471.
5. L. Steen and J. Seeback, *Counter Examples in Topology*, Holt, Rinehart and Winston, Inc., London, 1970.
6. F. D. Tall, *Set-theoretic consistency results and topological theorems*, Dissertiones Mathematicae, CXLVIII (1977) 1–53.
7. S. Willard, *General Topology*, Addison-Wesley, London, 1970.

Hasan Z. Hdeib
Department of Mathematics
Yarmouk University
Irbid
Jordan

Current address:
Department of Mathematics
Kuwait University
P. O. Box 5969
Kuwait 13060

THE MATH. HERITAGE OF C.F. GAUSS (pp. 349-357)
edited by George M. Rassias
©1991 World Scientific Publ. Co. Singapore

THE MAXWELL CONDITION
IN FRIEDMANN COSMOLOGY

C. Denson Hill

We propose that an additional "working hypothesis" (analogous to the Copernican principle) be introduced into cosmology. We call it the *Maxwell Condition*. For the standard Friedmann cosmological models, it then follows that $\rho \leq \rho_{\mathrm{crit}}$ (open expanding universe).

1. Introduction

From an observational point of view the universe appears on a large scale to be isotropic, homogeneous, matter-dominated, and to have negligible pressure. Thus it can be described by one of the Friedmann models in which the perfect cosmic fluid is thought of as a dust having zero pressure, spatially constant density $\rho = \rho(t)$, and whose particle world lines are the geodesic time lines. But the interstellar vaccuum is filled with electromagnetic radiation. How are we to take this into account?

In any spacetime X, Maxwell's equations for the electromagnetic field tensor F are commonly written in terms of the covariant derivative:

$$\left.\begin{array}{l} F_{[ij;k]} = 0 \\[2ex] F^{ij}{}_{;j} = 4\pi J^i\,; \end{array}\quad (c=1)\right\} \tag{1}$$

and as a compatibility condition, the charge-current vector density J is required to satisfy the continuity equation

$$J^i{}_{;i} = 0\,, \tag{2}$$

which is the condition of conservation of charge. If we take signature $(-++ +)$ and choose a Minkowski orthonormal coframe basis, then the covariant components F_{ij} of F are related to the usual electric field $\mathbf{E}$ and magnetic field $\mathbf{B}$ by

$$F_{ij} = \begin{bmatrix} 0 & -E_1 & -E_2 & -E_3 \\ E_1 & 0 & B_3 & -B_2 \\ E_2 & -B_3 & 0 & B_1 \\ E_3 & B_2 & -B_1 & 0 \end{bmatrix}.$$

Likewise the contravariant components J^i of J are related to the usual charge density σ and current density $\mathbf{j} = (j^1, j^2, j^3)$ by

$$J^i = (\sigma, j^1, j^2, j^3).$$

When X is Minkowski spacetime, the first set of equations in (1) corresponds to

$$\operatorname{div} \mathbf{B} = 0 \quad \text{and} \quad \operatorname{curl} \mathbf{E} + \mathbf{B}_t = 0, \tag{3}$$

the second set of equations in (1) corresponds to

$$\operatorname{div} \mathbf{E} = 4\pi\sigma \quad \text{and} \quad \operatorname{curl} \mathbf{B} - \mathbf{E}_t = 4\pi j, \tag{4}$$

and (2) is usually written as

$$\sigma_t + \operatorname{div} \mathbf{j} = 0. \tag{5}$$

2. The Maxwell Condition

Has anyone ever observed a charge-current $J = (\sigma, \mathbf{j})$, that satisfies the conservation of charge condition (2), to which there does not exist at least one electromagnetic field $F = (\mathbf{E}, \mathbf{B})$ satisfying Maxwell's equations (1)? Is the solvability of the system (1), (2) in the spacetime X in which we actually live not a subconscious aspect of Maxwell theory? Physicists have often argued that the partial differential equations which arise in mathematical physics should always have solutions, since it would be *physically unreasonable* if they did not. Mathematicians who work in several complex variables or in differential topology, on the other hand, are familiar with the following: To insist that an overdetermined system of partial differential equations should always have a solution, for every choice of the inhomogeneous term which satisfies the necessary compatibility conditions, is to impose some constraint on the geometrical nature of the underlying space. This is essentially a type of *cohomological condition* on the space under consideration. Thus we are lead to a rough formulation of the *Maxwell condition*:

> Given a J which is defined and satisfies (2) in X, there should exist at least one corresponding F which is defined and satisfies (1) in X.

This condition is to be regarded as a constraint on the geometrical nature of a physically realistic spacetime X.

It should be emphasized that our intention is for this condition to hold not just for the single J of our actual spacetime X, but rather that it should hold for an entire reasonable class of conceivable J's. The point of the present note is to simply read off what the Maxwell condition implies for the standard Friedmann cosmological models.

3. The Friedmann Models

In these models the spacetimes X are all diffeomorphic to

$$X \simeq \mathbb{R} \times M \,,$$

where M is a three-dimensional manifold with a fixed Riemannian metric dl^2 of constant sectional curvature K_M equal to either 1, 0, or -1. No assumptions are made *a priori* on the global topological nature of M. The Lorenztian metric on X has the form

$$ds^2 = -dt^2 + R(t)^2 dl^2 \,,$$

where the variable t now ranges over some open interval.

The function $R(t)$ is determined via Einstein's equations (in which one takes the cosmological constant $\Lambda = 0$), and the assumption of a perfect cosmic fluid having zero pressure, as the solution of an ordinary differential equation: the *Friedmann equation*

$$\dot{R}^2 + K_M = \frac{8\pi\kappa}{3}\rho R^2 \,. \tag{6}$$

The relation

$$\rho(t)R^3(t) = \text{constant} \tag{7}$$

is also derived from the conservation of mass-energy condition $T^{ij}{}_{;j} = 0$ (see
[5]). It follows from (6) and (7) that a specific one of the Friedmann models
is determined completely by the choice of two real parameters. The two
parameters are usually taken to be the *Hubble constant*

$$H_0 = (\dot{R})_0/R_0 \, ,$$

and the *deceleration parameter*

$$q_0 = -(\ddot{R})_0 R_0/(\dot{R})_0^2 \, ,$$

where $R_0 = R(t_0)$ is the present value of the scale factor.

In fact by using (7) and differentiating (6) one obtains

$$\rho_0 = \frac{3H_0^2}{4\pi\kappa} q_0 \, , \tag{8}$$

where $\rho_0 = \rho(t_0)$ is the present mean density of matter in the universe.
Hence if the values of H_0, q_0 are prescribed, then ρ_0 and R_0 are uniquely
determined by (8) and (6). The same calculation allows one to solve for
K_M:

$$K_M = H_0^2 R_0^2 (\Omega - 1) \, ,$$
$$\Omega = 2q_0 = \rho_0/\rho_{\text{crit}} \, ,$$

and to see that the critical value $q_0 = \frac{1}{2}$ corresponds to

$$\rho_{\text{crit}} = \frac{3H_0^2}{8\pi\kappa} \, . \tag{9}$$

The significance of these quantities is that $\Omega < 1, \Omega = 1, \Omega > 1$ correspond
to $K_M = -1, 0, 1$ respectively.

It is well known that if $\Omega \leq 1$ the Friedmann universe expands forever;
whereas if $\Omega > 1$, it expands and then contracts. For further details see [4],
[5].

We should point out however that all of the Friedmann spacetimes X
are solutions of the Einstein field equations

$$\mathbf{R}_{ij} - \frac{1}{2} g_{ij} \mathbf{R} = 8\pi\kappa T_{ij} \, , \tag{10}$$

with a stress-energy-momentum tensor T whose covariant components T_{ij}, with respect to a Minkowski orthonormal coframe basis, are all zero except for $\rho(t)$ in the upper left-hand corner.

4. What is the Topology of the Universe?

Up to this point the manifold M is any three dimensional manifold having constant sectional curvature K_M equal to $-1, 0$, or 1. We assume that M is oriented and connected. Then for any such Friedmann spacetime X, all we know about its homology groups is that $H_0(X, \mathbb{Z}) = \mathbb{Z}$ and $H_4(X, \mathbb{Z}) = 0$, since M is a deformation retract of X. But what about $H_i(X, \mathbb{Z})$ for $1 \leq i \leq 3$?

Without some additional assumptions on M nothing more can be said. A natural assumption to make is that M is geodesically complete:

Proposition 1. Let $X \simeq \mathbb{R} \times M$ be any such Friedmann spacetime $(\Lambda = 0)$ such that:
  (a) M is geodesically complete, and
  (b) X satisfies the Maxwell condition.
Then
  (1) $\rho(t) \leq \rho_{\mathrm{crit}}(t)$,
  (2) M is not compact, and
  (3) $H_j(X, \mathbb{Z}) = 0$ for $j > 2$.

Consider the case where $K_M = 0$ ($\Omega \equiv 1$), for example: The proposition does not rule out the possibility that M might be $\mathbb{R} \times T^2$, where T^2 is the flat torus; nor does it rule out similar M with $K_M = -1$.

In order to demonstrate the proposition, we first write (1), (2) in terms of differential forms: The electromagnetic field tensor F, in its covariant version, is to be regarded as a 2-form. In terms of local coordinates,

$$F = F_{ij}\, dx^i \wedge dx^j .$$

The covariant version J of the charge-current density is then a 1-form

$$J = J_k\, dx^k .$$

It is well known that Maxwell's equations can also be written as

$$\left. \begin{array}{l} dF = 0 \\ \delta F = 4\pi J , \end{array} \right\} \qquad (1')$$

and the continuity equation becomes

$$\delta J = 0 \,, \qquad\qquad (2')$$

where d is exterior differentiation and $\delta = - *d*$ is its formal adjoint. Here $*$ is the Hodge $*$-operator ($** = -1$); see [4], [2].

The first equation in $(1')$ does not depend on the metric at all. The next two equations are better written in their manifestly conformally invariant form as

$$d * F = 4\pi * J \,, \qquad\qquad (1'')$$
$$d * J = 0 \,. \qquad\qquad (2'')$$

These equations say that $*J$ is a closed 3-form in X, and that it is exact whenever there is a 2-form $*F$ which solves $(1'')$ in X. Therefore the Maxwell condition on X implies in particular the vanishing of the deRham cohomology group

$$H^3_{\text{deRham}}(X) = 0 \,. \qquad\qquad (11)$$

Now $H_3(M, \mathbb{Z})$ has no torsion since it is a subgroup of a free abelian group. Hence

$$H_3(X, \mathbb{Z}) \cong H_3(M, \mathbb{Z}) \subset H_3(M, \mathbb{R}) \cong$$
$$H^3_{\text{sing}}(M, \mathbb{R}) \cong H^3_{\text{sing}}(X, \mathbb{R}) \cong$$
$$H^3_{\text{deRham}}(X) = 0 \,,$$

using duality and deRham's theorem, which establishes (3).

If M were compact it would represent a nontrivial cycle in $H_3(M, \mathbb{Z})$; hence (2). We have used only the Maxwell condition up to this point.

Finally if for some t_0, one has $\rho(t_0) > \rho_{\text{crit}}(t_0)$, then $\Omega > 1$ so $K_M = 1$, and it follows that M must be compact if one assumes that M is geodesically complete. This contradiction yields (1).

Remark. The conclusion (11) is rather insensitive to one's precise implementation of the Maxwell condition: It does not matter whether J and F are required to have C^∞ coefficients, distribution ccoefficients (currents in the sense of deRham), or even hyperfunction coefficients, because by the abstract deRham theorem any fine or flabby resolution of the constant sheaf $\mathbb{R}$ will do.

It is interesting to point out that we have not yet ruled out the possibility of a nontrivial $H^2_{\text{deRham}}(X)$. That a given electromagnetic field F represent the zero class in this cohomology group is the necessary and sufficient condition for the existence of a four-vector potential

$$F_{ij} = A_{j;i} - A_{i;j} \quad \text{in } X \,.$$

Also if two potentials A and A' represent the same F, then they differ by a change of gauge in X if and only if the difference $A' - A$ represents the zero class in $H^1_{\text{deRham}}(X)$, and we haven't ruled out the possibility that this group might be nontrivial either.

5. Influence of the Copernican Principle

The Friedmann models already incorporate a weak version of the Copernican principle: namely that M is locally isotropic about each point. That means each point P of M has a neighborhood in M which is invariant (leaving P fixed) under a local three-parameter group of isometries (isomorphic to the orthogonal group $O(3)$). In fact this is the physical assumption which (by Schur's theorem) dictates that M should have constant curvature in these models. This weak isotropy principle is consistent with the intuitive notion that the spatial universe M should (for co-moving observers) look locally near any point just like that local region we are able to see. Any more global hypothesis than that is of a fundamentally different nature.

By the *Copernican principle* here we mean the stronger hypothesis of global isotropy about each point: for each $P \in M$, the local isometries extend to isometries of all of M. It is easy to see that this stronger principle implies that M is geodesically complete; which in turn implies that M is a quotient space with universal cover either: $\mathcal{H}^3$ (hyperbolic three-space) if $K_M = -1$, $\mathbb{R}^3$ if $K_M = 0$, or S^3 if $K_M = 1$.

It is interesting to note that the Copernican principle alone does not rule out the possibility that M might, for example, be the orientable but not simply connected compact manifold $\mathbb{RP}^3$.

Proposition 2. Let $X \simeq \mathbb{R} \times M$ be any Friedmann spacetime ($\Lambda = 0$) such that:

(i) M satisfies the Copernican principle, and
(ii) X satisfies the Maxwell condition.

Then $\rho \leq \rho_{\text{crit}}$, and X is diffeomorphic to $\mathbb{R}^4$.

The demonstration of the proposition shows that (i) can be replaced by the slightly weaker assumption that M is geodesically complete and globally isotropic about just one point.

In view of proposition 1, it will suffice to show that M is simply connected: If $\pi_1(M) \neq 0$ then there must, on the one hand be a closed geodesic at P; see [3]. But since M is not compact, there must be an infinite geodesic ray eminating from P; it can be reflected, by hypothesis, to give an infinite geodesic line passing through P which does not intersect itself. By global isotropy one can rotate that geodesic line into any other geodesic passing through P. So on the other hand, there can be no closed geodesic at P. Hence $\pi_1(M) = 0$ and the proposition is established. Thus M is either $\mathcal{H}^3(K_M = -1)$ or else $\mathbb{R}^3(K_M = 0)$.

6. Comments

We have not incorporated the Maxwell stress-energy-momentum tensor

$$E_{ij} = \frac{1}{4\pi}\left[F_{ik}F_j^k - \frac{1}{4}g_{ij}F^{rs}F_{rs}\right],$$

for the electromagnetic field into the T_{ij} on the right-hand side of (10). That is because our point of view here is to regard the purely gravitational effects as dominating insofar as they determine, on a large scale, that space is curved; whereas electromagnetic effects of a much smaller order of magnitude are allowed to act to constrain its global topology.

One way to look at the influence of the Copernican principle on these Friedmann models (when it is added to the Maxwell condition) is that it is the condition which insures:

1^0 Every electromagnetic field F has a four-vector potential A in X; and

2^0 If two potentials A and A' represent the same field F, then they differ by exactly a change of gauge in X.

It is interesting to compare our discussion with the paper of Gott, Gunn, Schramm and Tinsley [1].

If M is in fact compact, it follows via Gauss' law that

$$\int_M *J = 0, \tag{12}$$

for any J which satisfies (1″). Indeed this condition (12) that the universe should have "total charage zero" is built into many cosmolgical theories, such as the geometro-dynamical ones where charged particles (in pairs) are replaced by topological handles ("wormholes"). Condition (12) also follows if one regards the field F rather than the source J as the fundamental quantity (then the only J's which occur are those defined by (1″)). Thus our point of view is apt to be rejected by a majority of professional cosmologists who desire a bound universe.

Our interest in investigating the consequences of the Maxwell condition for these simple Friedmann models stems rather from a purely mathematical analogy: Maxwell's equations on a spacetime X are anologous to the inhomogeneous Cauchy-Riemann equations on a complex manifold X. The Maxwell condition is the analogue of the vanishing of the Dolbeault cohomology groups, and the latter condition characterizes the (necessarily noncompact) Stein manifolds.

References

1. J. R. Gott III, J. E. Gunn, D. N. Schramm and B. M. Tinsley, *An unbound universe?*, Astrophysical J. **194** (1974), 543–553.
2. W. V. D. Hodge, *The Theory and Applications of Harmonic Integrals*, Cambridge Univ. Press, 1959.
3. W. Klingenberg, *Lectures on Closed Geodesics*, Springer, 1978.
4. C. W. Misner, K. S. Thorne and J. A. Wheeler, *Gravitation*, Freeman, 1973.
5. S. Weinberg, *Gravitation and Cosmology*, Wiley, 1972.

C. Denson Hill
Mathematics Department
SUNY at Stony Brook
Stony Brook, NY 11794
USA

SPACE-TIME COMPACTIFICATION AND RIEMANNIAN SUBMERSIONS

Stere Ianus and Mihai Visinescu

The purpose of this paper is to consider the Kaluza-Klein theories combined with the Riemannian submersion. The compactification of the extra dimensions is produced spontaneously by a scalar sector in the form of a nonlinear sigma model. It is shown that the existence of some vertical Killing vectors on the extra space can generate massless gauge bosons. Explicit examples are constructed taking for the extra dimensional space an even dimensional generalized Hopf manifold or an odd dimensional manifold of cosymplectic type.

1. Introduction

In the last time, space-time with more than four dimensions is frequently considered in many attempts for a unified theory including gravity such as supergravity [6] and superstring [9]. Following Kaluza and Klein [14] [15] the extra dimensions are compactified to a small size, apparently beyond our ability of experimental detection.

In many physically interesting models in $(4+m)$ dimensions, the compactification of the extra m dimensions is produced spontaneously by the non-trivial vacuum configuration of an antisymmetric tensor field [5].

Another mechanism for space-time compactification was proposed by Omero and Percacci [17] and Gell-Mann and Zwiebach [7]. The compactification of the extra dimensions is induced by a scalar sector in the form of a nonlinear sigma model.

The compactified space becomes isomorphic to the manifold in which the scalar fields take values and the four dimensional space has no cosmological term at the classical level. Unfortunately, all symmetries in the extra dimensions are broken by the scalars and the model gives no massless gauge vector bosons of the Kaluza-Klein type.

In a recent paper [10] it was investigated the possibility of extending the model, assuming that the internal space can be larger than the manifold in which the fields take values.

The general solution of the model can be expressed in terms of harmonic maps satisfying the Einstein equations. It was shown that a very general class of solutions is given by Riemannian submersions from the extra dimensional space onto the space in which the scalar fields take values. Investigating the isometries of the metric of the extra m dimensional space we found that the gauge fields associated with the vertical Killing vectors are massless. Explicit examples are constructed taking for the extra dimensional space an even dimensional generalized Hopf manifold [11] or an odd dimensional manifold of cosymplectic type [12].

The present paper is organised as follows:

In Sec. 2 we discuss general aspects of dimensional reduction induced by nonlinear scalar dynamic. In Sec. 3 we construct an explicit solution with a massless gauge field associated with a vertical Killing vector in a generalized Hopf manifold [24]. In Sec. 4 the extra space is chosen to be odd dimensional of cosymplectic type [2] and a massless gauge field is associated with the characteristic vector field of the cosymplectic manifold. The paper ends with some concluding comments.

2. Space-time Compactification Induced by a General Nonlinear Sigma Model

The model we shall discuss consists of Einstein gravity in $(4+m)$ dimensions coupled to a nonlinear sigma model. The action contains the usual gravitational part without a cosmological constant and a term corresponding to the generalized sigma model in curved space:

$$S = \frac{1}{2} \int_{M^{4+m}} \left[-\frac{R}{2} + \frac{g^{ij}}{\lambda^2} h_{ab}(\phi) \frac{\partial \phi^a}{\partial z^i} \frac{\partial \phi^b}{\partial z^j} \right] dv \qquad (2.1)$$

where dv is the volume element of the orientated pseudo-Riemannian manifold M^{4+m}. We parametrise $(4+m)$-dimensional space-time M^{4+m} by local coordinates $z^i = (x^\mu, y^r), i = 1, 2, \ldots, 4+m; \mu = 0, 1, 2, 3; r = 5, 6, \ldots, 4+m$. The scalar fields $\phi^a(z)$ are thought of as coordinates of a n-dimensional compact space B with the metric h_{ab}, latin letters from the beginning of the alphabet taking values $5, 6, \ldots, 4 + n$. We take $\mathrm{diag}\,(g_{ij}) = (+ - \ldots -), S_{ij} = R^k_{ikj}, R = g^{ij} S_{ij}$ and λ^2 is a constant giving the strength of the self-coupling of the scalar fields.

The corresponding Euler-Lagrange equations are the Einstein equations

$$R_{ij} = \frac{2}{\lambda^2} h_{ab} \frac{\partial \phi^a}{\partial z^i} \frac{\partial \phi^b}{\partial z^j} \tag{2.2}$$

and the equations of the harmonic maps [4]

$$\tau(\phi)^c = g^{ij} (\nabla(d\phi))^c_{ij} = \Delta \phi^c + {}^B\Gamma^c_{ab} \frac{\partial \phi^a}{\partial z^i} \frac{\partial \phi^b}{\partial z^j} g_{ij} = 0 \,. \tag{2.3}$$

In Eq. (2.3) $\tau(\phi)$ is the tension field of ϕ, Δ is the Laplace-Beltrami operator on the manifold M^{4+m} and ${}^B\Gamma$ is the Cristoffel symbol of the Levi-Civita connection of h.

The stress-energy tensor of the map $\phi : (M^{4+m}, g) \rightarrow (B, h)$ is [1]

$$(S_\phi)_{ij} = \frac{1}{2} g^{kl} h_{ab} \frac{\partial \phi^a}{\partial z^k} \frac{\partial \phi^b}{\partial z^l} g_{ij} - h_{ab} \frac{\partial \phi^a}{\partial z^i} \frac{\partial \phi^b}{\partial z^j} \,. \tag{2.4}$$

The divergence of the stress-energy tensor S_ϕ is

$$\mathrm{div}\, S_\phi = -\langle \tau(\phi), d\phi \rangle \tag{2.5}$$

and consequently if ϕ is harmonic, then $\mathrm{div}\, S_\phi = 0$.

But in our case S_ϕ is divergence free [1] [10] and we come to the conclusion that ϕ is harmonic if the map is a differentiable submersion almost everywhere.

With this observation the study of the coupled Eqs. (2.2) and (2.3) is reduced to the search of the submersions $\phi : M^{4+m} \rightarrow B$ satisfying the Einstein equations (2.2), while Eq. (2.3) is automatically verified. For this class of solutions we have $\dim M^{4+m} = 4 + m \geq \dim B = n$ and $\mathrm{rank}\, \phi = n$. Probably, there are solutions of Eqs. (2.2) and (2.3) which are not submersions, but the rank of them is not maximal.

Assuming that ϕ is a Riemannian submersion it is possible to choose for any point $z \in M^{4+m}$ a local coordinate system (z^i) and a local coordinate system around $\phi(z) \in B$ such that

$$\phi^a(z) = y^a \quad , \quad a = 5, 6, \ldots, 4 + n \,. \tag{2.6}$$

In the above local charts we get

$$S_{ab} = \frac{2}{\lambda^2} h_{ab} \quad , \quad a, b = 5, 6, \dots, 4 + n \tag{2.7}$$

and the other components of the Ricci tensor vanish.

From Eq. (2.7) we observe that the model admits a background solution $M^{4+m} = M^4 \times M^m$ where M^4 is a flat (Minkowski) space

$$R_{\mu\nu} = 0 \qquad \mu, \nu = 0, 1, 2, 3 \,. \tag{2.8}$$

The extra space M^m rolls up to form the manifold B which we choose to be compact. In what follows we shall assume that the submersion $\phi : M^m \to B$ is independent of the spacetime coordinates x^μ (μ=0,1,2,3).

Therefore we shall suppose that we know a "background" metric $\bar{g}^{ij}$ solution of Eq. (2.2). In particular for $\bar{g}^{\mu\nu}(\mu, \nu = 0, 1, 2, 3)$ we can take any solution of the Einstein equation in vacuum (2.8) without a cosmological term.

As usual in the standard Kaluza-Klein theory, the spin-1 gauge bosons arise from the expansion of the metric tensor [20]

$$g^{rs}(z) = \bar{g}^{rs}(z) + A^{\mu\alpha}(x) A_\mu^\beta(x) X_\alpha^r(y) X_\beta^s(y) + \dots \tag{2.9}$$
$$r, s = 5, 6, \dots, 4 + m$$

where X_α are a set of Killing vectors of the manifold M^m. Including the above fluctuations into the action (2.1) we get an additional term representing the interaction between the gauge fields A_μ and the scalar fields ϕ^a:

$$\frac{1}{2\lambda^2} \int_{M^4} A^{\mu\alpha}(x) A_\mu^\beta(x) dv_1 \cdot \int_{M^m} h_{ab} X_\alpha^r(y) X_\beta^s(y) \frac{\partial\phi^a}{\partial y^r} \frac{\partial\phi^b}{\partial y^s} dv_2 \tag{2.10}$$

where dv_1 and dv_2 are the volume elements for the manifolds M^4 and respectively M^m.

In general this term is not zero which means that the gauge fields acquire masses. The necessary condition in order that a Killing vector X^α of M^m yields a massless gauge boson is [10]:

$$X^\alpha \phi^a = 0 \quad \text{for } a = 5, 6, \dots, 4 + n \,. \tag{2.11}$$

Writing a vector field X^α on M^m in a local chart

$$X = \sum_{r=5}^{4+m} X^{\alpha r} \frac{\partial}{\partial y^r} \tag{2.12}$$

we get that a vector field vanishes on the submersion map (Eq. (2.11)) if it is vertical

$$X^\alpha = \sum_{r=5+n}^{4+m} X^{\alpha r} \frac{\partial}{\partial y^r}. \tag{2.13}$$

Let us remark that in references [7] [17] the scalar fields have to comprise the same space as the internal coordinates $(m = n)$, ϕ is the identity map and consequently the fibres of the submersion are discrete sets without vertical Killing vectors.

A way to get massless gauge bosons was proposed in reference [19] including in the action (2.1) a higher derivative term and choosing appropriately some arbitrary constants.

In addition a cosmological term is added which reduces the initial worth of the model.

Another possibility is to use a larger extra dimensional space than the space in which the scalar fields take values. In this manner the main features of the spontaneous compactification induced by nonlinear dynamics are preserved and the problem is reduced to the search of some bundle manifolds with vertical Killing vectors and with suitable properties, adequate for the physical purposes.

In the next two chapters we shall present some definite examples of space-time compactification induced by a scalar sector in the form of a nonlinear sigma model. We shall point out the existence of some vertical Killing vectors on the extra space which can generate massless gauge bosons.

3. An Example of Extra Even Rimensional Space

The purpose of this section is to construct an explicit example of submersion $\phi : M^m \rightarrow B$ with a vertical Killing vector taking for the extra dimensional space a generalized Hopf manifold [23] [24].

In the beginning let us consider a compact differentiable manifold M of class C^∞ on which is given a vector field ξ of class C^∞, regular [18] and without singularities. This vector field generates a one-dimensional distribution $\{\xi\}$ and a group of global diffeomorphisms of M [16]. Let B be the set of maximal orientated integral curves of the distribution $\{\xi\}$ and $\pi : M \to B$ the natural projection. B with the natural topology is a differentiable manifold and the map π is a differentiable submersion of class C^∞ [18]. Let us denote by g a metric on M for which ξ is a Killing vector field. Therefore the group of global diffeomorphisms generated by ξ is a group of isometries of the metric g.

Let us denote by $\mathcal{Y}$ the 1-form dual to ξ with respect to g

$$\eta(x) = g(X, \xi), \text{ for any } X \in \mathcal{X}(M), \tag{3.1}$$

$\mathcal{X}(M)$ being the Lie algebra of vector fields on M.

Now we shall impose a quite restrictive, condition on ξ assuming that this vector field is parallel with respect to the Levi-Civita connection ∇ associated with metric g:

$$\nabla_X \xi = 0 \quad \text{for any } X \in \mathcal{X}(M) \tag{3.2}$$

which means that ξ is a Killing vector.

Taking into account this condition we obtain that the 1-form η is closed

$$d\eta = 0 \tag{3.3}$$

since from Eqs. (3.1) and (3.2) we have

$$d\eta(X, Y) = g(\nabla_X \xi, Y) - g(\nabla_Y \xi, X) = 0$$
$$\text{for any} X, Y \in \mathcal{X}(M). \tag{3.4}$$

We can define the distribution H

$$H = \{X \in TM / \eta(X) = 0\} \tag{3.5}$$

and in every point $x \in M$ the fibre H_x is a subspace of codimension one in $T_x M$ (tangent space of M at x) orthogonal to ξ. Therefore we have the orthogonal decomposition

$$TM = \{\xi\} \oplus H \tag{3.6}$$

and ξ is a vertical vector field of the submersion $\pi : M \to B$.

The metric g on M induces on B a metric g' in a natural way. Before we define it let us make the following notation. For any $X' \in T_{x'}B$ let us denote by X^h the unique vector from $H_x(x \in M, \pi(x) = x')$ such that $\pi_* X^h = X'$ where π_* is the derivative map $\pi_* : TM \to TB$.

We call X^h the horizontal lift of the vector X' through the submersion π and we shall use the same notation for the horizontal lift of any vector field $X' \in \mathcal{X}(B)$.

Using this convention we can define the metric g' on B by

$$g_{x'}(X', Y') = g_x(X^h, Y^h) \text{ for any } X', Y' \in T_{x'} B. \tag{3.7}$$

With these preparations we are ready to evaluate the Ricci tensor of the metric g on M:

$$S(Y, Z) = \sum_{i=1}^{m} g(R(E_i, Y)Z, E_i) \tag{3.8}$$

where $\{E_i\}$ is a local orthogonal base of vector fields on M. In particular we get [11]

$$\begin{aligned}
S(\xi, \xi) &= 0 \\
S(\xi, Y) &= S(Y, \xi) = 0 \\
S(Y^h, Z^h) &= \pi^* S(Y', Z').
\end{aligned} \tag{3.9}$$

Finally, if S is Einstein $(S = c \cdot g')$ then

$$S(Y^h, Z^h) = c\pi^* g'(Y', Z'). \tag{3.10}$$

We remark that Eqs. (3.9), (3.10) are in agreement with the Eq. (2.7) choosing appropriately the constant c.

We shall end this section with a concrete realization of the model presented in Sec. 2, taking for the extra space a generalized Hopf manifold. A Hermitian manifold (M, J, g) is called a generalized Hopf manifold (g.H.m.) [23] [24] if the fundamental 2-form F of the Hermitial metric g given by $F(X, Y) = g(X, JY)$ satisfies

$$dF = \eta \wedge F \tag{3.11}$$

J is the complex structure of the manifold and η is a closed non-vanishing 1-form called the Lee form of g.H.m. (M.J.g) which is parallel with respect to the Levi-Civita connection ∇ of g,

$$\nabla \eta = 0. \tag{3.12}$$

The dual vector field ξ associated with η is also parallel:

$$\nabla\xi = 0. \tag{3.13}$$

As stated above (Eq. (3.2)) the Lee vector field ξ is a Killing vector.

Using the closed form η we can write locally $\eta = d\sigma$ and we can define a new metric $(e^{-\sigma}g)$ conformal equivalent to the metric g of M. The Weyl connection $\widetilde{\nabla}$ is defined by [18] [22].

$$\widetilde{\nabla}_X(e^{-\sigma}g) = 0 \quad \text{for any } X \in \mathcal{X}(M). \tag{3.14}$$

A g.H.m. is called $\mathcal{P}_0 K$-manifold [10] if the curvature tensor $\tilde{R}$ associated with the Weyl connection $\widetilde{\nabla}$ is zero. A standard evaluation gives the Ricci tensor

$$S(Y,Z) = 2(m-1)\{g(Y,Z) - \eta(Y)\eta(Z)\} \tag{3.15}$$

where $2m$ is the real dimension of M. In particular Eq. (3.9) becomes

$$\begin{aligned} S(\xi,\xi) &= S(\xi,Y) = 0 \\ S(Y,Z) &= 2(m-1)g(Y,Z) \end{aligned} \tag{3.16}$$

where Y, Z are vector fields orthogonal on ξ and therefore in the horizontal distribution H.

We mention that any Hopf manifold [16] and more generally any manifold local analytic homotetic with a Hopf manifold is a $\mathcal{P}_0 K$-manifold [24].

In this case the base manifold of the fibration has constant curvature.

In conclusion, the example described above satisfies all the requirements of the model presented in Sec. 2, identifying the mapping $\phi : M^{2m} \rightarrow B$ with the submersion $\pi : M \rightarrow B$. Choosing for M a g.H.m. of $\mathcal{P}_0 K$ type we are able to fulfil the Euler Lagrange equations of the nonlinear sigma model in curved space. In addition we make manifest a vertical Killing vector field indispensable for the generation of a massless gauge boson.

4. An Example of Extra Odd Rimensional Space with Cosymplectic Structure

The almost contact manifolds represent the natural analogue in odd dimensions of the almost complex manifolds. An almost contact manifold M is a $(2m+1)$ dimensional C^∞ manifold for which the structural group of the tangent bundle TM is the reducible to $U(n) \times 1$ [21]. On an almost contact manifold there exists a tensor field φ of type (1,1), a vector field ξ and a 1-form η satisfying

$$\eta(\xi) = 1$$
$$\varphi^2 = -I + \eta \otimes \xi \tag{4.1}$$

where I denotes the identity transformation. The vector field ξ is called the characteristic vector field of the almost contact manifold. An almost contact manifold admits a Riemannian metric g such that

$$g(\varphi X, \varphi Y) = g(X, Y) - \eta(X)\eta(Y), \text{ for any } X, Y \in TM \tag{4.2}$$

$$g(X, \xi) = \eta(X). \tag{4.3}$$

Let us denote by H the canonical distribution of the almost contact manifold M defined by

$$H = \{X \in TM | \eta(X) = 0\}. \tag{4.4}$$

We can observe that we have the orthogonal decomposition with respect to the metric g

$$TM = \{\xi\} \oplus H \tag{4.5}$$

where $\{\xi\}$ is the distribution generated by ξ.

The geometrical structure $(\varphi, \xi, \eta, g\}$ with the properties (4.1), (4.2), (4.3) is called an almost contact metric structure and a manifold with such a structure is called an almost contact metric manifold.

The class of manifolds of odd dimensions admitting a geometric structure (φ, ξ, η, g) with the above properties is very large [3] [21]. D. Blair [2] defined a cosymplectic manifold to be an almost contact metric manifold which satisfies the conditions:

$$d\eta = 0 \ , \ d\phi = 0 \ , \ N_\varphi = 0 \tag{4.6}$$

where ϕ is the 2-form defined by

$$\phi(X, Y) = g(X, \varphi Y) \tag{4.7}$$

and N_φ is the Nujenhuis tensor associated to φ. The cosymplectic manifolds represent the natural analogue in odd dimensions of the Kähler manifolds.

By covariant derivation from Eq. (4.3) we get

$$(\nabla_X \eta)Y = g(\nabla_X \xi, Y) \quad \text{for any } X, Y \in \mathcal{X}(M) \tag{4.8}$$

and consequently ξ is a vector field parallel with respect to ∇:

$$\nabla \xi = 0 \tag{4.9}$$

which implies that ξ is a Killing vector.

In what follows we shall assume that ξ is a regular vector field [18]. In this case the quotient space $B = M/\xi$ is a differentiable (Hausdorff) manifold and the canonical map $\pi : M \to B$ is a differentiable submersion. As in the previous chapter we shall define a metric g' on B in a canonical way such that $\pi : M \to B$ becomes a Riemannian submersion:

$$g_x(X^h, Y^h) = g'_{x'}(\pi_* X^h, \pi_* Y^h) \tag{4.10}$$

where X^h and Y^h are the horizontal lifts of the fields X' and Y' from B, i.e., $\pi_* X^h = X', \pi_* Y^h = Y'$. In Eq. (4.10) $x \in M, x' \in B$, $\pi(x) = x'$ and π_* is the derivative map, $\pi_* : TM \to TB$.

Let us consider a local orthogonal base of vector fields on $M\{\xi, E_i^h\}$, $1 \leq i \leq 2m$ such that E_i^h are the horizontal lifts of some orthonormal vector fields on B.

Using the same procedure as in Sec. 3 we can evaluate the Ricci tensor on the manifold M [12]

$$S(\xi, \xi) = 0 \tag{4.11}$$

$$S(\xi, Y) = S(Y, \xi) = 0 \quad \text{for any } Y \in H \tag{4.12}$$

$$S(Y^h, Z^h) = \sum_{i=1}^{2m} g(R(E_i^h, Y^h)Z^h, E_i^h). \tag{4.13}$$

In the theory of almost contact metric manifolds and in particular for the cosymplectic manifolds there is the notion of η-Einstein manifold [21]. A cosymplectic manifold is η-Einstein if the Ricci tensor satisfies the relation

$$S(X, Y) = ag(X, Y) + b\eta(X)\eta(Y) \tag{4.14}$$

where a and b are constants.

Using these preparatives we shall construct a definite example of space-time compactification taking for the extra dimensional space a η-Einstein cosymplectic manifold. More precisely we shall assume that the characteristic vector field ξ of the cosymplectic manifold is regular and we shall identity the sigma field with the submersion $\pi : M \to B$ described above.

Using the η-Einstein property of the manifold we get for the Ricci tensor

$$\begin{aligned}
S(X^h, Y^h) &= ag(X^h, Y^h) \\
S(\xi, X^h) &= S(\xi, \xi) = 0 \, .
\end{aligned} \tag{4.15}$$

Therefore in the above mentioned local base $\{\xi, E_i^h\}$ we have

$$\begin{aligned}
S_{5+2m,5+2m} &= S(\xi, \xi) = 0 \\
S_{4+i,5+2m} &= S_{5+2m,4+i} = S(\xi, E_i^h) = 0 \\
S_{4+i,4+j} &= S(E_i^h, E_j^h) = ag(E_i^h, E_j^h) \\
&= ag'(E_i, E_j) \circ \pi = ag'_{ij} \circ \pi \, ; \quad 1 \le i, j \le 2m \, .
\end{aligned} \tag{4.16}$$

Comparing Eqs. (2.7) and (4.16) we have a relation between the constant λ^2 and the constant a occuring in Eq. (4.14). The gauge field associated with the characteristic vector field ξ is massless.

To give an example of η-Einstein cosymplectic manifold we shall consider the class of cosymplectic manifolds of constant φ-holomorphic sectional curvature [3] which are the natural analogues of the Kähler manifolds of constant holomorphic curvature.

The curvature tensor of a cosymplectic manifold of constant φ-holomorphic sectional curvature has the form:

$$\begin{aligned}
R(X, Y)Z = \frac{c}{4}\{ &g(\varphi Y, \varphi Z)X - g(\varphi X, \varphi Z)Y \\
&+ g(\varphi Y, Z)\varphi X - g(\varphi X, Z)\varphi Y + 2g(X, \varphi Y)\varphi Z \\
&+ \eta(Y)g(X, Z)\xi - \eta(X)g(Y, Z)\xi \}
\end{aligned} \tag{4.17}$$

where c is a constant. From the last relation it results that the Ricci tensor of a cosymplectic manifold of constant φ-holomorphic sectional curvature is

$$\begin{aligned}
S(X, Y) &= \frac{(m+1)c}{2}g(\varphi X, \varphi Y) \, , \text{ for any } X, Y \in H \\
S(\xi, \xi) &= 0 \, , \, S(\xi, X) = 0 \text{ for any } X \in H \, .
\end{aligned} \tag{4.18}$$

5. Further Results and Concluding Remarks

The purpose of this paper is to present new explicit examples of space-time compactification triggered by a scalar section in the form of a nonlinear sigma model.

Concerning the drawback related to the absence of the massless gauge bosons in previous studies of the model, we point out the possibility of removing it. Using a larger extra dimensional space than the space in which the scalar fields take values there is room for vertical Killing vectors vanishing on the submersion map. In this way the problem is reduced to the search of some fibre bundles manifolds with vertical Killing vectors and with suitable properties, adequate for the physical purpose.

We mention that there are some possible extensions of the model [12], [13]. For example it is possible to generate more than one massless gauge field choosing the extra space with adequate properties. Unfortunately the massless gauge fields will result with an Abelian structure. In general, in the space-time compactification induced by a nonlinear sigma model, if the extra space is Riemannian, compact, without boundary, the massless gauge fields admit only Abelian symmetries [25].

In closing we remark that the sigma model coupled to gravity is very flexible. It involves many sorts of space-time manifolds with interesting geometrical structures, deserving further study.

References

1. P. Baird and J. Eells, Lecture Notes in Mathematics, Vol. 894, Springer, Berlin, 1980.
2. D. E. Blair, *The theory of quasi-Sasakian structures*, J. Diff. Geometry 1 (1967) 33.
3. D. E. Blair, *Contact Manifolds in Riemannian Geometry*, Lecture Notes in Mathematics, Vol. 509, Springer, Berlin, 1976.
4. J. Eells and L. Lemaire, *A report on harmonic maps*, Bull. London Math. Soc. **10** (1978) 1.
5. P. G. O. Freund and M. A. Rubin, *Dynamics of dimensional reduction*, Phys. Lett. **97B** (1980) 233.
6. S. J. Gates, M. Grisaru, M. Rocek and W. Siegel, *Superspace*, Benjamin/Cummings, 1983.
7. M. Gell-Mann and B. Zwiebach, *Spacetime compactification indiced by scalars*, Phys. Lett. **141B** (1984) 333.
8. S. Goldberg, *Curvature and Homology*, Academic Press, New York, 1962.

9. M. B. Green, J. H. Schwarz and E. Witten, *Superstring Theory*, Cambridge University Press, 1987.

10. S. Ianus and M. Visinescu, *Spontaneous compactification induced by nonlinear scalar dynamics, gauge fields and submersions*, Class. Quantum Grav. **3** (1986) 889.

11. S. Ianus and M. Visinescu, *Kaluza-Klein theory with scalar field and generalised Hopf manifolds*, Class. Quantum Grav. **4** (1987) 1317.

12. S. Ianus and M. Visinescu, *Space-time compactification induced by scalar fields and cosymplectic structures*, preprint Central Institute of Physics, Bucharest, FT-336 (1988), to be published.

13. S. Ianus and M. Visinescu, in preparation.

14. Th. Kaluza, *On the problem of unity in physics*, Sitz. Preuss. Akad. Wiss. K1 (1921) 966.

15. O. Klein, *Quantentheorie und fünfdimensionale Relativitatstheorie*, Z. Phys. **37** (1926) 895.

16. S. Kobayashi and K. Nomizu, *Foundation of Differential Geometry*, Interscience, New York, vol. I (1963), vol. II (1969).

17. C. Omero and R. Percacci, *Generalized nonlinear sigma models in curved space and spontaneous compactification*, Nucl. Phys. **B165** (1980) 351.

18. R. Palais, *A global formulation of the Lie theory of transformation groups*, Mem. Amer. Math. Soc. No. 22 (1957).

19. R. Parthasarathy, *On space-time compactification induced by a general nonlinear sigma model*, Phys. Lett. **B181** (1986) 91.

20. A. Salam and J. Strathdee, *On Kaluza-Klein theory*, Ann. of Physics **141** (1982) 316.

21. S. Sasaki, *Almost contact manifold*, Lecture Notes, Mathematical Institute, Tohoku University, 1965.

22. I. Vaisman, *On locally conformal almost Kähler manifolds*, Israel J. of Math. **24** (1976) 338.

23. I. Vaisman, *Locally conformal Kähler manifolds with parallel Lee form*, Rendiconti di Matematica **12** (1979) 263.

24. I. Vaisman, *Generalized Hopf manifolds*, Geometriae Dedicata **13** (1982) 231.
25. M. Visinescu, *Massless Kaluza-Klein gauge fields and space-time compactification induced by scalars*, Europhysics Lett. **4** (1987) 767.

Stere Ianus
Department of Mathematics
University of Bucharest
Str. Academiei 14, Bucharest
Romania

Mihai Visinescu
Department of Fundamental Physics
Institute for Physics and Nuclear Engineering
P. O. Box MG 6
Măgurele, Bucharest
Romania

THE MATH. HERITAGE OF C.F. GAUSS (pp. 372-378)
edited by George M. Rassias
©1991 World Scientific Publ. Co. Singapore

INDEPENDENCE AND SCRAMBLED SETS
FOR CHAOTIC MAPPINGS

A. Iwanik

A notion of independence is applied to obtain large scrambled sets for chaotic mappings of the interval. Under mixing conditions, scrambled sets are obtained as independent sets of transitive points. Thus several known results on scrambled sets are strengthened and simplified.

Let f be a continuous mapping of the unit interval I into itself. according to Li and Yorke [7], f is called chaotic if there exists an uncountable subset S of I such that for any $x, y \in S, x \neq y$, and any f-periodic point p

(1) $\lim \sup |f^n(x) - f^n(y)| > 0$

(2) $\lim \inf |f^n(x) - f^n(y)| = 0$

(3) $\lim \sup |f^n(x) - f^n(p)| > 0$.

Any such set S is called scrambled.

The size of a scrambled set can be viewed as the degree of chaos. Although scrambled sets are typically small for both measure and category (see [8]), there exist mappings that offer very strong chaotic behavior. In fact several authors attempted to construct possibly large scrambled sets for specific mappings. Scrambled sets of positive Lebesgue measure ([6], [13]) and even of measure one ([9], [2], cf. [12]) have been constructed. They can never be residual ([3], cf. [2]), but assuming the continuous hypothesis there exist scrambled sets which are of second category in every interval ([2]).

The aim of this paper is to propose a new method for constructing large scrambled sets. Our approach is based on a notion of independence introduced in [4]. A slightly modified version of results of [4] shows that certain mixing conditions are sufficient for the existence of large totally independent sets, which are even more chaotic than scrambled sets. In particular, assuming the continuum hypothesis we are able to construct for the "tent" function $t(x) = 1 - |2x - 1|$ a scrambled set (in fact totally independent) which is of second category in every interval and at the same time has outer measure one (Corollary). We also obtain a simple construction of a scrambled set of full measure (Theorem 3).

1. Totally Independent Sets

Consider a flow (X, T) where X is a dense-in-itself Polish space and $T : X \to X$ is a continuous mapping. A point x in X is called transitive if its orbit

$$\{T^j x : j \geq 0\}$$

is dense in X. If there exists a transitive point (x, y) for the product flow $(X, T)^2 = (X \times X, T \times T)$ then (X, T) is called topologically weakly mixing. More generally, a subset E of X is said to be independent ([4]) if for any finite collection $x_1, \dots, x_n$ of distinct points of E the point $(x_1, \dots, x_n)$ is transitive in the product flow $(X, T)^n$. We introduce a yet stronger notion:

Definition. A subset E of X is called totally independent if for any $r > 0$ the set E is independent for the flow (X, T^r).

In other words, E is totally independent if for any distinct $x_1, \dots, x_n$ in E and every $r > 0$ the orbit

$$\{(T^{rj} x_1, \dots, T^{rj} x_n) : j \geq 0\}$$

is dense in X^n.

2. Independence Implies Chaos

For the rest of this paper f denotes a continuous mapping of I into itself. We recall that S is called δ-scrambled ($\delta > 0$) if it is scrambled and the right hand sides of the inequalities (1), (3) are replaced by δ ([5], [14]).

Assume that $X \subset I$ is f^m-invariant for some $m > 0$, i.e., $f^m(X) \subset X$.

Lemma 1. Let E be totally independent for the flow (X, f^m). Then for any $\delta < (\text{diam } X)/2$ the set E is δ-scrambled.

Proof. Since for any $x, y, \in E, x \neq y$, the f^m-orbit of (x, y) is dense in $(X \times X)$, we have

$$\limsup |f^n(x) - f^n(y)| = \operatorname{diam} X.$$

The condition (2) is also clear. Now fix $r > 0$ with $f^{mr}(p) = p$. Since $\{f^{mrj}(x) : j \geq 0\}$ is dense in X, we have

$$\limsup_n |f^n(x) - f^n(p)| \geq \limsup_j |f^{mrj}(x) - p| \geq (\operatorname{diam}\ X)/2.$$

Note that if E is only assumed to be independent then (1) and (2) still hold true, and (3), with $\delta = 0$, fails for at most one point x, whence $E \backslash \{x\}$ is scrambled.

It may happen that diam $X = 1$ in Lemma 1, in which case E is "extremally scrambled" in the sense of Kan [6]. It is so for the "tent" function (see comments following Corollary).

3. Mixing Implies Independence

The existence of independent sets for topologically weakly mixing flows and strongly mixing dynamical systems is ensured by Theorems 1 and 3 of [4]. To obtain similar results for totally independent sets, only minor change of the arguments is required. More specifically, in Theorem 1, when applying Mycielski's theorem [11], we now have to consider the relations R_n^r for each $T^r, r > 0$, but otherwise the proof remains the same as the number of relations is still countable. The modification of transfinite induction required for the proof of Theorem 2 is obvious. Thus we have for any flow (X, T) of Sec. 1,

Theorem 1. Let (X, T^r) be topologically weakly mixing for every $r > 0$. Then for every countable neighborhood basis $U_1, U_2, \ldots$ there exists a totally independent set E which is a union of perfect sets C_n such that $C_n \subset U_n$ for $n = 1, 2, \ldots$

Theorem 2. Let μ be a T-invariant probability (Borel) measure on X such that supp $\mu = X$ and the dynamical system (X, μ, T) is strongly mixing. Assuming the continuum hypothesis there exists a totally independent set F such that $F \cap B \neq \emptyset$ whenever B is a Borel subset of either second category or positive μ measure.

Remarks. 1. The assumption of Theorem 1 is weaker than the assumption of Theorem 2 since if T is strongly mixing then so is T^r, hence $T^r \times T^r$ is ergodic, and transitive points exist in $(X, T^r)^2$ $(r > 0)$.

2. The strong mixing assumption in Theorem 2 can be relaxed to mild mixing and the continuum hypothesis can be weakened by only assuming that the covering numbers for both measure and category are equal to the continuum ([4]).

3. It is shown in [4] that if μ is T-invariant and F is independent that $\mu_*(F) = 0$. Therefore we have $\mu_*(F) = 0$ and $\mu^*(F) = 1$ in Theorem 2.

4. By the definition of totally independent sets the converse of Theorem 1 is also valid.

For mappings of the interval we have the following corollary.

Corollary. Let $m > 0$ and μ be an f^m-invariant probability measure on I. Assume that the dynamical system (I, μ, f^m) is strongly mixing. Then there exist f^m-totally independent sets E, F for the flow (supp μ, f^m) such that

(a) E is an F_σ set and is uncountable in every interval of positive μ measure,

(b) assuming the continuum hypothesis F is of second category (relative to supp μ) in every interval of positive μ measure, and in addition $\mu_*(F) = 0, \mu^*(F) = 1$.

Proof. The set $X = \text{supp } \mu$ is dense-in-itself since μ is nonatomic by mixing property. The assumptions of Theorem 2 (and Theorem 1) are now satisfied with $T = f^m$.

In particular the assumption of Corollary is fulfilled by the "tent" function $t(x)$ with $m = 1$ and μ equal to Lebesgue measure λ. Indeed, t is measure theoretically conjugate with the 0–1 unilateral Bernoulli shift

(1/2, 1/2) via the conjugacy

$$\varphi(x_0, x_1, \dots) = \sum_{n=0}^{\infty} 2^{-n-2}(1 - (-1)^{x_0 + x_1 + \dots + x_n})$$

so (I, λ, t) is strongly mixing. As a consequence of Lemma 1 we obtain a scrambled set which is of second category in every interval and of outer measure one (under the continuum hypothesis or the weaker assumption of Remark 2). Sets satisfying each of these two properties individually were constructed in [2] and [12], respectively.

4. Positive Entropy Implies Mixing

The following lemma is known (see [8] and [1], p.577).

Lemma 2. If f has positive entropy then there exists an $m > 0$ and an f^m-invariant probability measure μ such that the dynamical system (I, μ, f^m) is conjugate to Bernoulli shift, hence is strongly mixing.

As a consequence of Lemma 2, Corollary, and Lemma 1 we obtain a result of Janková and Smítal ([5], Thm.1): Positive entropy implies the existence of a perfect scrambled set (by [8], positive entropy is equivalent to the existence of cycles divisible by an odd prime, the original assumption in [5]).

5. Totally Independent Set of Full Measure

Misiurewicz [9] constructed a scrambled set of full Lebesgue measure for a mapping conjugate to the "tent" function. Another construction of a scrambled set of measure one was given by Bruckner and Hu [2]. By using some ideas of [9], but avoiding calculations, we construct a totally independent set (hence "extremally scrambled") of full measure. The assumption of Theorem 3 is clearly satisfied for $f = t$.

Theorem 3. Let f^r be topologically weakly mixing for every $r > 0$. Then f is topologically conjugate to a mapping g having a totally independent set E of Lebesgue measure one.

Proof. Let $U_1, U_2, \ldots$ be a countable neighborhood basis for I. By Theorem 1 there exists a totally independent set $D = C_1 \cup C_2 \cup \ldots$ such that $C_n \subset U_n$ and each C_n is a perfect set. Let ν_n be a nonatomic probability measure concentrated on C_n. We set

$$\nu = \sum_{n>0} 2^{-n} \nu_n$$

so ν is a nonatomic probability measure and $\nu(D) = 1$. The distribution function

$$\psi(x) = \nu[0, x]$$

is a homeomorphism of I onto itself, so $g = \psi f \psi^{-1}$ is conjugate to f. This implies that $E = \psi(D)$ is g-totally independent. Moreover $\nu \circ \psi^{-1} = \lambda$, so

$$\lambda(E) = \nu(\psi^{-1}(E)) = \nu(D) = 1.$$

References

1. L. Block, *Homoclinic mappings of the interval*, Proc. Amer. Math. Soc. **72** (1978) 576–580.
2. A. M. Bruckner and Thakyin Hu, *On scrambled sets for chaotic functions*, Trans. Amer. Math. Soc. **301** (1987) 288–297.
3. T. Gedeon, *There are no chaotic mappings with residual scrambled sets*, Bull. Austral. Math. Soc. **36** (1987) 411–416.
4. A. Iwanik, *Independent sets of transitive points*, in: Dynamical Systems and Ergodic Theory, Banach Center Publications, vol. 23, PWN — Polish Scientific Publishers, Warszawa, to appear in 1989.
5. K. Janková and J. Smítal, *A characterization of chaos*, Bull. Austral. Math. Soc. **34** (1986) 283–292.
6. I. Kan, *A chaotic function possessing a scrambled set of positive Lebesgue measure*, Proc. Amer. Math. Soc. **92** (1984) 45–49.
7. T. Y. Li and J. A. Yorke, *Period three implies chaos*, Amer. Math. Monthly **82** (1975) 985–992.
8. M. Misiurewicz, *Horseshoes for mappings of the interval*, Bull. Acad. Polon. Sci. Sér. Sci. Math. **27** (1979) 167–169.
9. M. Misiurewicz, *Chaos almost everywhere*, in: Iteration Theory and Its Functional Equations, Ed. R. Liedl et al., Lecture Notes in Math. 1163, Springer (1985) 125–130.

10. Ivan Mizera, *Continuous chaotic functions of the interval have generically small scrambled sets*, Bull. Austral. Math. Soc. **37** (1988) 89–92.
11. J. Mycielski, *Independent sets in topological algebras*, Fund. Math. **55** (1964) 139–147.
12. J. Smítal, *A chaotic function with some extremal properties*, Proc. Amer. Math. Soc. **87** (1983) 54–56.
13. J. Smítal, *A chaotic function with a scrambled set of positive Lebesgue measure*, Proc. Amer. Math. Soc. **92** (1984) 50–54.
14. J. Smítal, *Chaotic functions with zero topological entropy*, Trans. Amer. Math. Soc. **297** (1986) 269–282.

A. Iwanik
Institute of Mathematics
Technical University of Wrocław
50-370 Wrocław, Poland

THE MATH. HERITAGE OF C.F. GAUSS (pp. 379-404)
edited by George M. Rassias
©1991 World Scientific Publ. Co. Singapore

A RECENT MODIFICATION OF ITERATIVE METHODS FOR SOLVING NONLINEAR PROBLEMS

A. J. Jerri

Modifications to the various iterative schemes are very familiar to numerical methods. Eight years ago, by chance, and for a very different purpose we were led to a modification to an iterative method that greatly improved the convergence of the solution to a nonlinear chemical reaction problem. This started with the iteration of the nonlinear spectral (Fourier transform) representation, then moved to the Green's function integral representation of the given nonlinear boundary and sometimes initial value problems. The better convergence of the present modification to these iterative schemes, for nonlinear problems, is based on better contraction as supported clearly by a fixed point theorem. The present method was then illustrated with success for other problems such as variable velocity waves, evolution equations, chemical concentration in catalyst pellets of different geometries, elastic beam on nonlinear foundation, nonlinear dynamics, and most recently for an important problem like the Poisson-Boltzman (P–B) equation. For the P–B problem the present modification was tried for the Green's function integral representation as well as the successive overrelaxation (SOR) method with a, surprisingly, superior advantage in speed and storage for the latter. Presently we are investigating the application of this modification to other similar relaxation schemes like the Jacobi, the weighted Jacobi, and the Gauss Seidel methods. The most recent tentative results point in the same direction of success and can also be supported by a better contraction argument via the fixed point theorem.

Introduction

1.1. *An Overview — The Application of the Modified Iterative Method*

It was almost eight years ago where by chance, and for an entirely

---

On leave from the Department of Mathematics and Computer Science, Clarkson University, Potsdam, New York 13676.

different purpose, we were led to introducing a modification to an iterative scheme [1] for solving what we presented as a nonlinear integral equation representation of a chemical reaction in a planar catalyst pellet [2,3]. We will use this simple nonlinear boundary value problem for the chemical concentration $y(x)$ (with φ as the Theile modulus),

$$\frac{d^2y}{dx^2} = \varphi^2 y^2 \,,\, 0 < x < 1 \,, \tag{1}$$

$$y'(0) = 0 \tag{2}$$

$$y(1) = 1 \,, \tag{3}$$

as our model example to investigate and illustrate clearly the application and the good convergence of our present modification to the various important iterative schemes of the nonlinear integral representation as well as the main relaxation schemes of finite difference. The initial numerical computations showed very definitely that the usual — we term it "direct" versus the present "modified" — iteration of the nonlinear integral representation of the above problem (among others to follow) does not converge for all φ. However, through a simple modification to the iterative process, to be discussed in parts 2 and 3, we were able to obtain convergence for a larger range of φ-values [1,4]. Such good results were soon to be explained in terms of a better contraction property, of the new modified form (mapping) of the iteration, as it is supported by the fixed point theorem [5, 6, 7]. This important convergence question [5,8] will be discussed in Sec. 1.2 C for the modified iteration of the nonlinear integral representation of part 2, and in Part 3 for the modification of the (finite difference) successive overrelaxaion (SOR) [9] and other relaxation schemes [10].

The method was also applicable to variable coefficient terms, like $f(x)y(x)$, instead of the nonlinear term $y^2(x)$ in (1). This led us to consider the problem of hyperbolic (wave) equation in $u(x,t)$ with variable velocity $c(x)$ [4],

$$\frac{\partial u}{\partial t} + c(x)\frac{\partial u}{\partial x} = g(x,t)\,,\, 0 < x < 1\,. \tag{4}$$

The computations showed excellent agreement with the exact solution $u(x,t) = \cos(\pi x e^t)$ of the particular problem [11],

$$\frac{\partial u}{\partial t} - x\frac{\partial u}{\partial x} = 0\,,\, -1 < x < 1\,; \qquad t > 0 \tag{5}$$

$$u(-1,t) = u(1,t)\,, \tag{6}$$

$$u(x,0) = \cos \pi x\,, \qquad -1 < x < 1\,. \tag{7}$$

Furthermore, our results compared well with the results we obtained by using the pseudospectral method [12], a typical numerical method for such problems.

The next stage of the development was the application of this modified iterative scheme to a) more of the relevant nonlinear boundary value problems and b) for nonlinear waves:

A. *Nonlinear boundary value problems*

The first direction was to illustrate the modified iteration scheme for more relevant nonlinear or variable coefficients boundary value problems [8, 9, 13] with emphasis on the convergence question [5, 8, 6]. The second direction, which we will cover in the next section B, was to use the modified iteration for solving nonlinear waves problems [14, 15]. The nonlinear boundary value problems involved the illustration of the method to solve for the concentration $u(r)$ in a cylindrical catalyst pellet [8],

$$\frac{1}{r}\frac{d}{dr}\left[r\frac{du}{dr}\right] = \varphi^2 f(u), \quad 0 < r < 1 \tag{8}$$

$$u_r(0) = 0 \tag{9}$$

$$u(1) = 1 \tag{10}$$

and also for that of a spherical shape pellet,

$$\frac{1}{r^2}\frac{d}{dr}\left[r^2\frac{du}{dr}\right] = \varphi^2 f(u), \quad 0 < r < 1 \tag{11}$$

$$u_r(0) = 0 \tag{12}$$

$$u(1) = 1. \tag{13}$$

For the nonlinear transform (spectral) representation of the cylindrical pellet (8)–(10) we used the finite Hankel (Bessel) transform as compared to the finite Fourier transform for the planar pellet (1)–(3). For the three geometries, our results compared very well with those of the shooting method, whenever the latter was applicable [16]. The modified iteration was done on both the integral transform nonlinear integral equation in the spectral space as well as the (Green's function) integral representation in the physical space. The next application was for solving a very important nonlinear equation, namely, the Poisson-Boltzman (P–B) equation [9] in the potential $\psi(\eta, \theta)$ between two charged spherical particles suspended in an electrolytic solution [17],

$$\nabla^2 \psi = \sinh \psi \tag{14}$$

The problem here is expressed in the bispherical coordinates (η, θ) (see [17], [18]). The application of the modified iterative method to the integral transform (spectral) representation and the (Green's function) integral representation of the above boundary value problems, is covered in Sec. 1.2 A and Sec. 1.2 B respectively.

In addition to the iteration of the nonlinear integral representation, we have recently applied the modification to the iteration in a relaxation scheme, namely, the successive overrelaxation (SOR) method [19, 20, 21] to solve the Poisson-Boltzman equation (14), and with surprisingly good results [9]. This is in the sense that while the "direct" SOR method does not work for this particular problem, and the modified iteration for the integral representation is limited in its convergence, our "modification" of the SOR does converge fast. In the absence of an exact solution for the complete problem in (14) that we considered, we had to repeat previous finite element computations [22] on the same machine to make a precise comparison with the present modification of the various iterative schemes. The modified SOR surpassed both methods in simplicity and speed; it is about 100 times faster than the modified iteration of the (Green's function) integral representation, with the latter being still simpler and faster than the finite element method [9]. Encouraged by such excellent results of the modified SOR, we turned our attention very recently to investigating the modification of other well known relaxation schemes, namely, the Jacobi, the weighted Jacobi and the Gauss-Seidel methods [10, 21]. The latest tentative results point in the same direction of success. Furthermore, and as it is the case with our modification of the earlier iterative schemes, this research is guided by a better contraction for its sure convergence as supported by the fixed point theorem. This subject of modifying the SOR and other relaxation schemes is covered in part 3.

Another nonlinear problem, that involves fourth order differential equation, was that describing (in dimensionless form) the deflection $y(x)$ of an infinite elastic beam on generally nonelastic foundation [13]:

$$\frac{d^4 y}{dx^4} = -y^n(x), \quad 0 < x < \infty \tag{15}$$

$$y'(0) = 0 \tag{16}$$

$$y(\infty) = 0 \tag{17}$$

$$y''(0) = 1, \tag{18}$$

$$y'''(\infty) = 0. \tag{19}$$

For this problem, a cosine Fourier transform was used to reduce the boundary value problem (15)–(19) to a nonlinear integral equation in the transform (spectral) space. The modified iteration gave good agreement (for $n = 2$, and n close to 1) with the available numerical results [23], especially for x not very far from the important location $x = 0$. The numerical results we compared to were based on quasilinearlization method [24].

B. *Nonlinear evolution equations*

The next, or parallel stage, in our effort of demonstrating the success of the modified iteration was its aplication to solving some very basic evolution equations [25]. For such (nonlinear) evolution equations in $u(x,t)$, the Fourier transform (spectral) representation was employed on the x-spatial variation with a forward finite difference in time [14]. Of course, the fast Fourier transform algorithm (FFT) was used for computing the Fourier transform and its inverse. The first equation considered was (understandably!) the *Korteweg-de Vries (KdV) equation* in $u(x,t)$,

$$u_t + 6uu_x + u_{xxx} = 0 \qquad -\infty < x < \infty; t > 0 \tag{20}$$

along with the solitary initial condition

$$u(x,0) = 2k^2 \text{sech}^2 kx \tag{21}$$

as well as a nonsolitary condition like

$$u(x,0) = -\text{sech}\, x \, . \tag{22}$$

The Fourier transform of this problem (20)–(21) results in an integro-differential equation, where the modified iteration was used with results that compared well with other known methods [26, 12, 27] including the pseudospectral method [12].

Our next example was the following *Benjamin-Ono (B-O) equation*:

$$u_t + 2uu_x + H\{u_{xx}\} = 0 \tag{23}$$

where $H\{f\}$ is the Hilbert transform of f,

$$H\{f\} = \frac{P}{\pi} \int_{-\infty}^{\infty} \frac{f(t)}{(t - x)} dt \, ; \tag{24}$$

P stands for "principal value". We note that this problem has the same nonlinear term uu_x as that of the KdV (20), but in addition it involves $H\{u_{xx}\}$, the Hilbert transform of u_{xx} in comparison to the simpler constant coefficient term u_{xxx} of the KdV. This signifies the importance of illustrating the present modified iterative method with such equation that involves both nonlinear as well as "variable coefficient" terms in a nonlinear wave equation. We emphasize that these are the main features of the present method which we have already illustrated in the case of nonlinear or variable coefficient chemical reactions [1, 8] and a variable velocity wave equation [4]. Here we also used Fourier transform for the resulting nonlinear integral equation, where we should mention the following essential Fourier transform $F\{H\{f\}\}$ of the Hilbert transform $H\{f\}$ of f,

$$F\{H\{f\}\} = -i\sqrt{2\pi}\,\mathrm{sgn}\,\lambda\, F(\lambda) \tag{25}$$

where

$$\mathrm{sgn}\,\lambda = \begin{cases} -1, & \lambda < 0 \\ 1, & \lambda > 0 \,. \end{cases} \tag{26}$$

To check our numerical results of the modified spectral iterative method, we compared to the following exact solution [25, p. 212]

$$u(x,t) = \frac{2c_1}{1 + c_1^2(x - c_1 t)^2} \tag{27}$$

where $c_1 \sim o(1)$.

The next evolution equation we considered was the nonlinear Schrodinger equation:

$$iu_t(x,t) = u_{xx} + 2|u|^2 u \tag{28}$$

where its main difference from the KdV equation (20) is its cubic nonlinear term.

The numerical results for the above three nonlinear waves problems (20), (23) and (28) compared well with either some exact solutions or with the results of leading methods [12, 25, 26]. With that we felt encouraged to apply the present method to two dimensional nonlinear waves, namely the following *Kadomtsev-Petviashvilli (K-P) equation* in $u(x,y;t)$,

$$(u_t + 6uu_x + u_{xxx})_x = 3u_{yy} \,. \tag{29}$$

For this two dimensional problem, we need the double Fourier transform, where for its computations in the iterative process we used the double fast Fourier transform (FFT). As a simple illustration we compared the results of the present modified iterative method with the following exact solution found in [25, p. 198],

$$u(x,y,t) = \frac{4[-(x' + p_r y')^2 + p_i^2 y'^2 + 3p_i^{-2}]}{[(x' + p_r y')^2 + p_i^2 y'^2 + 3p_i^{-2}]^2} \tag{30}$$

with

$$x' = x - (p_r^2 + p_i^2)t \tag{31}$$

$$y' = y + 2p_r t. \tag{32}$$

As a special case, we took $\alpha = -1, p_r = 0, p_i = 0.125$. The computations were done for $t = 0.01$ with $\Delta t = 0.005, \Delta x = 0.816$ for the 50 points in the domain $(-20, 20)$ of the variable $x, \Delta y = 1.04$ for the 26 points in the domain of the variable y. The maximal error was of the order of 10^{-5}.

Our present effort is directed towards a two dimensional generalization of the KdV for $u(x, r; t)$ in cylindrical coordinates, which we term the *cylindrical Kadomtsev-Petviashvilli (CKP) equation* [25]

$$[u_t + uu_x + \frac{1}{2}u_{xxx}]_x + \frac{1}{2}(1/r)(ru)_r = 0. \tag{33}$$

Even though the KP is the most common two dimensional generalization (of the KdV) in $u(x, y; t)$ with soliton solutions, it is reported experimentally [28] that a two dimensional ion-acoustic solution was observed as a disc shaped object, which has axial symmetry. This led us to think that perhaps the equation governing the observed object [28] is not the KP equation (29) but a cylindrical generalization to two-dimensions (of the KdV) with axial symmetry, namely the CKP in (33). The CKP equation (33) is derived from the plasma fluid equations [15] using the reductive perturbation method under the assumption of a small perturbation in the radial direction. As this equation is not integrable, numerical methods must be used to follow the evolution of initial profiles. As most numerical methods which are used to study nonlinear wave equations do not employ cylindrical coordinates, we found it necessary to generalize a basic one, namely, the pseudospectral method [12] to cylindrical geometry. This is done for the comparison with our modified iteration of the spectral representation. For the two methods, compatible transforms consisting of the combination of Fourier and Hankel transforms are used for the axial and radial variations in (33), respectively. The application of the modified iterative method, to these nonlinear waves' integral transform (spectral) representation, is covered in Sec. 2.1.

We have also applied the modified iteration in an attempt for predicting the period — doubling solutions of nonlinear dynamics, where the convergence was better than other approaches of its kind. We use the method for constructing a sequence of periodic solutions to the example of the nonlinear Mathiew equation [30],

$$\frac{d^2 x}{dt^2} + (1 + 2q \cos 2t)x - x^3 = 0, \ 0 < t < L \tag{34}$$

$$\dot{x}(0) = 0 \tag{35}$$

$$\dot{x}(L) = 0. \tag{36}$$

The initial results show an improvement over the (spectral) — iterative variational approach [31, 32, 33] in that it has better convergence properties and requires fewer or no "relocation" parameters of these methods. In addition to applying the present modified iteration to the previously used spectral representation of the above Mathiew problem (34)–(36), we also applied it to its corresponding nonlinear integral equation representation in the physical space,

$$x(t) = \int_0^L G(t,\tau)[(1 + 2q \cos 2\tau)x(\tau) - x^3(\tau)]d\tau + \frac{1}{L}\int_0^L x(\tau)d\tau, \quad (36)$$

where $G(t,\tau)$ is the (modified) Green's function for this problem (34)–(36) that can be derived [34] (with a consistency condition $\int_0^L F(x(t),t)dt = 0$) as

$$G(t,\tau) = \frac{L}{3} + \frac{1}{2L}(t^2 + \tau^2) - \begin{cases} \tau, & t < \tau \\ t, & t > \tau \end{cases}. \tag{37}$$

This integral representation (36)–(37) is used to solve the above Mathiew problem (34)–(36) for the first time. The preliminary computations show very clearly that this (Green's function) representation has besides its good convergence, the added advantage that it does not require the "relocation" parameters of the previous spectral representation of the problem [33, 32, 31, 30].

1.2. *Preliminaries and the Origin of the Present Method*

A. *The spectral representation*

To simplify the notation, and especially that of the convolution product of the Fourier transforms, we will use the (infinite) Fourier integral representation. Since many problems we deal with are defined on the finite domain, we will be sure to make the appropriate adjustments that are compatible with the given boundary conditions. After all, whether we have infinite, finite or discrete domain, all the Fourier analysis computations of present day are approximated by the discrete Fourier transform in the way to utilizing its efficient algorithm, the FFT.

Consider the (exponential) Fourier transform of $f(x)$,

$$F(\lambda) = \int_{-\infty}^{\infty} f(x)e^{-i\lambda x}dx \tag{38}$$

and it inverse

$$f(x) = \frac{1}{2\pi} \int_{-\infty}^{\infty} F(\lambda)e^{i\lambda x}dx \,. \tag{39}$$

The convolution theorem states that if $F(\lambda)$ and $G(\lambda)$ are the respective Fourier transform of $f(x)$ and $g(x)$, then the Fourier transform of the product $f(x)g(x)$ is the following convolution product

$$(F^*G)(\lambda) = \frac{1}{2\pi} \int_{-\infty}^{\infty} F(\lambda - \mu)G(\mu)d\mu \,. \tag{40}$$

We note that this product is commutative i.e., $F^*G = G^*F$.

We remind here, that in case of the finite domain, this convolution integral would, through the Fourier series representation, become a summation [1].

We will use, as our first simple example the problem of the concentration pellet as in (1)–(3).

Since this problem is defined on a finite domain we should use a Fourier series where the transform is the Fourier coefficient (or finite Fourier transform). In this case if we are to stay with the above simple forms, we will drop the infinite limits of the integrals and add a term $B(\lambda)$ to the transform $-\lambda^2 Y(\lambda)$ of $\frac{d^2y}{dx^2}$ to cover the boundary conditions, i.e., the formal Fourier transform of (1) becomes

$$\begin{aligned} -\lambda^2 Y(\lambda) + B(\lambda) &= \varphi^2 \cdot (Y^*Y)(\lambda) \\ &= \frac{\varphi^2}{2\pi} \int Y(\lambda - \mu)Y(\mu)d\mu \end{aligned} \tag{41}$$

           A. J. Jerri

where $Y(\lambda)$ is the Fourier transform of $y(x)$. In the present treatment we look at this Eq. (41) as a nonlinear integral (or summation equation in $Y(\lambda)$), which can be written in symbolic terms as

$$Y = T(Y),\tag{42}$$

a nonlinear mapping T with a fixed point Y as our desired solution of (41).

The existing successive approximations, if attempted here, would be written as

$$Y^{(n+1)} = T(Y^{(n)})$$
$$-\lambda^2 Y^{(n+1)}(\lambda) + B(\lambda) = \frac{\varphi^2}{2\pi} \int Y^{(n)}(\lambda - \mu) Y^{(n)}(\mu) d\mu \tag{43}$$

where $Y^{(n)}$ stands for the input estimate and $Y^{(n+1)}$ for the output. Of course, in the final analysis we will inquire about whether this mapping T is contractive to guarantee the existence of its fixed point $Y, Y = T(Y)$ as the desired unique solution of (42). Our computations show that this very familiar iterative process (43) is slowly convergent, and is very sensitive to the magnitude of $\varphi^2 > 1$. Indeed, for the particular boundary conditions

$$y'(0) = 0 \tag{2}$$
$$y(1) = 1 \tag{3}$$

where the Fourier cosine series is used, we observed that this iterative process ceases to converge beyond $\varphi = 1.35$. This can be explained in terms of the mapping (43) being no longer contractive. In contrast, the modified iteration process we are about to introduce converges even for $\varphi = 50$ or more, where we now have a valid contraction argument to assure the existence of the unique solution $Y(\lambda)$, [8, 5].

The origin of the present iterative method was as follows: We start with the first approximation $Y^{(1)}(\lambda)$, which can be taken as the solution of the linearization of (41), or the following equivalent approximation that replaces $Y(\lambda - \mu)$ inside the convolution integral of (41) by the first term $Y(\lambda)$ of its Taylor series expansion (45) about $\mu = 0$,

$$-\lambda^2 Y(\lambda) + B(\lambda) = \frac{\varphi^2}{2\pi} \int Y(\lambda - \mu) Y(\mu) d\mu$$
$$= \frac{\varphi^2}{2\pi} Y(\lambda) \int Y(\mu) d\mu = \frac{\varphi^2}{2\pi} Y(\lambda) y(0),$$

$$-\lambda^2 Y(\lambda) + B(\lambda) = \frac{\varphi^2}{2\pi} y(0) Y^{(1)}(\lambda),$$

$$Y^{(1)}(\lambda) = \frac{B(\lambda)}{\lambda^2 + \frac{\varphi^2}{2\pi} y(0)} \tag{44}$$

after using the first term of the following *Taylor series* to approximate $Y(\lambda = \mu)$ in (44),

$$Y(\lambda - \mu) = Y(\lambda) - \mu Y'(\lambda) + \frac{\mu^2}{2} Y''(\lambda) + \dots, \tag{45}$$

and where $y(0)$ is obtained from (39) with $x = 0$.

The second approximation $Y^{(2)}(\lambda)$ is the first step of the iteration, where we divide and multiply by $Y^{(1)}(\lambda - \mu)$ inside the convolution integral of (41),

$$-\lambda^2 Y(\lambda) + B(\lambda) = \frac{\varphi^2}{2\pi} \int \frac{Y(\lambda - \mu)}{Y^{(1)}(\lambda - \mu)} Y^{(1)}(\lambda - \mu) Y(\mu) d\mu. \tag{41}$$

We then approximate (41) by using the first term of the Taylor series expansion of $\frac{Y(\lambda-\mu)}{Y^{(1)}(\lambda-\mu)}$ about $\mu = 0$,

$$\frac{Y(\lambda - \mu)}{Y^{(1)}(\lambda - \mu)} = \frac{Y(\lambda)}{Y^{(1)}(\lambda)} - \frac{d}{d\lambda}\left[\frac{Y(\lambda)}{Y^{(1)}(\lambda)}\right] + \dots \tag{46}$$

to have an equation,

$$-\lambda^2 Y(\lambda) + B(\lambda) = \frac{\varphi^2}{2\pi} \frac{Y(\lambda)}{Y^{(1)}(\lambda)} \int Y^{(1)}(\lambda - \mu) Y(\mu) d\mu \tag{47}$$

in the second approximation $Y^{(2)}(\lambda)$ as the output $Y(\lambda) = Y^{(2)}(\lambda)$ outside the integral. The estimate for the input $Y(\mu)$ inside the integral (47) takes the previous value which in this case is $Y^{(1)}(\mu)$,

$$Y^{(2)}(\lambda) = \frac{B(\lambda)}{\lambda^2 + \frac{\varphi^2}{2\pi} \frac{1}{Y^{(1)}(\lambda)} \int Y^{(1)}(\lambda - \mu) Y^{(1)}(\mu) d\mu}. \tag{48}$$

This process is continued to $Y^{(m+1)}(\lambda)$,

$$Y^{(m+1)}(\lambda) = \frac{B(\lambda)}{\lambda^2 + \frac{\varphi^2}{2\pi} \frac{1}{Y^{(m)}(\lambda)} \int Y^{(m)}(\lambda - \mu) Y^{(m)}(\mu) d\mu}. \tag{49}$$

At the time [4] we considered this to be a somewhat new or "modified" iteration (which we use to call "successive divisions" approximations) that represents the mapping

$$Y^{(m+1)} = S(Y^{(m)}) \tag{50}$$

towards the solution Y of

$$Y = S(Y), \tag{51}$$

which is our nonlinear equation (41).

It is this "modified" mapping $S(Y)$ (of the successive divisions process) when compared with that of the direct iteration $T(Y)$ of (43), that showed fast convergence, and in particular, being less sensitive to φ^2 of (41). Indeed we can look at $S(Y^{(m)})$ in (50) to introduce a narrower window to help the contraction, or that it operates with a spirit of continued fractions approximations. Some experts have described this process as one with a sort of "dynamic" Green's function that is kept updated along with the approximation of the nonlinear term.

In the case of variable coefficients, for example, if $y^2(x)$ in (1) is replaced by $g(x)y(x)$, the same method can be employed to have

$$\begin{aligned}
-\lambda^2 Y^{(n+1)}(\lambda) + B(\lambda) &= \frac{\varphi^2}{2\pi} \int_{-\infty}^{\infty} G(\lambda - \mu)Y(\mu)d\mu \\
&= \frac{\varphi^2}{2\pi} \int_{-\infty}^{\infty} Y(\lambda - \mu)G(\mu)d\mu
\end{aligned} \tag{52}$$

where $G(\lambda)$ is the Fourier transform of $g(x)$. The modified iterative process would result in

$$Y^{(m+1)}(\lambda) = \frac{B(\lambda)}{\lambda^2 + \frac{\varphi^2}{2\pi} \frac{1}{Y^{(m)}(\lambda)} \int Y^{(m)}(\lambda - \mu)Y^{(m)}(\lambda - \mu)G(\mu)d\mu}. \tag{53}$$

The existing successive approximations method for solving integral equations normally starts with an input estimate $Y^{(n)}$ inside the integral in (41) to result in an output $Y^{(n+1)}$ as in (43)

$$\begin{aligned}
-\lambda^2 Y^{(n+1)}(\lambda) + B(\lambda) &= \frac{\varphi^2}{2\pi} \int Y^{(n)}(\lambda - \mu)Y^{(n)}(\mu)d\mu \\
Y^{(n+1)} &= T(Y^{(n)}).
\end{aligned} \tag{43}$$

There is a very important property of such a mapping (or iterative process) that would, according to Banach's fixed point theorem, guarantee the existence of a unique solution for the Eqs. (41) or (42). This property being that the mapping is contractive, which means that for the two distinct inputs $Y^{(n-1)}$ and $Y^{(n)}$ in (43) we should have

$$|T(Y^{(n)}) - T(Y^{(n-1)})| \leq \alpha |Y^{(n)} - Y^{(n-1)}|, \qquad 0 \leq \alpha < 1. \tag{54}$$

Here we are assuming the use of real valued functions and the Euclidean distance $|x - y|$ between x and y. α here is the *Lipschitz parameter* of the mapping $T(Y)$ and can often be found from

$$\alpha = \max \left| \frac{\partial T(Y)}{\partial Y} \right|. \tag{55}$$

When we first tried this iterative process (43) numerically, convergence was possible only for φ below about 1.35. This result can now be explained in light of the proof we have developed recently on the basis that the above mapping (43) ceases to be contractive about this φ value; hence, there is no obvious guarantee for the existence of a fixed point for such mapping or, in other words, a solution to our nonlinear integral equation (41). On the other hand we showed [1, 5, 8] that the modified iteration in (49) or (50) remains contractive for a much wider range of φ, which explains the successful numerical results we had obtained for $\varphi = 50$ or more in [1]. This good contractive property can now be explained as a consequence of the "new form" of the mapping for the modified iterative process as shown in (49)–(50). Here we have

$$\alpha_s = \max \left| \frac{\partial S(Y)}{\partial Y} \right| \tag{56}$$

which happened to have a larger domain of convergence than its corresponding direct iteration one,

$$\alpha_d = \max \left| \frac{\partial T(Y)}{\partial Y} \right| \tag{57}$$

of (43) (or (42)). Even though the proof of this good contraction for $S(Y)$ was first done [5] for the above Fourier (spectral) representation, it is now more simple and compact to show for the (Green's function) integral representation in the physical space of the same problem. This is also true for the actual numerical iteration.

B. *The (green's function) integral representation in the physical space*

The details, illustrations and proof of the modified-iterative method were first done for the above spectral representation (41), associated with the quadratic nonlinearity $\varphi^2 y^2(x)$. For the same nonlinear problem (1)–(3), with nonlinear term $\varphi^2 f(y)$, instead of $\varphi^2 y^2$, we can use the Green's function

$$G(x,t) = \begin{cases} 1-t, & x \leq t \\ 1-x, & x \geq t \end{cases} \tag{58}$$

to write the nonlinear integral equation representation in $y(x)$ of the physical space.

$$y(x) = 1 - \varphi^2 \int_0^1 G(x,t) f(y(t)) dt \equiv 1 - \varphi^2 K(y) \equiv T_d(y) \tag{59}$$

where the 1 here is for $y_h = 1$, the solution associated with the homogeneous version for the differential equation (1).

We can see here how simple it is to apply this method to the more general nonlinearity $f(x)$. In the next section we will illustrate the modified iteration for this representation (59), compare with the other methods, and discuss the dependence of the contraction property of its mapping $T_d(y)$ on the particular nonlinearity $f(y)$ and the Green's function.

C. *On the convergence of the modified iterative method*

As we have just indicated the modified iterative method can also be applied to other representations or mappings besides the above (spectral) nonlinear integral equation representation (41). In particular, for the Green's function integral representation in (59) we can have, according to (55), a direct iteration contraction factor

$$a_d = \max \left| \frac{\partial}{\partial y} T_d(y) \right|, \tag{57}$$

where the maximum is taken over the interval $x \in (0,1)$,

$$\frac{\partial}{\partial y} T_d(y) = -\varphi^2 \frac{\partial}{\partial y} \int_0^1 G(x,t) y^2(t) dt \equiv -\varphi^2 \frac{\partial}{\partial y} K(y) \tag{60}$$

$$= -\varphi^2 \frac{\int_0^1 G(x,t)[2y(t) + \eta(t)]\eta(t)dt}{\eta(x)} \, , \quad \eta(t) \equiv \Delta y(t)$$

$$= -2\varphi^2 \frac{\int_0^1 G(x,t)y(t)\eta(t)dt}{\eta(x)} \, , \tag{61}$$

after neglecting the η^2 term inside the integral. On the other hand, if we multiply on the left of the integral in (59),

$$y(x) = 1 - \varphi^2 \frac{y(x)}{y(x)} \int_0^1 G(x,t)f(y(t))dt \tag{59a}$$

we can write (59) in the following equivalent form (for $f(y) = y^2$),

$$y(x) = \frac{1}{1 + \varphi^2 \frac{\int_0^1 G(x,t)y^2(t)dt}{y(x)}} = \frac{1}{1 + \varphi^2 \frac{K(y)}{y(x)}} \equiv T_s(y) \tag{62}$$

the mapping associated with the present modified iterations, where

$$\alpha_s = \max \left| \frac{\partial T_s}{\partial y} \right| = \max \left| \frac{1}{\left[1 + \varphi^2 \frac{K(y)}{y(x)}\right]^2} \frac{y(x)\frac{\partial K(y)}{\partial y} - K(y)}{y^2(x)} \right|$$

$$= \max \left| \frac{\varphi^2 \frac{\partial K(y)}{\partial y} - \frac{\varphi^2}{y(x)}K(y)}{1 + \frac{\varphi^2}{y(x)}K(y)} \right| . \tag{63}$$

Here, we see clearly the term $\varphi^2[\partial K(y)/\partial y]$ of α_d in (60) and the additional term of this mapping $[\varphi^2/y(x)]K(y)$ that, as we shall see in Fig. 1, causes the larger domain of α_s, compared to α_d.

Indeed if we consider the more general problem

$$y(x) = f(x) + T(y) \tag{64}$$

and its present modified mapping

$$y(x) = \frac{f(x)}{1 - \frac{T(y)}{y}} \equiv T_s(y) \, , \tag{65}$$

we have $\alpha_d = \max \left| \frac{\partial T}{\partial y} \right|$ and $\alpha_s = \max |\partial T_s / \partial y|$,

$$\frac{\partial T_s}{\partial y} = -\frac{f(x)}{\left[1 - \frac{T(y)}{y}\right]^2} \cdot \frac{-\left[y\frac{\partial T}{\partial y} - T(y)\right]}{y^2}$$

$$\frac{\partial T_s}{\partial y} = \frac{\frac{\partial T}{\partial y} - \frac{T(y)}{y}}{1 - \frac{T(y)}{y}} = \frac{\frac{\partial T}{\partial y} - \beta(x)}{1 - \beta(x)} \, , \qquad \beta = \frac{T(y)}{y} \, . \tag{66}$$

In Fig. 1 we illustrate the larger domain, (shaded area) in $\beta = T(y)/y$ versus $\partial T/\partial y$ for $|\alpha_s| > 1$ compared to the interval $(-1, 1)$ for $|\alpha_d| < 1$.

For the region of validity of $|\alpha_s| < 1$, we must note that it is easier for the present method to work when $|\alpha_d|$ exceeds one on the left of $\alpha = -1$. This means that we would only seek the problems that have β below the line $\beta = (\alpha + 1)/2$ of Fig. 1. This is in contrast to $|\alpha_d|$ exceeding one from the right of $\alpha = 1$, where in this case the sought β had to change character to stay in the shaded region. This explains the success of this method for the particular examples with quadratic nonlinearties (41) and the special case of (59) where $\partial T/\partial y$ is negative. This means that if we change $\varphi^2 y^2$ to $-\varphi^2 y^2$ in the original problem (1) then α_d would exceed 1 on the right of $\alpha_d = 1$ and the present method would have trouble, which is verified numerically. However, for very small values of φ around $\varphi = 0.1$, the α_d would be to the left of $\alpha_d = 1$ and away from this difficulty. This was verified for the nonlinearity in the equation $\frac{d^2 y}{dx^2} + \frac{1}{x}\frac{dy}{dx} = -\varphi^2(1+\beta-y)e^{\gamma(y-1)/y}, 0 < x < 1$ of the first order Arrenhenius kinetic problems in cylindrical pellet with $\beta = \frac{1}{2}, \gamma = 50$. These results were also verified by comparing very well with those in [Figure 10, in [35]] for $\varphi = 0.1$.

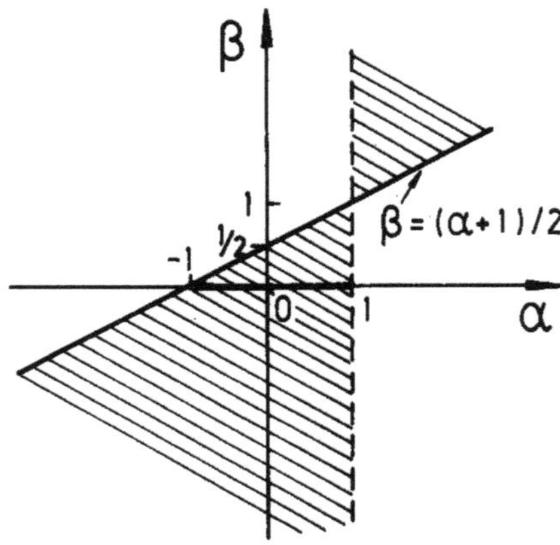

Fig. 1. Plot of the domain (shadowed area) for the modified iteration (T_s) contraction $|\alpha| < 1$ of the direct iteration. (From Jerri [8], Courtesy Chem. Eng. Comm., 1987 Vol. 52, see Ref. [8]).

2. Modifying the Iteration for the Nonlinear Integral Representation

2.1. *Nonlinear Waves*

With our introduction of modifying the spectral representation for the nonlinear chemical reaction problem (1)–(3) in Sec. 1.2A, we shall concentrate here on an example from nonlinear waves.

We will first have a look at the resulting nonlinear integro-differential equation (71) of the KdV equation (20) in the Fourier (spectral) space, then follow it by a more detailed analysis of the Benjamin-Ono (B–O) equation (23).

Consider the Fourier transform of $u(x,t)$

$$F\{u\} = U(\lambda,t) = \int_{-\infty}^{\infty} u(x,t)e^{-i2\pi\lambda x}dx \qquad (67)$$

and its inverse

$$F^{-1}\{U(\lambda,t)\} = u(x,t) = \int_{-\infty}^{\infty} U(\lambda,t)e^{i2\pi\lambda x}dx . \qquad (68)$$

In the case of the KdV equation (20), we Fourier transform it to have:

$$\frac{dU(\lambda,t)}{dt} = 8i\pi^3\lambda^3 U(\lambda,t) - 6F\{uu_x\}, \qquad (69)$$

then use the convolution theorem (40) for the nonlinear term $F\{uu_x\}$,

$$F\{uu_x\} = \int_{-\infty}^{\infty} U(\lambda-\mu,t)[2\pi i\mu U(\mu,t)]d\mu \qquad (70)$$

to obtain the following nonlinear integro-differential equation in $U(\lambda,t)$:

$$\frac{dU(\lambda,t)}{dt} = 8i\pi^3\lambda^3 U(\lambda,t) - 12\pi i \int_{-\infty}^{\infty} (\lambda-\mu)U(\lambda-\mu,t)U(\mu,t)d\mu . \qquad (71)$$

The (direct) iteration would start with an estimate $U_n(\lambda,t)$ inside the integral of (71), and we consider the $U(\lambda,t)$ outside the integral in (71) as the successive estimate, or output, $U_{n+1}(\lambda,t)$. The present modification on this iteration process would look like as if we multiply and divide the integral of (71) by $U(\lambda,t)$,

$$\frac{dU(\lambda,t)}{dt} = 8i\pi^3\lambda^3 U(\lambda,t) - 12\pi i\frac{U(\lambda,t)}{U(\lambda,t)} \int_{-\infty}^{\infty} (\lambda-\mu)U(\lambda-\mu,t)U(\mu,t)d\mu$$

$$(72)$$

then, divide both sides of (72) by $U(\lambda, t)$, and integrate with respect to t to obtain

$$
\begin{aligned}
U(\lambda, t) =\, & U(\lambda, 0) \exp\Big\{8i\pi^3\lambda^3 t - 12\pi i \int_0^t \frac{1}{U(\lambda, t)} \\
& \cdot \Big(\int_{-\infty}^{\infty} (\lambda - \mu)U(\lambda - \mu, t)U(\mu, t)d\mu\Big)\, dt\,.
\end{aligned} \tag{73}
$$

For $U(\lambda, 0)$ we use

$$
U(\lambda, 0) = 4\pi^2 \lambda \operatorname{csch} \frac{\pi^2}{k}\lambda\,, \tag{74}
$$

the Fourier transform of the solitary initial condition (21). The iterative process, including the use of the fast Fourier transform (FFT) for computing the inner Fourier convolution integral, and the time integration over the interval $(0, t)$ in (73) is discussed in some details in [14].

The Benjamin-Ono (B-O) equation

Now, we turn to the Bengamin-Ono equation (23) and Fourier transform it, using the identity (25) for the $H\{uu_x\}$ term and the convolution product of (40) for the Fourier transform of the nonlinear term uu_x, to have the following nonlinear integro-differential equation in $U(\lambda, t)$ the Fourier transform of $u(x, t)$,

$$
\begin{aligned}
\frac{dU}{dt}(\lambda, t) =\, & - 2(U^* 2\pi i\lambda U) + i\sqrt{2\pi}(\operatorname{sgn}\lambda)(-4\pi^2\lambda^2 U)\,, \\
\frac{dU}{dt} =\, & - 4\pi i \int_{-\infty}^{\infty} (\lambda - \mu)U(\lambda - \mu, t)U(\mu, t)d\mu \\
& - 4\pi^2\lambda^2 i\sqrt{2\pi}(\operatorname{sgn}\lambda)U(\lambda, t)\,.
\end{aligned} \tag{75}
$$

The iterative method for such a nonlinear integral equation, as we have indicated very briefly for the KdV equation in (73), starts with dividing and multiplying inside the integral in (75) by a first approximate of the solution $U^{(1)}(\lambda - \mu, t)$. Then, the quotient $\frac{U(\lambda-\mu,t)}{U^{(1)}(\lambda-\mu,t)}$ inside the integral is approximated by $\frac{U(\lambda,t)}{U^{(1)}(\lambda,t)}$, the first term of its Taylor series expansion about $\mu = 0$ to have

$$
\begin{aligned}
\frac{dU}{dt} =\, & - 4\pi i \frac{U(\lambda, t)}{U^{(1)}(\lambda, t)} \int_{-\infty}^{\infty} (\lambda - \mu)U^{(1)}(\lambda - \mu, t)U(\mu, t)d\mu \\
& - 4\pi^2\lambda^2 i(\sqrt{2\pi}(\operatorname{sgn}\lambda)U(\lambda, t))
\end{aligned} \tag{76}
$$

as an approximation to the integral equation (75) with a temporary kernel $U^{(1)}(\lambda, t)$. If we divide both sides of (76) by $U(\lambda, t)$, then integrate with respect to t we have:

$$
\begin{aligned}
U(\lambda, t) =& U(\lambda, 0) \exp\{-4\pi^2 t\lambda^2 i\sqrt{2\pi}\, \text{sgn}\,\lambda - 4\pi i \int_0^t \frac{1}{U^{(1)}(\lambda, t)} \\
& \times \int_{-\infty}^{\infty} U^{(1)}(\lambda - \mu, t)\mu U(\mu, t)d\mu dt\}
\end{aligned}
\tag{77}
$$

where $U(\lambda, 0) = F\{u(x, 0)\}$.

The computations start with the $U(\lambda, t - \Delta t)$ value (of the previous time step) for $U(\lambda, t)$ inside the integral of (77) to give a new approximation as $U(\lambda, t)$ on the left side of (77). This process is repeated (at time t) until $U(\lambda, t)$ approached the solution to this integral equation with the approximate (temporary) kernel $U^{(1)}(\lambda, t)$. Next, we update the kernel to the above solution and repeat the process until we get a solution which in turn will be used to update the latter temporary kernel. This continues until the iterates as inputs and outputs, and the kernel of (77) approach the same value, which is what we term the solution to the original exact integro-differential equation (75).

The convolution product in (77) is evaluated via the fast Fourier transform (FFT) and for the marching in time. We split the interval $(0, t)$ to $(0, t - \Delta t)$ and $(t - \Delta t, t)$, where the first interval transferred from the last time interval, and where for the second interval we use a 5-point integration formula.

2.2. *Nonlinear Boundary Value Problems*

Here we will illustrate the modified iteration for the (Green's function) integral representation of the important problem of Poisson-Boltzmann equation (14). In part 3 we use the same problem to illustrate the success of the modified iteration to the successive overrelaxation (SOR) method.

The Poisson-Boltzmann equation

In bispherical coordinates (η, θ) the Poisson-Boltzmann equation (14) becomes [17, 18, 22]

$$
\left(\frac{\cosh\eta - \cos\theta^2}{\alpha}\right)\left[\frac{\partial^2\psi}{\partial\eta^2} - \left(\frac{\sinh\eta}{\cosh\eta - \cos\theta}\right)\frac{\partial\psi}{\partial\eta} + \frac{\partial^2\psi}{\partial\theta^2} + \left[\cot\theta - \frac{\sin\theta}{\cosh\eta - \cos\theta}\right]\frac{\partial\psi}{\partial\eta}\right] = \sinh\psi
\tag{78}
$$

employing the φ-independence. The boundary conditions of fixed potentials on the surfaces ($\eta = \eta_1$ and $\eta = \eta_2$) of the two spheres are given as

$$\begin{aligned}
\psi = \psi_1 \,, \ \eta = \eta_1 \,, \ 0 < \theta < \pi \\
\psi = \psi_2 \,, \ \eta = \eta_2 \,, \ 0 < \theta < \pi \,.
\end{aligned} \tag{79}$$

Equation (78) is a nonlinear partial diffrential equation in ψ, where the integral representation is

$$\psi(\eta,\theta) = \psi_L(\eta,\theta) - \frac{1}{2}\int G(\eta,\eta';\theta,\theta')\sinh\psi(\eta',\theta')dA' \tag{80}$$

where ψ_L is the solution to the (homogeneous) Laplace equation in (78) with the inhomogeneous boundary conditions given in Eq. (79), and $G(\eta,\eta';\theta,\theta')$ is the Green's function, which satisfies

$$\nabla^2 G(\eta,\eta';\theta,\theta') = -4\pi\delta(\eta - \eta')\delta(\theta - \theta') \,, \tag{81}$$

and is given in [9, 17].

The final solution can be obtained from solving the following (nonlinear) integral equation [17] in ψ, as given in (82)–(83):

$$\psi(\eta,\theta) = \psi_L(\eta,\theta) - \frac{1}{2}\int_{\eta_1}^{\eta_2} d\eta' \int_0^{\pi} d\theta' \, G(\eta,\eta';\theta,\theta')J(\eta',\theta')\rho(\eta',\theta') \,, \tag{82}$$

where $\rho(\eta',\theta')$ is the charge density.

$$\rho(\eta',\theta') = \sinh\psi(\eta',\theta') \,, \tag{83}$$

and the Jacobian $J(\eta',\theta')$ is given by

$$J(\eta',\theta') = \frac{a^3 \sin\theta}{(\cosh\eta' - \cos\theta')^3} \,. \tag{84}$$

The problem now reduces to solving the integral equation (82)–(84) for the potential ψ. Earlier researchers [17] had proposed a direct iteration scheme, which can be symbolically written as:

$$\psi_{n+1} = \psi_L + T(\psi_n) \,, \ \psi_0 = \psi_L \,, \tag{85}$$

where

$$T(\psi) = -\frac{1}{2}\int\int dA'G \sinh\psi \,, \tag{86}$$

as given in (82) with $dA' = J(\eta', \theta')d\eta' d\theta'$. This is implemented by starting with ψ_0 to obtain ψ_1, then using ψ_1 to obtain ψ_2. This process continues by getting ψ_n from ψ_{n-1} until the difference of the two is less than some prescribed small number. We have tried this for Eqs. (82)–(84) and (as we had expected from theory) found it not to converge for the separations of interest. Before we mention the result of the present modified iteration we may note that the symbolic form of the integral equation (82) (with (83) and (84)) is

$$\psi = H + T(\psi). \tag{87}$$

The modified iterative method can be looked at as if we were to multiply $T(\psi)$ in (87) by $1 = \frac{\psi}{\psi}$; i.e.,

$$\psi = H + \frac{\psi}{\psi}T(\psi). \tag{88}$$

Now label the nth and $(n-1)$st iterates of the solution by ψ_n and ψ_{n-1}, respectively and write (88) in the form of a recursion relation:

$$\psi_n = H + \frac{\psi_n}{\psi_{n-1}}T(\psi_{n-1}). \tag{89}$$

Solving for ψ_n, we obtain the scheme for the modified iteration

$$\psi_n = \frac{H}{1 - \frac{T(\psi_{n-1})}{\psi_{n-1}}}. \tag{90}$$

So symbolically the present modified iteration for (85) is

$$\psi_n = \frac{\psi_L}{1 - \frac{T(\psi_{n-1})}{\psi_{n-1}}} \tag{91}$$

where we need to use the known [9, 17] homogeneous term ψ_L, the Green's Function from [17], and the integral in T is the double integral of Eq. (82). When this modified scheme (91) was used instead of the direct scheme (85), convergence was attained for the same parameters of interest [17, 22]. This was compared with repeating a finite element method computations for doing the complete problem [22], since other methods make the same approximations in the geometry of the problem.

3. Modifying the Relaxation Schemes for Solving Nonlinear Boundary Value Problems

Here we will illustrate the success of modifying the successive over-relaxation (SOR) scheme as a typical finite difference relaxation scheme as compared to all the preceding integral representations of the boundary value problem. For an example we choose, again, our first successful example of the Poisson-Boltzmann equation as in (78). Due to the limitation of space we will leave the modification of other relaxation schemes to our future research [10].

The Modified Successive Overrelaxation (SOR) for the Poisson-Boltzmann equation

It is well known that one method for numerically finding solutions of elliptic boundary value problems is the method of successive overrelaxation, or SOR [20, 19, 21]. In this section we review this method and apply it to our problem of finding the solution to the two sphere problem. We have found that the SOR method does not converge very well for the Poisson-Boltzmann equation. As this is also an iterative method, we attempted our modification on its iterative process as in the form of Eq. (89). This modification gave a scheme that converges and is more efficient than the methods of the last section. In particular it is much faster than our modification on the iteration of the Green's function integral representation. The latter suffers from the slowly convergent double Fourier series for its Green's function. Also, the modified SOR is much simpler to implement in addition to using much less storage, when compared to that needed for the Green's function.

We begin by writing the first order derivatives in Eq. (78) as forward differences and the second order derivatives as central differences. This leads to the five point scheme

$$a_{ij}\psi_{i+1,j} + b_{ij}\psi_{i-1,j} + c_{ij}\psi_{i,j+1} + d_{ij}\psi_{i,j-1} + e_{ij}\psi_{ij} = f_{ij}, \qquad (92)$$

where the coefficients a_{ij} through e_{ij} are determined from the coefficients of the differential equation, and the f_{ij} is proportional to the source, $\sinh\psi_{ij}$. One can solve for ψ_{ij} from this equation and obtain

$$\psi_{ij} = \frac{1}{e_{ij}}\left[f_{ij} - a_{ij}\psi_{i+1,j} - b_{ij}\psi_{i-1,j} - c_{ij}\psi_{i,j+1} - d_{ij}\psi_{i,j-1}\right]. \qquad (93)$$

We could set up an iterative scheme for this equation by evaluating the right hand side at $\psi_{ij}^{(n)}$ to obtain the left side as $\psi_{ij}^{(n+1)}$. This is known as the Jacobi method [20]. We could also use the new iterates on the left side of (93) as soon as they are known. This leads to the Gauss-Seidel method [20]. However, it is well knwon that the SOR method converges faster than either of these methods. So, we will apply our modification to this method, though it is also possible to modify the former methods in the same manner [10].

The successive overrelaxation sets up an iterative procedure, such that the new iterate, $\psi_{ij}^{(n+1)}$ is given as a weighted average of the old iterate $\psi_{ij}^{(n)}$ and the value obtained in (93) [20]. The value of ψ_{ij} obtained on the right side of (93) will be denoted as ψ_{ij}^* which we compute by evaluating the left side using the old iterates $\psi_{ij}^{(n)}$. Namely, we have for the weighted average

$$\psi_{ij}^{(n+1)} = \omega\psi_{ij}^* + (1-\omega)\psi_{ij}^{(n)}, \tag{94}$$

or, rewriting (94) in a more explicit form:

$$\begin{aligned}
\psi_{ij}^{(n+1)} =\psi_{ij}^{(n)} - \frac{\omega}{e_{ij}}&\left[a_{ij}\psi_{i+1,j}^{(n)} + b_{ij}\psi_{i-1,j}^{(n)} + c_{ij}\psi_{i,j+1}^{(n)}\right.\\
&\left.+ d_{ij}\psi_{i,j-1}^{(n)} + e_{ij}\psi_{ij}^{(n)} - f_{ij}^{(n)}\right].
\end{aligned} \tag{95}$$

The weight ω is known as the relaxation parameter for this scheme.

Since this is basically an iterative scheme, we have applied our modification from Eq. (90) with very good success. Our modification, as was done for other iterative schemes [1, 4, 6, 8, 14] amounts here to multiplying the nonhomogeneous term $f_{ij}^{(n)}$ in (95) by $\psi_{ij}^{(n+1)}/\psi_{ij}^{(n)}$ to have

$$\begin{aligned}
\psi_{ij}^{(n+1)} = \psi_{ij}^{(n)} - \frac{\omega}{e_{ij}}&\left[a_{ij}\psi_{i+1,j}^{(n)} + b_{ij}\psi_{i-1,j}^{(n)} + c_{ij}\psi_{i,j+1}^{(n)} + d_{ij}\psi_{i,j-1}^{(n)}\right.\\
&\left.+ e_{ij}\psi_{ij}^{(n)} - \frac{\psi_{i,j}^{(n+1)}}{\psi_{ij}^{(n)}}f_{ij}^{(n)}\right].
\end{aligned} \tag{96}$$

Solving for the $(n+1)$st iterate leads to the present modification of the SOR scheme:

$$\begin{aligned}
\psi_{ij}^{(n+1)} = \psi_{ij}^{(n)} - \frac{\omega}{e_{ij}}&\left[a_{ij}\psi_{i+1,j}^{(n)} + b_{ij}\psi_{i-1,j}^{(n)} + c_{ij}\psi_{i,j+1}^{(n)} + d_{ij}\psi_{i,j-1}^{(n)}\right.\\
&\left.+ e_{ij}\psi_{ij}^{(n)}\right]\Bigg) \Bigg/ \left(1 + \frac{\omega f_{ij}^{(n)}}{e_{ij}\psi_{ij}^{(n)}}\right).
\end{aligned} \tag{97}$$

This scheme is found to converge quickly to the solution of the Poisson-Boltzmann equation between two spheres (particles) in an electrolyte. Indeed it is about 100 times as fast as our first modification of the iteration for the Green's function integral representation in (86). It is even faster than the leading method for solving such a problem, namely, the finite element method [22].

As we mentioned earlier our research is continuing for modifying other relaxation schemes [10], and applying the present modification to more of the relevant applied nonlinear problems. Most recently Oswald [36] showed serious interest in the method and did some computations for only one dimensional problems to compare with other well known methods.

Acknowledgement

The author would like to thank his colleague Prof. R. L. Herman for a number of discussions, and for using our cooperate results, Dr. K. W. Tse for helping with the initial numerical computations. Thanks are also due Professor M. Z. Nashed and R. Vichnevetsky for their very early support of the idea, Professor A. Fokas for his enthusiasm about writing this paper, and Professor M. J. Ablowitz for early helpful comments about the method. Mrs. S. A. Khan deserves thanks for typing the manuscript with patience and care. I am indeleted to my wife Suad for her continue support and patience and Profs. M. Z. Nashed and R. Vichnevetsky for their initial support of the idea.

References

1. A. J. Jerri, *Application of the transformation-iterative method to nonlinear concentration boundary value problems*, Chem. Eng. Commun. **23** (1983) 101–113.
2. D. D. Do and R. H. Weiland, *Self-Poisoning of Catalyst Pellets*, Alch. E. Symposium Series, pp. 36–45.
3. D. D. Do and J. E. Bailey, *A formalism for the solution of problem involving chemical reaction and concentration-dependent diffusion coefficients*, Appl. Sci. Res. **37**, pp. 225–240.
4. A. J. Jerri, *A Transform-Iterative Method for Nonlinear or Variable Velocity Waves*, Advance in Computer Methods for Partial Differential Equations, ed. R. Vichnevetsky and R. S. Stepleman, IMACS Publ., 1984, pp. 313–317.

5. A. J. Jerri, *On the convergence of the recent variation on the iterative method for nonlinear problems*, Presented at the Congress on Math. and Appl., Alexandria University, Nov. 23–25, 1986.

6. A. J. Jerri, *Introduction to Integral Equations with Applications*, Marcel Dekker, New York, 1985.

7. L. Collatz, *Functional Analysis and Numerical Methods*, McGraw Hills, New York, 1966.

8. A. J. Jerri, R. L. Herman, and R. H. Weiland, *A modified iterative method for nonlinear chemical concentration in cylindrical and spherical pellets*, Chem. Eng. Commun. **52** (1987) 173–193.

9. A. J. Jerri, and R. L. Herman, *The solution of the Poisson Boltzmann equation between two spheres: modified iterative methods*, 1989, submitted.

10. A. J. Jerri, and R. L. Herman, *Modifying relaxation schemes for solving nonlinear boundary value problems*, to be submitted.

11. R. Vichnevetsky, and J. B. Bowles, *Fourier Analysis of Numerical Approximations of Hyperbolic Equations*, SIAM, Philadelphia, 1982.

12. B. Fornberg, and G. B. Whitham, *A numerical and theoretical study of certain nonlinear wave phenomena*, Phil. Trans. Roy. Soc., **289**, **373** (1978).

13. A. J. Jerri, *Integral transform-iterative method for elastic beam on nonlinear foundations*, to be submitted.

14. A. J. Jerri, and K. W. Tse, *A modified iterative method for the Kortewg de-Vries and other nonlinear equations*, Proceedings Appl. Conf., Cairo University, Cairo, Egypt, Jan. 3–6, 1987.

15. R. L. Herman and A. J. Jerri, *A Modified Iterative Methods for the Cylindrical Kadomtsev-Petviashvilli Equation for Two Dimensional Ion-Acoustic Waves*, to be submitted.

16. E. K. Blum, *Numerical Analysis and Computation-Theory and Practice*, Addison-Wesley, New York, 1972.

17. E. Barouch, *Double layer interaction between spheres with unequal surface potentials*, J. Chem. Soc., Faraday Trans. 1, **84** (1988) 3093–3095.

18. P. M. morse and H. Feshbach, *Methods of Theoretical Physics*, McGraw-Hill, New York, 1953.

19. G. E. Forsythe and W. R. Wason, *Finite Difference Methods for Partial Differential Equations*, John Wiley and Sons, New York, 1960.

20. W. H. Press, B. P. Flannery, S. A. Teukolsky, and W. T. Vettering, *Numerical Recipes, The Art of Scientific Computing*, Cambridge Univ. Press, Cambridge, 1986.

21. W. L. Briggs, *A Multigrid Tutorial*, SIAM, Phila. PA, 1987.

22. T. Ring, *Double-layer interaction energy for two unequal spheres*, J. Chem. Soc. Faraday Trans. 2, **78** (1982) 1513–1528.

23. G. M. Kurajian, and T. Y. Na, *Elastic beams on nonlinear continuous foundations*, ASME Winter Annual Meeting, Nov. 15–20, 1981, pp. 1–7.

24. T. Y. Na, *Computational Methods in Engineering Boundary Value Problems*, Academic Press, New York, 1979.

25. M. Ablowitz and H. Segur, *Solitons and the Inverse Scattering Transform*, SIAM, Philadelphia, 1981.

26. T. R. Taha and M. J. Ablowitz, Comp. Phys. **55** (1984) 231.

27. M. D. Kruskal, Private Communications to Ref. 26.

28. E. F. Gabl, J. M. Bulson, and K. E. Lonngren, *Excitation of a two-dimensional ion-acoustic soliton*, Phys. Fluids **27** (1984) 269–272.

29. A. J. Jerri, *Towards a discrete Hankel transform and its application*, Applicable Anal. **7** (1978) 97–109.

30. A. J. Jerri, R. L. Herman, T. Bountis and N. Mousa, *Modified iterative method for the determination of period doubling bifurcations in nonlinear dynamical systems*, to be submitted.

31. R. Helleman, in *Fundamental Problems in Statistical Mechanics*, ed. E. G. D. Cohen, V. 5, North Holland, 1981.

32. R. Helleman, and T. Bountis, in Lecture Notes in Physics, vol. 93, Springer, New York, 1978, p. 353.

33. N. Budinsky, *Stability of Periodic Orbits and Chaotic Behavior in Hamiltonian Systems*, Ph. D. Dissertation, Clarkson Univerisity, 1983.

34. R. P. Kanawal, *Linear Integral Equations*, Academic Press, New York, 1971, pp. 89–90.

35. A. K. Kapila, and B. J. Maatkowsky, *Reactive-diffuse systems with Arrhenius kinetics: Multiple solutions, ignition and extinction*, SIAM J. Appl. Math. **36** (1979) 373–389.

36. P. Oswald, Recent personal communications.

A. J. Jerri
Department of Mathematics
Kuwait University
P. O. Box 5969
13060 Safat
Kuwait

THE MATH. HERITAGE OF C.F. GAUSS (pp. 405-415)
edited by George M. Rassias
©1991 World Scientific Publ. Co. Singapore

EXPANSIONS OF SPECIAL CASES OF GAUSS' HYPERGEOMETRIC FUNCTIONS USING GENERALIZED CALCULUS

R. N. Kalia

We shall discuss here how some simple, yet elegant, expansion formulas involving special cases of Gauss' hypergeometric functions can be found by using four special cases of generalized Leibniz rule. First, we take a quick look backwards in the labyrinthine corridor of time to appreciate the contribution of Gauss in the development of hypergeometric series.

1. Historical Note

The chroniclers of the history of hypergeometric series may have to go back to the paper of Gauss [3] in 1812 where Gauss introduced capital Π in the designation of continued products. He introduced a closed relative of the gamma function.

$$\prod(k,z) = \frac{1 \cdot 2 \cdot 3 \ldots k}{(z+1)(z+2)\ldots(z+k)}k \, . \tag{1.1}$$

where k is a positive integer. He later arrived at $\Pi(z) = 1 \cdot 2 \cdot 3 \ldots z$. Further, he obtained $\Pi(-1/2) = \sqrt{\pi}$. He also denoted $d/dz(\log \Pi(z))$ by $\psi(z)$. See Cajori [2] for details.

In Gauss [3], it is not mentioned that "Euler had already introduced the defining equation, which one should assume, would have provided Gauss, too, with a natural way to introduce the function", opines Buhler [1].

In the above mentioned 1812 paper, Gauss discusses in Secs. 12–14 his investigations about the expansion of

$$\frac{F(\alpha, \beta + 1, \gamma + 1, x)}{F(\alpha, \beta, \gamma, x)}$$

in terms of continued fractions. According to Buhler [1] those expansions are "methodologically and substantially of no major interest." But recent contributions of M. E. H. Ismail and others cited in Ismail and Libis [4] have stimulated interest in the use of asymptotic methods in J-fractions. Ismail dwells on basic analogs in his systematic study in the subject.

Gauss probably developed his paper on the hypergeometric series as the series contains the elliptic functions on which he wrote extensively.

2. The Hypergeometric Function

The forward factorial or Pochhammer symbol is given by

$$(a)_k = \begin{cases} 1, & k = 0 \\ a(a+1)(a+2)\ldots(a+k-1), & k = 1, 2, \ldots, \end{cases} \tag{2.1}$$

and $\Gamma(a+k)/\Gamma(a) = (a)_k$ in terms of the gamma function $\Gamma(x)$.

The Gaussian Hypergeometric function is defined by

$$_2F_1(a, b; b; z) = \sum_{n=0}^{\infty} \frac{(a)_n (b)_n}{(c)_n} \frac{z^n}{n!}, c \neq 0, -1, -2, \ldots, \tag{2.2}$$

in terms of the Pochhammer symbol (2.1). The series (2.2) is absolutely convergent if $\mathrm{Re}(c - a - b) > 0$, conditionally convergent if $-1 < \mathrm{Re}(c - a - b) < 0$ and divergent if $\mathrm{Re}(c - a - b) \leq -1$ on the unit circle.

The generalized Hypergeometric function is defined by

$$_pF_q \begin{bmatrix} \alpha_1, \ldots, \alpha_p; \ z \\ \beta_1, \ldots, \beta_q \end{bmatrix} = \sum_{n=0}^{\infty} \frac{(\alpha_1)_n \ldots (\alpha_p)_n}{(\beta_1)_n \ldots (\beta_q)_n} \frac{z^n}{n!}$$

$$= \ _pF_q(\alpha_1, \ldots, \beta_1, \ldots, \beta_q; z) \tag{2.3}$$

where p, q are positive integers or zero (an empty product is interpreted as 1), and we assume that the variable z, the numerator parameters $\alpha_1, \ldots, \alpha_p,$

and the denominator parameters $\beta_1, \ldots, \beta_q$ take on complex values, provided that

$$\beta_j \neq 0, -1, -2, \ldots, j = 1, \ldots, q \ . \tag{2.4}$$

In what follows, we shall assume our parameters obey (2.4). Furthermore, if any of the numerator parameter is a negative integer or zero, the $_pF_q$ series terminates in view of $(-n)_k = 0, k > n$. In such a case the series will reduce to a generalized hypergeometric polynomial. The $_pF_q$ converges for $|z| < \infty$ if $p \leq q$, converges for $|z| < 1$ if $p = q + 1$, and diverges for all $z, z \neq 0$, if $p > q + 1$.

As one notes above, the function $_pF_q$ had not been defined for $p > q+1$. MacRobert and Meijer proposed their E and G functions to give a meaning to the hypergeometric function in the event when $p > q + 1$. We will refrain from reproducing the involved discussion on the evolution of E, G, and later H functions. An interested reader may refer to the works of Mathai and Saxena [7] and Srivastava and Manocha [11] and the extensive bibliographies therein.

We shall discuss below some very simple methods which encompass expansion formulas of incomplete gamma, incomplete beta, and related functions. Also, we shall see how the generalized derivatives lead us to such formulas which could otherwise only be obtained by using long drawn methods from the theory of integral transformation. However, one is reminded here that the integral transforms and fractional calculus are connected areas.

3. Generalized Calculus and Expansions of Incomplete Gamma Functions

The Gamma function is defined by the Euler integral.

$$\Gamma(z) = \int_0^\infty t^{z-1}e^{-t}dt, \quad \mathrm{Re}(z) > 0 \ . \tag{3.1}$$

The incomplete gamma function $\gamma(\alpha, z)$ and its complement $\Gamma(\alpha, z)$ are defined by

$$\gamma(\alpha, z) = \int_0^z t^{\alpha-1}e^{-1}dt, \quad \mathrm{Re}(\alpha) > 0, |\arg z| < \pi \ , \tag{3.2}$$

$$\Gamma(\alpha, z) = \int_0^\infty t^{\alpha-1}e^{-1}dt, \quad |\arg(z)| < \pi \ ; \tag{3.3}$$

so that

$$\gamma(\alpha, z) + \Gamma(\alpha, z) = \Gamma(\alpha) \ . \tag{3.4}$$

The 'entire incomplete gamma function', $\gamma^*(\alpha, z)$, is defined by means of the equations.

$$z^\alpha \gamma^*(\alpha; z) = \frac{\gamma(\alpha; z)}{\Gamma(\alpha)} = 1 - \frac{\Gamma(\alpha; z)}{\Gamma(\alpha)} \ , \tag{3.5}$$

where $\Gamma(\alpha)$ is the usual gamma function defined by (3.1). Conditions of validity and computational aspect to eight bit precision are given in Spanier and Oldham [10]. Connection with the error function is of great use in applications. The error function and its complement are related to the incomplete gamma functions as follows.

$$\gamma\left(\frac{1}{2}; x\right) = \sqrt{\pi} \ \mathrm{erf}(\sqrt{x}) \tag{3.6}$$

$$\Gamma\left(\frac{1}{2}; x\right) = \sqrt{\pi} \ \mathrm{erfc}(\sqrt{x}) \ . \tag{3.7}$$

The error functions are given by the following two equations.

$$\mathrm{erf}(x) = 2\pi^{-\frac{1}{2}} \int_0^x e^{-t^2} dt = 2x\pi^{-\frac{1}{2}} {}_1F_1\left(\frac{1}{2}; \frac{3}{2}; -x^2\right) \ , \tag{3.8}$$

where ${}_1F_1$ is the Kummer's confluent hypergeometric function (after E. E. Kummer (1810–1893)).

Also,

$$\mathrm{erfc}(x) = 2\pi^{-\frac{1}{2}} \int_x^\infty e^{-t^2} dt = 1 - \mathrm{erf}(x)$$
$$= (\pi x)^{-\frac{1}{2}} e^{-x^2/2} \ W_{-\frac{1}{4}, \frac{1}{4}}(x^2) \ ; \tag{3.9}$$

where $W_{k,m}(x)$ is Whittaker's function.

The fractional differential operator D_z^α where α may be a positive or negative integer, a rational number, or a complex number, is defined as

$$D_z^\mu\{z^\lambda\} = \frac{\Gamma(\lambda + 1)}{\Gamma(\lambda - \mu + 1)} z^{\lambda - \mu} \tag{3.10}$$

where μ is an arbitrary complex number, or a real number. For precise definitions and conditions of validity see Osler [8]. It is recommended by the participants of the New-Haven Conference on "Fractional Calculus and

Its Applications" to use $_0D_z^\alpha$ for D_z^α in (3.10) to indicate the particular definition using the integral to denote fractional differentiation where the zero in the symbol $_0D_z^\alpha$ denotes lower and the z, the upper, limits of integration. (See Ross, Bertram, ed. [9] for details).

Osler's [8] generalized Leibniz theorem with $g(z) = z$ is

$$D_z^\alpha u(z)v(z) = \sum_{n=-\infty}^{\infty} \binom{\alpha}{\delta+n} D_z^{\alpha-\delta-n} u(z) D_z^{\delta+n} v(z) \qquad (3.11)$$

where the generalized symbol

$$\binom{\alpha}{\delta+n} = \frac{\Gamma(\alpha+1)}{\Gamma(\alpha-\delta-n+1)\Gamma(\delta+n+1)} \qquad (3.12)$$

implies the right hand member of (3.12).

The special case of (3.11) for $\delta = 0$ is the well-known Liebniz theorem of the elementary calculus. One can employ (3.11) and its special cases to generate a host of known or hitherto unknown expansion formulas for special functions.

Case 1 ($\delta = 0$)

$$D_z^\alpha u(z)v(z) = \sum_{n=0}^{\infty} \frac{\Gamma(\alpha+1)D_z^{\alpha-n} u(z) D_z^n v(z)}{\Gamma(\alpha-n+1)n!} \qquad (3.13)$$

Case 2 ($v = 1$)

$$D_z^\alpha u(z) = \frac{\Gamma(\alpha+1)\sin(\alpha-\delta)\pi}{\pi} \sum_{n=-\infty}^{\infty} \frac{(-1)^n z^{n+\delta-\alpha}}{(\alpha-\delta-n)} \frac{D_z^{\delta+n} u(z)}{\Gamma(\delta+n+1)} \qquad (3.14)$$

The following generalized derivatives will be needed for the expansion formulas

$$D_z^{-\alpha} e^z = \frac{\gamma(\alpha,z)}{\Gamma(\alpha)} e^z, \quad \text{Osler [8] ,} \qquad (3.15)$$

$$D_x^\beta x^\alpha e^x \gamma^*(\alpha;x) = x^{\alpha-\beta} e^x \gamma^*(\alpha-\beta;x) \quad \text{Spanier [10] ;} \qquad (3.16)$$

α, β, positive or negative, integer or non integer.

$$D_z^\alpha \ln z = \frac{z^{-\alpha}}{\Gamma(1-\alpha)}[\ln z - \gamma - \psi(1-\alpha)] , \qquad (3.17)$$

410   R. N. Kalia

where $\psi(z) = \frac{d}{dx}\ln \Gamma(z)$, and γ is Euler Mascheroni constant.

$$D_z^\alpha z^p = \frac{\Gamma(p+1)}{\Gamma(p-\alpha+1)} z^{p-\alpha}, \quad p \neq -1, -2, \dots \qquad (3.18)$$

$$D_z^\alpha z^p \ln z = \frac{\Gamma(p+1)z^{p-\alpha}}{\Gamma(p-\alpha+1)} [\ln z + \psi(p+1) - \psi(p-\alpha+1)],$$
$$p \neq -1, -2, -3, \dots \qquad (3.19)$$

can be computed easily by the Osler's formulas (see Oslar [8]).

A. *Main Results*

In (3.11) take $u(z) = e^z, v(z) = z^\beta e^z \gamma^*(\beta; z)$. We obtain the expansion formula for the entire incomplete gamma function.

$$z^{-\delta}\gamma^*(\beta, z) = \frac{\sin \delta\pi}{\pi} \sum_{n=-\infty}^{\infty} \frac{(-1)^n \gamma(-\delta-n, z)}{\Gamma(-\delta-n)(\delta+n)} \gamma^*(\beta+\delta+n; z) z^n, \quad (3.20)$$

where δ is not an integer or zero.

Or, using (3.5), one obtains

$$\frac{\gamma(\beta; z)}{\Gamma(\beta)} = \frac{\sin \delta\pi}{\pi} \sum_{n=-\infty}^{\infty} \frac{(-1)^n \gamma(-\delta-n, z)\gamma(\beta+\delta+n, z)}{\Gamma(-\delta-n)(\delta+n)\Gamma(\beta+\delta+n)} \qquad (3.21)$$

where $\delta \neq 0$ and δ not an integer as before in last equation.

Now, setting in (3.14) $u(z) = z^\beta e^z \gamma^*(\beta, z), v(z) = 1$ and invoking (3.15) and (3.16) leads to

$$\gamma^*(\beta-\alpha; z) = \frac{\Gamma(\alpha+1)\sin(\alpha-\delta)\pi}{\pi} \sum_{n=-\infty}^{\infty} \frac{(-1)^n \gamma^*(\beta-\delta-n; z)}{(\alpha-\delta-n)\Gamma(\delta+n+1)},$$
$$(3.22)$$

where $\text{Re } \alpha > 0, \alpha \neq \delta$; and $\alpha - \delta$ not an integer.

A simplified expansion with a series from $n = 0$ to ∞ is obtained on taking $\delta = 0$ in (3.24).

To get yet one more expansion, set $u(z) = e^z$ in Case 2, Eq. (3.14), and using $\gamma(v; z) = z^v \gamma^*(v; z)\Gamma(v)$. One gets

$$z^{2(\alpha-\delta)}\gamma^*(\alpha; z) = \frac{\Gamma(\alpha+1)\sin(\alpha-\delta)}{\pi} \sum_{n=-\infty}^{\infty} \frac{(-1)^n z^{2n}\gamma^*(\delta+n; z)}{(\alpha-\delta-n)(\delta+n)} \qquad (3.23)$$

for Re $\alpha > 0, \alpha \neq \delta; \alpha - \delta$ not an integer.

The results (3.21–3.23) are believed to be new. It will not be surprising if the expansion formulas for extensions of gamma functions occur in the literature.

The expansions in this section have appeared in Kalia and Keith [6] where the delicate conditions on the various parameters were not indicated. It can be shown that the series are convergent by splitting the series.

4. Generalized Calculus and the Bilateral Expansions of Incomplete Beta and Elliptic Functions

In this section, some special cases of the generalized Leibniz Theorem are used to find bilateral expansion formulas for Elliptic functions of the first and second kind, incomplete Beta, and some elementary functions in terms of Hypergeometric functions. The method is similar to that of Kalia and Keith [6]. Some definitions and results used in this section follow.

The incomplete Beta function is defined by

$$B_z(a, b) : = a^{-1} z^a {}_2F_1(a, 1 - b; a + 1; z) \ . \tag{4.1}$$

The Elliptic functions $K(x)$ and $E(x)$ are defined by

$$K(z) = \int_0^{\pi/2} \frac{d\theta}{\sqrt{1 - x^2 \sin^2 \theta}} = \frac{1}{2}\pi {}_2F_1(1/2, 1/2; 1; x^2) \tag{4.2}$$

$0 < x < 1$, and

$$E(x) = \int_0^{\pi/2} \sqrt{1 - x^2 \sin^2 \theta} \ d\theta = \frac{1}{2}\pi {}_2F_1(-1/2, 1/2; 1; x^2) \ ; \tag{4.3}$$

$0 < x < 1$.

The function $E_1(x)$ is defined by

$$E_1(x) = \int_x^\infty \frac{\exp(-t)}{t} dt \ . \tag{4.4}$$

The fractional differential operator D_x^μ, where μ may be a positive or negative integer, a rational number, or a complex number, is defined by (3.10).

Two special forms of (3.11) are listed below for specific values of α, δ and v. These two formulas are of special interest to us as these can be employed judiciously with a table of fractional integrals or tables to generate a host of series expansions involving special and elementary functions.

Case 3 $(\alpha = 0)$

$$u(x)v(x) = \frac{\sin \delta\pi}{\pi} \sum_{n=-\infty}^{\infty} \frac{(-1)^n D_x^{-\delta-n} u(x) D_x^{\delta+n} v(x)}{\delta+n} \qquad (4.5)$$

Case 4 $(\alpha = 0, v \equiv 1)$

$$u(x) = \frac{\sin^2 \delta\pi}{\pi^2} \sum_{n=-\infty}^{\infty} \frac{\Gamma(\delta+n)x^{-\delta-n} D_x^{-\delta-n} u(x)}{\delta+n} \qquad (4.6)$$

Using (3.10) on (2.3), which is justifiable, we obtain after some manipulation

$$D_x^\mu \left\{ x^\lambda \, _pF_q \left[\begin{matrix} \alpha_1,\ldots,\alpha_p; \ x \\ \beta_1,\ldots,\beta_q \end{matrix} \right] \right\} = \sum_{k=0}^{\infty} \frac{\prod_{i=1}^{p}(\alpha_i)_k}{\prod_{i=1}^{q}(\beta_i)_k} \frac{\Gamma(\lambda+k+1)x^{\lambda-\mu+k}}{\Gamma(\lambda-\mu+k+1)k!} , \qquad (4.7)$$

or equivalently,

$$\left\{ D_x^\mu \, _pF_q \left[\begin{matrix} \alpha_1,\ldots,\alpha_p; \ x \\ \beta_1,\ldots,\beta_q \end{matrix} \right] \right\} = \frac{\Gamma(\lambda+1)x^{\lambda-\mu}}{\Gamma(\lambda-\mu+1)} \, _{p+1}F_{q+1}$$

$$\times \left[\begin{matrix} \alpha_1 \ldots,\alpha_p, \lambda+1; x \\ \beta_1,\ldots,\beta_q, \lambda+\mu+1 \end{matrix} \right] \qquad (4.8)$$

provided $\lambda > -1$ and $\lambda - \mu$ is not a negative integer.

When $\lambda = 0$, then

$$D_x^\mu \, _pF_q \left[\begin{matrix} \alpha_1,\ldots,\alpha_p; \ x \\ \beta_1,\ldots,\beta_q \end{matrix} \right] = \frac{x^{-\mu}}{\Gamma(1-\mu)} \, _{p+1}F_{q+1} \left[\begin{matrix} \alpha_1,\ldots,\alpha_p, 1; x \\ \beta_1,\ldots,\beta_q, 1-\mu \end{matrix} \right] \qquad (4.9)$$

provided μ is not a positive integer.

Our strategy would be to use (4.8) or (4.9), after throwing the special functions into the form involving a special case of $_pF_q$ and consequently use the cases 3 and 4 to derive the series expansions.

We will have to consider the expansions valid only when the order of the generalized derivative is not a positive integer. This restriction can be lifted with some modifications and we will study this later.

The expansion formulas follow in the next section.

Expansion Formulas

(i) Take $u = K(\sqrt{x})$ in case 4. We get

$$K(\sqrt{x}) = \frac{\sin^2 \delta\pi}{2\pi} \sum_{n=-\infty}^{\infty} \frac{{}_2F_2 \left[\begin{matrix} 1/2, 1/2; x \\ \delta + n + 1 \end{matrix} \right]}{(\delta + n)^2} \tag{4.10}$$

$\delta \neq 0, \delta$ not an integer; $|x| < \infty$.

(ii) Take $u = E(\sqrt{x})$, we get from case 4 above, the expansion for the elliptic function of the second kind

$$E(\sqrt{x}) = \frac{\sin^2 \delta\pi}{2\pi} \sum_{n=-\infty}^{\infty} \frac{{}_2F_1 \left[\begin{matrix} 1/2, -1/2; x \\ 1 + \delta + n \end{matrix} \right]}{(\delta + n)^2}, \tag{4.11}$$

$\delta \neq 0, \delta$ not an integer, $|x| < 1$.

(iii) Let $u = \arcsin(\sqrt{x})$ to get

$$\arcsin(\sqrt{x}) = \frac{\sin^2 \delta\pi}{2\pi^{3/2}} x^{1/2+\delta} \sum_{n=-\infty}^{\infty} \frac{\Gamma(\delta + n) x^n}{\delta + n} {}_2F_1 \left[\begin{matrix} 1/2, 1/2; x \\ 3/2 + \delta + n \end{matrix} \right] \tag{4.12}$$

δ not an integer; $|x| < 1$.

(iv) Taking $u = \text{arcsinh}(\sqrt{x})$ in case 4, we get

$$\text{arcsinh}(\sqrt{x}) = \frac{x^{1/2+\delta} \sin^2 \delta\pi}{2\pi^{3/2}} \sum_{n=-\infty}^{\infty} \frac{\Gamma(\delta + n) x^n {}_2F_1 \left[\begin{matrix} 1/2, 1/2; -x \\ 3/2 + \delta + n \end{matrix} \right]}{(\delta + n)\Gamma(3/2 + \delta + n)} \tag{4.13}$$

$\delta \neq 0, \delta$ not an integer; $|x| < 1$.

(v) Take $u = B_x(a, b)$, the incomplete beta function in case 4, to get

$$B_x(a, b) = \frac{\sin^2(\delta\pi) x^{a+\delta} \Gamma(a)}{\pi^2} \sum_{n=-\infty}^{\infty} \frac{\Gamma(\delta + n) {}_2F_1 \left[\begin{matrix} 1 - b, a; x \\ 1 + a + \delta + n \end{matrix} \right] x^n}{(\delta + n)\Gamma(1 + a + \delta + n)}; \tag{4.14}$$

414   *R. N. Kalia*

$\delta \neq 0, \delta$ not an integer; $|x| < 1$.

(vi) Taking $u = E_1(x) + \gamma + \ln x$, in case 4 where $E_1(x)$ is defined through (4.4) above and γ is Euler's constant we obtain

$$E_1(x) + \gamma + \ln x = \frac{x \sin^2 \delta\pi}{\pi^2} \sum_{n=-\infty}^{\infty} \frac{\Gamma(\delta+n)}{\delta+n} {}_2F_2 \begin{bmatrix} 1,1;-x \\ 2,1+\delta+n \end{bmatrix} , \quad (4.15)$$

$\delta \neq 0, \delta$ not an integer; $|x| < \infty$.

(vii) Lastly taking $u = K(\sqrt{x}), v = E(\sqrt{x})$ in case 3, we have the expansion formula for the product of Elliptic functions of the first and second kind as follows:

$$K(\sqrt{x})E(\sqrt{x}) = \frac{\pi}{4} \sum_{n=-\infty}^{\infty} \frac{{}_2F_1 \begin{bmatrix} 1/2,-1/2;x \\ 1+\delta+n \end{bmatrix} {}_2F_1 \begin{bmatrix} 1/2,1/2;x \\ 1-\delta-n \end{bmatrix}}{(\delta+n)} , \quad (4.16)$$

δ not an integer; $|x| < 1$.

More expansions and the integral analogs of the expansions can be found and will be dealt with elsewhere. Gauss would have loved to know the developed theory of generalized Calculus which was prophesied by Leibniz [5], in reply to a letter of L'Hospital. However, Gauss, the Prince of Mathematicians, developed complete investigations of most subjects he undertook and inspired great interest of the posterity in his hypergeometric functions; amongst others.

References

1. W. K. Buhler, *Gauss: A Biographical Study*, Springer-Verlag, New York, 1981.

2. Florian Cajori, *A History of Mathematical Notations*, Volume II, *Notations Mainly in Higher Mathematics*, the Open Court Publishing Company, Chicago-Illinois, 1929.

3. C. F. Gauss, *Disquisitions generales circa seriem infinitam: ...*, C. Gott. rec. 2, 1813 = Werke Band III, Gottingen, 1866.

4. M. E. H. Ismail and Carl A. Libis, *Contiguous relations, basic hypergeometric functions and orthogonal polynomials*, I. J. Math. Anal. Appl. **141** (1989) 349–372.

5. G. W. Leibniz, *Leibnizschen's Mathematische Schriffen*, 2 (1962) 301–302, George Olm, Hildesheim, Germany.

6. R. N. Kalia, Sandra Keith, *Fractional calculus and expansions of incomplete gamma functions*, Appl. Math. Lett. **3** (1990) 19–21.

7. A. M. Mathai and R. K. Saxena, *The H-Function with Applications in Statistics and Other Disciplines*, John Wiley and Sons, New York, 1978.

8. Thomas J. Osler, *Leibniz Rule for fractional derivatives generalized and an application to infinite series*, SIAM J. Appl. Math. **18** (1970) 658–674.

9. Bertram Ross, Ed., *Fractional Calculus and Its Applications*, Lecture Notes in Math. No. 457, 1975.

10. J. Spanier and K. B. Oldham, *Atlas of Functions*, Hemisphere Publishing Corporation, 1987.

11. H. M. Srivastava and H. L. Manocha, *A Treatise on Generating Functions*, Ellis Horwood Limited, New York, Chichester, 1984.

R. N. Kalia
Department of Mathematics
and Statistics
St. Cloud State University
720 South 4th Avenue
St. Cloud, MN 56301–4498
U.S.A.

THE MATH. HERITAGE OF C.F. GAUSS (pp. 416-426)
edited by George M. Rassias
©1991 World Scientific Publ. Co. Singapore

GAUGE-NATURAL OPERATORS TRANSFORMING CONNECTIONS TO THE TANGENT BUNDLE

Ivan Kolář

We determine all first order gauge-natural operators transforming connections on a principal fibre bundle $P \to M$ into connections on $TP \to TM$ in the case the structure group of P is the general linear group in an arbitrary dimension.

All manifolds and maps are assumed to be infinitely differentiable.

The tangent bundle TG of a Lie group G is also a Lie group with the multiplication

$$\left(\frac{d}{dt}\Big|_0 \gamma(t)\right)\left(\frac{d}{dt}\Big|_0 \delta(t)\right) = \frac{d}{dt}\Big|_0 (\gamma(t) \cdot \delta(t)) \tag{1}$$

provided the dot on the right hand side means the multiplication in G. Given a principal fibre bundle $p : P \to M$ with structure group G, the tangent bundle $Tp : TP \to TM$ has a canonical structure of a principal fibre bundle with structure group TG. Kobayashi pointed out, [3], that every connection Γ on P induces a connection on $TP \to TM$ as follows. Interpret Γ as the lifting map $\lambda(\Gamma) : P \oplus TM \to TP$ transforming the elements of TM into the horizontal vectors. If we construct the tangent map of $\lambda(\Gamma)$ and apply the canonical involutions ex_M and ex_P of TTM and TTP, then there is a unique connection $T\Gamma$ on $TP \to TM$ such that its lifting map $\lambda(T\Gamma)$ makes the following diagram commutative

---

AMS CLASSIFICATION: 53 C 05, 58 A 20

$$TP \oplus TTM \xrightarrow{\; T(\lambda(\Gamma)) \;} TTP$$

$$\mathrm{id}_{TY} \oplus \mathrm{ex}_M \Big\downarrow \qquad\qquad \Big\downarrow \mathrm{ex}_P \qquad\qquad (2)$$

$$TP \oplus TMM \xrightarrow[\; \lambda(T\Gamma) \;]{} TTP$$

Connection $T\Gamma$ will be called the tangent connection of Γ.

From the point of view of the recent theory of gauge-natural bundles and operators, see below, the fact that $T\Gamma$ is constructed geometrically can be expressed by saying that T is a gauge-natural operator. Taking into account our previous results about the gauge-natural operators, [4], we can pose the problem of finding all gauge-natural operators transforming connections on $P \to M$ into connections on $TP \to TM$. In order to apply our techniques from [4], we restrict ourselves to the first order operators and to the case where the structure group of P is the general linear group in an arbitrary dimension. Our main result is formulated in Sec. 4.

1. The general concept of a natural bundle over m-manifold was introduced by A. Nijenhuis as a modern reformulation of the classical concept of an arbitrary bundle of geometric objects, [7]. However, a disadvantage of the theory of natural bundles is that it does not cover the case of the connection bundle of any ("abstract") principal fibre bundle, which plays a basic role in the gauge theories of mathematical physics. For this reason D. J. Eck modified the idea of natural bundle as follows, [1]. Let $\mathcal{M}f_m$ be the category of m-dimensional manifolds and their local diffeomorphisms, let $\mathcal{F}M$ be the category of fibred manifolds and let B denote the base functor. Fix a Lie group G and define a category $\mathcal{P}_m(G)$, whose objects are principal G-bundles over m-manifolds and whose morphisms are the morphisms of principal G-bundles $f : P \to \bar{P}$ with the base map $Bf : BP \to B\bar{P}$ lying in $\mathcal{M}f_m$.

Definition 1. A gauge-natural bundle over m-manifolds is a functor $F : \mathcal{P}_m(G) \to \mathcal{F}M$ such that

(a) every $\mathcal{P}_m(G)$-object $p : P \to BP$ is transformed into a fibred manifold $q_P : FP \to BP$ over BP,

(b) every $\mathcal{P}_m(G)$-morphism $f : P \to \bar{P}$ is transformed into an $\mathcal{F}M$-morphism $Ff : FP \to F\bar{P}$ over Bf,

(c) for every open subset $U \subset BP$, the inclusion $i : P^{-1}(U) \hookrightarrow P$ is transformed into the inclusion $Fi : q_P^{-1}(U) \hookrightarrow FP$.

If we intend to point out the structure group G, we say that F is a G-natural bundle. We remark that Eck has assumed an additional continuity condition for F, but J. Slovák has recently deduced that this condition follows from the other axioms, [6].

Let $W^r P$ be the space of all r-jets $j^r_{(0,e)}\varphi$, where $\varphi : \mathbf{R}^m \times G \to P$ is a principal bundle morphism of $\mathcal{P}_m(G)$, $0 \in \mathbf{R}^m$, $e = $ the unit of G. The space $W^r P$ is a principal bundle over BP with structure group $W^r_m G$, which is the group of all r-jets $j^r_{(0,e)}\psi$, where $\psi : \mathbf{R}^m \times G \to \mathbf{R}^m \times G$ is a morphism of $\mathcal{P}_m(G)$ satisfying $B\psi(0) = 0$. Let $T^r_m G = J^r_0(\mathbf{R}^m, G)$ be endowed with the multiplication $(j^r_0 f(t))(j^r_0 g(t)) = j^r_0(f(t) \cdot g(t))$ and let G^r_m be the group of all invertible r-jets from $\mathbf{R}^m$ into $\mathbf{R}^m$ with source and target 0. The group $W^r_m G$ is equal to the semidirect product of G^r_m and $T^r_m G$ with respect to the following multiplication

$$(X, A)(Y, B) = (X \circ Y, (A \circ Y)B), \quad X, Y \in G^r_m, \quad A, B \in T^r_m G$$

where $\circ$ denotes the composition of jets, [2].

Every $\mathcal{P}_m(G)$-morphsim $f : P \to \bar{P}$ is extended into a principal bundle morphism $W^r f : W^r P \to W^r \bar{P}$ defined by the jet composition $W^r f(j^r_{(0,e)}\varphi)$
$= j^r_{(0,e)}(f \circ \varphi)$. Every smooth left action of $W^r_m G$ on a manifold S determines a G-natural bundle over m-manifolds transforming every $\mathcal{P}_m(G)$-object P into the fibre bundle associated to $W^r P$ with standard fibre S and every $\mathcal{P}_m(G)$-morphism f into $(W^r f, \mathrm{id}_S)$. The following result was deduced by D. J. Eck, [1], who used the continuity assumption, and completed by J. Slovák, [6].

Proposition 1. Every gauge-natural bundle is a fibre bundle associated to bundle W^r of a certain order r.

In such a case we shall say that the gauge-natural bundle in question has order r. (Another definition of the order of a gauge-natural bundle can be found in [4].)

In particular, consider the connection bundle $QP \to BP$ of P, which can be defined e.g., as the factor space $QP := J^1 P/G$, where $J^1 P$ means the first jet prolongation of fibred manifold $P \to BP$. Obviously, Q is a gauge-natural bundle and one finds easily that QP is a fibre bundle associated to W^1P. Its standard fibre is $\mathfrak{g} \otimes \mathbf{R}^{m*}$, where $\mathfrak{g}$ is the Lie algebra of the structure group G. The corresponding action of $W^1_m G$ on $\mathfrak{g} \otimes R^{m*}$ can be found in [4].

2. Let F, E and H be three G-natural bundles over m-manifolds and $\rho : H \to E$ be a natural transformation such that every $\rho_P : HP \to EP$ is a surjective submersion.

Definition 2. A gauge-natural (or G-natural) operator $A : F \to (H \to E)$ is a system of operators $A_P : C^\infty(FP \to BP) \to C^\infty(HP \to EP)$ for every $\mathcal{P}_m(G)$-object $p : P \to BP$ satisfying

(a) $A_{\bar{P}}(FfosoBf^{-1}) = HfoA_P(s)oEf^{-1}$ for every $\mathcal{P}_m(G)$-isomorphism $f : P \to \bar{P}$ and every section $s : BP \to FP$,

(b) $A_{p^{-1}(U)}(s|U) = (A_P s)|U$ for every open subset $U \subset BP$ and every section $s : BP \to FP$,

(c) every A_P is regular, i.e., A_P transforms every smoothly parametrized family of sections into a smoothly parametrized family.

Obviously, the k-th jet prolongation $J^k F$ of a G-natural bundle is a G-natural bundle as well. Consider the fibres $E_0 = E_0(\mathbf{R}^m \times G), H_0 = H_0(\mathbf{R}^m \times G), J^k_0 F = J^k_0 F(\mathbf{R}^m \times G)$ over $0 \in \mathbf{R}^m$ and denote by $\rho_0 : H_0 \to E_0$ the restriction of ρ. Quite analogously to Proposition 2 of [5] we deduce

Proposition 2. The k-th order G-natural operators $F \to (H \to E)$ are in a canonical bijection with the W^s_m-equivalent maps $f : J^k_0 F \times E_0 \to H_0$ satisfying $\rho_0 \circ f = pr_2$, where s is the maximum of the orders of $J^k F, E$ and H.

In our concrete problem we have two connection bundles $F = Q$ and $H = QT$, while $E = TB$ is the composition of the base functor with the tangent functor on $\mathcal{M}f_m$.

3. The construction of the tangent connection $T\Gamma$ by Kobayashi is one gauge-natural operator $Q \to (QT \to TB)$. The other operators can be characterized in terms of the so-called difference tensors. In general, consider the vector bundle LP associated to P by means of the adjoint action of G on its Lie algebra $\mathfrak{g}$. It is well-known that QP is an affine bundle with associated vector bundle $LP \otimes T^*BP$, i.e., the difference of two connections on P is a section of $LP \otimes T^*BP$. In particular, the difference of two connections on $TP \to TBP$ is a section of $L(TP) \otimes T^*TBP$.

One finds easily that the bundle projection $TG \to G$ is a group homomorphism and its kernel coincides with the abelian group $\mathfrak{g}$ (as a vector space). The zero section $G \to TG$ is a group homomorphism splitting the bundle projection. This implies that TG coincides with the semidirect product to G and $\mathfrak{g}$ with the following multiplication

$$(g_1, X_1)(g_2, X_2) = (g_1 g_2, \ \mathrm{ad}(g_2^{-1})(X_1) + X_2) \tag{3}$$

where ad means the adjoint action of G. This identifies the Lie algebra $\mathfrak{tg}$ of TG with $\mathfrak{g} \times \mathfrak{g}$ and a simple calculation gives the following formula for the adjoint action ad_{TG} of TG

$$\mathrm{ad}_{TG}(g, X)(Y, V) = (\mathrm{ad}(g)Y, \ \mathrm{ad}(g)([X, Y] + V)) \ . \tag{4}$$

Hence the subspace $0 \times \mathfrak{g} \subset \mathfrak{tg}$ is ad_{TG}-invariant, so that it defines a subbundle $K(TP) \subset L(TP)$. The injection $V \mapsto (0, V)$ induces a map $I_P : LP \to K(TP)$.

The curvature of a connection Γ on P can be interpreted as a linear morphism $\Lambda^2 TBP \to LP$. In general, having any linear base-preserving morphism $S : \Lambda^2 TBP \to LP$, we define a linear map $\mu_P(S) : TTBP \to L(TP)$ by

$$\mu_P(S)(Z) = I_P(S(\pi_1(Z) \wedge \pi_2(Z))), \quad Z \in TTBP \tag{5}$$

where $\pi_1 : TTBP \to TBP$ is the bundle projection and $\pi_2 : TTBP \to TBP$ is the tangent map of the bundle projection $TBP \to BP$.

4. From now on we assume $G = GL(n, \mathbf{R})$, where n is independent on the dimension m of BP. In [4] we have deduced that all $GL(n, \mathbf{R})$-natural operators $Q \to L \otimes \Lambda^2 T^*B$ are linearly generated by the curvature operator C and by the modified curvature operator D constructed as follows. The map $\mathfrak{gl}(n, \mathbf{R}) \to \mathfrak{gl}(n, \mathbf{R}), A \mapsto$ (trace A)Id is ad-invariant, so that it induces a linear morphism $\mathcal{H}_p : LP \to LP$. Then we define $D = (\mathcal{H} \otimes \Lambda^2 \mathrm{id}) \circ C : Q \to L \otimes \Lambda^2 T^*B$.

Hence $\Gamma \mapsto \mu_P(C\Gamma)$ and $\Gamma \mapsto \mu_P(D\Gamma)$ are two gauge-natural difference tensors. The analytic discussion of our problem, in which consists the proof of the following theorem, suggests that there are two other gauge-natural difference tensors $\tau(C)$ and $\tau(D)$ determined by C and D. However, the simplest geometric construction of the latter operators uses heavily their gauge-naturality. That is why we postpone the geometrical definition of both $\tau(C)$ and $\tau(D)$ after the proof of our Theorem. (This is in accordance with the discovery of our result: the analytical discussion gave the coordinate expression of a 4-parameter family of gauge-natural difference tensors and the result was interpreted geometrically.)

Theorem. All first order $GL(n, \mathbf{R})$-natural operators $Q \to (QT \to TB)$ form the 4-parameter family

$$T + k_1\mu(C) + k_2\mu(D) + k_3\tau(C) + k_4\tau(D), \quad k_1, k_2, k_3, k_4 \in \mathbf{R} , \qquad (6)$$

The proof requires several steps.

5. Let $x^i, y_q^p, \det y_q^p \neq 0$,

$$i, j, \ldots = 1, \ldots, m \quad p, q, \ldots = 1, \ldots, n$$

be the canonical coordinates on $P = \mathbf{R}^m \times GL(n, \mathbf{R})$. The action of $G = GL(n, \mathbf{R})$ on P is $(x^i, y_q^p)(z_q^p) = (x^i, y_r^p z_q^r)$, so that the equations of a connection Γ on P are

$$dy_q^p = \Gamma_{ri}^p(x) y_q^r dx^i . \qquad (7)$$

If $\xi^i = dx^i, \eta_q^p = dy_q^p, \zeta_q^p = dz_q^p$ are the additional coordinates on TP and TG, then (1) gives the following action of TG on TP

$$(x^i, \xi^i, y_q^p, \eta_q^p)(z_q^p, \zeta_q^p) = (x^i, \xi^i, y_r^p z_q^r, \eta_z^p z_q^r + y_r^p \zeta_q^r) .$$

Hence the equations of a connection on $TP \to TBP$ are

$$\begin{aligned}
dy_q^p &= A_{ri}^p y_q^r dx^i + B_{ri}^p y_q^r d\xi^i \\
d\eta_q^p &= C_{ri}^p y_q^r dx^i + D_{ri}^p y_q^r d\xi^i + A_{ri}^p \eta_q^r dx^i + B_{ri}^p \eta_q^r d\xi^i
\end{aligned} \qquad (8)$$

where $A_{qi}^p, B_{qi}^p, C_{qi}^p, D_{qi}^p$ are any smooth functions of x^i and ξ^i.

Since we are interested in the first order gauge-natural operators, Proposition 2 lets us know that we shall deal with the group $W_m^2 G$ of the 2-jets at $(0, e)$ of the transformations

$$\bar{x}^i = f^i(x), \quad \bar{y}_q^p = f_r^p(x) y_q^r, \quad f^i(0) = 0 .$$

The canonical coordinates on $W_m^2 G$ are $a_j^i = \partial_j f^i(0), a_{jk}^i = \partial_{jk}^2 f^i(0), a_q^p = f_q^p(0), a_{qi}^p = \partial_i f_q^p(0), a_{qij}^p = \partial_{ij}^2 f_q^p(0)$. Write $(\tilde{a}_j^i, \tilde{a}_{jk}^i)$ for the inverse element to $(a_j^i, a_{jk}^i) \in G_m^2$ and $\tilde{a}_q^p$ for the inverse matrix to a_q^p.

Let $\Gamma_{qi,j}^p$ be the induced coordinates on the standard fibre $J_0^1 Q$, i.e., $\Gamma_{qi,j}^p$ are the "formal" derivatives of the Christoffel's Γ_{qi}^p with respect to x^j. The equations of the action of $W_m^2 G$ on $J_0^1 Q$ can be found in [4]. However, we shall need the following change of coordinates

$$Y_{qij}^p = \Gamma_{qi,j}^p + \Gamma_{ri}^p \Gamma_{qj}^r . \tag{9}$$

Using [4] one evaluates easily the action of $W_m^2 G$ on $J_0^1 Q$ in the following form

$$\bar{\Gamma}_{qi}^p = a_r^p \Gamma_{sj}^r \tilde{a}_q^s \tilde{a}_i^j + a_{rj}^p \tilde{a}_q^r \tilde{a}_i^j \tag{10}$$

$$\bar{Y}_{qij}^p = a_r^p Y_{sk\ell}^r \tilde{a}_q^s \tilde{a}_i^k \tilde{a}_j^\ell + a_{r\ell}^p \Gamma_{sk}^r \tilde{a}_q^s \tilde{a}_i^k \tilde{a}_j^\ell + a_{rk}^p \Gamma_{s\ell}^r \tilde{a}_q^s \tilde{a}_i^k \tilde{a}_j^\ell$$
$$+ a_r^p \Gamma_{sk}^r \tilde{a}_q^s \tilde{a}_{ij}^k + a_{rk\ell}^p \tilde{a}_q^r \tilde{a}_i^k \tilde{a}_j^\ell + a_{rk}^p \tilde{a}_q^r \tilde{a}_{ij}^k . \tag{11}$$

Further we replace Y_{qij}^p by

$$S_{qij}^p = \frac{1}{2}(Y_{qij}^p + Y_{qji}^p), \quad R_{qij}^p = \frac{1}{2}(Y_{qij}^p - Y_{qji}^p) . \tag{12}$$

Then S_{qij}^p has the same transformation law as (11), while R_{qij}^p is a tensorial quantity (called the formal curvature tensor).

We have $E_0 = T_0 \mathbf{R}^m$ with coordinates ξ^i and the standard action $\bar{\xi}^i = a_j^i \xi^j$. The coordinates on $H_0 = Q_0(TP \to TBP), 0 \in \mathbf{R}^m$, are ξ^i together with $A_{qi}^p, B_{qi}^p, C_{qi}^p, D_{qi}^p$ from (8) and $\rho_0 : H_0 \to E_0$ means the projection into ξ^i. The action of $W_m^2 G$ on $A_{qi}^p, \ldots, D_{qi}^p$ can be deduced from (8), but we shall do it gradually in the course of the proof.

6. By Proposition 2, our problem is to find all $W_m^2 G$-maps $J_0^1 Q \times E_0 \to H_0$ satisfying $\xi^i = \xi^i$. Thus, at the beginning $A_{qi}^p, B_{qi}^p, C_{qi}^p, D_{qi}^p$ are any smooth functions of $\Gamma^p_{qi}, R^p_{qij}, S^p_{qij}, \xi^i$ and they should be simplified by the equivariancy conditions. In the course of such a procedure we shall need a representation lemma, the proof of which can be found in [6]. Consider $\mathbf{R}^n \otimes \mathbf{R}^{n*} \otimes \Lambda^2 \mathbf{R}^{m*}$ with typical element R^p_{qij} and $\mathbf{R}^m$ with typical element ξ^i.

Lemma 1. All $GL(m, \mathbf{R}) \times GL(n, \mathbf{R})$-equivariant maps $(\mathbf{R}^n \otimes \mathbf{R}^{n*} \otimes \Lambda^2 \mathbf{R}^{m*}) \times \mathbf{R}^m \to \mathbf{R}^n \otimes \mathbf{R}^{n*} \otimes \mathbf{R}^{m*}$ form the 2-parameter family

$$k_1 R^p_{qij}\xi^j + k_2 \delta^p_q R^r_{rij}\xi^j \quad k_1, k_2 \in \mathbf{R} \ . \tag{13}$$

7. From the first line of (8) we deduce the simplest transformation law

$$B^p_{qi} = a^p_r B^r_{sj} \tilde{a}^s_q \tilde{a}^j_i \ . \tag{14}$$

The equivariancy of $B^p_{qi}(\Gamma^p_{qi}, R^p_{qij}, S^p_{qij}, \xi^i)$ with respect to the subgroup

$$a^i_j = \delta^i_j, \quad a^p_q = \delta^p_q, \quad a^i_{jk} = 0, \quad a^p_{qi} = 0 \tag{15}$$

means $B^p_{qi}(\Gamma^p_{qi}, R^p_{qi}, S^p_{qij}, \xi^i) = B^p_{qi}(\Gamma^p_{qi}, R^p_{qij}, S^p_{qij} + a^p_{qij}, \xi^i)$. This implies that B^p_{qi} is independent on S^p_{qij}. Quite similarly, the equivariancy of B^p_{qi} with respect to the subgroup

$$a^i_j = \delta^i_j, \quad a^p_q = \delta^p_q, \quad a^i_{jk} = 0 \tag{16}$$

yields B^p_{qi} is independent on Γ^r_{sj}. Considering now the subgroup

$$a^i_{jk} = 0, \quad a^p_{qi} = 0, \quad a^p_{qij} = 0 \tag{17}$$

we can apply Lemma 1. This gives

$$B^p_{qi} = a R^p_{qij}\xi^j + b \delta^p_q R^r_{rij}\xi^j \ . \tag{18}$$

From the first line of (8) we next deduce

$$\bar{A}^p_{qi} = a^p_r A^r_{sj} \tilde{a}^s_q \tilde{a}^j_i - a^p_r B^r_{sm} \tilde{a}^s_q \tilde{a}^m_j a^j_{k\ell} \tilde{a}^\ell_i \xi^k + a^p_{rj} \tilde{a}^r_q \tilde{a}^j_i \ . \tag{19}$$

The equivariancy of $A^p_{qi}(\Gamma^p_{qi}, R^p_{qij}, S^p_{qij}, \xi^i)$ with respect to subgroup (15) yields in the same way as above that A^p_{qi} is independent on S^p_{qij}. The equivariancy on subgroup (16) means $A^p_{qi}(\Gamma^p_{qi}, R^p_{qij}, \xi^i) + a^p_{qi} = A^p_{qi}(\Gamma^p_{qi} + a^p_{qi}, R^p_{qij}, \xi^i)$. Together with (10) this implies that the difference $E^p_{qi} = A^p_{qi} - \Gamma^p_{qi}$ is independent on Γ^p_{qi}. Since the transformation law of E^p_{qi} on subgroup (17) is tensorial, Lemma 1 yields

$$A^p_{qi} = \Gamma^p_{qi} + cR^p_{qij}\xi^j + d\delta^p_q R^r_{rij}\xi^j \ . \tag{20}$$

The equivariancy of A^p_{qi} on the subgroup

$$a^i_j = \delta^i_j, \quad a^p_q = \delta^p_q, \quad a^p_{qi} = 0, \quad a^p_{qij} = 0 \tag{21}$$

then requires

$$(aR^p_{qj\ell}\xi^\ell + b\delta^p_q R^r_{rj\ell}\xi^\ell)a^j_{ik}\xi^k = 0 \ . \tag{22}$$

Hence $B^p_{qi} = 0$.

Now it suffices to find the transformation law of D^p_{qi} modulo the linear combinations of B^p_{qi}. The second line of (8) yields

$$D^p_{qi} = a^p_r D^r_{sj}\tilde{a}^s_q\tilde{a}^j_i + a^p_{sj}\tilde{a}^s_q\tilde{a}^j_i \quad (\text{mod } B^p_{qi}) \ . \tag{23}$$

Quite similarly to the previous step we obtain

$$D^p_{qi} = \Gamma^p_{qi} + eR^p_{qij}\xi^j + f\delta^p_q R^r_{rij}\xi^j \ . \tag{24}$$

8. To simplify the study of the most complicated case of C^p_{qi}, we shall consider the difference $A - T$, where A is the unknown operator in question and T is the known operator (2). Evaluating (2) one finds the following coordinate expression of T

$$A^p_{qi} = \Gamma^p_{qi}, \quad B^p_{qi} = 0, \quad C^p_{qi} = \Gamma^p_{qi,j}\xi^j, \quad D^p_{qi} = \Gamma^p_{qi} \ . \tag{25}$$

Hence the coordinates of $A - T$ are $G^p_{qi} = cR^p_{qij}\xi^j + d\delta^p_q R^r_{rij}\xi^j, 0, F^p_{qi} = C^p_{pi} - \Gamma^p_{qi,j}\xi^j, H^p_{qi} = eR^p_{qij}\xi^j + f\delta^p_q R^r_{rij}\xi^j$. From the second line of (8) we deduce the following transformation law of F^p_{qi} on the subgroup $a^i_j = \delta^i_j, a^p_q = \delta^p_q$

$$\bar{F}^p_{qi} = F^p_{qi} + a^p_{rj}\xi^j G^r_{qi} - G^p_{ri}a^r_{qj}\xi^j - H^p_{qk}a^k_{ij}\xi^j \ . \tag{26}$$

The equivariancy of $F^p_{qi}(\Gamma^p_{qi}, R^p_{qij}, S^p_{qij}, \xi^j)$ on subgroup (15) yields F^p_{qi} is independent on S^p_{qij}. Then (26) implies $H^p_{qi} = 0$ and we have

$$F^p_{qi}(\Gamma^p_{qi} + a^p_{qi}, R^p_{qij}, \xi^i) = F^p_{qi}(\Gamma^p_{qi}, R^p_{qij}, \xi^i) + a^p_{rj}\xi^j G^r_{qi} - G^p_{ri}a^r_{qj}\xi^j \ . \tag{27}$$

Differentiating (27) with respect to Γ^r_{sj}, we find that $\partial F^p_{qi}/\partial \Gamma^r_{sj}$ are independent on Γ^p_{qi}. Hence

$$F^p_{qi} = k^{psj}_{qri}(R^p_{qij}, \xi^i)\Gamma^r_{sj} + J^p_{qi}(R^p_{qij}, \xi^i) \,. \tag{28}$$

Then (27) yields

$$k^{psj}_{qri} = c(\delta^p_r R^s_{qik} - \delta^s_q R^p_{rik})\xi^j\xi^k \,. \tag{29}$$

Applying Lemma 1, we obtain $J^p_{qi} = k_1 R^p_{qij}\xi^j + k_2\delta^p_q R^r_{rij}\xi^j$.

Thus, the resulting 4-parameter family of $GL(n, \mathbf{R})$-natural operators has the coordinate expression

$$A^p_{qi} = \Gamma^p_{qi} + k_3 R^p_{qij}\xi^j + k_4\delta^p_q R^r_{rij}\xi^j \,, \quad B^p_{qi} = 0, \quad D^p_{qi} = \Gamma^p_{qi} \,,$$
$$C^p_{qi} = \Gamma^q_{qi,j}\xi^j + k_1 R^p_{qij}\xi^j + k_2\delta^p_q R^r_{rij}\xi^j + k_3(\Gamma^p_{rj}R^r_{qik} - R^p_{rik}\Gamma^r_{qj})\xi^j\xi^k \,. \tag{30}$$

9. Obviously, the coefficients by k_1 and k_2 in (30) correspond to $\mu(C)$ and $\mu(D)$ from (6).

To construct operator $\tau(C)$ geometrically, we interpret the curvature of a connection Γ on P as a map $\varphi : P \to \mathfrak{g}\otimes\Lambda^2 T^* BP$. An induced map $\tau\varphi : TP \to (\mathfrak{g}\times\mathfrak{g}) \otimes \Lambda^2 T^* BP$ can be defined as follows. Let $\omega(X) \in \mathfrak{g}$ be the value of the connection form of Γ at $X \in T_u P$. $\omega(X) = \frac{d}{dt}\big|_0\gamma(t)$ holds where $\gamma(t)$ is a curve on G. Then we set

$$\tau\varphi(X) = \frac{d}{dt}\bigg|_0(\mathrm{ad}(\gamma(t)) \otimes \mathrm{id})\varphi(u) \,.$$

For every $Z \in TTBP$ we take analogously to (5)

$$\tau\varphi(X)(\pi_1(Z)\wedge\pi_2(Z)) \,. \tag{31}$$

This defines a map $TP\oplus TTBP \to \mathfrak{g}\times\mathfrak{g}$, the coordinate expression of which coincides with the coefficient by k_3 in (30) at least one point of each fibre. It remains to prove that (31) corresponds to a section of $L(TP)\otimes T^* TBP$. By the foundations of the theory of associated fibre bundles, this follows from the fact that the coefficient by k_3 in (30) is equivariant.

Finally, we construct $\tau(D)$ in a quite analogous way.

References

1. D. J. Eck, *Gauge-natural bundles and generalized gauge theories*, Mem. Amer. Math. Soc. **33**, No 247 (1981).
2. C. Ehresmann, *Les prolongements d'un espace fibré différentiable*, CRAS Paris **240** (1955) 1755–1757.
3. S. Kobayashi, *Theory of connections*, Annali di Matematica Pura et Applicata **43** (1957) 119–185.
4. I. Kolář, *Some natural operators in differential geometry*, Proc. of Conf. Diff. Geom. and its Applications, Brno 1986, Dordrecht 1987, pp. 91–110.
5. I. Kolář, *Some natural operations with connections*, Journal of National Academy of Mathematics **5** (1987) 127–141.
6. I. Kolář, P. W. Michor, J. Slovák, *Natural Operations in Differential Geometry*, to appear.
7. A. Nijenhuis, *Natural bundles and their general properties*, Diff. Geom. in Honour of K. Yano, Kinokuniya, Tokyo (1972) 317–334.

Ivan Kolář
Mathematical Institute of the ČSAV
Brno branch
Mendelovo nám. 1
CS 66282 Brno
Czechoslovakia

THE MATH. HERITAGE OF C.F. GAUSS (pp. 427-431)
edited by George M. Rassias
©1991 World Scientific Publ. Co. Singapore

RETURNS UNDER BOUNDED NUMBER OF ITERATIONS

Zbigniew S. Kowalski

For any dynamical system $(X, \mathbb{B}, m, f)$ and for any set A of positive measure, $A \in \mathbb{B}$, the lower estimation of the quantity $\max \{ m(A \cap f^{-i_1} A \cap \ldots \cap f^{-i_r} A) : 1 \leq i_1 < i_2 < \ldots < i_r \leq k \}$ as the function of three variables k, r and $m(A)$ is obtained. It turns out that this estimation is the greatest lower bound with respect to the class of all dynamical systems.

Let $(X, \mathbb{B}, m)$ be a probabilistic measure space and let f be a measurable, measure preserving transformation from X to X i.e., $f^{-1}\mathbb{B} \subset \mathbb{B}$ and $m(f^{-1}A) = m(A)$ for every set $A, A \in \mathbb{B}$. Such a transformation f is called an endomorphism. The main purpose of this note is to find the greatest lower bound to the quantity

$$\max \{ m(A \cap f^{-i_1} A \cap \ldots \cap f^{-i_r} A) : 1 \leq i_1 < i_2 < \ldots < i_r \leq k \}$$

for $r, k \geq 1, m(A) = $ const., in the class of all dynamical systems. The following theorem establishes such an estimation.

Theorem. For an endomorphism f of the probabilistic space $(X, \mathbb{B}, m)$ and for any set $A, A \in \mathbb{B}, m(A) > 0$, and positive integers k, r

$$\max \{ m(A \cap f^{-i_1} A \cap \ldots \cap f^{-i_r} A) : 1 \leq i_1 < i_2 < \ldots < i_r \leq k \}$$
$$\geq \left(\left(\binom{I((k+1) \quad m(A))}{r+1} \right) + F((k+1)m(A)) \right.$$
$$\left. \times \binom{I((k+1) \quad m(A))}{r} \right) \binom{k+1}{r+1}^{-1} . \tag{1}$$

(1) holds with equality for some endomorphism and a set of measure equal to $m(A)$.

Here $I(a) = \max\{n : n \leq a, n \text{ is an integer }\}$ and $F(a) = a - I(a)$. For our purpose we assume that Newton symbol $\binom{a}{b} = 0$ for $a < b$.

Proof. Let $A \in \mathbb{B}$ and $m(A) > 0$. Let k and r be fixed positive integers. Let $c(k,r) = \max\left\{m(A \cap f^{-i_1}A \cap \ldots \cap f^{-i_r}A) : 1 \leq i_1 < i_2 < \ldots < i_r \leq k\right\}$. We will denote by α the family of sets $\{A, f^{-1}A, \ldots, f^{-k}A\}$ and by $A(\alpha)$ the partition of X generated by α, i.e.,

$$A(\alpha) = \{A^{\varepsilon_1} \cap (f^{-1}A)^{\varepsilon_2} \cap \ldots \cap (f^{-k}A)^{\varepsilon_{k+1}}$$
$$: \varepsilon_i \in \{0,1\} \quad i = 1, \ldots, k+1\} .$$

Here $A^1 = A$ and $A^0 = X - A$. Let $Y = \{0,1\}^S$ where $S = \{1, \ldots, k+1\}$. Denote by A_i the set

$$\{x : x \in Y \quad \text{and} \quad |\{j : (x)_j = 1\}| = i\} \quad \text{for} \quad i = 0, \ldots, k+1 .$$

Let $m_i = m\left(\bigcup_{x \in A_i} A_x\right), i = 0, \ldots, k+1$, where

$$A_x = A^{(x)_1} \cap (f^{-1}A)^{(x)_2} \cap \ldots \cap (f^{-k}A)^{(x)_{k+1}} .$$

Taking the above definitions into consideration we get the following equality

$$\bigcup\{f^{-i_0}A \cap f^{-i_1}A \cap \ldots \cap f^{-i_r}A : 0 \leq i_0 < i_1 < \ldots < i_r \leq k\}$$
$$= \bigcup\left\{A_x : x \in \bigcup_{i=r+1}^{k+1} A_i\right\} .$$

Here $f^{-0}A = A$. Hence by the measure preserving property of f we have

$$\binom{k+1}{r+1}c(k,r) \geq \sum_{0 \leq i_0 < \ldots < i_r \leq k} m\left(f^{-i_0}A \cap \ldots \cap f^{-i_r}A\right)$$
$$= \sum_{i=r+1}^{k+1} \binom{i}{r+1}m_i .$$

Let $T = \sum_{i=r+1}^{k+1} \binom{i}{r+1} m_i$. Then

$$c(k,r) \geq T \binom{k+1}{r+1}^{-1} . \tag{2}$$

We find the least value of the quantity T by the application of the ideas of the simplex method in linear programming ([1]) for our purposes. Let $L = I((k+1)m(A))$ and $\varepsilon = F((k+1)m(A))$. With regard to the previous definitions, we see that the following equalities hold:

$$L + \varepsilon = \sum_{i=1}^{k+1} i m_i \tag{3}$$

$$1 = \sum_{i=0}^{k+1} m_i \tag{4}$$

$$T = \sum_{i=r+1}^{k+1} \binom{i}{r+1} m_i \tag{5}$$

Using the equalities (3) and (4) we eliminate the variables m_L and m_{L+1} from the equality (5). Namely,

$$m_L = 1 - \varepsilon + \sum_{\substack{i=0 \\ i \neq L}}^{k+1} (i - L - 1) m_i , \quad m_{L+1} = \varepsilon + \sum_{\substack{i=0 \\ i \neq L+1}}^{k+1} (L - i) m_i ,$$

$$T = \binom{L}{r+1} + \binom{L}{r} \varepsilon + \sum_{i=0}^{r} \left(r \binom{L+1}{r+1} - i \binom{L}{r} \right) m_i$$

$$+ \sum_{\substack{i=r+1 \\ i \neq L, L+1}}^{k+1} \left(r \binom{L+1}{r+1} - i \binom{L}{r} + \binom{i}{r+1} \right) m_i . \tag{6}$$

We will show that in (6) the coefficients before m_i are non-negative for $i = 0, \ldots, k+1$. By $\binom{L+1}{r+1} \geq \binom{L}{r}$ we get

$$r \binom{L+1}{r+1} - i \binom{L}{r} \geq 0 \quad \text{for} \quad i = 0, \ldots, r . \tag{7}$$

Next we prove that

$$r \binom{L+1}{r+1} - i \binom{L}{r} + \binom{i}{r+1} \geq 0 \quad \text{for} \quad i = r+1, \ldots , \tag{8}$$

(by the induction on r). For $r = 1$ we have

$$(L - i)(L + 1 - i) \geq 0 .$$

Applying the induction assumption for r we obtain

$$(r+1)\binom{L+1}{r+2} - i\binom{L}{r+1} + \binom{i}{r+2}$$

$$= (r+1)(L-r)(r+2)^{-1}\binom{L+1}{r+1} + (i-r-1)(r+2)^{-1}\binom{i}{r+1}$$

$$- i(L-r)(r+1)^{-1}\binom{L}{r}$$

$$\geq \binom{L+1}{r+1}(r+1)(L-r)(r+2)^{-1} + \left[i\binom{L}{r} - r\binom{L+1}{r+1}\right]$$

$$\times (i-r-1)(r+2)^{-1} - i(L-r)(r+1)^{-1}\binom{L}{r}$$

$$= ((r+1)i^2 - (2L(r+1)+r+1)i + L(L+1)(r+1))\binom{L+1}{r+1}$$

$$\cdot [(r+2)(L+1)]^{-1} .$$

In order to show that the last expression is non-negative it is sufficient to prove that

$$i^2 - (2L+1)i + L(L+1) \geq 0 .$$

But here $i \neq L, L+1$, which finishes the proof of the inequality (8). From (7) and (8) we conclude that the coefficients before m_i, in (6), are non-negative. Therefore T attains the minimum value for $m_i = 0, i \neq L$ and $i \neq L+1$. Here $T_{\min} = \binom{L}{r+1} + \binom{L}{r}\varepsilon$. Thus by (2)

$$c(k,r) \geq \left(\binom{L}{r+1} + \varepsilon\binom{L}{r}\right)\binom{k+1}{r+1}^{-1} . \tag{9}$$

In order to finish the proof of the Theorem we define the dynamical system (Z, μ, g) which realizes the equality in (9) as follows: $Z = A_L \cup A_{L+1} \subset Y$

$$\mu(x) = \begin{cases} (1-\varepsilon)|A_L|^{-1} & \text{if } x \in A_L \\ \varepsilon|A_{L+1}|^{-1} & \text{if } x \in A_{L+1} \end{cases}$$

and $(g(x))_i = (x)_{i-1}$ for $i > 1$ and $(g(x))_1 = (x)_{k+1}$. Here

$$|A_L| = \binom{k+1}{L}, |A_{L+1}| = \binom{k+1}{L+1} \text{ and}$$

$$g(A_L) = A_L , g(A_{L+1}) = A_{L+1} .$$

The set $B, B = \{x : x \in Z \text{ and } (x)_1 = 1\}$, has the measure

$$\mu(B) = \binom{k}{L-1}\binom{k+1}{L}^{-1}(1-\varepsilon) + \binom{k}{L}\binom{k+1}{L+1}^{-1}\varepsilon$$
$$= (L+\varepsilon)(k+1)^{-1} = m(A)$$

and

$$\mu\left(B \cap g^{-i_1}B \cap \ldots \cap g^{-i_r}B\right) = \binom{k-r}{L-r-1}\binom{k+1}{L}^{-1}(1-\varepsilon)$$
$$+ \binom{k-r}{L-r}\binom{k+1}{L+1}^{-1}\varepsilon$$
$$= \left(\binom{L}{r+1} + \varepsilon\binom{L}{r}\right)\binom{k+1}{r+1}^{-1}$$

for any sequence $1 \leq i_1 < i_2 < \ldots < i_r \leq k$.

References

1. G. B. Dantzig, *Linear Programming and Extensions*, Princeton, N.J. 1963.

Zbigniew S. Kowalski
Instytut Matematyki
Politechnika Wrocławska
50–370 Wrocław
Poland

THE MATH. HERITAGE OF C.F. GAUSS (pp. 432-453)
edited by George M. Rassias
©1991 World Scientific Publ. Co. Singapore

ON A STRUCTURAL SCHEME OF PHYSICAL THEORIES PROPOSED BY E. TONTI[a]

Ernesto A. Lacomb[b] and Fausto Ongay

A structural scheme of physical theories proposed by E Tonti is analyzed, using modern mathematical concepts, aiming to attract physicists to some recent developments in mathematics. At the same time the presentation avoids extreme generality or excessive abstractions. The merit of this scheme is that it emphasizes a systematic aspect of methodology in physics; this way some essential properties and equations of a given theory are easily read off. A few basic examples are studied, mainly electromagnetism and the linearized Einstein equation, and then the scheme is used to gain some insight into the structure of these theories.

1. Introduction

In a series of papers published around 1975 ([12]), E. Tonti developed a structural scheme trying to formalize the analogies among diverse physical theories, using some ideas of algebraic topology and differential geometry. These articles were preceded by that of Branin ([2]), published in 1966 and motivated by general ideas on electrical networks.

Tonti's work is essentially a wide recollection of laws or equations of physics, that he tries to reduce following several basic patterns in the form of commutative diagrams. The merit of his viewpoint lies mainly in the fact that it emphasizes a systematic aspect of methodology in physics.

The purpose of this work is to discuss this scheme in a not so wide context, but a more modern mathematical presentation, hence throughout

---

[a]Research partially supported by PRONAES (Mexico), grant C86-010260.
[b]Member of CIFMA (Mexico).

this work we use the language of modern differential geometry (manifolds, differential forms and fiber bundles). These notions are now an important tool in theoretical physics, e.g. in gauge theories, and there are several good accounts on them written for physicists ([9], [13], [14]). This analysis is carried out in Sec. 2, and then some examples are carefully discussed in Sec. 3.

A more original contribution is our discussion of reductions and extensions of theories in Sec. 4, in the context of our main representative examples: classical electromagnetism and the linear approximation to Einstein's equation. This way we describe some possible extensions and consequences of the method. In the last section we give a critique of the scheme.

Several other works related to this sort of ideas, although with different viewpoints, can be found in the literature ([6], [10], [11]).

2. Description of the Scheme

The starting point of this scheme is the following seemingly trivial observation, which in fact has very deep implications:

"Every measurement of a physical quantity is related to some region of space-time."

If we assume that we are dealing with a differentiable theory, that is, if we assume that space-time is a differentiable (C^∞) manifold, then the above remark says that the physical quantities are differentiable functions defined on submanifolds, possibly with boundary, of this manifold.

A further assumption, which is implicit in Tonti's works, is the superposition principle. Using the language of differential geometry, these hypotheses may be expressed as a mathematically precise axiom as follows:

Physical quantities are sections of a vector bundle over a submanifold, possibly with boundary, of the space-time manifold.

Later on we will state two postulates for the structure scheme, but for the time being we accept the above remark as the mathematical description of physical quantities. Readers not familiar with the theory of vector bundles may think of sections as "generalized vector-valued functions."

According to Tonti and following his terminology, the equations of physics may be classified in three basic categories:

a) Topological equations.

b) Constitutive or phenomenological equations.

c) Equations of interaction between two theories.

For reasons described below, types b) and c) are also called metric equations. In this work however we shall restrict ourselves to topological and constitutive equations, because the equations of interaction are of a rather different nature.

To inscribe equations of class a) within the framework described above we must add a new hypothesis. Indeed, these equations are systematically obtained by a process of differentiation, which may be described as follows: if we let $B \to M$ denote the vector bundle whose sections are the physical quantities; then this differentiation is of one of the following two types:

i) B posses a differential operator; that is an operator d, acting on the sections of B, and such that $d^2 = 0$. In general one has to deal with vector-valued differential forms or even vector-bundle-valued forms and d is the usual exterior differential in B; to simplify the discussion we shall assume only vector valued forms, that is, forms whose coefficients are vectors rather than scalars. For those familiar with the theory of vector bundles, formally this means that B is of the form $B = \Lambda(M) \otimes V$, where $\Lambda(M)$ denotes the exterior bundle of T^*M, V is a trivial vector bundle whose rank (i.e., its fiber dimension) is not necessarily finite.

ii) B has an affine connection ∇ and differentiation is covariant derivation with respect to this connection. One may see [9] or [13] for a discussion of this concept; however, we shall not make use of this notion in the description of the scheme, restricting ourselves to the case described in i).

Thus, these topological equations are in fact differential equations although of a special type. It is also pertinent to point out that Tonti calls this differentiation process coboundary, but the term seems inappropriate, specially with the second kind, which is *not* in general a coboundary process. Since differentiation is a linear operation, these equations are always linear in the observable, that is, the section of the bundle, which is differentiated (although in some cases they may be equivalent to nonlinear equations in other variables).

Furthermore, the manifolds appearing in physics possess pseudo-Riemannian metrics; these metrics are the additional mathematical structure required for equations type b), hence the name metric equations. As a rule these metrics are linked to the physical properties of the media, so that they must be determined from experiment and moreover, the constitutive equations can also be nonlinear in contrast with the topological equations.

We also remark that if the manifold possesses a metric both differentiation processes are related, since the exterior differential may be computed using the standard Levi-Civita connection (see [13], Appendix B). This purely mathematical fact should also have physical implications.

Finally all the manifolds are assumed orientable, which means that the manifold admits a *global* volume form, but, as Tonti points out, one must not impose an *a priori* fixed orientation in M. Thus we are led to distinguish between what de Rham ([7]) called *odd* forms, which change sign under an orientation reversing change of coordinates, and *even* forms which do not change sign; the quantities related to even forms are called *configuration* variables and those related to odd forms are called *source* variables. Recall however that in an *oriented* manifold, that is, a manifold with a fixed orientation, both types of forms *may* be identified. We shall return to this point later on.

Now in any orientable n-dimensional manifold endowed with a pseudo-Riemannian metric there exists an operator $*$, called *Hodge's star operator* (see for instance [5] or [8]), giving a duality between even p-forms and odd $(n - p)$-forms; this operator plays a crucial role in the construction of constitutive equations.

We may now state the two basic postulates of the scheme:

P1: The domain of physical quantities are orientable submanifolds, possibly with boundary and endowed with pseudo-Riemannian metrics, of the space-time manifold.

P2: The physical quantities are vector-valued, possibly odd, differential forms and the differentiation is exterior differential on these forms.

These two postulates are more or less explicit in Tonti's works (see [12, (3)], p. 228 for a statement of P2). However, although P1 appears as a natural hypothesis, P2 is a subtler requirement. This is more so when we want to incorporate the general covariance principle. We will discuss this with more detail in the last section of this paper.

With the above conventions, the structure scheme can be summarized in the form of commutative diagrams of the sort described in Fig. 1. To fix ideas and simplify the writing, we will assume that M is an orientable 3 dimensional manifold endowed with a Riemannian metric g; thus the $*$-operator relates even p-forms and odd $(3 - p)$-forms.

It should be emphasized that once the above postulates are accepted, the form of the diagrams is a direct consequence of them, and this has already some consequences: From the relation $d^2 = 0$ satisfied by the ex-

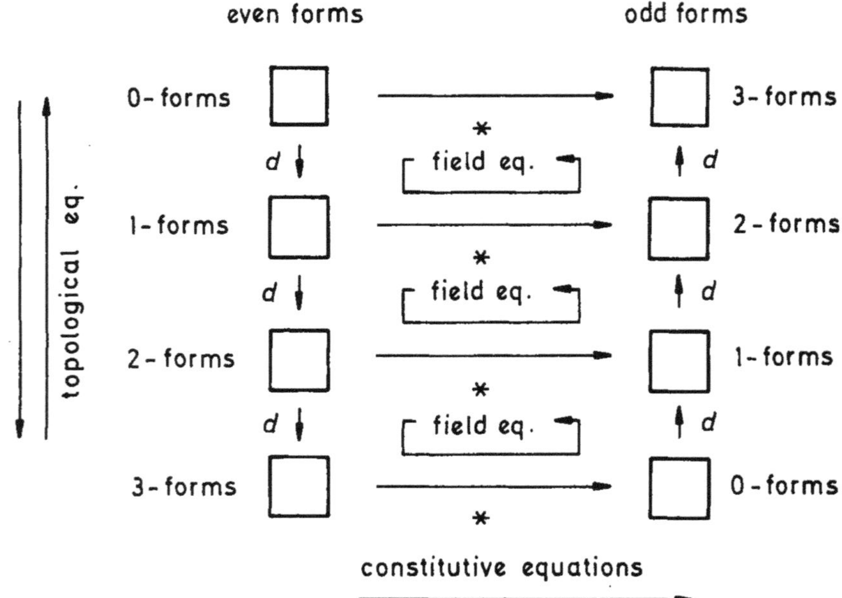

Fig. 1. Standard diagram in 3 dimensions.

terior differential it turns out that certain "physical quantities" appearing in Tonti's diagrams are either trivial or do not have a direct physical interpretation in a given theory; the importance of this observation will become evident when discussing concrete examples.

A key role in the field equations of a given theory is played by the Laplace-Beltrami operator, which generalizes the laplacian and which is defined as follows: recall that the adjoint δ of d is the operator defined on p-forms as $\delta = (-1)^{p(n-p)} * d*$, then the Laplace-Beltrami operator is defined as $\Delta = d\delta + \delta d$. Since the $*$ operator is used, the Laplace-Beltrami operator depends on the metric of the manifold. We also remark that linear constitutive equations (or their linear approximations) automatically yield a field equation which is a Poisson equation.

3. Basic Examples

In this section we discuss the application of the method to some representative examples. The notation for the different physical magnitudes is as much as possible as the one used by Tonti in [12, (2)], to make comparisons easier.

a) Dynamics of a classical particle in a conservative force field

In this example the equation to deduce is Netwon's second law $f = ma$. The base manifold here is the time axis, so that the diagram is in dimension 1, and associated to position, velocity, linear momentum and force we have vector valued differential forms of rank 3. The exterior differential is then identified with the usual derivative of curves in $\mathbb{R}^3$ by the choice of a basis for the time axis (which in this case is equivalent to choosing a metric and an orientation).

The constitutive equations are the one relating momentum with velocity plus the force law; the resulting field equation is precisely Newton's law. So, Tonti's diagram related with this equation is that of Fig. 2.

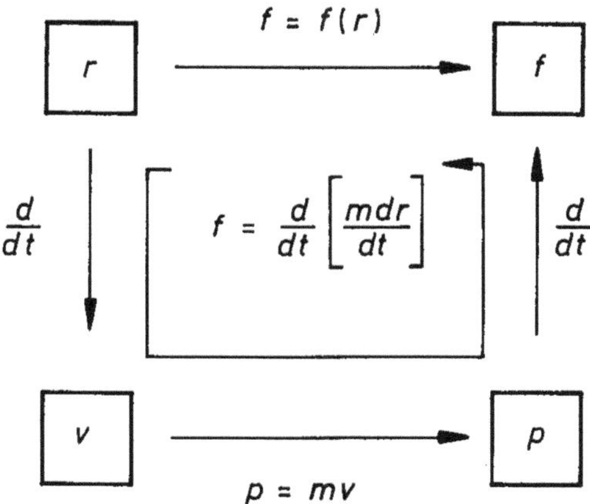

Fig. 2. Tonti's diagram for Newton's second law.

b) Classical electromagnetism

Tonti's method is perhaps best shown in this case so, although there are several excellent accounts of it ([4], pp. 44–48 or [8]), we will give a rather complete discussion.

The manifold in question here is $M = \mathbb{R}^4$ with the Minkowski metric g_1, which in the global coordinates (t, x, y, z) is given by

$$ds = dt^2 - (dx^2 + dy^2 + dz^2) . \tag{1}$$

The vector bundle is just $B = \Lambda(M)$, so that the physical quantities are usual differential forms and if $\chi : M \to \mathbb{R}$ is a differentiable function recall that the differential of χ is

$$d\chi = \partial_t \chi \, dt + \partial_x \chi \, dx + \partial_y \chi \, dy + \partial_z \chi \, dz \ , \tag{2}$$

where $\partial_t = \frac{\partial}{\partial t}$, etc.

The Hodge $*$ operator is not the same as in the euclidean case since g_1 is different from the euclidean metric. For our purposes it is enough to recall that this operator, associated to the metric g_1 and the orientation given by $dt \wedge dx \wedge dy \wedge dz$, takes on 2-forms the following expression:

$$*(dx \wedge dy) = -dz \wedge dt, \quad *(dx \wedge dt) = dy \wedge dz$$
$$*(dx \wedge dz) = dy \wedge dt \ \text{ and } \ *^2 = - \text{ identity} \tag{3}$$

The electric field is a (even) 1-form $E = E_1 dx + E_2 dy + E_3 dz$, while the magnetic field is a 2-form $B = B_1 dy \wedge dz + B_2 dz \wedge dx + B_3 dx \wedge dy$. This allows us to define the four dimensional Maxwell 2-form

$$F = E \wedge dt + B \tag{4}$$

representing the classical electromagnetic field. Applying the exterior differential we get

$$dF = \ \text{rot}(E) \wedge dt + \ \text{div}(B) \wedge dx \wedge dx \wedge dz + \partial_t B \wedge dt \tag{5}$$

where

$$\begin{aligned} \text{rot}(E) &= (\partial_x E_2 - \partial_y E_1) dx \wedge dy \\ &+ (\partial_x E_3 - \partial_z E_1) dx \wedge dz \\ &+ (\partial_y E_3 - \partial_z E_2) dy \wedge dz \end{aligned} \tag{6}$$

and

$$\text{div}(B) = \partial_x B_1 + \partial_y B_2 + \partial_z B_3 \ . \tag{7}$$

Hence, Faraday's law together with the absence of magnetic monopoles are equivalent to $dF = 0$, which is known as homogeneous Maxwell equation. Since M is contractible, which means that it may be smoothly deformed into a point, this equation is equivalent to $F = dA$, for some 1-form, A, known as the Maxwell 4-potential.

The electric displacement D and the magnetic field strength H are identified with 2- and 1-forms respectively. If we denote by $\star$ the Hodge star operator associated to the euclidean metric in dimension 3, and assuming the medium to be isotropic, these fields are simply described by $D = \varepsilon \star E$ and $H = \mu^{-1} \star B$, where the scalars dielectric permitivity ε and magnetic permeability μ are functions of the medium.

These constitutive equations may be summarized in a single 4-dimensional equation, by a slight change of the metric on M with the corresponding change in the $*$-operator. This can be described as follows: writing $c = (\varepsilon\mu)^{-1/2}$ and modifying the metric g_1 by taking in general g_c defined by

$$ds^2 = c^2 dt^2 - (dx^2 + dy^2 + dz^2) . \tag{8}$$

Now define the Maxwell dual field G, which is an odd 2-form which in coordinates collects together D and H as

$$G = D - H \wedge dt . \tag{9}$$

Then the constitutive equation of classical electromagnetism becomes

$$G = (\varepsilon/\mu)^{1/2} * F . \tag{10}$$

The above defined function c of the medium has units of speed: it is of course just the speed of propagation of electromagnetic signals in the medium. This can be concluded from the field equations which we will briefly discuss below.

The charge density ρ and the current density j are put together in the 1-form J, called the 4-current:

$$J = -\rho dt + j_1 dx + j_2 dy + j_3 dz \tag{11}$$

and Ampère's and Gauss' laws can be written as $dG = 4\pi * J$. This is the so-called inhomogeneous Maxwell equation.

In this case the operator Δ applied to functions is modulo constants equivalent to the D'Alembert operator, denoted by $\Box$, and given in coordinates by

$$\Box \psi = \frac{1}{c^2} \frac{\partial^2 \psi}{\partial t^2} - \frac{\partial^2 \psi}{\partial x^2} - \frac{\partial^2 \psi}{\partial y^2} - \frac{\partial^2 \psi}{\partial z^2} . \tag{12}$$

This inhomogeneous Maxwell equation can be rewritten as

$$\delta F = (\mu/\varepsilon)^{1/2} 4\pi J, \quad \text{and hence}$$
$$\Delta F = (d\delta + \delta d)F = d\delta F = (\mu/\varepsilon)^{1/2} 4\pi dJ \tag{13}$$

so that the electromagnetic field satisfies a Poisson equation. If charges and currents are absent (or more generally if J is a closed form, i.e., $dJ = 0$), we obtain the Laplace equation associated to g_c. If c is a constant this is equivalent to a wave equation for each component of F, with propagation speed equal to c. It is interesting to remark that if A is chosen so that $\delta A = 0$ (null divergence or Lorentz gauge condition), then this 4-potential also satisfies the wave equation, with the same propagation speed c, provided that $J = 0$.

We have seen that F is a closed form, while $G = (\varepsilon/\mu)^{1/2} * F$ is not; however, the equation $d^2 G = 0$ is also important in physics: indeed, $d^2 G = 0$ is equivalent to $d * J = 0$, which in coordinates becomes

$$\frac{\partial \rho}{\partial t} + \nabla \cdot j = 0 \tag{14}$$

which is the so-called continuity equation, expressing charge conservation. This continuity equation is the archetype of conservation laws stemming from the relationship $d^2 = 0$.

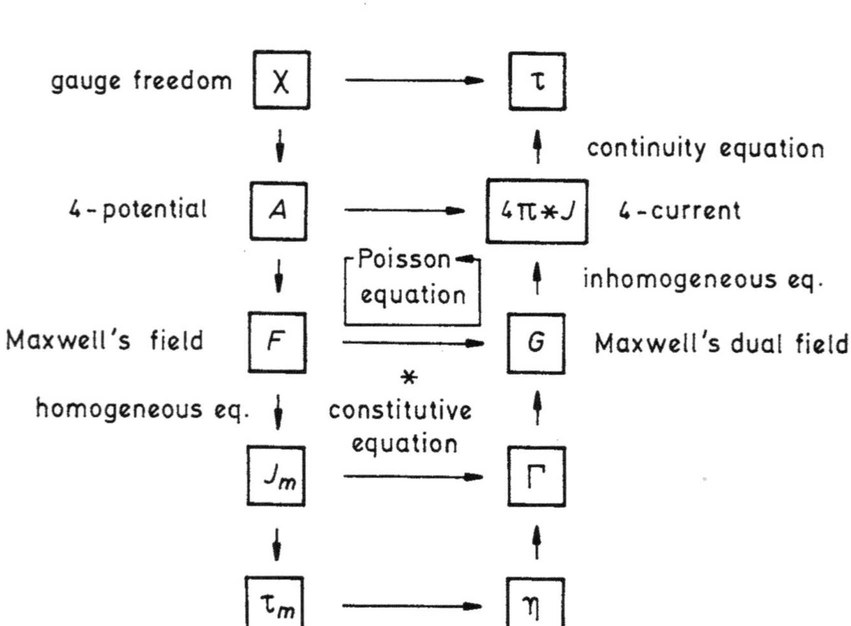

Fig. 3. Tonti's diagram for classical eletromagnetism.

Tonti's diagram for classical electromagnetism is as in Fig. 3.

Recall that not all the indicated magnitudes have a sound physical interpretation.

c) Linearized Einstein equation

The linear approximation to Einstein's equation is quite similar to, although somewhat more complicated than, classical electromagnetism.

Once again the base manifold is $M = \mathbb{R}^4$ with a pseudo-metric g of index 3, referred to as a Lorentz metric, where the index of the metric may be defined as the number of minuses appearing in the coordinate expression of the metric when it is written in diagonal form; for instance the euclidean metric has index 0 while the metrics g_c have index 3. The explicit expression of the metric is however unknown since it is determined through Einstein's equation.

Before showing how postulates P1 and P2 are satisfied, let us recall Einstein's linearized equation in the more familiar tensor notation; we follow here the presentation of Wald [13].

Denote by g_{ab} the components of the metric tensor, and assume they have the form

$$g_{ab} = h_{ab} + \gamma_{ab} \ . \tag{15}$$

Here h_{ab} are the components of the metric g_c and γ_{ab} is a small peturbation. We will discuss the meaning of smallness in this context in somewhat greater length in the next section, where more explicit calculations are done, but a more complete account of this point may be found in [13]. Einstein's equation can then be obtained directly in terms of the perturbation: indeed computing the Christoffel symbols and Ricci tensor to first order, we get as Einstein tensor

$$G_{ab} = \partial^c \partial_{(b} \gamma_{a)c} - \frac{1}{2} \partial_a \partial_b \gamma - \frac{1}{2} h_{ab} (\partial^c \partial^d \gamma_{cd} - \partial^c \partial_c \gamma) \tag{16}$$

where $\gamma = h^{ac} \gamma_{ca}$. Parentheses enclosing some indices indicate a symmetrical sum with respect to those indices. Notice also that in the computations the flat metric tensor h_{ab} is used to raise and lower indices.

The above expression can be simplified by a judicious choice of the perturbation: first of all we can rewrite the Einstein tensor in terms of a new tensor defined by

$$\phi_{ab} = \gamma_{ab} - \frac{1}{2} h_{ab} \gamma \ . \tag{17}$$

Indeed, γ_{ab} can be recovered from ϕ_{ab} by contracting with h^{ab} (recall that this tensor is defined by the condition $h^{ab}h_{bc} = \delta^a_c$) to get $\phi = \frac{1}{2}\gamma$. Then Einstein's equation becomes

$$G_{ab} = -\frac{1}{2}\partial^c\partial_c\phi_{ab} + \partial^c\partial_{(b}\phi_{a)c} - h_{ab}\partial^c\partial^d\phi_{cd} = 8\pi T_{ab} \tag{18}$$

where T_{ab} denotes the energy-momentum tensor.

On the other hand, we can make a gauge transformation by means of an "infinitesimal transformation" of the base manifold M, corresponding to the fact that it has no natural coordinate system, which is indeed the mathematical statement of Einstein's equivalence principle. More precisely, we can perform on ϕ_{ab} a transformation of the type

$$\phi_{ab} \rightarrow \phi_{ab} + \partial_{(a}\xi_{b)} \tag{19}$$

without changing Einstein's equations; here ξ_a is just a vector field on M and we must symmetrize to preserve the symmetric nature of the metric. We can then choose a tensor ϕ_{ab} satisfying $\partial^b\phi_{ab} = 0$, which is the equivalent of the previously described Lorentz gauge condition of the electromagnetism. With these choices a straightforward computation shows that only one term of the Einstein tensor remains, so that Einstein's equation simply becomes

$$\partial^c\partial_c\phi_{ab} = -16\pi T_{ab} \tag{20}$$

which is a Poisson equation.

We can now easily describe this equation by a Tonti diagram as follows:

Since the base manifold is contractible, any fiber bundle is trivial; this allows us to consider tensors as tensor-valued differential forms by raising one of the indices with the metric tensor h^{ab}. In particular the tensors ϕ_{ab} and T_{ab} are identified with vector-valued 1-forms denoted by Φ and T respectively.

Moreover, associated to the metric g_c there is a well defined exterior differential and a Hodge $*$ operator as described in the previous example. In these terms the operator ∂^a is the adjoint of ∂_a, so that in intrinsic notation and due to our choice of gauge, the form Φ satisfies $\delta\Phi = 0$. Einstein's equation then becomes

$$\delta d\Phi = -16\pi T \tag{21}$$

or more briefly

$$\Delta\Phi = -16\pi T \ . \tag{22}$$

Taking into account all of the above, Tonti's diagram for this case is depicted in Fig. 4. Once again, not all the quantities have a natural physical interpretation. We also remark that this analysis of the linear case does not apply directly to the general Einstein equation (see [14]).

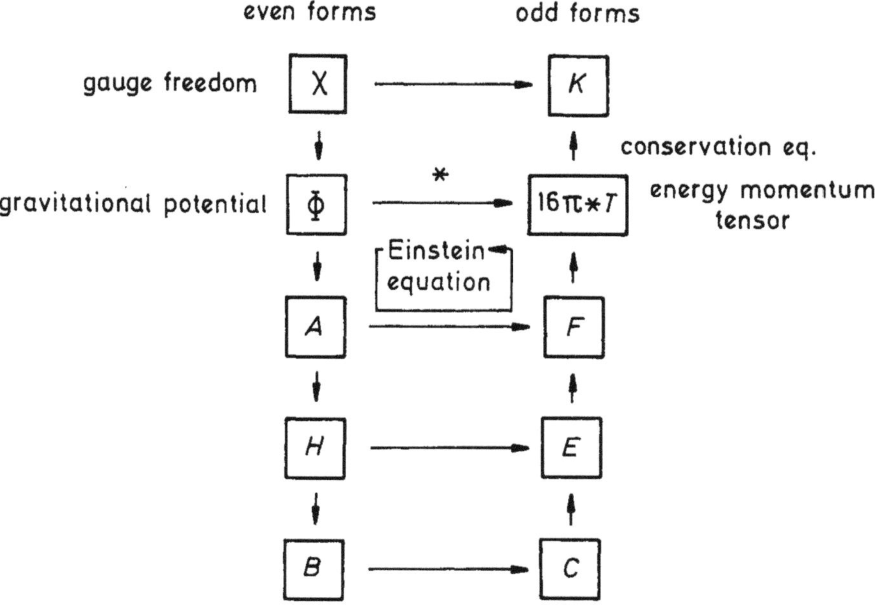

Fig. 4. Tonti's diagram for the linearized Einstein equation.

4. Extension and Reduction of Theories

We may summarize the discussion of Sec. 2, by saying that a physical theory is determined by a base manifold M, a vector bundle $B \to M$ and the constitutive relations of the theory. This formal framework will be used in this section to show how "formal manipulations" of Tonti's diagrams, suggested by usual techniques of algebraic topology and differential geometry, furnish a more or less systematic method to analyze related physical theories. We shall discuss two processes, which in some sense are dual to each other, which will be called reduction and coupling of physical theories.

4.1. *Reductions of a Physical Theory*

We may heuristically define a reduction of a physical theory as a combination of some of the following steps:

i) Choice of a submanifold or a quotient manifold of the base manifold. Recall that a quotient manifold is a manifold obtained by identifying some parts of the original manifold to points. In the cases we shall consider the base manifold is $\mathbb{R}^4$, and by identifying points differing by the time coordinate we get a manifold which is "identical" to $\mathbb{R}^3$. This new manifold will be the base manifold of the restricted theory.

ii) Determination of a subbundle or a quotient bundle of the bundle B. This will determine the physical quantities of the restricted theory.

iii) Performing some approximations, either on the geometry of the manifolds or on the constitutive equations.

As a rule, these choices are dictated by physical considerations.

Perhaps the best example of this process is the passage from electromagnetic theory to electrostatics. As already mentioned, in this example the base manifold is $M = \mathbb{R}^4$ with the Lorentz metric $\mathscr{g}_c$ and the bundle $B = \Lambda(M)$. M has a natural global system of coordinates (t, x, y, z) where the metric is diagonal, and a natural foliation, that is, a partition of the manifold into disjoint but isomorphic submanifolds, called the leaves of the foliation, given by $t = $ constant. As pointed out above, it is easy to see that both, the leaves of the foliation and the quotient manifold of M by t may be identified with $\mathbb{R}^3$, furnished with the euclidean metric; we shall denote these manifolds by $\widehat{M}$.

Electrostatics results from a projection of electromagnetism to $\widehat{M}$ for even forms and a restriction to a leaf of the foliation for odd forms, (recall that both manifolds may be identified). The electric differential forms E, D, ρ and ϕ, described before will then become identified with differential forms on $\widehat{M}$.

The fact that we must combine a projection and a restriction to recover electrostatics can be seen directly from the expressions of the electrostatic fields; however, we can get an idea as to why this is so if we consider the definitions of E and D in terms of line and surface integrals respectively: physically a line integral is associated to a displacement (i.e., motion along a trajectory) but the process is quite different for computing a surface integral.

Also, in order to project the forms to $\widehat{M}$, we must assume that the electric forms are static, that is, the time derivatives of their coefficients vanish. Then the absence of currents ($\partial_t \rho = 0$) allows us to neglect the magnetic fields and, by choosing an appropriate gauge, the magnetic potential may be taken as 0. In this gauge the 4-potential reduces to $A = \phi\, dt$, where now ϕ is time independent and, upon passing to the quotient, A becomes identified with the three dimensional 0-form ϕ, the electrostatic potential. A similar reasoning applied to the electric field E yields the electrostatic field form, also written E.

It is interesting to remark that this projection to a quotient manifold may be interpreted as taking a time average of the forms, which may be written symbolically as

$$\phi = \int A \Big(\int dt \Big)^{-1} . \tag{23}$$

This average process describes a physical way to relate both potentials but also shows that a similar approximation argument may be applied to quasi-static forms, that is forms with negligible time derivatives.

For odd forms the situation is simpler: since we are considering submanifolds in this case, every form that does not involve dt is automatically identified with a form in three dimensional space $\widehat{M}$. We also remark that there is no inconsistency with the interpretation of $\widehat{M}$ as a quotient manifold, because all the coefficients are assumed to be time independent.

These previous remarks are summarized in diagram 5, where A denotes the approximations performed and $\star$ denotes the usual $\ast$-operator in $\mathrm{I\!R}^3$. The middle columns are precisely those of Tonti's diagram of electrostatics.

A similar treatment will yield the diagram of magnetostatics. This theory is somehow dual to the previous one because we must now perform a projection down to a quotient for even forms and a restriction to a leaf for odd forms. Making approximations akin to those of the electrostatic case we get diagram 6, where the middle columns are those of magnetostatics.

We recall that not all the forms in these diagrams are physically meaningful. It is also obvious that some of the approximations appearing in the diagrams have no interpretation.

In both of these examples the restriction of the theory is obtained by combining a restriction to a submanifold and a projection down to a quotient manifold, but other possibilities appear in practice; for instance, to recover the diagram of linear electric conduction ([12, (2)], p. 160) one

E. A. Lacomba & F. Ongay

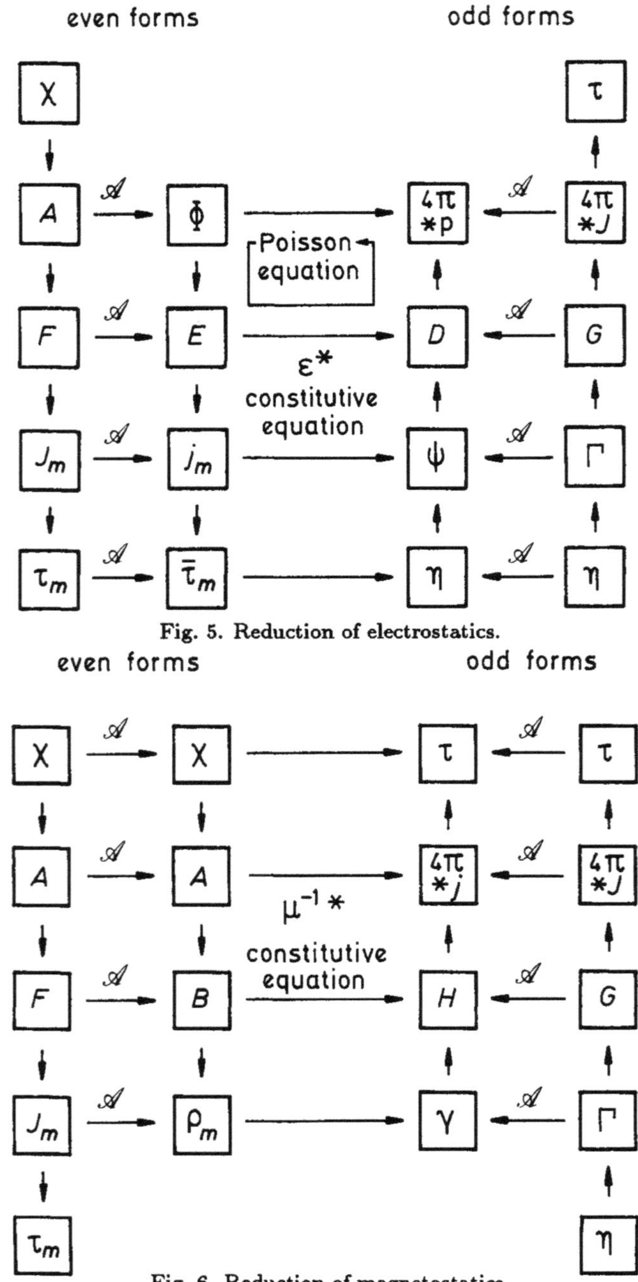

Fig. 5. Reduction of electrostatics.

Fig. 6. Reduction of magnetostatics.

must perform projections in both even and odd forms. We will not go into details of this case here, but instead we shall exploit the analogy between electromagnetism and linear gravitation to show how the use of Tonti's diagrams allows us to easily recover newtonian gravitation from Einstein's theory (see [3], p. 26–29, also [5] or [13]).

Let us consider the diagram of linear gravitation and make the newtonian approximations, namely a static universe, weak gravitational fields and $c \gg 1$. These hypotheses imply ([13]) that the metric tensor g may be written in the form

$$g_{00} \, dt^2 + \sum_{i,j=1}^{3} g_{ij} \, dx^i dx^j \tag{24}$$

where the metric coefficients are independent of t and $|g_{ij}|$ are of the order of 1, for $i, j = 1, 2, 3$ and negligible with respect to $g_{00} \simeq c^2$.

According to the discussion of the previous section we can assume that the metric tensor to first order can be written as

$$(c^2 + \phi_{00})dt^2 + \sum_{i,j=1}^{3} (\delta_{ij} + \phi_{ij})dx^i dx_j \tag{25}$$

where the components of the perturbation ϕ_{ab} are much smaller than those of the flat metric g_c. Then Einstein's equation reduces to a partial differential equation involving only spatial coordinates for the component ϕ_{00} of the perturbation.

Since we already know that Newton's law of gravitation is similar in structure to Coulomb's law of electrostatics, we propose for the reduction a diagram similar to that of the reduction from electromagnetism to electrostatics (but we must point out that other possibilities do exist). With this in mind and using the same conventions as before, the diagram for the reduction is as in Fig. 7. In this diagram the middle columns represent the restricted theory where the base manifold is $\mathbb{R}^3$ equipped with a metric that is approximately the euclidean one. In this way the structure of the diagram suggests, as field equation of this restricted theory,

$$\nabla^2 \phi_{00} = 16\pi T_{00} \, , \tag{26}$$

where ∇^2 denotes the usual laplacian in $\mathbb{R}^3$. This equation coincides with the usual field equation of Newtonian gravitation

$$\nabla^2 \varphi = 4\pi \rho \, , \tag{27}$$

where φ is the newtonian gravitational potential and ρ the newtonian mass density, provided we make the identifications

$$\varphi = \frac{\alpha}{4}\phi_{00}; \quad \rho = \frac{1}{\alpha}T_{00} ; \tag{28}$$

and the usual choice for α is c^2 ([12, (2)], p. 177).

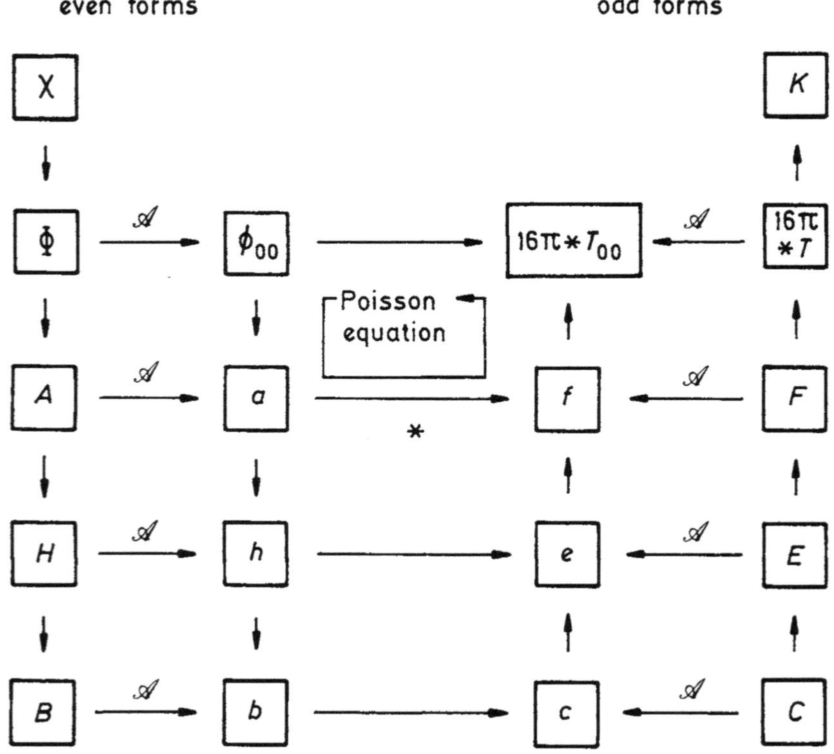

Fig. 7. Newtonian gravitation as a reduction of Einstein's equation.

This also justifies *a posteriori* our construction of the diagram of the reduction, since we actually know that configuration variables are measured in practice by taking time averages (for instance dropping a stone to measure the strength of gravity). However it would be interesting to analyze other reductions of Einstein's equation, as was done for electromagnetic theory.

On the other hand, in Newton's theory we have $\nabla\varphi = a$ as the equation of gravitational acceleration, where ∇ denotes the usual gradient. Reversing the approximation argument, the analog of this equation in Einstein's theory is

$$\frac{1}{2}\partial^i\phi_{00} = -\Gamma^i_{00} \, , \tag{29}$$

where Γ^a_{bc} denotes the Christoffel symbols of the metric g, and in this way one easily recovers the well known fact that gravitational acceleration is due to the fact that the metric is curved by the presence of massive objects.

4.2. *Coupling of Theories*

We shall now describe how to obtain electromagnetism as a "coupling" of electrostatics and magnetostatics, when we consider time-dependent fields. In this sense, electromagnetism may be considered as a very special extension of electrostatics or magnetostatics.

Keeping in mind the diagrams of these theories and also the dual nature of electrostatics and magnetostatics, the coupling is obtained by making a shift in the diagrams, whose effect is to put both constitutive equations at the same level. The key to perform this operation is given by the expressions of the 4-dimensional electromagnetic forms, $F = E \wedge dt + B$ and $G = D - H \wedge dt$. These equations show that the electric form E and the magnetic form H, which are defined in the 3-dimensional *quotient* manifold $\widehat{M}$, must be multiplied by dt to become forms defined in Minkowski space M. In technical language, we must *lift* the forms from the quotient manifold to the original manifold, but this may be intuitively understood as an "infinitesimal integration" of the forms, which is in a sense the inverse process to the time average used to pass to the quotient.

Diagram 8 summarizes these conclusions; it is clearly shown there the way to couple the columns of electrostatics and mangetostatics to get the corresponding columns of electromagnetism. The diagonal arrows marked ∂_t correspond to the additional part of the exterior differential added because of the time dependence of the fields. This coupling of the columns has a well defined physical counterpart, namely, that time-varying fields produce new fields that get coupled in a physical way with the previous fields.

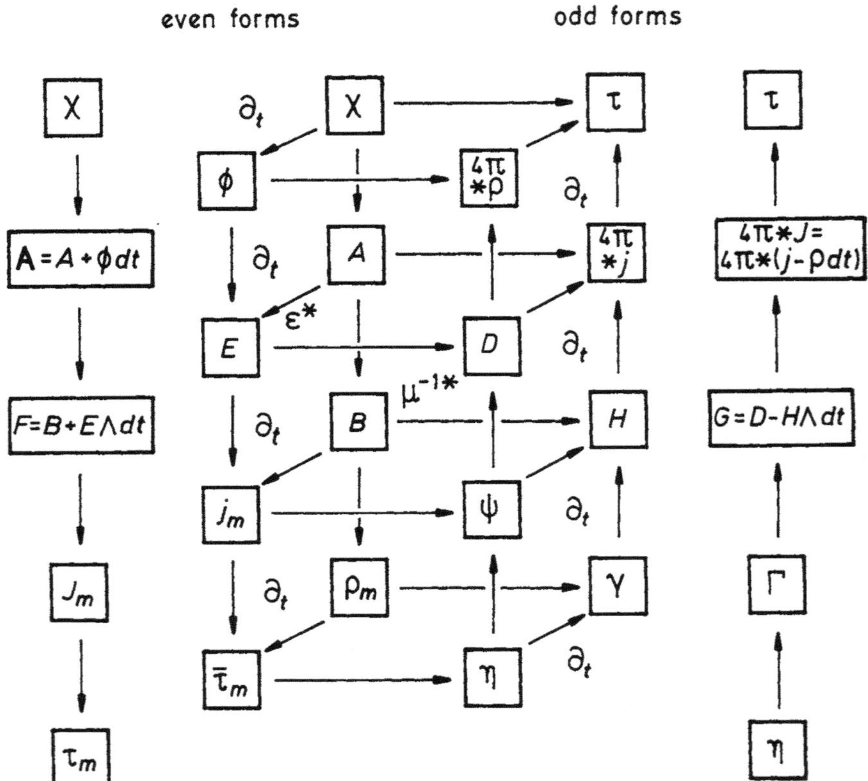

Fig. 8. Coupling of electrostatics and magnetostatics.

5. Conclusions

As mentioned, the key idea of Tonti's method is to exhibit in a formal way the analogies between different physical theories. This gives an algorithmic strategy to analyze the equations of physics. For instance, by using Tonti's diagrams one may gain understanding of a difficult theory by comparing it with a simpler or better known one. Likewise, some new aspects of a physical theory may be studied by trying to fill in the stages of the corresponding diagram.

Also, although this is not a new idea, the introduction of some notions of algebraic topology and differential geometry provides a unified language for the various physical theories, and also the means to exploit in the study of physics some well-known techniques of these mathematical branches. Such is the case of the reduction and coupling of theories discussed in this paper.

However the most important point about these diagrams is the fact that they point towards a systematic aspect of the methodology of physics, at both the theoretical and experimental levels, stemming from the mere fact that we are using mathematical models for physical theories. As a consequence, certain procedures and conservation laws are automatic within a given theory. Certainly, it will be impossible to reduce physics to a formal theory but realizing the existence of formal or formalizable aspects in science seems an important step towards a better understanding of it.

We must point out some deficiencies of Tonti's work. A most obvious critique is the old-fashioned mathematical language employed in his works which causes some notational inconveniences, and also some unnecessary limitations in both the comprehension of the ideas and the application of mathematical results. A good example of these shortcomings is the confusion arising in the definition of connection as a result of parallel transport without explaining that although neither parallelism nor connection are intrinsic notions in an arbitrary manifold, there is a canonical way to perform these operations once a Riemannian metric is given, using the Levi-Civita connection.

The problem of orientation, while fundamental because of the difference between configuration and source variables, is also unclear in Tonti's work. In particular the role of Hodge's *-operator is not mentioned, although it is systematically employed. (See for instance [12, (4)], p. 455, where a complete description of this operator is given, but regarding it as a little more than a notational trick).

On the other hand, since Tonti's papers deal only with local theories ([12, (2)], p. 57), the question arises as to the necessity of concepts which are only relevant when global problems are considered, as is the case of orientation. Recall that locally every manifold is orientable and as was mentioned, in an orientable manifold odd and even forms can *always* be identified, but not so in a nonorientable manifold; that is, nonorientability is a *global* problem. The intrinsic description of Tonti's diagrams given in this paper shows that global aspects may be incorporated into the scheme in a natural way, but the question still remains. However, there are at least two reasons for introducing global matters into this type of study:

On the one hand, there exist physical theories such as gravitation or analytical mechanics, which make considerations about the global structure of space-time. On the other hand, there are some experiments which show the existence of physical phenomena linked to a non trivial topology of the physical space where the experiment is performed; such is the case of the Aharanov-Bohm effect where one detects magnetic perturbations in electron scattering, in the absence of measurable magnetic fields, due to conductors which make the topology of space nontrivial. (See [1], for a theoretical discussion of some consequences of this experiment.)

Of course the analysis of physical theories in full generality is much more complicated and it remains to verify whether Tonti's scheme is still valid with this generality. For instance, in some cases it is not clear whether physical observables may be represented by vector-valued differential forms; for example, if the base space is not contractible, vector-bundle-valued differential forms may not be identified with vector-valued forms, so this more general theory may be required, and so on.

In a sense this shows why a postulate like P2 may not be enough to describe all physical quantities. But moreover, if we take into account the symmetry principle of modern physics, such as the general covariance principle of general relativity, we are forced to require invariance or equivariance properties of physical magnitudes which may not be satisfied by differential forms. While this type of restrictions may be incorporated into some theories, e.g. gauge theories, the formal structure of the full theory of gravitation seems to be of a more complicated nature.

Acknowledgement

The authors wish to thank Dr. O. A. Sánchez Valenzuela who read the manuscript and made some helpful comments.

References

1. A. Barut, Reports Math. Phys. **11** (1977) 415.
2. F. Branin, *The algebraic topological basis for network analysis*, Symp. on Generalized Networks, Brooklin Polyt. Press, N. Y., 1966.

3. P. A. M. Dirac, *General Theory of Relativity*, Wiley, 1975.
4. H. Flanders, *Differential Forms with Applications to the Physical Sciences*, Academic Press, 1963, pp. 44–48.
5. T. Frankel, *Gravitational Curvature*, W. H. Freeman, 1979, pp. 99–115.
6. C. W. Misner and J. A. Wheeler, Ann. of Phys. **2** (1957) 525.
7. G. de Rham, *Variétés Différentiables*, Hermann, Paris, 1955 pp. 17–27.
8. N. Schleifer, Am. J. Phys. **51** (1983) 1139.
9. B. Schutz, *Geometrical Methods of Mathematical Physics*, Cambridge Univ. Press, 1980.
10. M. Souriau, *La Structure des Systemes Dynamiques*, Dunod, Paris, 1970.
11. S. Sternberg, Springer Lect. Notes in Math. **676** (1978) 1. See also von Westenholz, op. cit. pp. 349–353.
12. E. Tonti, (1) *The algebraic topological structure of physical theories*, in Symmetry, Similarity and Group Theoretical Methods in Mechanics, eds. P. Glockner & M. Singh, American Academy of Mechanics, Calgary, 1974. (2) *On the formal structure of physical theories*, Quaderno dei Gruppi di Ricerca Matematica, CNR Italy, 1975. (3) Rendiconti del Seminario Matematico e Fisico de Milano **46** (1976) 163. (4) Appl. Math. Modelling **1** (1976) 37.
13. R. W. Wald, *General Relativity*, Univ. of Chicago Press, Chicago, 1984.
14. C. von Westenholz, *Differential Forms in Mathematical Physics*, 2nd edition, North Holland, 1983.

Ernesto A. Lacomba
Departamento de Matemáticas
Universidad Autónoma Metropolitana
Unidad Izbapalapa
09340 Mexico City, Mexico

Fausto Ongay
Centro de Investigación en Matemáticas
P.O. Box 402, 36000 Guanajuato
Mexico

THE MATH. HERITAGE OF C.F. GAUSS (pp. 454-457)
edited by George M. Rassias
©1991 World Scientific Publ. Co. Singapore

ARCHIMEDES VERSUS GAUSS:
THE CONSTRUCTION OF A REGULAR HEPTAGON

John F. Lamb, Jr.

Long ago, Archimedes devised a way to inscribe a regular heptagon in a given circle with straight edge and compass and a preliminary construction [1]. On the other hand, C. F. Gauss proved that a regular polygon of n sides can be constructed by compass and straight edge if and only if either:

$$n = 2^a$$

or

$$n = 2^a p_1 p_2 \ldots p_{r'}$$

where $p_1, p_2, \ldots, p_r$ are different prime numbers of the form

$$p_i = 2^{2^b} + 1$$

and a and b are non-negative integers [1].

Since $n = 7$ is not of this form, it should be impossible to construct a regular heptagon. After the preliminary construction, Archimedes' method uses only constructions that are possible with straight edge and compass. Therefore, there must be something "illegal" in the preliminary construction.

The preliminary construction is done as follows: Given a square ABEF, construct diagonal AE. Slide the straight edge along AB (extended) while passing through F until a point D is reached on AB so that triangle FEG and triangle BDH have the same area where G is the point where the straight edge meets AE and H is the point where the straight edge meets BE (See Fig. 1) [1]. Since the areas of triangles BDH and FEG are the same, FE·GK=BD·BH. From this basic relationship two other relationships needed in the Archemedean Heptagon Construction Method can be derived. They are

$$AB \cdot AC = BD^2$$

and

$$CD \cdot CB = AC^2$$

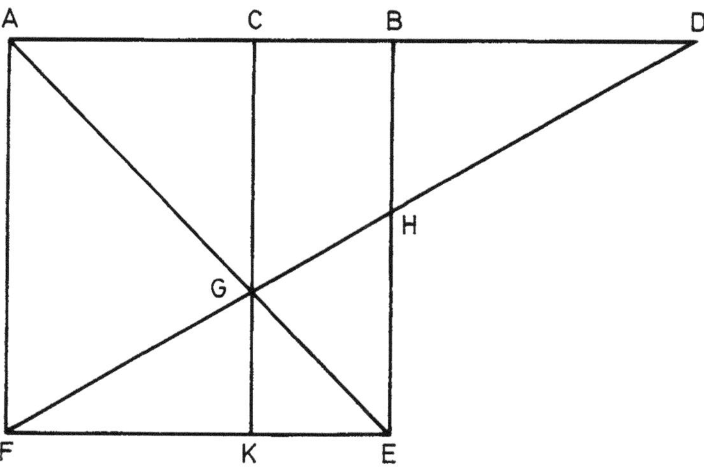

Fig. 1.

Everything in the above process follows the rules except the neusis construction where we slide the straightedge along AB. Can BD actually be constructed with straight edge and compass so that DF can be drawn? In other words, if we assume the lengths of the sides of the square are one unit, can we construct a number for the length of BD?

For easier calculations, let GK=x, HB = y and BD=z. Thus, KE=x since GKL is an isosceles right triangle, so CB= x. Also, CG=1−x, FK =1 − x and HE = 1 − y. Note that the following triangles are similar to each other: FKG, DBH and DCG.

Using triangles DCG and FKG, we get

$$(x + z)/(1 - x) = (1 - x)/x$$

so

$$x^2 + xz = (1 - x)^2 = 1 - 2x + x^2,$$

so

$$xz = 1 - 2x,$$

so

$$x(z + 2) = 1.$$

Thus

$$x = 1/(z + 2). \tag{1}$$

Using triangles DBH and FKG, we get

$$z/y = (1 - x)/x,$$

so

$$xz = y - xy.$$

The equal area condition means $yz = x$, so $y = z/x$. Substituting for y yields

$$xz = x/z - x^2/z.$$

Thus

$$xz^2 = x - x^2.$$

Dividing through by x yields

$$z^2 = 1 - x. \tag{2}$$

Equating (1) and (2), we get

$$1 - z^2 = 1/(z + 2),$$

so

$$1 = z - z^3 - 2z^2 + 2,$$

so

$$z^3 + 2z^2 - z - 1 = 0.$$

Does this cubic equation have a constructible root? If so, it must be either a rational root or an irrational root of the form $a + b\sqrt{c}$ where a, b and c are rational numbers. Since the coefficient of z^3 is one and the constant term is one, the only rational roots possible are plus or minus one. Synthetic division shows that a remainder of one is obtained when these values are used, so all the roots must be irrational. If we assume that there is a root of the form $a + b\sqrt{c}$, then there is also a root of the form $a - b\sqrt{c}$. If we let r be the third root then $r + (a + b\sqrt{c}) + (a - b\sqrt{c}) = -3$ (the negative of the coefficient of z^2), so $r + 2a = -3$. Thus $r = -3 - 2a$, so r would be a rational root contradicting the fact that there are none. Hence, none of the roots of the cubic equation in z are constructible, so z is not constructible using compass and straight edge.

This result leads one wonder how Archimedes went about finding the length z of BD. The Greeks often solved problems by intersecting conic sections [1]. Perhaps Archimedes had this in mind because we have $x = 1 - z^2$ (a parabola) and $x = 1/(z + 2)$ (a hyperbola). These conic sections intersect at three points. The z-coordinates are approximately $-2.2474, -.55498$ and $.80194$. When the positive value is used, we get $x = .35689$ and $y = .44505$. Substituting them in the equation $yz = x$ shows that they are approximately correct, so intersecting these conic sections would give a good approximation for the length of BD to use in the rest of the construction process.

Reference

1. Aaboe, Asge, *Episodes From the Early History of Mathematics*, New Mathematical Library, Vol. 13, Washington, D.C., Mathematical Association of America, 1964.

John F. Lamb, Jr.
Department of Mathematics
East Texas State University
Commerce, Texas 75428
USA

THE MATH. HERITAGE OF C.F. GAUSS (pp. 458-478)
edited by George M. Rassias
©1991 World Scientific Publ. Co. Singapore

SOME PROPERTIES OF COVARIANT OPERATORS IN GAUGE THEORIES

*Kishore B. Marathe** and *Giovanni Giachetta*

Key-Words: Jet spaces, Principal Connecetions, Gauge Fields, Gauge Transformations, Lagrangian Formalism, Covariant and Gauge Invariant Operators, Conservation Laws

We give a geometrical formulation of classical theories of gauge fields and their associated fields using the formalism of principal connections and jet bundles. We define covariant and gauge invariant operators and discuss their properties. The relation of these operators to the conservation laws associated with the classical Lagrangian formalism is also discussed.

1. Introduction

The publication of Gauss' "Disquisitiones generales cicra superficies curvas" in 1827 (an earlier version prepared in 1825 was not published until 1901) marked the creation of Differential Geometry (of surfaces) as a new field of study in geometry. In this work Gauss introduced the notion of local coordinates on a surface and thus inaugurated the intrinsic method for studying the geometrical and topological properties of a surface. In 1822 Gauss had used curvilinear coordinates to solve the fundamental problem of cartography: to map one surface onto another so that the image is similar to the original in the smallest parts. This work together with his work in geodesy and astronomy greatly influenced his work in geometry. After

---

Mathematics Subject Classification: 81E13, 83E05

*Permanant address: Department of Mathematics, Brooklyn College of CUNY, Brooklyn, N. Y. 11210, U.S.A.

introducing the first and second fundamental forms, Gauss proposed a measure for the curvature $\kappa(p)$ of the surface S at a point $p \in S$ by using the spherical image of a neighborhood of p. He then proved his famous "theorema egregium" which states that the curvature depends only on the first fundamental form. Gauss defined the integral curvature $\kappa(\Sigma)$ of a bounded surface Σ to be $\int_\Sigma \kappa d\sigma$. He computed $\kappa(\Sigma)$ when Σ is a geodesic triangle to prove his celebrated theorem:

$$\kappa(\Sigma) = A + B + C - \pi, \tag{1}$$

where, A, B, C are the angles of the geodesic triangle Σ. Gauge was aware of the significance of Eq. (1) in relation to the study of the Euclidean parallel postulate. He was interested in surfaces of constant curvature and mentioned a surface of revolution of constant negative curvature, namely, a pseudosphere. The geometry of the pseudosphere turns out to be the non-Euclidean geometry of Lobachevaski-Bolyai. Gauss had discovered several propositions of non-Euclidean geometry. But true to his motto "Pauca sed matura", he did not publish his results. Equation (1) is a special case of the well-known Gauss-Bonnet theorem. When applied to a compact surface Σ the Gauss-Bonnet theorem states that

$$\int_\Sigma \kappa d\sigma = 2\pi\chi(\Sigma), \tag{2}$$

where $\chi(\Sigma)$ is the Euler characteristic of Σ. The left-hand side of the above Eq. (2) is obtained by using the differential structure of Σ, while the right-hand side depends only on the topology of Σ. Thus Eq. (2) is a relation between geometric and topological invariants of Σ. This result admits far reaching generalizations. In particular, it can be regarded as a prototype of an index theorem. Gauss-Bonnet theorem was generalized to Riemannian polyhedra by Allendoerffer and Weil and to arbitrary manifolds by Chern. In this generalization the Gaussian integral curvature is replaced by an invariant formed from the Riemann curvature and forms the starting point of the theory of characteristic classes.

Gauss' idea of studying the geometry of a surface without leaving it, i.e., by means of measurements made on the surface itself is of fundamental importance in modern differential geometry and its applications to physical theories. We are forced to study the geometry of the three dimensional physical world by the intrinsic method, i.e., without leaving it. This idea was already implicit in Riemann's work which extended Gauss' intrinsic method to the study of manifolds of arbitrary dimension. This work together with the work of Ricci and Levi-Civita provided the foundation for Einstein's theory of general relativity. A differential geometric formulation of Einstein's equations and their generalization is given in [MA 7], [MA 8], [MA 9].

Gauss practiced mathematics in its broadest sense. His contributions range from the purest of pure mathematics (number theory) to computational mathematics. His work has had a profound influence on the development of modern mathematics and it will continue to inspire the future practitioners of this great art and science. We are happy to dedicate this paper to the memory of Gauss, "The Prince of Mathematicians".

2. Principal Connections and Yang-Mills Fields

The theory of gauge fields and their associated fields, such as the Yang-Mills-Fields, was developed by physicists to explain and unify the fundamental forces of nature. The theory of connections in a principal fiber bundle and characteristic classes was developed by mathematicians, during approximately the same period, but the fact that they are closely related was not noticed for many years. Since then substantial progress has been made in understanding this relationship and in applying it successfully to problems in both physics and mathematics. For further details and notation see [MA 10].

We begin with a definition of connection in an arbitrary fiber bundle and then consider the special case of principal connections. Let E be a fiber bundle over B with projection $p : E \to B$. Let VE denote the vertical vector bundle over E. VE is a subbundle of the tangent bundle TE, the fiber $V_u E, u \in E$ being the tangent space to the fiber $E_{p(u)}$ of E passing through u. Let $p^*(TB)$ be the pull back of the tangent bundle of B to E. Then we have the following short exact sequence of vector bundles and morphisms over E:

$$0 \longrightarrow VE \xrightarrow{i} TE \xrightarrow{a} p^*(TB) \longrightarrow 0 \tag{3}$$

where i is the injection of the vertical bundle into the tangent bundle, and a is defined by $(e, X_e) \mapsto (e, p_*(X_e))$. We define a connection on E to be a splitting of the above exact sequence, i.e., a vector bundle morphism

$c : p^*(TB) \to TE$ such that $a \circ c = \mathrm{id}_{p^*(TB)}$. Let $b \in B$ and let $e \in p^{-1}(b)$. Then the splitting c induces an injection $\hat{c}_e : T_b(B) \to T_e(E), X \mapsto c(e, X)$. We call $\hat{c}_e(X)$ the *horizontal lift* of the tangent vector X to $e \in E$. We call $\hat{c}_e(T_b(B))$ the *horizontal space at $e \in E$* and denote it by $H_e E$. The spaces $H_e E$ are the fibers of the *horizontal bundle HE* and we have the decomposition

$$TE = HE \oplus VE.$$

Alternatively, a connection may be defined as a section of the first jet bundle $J_1(E)$ over E. The definitions of covariant derivative, covariant differential and curvature can be formulated in this general context. An introduction to this approach and its physical applications is given in [MA 4], [MA 5], [MA 6], [MO 1].

Let M be an m-dimensional manifold and $P(M, G)$ a principal bundle with the gauge group G as its structural group. If $E = P(M, G)$, then we can recover the usual definition of connection as follows. The action of G on P extends to all the vector bundles in the exact sequence (3) to give the following short exact sequence of vector bundles over M:

$$0 \longrightarrow VP/G \overset{i}{\longrightarrow} TP/G \overset{\hat{a}}{\longrightarrow} \pi^*(TM)/G \longrightarrow 0$$

We can rewrite the above sequence as follows:

$$0 \longrightarrow \mathrm{ad}(P) \overset{i}{\longrightarrow} T_G P \overset{\hat{a}}{\longrightarrow} TM \longrightarrow 0 \tag{4}$$

A connection on P is then defined as a splitting of the short exact sequence (4). This splitting induces a splitting of the sequence (3) and we recover the definition of the connection given earlier. We call such a connection a principal connection. This approach helps to clarify the role of the structure group in the usual definition of connection in a principal bundle. Given a principal connection we can define a 1-form ω on P with values in the Lie algebra $\mathbf{g}$ which characterizes the connection. It is called the connection 1-form and is traditionally used to define the principal connection.

In physical applications one is usually interested in a fixed Lie group G called the *gauge group* which represents an internal or local symmetry of the field. The base manifold M of the principal bundle $P(M, G)$ is usually the space-time manifold or its Euclidean version, i.e., a Riemannian manifold of dimension 4. But in some physical applications, such as superspace,

Kaluza-Klein and string theories, the base manifold can be an essentially arbitrary manifold.

A connection in P is called a *gauge connection*. The connection 1 form ω is called the *gauge connection form* or simply the *gauge connection*. A *global gauge* or simply a *gauge* is a section $s \in \Gamma(P)$. The gauge potential A on M in gauge s is given by $A = s^*(\omega)$. A global gauge and hence the gauge potential on M exists if and only if the bundle P is trivial. A *local gauge* is a section of the bundle $P(M,G)$ restricted to some open subset $U \subset M$. Local gauges defined for the local representations $(U_i, \psi_i)_{i \in I}$ of P, always exist. Let $t \in \Gamma(U_i, P)$ be a local gaue; then the 1-form $t^*\omega \in \Lambda^1(U_i, \mathbf{g})$ is called the *gauge potential* in the local gauge t and is denoted by A_t. If the local gauge t is given we often denote A_t by A and call it a *local gauge potential*. A connection ω on P defines a covariant derivative

$$\nabla^\omega : \Lambda^0(M, \mathrm{ad}\, P) \to \Lambda^1(M, \mathrm{ad}\, P).$$

We denote by the same symbol the extension of the covariant derivative

$$\nabla^\omega : \Lambda^k(M, \mathrm{ad}(P)) \to \Gamma(T^*M \otimes \Lambda^k T^*M \otimes \mathrm{ad}\,(P)).$$

The corresponding covariant exterior derivative is denoted by d^ω

$$d^\omega : \Lambda^k(M, \mathrm{ad}\,(P)) \to \Lambda^{k+1}(M, \mathrm{ad}\,(P)),$$

and its adjoint is denoted by δ^ω

$$\delta^\omega : \Lambda^k(M, \mathrm{ad}\,(P)) \to \Lambda^{k-1}(M, \mathrm{ad}\,(P)).$$

Let $\Omega = d^\omega \omega$ be the curvature 2-form of ω with values in the Lie algebra $\mathbf{g}$. We call Ω the *gauge field* on P. Although this terminology is fairly standard, we would like to warn the reader that sometimes, in the physics literature, our gauge potential is called the gauge field and our gauge field is called the field strength tensor. As is well known there exists a unique 2-form F_ω on M with values in the Lie algebra bundle $\mathrm{ad}(P)$ associated to the curvature 2-form Ω. The 2-form $F_\omega \in \Lambda^2(M, \mathrm{ad}\,(P))$ is called the *gauge field* on M corresponding to the gauge connection ω. The gauge field F_ω is globally defined on M, even though, in general, there is no corresponding globally defined gauge potential on M. If we are given a local gauge potential $A_t \in \Lambda^1(U_i, \mathbf{g})$, and a local trivialization of ad P over U_i, then on U_i we have the relations

$$t^*\omega = A_t \text{ and } F_\omega = d^\omega A_t \,. \tag{5}$$

We denote by $\mathcal{A}(P)$ the *space of gauge connections* on P defined by

$$\mathcal{A}(P) := \{\omega \in \Lambda^1(P, \mathbf{g}) |\ \omega \text{ is a connection on } P\} \,. \tag{6}$$

If P is fixed we will denote $\mathcal{A}(P)$ simply by $\mathcal{A}$ and a similar notation will be followed for other related spaces. The space $\mathcal{A}$ is an affine space of the underlying vector space $\Lambda^1(M, \text{ad } P)$. Indeed from the definition of $\mathcal{A}$, it follows that $\omega_1, \omega_2 \in \mathcal{A}$ implies that $\omega_1 - \omega_2$ is horizontal and therefore defines a unique 1-form on M with values in the bundle ad P. Then we have that, for a fixed connection α

$$\mathcal{A} \cong \{\alpha + \pi^* A | A \in \Lambda^1(M, \text{ad } P)\} \,. \tag{7}$$

In our work we need to consider the k jet extensions $J_k\mathcal{A}$ of the space of gauge connections $\mathcal{A}$ regarded as fiber bundles over M. We define the space $\Lambda^i(J_k\mathcal{A})$ of *horizontal i-forms* on $J_k\mathcal{A}$ as the space of bundle morphisms $J_k\mathcal{A} \to \Lambda^i(T^*M)$. Let $E(M, V_r, r, P)$ be a vector bundle associated to P by the representation $r : G \to \text{End}(V_r)$. Then we define the space $\Lambda^i(J_k\mathcal{A}; E)$ of *horizontal E-value i-forms* on $J_k\mathcal{A}$ as the space of bundle morphisms $J_k\mathcal{A} \to \Lambda^i(T^*M) \otimes E$. If $E = \text{ad}(P)$, we call these forms the *adjoint, horizontal* forms. For example, by considering the canonical affine splitting of $J_1\mathcal{A}$ over $\mathcal{A}$ we obtain an adjoint, horizontal 2-form $\hat{F}$. We call $\hat{F}$ the *unversisal curvature form*. We denote by $d^\sharp$ the jet shift exterior differential ([MA 1], [MA 2])

$$d^\sharp : \Lambda^i(J_k\mathcal{A}\,; E) \to \Lambda^{i+1}(J_{k+1}\mathcal{A}\,; E) \,.$$

We note that $\mathcal{A}$ is an affine space modelled after the vector space $\Lambda^1(M, \text{ad } P) = T^*M \otimes \text{ad } P$. Thus the tangent space $T_\alpha\mathcal{A}$ is isomorphic to $\Lambda^1(M, \text{ad } P)$ and we identify the two spaces. If $\langle,\rangle_\mathbf{g}$ is an invariant inner product on $\mathbf{g}$, we have a natural inner product defined on $T_\alpha\mathcal{A}$ as follows

$$\langle A, B \rangle_\alpha = \int_M g^{ij} \langle A_i, B_j \rangle_\mathbf{g} \,, \quad \forall A, B \in T_\alpha\mathcal{A} \tag{8}$$

where g^{ij} are the components of the metric tensor g on M with respect to a base dx^i of $T_x^* M$ and $A = A_i dx^i, B = B_i dx^i$. We observe that an invariant inner product always exists for semisimple Lie algebras and is given by a multiple of the Killing form K on $\mathbf{g}$ given by

$$K(x, y) = -\mathrm{Tr}\,(\mathrm{ad}_x\, , \mathrm{ad}_y)\,.$$

The inner product defined in (8) can be easily extended to $\Lambda^k(M, \mathrm{ad}\, P)$. We now assume that metrics are chosen on M and the bundles over M and the norm is defined on sections of these bundles as an L^2-norm if M is not compact. The *Yang-Mills Lagrangian* $\mathcal{L}_{YM}$ is defined by

$$\mathcal{L}_{YM}(\omega) = \frac{1}{8\pi^2}\|F_\omega\|^2\,. \tag{9}$$

The *Yang-Mills action* $\mathcal{A}_{YM}$ is defined by

$$\mathcal{A}_{YM}(\omega) = \frac{1}{8\pi^2}\int_M \|F_\omega\|^2\,. \tag{10}$$

The corresponding pure *Yang-Mills equation* is

$$\delta^\omega F_\omega = 0 \tag{11}$$

which is equivalent to

$$d^\omega * F_\omega = 0\,. \tag{12}$$

Note that in this case there is only one nontrivial Bianchi identity

$$d^\omega F_\omega = 0\,. \tag{13}$$

This identity is a consequence of the Cartan structure equations and expresses the fact that locally, F is derived from a potential. It is customary to consider the pair (11) and (13) or (12) and (13) as the Yang-Mills equations. This is consistent with the fact that they reduce to the Maxwell equations for the electromagnetic field F when the gauge group G is $U(1)$ and M is the Minkowski space. A non-trivial example of a Yang-Mills field is provided by the Dirac Monopole.

Example. (*The Dirac Monopole*) Let $S^3(S^2, U(1))$ be the principal $U(1)$-bundle over S^2 determined by the Hopf fibration of S^3. Let μ denote the connection 1-form of the canonical connection on this bundle and let F_μ be the corresponding gauge field on S^2. In this case there is no globally defined gauge potential on S^2. We need at least two charts to cover S^2 and therefore, at least two local potentials which give rise to a single globally defined gauge field. This field can be shown to be equivalent to the Dirac monopole field. The Dirac monopole quantization condition corresponds to the classification of principal $U(1)$-bundles over S^2. These are classified by $\pi_1(U(1)) \cong \mathbf{Z}$. In general, the principal G-bundles over S^2 are classified by $\pi_1(G)$. Thus $\pi_1(SU(2)) = \mathrm{id}$ implies that there is a unique $SU(2)$-monopole on S^2 and $\pi_1(SO(3)) = \mathbf{Z}_2$ implies that there are two inequivalent $SO(3)$-monopoles on S^2.

Gauge potentials and gauge fields acquire physical significance only after one postulates the field equations to be satisfied by them. These equations and their consequences must then be subjected to suitably devised experiments for verification. On more than one occasion a theory was abandoned, when its predictions seemed to contradict an experimental result and later this experiment or its conclusions turned out to be incorrect and the abandoned theory turned out to be correct. In any case there is no natural mathematical method for assigning field equations to gauge fields. Thus the Riemann curvature of a space-time manifold M is the gauge field corresponding to the gauge potential given by the Levi-Civita connection on the orthonormal frame bundle of M, but it does not describe the gravitational field until it is subjected to Einstein's field equations. If instead it satisfies Yang-Mills equations, then it describes a classical Yang-Mills field. This aspect of gauge fields is already evident in the following remark of Yang [YA 1]: "The electromagnetic field is a gauge field. Einstein's gravitational theory is intimately related to the concept of gauge fields, although to *identify* the gravitational field as a gauge field is not an absolutely straightforward matter." However, a study of physically interesting field equations such as Maxwell's equations of electro-magnetic field and their quantization indicates some describe features for the gauge field equations. One of these features is gauge invariance of the field equations. This requirement is formulated in terms of the group of gauge transformations which acts on the various fields involved. In the following paragraph we give a mathematical formulation of this group.

3. Generalized Gauge Transformations

The group $\mathrm{Diff}(P)$ of the diffeomorphisms of P is too large to serve as a group of gauge transformations, since it mixes up the fibers of P. The requirement that fibers map to fibers may be expressed by the condition that the following diagram commutes.

$$
\begin{array}{ccc}
P & \xrightarrow{\;f\;} & P \\
{\scriptstyle \pi}\downarrow & & \downarrow{\scriptstyle \pi} \\
M & \xrightarrow[f_M]{} & M
\end{array}
$$

When this condition is satisfied, we say that $f \in \mathrm{Diff}(P)$ covers $f_M \in \mathrm{Diff}(M)$. The set $\mathrm{Diff}_M(P)$ of all $f \in \mathrm{Diff}(P)$ such that f covers some $f_M \in \mathrm{Diff}(M)$, is a group called the *group of generalized gauge transformations*. We note that the fiber preserving property of f completely determines the diffeomorphism f_M. The group of generalized gauge transformations seems to be the appropriate group for considering covariance and gauge invariance of Lagrangians and operators corresponding to coupled fields. We define the group $\mathcal{G}$ to be the subgroup $\mathrm{Aut}(P) \subset \mathrm{Diff}_M(P)$ of principal bundle automorphisms of P. Thus

$$
\mathcal{G} := \mathrm{Aut}\,(P) = \{f \in \mathrm{Diff}_M(P) | f_M = \mathrm{id}_M\} . \tag{14}
$$

Then $\mathcal{G}$ is a normal subgroup of $\mathrm{Diff}_M(P)$. We call $\mathcal{G}$ the *group of gauge transformations*. From the definition (8) it is clear that the group $\mathcal{G}$ maps each fiber of P into itself and we have the following exact sequence of groups.

$$
1 \longrightarrow \mathcal{G} \xrightarrow{\;i\;} \mathrm{Diff}_M(P) \xrightarrow{\;j\;} \mathrm{Diff}(M) \longrightarrow 1 ,
$$

where i denotes the inclusion map and j is defined by

$$
j(f) = f_M , \quad \forall f \in \mathrm{Diff}_M(P) .
$$

In several applications one is interested in splitting the above exact sequence or in constructing an extension of $\mathrm{Diff}(M)$ by $\mathcal{G}$. In particular, we often want to lift the action of $\mathrm{Diff}(M)$ to some subgroup of $\mathrm{Diff}_M(P)$. Additional geometric structures may also be involved in this process. For example, if $P = L(M)$, the bundle of frames of M, then it is a principal bundle but carries the additional structure given by the soldering form θ and we have the following proposition.

Proposition 1. There exists a natural lift $\lambda : \mathrm{Diff}(M) \to \mathrm{Diff}(L(M))$ which splits the exact sequence of groups

$$1 \longrightarrow \mathcal{G} \xrightarrow{\ i\ } \mathrm{Diff}_M(L(M)) \xrightarrow{\ j\ } \mathrm{Diff}(M) \longrightarrow 1 \,.$$

Furthermore, $f \in \mathrm{Diff}(L(M))$ is the natural lift of a diffeomorphism $f_M \in \mathrm{Diff}(M)$ (i.e., $f = \lambda(f)$) if and only if f leaves the soldering form invariant, i.e., $f^*\theta = 0$.

When M is a four dimensional Lorentz manifold, connections on the frame bundle $L(M)$ play the role of geravitational potentials. Action functionals involving connections and metrics on M form the starting point of gauge theories of gravitation.

The group $\mathcal{G}$ acts on the space of gauge connections $\mathcal{A}(P)$ by pulling back the connection form, i.e.,

$$(f,\omega) \mapsto f \cdot \omega = f^*\omega \,, \quad f \in \mathcal{G}, \, \omega \in \mathcal{A}(P) \,. \tag{15}$$

This action extends to the associated bundles and various jet spaces. We shall also be interested in the local (infinitesimal) version of these actions. We say that connections $\alpha, \beta \in \mathcal{A}$ are *gauge equivalent* if there exists a gauge transformation $f \in \mathcal{G}$ such that $\beta = f \cdot \alpha$. From the definition of the action of $\mathcal{G}$ on $\mathcal{A}$ given above, it follows that each equivalence class of gauge equivalence connections is an orbit of $\mathcal{G}$ in $\mathcal{A}$. The orbit space $\mathcal{O} = \mathcal{A}/\mathcal{G}$ thus represents gauge inequivalent connections and is called the *moduli space of gauge connections* on $P(M,G)$.

A physical interpretation of a gauge transformation $f \in \mathcal{G}$ is that f is a local (pointwise) change of gauge over M. For this reason it is sometimes called a local gauge group, but we will not use this terminology. There are several alternative definitions of the group of gauge transformations. We have collected together the most frequently used definitions in the following theorem.

Theorem 2. There exists a one-to-one correspondence between each pair of the following three sets:

(i)   the group of gauge transformations $\mathcal{G}$,

(ii)   the set $\mathcal{F}_G(P, G)$ of all functions $f : P \to G$ such that f is G-equivariant, with respect to the adjoint action of G on itself,

(iii) the set of sections $\Gamma(P \times_{Ad} G)$ of the associated bundle $(P \times_{Ad} G)$ over M.

In view of this theorem we use any one of the three representations above for the group of gauge transformations as needed. For example, regarding $\mathcal{G}$ as the space sections of $(P \times_{Ad} G)$, the bundle of groups (not a principal G-bundle), we can show that a suitable Sobolev completion of $\mathcal{G}$ (also denoted by $\mathcal{G}$) is a Banach Lie group (i.e., $\mathcal{G}$ is a Banach manifold with smooth group operations). Let ad denote the adjoint action of the Lie group G on its Lie algebra **g**. Let $E(M, \mathbf{g}, \mathrm{ad}\, P)$ be the associated vector bundle with fiber type **g** and action ad, the adjoint action of G on **g**. Recall that this bundle is a bundle of Lie algebras denoted by $P \times_{ad} \mathbf{g}$ or ad P. We denote $\Gamma(\mathrm{ad}\, P)$ by $\mathcal{LG}$; it is a Lie algebra under pointwise bracket and pointwise exponential map to $\mathcal{G}$. We will show in the next section that a suitable Sobolev completion of $\mathcal{LG}$ is a Banach Lie algebra which is the Lie algebra of the infinite dimensional Banach Lie group $\mathcal{G}$.

4. Associated Fields and Coupled Equations

Let $E(M, F, r, P)$ be a vector bundle associated to the principal bundle $P(M, G)$. We call a section $\phi \in \Gamma(E)$ an *E-potential* (or a generalized Higgs potential) and $d^\omega \phi$ an *E-field* (or a generalized Higgs field) associated to the gauge connection ω. In physical applications $d^\omega \phi$ represents the interaction of gauge and matter fields by minimal coupling. An example of this is given later. If $E = \mathrm{ad}(P) := P \times_{ad} \mathbf{g}$ then ϕ is called the *Higgs potential* and $d^\omega \phi$ the *Higgs field* associated to the gauge connection ω. We call ad (P) the *Higgs bundle*. In general there are several fields that can be defined on bundles associated with a given manifold. For example, on a Lorentz 4-manifold the Levi-Civita connection is interpreted as representing a gravitational potential. Recall that the Levi-Civita connection is the unique torsion-free connection defined on the bundle of orthonormal frames $\mathcal{O}(M)$ of M. In general if M is a pseudo-Riemannian manifold, we can define the space of linear connections $\mathcal{A}(\mathcal{O}(M))$ on M by

$$\mathcal{A}(\mathcal{O}(M)) := \{\alpha \in \Lambda^1(\mathcal{O}(M), \mathrm{so}\,(m)) | \alpha \text{ is a connection on } \mathcal{O}(M)\}\,.$$

Generalized theories of gravitation often use this $\mathcal{A}(\mathcal{O}(M))$ as their configuration space. If M is a spin manifold and $S(M)(M, \mathrm{Spin}(m))$ the Spin (m)-principal bundle, one can consider the space $\mathcal{A}(S(M))$ of spin connections on the bundle $S(M)$, defined by

$$\mathcal{A}(S(M)) := \{\beta \in \Lambda^1(S(M), \mathrm{Spin}(m)) | \beta \text{ is a connection on } S(M)\}.$$

For a given signature (p, q) we may consider the space $\mathcal{RM}_{(p,q)}(M)$ of all pseudo-Riemannian metrics on M of signature (p, q). The space $\mathcal{RM}_{(m,0)}(M)$ of Riemannian metrics on M is denoted simply by $\mathcal{RM}(M)$. Recall that there is a canonical principal GL $(m, \mathbf{R})$-bundle over M, namely $L(M)$ the bundle of frames of M. If $\rho : \mathrm{GL}(m, \mathbf{R}) \to \mathrm{End}\, V$ is a representation of $\mathrm{GL}(m, \mathbf{R})$ on V and $E(M, V, \rho, L(M))$ is the corresponding associated bundle of $L(M)$, then we denote by $\mathcal{W}$ the space of E-potentials $\Gamma(E)$, i.e.,

$$\mathcal{W} = \Gamma(E(M, V, \rho, L(M))).$$

Thus we see that we have an array of fields on a given base manifold M and we must specify the equations governing the evolution and interactions of these fields and study their physical meaning. There is no standard procedure for doing these things. In many physical applications one obtains the coupled field equations of interacting fields as the Euler-Lagrange equations of a variational problem with the Lagrangian constructed from the fields. For any given problem the Lagrangian is chosen subject to certain invariance or covariance requirements related to the symmetries of the fields involved. We now discuss three general conditions for Lagrangians that are frequently imposed in physical theories.

We assume that the field equations are the variational equations of an action integral defined by a Lagrangian L on the configuration space with values in $\Lambda^4(M)$. When a fixed volume form such as the metric volume form is given we regard L as a real valued function. We shall use any one of these conventions without comment. The action $\mathcal{E}$ is given by

$$\mathcal{E}(g, \phi, \omega, \psi) = \int_M L(g, \phi, \omega, \psi).$$

There are various groups associated with the geometric structures involved in the construction of these fields, which have natural actions on them. We note the following exact sequence of groups:

$$0 \to \mathcal{G} \to \mathrm{Diff}_M(P) \to \mathrm{Diff}(M).$$

We shall require the Lagrangian to satisfy the following conditions:
i) Naturality, ii) Local regularity, iii) Conformal invariance.

i) *Naturality*: Naturality with respect to the group of generalized gauge transformations is defined as follows. Let $F \in \mathrm{Diff}_M(P)$ be a generalized transformation covering $f \in \mathrm{Diff}(M)$. Then by naturality with respect to $\mathrm{Diff}_M(P)$ we mean that

$$L(f^*g, f_\rho^*\phi, F^*\omega, F_r^*\psi) = f^*L(g, \phi, \omega, \psi), \tag{16}$$

where f_ρ^* is the induced action of f on $\mathcal{W}$, and F_r^* is the action induced by the generalized gauge transformation F on $\mathcal{H}$. In the absence of the principal bundle P this condition reduces to naturality with respect to $\mathrm{Diff}(M)$ and is Einstein's condition of general covariance of physical laws derived from the Lagrangian formalism. Further in the absence of $\mathcal{W}(M)$ this condition corresponds to the covariance of gravitational field equations when the Lagrangian is taken to be the standard Einstein-Hilbert Lagrangian. If we require naturality with respect to the group $\mathcal{G}$ then the condition (16) is precisely the principle of gauge invariance introduced by Weyl. Since in this case $f = \mathrm{id}$ the condition (16) becomes

$$L(g, \phi, F^*\omega, F_r^*\psi) = L(g, \phi, \omega, \psi).$$

The concept of natural tensors on a Riemannian manifold was introduced in [EP 1] and was extended to oriented Riemannian manifolds in [ST 1] where a functional formulation of naturality is given and a complete classification of natural tensor fields is given under some regularity conditions.

ii) *Local regularity*: Given any coordinate chart on M and a local gauge we can express the various potentials and fields with respect to induced bases. We require that in this system the Lagrangian be expressible as a universal polynomial in

$$(\det g)^{-1/2}, (\det h)^{-1/2}, g_{ij}, \partial^{|\alpha|}g_{ij}/\partial x^\alpha, \phi_{|\beta|}, \omega_{|\gamma|}, \psi_{|\delta|} \cdots$$

where $\alpha, \beta, \gamma, \delta \ldots$ are suitable multi-indices (i.e., in the coefficients and derivatives of the potentials and fields in the induced bases). In physical applications one often restricts the order of derivatives that can occur to 2. For example in gravitation one considers natural tensors satisfying the conditions that they contain derivatives up to order 2 and depend linearly on the second order derivatives. Then it is well known that such tensors can be expressed as

$$c_1 R^{ij} + c_2 g^{ij} S + c_3 g^{ij} \, ,$$

where R^{ij} are the components of the Ricci tensor and S is the scalar curvature. Einstein's equations with or without the cosmological constant involve the above combination with suitable value of the constants c_1, c_2, c_3. Applying the classification theorem of [ST 1] to $SO(4)$ actions on the metric and gauge fields, we get the following general form for the Lagrangian:

$$
\begin{aligned}
L(g,\omega) =\,& c_1 S^2 + c_2 \|K\|^2 + c_3 \|W^+\|^2 + c_4 \|W^-\|^2 \\
& + c_5 \|F_\omega \wedge F_\omega\| + c_6 \|F_\omega \wedge (*F_\omega)\| ,
\end{aligned}
\tag{17}
$$

where S, K, W^+, W^- are the $SO(4)$-invariant components of the Riemannian curvature and F_ω is the gauge field of the gauge potential ω. For a suitable choice of constants in the above Lagrangian we obtain various topological invariants of M and P as well as the pure Yang-Mills action. For example the first Pontryagin class of M is given by

$$p_1(M) = \frac{1}{4\pi^2} \int_M (\|W^+\|^2 - \|W^-\|^2) \, .$$

The first Pontryagin class of P is given by

$$p_1(P) = \frac{1}{8\pi^2} \int_M (\|F_\omega^+\|^2 - \|F_\omega^-\|^2) \, ,$$

which turns out to be the instanton number or the topological quantum number associated to the bundle P.

To satify the conditions of naturality and local regularity for fields coupled to gauge fields physicists often start with ordinary derivatives of associated fields and the coupling is achieved by replacing these by gauge covariant derivatives in the Lagrangian. This is called the principle of minimal coupling (or interaction). These two requirements can also be formulated by taking the Lagrangian to be defined on sections of suitable jet bundles on the space of connections. Using this approach a generalization of the classical theorem of Utiyama has been obtained in [GA 1] and [MA 3].

iii) *Conformal invariance*: The condition of conformal invariance of the Lagrangian may be expressed as follows

$$L(e^{2\sigma}g, \phi, \omega, \psi) = L(g, \phi, \omega, \psi) \, , \quad \forall \sigma \in \mathcal{F}(M) \, .$$

In general, Lagrangians satisfying the conditions of naturality and regularity need not satisfy the condition of conformal invariance. This condition is often used to select parameters such as the dimension of the base space and the rank of the representation. A particular case of (17) is the Yang-Mills Lagrangian with action

$$\mathcal{E} = \frac{1}{8\pi^2} \int_M F_\omega \wedge *F_\omega \, .$$

It is an example of Lagrangian that is conformally invariant only if the dimension of M is 4.

A large number of Lagrangians satisfying the naturality and regularity requirements are used in the physics literature. They are broadly classified into bosonic Lagrangians and fermionic Lagrangians, depending on the absence or presence of spin structures and their associated fields. For example a bosonic Lagrangian is given by

$$L_{\text{boson}}(g,\omega,\psi) = \|F_\omega\|^2 + \|\nabla\psi\|^2 + \frac{1}{6}S\|\psi\|^2 - V(\psi) \, ,$$

where V is the potential function which is taken to be a gauge invariant polynomial of degree ≤ 4 on the fibers of E. If M is a spin manifold and if Σ is a bundle associated to the spin bundle, then we define an E-valued spinor to be a section of $\Sigma \otimes E$. The fermion Lagrangian is defined by

$$L_{\text{fermion}}(g,\omega,\xi) = \|F_\omega\|^2 + \langle \mathcal{D}(\xi),\xi \rangle \, ,$$

where $\xi \in \Gamma(\Sigma \otimes E)$ and $\mathcal{D}$ is the Dirac operator on E-valued spinors. Several important properties of coupled field equations are studied in [PA 1].

5. Covariant and Gauge Invariant Operators

Of the three conditions for Lagrangians discussed in the previous section we now consider in detail some applications of the first condition. Further applications of this condition as well as those of the other conditions will be taken up in a later work. In what follows we restrict ourselves to a fixed, 4-manifold M as the base manifold, but the discussion can be easily extended to apply to an arbitrary base manifold. In most physical applications, M will be a space-time manifold and we will be concerned with the space of Lorentz metrics. Let $P(M, G)$ be a principal bundle over M whose structure group G carries a bi-invariant metric h. For example if G is a semisimple Lie group, then a suitable multiple of the Killing form on $\mathbf{g}$ provides a bi-invariant metric on G.

We want to consider coupled field equations for a metric $g \in \mathcal{RM}(M)$, a connection $\omega \in \mathcal{A}(P)$ and a generalized Higgs potential $\psi \in \mathcal{H} = \Gamma E(M, V_r,$

$r, P)$, where $r : G \to \mathrm{End}(V_r)$ is a representation of the gauge group G. Thus our *configuration space* is defined by

$$\mathcal{C} := \mathcal{RM} \times \mathcal{A} \times \mathcal{H}.$$

We note that $\mathcal{RM}$ is an open submanifold of $\vee^2 TM$, the symmetric tensor product of $TM, \mathcal{A}$ is an affine space modelled on the vector space $T^*M \otimes \mathrm{ad}(P)$ and $\mathcal{H}$ is a finite dimensional vector bundle over M. Thus the space $\mathcal{C}$ is a finite dimensional fiber bundle over M. We are interested in considering Lagrangians and Euler-Lagrange type operators defined on the sections of the configuration space and their jet extensions. In order to avoid topological restrictions, we consider only infinitesimal (local) actions, i.e., we consider the local version of Eq. (16) in defining covariance and gauge invariance. In what follows all considerations are local unless otherwise stated. Let ξ be a section of $T_G P$ over M and let u_ξ be the associated vector field on $\mathcal{C}$. We note that ξ determines a vector field u over M and that u_ξ projects onto this vector field. Let $J_k \mathcal{C}$ be the k-th jet extension of $\mathcal{C}$. Then u_ξ can be lifted to a vector field on $J_k \mathcal{C}$ which we denote by the same symbol. This extension is needed in order to define covariance and gauge invariance of Lagrangians and operators defined on these jet spaces. Let $\mathcal{L} : J_1 \mathcal{C} \to \Lambda^m T^* M$ be a Lagrangian density. We say that L is *covariant* if

$$L_{u_\xi} \mathcal{L} = 0, \ \forall \xi : M \to T_G P. \tag{18}$$

We say that $\mathcal{L}$ is *gauge invariant* if

$$L_{u_\xi} \mathcal{L} = 0, \ \forall \xi : M \to \mathrm{ad}(P) \subset T_G P. \tag{19}$$

A morphism $\mathcal{E} : J_2 \mathcal{C} \to \Lambda^m T^* M \otimes V^* \mathcal{C}$ over M is said to be a *covariant operator* if

$$L_{u_\xi} \mathcal{E} = 0, \ \forall \xi : M \to T_G P. \tag{20}$$

We say that $\mathcal{E}$ is a *gauge invariant operator* if

$$L_{u_\xi}\mathcal{E} = 0\,, \quad \forall \xi : M \to \mathrm{ad}\,(P) \subset T_G P\,. \tag{21}$$

A morphism $\mathcal{E} : J_2 C \to \Lambda^m T^* M \otimes V^* C$ over M is called an operator of *Euler-Lagrange type* if it is associated to a sheaf of local Lagrangian densities $\mathcal{L}$. We observe that if $\mathcal{L}$ is covariant (gauge invariant) then the associated Euler-Lagrange operator $\mathcal{E}$ is also covariant (gauge invariant) but the converse is not true. In general, we can decompose the operator $\mathcal{E}$ into three parts corresponding to the product decomposition of the configuration space as follows:

$$\mathcal{E} = \mathcal{E}_1 \oplus \mathcal{E}_2 \oplus \mathcal{E}_3\,, \tag{22}$$

where

$$\mathcal{E}_1 : J_2 C \to \Lambda^m T^* M \otimes \mathsf{V}^2 T^* M\,, \tag{23}$$

$$\mathcal{E}_2 : J_2 C \to \Lambda^m T^* M \otimes TM \otimes \mathrm{ad}^*(P)\,, \tag{24}$$

and

$$\mathcal{E}_3 : J_2 C \to \Lambda^m T^* M \otimes \mathcal{H}^*\,. \tag{25}$$

It follows that $\mathcal{E}$ is covariant (gauge invariant) if and only if each one of $\mathcal{E}_1, \mathcal{E}_2, \mathcal{E}_3$ is covariant (gauge invariant).

In general, one can consider Lagrangians defined on the configuration space obtained by taking various jet bundles over the components of C, i.e., we can take as configuration space Q the space

$$Q := J_i(\mathcal{RM}) \times J_j(\mathcal{A}) \times J_k(\mathcal{H})\,. \tag{26}$$

The choice of the particular jet extensions is determined by the specific physical application. One often begins with some choice and then modifies it as additional fields or invariance requirements are imposed. In what follows we choose

$$Q := (\mathcal{RM}) \times J_1(\mathcal{A}) \times J_1(\mathcal{H})\,. \tag{27}$$

This choice allows us to consider the energy-momentum tensor of the interaction between gauge and matter fields by regarding the metric on M as rigid. Let

$$\mathcal{L} : (\mathcal{RM}) \times J_1(\mathcal{A}) \times J_1(\mathcal{H}) \to \Lambda^m T^* M\,, \tag{28}$$

be a Lagrangian density and let

$$\mathcal{E} : J_1(\mathcal{RM}) \times J_2(\mathcal{A}) \times J_2(\mathcal{H}) \to \Lambda^m T^* M \otimes V^* \mathcal{C}, \qquad (29)$$

be the associated Euler-Lagrange operator. Then $\mathcal{E}$ decomposes according to the scheme indicated in Eqs. (22) to (25). We note that in this case the component $\mathcal{E}_1$ is the pullback of operator $\tau : \mathcal{Q} \to \Lambda^{m-1} T^* M \otimes T^* M$. This operator τ is the energy-momentum tensor corresponding to the variation of the metric and is called the *metric energy-momentum tensor*. We denote by β the composition of $\mathcal{E}_2$ with the canonical contraction. Thus

$$\beta : J_1(\mathcal{RM}) \times J_2(\mathcal{A}) \times J_2(\mathcal{H}) \to \Lambda^{m-1} T^* M \otimes \mathrm{ad}^*(P). \qquad (30)$$

We denote the third component $\mathcal{E}_3$ by η. Thus

$$\eta : J_1(\mathcal{RM}) \times J_2(\mathcal{A}) \times J_2(\mathcal{H}) \to \Lambda^m T^* M \otimes \mathcal{H}^* . \qquad (31)$$

We now define the interaction form γ

$$\gamma : \mathcal{A} \times J_1(\mathcal{H}) \to T^* M \otimes \mathcal{H} \text{ by } \gamma(A, j_1 \phi) = d^\omega \phi , \qquad (32)$$

where, ω is the connectioin 1-form corresponding to the potential A. Our definition of γ illustrates the usual minimal coupling condition. We denote by the same letter γ the pull back of γ to $\mathcal{Q}$. By taking the fiber derivative of the Lagrangian with respect to the first order jets we obtain the partial momentum maps μ and ν as follows:

$$\mu : \mathcal{Q} \to \Lambda^{m-1} T^* M \otimes TM \otimes \mathrm{ad}^*(P) , \qquad (33)$$

$$\nu : \mathcal{Q} \to \Lambda^{m-1} T^* M \otimes \mathcal{H}^* . \qquad (34)$$

We note that while ν is a horizontal form, μ is not, in general, horizontal. In fact, a necessary condition for the Lagrangian to be gauge invariant is that μ defines a horizontal form. We denote by $\mathcal{I}$ and $\mathcal{F}$ the current and the force, respectively (see [MA 1], [MA 2] for further details). The full energy-momentum tensor

$$T : \mathcal{Q} \to \Lambda^{m-1} T^* M \otimes T^* M , \qquad (35)$$

is defined by

$$T = i \circ \mathcal{L} - \hat{F} \cdot \mu - \gamma \cdot \nu , \qquad (36)$$

where i is the cannonical inclusion, $\hat{F}$ is the universal connection and $\cdot$ denotes the obvious contractions. We now state several results on gauge invariance and covariance of Lagrangians $\mathcal{L}$ and associated Euler-Lagrange operators $\mathcal{E}$ as defined on the spaces indicated in Eqs. (28) and (29).

Theorem 3. Lagrangian density $\mathcal{L}$ is gauge invariant if and only if the following three conditions are satisfied.

$$\mu \in \Lambda^{m-2}(Q; \text{ad}^*(P)), \tag{37.1}$$

$$\beta = d^{\natural}\mu - \mathcal{I}, \quad \eta = -d^{\natural}\nu + \mathcal{F}, \tag{37.2}$$

and

$$d^{\natural}\beta = r^*\eta, \tag{37.3}$$

where $d^{\natural}$ is the jet shift differential and r^* denotes the pullback induced by the defining representation r of the associated bundle E.

Condition (37.3) can be interpreted as internal charge conservation. This result includes that of Horndeski [HO 1] as a special case. We note that the conditions of this theorem apply to the Yang-Mills Lagrangian $\mathcal{L}_{YM}$ defined in (9). In addition, the Yang-Mills Lagrangian is conformally invariant and satisfies further local conditions. We propose to undertake the study of conformally invariant Lagrangians and operators in future.

Theorem 4. Let $\mathcal{L}$ be a gauge invariant Lagrangian. Then $\mathcal{L}$ is covariant if and only if the following two conditions are satisfied.

$$\tau = T, \tag{38.1}$$

and

$$d^{\natural}\tau = \hat{F} \cdot \beta + \gamma \cdot \eta, \tag{38.2}$$

where $\cdot$ denotes the obvious contractions.

Theorem 5. The following three conditions are equivalent:
(i) $\mathcal{E}$ is gauge invariant;
(ii) β, η, τ are gauge invariant;
(iii) $d^{\natural}\beta - r^*\eta$ projects onto M, i.e., defines an $\text{ad}^*(P)$-valued m-form on M. Moreover, if the gauge group G is abelian or semisimple, then condition iii) above can be replaced by condition (iii)' below.
(iii)' $d^{\natural}\beta - r^*\eta = 0$.

The proofs of the above theorems involve extensive computation using local coordinates to express the various quantities involved and will be given elsewhere.

Acknowledgement

One of the authors (KBM) would like to thank the Consiglio Nazionale delle Ricerche, Italy for a visiting professorship at the Universitá di Camerino, during the Fall 89 term, when this work was completed. The work of (LM) and (GG) was supported by Ministero della Pubblica Istruzione, Italy (national and local funds).

References

[EP 1]   D. B. A. Epstein, *Natural tensors on Riemannian manifolds*, J. Diff. Geom. **10** (1975) 631–645.

[GA 1]   P. Garcia, *Gauge algebras, curvature and symplectic structure*, J. Diff. Geom. **12** (1977) 209–227.

[HO 1]   G. W. Horndeski, *Gauge invariance and charge conservation in non-abelian gauge theories*, Arch. Rational Mech. **75** (1981) 211–227.

[LE 1]   T. D. Lee and C. N. Yang, *Conservation of heavy particles and generalized Gauge transformations*, Phys. Rev. **98** (1955) 1501.

[MA 1]   L. Mangiarotti, *Jet shift differentials in classical Gauge theories*, in Proc. Differential Geometry and its Applications, Brno, 1986, Purkyně University, Brno, 1987, pp. 225–234.

[MA 2]   L. Mangiarotti, *Local Gauge invariant Lagrangians*, in Proc. VII Italian Conf. on General Relativity and Gravitational Physics, Rapallo, 1986, World Scientific, Singapore, 1987, pp. 227–239.

[MA 3]   L. Mangiarotti and M. Modugno, *Some results on the calculus of variations on jet spaces*, Ann. Inst. H. Poincaré **39** (1983) 29–43.

[MA 4]   L. Mangiarotti and M. Modugno, *Graded Lie algebras and connections on a fibred space*, J. Math. Pur. et Appl. **63** (1984) 111–120.

[MA 5]   L. Mangiarotti and M. Modugno, *Fibered Spaces, jet spaces and connections for field theories*, in Proc. Int. Meeting Geometry and Physics, Florence, 1982, Pitagora Editrice, Bologna, 1983, pp. 135–165.

[MA 6]   L. Mangiarotti and M. Modugno, *On the geometric structure of gauge theories*, J. Math. Phys. **26** (1985) 1373–1379.

[MA 7]   K. B. Marathe, *Spaces admitting gravitational fields*, J. Math. Phys. **14** (1973) 228–233.

[MA 8]   K. B. Marathe, *The mean curvature of gravitational fields*, Physica **114A** (1982) 143–145.

[MA 9]   K. B. Marathe, *Generalized gravitational instantons*, in Proc. Coll. on Diff. Geom., Debrecen (Hungary) 1984, Colloquia Math Soc. J. Bolyai, Hungary, 1987, pp. 763–775.

[MA 10]   K. B. Marathe and G. Martucci, *Geometry of Gauge theories*, J. Geo. and Phy. (1) **1** (1989) 1–106.

[MO 1]   M. Modugno, *Systems of vector valued forms on a fibred manifold and applications to gauge theories*, in Lect. Notes in Math. 1251, Springer-Verlag, New York, 1987, pp. 238–264.

[PA 1]   T. Parker, *Gauge theories on four dimensional Riemannian manifolds*, Comm. Math. Phys. **85** (1982) 563–602.

[ST 1]   P. Stredder, *Natural differential operators on Riemannian manifolds and representations of the orthogonal and special orthogonal groups*, J. Diff. Geo. **10** (1975) 657–660.

[YA 1]   C. N. Yang, *Some concepts in current elementary particle physics*, in The Physicist's Conception of Nature, ed. J. Mehra, Reidel, Dordrecht, Holland, 1972, pp. 447–453.

Kishore B. Marathe and Giovanni Giachetta
Dipartimento di Matematica e Fisica
Università di Camerino
Camerino 62032 (MC)
Italy

THE MATH. HERITAGE OF C.F. GAUSS (pp. 479-504)
edited by George M. Rassias
©1991 World Scientific Publ. Co. Singapore

A SIGNAL DISCRIMINATOR

Frank McNolty, William Sherwood and Jean Mirra

The Carl Friedrich Gauss least squares method (1821) and a quasi-ridge modification of that procedure are uniquely applied to the continuous variable domain by a bank of parallel matched filters. The resulting signal processor is useful in identifying which, if any, of several candidate signals are present at the sensor input. The signals may be present in any combination or not at all, but are always accompanied by additive colored noise. The main topics of the paper focus upon: biased and unbiased estimation of the signal amplitudes, the admissible region for the quasi-ridge parameter k in which the ridge total mean square error TMSE* is less than the least squares TMSE, the ratio TMSE*/TMSE as a function of correlation and signal amplitude, the selection of a suitable value of k to provide a TMSE* smaller than TMSE, the effects of mismatched filters, the conditions for realizable matched filters and nondeterministic noise, and the expected values and covariances of the signal amplitude estimates. It is shown that the shrinkage parameter k need not be chosen with pinpoint accuracy in order to enhance signal estimation and discrimination.

1. Introduction

One could easily argue that Gauss's method of least squares is the most useful of all mathematical techniques. During the nearly 170 years since its inception, the least squares approach has been applied, either explicitly or implicitly, in numerous problem areas. It is a tractable, useful, and aesthetic device with wide appeal to engineers, mathematicians, physicists, and statisticians. In our own problem, Gauss's method will be modified by uniquely introducing the ridge shrinkage parameter k into the realm of continuous variables. We are reluctant to refer to our analysis as ridge estimation, since a known rather than an estimated covariance matrix is used. Accordingly, appropriate key-word phrases might be: quasi-ridge or modified least squares. The assumption of Gaussian noise also admits another descriptor, namely modified maximum likelihood.

The problem addressed in this paper may be concisely defined as follows: The input to a sensor is denoted by a continuous function of time $\nu(t)$. This input might be a random voltage and consists of

$$\nu(t) = A_1 s_1(t - t_0) + A_2 s_2(t - t_0) + \ldots + A_p s_p(t - t_0) + n(t) \qquad (1.1)$$

where $n(t)$ is a colored, Gaussian noise and is at least wide-sense stationary. The signals $s_i(t - t_0)$ may appear in any combination or not at all. Their functional forms and (for the present) their arrival times t_0 are assumed to be known.

Now having received $\nu(t)$, what is a good way to decide which, if any, of the signals are present (the noise $n(t)$ is always present)? Secondly, how can one make the best estimates of the signal amplitudes A_i using linear filtering operations? What if the shape of $s_i(t)$ is somewhat different than anticipated?

If the choice were, say, $\nu(t) = s(t) + n(t)$ versus $n(t)$ alone, the analysis would fall neatly into the well-known detection problem. In our case, however, we are asked to identify each of the several possible signals which might accompany $n(t)$ at the input. This is another step beyond detection, i.e., discrimination. In this paper we propose to perform discrimination and detection jointly through the device of estimating the signal amplitudes as a function of the matched filter outputs.

2. Matched Filters

The matched filter provides a means for maximizing the peak signal-to-average-noise power ratio $(S/N)_p$ at a particular instant of time $t = \Delta$. In a discrimination scenario, it can also be useful as a clutter rejection filter. In terms of maximizing $(S/N)_p$, it is the optimum linear or nonlinear filter when the noise is Gaussian.

A typical input-output configuration for a matched filter is shown below: Here $c(t), s(t)$ are the input and output signals and $n(t)$ is the accompanying additive, mean zero, colored noise assumed to be at least wide-sense stationary.

The transfer function for the matched filter of Fig. 1 is defined as

$$H(f) = KC^*(f) \exp(-j2\pi f \Delta)/N_c(f) \qquad (1.2)$$

$$c(t) \longrightarrow \boxed{\begin{array}{c} \text{MATCHED} \\ \text{FILTER} \\ H(f) \qquad h(t) \end{array}} \longrightarrow s(t)$$

$$n(t) \longrightarrow \qquad \qquad \longrightarrow n_m(t)$$

Fig. 1. Matched Filter

where

$$C(f) = \int_{-\infty}^{\infty} c(t)\exp(-j2\pi ft)dt \,. \tag{1.3}$$

$N_c(f)$ is the noise power spectrum at the input and Δ is the realizability delay.

Some convenient general relationships for the matched filter (Eq. (1.2)) are given below. For instance, the noiseless signal voltage $s(t)$ out of the filter is given by

$$\begin{aligned}
s(t) &= \int_{-\infty}^{\infty} S(f)\exp(j2\pi ft)df \\
&= (2\pi K)^{-1} \cdot \int_{-\infty}^{\infty} N_m(\omega)\exp[-j\omega(\Delta - t)]d\omega \\
&= \int_{-\infty}^{\infty} C(f)H(f)\exp(j2\pi ft)df \\
&= K \cdot \int_{-\infty}^{\infty} |C(f)|^2[N_c(f)]^{-1}\exp[-j2\pi f(\Delta - t)]df
\end{aligned} \tag{1.4}$$

where $N_m(\omega)$ is the power spectrum of the output noise $n_m(t)$, as usual $\omega = 2\pi f$, and we will use ω and f interchangeably in the sequel.

The autocorrelation function of $n_m(t)$ out of the matched filter is

$$\begin{aligned}
R_m(\tau) &= (1/2\pi) \cdot \int_{-\infty}^{\infty} N_m(\omega)\exp(j\omega\tau)d\omega \\
&= (K/2\pi) \cdot \int_{-\infty}^{\infty} S(\omega)\exp[j\omega(\Delta + \tau)]d\omega \\
&= Ks(\Delta + \tau) \,.
\end{aligned} \tag{1.5}$$

The matched filter transfer function may also be expressed as

$$H(f) = [KC(f)]^{-1}N_m(f)\exp(-j2\pi f\Delta) \,. \tag{1.6}$$

Additionally,

$$N_m(f) = K^2|C(f)|^2/N_c(f) = N_c(f)|H(f)|^2 \tag{1.7}$$

and

$$\sigma_n^2 = R_m(0) = Ks(\Delta) = \int_{-\infty}^{\infty} N_m(f)df = K^2 \left(\frac{S}{N}\right)_p$$
$$= E[n_m^2(t)] = K^2[s(\Delta)]^2/R_m(0) \,. \tag{1.8}$$

When

    (a) $h(t) = 0$ for $t < 0$

and

    (b) $\int_0^{\infty} |h(t)|dt < \infty$, i.e., $h(t) \in L^1$

the matched filter (Eq. (1.2)) is said to be realizable. The quantity Δ appearing in the preceding expressions has been defined as a realizable delay, however, the corresponding filter is not necessarily realizable. For a signal of finite duration accompanied by wide-sense stationary white noise with two-sided spectrum $N_0/2$, the filter can be made realizable through a proper choice of Δ, and in this case the maximum $(S/N)_p$ at $t = \Delta$ becomes $2E/N_0$ where E is the signal energy.

If $N_c(f)$ of Eq. (1.2) is an arbitrary noise spectrum, one can obtain the realizable matched filter in one of two ways:

    (1) by solving an integral equation [1] involving $R_n(\tau)$, $h(t)$, and $s(t)$ where $R_n(\tau)$ is the autocorrelation function of the input noise $n(t)$, or

    (2) by constructing a realizable whitening filter which then precedes the desired white-noise matched filter.

For various reasons, we will proceed by means of method 2, which will yield the same solution for $h(t)$ as does method 1.

The filtering configuration is shown in Fig. 2.

$c(t) \longrightarrow$	WHITENING FILTER $H_w(f) \qquad h_w(t)$	$\xrightarrow{a(t)}$	MATCHED FILTER $H_1(f) \qquad h_1(t)$	$\longrightarrow s(t)$
$n(t) \longrightarrow$		$\xrightarrow{n_1(t)}$		$\longrightarrow n_m(t)$

Fig. 2. Two-Stage Matched Filter $H_m(\omega)$.

In this procedure, we must still choose $H_w(f)$ carefully [1] so that $h_w(t)$ is realizable. We write

$$N_c(f) \cdot |H_w(f)|^2 = 1 \,, \text{ i.e.,}$$
$$H_w(f) = 1/\sqrt{N_c(f)}$$

where now $[N_c(f)]^{1/2}$ is a symbolic notation for $N_c^+(\omega)$, which is a function having all of its poles and zeros in the upper-half complex plane. Such a function has an inverse Fourier transform $h_w(t) = 0$ for $t < 0$.

If $N_c(f)$ is amenable to the spectrum factorization

$$N_c(\lambda) = N_c^+(\lambda)N_c^-(\lambda) \tag{1.9}$$

where all of the poles and zeros of $N_c^-(\lambda)$ lie in the lower-half complex plane ($\lambda = \omega + j\sigma$), then $H_w(\omega)$ will be a realizable filter having output noise spectrum $N_1(\omega) = 1$.

The condition for the existence of this factorization of a square-integrable function $N_c(\omega)$ is the Paley-Wiener [2] criterion

$$\int_{-\infty}^{\infty} \frac{|\ln N_c(\omega)|}{1 + \omega^2} d\omega < \infty \tag{1.10}$$

and the solution for the corresponding two-stage realizable matched filter $H_m(\omega)$ is then [1] shown to be

$$\begin{aligned}
H_m(\omega) =& [2\pi N_c^+(\omega)]^{-1} \cdot \int_0^{\infty} \exp(-j\omega t) \\
& \times \int_{-\infty}^{\infty} C(-u)[N_c^-(u)]^{-1} \exp[ju(t - \Delta)] du\, dt\,.
\end{aligned} \tag{1.11}$$

Having legitimatized the matched filter of Fig. 2, we must now ask whether or not the output noise $n_m(t)$ is truly random. Doob [3] tells us that a wide-sense stationary random process $\{n_m(t)\}$ will have a mean-square prediction error > 0 if and only if

$$\int_{-\infty}^{\infty} \frac{|\ln N_m(\omega)|}{1 + \omega^2} d\omega < \infty \tag{1.12}$$

which is the same type of condition required for the factorization of Eq. (1.9). The noise spectrum out of the realizable filter $H_1(\omega)$ is then

$$1 \cdot |H_1(\omega)|^2 = N_m(\omega) \tag{1.13}$$

so that Eq. (1.12) becomes

$$\int_{-\infty}^{\infty} \frac{|\ln N_m(\omega)|}{1 + \omega^2} d\omega = 2 \int_{-\infty}^{\infty} \frac{|\ln |H_1(\omega)||}{1 + \omega^2} d\omega < \infty \tag{1.14}$$

and the last inequality holds, because of the necessary and sufficient Wiener [4] condition for realizable filters where $H_1(\omega) \in L^2$. That is, $\{n_m(t)\}$ is a nondeterministic process. Note that $f(\omega) \in L^1$ implies $f(\omega) \in L^2$ provided that there are no singularities at the origin. From Doob's criterion, one can show that band-limited noise and processes having a Gaussian power spectrum fail to meet the condition for nondeterministic processes.

3. Signal Discrimination

A. *Introduction*

In certain applications, a sensor is confronted with the problem of distinguishing between different types of input signals. The initial approach to the problem will exploit the radar resolution techniques of Helstrom [5] and Nilsson [6]. We then apply a quasi-ridge method of discrete multivariate analysis to the continous signal domain. This application of ridge to the continuous case provides additional insight into the role of biased estimation in the more unwieldly discrete multivariate regime.

From a mathematical viewpoint, it would be expedient to present the analysis entirely in terms of matrix algebra; however, some physical applications might involve no more than two, three, or four types of signals. Thus, much is to be gained by frequently displaying the results in scalar form. For the present, we arbitrarily assume that the sensor input consists of three signals

$$v_1(t) = A_1 s_1(t - t_0) + A_2 s_2(t - t_0) + A_3 s_3(t - t_0) + n(t) \qquad (2.1)$$

where the A_i are signal amplitudes, t_0 is the arrival time, and $n(t)$ is the stationary, mean zero, Gaussian random noise which has a colored power spectrum $N(\omega)$. In our mathematical model, the input $v_1(t)$ is passed through a noise whitener $H_w(\omega)$ and then through a bank of three matched filters in parallel. The input to the ith matched filter $H_i(\omega)$ is denoted by

$$v(t) = A_1 a_1(t - t_0) + A_2 a_2(t - t_0) + A_3 a_3(t - t_0) + n_1(t) \qquad (2.2)$$

where $a_i(t - t_0)$ is the $H_w(\omega)$ distorted version of $s_i(t - t_0)$. Again, it is worthwhile to point out that the signals $s_i(t - t_0)$ may be present in any combination or not at all.

Figure 3 describes the filtering operation on the individual signal and noise components.

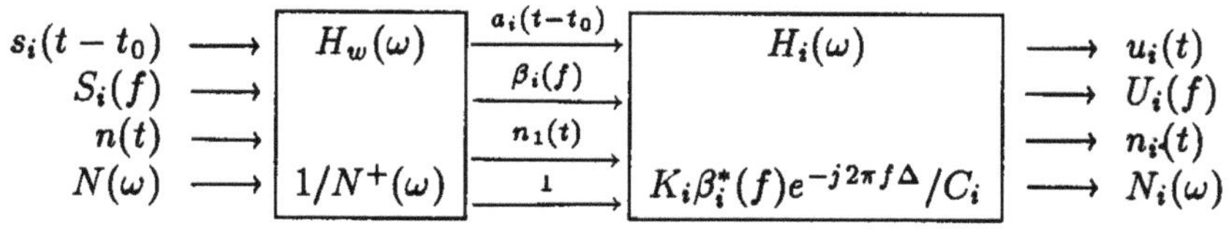

$s_i(t - t_0) \longrightarrow$ | $H_w(\omega)$ $\dfrac{a_i(t-t_0)}{}$ | $H_i(\omega)$ $\longrightarrow u_i(t)$

$$
\begin{array}{lll}
s_i(t - t_0) \longrightarrow & & \longrightarrow u_i(t) \\
S_i(f) \longrightarrow & \boxed{H_w(\omega)} \xrightarrow{\ a_i(t-t_0)\ } \boxed{H_i(\omega)} & \longrightarrow U_i(f) \\
n(t) \longrightarrow & \xrightarrow{\ \beta_i(f)\ } & \longrightarrow n_i(t) \\
N(\omega) \longrightarrow & \boxed{1/N^+(\omega)} \xrightarrow{\ n_1(t)\ }\xrightarrow{\ 1\ } \boxed{K_i\beta_i^*(f)e^{-j2\pi f\Delta}/C_i} & \longrightarrow N_i(\omega)
\end{array}
$$

Fig. 3. Input-Output Relationships for $H_w(\omega)$ and $H_i(\omega)$.

The following relationships are implied by Fig. 3.

$$S_i(f) = \int_{-\infty}^{\infty} s_i(t - t_0)\exp(-j2\pi ft)dt \tag{2.3}$$

$$\beta_i(f) = S_i(f)/N^+(f) = \int_{-\infty}^{\infty} a_i(t - t_0)\exp(-j2\pi ft)dt \tag{2.4}$$

where again the subscript corresponds to the ith signal. Also,

$$
\begin{aligned}
u_i(t) &= \int_{-\infty}^{\infty} U_i(f)\exp(j2\pi ft)df = \int_{-\infty}^{\infty} \beta_i(f)H_i(f)\exp(j2\pi ft)df \\
&= (k_i/c_i)\cdot \int_{-\infty}^{\infty} |\beta_i(f)|^2 \exp[-j2\pi f(\Delta - t)]df \\
&= (c_i/2\pi k_i)\cdot \int_{-\infty}^{\infty} N_i(\omega)\exp[-j\omega(\Delta - t)]d\omega .
\end{aligned}
\tag{2.5}
$$

The quantities c_i^2, λ_{ij}, and $R_i(t)$ are defined as follows

$$c_i^2 = \int_0^{\Delta} a_i^2(t - t_0)dt \tag{2.6}$$

$$\lambda_{ij} = (1/c_ic_j)\int_0^{\Delta} a_i(t - t_0)a_j(t - t_0)dt \tag{2.7}$$

$$R_i(\Delta) = \int_0^{\Delta} a_i(t - t_0)v(t)dt \tag{2.8}$$

$$R_i(t) = \int_0^{t} a_i(\Delta - \tau - t_0)v(t - \tau)d\tau \tag{2.9}$$

where $R_i(\Delta)$ is not the output of the ith matched filter at time $t = \Delta$. The actual output would be $(k_i/c_i)R_i(\Delta)$. We note that the signal $a_i(u)$ is all in by some time interval no larger than $\Delta - t_0$. The impulse-response function corresponding to $H_i(\omega)$ is given by

$$
\begin{aligned}
h_i(\tau) &= (k_i/c_i)a_i(\Delta - \tau - t_0) \quad \text{for } \tau \geq 0 \\
&= 0 \qquad\qquad\qquad\quad \text{for } \tau < 0 \,.
\end{aligned}
\tag{2.10}
$$

B. *Signal Amplitude Estimation*

In order to provide estimates for the signal amplitudes of Eq. (2.2), three additional quantities must also be defined

$$
R_0 = \begin{pmatrix} c_1^2 & c_1c_2\lambda_{12} & c_1c_3\lambda_{13} \\ c_1c_2\lambda_{12} & c_2^2 & c_2c_3\lambda_{23} \\ c_1c_3\lambda_{13} & c_2c_3\lambda_{23} & c_3^2 \end{pmatrix}
\tag{2.11}
$$

$$
\mathbf{R}(\Delta) = \begin{bmatrix} R_1(\Delta) \\ R_2(\Delta) \\ R_3(\Delta) \end{bmatrix} = \begin{pmatrix} R_1 \\ R_2 \\ R_3 \end{pmatrix} \,.
\tag{2.12}
$$

And the vector of maximum likelihood estimates is denoted by

$$
\hat{\mathbf{A}} = \begin{pmatrix} \hat{A}_1 \\ \hat{A}_2 \\ \hat{A}_3 \end{pmatrix} \,.
\tag{2.13}
$$

Although the $a_i(t)$ are not random, R_0 has some properties of a covariance matrix: symmetric, nonsingular, positive definite, $-1 \leq \lambda_{ij} \leq 1$. Perhaps we should refer to R_0 as a quasi-covariance matrix.

By using the Gauss straightforward least-squares approach, it can be shown that for the general P dimensional case

$$
\hat{\mathbf{A}} = R_0^{-1} \cdot \mathbf{R}(\Delta)
\tag{2.14}
$$

$$
E(\hat{\mathbf{A}}) = R_0^{-1} \cdot E[\mathbf{R}(\Delta)] = \mathbf{A} \,, \; \text{Cov}\,(\hat{\mathbf{A}}) = R_0^{-1}
\tag{2.15}
$$

$$
\text{Cov}\,[\mathbf{R}(\Delta)] = R_0 \,.
\tag{2.15a}
$$

An important property of Eq. (2.14) is that

$$
\begin{aligned}
E(\hat{A}_i) &= A_i \quad \text{when signal } a_i(t) \text{ is present} \\
&= 0 \qquad\quad \text{otherwise} \,.
\end{aligned}
\tag{2.16}
$$

The property Eq. (2.16) holds true regardless of which other signals are present or absent. In scalar form, for the case of three signals

$$\hat{A}_1 = (1/\beta c_1)[R_1(1 - \lambda_{23}^2)/c_1 - R_2(\lambda_{12} - \lambda_{13}\lambda_{23})/c_2$$
$$- R_3(\lambda_{13} - \lambda_{12}\lambda_{23})/c_3] \tag{2.17}$$

$$\hat{A}_2 = (1/\beta c_2)[R_2(1 - \lambda_{13}^2)/c_2 - R_1(\lambda_{12} - \lambda_{13}\lambda_{23})/c_1$$
$$- R_3(\lambda_{23} - \lambda_{12}\lambda_{13})/c_3] \tag{2.18}$$

$$\hat{A}_3 = (1/\beta c_3)[R_3(1 - \lambda_{12}^2)/c_3 - R_1(\lambda_{13} - \lambda_{12}\lambda_{23})/c_1$$
$$- R_2(\lambda_{23} - \lambda_{12}\lambda_{13})/c_2] \tag{2.19}$$

where

$$\beta = 1 + 2\lambda_{12}\lambda_{13}\lambda_{23} - \lambda_{12}^2 - \lambda_{13}^2 - \lambda_{23}^2 \,.$$

Additional scalar relationships corresponding to Eq. (2.15) are

$$\text{Var}\,(\hat{A}_j) = (1/\beta c_j^2)(1 - \lambda_{ik}^2)\,, \quad j \neq i, k\,; \ i \neq k \tag{2.20}$$

$$\text{Cov}\,(\hat{A}_i, \hat{A}_k) = (1/\beta c_i c_k)(\lambda_{ij}\lambda_{jk} - \lambda_{ik})\,; \quad j \neq i, k\,; \ i \neq k\,. \tag{2.20a}$$

For instance, the simple correlation between $\hat{A}_1$ and $\hat{A}_2$ is written as

$$\rho_{12} = (\lambda_{13}\lambda_{23} - \lambda_{12})[(1 - \lambda_{23}^2)(1 - \lambda_{13}^2)]^{-1/2} \tag{2.21}$$

and Eq. (2.21) is actually the partial correlation coefficient for $a_1(t)$ and $a_2(t)$ holding $a_3(t)$ fixed. The quantity $\hat{A}_i$ will often be referred to in the sequel as the output of the ith channel.

It is worthwhile to point out that Eq. (2.20) is the variance of the jth channel regardless of which signals $a_i(t)$, if any, are present.

In the case of two channels, the scalar signal-amplitude estimates are

$$\hat{A}_1 = \frac{1}{c_1(1 - \lambda_{12}^2)}\left[\frac{R_1(\Delta)}{c_1} - \frac{\lambda_{12}R_2(\Delta)}{c_2}\right] \tag{2.22}$$

$$\hat{A}_2 = \frac{1}{c_2(1 - \lambda_{12}^2)}\left[\frac{R_2(\Delta)}{c_2} - \frac{\lambda_{12}R_1(\Delta)}{c_1}\right]\,. \tag{2.23}$$

We now ask what changes occur in the property Eq. (2.16) when the filters are mismatched to the input signals. That is, instead of having an input

$$v(t) = A_1 a_1(t - t_0) + A_2 a_2(t - t_0) + n_1(t) \tag{2.24}$$

we have, say,

$$v(t) = A_1 a_1(t - t_0) + A_3 a_3(t - t_0) + n_1(t). \qquad (2.25)$$

In this case

$$E(\hat{A}_1) = A_1 + \frac{A_3 c_3(\lambda_{13} - \lambda_{12}\lambda_{23})}{c_1(1 - \lambda_{12}^2)} = A_1 + B_1 \qquad (2.26)$$

$$E(\hat{A}_2) = \frac{A_3 c_3(\lambda_{23} - \lambda_{12}\lambda_{13})}{c_2(1 - \lambda_{12}^2)} = B_2 \qquad (2.27)$$

where the B_i may be regarded as bias terms. Of course, as $a_3(t - t_0)$ more closely resembles $a_2(t - t_0)$, then $B_1 \to 0$ and $B_2 \to A_2$.

There is considerable flexibility in the interpretation of mismatched signals. For instance, $a_3(t)$ might indeed be $a_2(t)$, but with a delayed time of arrival. That is, $\lambda_{13} = \lambda_{23} = 0$ and $E(\hat{A}_1) = A_1, E(\hat{A}_2) = 0$, which is the desired result.

When the mismatching is in $a_1(t)$, then

$$v(t) = A_3 a_3(t - t_0) + A_2 a_2(t - t_0) + n_1(t) \qquad (2.28)$$

and we obtain

$$E(\hat{A}_1) = B_1 \qquad (2.29)$$

$$E(\hat{A}_2) = A_2 + B_2. \qquad (2.30)$$

If both filters are mismatched

$$v(t) = A_3 a_3(t - t_0) + A_4 a_4(t - t_0) + n_1(t) \qquad (2.31)$$

then

$$E(\hat{A}_1) = \frac{1}{c_1(1 - \lambda_{12}^2)} \sum_{i=3}^{4} A_i c_i(\lambda_{1i} - \lambda_{12}\lambda_{2i}) \qquad (2.32)$$

$$E(\hat{A}_2) = \frac{1}{c_2(1 - \lambda_{12}^2)} \sum_{i=3}^{4} A_i c_i(\lambda_{2i} - \lambda_{12}\lambda_{1i}). \qquad (2.33)$$

Appendix A discusses mismatching in the case of three and four channels. In all cases, the built-in variances are unaffected by the mismatched filters.

C. *Ridge Estimation Model*

We write the squared distance from $\hat{\mathbf{A}}$ to the vector of true coefficients $\mathbf{A}$ as [7]

$$L_1^2 = (\hat{\mathbf{A}} - \mathbf{A})'(\hat{\mathbf{A}} - \mathbf{A}) \qquad (2.34)$$

and then, letting $c_i = 1$ for brevity, total mean square error is

$$E(L_1^2) = \sum_{i=1}^{P} 1/\mu_i \tag{2.35}$$

where the μ_i are the eigenvalues of R_0. Also,

$$\text{Var}\,(L_1^2) = 2\sum_{i=1}^{P} 1/\mu_i^2 \,. \tag{2.36}$$

In the two-channel case,

$$E(L_1^2) = \text{Tr}\,(R_0^{-1}) = 1/\mu_1 + 1/\mu_2 = 2/(1 - \lambda_{12}^2) \tag{2.37}$$

$$\text{Var}\,(L_1^2) = 2\text{Tr}\,(T_0^{-2}) = 2(1/\mu_1^2 + 1/\mu_2^2) \tag{2.38}$$

when λ_{12}^2 is close to unity, one of the eigenvalues (μ_1 or μ_2) will be small so that the expected value of L_1^2 will be large as will the variance of L_1^2.

In discrete multivariate analysis, it is known [7, 8, 9] that the ridge technique can provide estimates which have a smaller total mean square error than the unbiased least squares or maximum likelihood estimates. The authors are, therefore, motivated to introduce the Hoerl-Kennard ridge methodology into the present continuous-signal application.

Denoting the ridge vector estrimate by $\hat{\mathbf{A}}^*$, we write

$$\hat{\mathbf{A}}^* = (R_0 + kI)^{-1} \cdot \mathbf{R}_0\hat{\mathbf{A}} = [I + kR_0^{-1}]^{-1}\hat{\mathbf{A}} \tag{2.39}$$

where k is sometimes referred to as a shrinkage parameter. Equation (2.39) is the general P dimensional estimate, while in two dimensions, the scalar ridge estimates are

$$\hat{A}_1^* = \frac{1}{(1+k)^2 - \lambda^2}[(1+k)R_1(\Delta) - \lambda R_2(\Delta)] \tag{2.40}$$

$$\hat{A}_2^* = \frac{1}{(1+k)^2 - \lambda^2}[(1+k)R_2(\Delta) - \lambda R_1(\Delta)] \,. \tag{2.41}$$

Alternatively, from Eqs. (2.22) and (2.23), one can write

$$\hat{A}_1^* = \frac{1}{(1+k)^2 - \lambda^2}[\hat{A}_1(1 - \lambda^2) + k(\hat{A}_1 + \lambda\hat{A}_2)] \tag{2.42}$$

$$\hat{A}_2^* = \frac{1}{(1+k)^2 - \lambda^2}[\hat{A}_2(1 - \lambda^2) + k(\hat{A}_2 + \lambda\hat{A}_1)] \tag{2.43}$$

where, in two dimensions, it is convenient to drop the subscripts on λ. It is not difficult to show that

$$E(\hat{A}_1^*) = (y - \lambda^2)A_1/x + k\lambda A_2/x \tag{2.44}$$

$$E(\hat{A}_2^*) = (y - \lambda^2)A_2/x + k\lambda A_1/x \tag{2.45}$$

where $y = 1 + k$ and $x = (1 + k)^2 - \lambda^2$; thus, when the input to the two-channel ridge system is

$$v(t) = A_1 a_1(t - t_0) + A_2 a_2(t - t_0) + n_1(t) \tag{2.46}$$

the total mean square error becomes

$$
\begin{aligned}
\text{TMSE}^* =\; & 2[y^2 - \lambda^2(y + k)]/x^2 + A_1^2 + A_2^2 \\
& + (y^2 + \lambda^2)[4\lambda A_1 A_2 + (1 + \lambda^2)(A_1^2 + A_2^2)]/x^2 \\
& - 4\lambda y[\lambda(A_1^2 + A_2^2) + (1 + \lambda^2)A_1 A_2]]/x^2 \\
& - 2[y(A_1^2 + 2\lambda A_1 A_2 + A_2^2) - \lambda(\lambda A_1^2 + 2A_1 A_2 + \lambda A_2^2)]/x \, .
\end{aligned}
\tag{2.47}
$$

Expression (2.47) is in the form

$$
\begin{aligned}
\text{TMSE}^* =\; & \text{total variance of the ridge estimates} \\
& + \text{total squared bias of the ridge estimates} \\
=\; & V(k) + B^2(k) \, .
\end{aligned}
$$

When $A_1 = A_2$, expression (2.47) reduces to

$$\text{TMSE}^* = 2[y^2 - \lambda^2(y + k)]/x^2 + 2A_1^2[x - (1 + \lambda)(y - \lambda)]^2/x^2 \tag{2.48}$$

and we recall that for the maximum likelihood estimator

$$\text{TMSE} = \frac{2}{1 - \lambda^2} \, .$$

It is easy to show that

$$\text{TMSE}^* \big|_{k=0} = \text{TMSE}$$

It is also seen from Eqs. (2.44) and (2.45) that for $k = 0$

$$E(\hat{A}_1^*) = A_1 \text{ and } E(\hat{A}_2^*) = A_2$$

while for $\lambda = 0$

$$E(\hat{A}_1^*) = A_1/(1 + k), \quad E(\hat{A}_2^*) = A_2/(1 + k) \, .$$

In addition, we have

$$\lim_{k \to \infty} \text{TMSE}^* = A_1^2 + A_2^2 = \mathbf{A}'\mathbf{A} \, .$$

Generally for the case of P channels

$$E(\hat{\mathbf{A}}) = (R_0 + kI)^{-1} \cdot E[R(\Delta)] = (I + kR_0^{-1})^{-1} \cdot \mathbf{A} \tag{2.49}$$

$$\text{Cov}(\hat{\mathbf{A}}^*) = (R_0 + 2kI + k^2 R_0^{-1})^{-1} \tag{2.50}$$

where -1 always denotes the matrix inverse.

$$E[L_1^2(k)] = \sum_{i=1}^{P} \mu_i/(\mu_i + k)^2 + k^2 \mathbf{A}'(R_0 + kI)^{-2}\mathbf{A}$$

$$= \sum_{i=1}^{P} \mu_i/(\mu_i + k)^2 + k^2 \sum_{i=1}^{P} a_i^2/(\mu_i + k)^2$$

where $\boldsymbol{\alpha} = P\mathbf{A}, R_0 = P'\Lambda P, PP' = P'P = I, P' = P^{-1}, PR_0P' = \Lambda$, and the columns of P' are the normalized eigenvectors of R_0.

4. Graphical Results

Figures 4 and 5 provide some typical examples showing the relationship of TMSE*, TMSE, $V(k)$, and $B^2(k)$. The least squares TMSE is shown as a horizontal dashed line running across the graph (roughly at 5.65 in Fig. 4), while the dashed curve representing TMSE* has a fishnet appearance and is the result of $V(k)$ and $B^2(k)$. It is important to note that the derivative of $V(k)$ is negative from the outset (i.e., as $k \to 0^+$), while the derivative of $B^2(K)$ is flat as $k \to 0^+$; thus, we need only impose a very weak boundedness condition on A_i to be assured of a minimum.

The authors have defined two quantities, R and d, which are useful in characterizing the behavior of TMSE*. The quantity R is defined as

$$R = \frac{\text{TMSE}^*(k_0)}{\text{TMSE}}$$

where k_0 is the value of k for which TMSE* takes on its minimum value. In other words, R provides a comparison of TMSE* with TMSE when k has been chosen optimally. Figure 6 portrays R versus λ (for fixed A_i), and it is seen that the ridge technique becomes increasingly effective as λ increases. Note that as λ increases, the eigenvalues of R_0 become smaller so that the expected value of the squared distance of the MLE vector from the true vector becomes larger. In contrast, the ridge vector is the shortest vector commensurate with a specified increase in the minimum residual sum of squares.

Figure 7 provides a graph of R versus A_i (for fixed λ) which shows that the ridge technique becomes less effective as A_i becomes very large. We recall from expression (2.47) that $B^2(k)$ could actually be written as $B^2(k, A_i)$, which is an increasing function of A_i. Thus, as A_i increases (for a specified λ), the bias increases, which in turn raises the minimum value

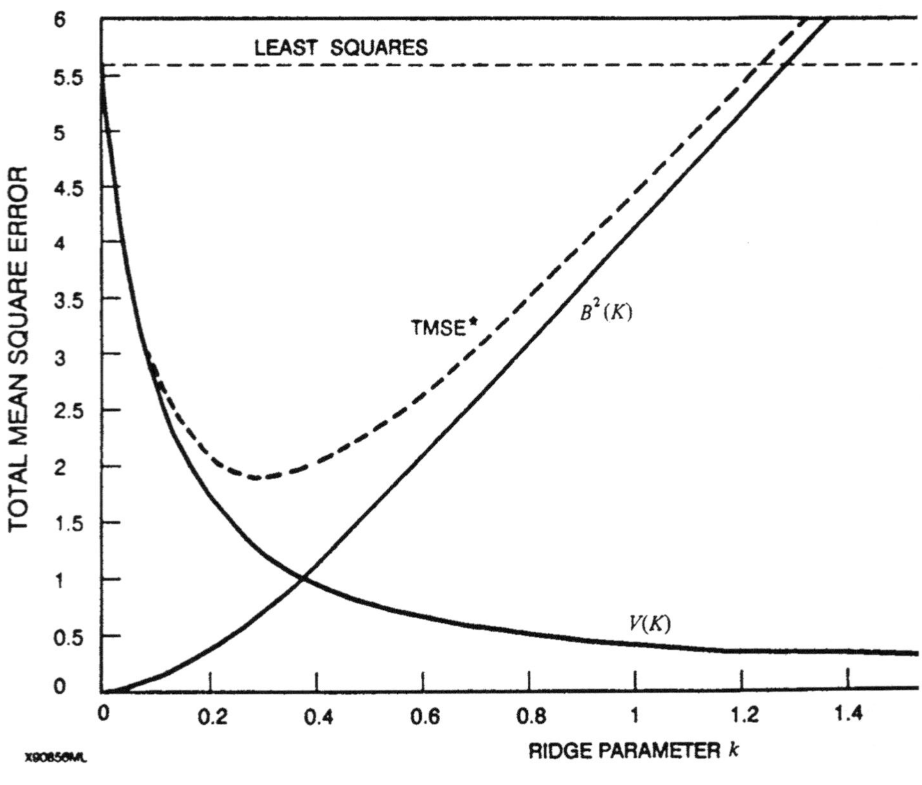

Fig. 4. Continuous Ridge Method TMSE*, $V(k)$, and $B^2(k)$ Ver:

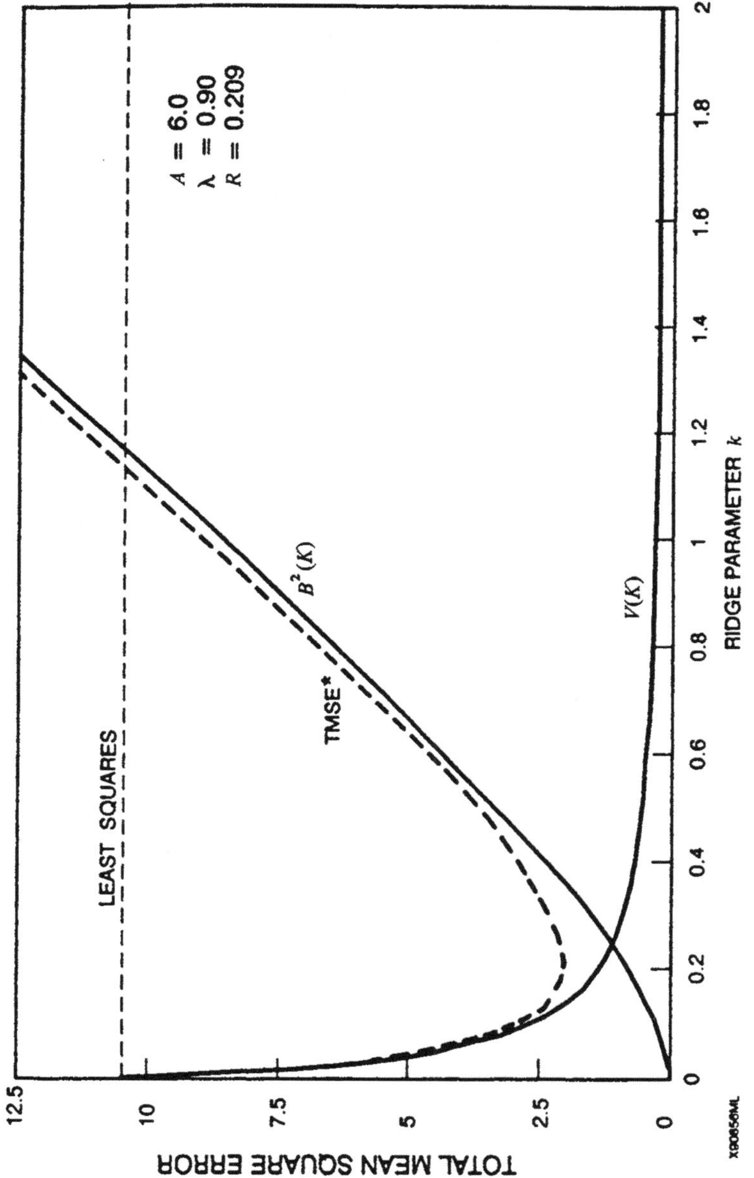

Fig. 5. Continuous Ridge Method TMSE*, $V(k)$, and $B^2(k)$ Versus k.

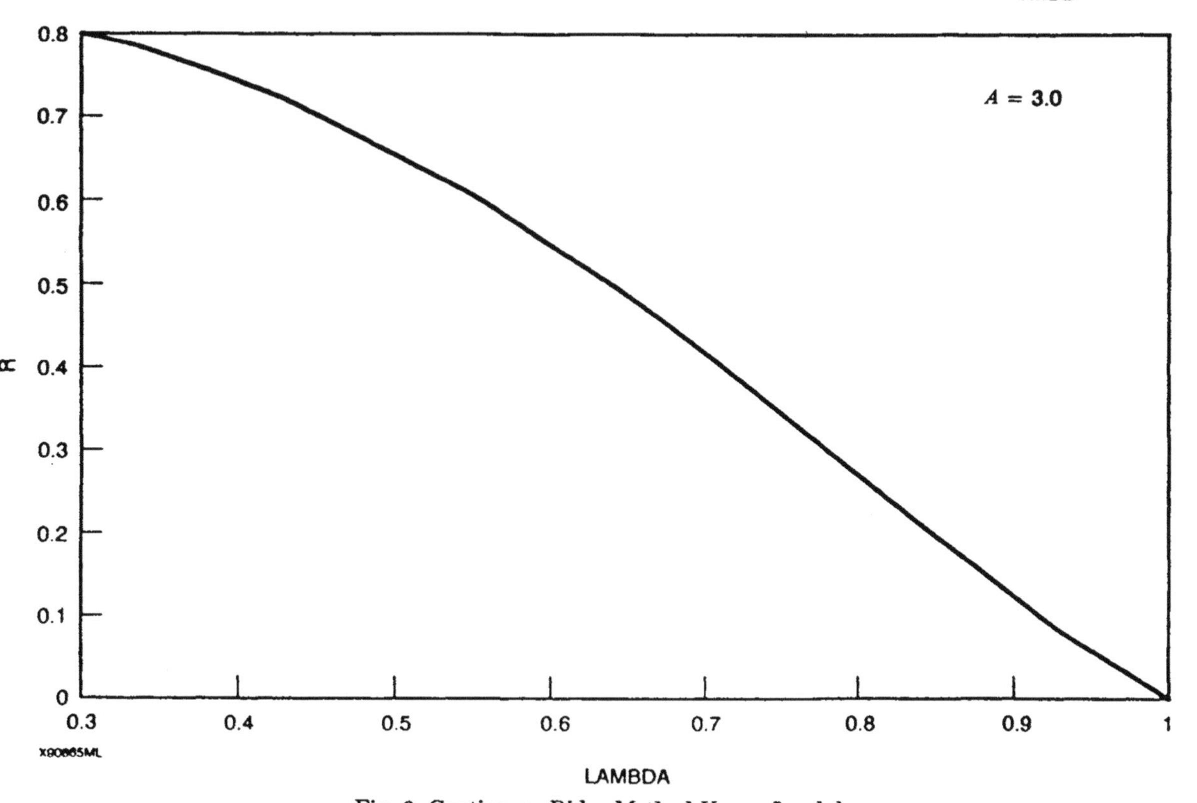

Fig. 6. Continuous Ridge Method Versus Lambda.

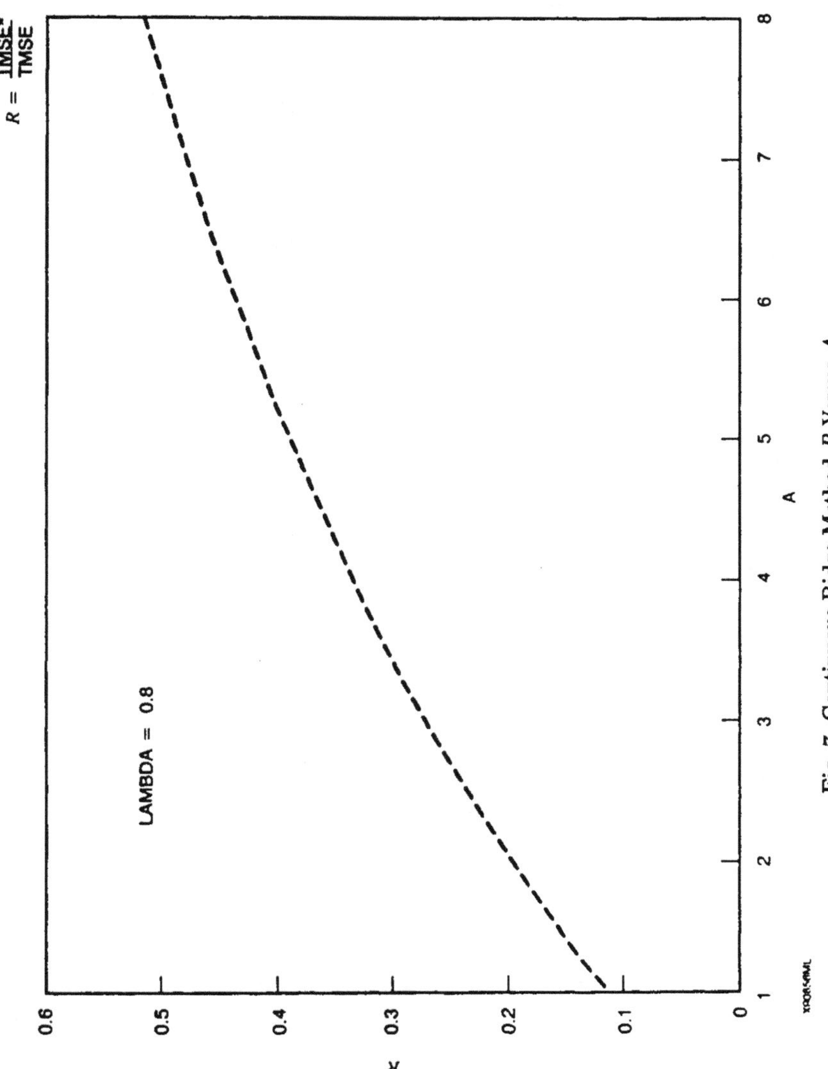

Fig. 7. Continuous Ridge Method *R* Versus *A*.

of TMSE* and also moves this minimum value (or equivalently k_0) closer to the origin.

The quantity d defines the admissible region from which one may select a k to obtain a smaller TMSE than that provided by the least squares estimator. The length (d) of the admissible region is measured along the horizontal TMSE line (least squares) from the origin to the point of intersection with the dashed TMSE* curve in Figs. 4 and 5. Any k in this region will lead to a reduction in the total mean square error. Figure 8 shows how the width of the admissible region increases as λ increases. That is, as λ increases, it becomes easier to select a value of k which will yield a smaller TMSE* than that given by least squares. If we consider $k = 1.0$ to be the practical upper limit for the ridge parameter, then Fig. 8 tells us that any $k > 0$ will yield a smaller total mean square error for $0 < \lambda < 0.84$. In applications, we would anticipate values of λ between, for example, $0.1 < \lambda < 0.80$.

Figure 9 describes the behavior of the optimum k_0 versus λ for a fixed A_i. As λ increases, so does k_0 and, in addition, the TMSE* curve becomes flatter in a neighborhood around the minimum for increasing values of λ. Again, this behavior means that one need not pinpoint the selection of k. Thus, we will simply substitute the least squares estimates $\hat{A}_1$ and $\hat{A}_2$ into Eq. (2.47) and solve for the k_0 which yields a minimum TMSE*. Expressions (2.42) and (2.43) then yield $\hat{A}_1^*$ and $\hat{A}_2^*$ from $k_0, \hat{A}_1$ and $\hat{A}_2$. For the three, four, $\dots$, P channel cases, expressions similar to Eq. (2.47) would be used. It is worthwhile to point out again that the quantities λ_{ij} are known, unlike the discrete multivariate ridge method in which the covariance matrix is estimated from the data.

Figure 10 shows the schematic diagram for a four-channel implementation of the least squares amplitude estimation.

In effect, the quantities $\hat{A}_1$ and $\hat{A}_2$ provide the initial values for $\hat{A}_1^*$ and $\hat{A}_2^*$, but in order to use an $\hat{A}_i$ for this purpose, one must first decide whether the output of the ith channel is due to noise alone or to signal plus noise. That is, we have a typical N versus $S + N$ thresholding situation arising from the mean zero, Gaussian sensor noise (assumed stationary). An automatic method of detection in the case of two channels is to use a confirmation-inhibit strategy in the decision process. A first-channel T_1 threshold crossing indicates the possible presence of $a_1(t)$ and/or $a_2(t)$, and it activates an output quantity $Q_2 = I_2 - \hat{A}_2$ in the second channel. When $a_1(t)$ (but not $a_2(t)$) is present, $\hat{A}_2$ will dip toward zero at time t_0 and Q_2

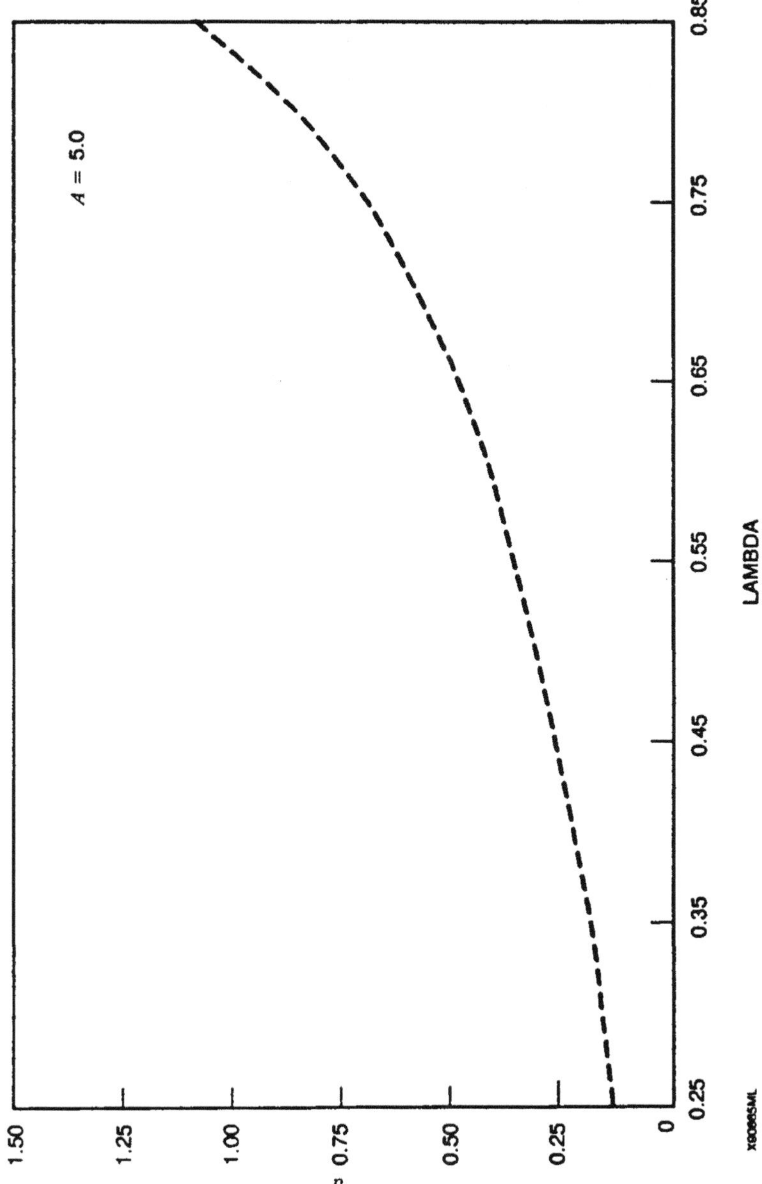

Fig. 8. Continuous Ridge Method d Versus Lambda.

F. McNolty et al.

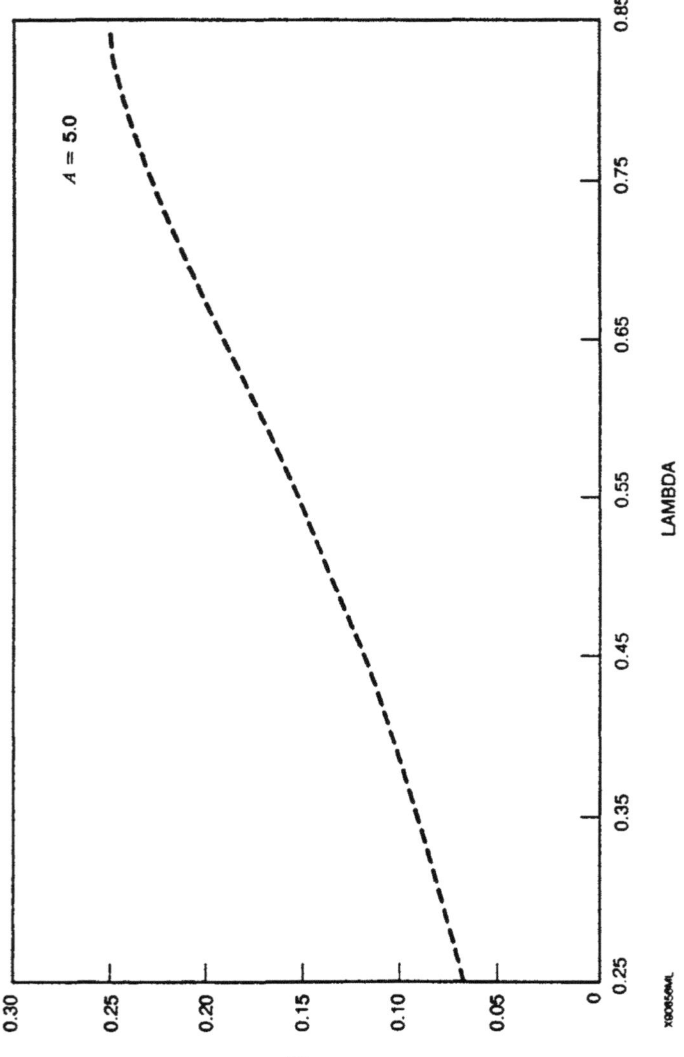

Fig. 9. Continuous Ridge Method k_0 Versus Lambda.

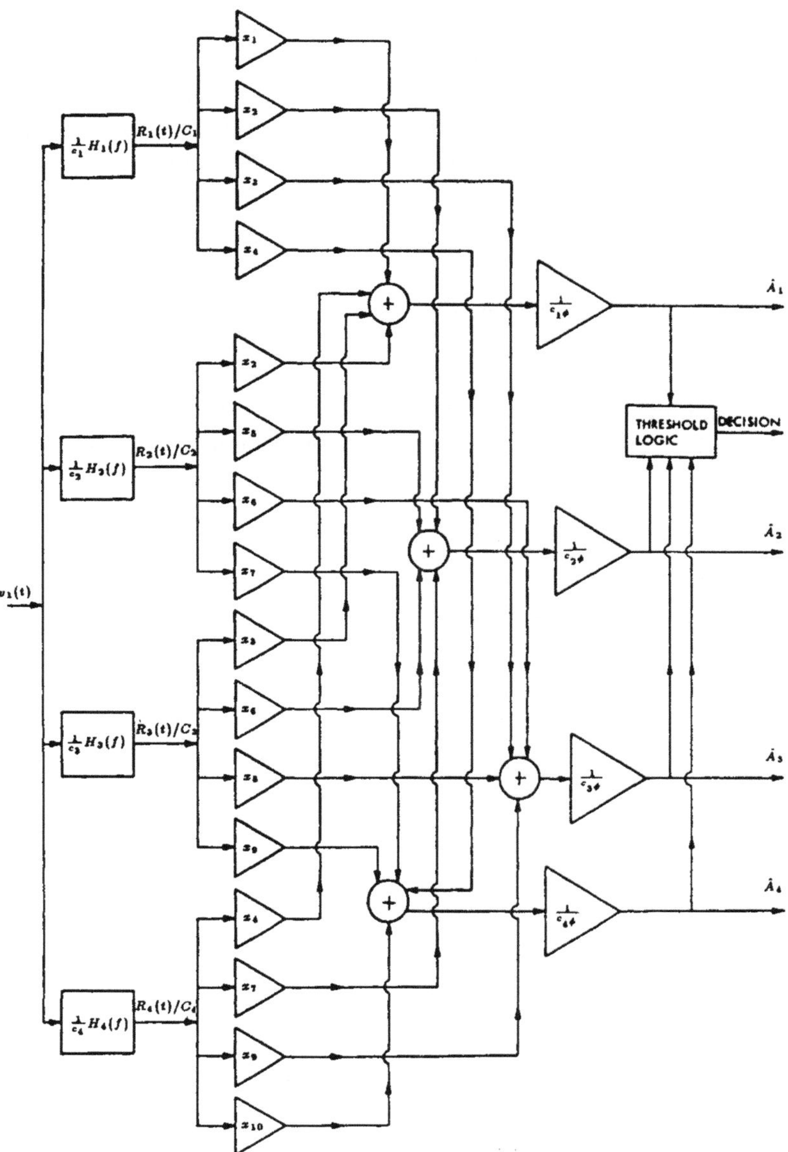

Fig. 10. Four-Channel Implementation of Least Squares Estimation.

will tend to cross a confirm threshold, thus confirming the presence of $a_1(t)$ and the absence of $a_2(t)$. Of course, the T_1 crossing indicates that sensor noise only is not present at the first channel output. The output Q_2, from the second channel, is active only as long as $\hat{A}_1$ is above T_1.

A second-channel T_2 threshold crossing indicates the possible presence of $a_1(t)$ and/or $a_2(t)$, and it activates an output quantity $Q_1 = I_1 - \hat{A}_1$ in the first channel. When $a_2(t)$ (but not $a_1(t)$) is present, $\hat{A}_1$ will dip toward zero at time t_0 and Q_1 will tend to cross an inhibit threshold, inhibiting the decision to declare $a_1(t)$ present. The Q_1 threshold is active only as long as $\hat{A}_2$ is above T_2. Again, the T_2 crossing indicates that sensor noise only is not present in the second channel output. The T_1 and T_2 threshold crossings can occur simultaneously, which activates both the Q_2 (confirm) and Q_1 (inhibit) thresholds. When both $\hat{A}_1$ and $\hat{A}_2$ remain above their thresholds together, the decision is for both $a_1(t)$ and $a_2(t)$ present.

Appendices

A. *Mismatched Filters for Three Channels*

In the three-channel case when all three filters are mismatched, we rewrite Eq. (2.2) in the form

$$v(t) = A_4 a_4(t - t_0) + A_5 a_5(t - t_0) + A_6 a_6(t - t_0) + n_1(t) \tag{A.1}$$

and from Eqs. (2.17), (2.18), and (2.19), the expected values of the least squares estimates become

$$E(\hat{A}_1) = \frac{1}{\beta c_1} \sum_{i=4}^{6} A_i c_i \left[(1 - \lambda_{23}^2)\lambda_{1i} - (\lambda_{12} - \lambda_{13}\lambda_{23})\lambda_{2i} \right.$$
$$\left. - (\lambda_{13} - \lambda_{12}\lambda_{23})\lambda_{3i} \right] \tag{A.2}$$

$$E(\hat{A}_2) = \frac{1}{\beta c_2} \sum_{i=4}^{6} A_i c_i \left[(1 - \lambda_{13}^2)\lambda_{2i} - (\lambda_{12} - \lambda_{13}\lambda_{23})\lambda_{1i} \right.$$
$$\left. - (\lambda_{23} - \lambda_{12}\lambda_{13})\lambda_{3i} \right] \tag{A.3}$$

$$E(\hat{A}_3) = \frac{1}{\beta c_3} \sum_{i=4}^{6} A_i c_i \left[(1 - \lambda_{12}^2)\lambda_{3i} - (\lambda_{13} - \lambda_{12}\lambda_{23})\lambda_{1i} \right.$$
$$\left. - (\lambda_{23} - \lambda_{12}\lambda_{13})\lambda_{2i} \right]. \tag{A.4}$$

Again, the mismatching may be interpreted in a variety of ways. For instance, if $a_4(t), a_5(t), a_6(t)$ are identical to $a_1(t)$, then one obtains

$$E(\hat{A}_1) = A_4 + A_5 + A_6\,, \quad E(\hat{A}_2) = E(\hat{A}_3) = 0\,.$$

Expressions (2.32), (2.33), (A.2), (A.3), and (A.4) are extremely useful in analyzing the system outputs. Considering this utility, the term "mismatching" is somewhat of a misnomer.

B. *Mismatched Filters for Four Channels*

For the four-channel system with all filters "mismatched", the input is expressed as

$$v(t) = A_5 a_5(t-t_0) + A_6 a_6(t-t_0) + A_7 a_7(t-t_0) + A_8 a_8(t-t_0) + n_1(t) \quad \text{(B.1)}$$

and one can show that

$$E(\hat{A}_1) = \frac{1}{c_1\phi} \sum_{i=5}^{8} A_i c_i [x_1\lambda_{1i} + x_2\lambda_{2i} + x_3\lambda_{3i} + x_4\lambda_{4i}] \qquad \text{(B.2)}$$

$$E(\hat{A}_2) = \frac{1}{c_2\phi} \sum_{i=5}^{8} A_i c_i [x_2\lambda_{1i} + x_5\lambda_{2i} + x_6\lambda_{3i} + x_7\lambda_{4i}] \qquad \text{(B.3)}$$

$$E(\hat{A}_3) = \frac{1}{c_3\phi} \sum_{i=5}^{8} A_i c_i [x_3\lambda_{1i} + x_6\lambda_{2i} + x_8\lambda_{3i} + x_9\lambda_{4i}] \qquad \text{(B.4)}$$

$$E(\hat{A}_4) = \frac{1}{c_4\phi} \sum_{i=6}^{8} A_i c_i [x_4\lambda_{1i} + x_7\lambda_{2i} + x_9\lambda_{3i} + x_{10}\lambda_{4i}] \qquad \text{(B.5)}$$

where

$$x_1 = 1 + 2\lambda_{23}\lambda_{24}\lambda_{34} - \lambda_{23}^2 - \lambda_{24}^2 - \lambda_{34}^2$$

$$x_2 = -\lambda_{12} + \lambda_{13}\lambda_{23} + \lambda_{14}\lambda_{24} - \lambda_{14}\lambda_{23}\lambda_{34} - \lambda_{13}\lambda_{24}\lambda_{34} + \lambda_{12}\lambda_{34}^2$$

$$x_3 = -\lambda_{13} + \lambda_{12}\lambda_{23} + \lambda_{14}\lambda_{34} - \lambda_{12}\lambda_{24}\lambda_{34} - \lambda_{14}\lambda_{24}\lambda_{23} + \lambda_{13}\lambda_{24}^2$$

$$x_4 = -\lambda_{14} + \lambda_{12}\lambda_{24} + \lambda_{13}\lambda_{34} - \lambda_{12}\lambda_{23}\lambda_{34} - \lambda_{13}\lambda_{23}\lambda_{24} + \lambda_{14}\lambda_{23}^2$$

$$x_5 = 1 + 2\lambda_{13}\lambda_{14}\lambda_{34} - \lambda_{13}^2 - \lambda_{14}^2 - \lambda_{34}^2$$

$$x_6 = -\lambda_{23} + \lambda_{12}\lambda_{13} + \lambda_{24}\lambda_{34} - \lambda_{12}\lambda_{14}\lambda_{34} - \lambda_{13}\lambda_{14}\lambda_{24} + \lambda_{14}^2\lambda_{23}$$

$$x_7 = -\lambda_{24} + \lambda_{12}\lambda_{14} + \lambda_{23}\lambda_{34} - \lambda_{12}\lambda_{13}\lambda_{34} - \lambda_{13}\lambda_{14}\lambda_{23} + \lambda_{13}^2\lambda_{24}$$

$$x_8 = 1 + 2\lambda_{12}\lambda_{14}\lambda_{24} - \lambda_{12}^2 - \lambda_{14}^2 - \lambda_{24}^2$$

$$x_9 = -\lambda_{34} + \lambda_{13}\lambda_{14} + \lambda_{23}\lambda_{24} - \lambda_{12}\lambda_{13}\lambda_{24} - \lambda_{12}\lambda_{14}\lambda_{23} + \lambda_{12}^2\lambda_{34}$$

$$x_{10} = 1 + 2\lambda_{12}\lambda_{13}\lambda_{23} - \lambda_{12}^2 - \lambda_{13}^2 - \lambda_{23}^2$$

and

$$\phi = 1 - \lambda_{12}^2 - \lambda_{13}^2 - \lambda_{14}^2 - \lambda_{23}^2 - \lambda_{24}^2 - \lambda_{34}^2 + 2\lambda_{12}\lambda_{13}\lambda_{23} + 2\lambda_{12}\lambda_{14}\lambda_{24}$$
$$+ 2\lambda_{13}\lambda_{14}\lambda_{34} + 2\lambda_{23}\lambda_{24}\lambda_{34} + \lambda_{12}^2\lambda_{34}^2 + \lambda_{13}^2\lambda_{24}^2 + \lambda_{14}^2\lambda_{23}^2$$
$$- 2\lambda_{12}\lambda_{13}\lambda_{24}\lambda_{34} - 2\lambda_{13}\lambda_{14}\lambda_{23}\lambda_{24} - 2\lambda_{12}\lambda_{14}\lambda_{23}\lambda_{34}\,.$$

Eqs. (B.2) through (B.5) are obtained by means of the scalar amplitude estimates

$$\hat{A}_1 = (1/c_1\phi)[R_1 x_1/c_1 + R_2 x_2/c_2 + R_3 x_3/c_3 + R_4 x_4/c_4] \qquad (B.6)$$
$$\hat{A}_2 = (1/c_2\phi)[R_1 x_2/c_1 + R_2 x_5/c_2 + R_3 x_6/c_3 + R_4 x_7/c_4] \qquad (B.7)$$
$$\hat{A}_3 = (1/c_3\phi)[R_1 x_3/c_1 + R_2 x_6/c_2 + R_3 x_8/c_3 + R_4 x_9/c_4] \qquad (B.8)$$
$$\hat{A}_4 = (1/c_4\phi)[R_1 x_4/c_1 + R_2 x_7/c_2 + R_3 x_9/c_3 + R_4 x_{10}/c_4]\,. \qquad (B.9)$$

C. *Effects of Delayed Arrival Times*

It is well known that a matched filter need not be perfectly matched to a given input in order to effectively enhance detection; nevertheless, it is worthwhile to quantitatively describe the effects of a delayed arrival time for one or more signals. For this purpose, we consider a two-channel system in which the input corresponding to expression (2.2) is now written as

$$v(t) = A_1 a_1(t - t_0) + A_2 a_2(t - t_0 - \epsilon) + n_1(t) \qquad (C.1)$$

where the signal $a_2(t)$ follows slightly behind the signal $a_1(t)$. In Eq. (C.1) we are considering a worse case where the quantity $\epsilon > 0$ is assumed to be small; otherwise, we would simply resolve $\nu(t)$ into two separate signals and there would be no need for discrimination.

It can be shown from expressions (2.8) and (2.9) that the expected values of the two-channel estimators are as follows.

At time $t = \Delta$,

$$E(\hat{A}_1) = A_1 - A_2 \cdot [\delta(\epsilon)/c_1 - \lambda\gamma(\epsilon)/c_2][c_1(1 - \lambda^2)]^{-1} \qquad (C.2)$$
$$E(\hat{A}_2) = A_2 - A_2 \cdot [\gamma(\epsilon)/c_2 - \lambda\delta(\epsilon)/c_1][c_2(1 - \lambda^2)]^{-1} \qquad (C.3)$$

and at time $t = \Delta + \epsilon$,

$$E(\hat{A}_1) = A_1 - A_1 \cdot [\beta(\epsilon)/c_1 - \lambda\delta(\epsilon)/c_2][c_1(1 - \lambda^2)]^{-1} \qquad (C.4)$$
$$E(\hat{A}_2) = A_2 - A_1 \cdot [\delta(\epsilon)/c_2 - \lambda\beta(\epsilon)/c_1][c_2(1 - \lambda^2)]^{-1} \qquad (C.5)$$

where

$$\delta(\epsilon) = \int_0^\Delta a_1(t - t_0)[a_2(t - t_0) - a_2(t - t_0 - \epsilon)]dt \qquad (\text{C.6})$$

$$\beta(\epsilon) = \int_0^\Delta a_1(t - t_0)[a_1(t - t_0) - a_1(t - t_0 - \epsilon)]dt \qquad (\text{C.7})$$

$$\gamma(\epsilon) = \int_0^{\Delta+\epsilon} a_2(t - t_0)[a_2(t - t_0) - a_2(t - t_0 - \epsilon)]dt \qquad (\text{C.8})$$

and

$$\lambda = \lambda_{12} = (1/c_1 c_2) \cdot \int_0^\Delta a_1(t - t_0)a_2(t - t_0)dt .$$

Expressions (C.2) through (C.5) apply to situations in which $0 < \lambda < 1$, since for $\lambda = 0$ we would not need to subtract a portion of the second signal and for $\lambda = 1$ the two signals are *a priori* indistinguishable in terms of pulse shape.

When $\epsilon = 0$, then $\delta(0) = \beta(0) = \gamma(0) = 0$ and $E(\hat{A}_1) = A_1, E(\hat{A}_2) = A_2$.
Also

$$\lim_{\epsilon \to \infty} \delta(\epsilon) = \lambda c_1 c_2 \,, \ \lim_{\epsilon \to \infty} \beta(\epsilon) = c_1^2 \,, \ \lim_{\epsilon \to \infty} \gamma(\epsilon) = c_2^2$$

so that at time $t = \Delta$

$$\lim_{\epsilon \to \infty} E(\hat{A}_1) = A_1 \,, \ \lim_{\epsilon \to \infty} E(\hat{A}_2) = 0$$

and at time $t = \Delta + \epsilon$

$$\lim_{\epsilon \to \infty} E(\hat{A}_1) = 0 \,, \ \lim_{\epsilon \to \infty} E(\hat{A}_2) = A_2 \,.$$

The expected value of, for example, $\hat{A}_1^2$ at $t = \Delta$ is given by

$$\begin{aligned}
E(\hat{A}_1^2) = &A_1^2 + [c_1^2(1 - \lambda^2)]^{-1} + A_2^2[\theta(\epsilon)/c_1 - \lambda\phi(\epsilon)/c_2]^2 \cdot [c_1^2(1 - \lambda^2)^2]^{-1} \\
&+ 2A_1 A_2[\theta(\epsilon) - \lambda c_1 \phi(\epsilon)/c_2] \cdot [c_1^2(1 - \lambda^2)]^{-1}
\end{aligned}$$
$$(\text{C.9})$$

where $\phi(\epsilon) = c_2^2 - \gamma(\epsilon)$ and $\theta(\epsilon) = \lambda c_1 c_2 - \delta(\epsilon)$, and the second term of Eq. (C.9) is the variance of $\hat{A}_1$ when $\epsilon = 0$. In expressions (C.2) through (C.9), we allowed the sensor to read out the filters at the preferred times $t = \Delta$ or $t = \Delta + \epsilon$. No errors were incurred in attaining these readout times, and system degradation was reflected in the bias terms of Eqs. (C.2) through (C.5) and also in the increased variance.

References

1. John B. Thomas, *An Introduction to Statistical Communication Theory*, New York, John Wiley and Sons, Inc., 1969.
2. R. E. Paley and Norbert Wiener, *Fourier Transforms in the Complex Domain*, Chicago, American Mathematical Society Colloquium Publication, No. 10, 1934.
3. J. L. Doob, *Stochastic Processes*, New York, John Wiley and Sons, Inc., 1953.
4. Norbert Wiener, *Extrapolation, Interpolation and Smoothing of Stationary Time Series*, New York, John Wiley and Sons, Inc., 1949.
5. Carl Helstrom, *Statistical Theory of Signal Detection*, Oxford, Pergamon Press, 1960.
6. N. J. Nilsson, *On the optimum range resolution of radar signals in noise*, Trans. IRE, Vol. IT-7, (Oct 1961) 245–253.
7. A. E. Hoerl and R. W. Kennard, *Ridge regression: Biased estimation for nonorthogonal problems*, Technometrics, **12** (Feb 1970) 55–67.
8. D. W. Marquardt and R. D. Snee, *Ridge regression in practice*, The American Statistician (1) **29** (Feb 1975) 3–20.
9. J. Mirra, F. McNolty, E. Hansen, and M. Brusato, *Ridge detection of culprit variables*, 1983 Annual Proceedings of Reliability and Maintainability, pp. 212–220.

Frank McNolty, William Sherwood and Jean Mirra
Lockheed Missiles & Space Company, Inc.
P.O. Box 3504
Sunnyvale, California 94088-3504
U.S.A

THE MATH. HERITAGE OF C.F. GAUSS (pp. 505-516)
edited by George M. Rassias
©1991 World Scientific Publ. Co. Singapore

POWERS OF 2, CONTINUED FRACTIONS, AND THE CLASS NUMBER ONE PROBLEM FOR REAL QUADRATIC FIELDS $Q(\sqrt{d})$, WITH $d \equiv 1 \pmod 8$

R. A. Mollin[1] *and H. C. Williams*[2]

Herein we investigate the connection between the class number one problem for real quadratic fields $Q(\sqrt{d})$ where $d \equiv 1 \pmod 8$, and the continued fraction expansion of $w = (1+\sqrt{d})/2$. In particular, we classify precisely those $d \equiv 1 \pmod 8$ for which all of the Q_i's (see Sec. 1) in the continued fraction expansion of w, are powers of 2. Using this approach we find, for example, a lower bound on d (in terms of the period of w) beyond which the class number $h(d) > 1$. Other connections with the class number one problem are also elucidated. This continues from the work of the authors of [5]–[13].

Introduction

In this paper we investigate real quadratic fields $Q(\sqrt{d})$ where $d \equiv 1 \pmod 8$. In particuar for these fields, we look at the continued fraction expansion of $w = (1 + \sqrt{d})/2$ and completely classify such d for which all of the Q_i's (see Sec. 1) are powers of 2. We then investigate the relationship between this phenomenon and the class number one problem for $Q(\sqrt{d})$ when $d \equiv 1 \pmod 8$. Specifically, we pose a conjecture to the effect that when all Q_i's are powers of 2 then there are exactly five such fields of class number one; viz. for $d \in \{17, 41, 113, 353, 1217\}$. We are able to prove this conjecture (using techniques of [5]–[7]) for all except possibly one value;. i.e., that the conjecture holds except for possibly one remaining value.

In the last section we investigate the case where all the Q_i's are *not*

---

[1]This author's research is supported by NSERC Canada Grant ♯A8484.

[2]This author's reserach is supported by NSERC Canada Grant ♯A7649.

powers of 2. Here we are able to give an explicit lower bound on d (in terms of the period of w) such that the class number $h(d) > 1$ beyond the bound.

1. Notation and Preliminaries

Let $d \equiv 1 \pmod 4$ be a square-free and positive integer. If $w = (1 + \sqrt{d})/2$ then let $w = \langle a, \overline{a_1, a_2, \ldots, a_k} \rangle$ denote the continued fraction expansion of w, hence with *period k*. Then $a_0 = a = [w]$, (where $[x]$ denotes the greatest integer less than or equal to x), and $a_i = [(P_i + \sqrt{d})/Q_i]$ for $i \geq 1$ where $P_0 = 1, Q_0 = 2$ and

$$P_{i+1} = a_i Q_i - P_i \quad \text{for } i \geq 0 \tag{1.1}$$

$$Q_{i+1} Q_i = d - P_{i+1}^2 \quad \text{for } i \geq 0. \tag{1.2}$$

Thus:

$$a_i < a \quad \text{for } k > i \geq 1 \tag{1.3}$$

$$P_i < \sqrt{d} \quad \text{for } i \geq 0 \tag{1.4}$$

$$Q_i < 2\sqrt{d} \quad \text{for } i \geq 0 \tag{1.5}$$

$$-1 < (P_i - \sqrt{d})/Q_i < 0 \quad \text{for } i \geq 0. \tag{1.6}$$

Since $d \equiv 1 \pmod 4$ then

$$w = \langle a, \overline{a_1, a_2, \ldots, a_{k-1}, 2a - 1} \rangle \tag{1.7}$$

where $a_i = a_{k-i}$ for $1 \leq i \leq k - 1$.

The ring of integers in $K = Q(\sqrt{d})$ is denoted by $\mathcal{O}_K$, principal ideals generated by an element a are denoted (a), and $\mathcal{O}_K$-primes are denoted by upper case script letters $\mathcal{P}, \mathcal{R}$, etc. We refer the reader to [9] or [14]–[15] for details of the theory of reduced ideals used herein.

2. Q_i's as Powers of 2 and $d \equiv 1 \pmod 8$

Throughout this section d will be a positive square-free integer with $d \equiv 1 \pmod 8$. The notation and definitions of Sec. 1 are in force throughout as well. The main goal of this section is to classify such d for which *no* Q_i has an odd prime factor.

Lemma 2.1. (a) If $d = (2a - 1)^2 + 8r$ and $Q_1 = 4r$ and so Q_1 has an odd prime factor whenever r has one.

(b) If $d = (2a - 3)^2 + 8r$ where r is divisible by an odd prime and the $\mathcal{O}_K$-primes above 2 are principal then Q_i has an odd prime factor for some i with $0 < i < k$.

Proof. (a) is obvious.

(b) Since the $\mathcal{O}_K$-ideals above 2 are principal and it is possible to find a reduced such ideal, (see for example [9, Theorem 2.1] or [15]), then $Q_i = 4$ for some i with $0 < i < k$, since it is well-known that the norm of a principal reduced ideal is some $Q_i/2$, (see for example [4, Proposition 2, p. 169] or [8, proof of Theorem 1.3]. Now, by [8, Theorem 4.1] we must have $a_i = a - 1$. Since $[(P_i + \sqrt{d})/4] = a - 1$ then $P_i \geq 2a - 3$. However, $P_i < \sqrt{d}$; whence $P_i = 2a - 1$ or $2a - 3$, keeping in mind that P_i must be odd since Q_i is even (and that $[\sqrt{d}] = 2a$ or $2a - 1$).

If $P_i = 2a - 3$ then $Q_{i-1} = 2r$ since $4 = Q_i = (d - P_i^2)/Q_{i-1} = (d - (2a - 3)^2)/Q_{i-1} = 8r/Q_{i-1}$; whence Q_{i-1} has an odd prime factor. If $P_i = 2a - 1$ then $P_{i+1} = a_i Q_i - P_i = 2a - 3$. Hence, $Q_{i+1} = (d - P_{i+1}^2)/Q_i = (d - (2a - 3)^2)/4 = 2r$. Therefore, Q_i has an odd prime factor.

QED

Remark 2.1. Lemma 2.1(b) fails if the $\mathcal{O}_K$-primes above 2 are not principal. For example, if $d = 257$ then $a = 8$ so $(2a - 3)^2 + 8r = 13^2 + 8 \cdot 11$. However, it is easily checked that $k = 3, Q_0 = Q_3 = 2$ and $Q_1 = Q_2 = 16$. Here $h(257) = 3$ and the $\mathcal{O}_K$-primes above 2 are not principal, in particular.

Now we examine those d for which the hypotheses of Lemma 2.1 fail.

Lemma 2.2. Suppose that both $d - (2a - 1)^2$ and $d - (2a - 3)^2$ are powers of 2, and that $[\sqrt{d}] = 2a - 1$. Then $d = (2^m - 2^n + 1)^2 + 2^{n+2}$ with $m > n \geq 1$, and $[\sqrt{d}] = 2^m - 2^n + 1$. All Q_i's are powers of 2 if and only if $k = 2j + 1 = (m + n)/(m - n); j \geq 1; (j + 1)n = jm$ with j properly dividing n.

Proof. We have that $d - [\sqrt{d}]^2 = 2^s$ and $d - ([\sqrt{d}] - 2)^2 = 2^t$ with $t > s \geq 3$. Also, $2^s + 4[\sqrt{d}] - 4 = 2^t$, so $[\sqrt{d}] = 2^{t-2} - 2^{s-2} + 1$. Therefore, $d = (2^{t-2} - 2^{s-2} + 1) + 2^s$. Set $n = s - 2$ and $m = t - 2$ to get the first statement; viz., that $d = (2^m - 2^n + 1)^2 + 2^{n+2}$ with $m > n \geq 1$ and

$[\sqrt{d}] = 2^m - 2^n + 1$. If $k = 1$ then $n = 0$, a contradiction and if $k = 2$ then $m = n$, a contradiction.

Now we calculate for $k > 2$ that:

i	0	1	2
P_i	1	$2^m - 2^n + 1$	$2^m - 2^n - 1$
Q_i	2	2^{n+1}	2^{m-n+1}
a_i	$2^{m-1} - 2^{n-1} + 1$	$2^{m-n} - 1$	$2^n - 1 (\text{if } m > 2n)$ $2^n - 2^{2n-m}(\text{if } m \leq 2n)$

<u>Case 1</u>: $m > 2n$. From well-known recurrence relations we have:

$$Q_3 = Q_1 - a_2^2 Q_2 + 2a_2 P_2 = 2(2^n + 2^m - 2^{m-n} - 2^{2n} + 1);$$

whence $Q_3/2$ is odd.

<u>Case 2</u>: $m \leq 2n$. We calculate:

i	3	4
P_i	$2^m - 2^n + 1$	$2^m - 2^n - 1$
Q_i	2^{2n-m+1}	$2^{2m-2n+1}$
a_i	$2^{2m-2n} - 2^{m-n}$	$2^{2n-m} - 2^{3n-2m} \text{ if } 3n \geq 2m$ $2^{2n-m} - 1 \quad\;\; \text{ if } 3n < 2m$

Now if $3n < 2m$ then it is straightforward to show that $Q_5 = 2(1 + 2^n - 2^{3n-m} - 2^{2m-2n} + 2^m)$ whence $Q_5/2$ is odd. Thus we assume $3n \geq 2m$, and we find by induction that

$$P_{2i+1} = 2^m - 2^n + 1 \quad \text{for } i \geq 0$$
$$P_{2i} = 2^m - 2^n - 1 \quad \text{for } i \geq 1$$
$$Q_{2i+1} = 2^{(i+1)n-im+1} \quad \text{for } i \geq 0$$

$$Q_{2i} = 2^{im-in+1} \quad \text{for } i \geq 1$$

and $k = 2j + 1$ where $(j + 1)n = jm; j \geq 1$ and j divides n. If $j = n$ then one easily shows that

$$d = (2^n + 1)^2 + 2^{n+2} \quad \text{and}$$
$$(2^n + 2)^2 < d; \quad \text{whence } [\sqrt{d}] = 2a,$$

a contradiction. This completes the proof.

QED

Corollary 2.1. If $d - (2a - 1)^2$ and $d - (2a - 3)^2$ are powers of 2, $[\sqrt{d}] = 2a - 1$ and the $\mathcal{O}_K$-primes above 2 are principal then not all the Q_i's are powers of 2.

Proof. If all Q_i's are powers of 2 then for some i we must have $Q_i = 4$; whence we must have $(i + 1)n - im = 1$. Since $k = 2j + 1$ then $j = i + x$ with $x \geq 1$. Since $(j + 1)n = jm$ then $(i + x + 1)n = (i + x)m$. Thus $x = 1, m = n + 1$, and $j = n$. Thus $k = 2n + 1$ and $d = (2^n + 1)^2 + 2^{n+2}$; whence $[\sqrt{d}] = 2a = 2^n + 2$ as in Lemma 2.2.

QED

Example 2.1. Illustrations of Lemma 2.2 for $k = 5$ are $d \in \{2465, 201857, 14754305\}$. When $k = 3$ then $m = 2n$ and we have as examples $d \in \{185, 3281, 58145, 986177\}$.

Lemma 2.3. Suppose that $[\sqrt{d}] = 2a$ and both $d - (2a - 1)^2$ and $d - (2a - 3)^2$ are powers of 2. Then $d = (2^m + 1)^2 + 2^{m+2}, m \geq 1, k = 2m + 1$ and all Q_i's are powers of 2.

Proof. We have that:

$$d - ([\sqrt{d}] - 1)^2 = 2^s \quad \text{and} \quad d - ([\sqrt{d}] - 3)^2 = 2^t \quad \text{with } s, t \geq 3 \text{ and } t > s.$$

Thus, $[\sqrt{d}] = 2^{t-2} - 2^{s-2} + 2$ and $d = (2^{t-2} - 2^{s-2} + 1)^2 + 2^s$.

<u>Claim:</u> $s = t - 1$.

$$[\sqrt{d}]^2 = (2^{t-2} + 2^{s-2} + 1)^2 + 2(2^{t-2} - 2^{s-2} + 1) + 1$$
$$< (2^{t-2} - 2^{s-2} + 1)^2 + 2^s \text{ since } [\sqrt{d}]^2 < d.$$

Thus, $2^{t-1} - 2^{s-1} + 3 < 2^s$; whence

$$3 + 2^{t-1} < 3 \cdot 2^{s-1}. \qquad (*)$$

If $s \neq t - 1$ then $s \leq t - 2$ whence; $2^{t-1} > 3 \cdot 2^{t-3} \geq 3 \cdot 2^{s-1} > 2^{t-1} + 3$ where the last inequality is $(*)$. Hence this contradiction establishes the claim. Set $m = s - 2$ and we get that $d = (2^m + 1)^2 + 2^{m+2}$ for $m \geq 1$. It remains to show that $k = 2m + 1$ and all Q_i's are powers of 2. We do this by the explicit calculation of the continued fraction expansion of w. (Note that $[\sqrt{d}] = 2^m + 2$). It is easily shown by induction that:

$$P_{2i+1} = 2^m + 1 \quad \text{for} \quad k \geq i \geq 0;$$
$$P_{2i} = 2^m - 1 \quad \text{for} \quad k > i \geq 1;$$
$$Q_{2i+1} = 2^{m-i+1} \quad \text{for} \quad k \geq i \geq 0;$$
$$Q_{2i} = 2^{i+1} \quad \text{for} \quad k > i \geq 1;$$
$$a_{2i+1} = 2^i \quad \text{for} \quad k > i > 0;$$
$$a_{2i} = 2^m + 1 \quad \text{for} \quad k > i > 0.$$

Hence $k = 2m + 1$.

QED

Remark 2.2. We observe from the proof of Lemma 2.3 that $Q_2 = 4$ for such d as given in the hypothesis. It follows that the $\mathcal{O}_K$-primes above 2 are principal necessarily.

The following table illustrates Lemma 2.3. We include the class number for reference later.

m	k	d	$h(d)$
1	3	17	1
2	5	41	1
3	7	113	1
4	9	353	1
5	11	1217	1
6	13	4481	3
7	15	17153	2
8	17	67073	3
9	19	265217	6
10	21	10485761	?

Table 2.1.

Remark 2.3. All class numbers in Table 2.1 and in fact all class numbers throughout the paper, are taken from an unpublished table of class numbers of real quadratic fields for all values of d up to one million. This table was compiled by the second author (under the nearly continuous prompting of the first author).

We now have our first main result.

Theorem 2.1. If $d \equiv 1 \pmod{8}$ and both $d-(2a-1)^2$ and $d-(2a-3)^2$ are powers of 2 then all Q_i's are powers of 2 if and only if either:

(a) $d = (2^m + 1)^2 + 2^{m+2}; m \geq 1$ and, $k = 2m + 1$; or

(b) $d = (2^m - 2^n + 1)^2 + 2^{n+2}$ with $m > n \geq 1; k = 2j + 1 = (m + n)/(m - n)$; some $j \geq 1$ and $(j + 1)n = jm$ with j properly dividing n.

Remark 2.4. Note that in Theorem 2.1(a) we necessarily have $[\sqrt{d}] = 2^m + 2 = 2a$ and the $\mathcal{O}_K$-primes above 2 are principal. This follows from Lemma 2.3. On the other hand in Theorem 2.1(b) then $\mathcal{O}_K$-primes above 2 are *not* principal and $[\sqrt{d}] = 2^m - 2^n + 1$ necessarily, from Lemma 2.2.

Conjecture 2.1.[3] The values $d \in \{17, 41, 113, 353, 1217\}$ are all those $d \equiv 1 \pmod 8$ having $h(d) = 1$ and all Q_i's as powers of 2.

Remark 2.5. Using the techniques of [5]–[7] it can be shown that Conjecture 2.1 holds *with possibly only one more value remaining.*

Remark 2.1 shows us that we missed some values because of the assumption in Lemma 2.1(b). Therefore, to complete the classification we consider the case where the $\mathcal{O}_K$-primes above 2 are *not* principal.

Theorem 2.2. If $d \equiv 1 \pmod 8$ and both $d-(2a-1)^2$ and $d-(2a-3)^2$ are powers of 2 and the $\mathcal{O}_K$-primes above 2 are *not* principal then all Q_i's are powers of 2 if and only if $d = (2a-1)^2 + 2^{f+2}$ where $2a-1 = 2^f b + s$ (with $a_1 = b$), $0 < s < 2^f$ and $bs+1 = 2^{2f/(k-1)}$ with $f \equiv 0 \pmod{(k-1)/2}; k > 1$ odd and $2f \neq k - 1$.

Proof. By virtue of Lemma 2.1(a) we may assume that $d = (2a-1)^2 + 2^{f+2}$ with $f \geq 2$. Thus, $Q_1 = 2^{f+1}$. Clearly $2a - 1 = 2^f b + s$ where $0 < s < 2^f$. Therefore, $P_2 = 2^f b - s$ and $Q_2 = 2(bs + 1)$ where $bs + 1 = 2^t$ with $t \geq 2$. Therefore $a_2 = 2^{f-t}b; P_3 = 2^f b + s; Q_3 = 2^{f+1-t}; a_3 = 2^t b + i$ where $s = 2^{f-t}i + j$ with $i \geq 0; 0 \leq j < 2^{f-t}$. Hence $P_4 = 2^f b - j + 2^{f-t}i$ and $Q_4 = 2^{t+1}bj + 2ij + 2^{t+1}$. Since $Q_4 = 2^u$ for $u \geq 3$ then:

<u>Case 1</u>: If $u < t + 1$ then $1 = 2^{t+1-u}bj + (ij/2^{u-1}) + 2^{t+1-u}$. If $j \neq 0$ then the right-hand side of the equality is at least 4, a contradiction. Thus, $j = 0$ which forces $u = t + 1$, a contradiction.

<u>Case 2</u>: $u \geq t+1$. Thus, $2^{u-t-1} = bj + (ij/2^t) + 1$. If $ij \neq 0$ then 2^t divides i, whence, $s \geq 2^f + j > 2^f$, a contradiction. If $j = 0$ then $f = t$ since $2a - 1 = 2^f b + s = 2^f b + 2^{f-t}i$ which forces $Q_3 = 2$; whence $k = 3$ and the result follows. So we assume $j \neq 0$. If $i = 0$ then $2^{u-t-1} = bj + 1 = bs + 1 = 2^t$; whence $u = 2t + 1$.

---

[3]Note that although Shewks found only these values up to a large bound in [*On Gauss' class number problems*, Math. Comp. **23** (1969) 151–163], he made no comment or conjecture to the effect that these were all of them.

Continuing the above process we see that the following occurs:

$$P_{2i+1} = 2^f b + s \quad \text{for} \quad i = 0, 1, \ldots, (k-1)/2$$
$$P_{2i} = 2^f b - s \quad \text{for} \quad i = 1, 2, \ldots, (k-1)/2$$
$$Q_{2i} = 2^{it+1} \quad \text{for} \quad i = 1, 2, \ldots, (k-1)/2$$
$$Q_{2i+1} = 2^{f+1-it} \quad \text{for} \quad i = 0, 1, \ldots, (k-1)/2$$
$$a_{2i} = b \cdot 2^{f-it} \quad \text{for} \quad i = 1, 2, \ldots, (k-1)/2$$
$$a_{2i+1} = b \cdot 2^{it} \quad \text{for} \quad i = 0, 1, \ldots, (k-1)/2$$

Since $s \neq 0$ then $P_i \neq P_{i+1}$ for all i so k must be odd. Moreover, $Q_k = 2$ implies that $f + 1 - (k-1)t/2 = 1$; i.e., $t = 2f/(k-1)$.

Finally since the $\mathcal{O}_K$-primes above 2 are not principal then $it + 1 \neq 2$ for any i and $f + 1 - it \neq 2$ for any i. Hence $2f \neq k - 1$.

QED

Remark 2.6. In [1]–[2] L. Bernstein was interested in the forms of d in this section. However, he was interested in them from the perspective of the continued fraction expansion of $\sqrt{d}$ rather than $(1 + \sqrt{d})/2$. Moreover he was interested in them from the perspective of calculating the fundamental of $Q(\sqrt{d})$. C. Levesque et al. also demonstrated some similar interest in [3].

3. Q_i's Having an Odd Prime Factor and $h(d) = 1$.

As in Sec. 2, d is a positive square-free integer with $d \equiv 1 \pmod 8$ throughout. Herein, we wish to find a lower bound on d (based on k) beyond which we know that $h(d) > 1$. We do this in terms of the odd primes dividing Q_i. We need a preliminary result first.

Lemma 3.1. Let $d - x^2 = 2^c t, c \geq 3, x > 0, \sqrt{d} - x < 2^{c-1}$ and $\sqrt{d} + x > 2^{c-1}$ then $\mathcal{J} = [t, (x + \sqrt{d})/2]$ is a reduced ideal in $\mathcal{O}_K$.

Proof. That $\mathcal{J}$ is an ideal follows from the facts elucidated in [14] (see also [9]). Set $\phi = (x + \sqrt{d})/2t$. From [9, Theorem 2.2] we have that $\mathcal{J}$ is reduced if and only if there exists a $\beta \in \mathcal{J}$ such that $\beta > t$ and $-t < \overline{\beta} < 0$, where $\overline{\beta}$ is the conjugate of β. Thus, it suffices to show that

$-1 < \overline{\phi} < 0$ and that $\phi > 1$. We have that $(\sqrt{d} - x)\phi = 2^{c-1}$; whence $\phi > 1$, since $\sqrt{d} - x < 2^{c-1}$. Moreover $\overline{\phi} = -2^{c-1}/(\sqrt{d} + x)$ which implies that $-1 < \overline{\phi} < 0$.

QED

Remark 3.1. For example, when $c = 3$ then possible values of x in Lemma 3.1 are $[\sqrt{d}], [\sqrt{d}] - 1, [\sqrt{d}] - 2$ and $[\sqrt{d}] - 3$. Moreover, exactly two of these will generate reduced ideals; i.e., if $[\sqrt{d}]$ is even then, since x must be odd in Lemma 3.1, only $[\sqrt{d}] - 1$ and $[\sqrt{d}] - 3$ are allowable values of x. Similarly, if $[\sqrt{d}]$ is odd then x is allowed to be only $[\sqrt{d}]$ or $[\sqrt{d}] - 2$.

Lemma 3.2. Suppose that not all Q_i's are powers of 2. Then there exists an odd prime p dividing Q_i for some i such that the $\mathcal{O}_K$-primes above p are reduced.

Proof. If p divides Q_i and $p < \sqrt{d}/2$ then the ideals above p are reduced by [9, Corollary 2.2].

Thus we assume for the remainder of the proof that $p > \sqrt{d}/2$ whenever an odd prime p divides some Q_i. If p divides Q_i and $Q_i \neq 2p$ for some i then $Q_i \geq 4p$. Since $2\sqrt{d} > Q_i$ by (1.5) then $p < \sqrt{d}/2$, a contradiction. Therefore we may assume that $Q_i = 2p$ for any i where Q_i is divisible by an odd prime p. By (1.2), $d = P_i^2 + 2pQ_{i-1}$. Since $d \equiv 1 \pmod{8}$ then 4 must divide Q_{i-1} in which case no odd prime can divide Q_{i-1} by assumption. Thus, $Q_{i-1} = 2^f$ and $d = P_i^2 + 2^{f+1}p$ where $f \geq 2$. Now, if $\sqrt{d} - P_i \geq 2^f$ then $2^f(\sqrt{d} + P_i) \leq (\sqrt{d} - P_i)(\sqrt{d} + P_i) = 2^{f+1}p$; i.e., $(\sqrt{d} + P_i)/2p \leq 1$. However, $a_i = [(\sqrt{d} + P_i)/2p] < (\sqrt{d} + P_i)/2p$, contradicting the fact that $a_i \geq 1$. On the other hand, if $\sqrt{d} + P_i \leq 2^f$ then $2^f(\sqrt{d} - P_i) \geq (\sqrt{d} + P_i)(\sqrt{d} - P_i) = 2^{f+1}p$; whence $p \leq (\sqrt{d} - P_i)/2 < \sqrt{d}/2$, contradicting our initial assumption. Hence we have shown that the hypothesis of Lemma 3.1 is satisfied; whence, the ideal $[p, (P_i + \sqrt{d})/2]$ is reduced.

QED

The following improves [8, Theorem 2.1] for the $d \equiv 1 \pmod{8}$ case.

Theorem 3.1. Let $d \equiv 1 \pmod 8$ such that not all Q_i's are powers of 2. Thus, if $d > 2^{k+1}$ then $h(d) > 1$.

Proof. Assume $h(d) = 1$.

Since $d > 2^{k+1}$ then $2^{\frac{k-1}{2}} < \frac{\sqrt{d}}{2}$. Hence $2^i < \sqrt{d}/2$ for $i = 0, 1, \ldots, m$ where $m = \left[\frac{k-1}{2}\right]$. Since $(2) = \mathcal{R}\overline{\mathcal{R}}$ in $\mathcal{O}_K$ then $\{1, \mathcal{R}^i, \overline{\mathcal{R}}^i\}_{i=1}^{m}$ provides $2m + 1$ distinct reduced ideals. By Lemma 3.2 there is an odd prime p with $(p) = \mathcal{P}\overline{\mathcal{P}}$ and $\mathcal{P}, \overline{\mathcal{P}}$ reduced. Thus we have a total of $2m+3$ distinct reduced ideals. Since k counts precisely the number of reduced ideals equivalent to a given one (in this case, equivalent to the principal class), (e.g. see [15, pp. 415–416]), then, $k \geq 2m + 3 = 2[(k-1)/2] + 3 > 2(k-3)/2 + 3 = k$, a contradiction.

QED

References

1. L. Bernstein, *Fundamental units and cycles* I, J. Number Thy. **8** (1976) 446–491.

2. L. Bernstein, *Fundamental units and cycles in the period of real quadratic number fields*, Part II, Pacific J. Math. **63** (1976) 63–78.

3. C. Levesque and G. Rhin, *A few classes of periodic continued fractions*, preprint #85-5, Université Laval.

4. S. Louboutin, *Continued fractions and real quadratic fields*, J. Number Thy. **30** (1988) 167–176.

5. R. A. Mollin and H. C. Williams, *A conjecture of S. Chowla via the generalized Riemann hypothesis*, Proc. Amer. Math. Soc. **102** (1988) 794–796.

6. R. A. Mollin and H. C. Williams, *On prime-valued polynomials and class numbers of real quadratic fields*, Nagoya Math J. **112** (1988) 143–151.

7. R. A. Mollin and H. C. Williams, *Solution of the class number one problem for real quadratic fields of extended Richaud-Degert type (with one possible exception)*, in Number Theory (ed. R. A. Mollin) Walter de Gruyter (1990) 417–425.

8. R. A. Mollin and H. C. Williams, *Real quadratic fields of class number one and related continued fractions*, to appear.

9. R. A. Mollin and H. C. Williams, *Class number one for real quadratic fields, continued fractions and reduced ideals*, in Number Theory and Applications (ed. R. A. Mollin) Kluwer Academic Pub. (1989) 481–496.

10. R. A. Mollin and H. C. Williams, *Computation of the class number of a real quadratic field*, to appear in Advances in the Theory of Computation and Computational Mathematics.

11. R. A. Mollin and H. C. Williams, *Continued fractions of period five and real quadratic fields of class number one*, to appear in Acta. Arith.

12. R. A. Mollin and H. C. Williams, *Prime-producing quadratic polynomials and real quadratic fields of class number one*, in Number Theory (ed. J. M. De Koninck and C. Levesqve) Walter de Gruyter (1989) 654–663.

13. R. A. Mollin and H. C. Williams, *Quadratic non-residues and prime-producing polynomials*, Canad. Math. Bull. **32** (1989) 474–478.

14. H. C. Williams, *Continued fractions and number theoretic omputations*, Rocky Mtn. J. Math. **15** (1985) 621–655.

15. H. C. Williams, and M. C. Wunderlich, *On the parallel generation of the residues for the continued fraction factoring algorithm*, Math. Comp. **177** (1987) 405–423.

R. A. Mollin
Department of Mathematics & Statistics
University of Calgary
Calgary, Alberta
T2N 1N4 Canada

H. C. Williams
Computer Science Department
University of Manitoba
Winnipeg, Manitoba
R3T 2N2 Canada

THE MATH. HERITAGE OF C.F. GAUSS (pp. 517-525)
edited by George M. Rassias
©1991 World Scientific Publ. Co. Singapore

THE VALIDITY OF GAUSSIAN ELECTRODYNAMICS

*Parry Moon, Domina Eberle Spencer,
Shama Y. Uma and Philip J. Mann*

At low velocities, there are three expressions for the force between current elements: the equation suggested by Ampère (1823), the classical equation first suggested by Grassmann (1845) and the relative velocity formulation introduced by Moon, Spencer, Uma and Mann (1987) which incorporates Gauss[1] electrodynamic concept. The first and third equations satisfy Newton's third law; the classical equation does not. The paper shows that for forces on a current element produced by a closed circuit, only the third equation can produce tangential forces. The validity of Gaussian electrodynamics must be determined by careful analysis of the many experiments already reported which appear to indicate the existence of tangential forces.

A previous paper [1] has shown that for moving charges in current elements at steady low velocities, there are three important equations for the force between current elements, Fig. 1. The first was suggested by Ampère [2] in 1823

$$d^2\mathbf{F}_A = -\frac{I_1 I_2}{4\pi\epsilon_0 c^2 r^2}\mathbf{a}_r[2(\mathbf{ds}_1 \cdot \mathbf{ds}_2) - 3(\mathbf{ds}_1 \cdot \mathbf{a}_r)(\mathbf{ds}_2 \cdot \mathbf{a}_r)]. \qquad (1)$$

This equation was derived by Ampère himself based on his own experiments. It is the equation which was also derived by Gauss from his unpublished electrodynamic equation. It was also derived by Weber from the first published electrodynamic equation which was equivalent to the Gauss equation for constant currents. This equation was used by Bush [3] and by Moon and Spencer [4]. More recently it has been used by Graneau [5], Pappas [6], Wesley [7] and Phipps [8]. Despite all the illustrious scientists who have thought highly of this equation it has one very serious fault. It can be derived from a force field that is expressible as the gradient of a scalar potential. Such a force field would appear to be inconsistent with Faraday's experiments on induction.

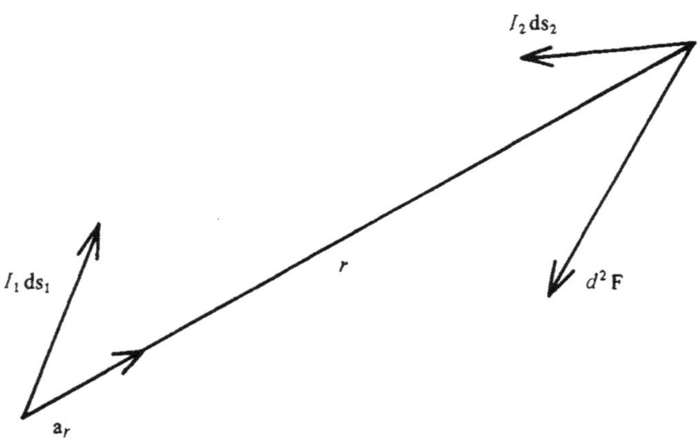

Fig. 1. The force between current elements.

Classical physics, as taught today employs an equation first suggested by Grassmann [9] in 1845

$$d^2\mathbf{F}_C = \frac{I_1 I_2}{4\pi\epsilon_0 c^2 r^2}\mathbf{ds}_2 \times (\mathbf{ds}_1 \times \mathbf{a}_r)\,. \tag{2}$$

This equation can be derived from a scalar and a vector potential, if the velocity in the numerator of the vector potential is the velocity of the source at the instant of emission in the coordinate system in which the force vector is expressed. This follows the Gaussian ideas in one way: it is derived from retarded potentials. But this equation violates the general principles for an electrodynamic force equation laid down by Gauss: it is defined in terms of velocities relative to the coordinate system rather than the velocity of the source relative to the receiver. This expression was used by Lorentz [10] in 1892.

The third possible equation was introduced by Moon, Spencer, Uma and Mann [11] in 1987. Following the ideas of Gauss, Weber and Riemann, it is expressed in terms of relative velocity between source and receiver. It follows from retarded scalar and vector potentials [12].

$$d^2\mathbf{F}_R = \frac{I_1 I_2}{8\pi\epsilon_0 c^2 r^2}[\mathbf{ds}_1 \times (\mathbf{ds}_2 \times \mathbf{a}_r) + \mathbf{ds}_2 \times (\mathbf{ds}_1 \times \mathbf{a}_r)]\,. \tag{3}$$

This equation is in harmony with the two important ideas first suggested by Gauss that the electrodynamic equation should include retardation and should be expressed in terms of relative velocity. It is the equation Gauss himself might have derived if he had worked on the subject long enough to realize that both a scalar and a vector potential were necessary. For this reason we propose to call this approach Gaussian Electrodynamics.

A disquieting characteristic of the classical equation, Eq. (2) for the force between current elements is that it does not satisfy Newton's third law. That is, the force exerted by $I_1 \mathbf{ds}_1$ on $I_2 \mathbf{ds}_2$ is not equal and opposite to the force exerted by $I_2 \mathbf{ds}_2$ on $I_1 \mathbf{ds}_1$. Both the Ampère equation and the proposed equation based on Gauss' principles are symmetric in $\mathbf{ds}_1$ and $\mathbf{ds}_2$. Consequently both Eqs. (1) and (3) satisfy Newton's third law.

In order to study the question of how the validity of Eqs. (1), (2) and (3) can be tested we wish to compare the three equations, remembering that current elements themselves are fictitious quantities. This paper attempts to produce criteria for comparison which can be tested experimentally.

1. Three Expressions for $d^2\mathbf{F}$

It will be convenient to express, $d^2\mathbf{F}_A, d^2\mathbf{F}_C$ and $d^2\mathbf{F}_R$ in terms of rectangular coordinates so that normal and tangential components can be readily identified, Fig. 2. Suppose that the current element on which the force acts is at the origin of the coordinate system and in the z-direction, then

$$\mathbf{ds}_2 = \mathbf{a}_z ds_2 . \tag{4}$$

The source is at distance r,

$$r = \left[x^2 + y^2 + z^2 \right]^{1/2} . \tag{5}$$

The unit vector from $\mathbf{ds}_1$ to $\mathbf{ds}_2$ is $\mathbf{a}_r$ and

$$\mathbf{a}_r = -\frac{x}{r}\mathbf{a}_x - \frac{y}{r}\mathbf{a}_y - \frac{z}{r}\mathbf{a}_z . \tag{6}$$

The source current element $\mathbf{ds}_1$ is in an arbitrary direction

$$\mathbf{ds}_1 = \mathbf{a}_x dx + \mathbf{a}_y dy + \mathbf{a}_z dz . \tag{7}$$

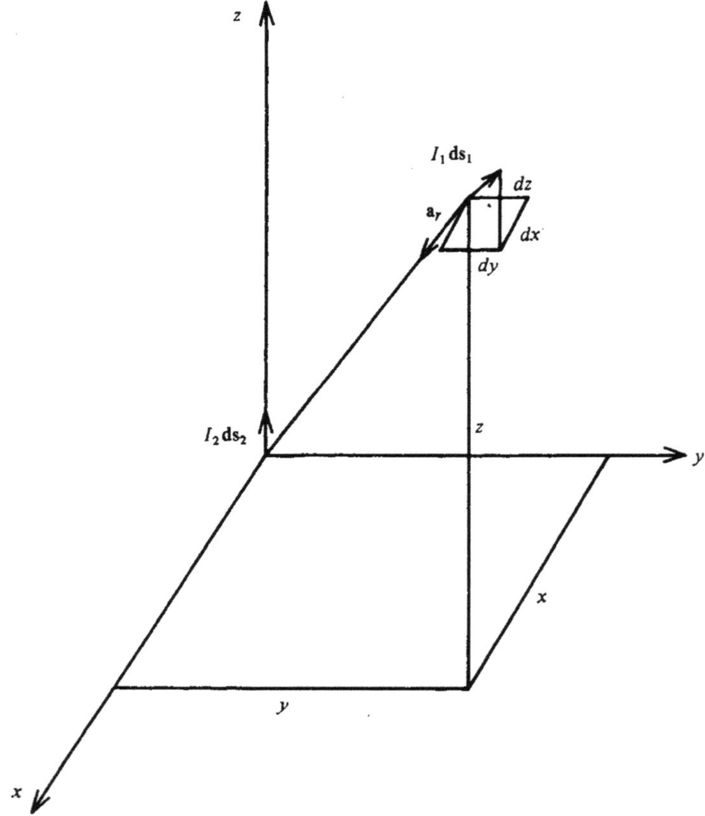

Fig. 2. Current elements in rectangular coordinates.

Before introducing these equations, it is convenient to recall [13] the vector identities

$$\mathbf{ds}_2 \times (\mathbf{ds}_1 \times \mathbf{a}_r) = (\mathbf{ds}_2 \cdot \mathbf{a}_r)\mathbf{ds}_1 - (\mathbf{ds}_1 \cdot \mathbf{ds}_2)\mathbf{a}_r$$
$$\mathbf{ds}_1 \times (\mathbf{ds}_2 \times \mathbf{a}_r) = (\mathbf{ds}_1 \cdot \mathbf{a}_r)\mathbf{ds}_2 - (\mathbf{ds}_1 \cdot \mathbf{ds}_2)\mathbf{a}_r \ . \tag{8}$$

Introducing Eq. (8) into Eqs. (2) and (3)

$$d^2\mathbf{F}_C = \frac{I_1 I_2}{4\pi\epsilon_0 c^2 r^2}[(\mathbf{ds}_2 \cdot \mathbf{a}_r)\mathbf{ds}_1 - (\mathbf{ds}_1 \cdot \mathbf{ds}_2)\mathbf{a}_r] \tag{9}$$

and

$$d^2\mathbf{F}_R = \frac{I_1 I_2}{8\pi\epsilon_0 c^2 r^2}[(\mathbf{ds}_2 \cdot \mathbf{a}_r)\mathbf{ds}_1 + (\mathbf{ds}_1 \cdot \mathbf{a}_r)\mathbf{ds}_2 - 2(\mathbf{ds}_1 \cdot \mathbf{ds}_2)\mathbf{a}_r] \ . \tag{10}$$

From Eqs. (4) and (7)

$$\mathbf{ds}_1 \cdot \mathbf{ds}_2 = ds_2 dz \,. \tag{11}$$

From Eqs. (6) and (7)

$$\mathbf{ds}_1 \cdot \mathbf{a}_r = \frac{-x\, dx - y\, dy - z\, dz}{r} \tag{12}$$

and from Eqs. (4) and (6),

$$\mathbf{ds}_2 \cdot \mathbf{a}_r = -\frac{ds_2 z}{r} \,. \tag{13}$$

Substitution into Eq. (1) gives

$$d^2\mathbf{F}_A = \frac{I_1 I_2 ds_2}{4\pi\epsilon_0 c^2}\left[\mathbf{a}_x\left[\frac{2x\, dz}{r^3} - \frac{3xz}{r^5}(x\, dx + y\, dy + z\, dz)\right]\right.$$
$$+ \mathbf{a}_y\left[\frac{2y\, dz}{r^3} - \frac{3yz}{r^5}(x\, dx + y\, dy + z\, dz)\right] \tag{14}$$
$$\left. + \mathbf{a}_z\left[\frac{2z\, dz}{r^3} - \frac{3z^2}{r^5}(x\, dx + y\, dy + z\, dz)\right]\right].$$

Substitution into Eq. (9) gives

$$d^2\mathbf{F}_C = \frac{I_1 I_2 ds_2}{4\pi\epsilon_0 c^2}\left[\mathbf{a}_x\left[\frac{x\, dz - z\, dx}{r^3}\right] + \mathbf{a}_y\left[\frac{y\, dz - z\, dy}{r^3}\right]\right]. \tag{15}$$

Substitution into Eq. (10) gives,

$$d^2\mathbf{F}_R = \frac{I_1 I_2 ds_2}{8\pi\epsilon_0 c^2}\left[\mathbf{a}_x\left[\frac{2x\, dz - z\, dx}{r^3}\right] + \mathbf{a}_y\left[\frac{2y\, dz - z\, dy}{r^3}\right]\right.$$
$$\left. - \mathbf{a}_z\left[\frac{x\, dx + y\, dy}{r^3}\right]\right]. \tag{16}$$

2. The Tangential Component of $d^2\mathbf{F}$

The tangential component of $d^2\mathbf{F}$ is simply the z-component. So from Eq. (14)

$$d^2 F_{Az} = \frac{I_1 I_2 ds_2}{4\pi\epsilon_0 c^2}\left[\frac{2zdz}{r^3} - \frac{3z^2}{r^5}(x\, dx + y\, dy + z\, dz)\right]. \tag{17}$$

From Eq. (15)

$$d^2 F_{Cz} = 0 \tag{18}$$

and from Eq. (16)

$$d^2 F_{Rz} = \frac{I_1 I_2 ds_2}{8\pi\epsilon_0 c^2}\left[\frac{-x\,dx - y\,dy}{r^3}\right].\tag{19}$$

It will be convenient, see Moon and Spencer [4], to express the three tangential forces in the form,

$$d^2 F_z = \frac{I_1 I_2 ds_2}{4\pi\epsilon_0 c^2}\left[P_x dx + P_y dy + P_z dz\right].\tag{20}$$

Comparing Eq. (20) with Eq. (17)

$$P_{Ax} = -\frac{3xz^2}{r^5}, \ \ P_{Ay} = -\frac{3yz^2}{r^5}, \ \ P_{Az} = \frac{2z}{r^3} - \frac{3z^3}{r^5}.\tag{21}$$

Since the Grassmann force, Eq. (18) is always perpendicular to $\mathbf{ds}_2$,

$$P_{Cx} = P_{Cy} = P_{Cz} = 0.\tag{22}$$

Comparison with Eq. (19) gives

$$P_{Rx} = -\frac{x}{2r^3}, \ \ P_{Ry} = -\frac{y}{2r^3}, \ \ P_{Rz} = 0.\tag{23}$$

A source $\mathbf{ds}_1$ must be part of a closed circuit or no current can flow. According to Ampère's third experiment [2], the tangential force on ds_2 produced by integrating ds_1 about a closed circuit C is zero or

$$\begin{aligned}
dF_z = \oint_C d^2 F_z = 0 &= \frac{I_1 I_2 ds_2}{4\pi\epsilon_0 c^2}\oint P_x dx + P_y dy + P_z dz \\
&= \frac{I_1 I_2 ds_2}{4\pi\epsilon_0 c^2}\oint_C \mathbf{P}\cdot\mathbf{ds}_1.
\end{aligned}\tag{24}$$

But by Stoke's Theorem [14],

$$\oint_C \mathbf{P}\cdot\mathbf{ds}_1 = \int_S \text{curl } \mathbf{P}\cdot\mathbf{dA},\tag{25}$$

where S is any surface bounded by C. Since this equation must hold for *all* closed circuits, according to Ampère's third experiment,

$$\int_S \text{curl } \mathbf{P}\cdot\mathbf{dA} = 0.\tag{26}$$

Therefore, the necessary and sufficient condition for the validity of Ampère's third experiment is

$$\text{curl } \mathbf{P} = 0. \tag{27}$$

For the Ampère equation,

$$\text{curl } \mathbf{P}_A = \begin{vmatrix} \mathbf{a}_x & \mathbf{a}_y & \mathbf{a}_z \\ \frac{\partial}{\partial x} & \frac{\partial}{\partial y} & \frac{\partial}{\partial z} \\ -\frac{3xz^2}{r^5} & -\frac{3yz^2}{r^5} & \frac{2z}{r^3} - \frac{3xz^2}{r^5} \end{vmatrix} \equiv 0.$$

Therefore,

$$\text{curl } \mathbf{P}_A \equiv 0 \tag{28}$$

and *no closed circuit can produce any tangential component of force along* ds_2 in accordance with Ampère's third experiment. Likewise for the Grassmann equation

$$\text{curl } \mathbf{P}_C = 0 \tag{29}$$

so the classical electrodynamic theory is again in agreement with Ampère's third experiment. However, for the relative velocity formulation advocated by Gauss, Weber and Riemann,

$$\text{curl } \mathbf{P}_R = \begin{vmatrix} \mathbf{a}_x & \mathbf{a}_y & \mathbf{a}_z \\ \frac{\partial}{\partial x} & \frac{\partial}{\partial y} & \frac{\partial}{\partial z} \\ -\frac{x}{2r^3} & -\frac{y}{2r^3} & 0 \end{vmatrix} = \frac{3z}{2r^5}(-\mathbf{a}_x y + \mathbf{a}_y x). \tag{30}$$

Since curl $\mathbf{P}_R \neq 0$, the relative velocity formulation of Eq. (3) is inconsistent with Ampère's third experiment.

Either the relative velocity formulation employing retarded potentials φ and A is inconsistent with experiment or there exist some closed circuits for which the tangential force produced by a closed circuit is different from zero. Such circuits may have been overlooked by Ampère.

3. Conclusions

The paper proposes a simple experimental test that will prove which of the low velocity electrodynamic equations is correct. Is it possible that Ampère did not try all possible configurations of current carrying conductors? If a single configuration of current carrying conductors can be found which produces a tangential force, the validity of the relative velocity formulation will have been proved.

Many experimentalists have believed that they had demonstrated the existence of tangential forces. Hering [15], Graneau [5], Pappas [6] and Phipps [8] demonstrate the existence of tangential forces in their experiments. Yet classical physicists have not yet faced the possibility that tangential forces may actually exist. Further study is needed of all previous experiments. New experiments should be performed.

But there is a strong possibility that Gauss was right in his hunch that the basic electrodynamic force equation can be defined in terms of the relative velocity of source and receiver.

References

1. D. E. Spencer and S. Y. Uma, *Gauss and the electrodynamic force*, The Mathematical Heritage of C. F. Gauss, 1990, p.

2. A. M. Ampère, *Mémoire sur la théorie mathématique des phénomènes électrodynamiques, uniquement déduite de l'expérience*, Mém. de l' Acad. des Sci. **6** (1823) p. 175.

3. V. Bush, *The force between moving charges*, J. Math. Phys. (3) **5** (1926) p. 129.

4. P. Moon and D. E. Spencer, *Interpretation of the Ampère force*, J. Franklin Inst. **257** (1954), p. 203; *The Coulomb force and the Ampère force*, J. Franklin Inst. **257** (1954), p. 305.

5. P. Graneau, *Ampère–Neumann Electrodynamics of Metals*, Hadronic Press, Nonantum, Mass., 1985.

6. P. T. Pappas, *The original Ampère force and Biot–Savart and Lorentz forces*, Nuovo Cimento **76B** (1983), p. 189.

7. J. P. Wesley, *Weber electrodynamics extended to include radiation*, Speculations in Science and Technology (1) **10** (1986), p. 47.

8. T. E. Phipps and T. E. Phipps, Jr., *Observation of Ampère forces in mercury*, to be published.

9. H. Grassmann, *Neue Theorie der Elektrodynamik*, Ann. d. Phys. **64** (1845), p. 1.

10. H. A. Lorentz, *The Theory of Electrons*, G. E. Stechert & Co., New York, N. Y., 1906.

11. P. Moon, D. E. Spencer, S. Y. Uma and P. J. Mann, *The Riemann Force*, International Congress on Relativity and Gravitation, Munich, West Germany, April, 1988.

12. A. S. Mirchandaney, D. E. Spencer, S. Y. Uma and P. J. Mann, *The theory of retarded potentials*, Hadronic J. **2** (1988), p. 231; *The electrodynamic fields of a moving charge*, to be published in Hadronic J.

13. P. Moon and D. E. Spencer, *Vectors*, D. Van Nostrand, Princeton, N. J., 1965, p. 153.
14. P. Moon and D. E. Spencer, *Vectors*, D. Van Nostrand, Princeton, N. J., 1965, p. 232.
15. C. Hering, *Electromagnetic forces: a search for more rational fundamentals: a proposed revision of the laws*, J. A. I. E. E. **42** (1923), p. 139.

Parry Moon
Massachusetts Institute of Technology
Cambridge, MA 02139
USA

Domina Eberle Spencer
University of Connecticut
Storrs, CT 06268
USA

Shama Y. Uma
Bridgewater State College
Bridgewater, MA 02325
USA

Philip J. Mann
University of Connecticut
Storrs, CT 06268
USA

THE MATH. HERITAGE OF C.F. GAUSS (pp. 526-543)
edited by George M. Rassias
©1991 World Scientific Publ. Co. Singapore

DIMENSIONS, FRACTALS, AND SPHERE PACKING

C. Musès

Introduction

The following sections all honour Carl Friedrich Gauss in their way. First, because of his genius in number theory; and second, because of his pioneering into higher dimensions through his daring work on biquadratic residues and higher geometry.

Fractals Express Kinkiness, Not Dimensions

1.0. The essence of fractals lies not in dimension theory, but in the theory of differentiation, and more precisely in the important theorem that a finite area can be enclosed by a closed boundary curve that is nowhere differentiable.

This theorem may be enlarged to embrace nowhere differentiable closed surfaces that bound a finite, calculable volume; and may be generalized to the fact that a nowhere differentiable n-manifold can be closed upon itself, so as to contain infinite and calculable content of $(n + 1)$-space.

But the dimension of the enclosing boundary for any $n + 1$ space is still n. We speak here of true or topological dimension, which is an invariant to all transformations within it. The term "dimension" should not be applied to fractals, since their so-called "Hausdorff dimension" implies only their degree of "kinkiness."

It is, above all, a great mistake to speak of fractals as "fractional dimensions" since they have nothing to do with true fractional dimensions at all.

It is easy to show that the fractal measure of kinkiness is not a dimension in any proper sense since a line of 1-dimensional manifold can have a Hausdorff or fractal measure greater than 2; thus providing a proof by contradiction of the false assertion that fractals are fractional dimensions.

1.1. The following examples and counterexamples illustrate this important point.

2.0. Fractals and Noninteger Dimensions

Benoit Mandelbrot [1] has introduced a useful measure he calls "fractals" in connection with derivative-less and plane filling curves and volume filling surfaces. Fractals relate to Hausdorff's concept, and not to bonafide dimensions as will soon be clear.*

Though he specifically cautions that his fractals are measures of content (actually, as will appear, they are not that either) and are not dimensions in any geometric/topological sense, he still is equivocal in his conclusions as to what his fractals actually measure, and this has led to the erroneous description of them as "noninteger dimensions" in even otherwise well-written popular expositions [2], indeed, in Mandelbrot's own ambiguous title. True fractional dimensions are a very different story [3].

Actually, Mandelbrot's fractals deal *not* with the dimension, but with the "kinkiness" of a space-filling or derivative-less curve or surface. The following examples and counter-examples illustrate this fundamental point.

If we make the Koch snowflake curve the profile curve of a multiply bent plane surface instead of merely a bent line, it should, according to Mandelbrot's fractal reasoning have a dimension between 2 and 3. We start in Fig. 1 (figures not exactly to scale): where each unit is now a unit area instead of a unit length. Since the fractal measure of the curve was 1.26 the fractal measure of the 2-dimensional surface will be $1.26(2) = 2.52$. So far so good, and note that we need a 3-space to contain this curved surface, as should be the case.

---

*Hausdorff's measure is not the actual dimension, which is a topological invariant (Brouwer).

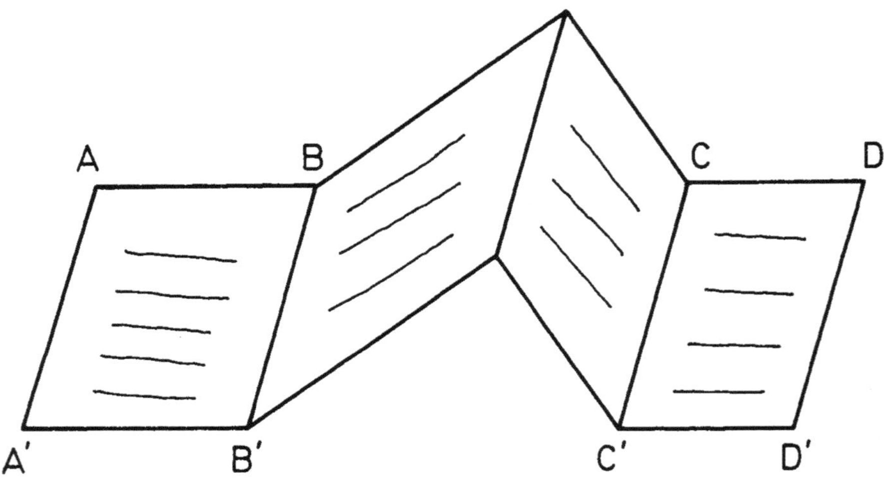

Fig. 1.

If we now make each unit a *cube* instead of a square, and form a "snowflake solid" (in a containing 4-space) we have a fractal measure of $1.26(3) = 3.78$ for its volume – also acceptable in the sense that it lies between 3 and 4. But this is too close to 4 to be realistic in any truly dimensional sense.

Continuing, we now calculate the fractal measure of such a dimensionally extended Koch figure's 4-space content at the next stage, in which each unit is a 4-cube, with these 4-cubes disposed in a containing 5-space. Now the fractal *exceeds* 5 dimensions, since $(1.26)(4) = 5.04$. Clearly, this is an inadmissible result for the dimension of a figure which by Mandellbrot's thesis is supposed to lie between 4 and 5 dimensions. Thus fractals, whatever they measure, do not measure dimension.

A second and independent counterexample that Mandelbrot's fractals measure "kinkiness" rather than dimensions is to construct a derivativeless curve (in this respect like the snowflake and other self-similar curves) which involves very close hairpin turns that in ordinary drawing amount to retracings. So we define our curve thus (Fig. 2):

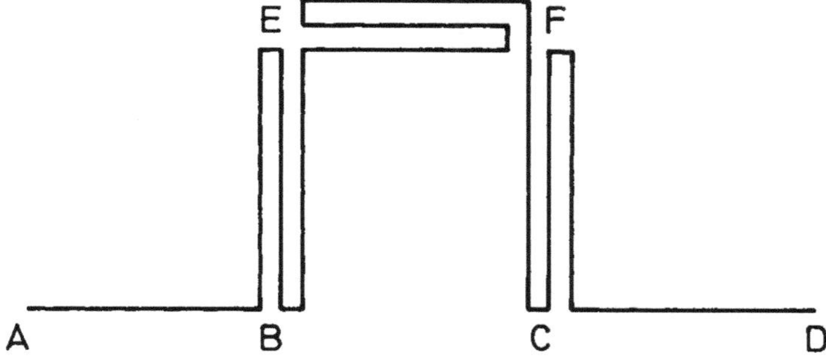

Fig. 2.

Now the fractal measure is $\log[1+3(3)+1]/\log 3 = \log 11/\log 3$, where any base may be used, here log denoting the logarithm to the base e. This fractal already exceeds 2, which is inadmissible if fractals are claimed to measure dimension: the dimension of no line can reach 2.

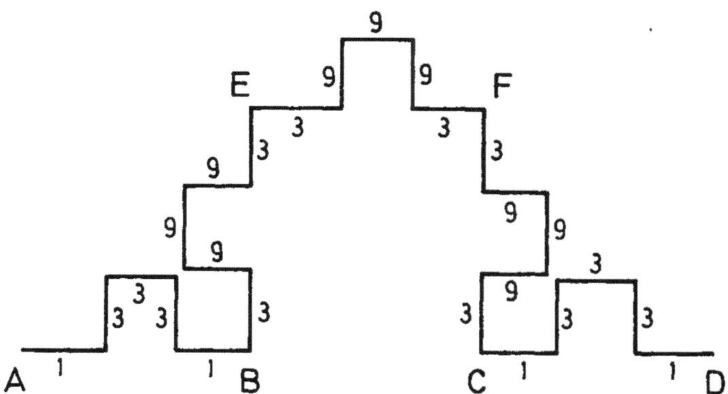

Fig. 3a.

The second stage is shown schematically in **Fig. 3a**, with detail of one of the second order $(3-9-9-9-3)$ loops in **Fig. 3b.** where the numbers indicate the number of retracings in between two successive points. Here we have a fractal measure amounting to $[\log \equiv \log e] : \log[2(11)+3(33)]/\log 9 = \log 121/\log 9$, which is again $\log 11/\log 3$. Thus the fractal measure as in the ordinary snowflake curve remains fixed, but it is $\log 11/\log 3 = \log_3 11 = 2.1826$ to four decimal places, which obviously is not a dimension between

1 and 2.* This discrepancy demonstrates that *though Mandelbrot's fractals may be useful in other respects, they do not measure dimensionality.* What they measure is order of bending or "kinkiness."

Fig. 3b.

2.1. Actual Fractional Dimensions

In conclusion, fractals are indeed a kinky concept, but they are not dimensions. The revivalists of space-filling curves, from Benoit Mandelbrot onward, have been too lax in confounding their "fractals" with bonafide fractional dimensions, of which practically nothing is known. As an interesting example stands the writer's theorem, that hyperspheres of unit radius *maximize* their content measure *not* in the fifth dimension (as previously thought) but in the dimension $5+ = 2(\Delta - 1)$, where $\psi(\Delta) = \log_e \pi$; and their (hyper) surface maximizes not in the seventh dimension but in the dimension $7+ = 2\Delta$. This theorem was first communicated to a friend and colleague, the hyperdimensional geometer Donald (H.S.M.) Coxeter, in August 1977. ψ is the log derivative of the gamma function. Note that it is actual (topological) dimensions rather than fractals, that bear on catastrophe and attractor theory and possess fascinating properties.

---

*If it be argued that we should here by rights use as the denominator, not 9 but infinity because of the vanishingly close hairpin turns, then by Mandelbrot's measure, the fractal is zero. Since no curve is of the zeroth dimension we again have a contradiction if we erroneously assume that fractals measure dimension per se.

3.0. Sphere Packings and Dimensions

The topologically unique nature of dimensions is nowhere better shown than in the geometry of higher-dimensional sphere packs, and we have already seen how hyperspheres furnished an important example of true fractional dimensions.

Every integer dimension has a characteristic (hyper) sphere packing of maximal density that can be extended throughout the entire space, thus creating a higher space tesselation of "tiling" – what Coxeter vividly calls a honeycomb [4].

Each positive integer dimension possesses a certain characteristic densest lattice or honeycomb, sections of which in successively lower dimensions each furnish the characteristic densest honeycomb for every lower dimension.

We often hear it said that non-Euclidean geometry is a totally different alternative to Euclidean and never the twain shall meet. Actually, however, any local non-Euclidean geometry in n dimensions can be embedded in Euclidean $(n + 1)$-space, just as the geometry of any curved space can be embedded in the geometry of flat k-dimensional space, where k is a measure of the dimensionality of the curvature of the given space. Note that even a line can be curved in 2, 3, 4, and even in n dimensions, where $4 < n < \infty$. Similarly, a surface can be curved in 3 or more dimensions; and the non-Euclidean geometry of a surface curved in k dimensions ($k \geq 3$) is included in the Euclidean geometry of flat k-space.

Each dimension has its own topological characteristics and in many ways the dimensions after the fourth are less exceptional than the first four, as the history of many theorems (e.g. Poincaré's conjecture) shows. Also, we note that of all regular polytopes, only three – the hypertetrahedron, the hyperoctahedron and the hypercube – exist in spaces of dimension > 4.

3.1. The First Eight Dimensions: A Remarkable Set

Yet dimensions continue to have unique characteristics. Thus, the first eight dimensions possess comparatively simple rules for sphere-packing centered about (hyper) tetrahedral symmetries and reflections. Interestingly, these eight dimension also correspond to the elementary hypernumber arithmetics that generate the hypercomplex algebras whose bases are composed of one real unit plus up to seven mutually perpendicular unit forms of

$\sqrt{-1}$. The (hyper) imaginary unit i_n, where $1 \le n \le 7$ are all such that $i_1, i_2 \Longrightarrow i_3$ and $i_1, i_2, i_3, i_4 \Longrightarrow i_5, i_6, i_7$; also $(\pm i_n)^4 = 1$.

There is a fascinating and very close relation between higher-dimensional sphere packing and some of these algebra-generating arithmetics (an arithmetic is always richer in theorems than an algebra).

The tangency number T_n for a given dimension n, sometimes fancifully called the kissing number, is the maximal number of equal (hyper) spheres that can fit around a central such sphere. Thus $T_1 = 2, T_2 = 6, T_3 = 12$, and so on. Now for $n = 1, 4$, or 8 it can be shown that $T_n = U_n$ where U_n is the number of units in the arithmetic (and hence the algebra) based on one real unit and $n - 1$ imaginary and hypercomplex units, as before explained. Thus $T_4 = 24$ and there are just 24 quaternion units; $T_8 = 240$ and there are just 24 unit octonions or Cayley numbers, and, of course, $T_1 = 2$ and there are 2 units in real arithmetic, namely ± 1.

The writer's finding, communicated to H.S.M. Coxeter in 1963 and kindly published by him in 1966 [5] shows how all the densest packings in spaces from one through eight dimensions have the same algebraic structures, reflecting underlyingly geometric and topological similarities. That finding is conveniently expressed as

$$T_n = n(\{2^{n-2}/3\} + n) \tag{1}$$

where $1 \le n \le 8$ and $\{R\} \equiv$ the largest integer $\le R$. For $2 \le n \le 8$ Eq. (1) can also be given in exact from depending on whether n be odd or even. Then Eq. (1) becomes

$$T_n = n\left(\frac{2^{n-2} + k}{3} + n\right) \tag{1a}$$

where $k = 2$ if n is even and $k = 1$ if n is odd.

3.2. Beyond Eight-Dimensional Space

It is probable that, as with so many theorems in number theory and related mathematics, 2 is an exception; and hence that since $T_{16} = 4320$, there are 4320 units in the 16-dimensional arithmetic formed of 8 mutually perpendicular, hypercomplex units whose square is -1 and 8 more hypercountercomplex units whose square is $+1$. We call the hypercountercomplex units $\varepsilon_1, \varepsilon_2$ etc., with $\varepsilon_0 \equiv 1$ and $\varepsilon_n^2 = 1$; and also we have another hypercomplex unit i_0, defined by $i_n \varepsilon_n = \varepsilon_n i_n = i_0$ and $i_0^2 = -1$. This 16-dimensional arithmetic and its algebra was first announced by the author [6]; and Kevin Carmody, a correspondent and a student of the author's work, later published very useful papers on it [7]. For a note on the physical applications of the author's countercomplex numbers and further references, see the *Journal of Mathematical Physics* [8].

In addition to the first 8, the first 24 dimensions also form a topologically distinct set. This interesting phenomenon is shown, for instance, by the fact that the densest sphere-packing density in dimension n is the same as that of $24 - n$, where $1 \leq n \leq 23$. Since $n = 12$ is the midpoint in this sequence of otherwise paired space, it is interesting that 12-space is characterized by *two* distinct kinds of lattice packing and that this alternative packing extends on both sides of 12, into 11-space and 13-space as well. As the author pointed out in a review of their book [9], J. H. Conway and N. J. A. Sloane made the fundamental error of equating the maximum packing density of 24-space with that of the point. Actually, the former is 1 while the latter is infinite. One must be very careful in n-dimensional geoemtry with the values $n = 0$ and $n = \infty$, for the point realm and the realm of infinite dimensionality are both extremely exceptional in relation to the topologies of spaces of finite dimension.

There is, however, some topological insights into 24-space afforded by the author's theorem that

$$T_n = 2n(2^{n/2} - 1) \tag{2}$$

where $n = 4m!$ and $m = 1, 2, 3$. Clearly, then, $n = 4, 8, 24$ and this formula thus structually relates spaces of 4, 8 and 24 dimensions. But there is no relation between 24-dimensional space and the point such as Conway and Sloane erroneously assert, probably following a false lead of John Leech whom they cite throughout the book. The review already cited [9] furnishes the detailed demonstration of the basic error in question.

4.0. Hypo- or Negadimensions

In papers and books on dimension theory to date, negadimensions are not discussed. Yet it is obvious from a generalization of Euler's basic topological theorem on polyhedral nets that at least the first negadimension pervades the entire geometry of positive dimensions. Euler's theorem can be written as

$$N_0 - N_1 + N_2 = 2 \tag{3}$$

where N_k is the number of k-dimensional elements in the polyhedral surface. But it is also clear that for any polyhedron whatsoever, $N_3 = 1$ since there is only one 3-dimensional element, namely the polyhedron itself.

Hence we can write Eq. (1) as

$$1 - N_0 + N_1 - N_2 + N_3 = 0 \tag{4}$$

where the 1 on the left is some N_k. Now k cannot exceed 3 and 0, 1, 2, and 3 have all been used. Hence $N_{-1} = 1$ and any polyhedron has exactly one element of dimension -1. The topology books have been forced (although miscalling it "the null dimension") to recognize this most fundamental of all negadimensions, most fundamental since

$$-1 > -k \tag{5}$$

where $k > 1$. But we can now write Eq. (3) as $\sum_{k=-1}^{n} (-1)^{k+1} N_k = 0$ for all closed figures of non-negative dimensions.* Thus for $n = 4$ we arrive at the well-known equation (see, for example, reference [4], p. 132, Sec. 7.65 and p. 165, Sec. 9.11)

$$N_0 - N_1 + N_2 - N_3 = 0 \tag{6}$$

whereas its more explicit and less opaque form is

$$N_{-1} - N_0 + N_1 - N_2 + N_3 - N_4 = 0 \tag{7}$$

where $N_{-1} = N_4 = 1$; and in general,

$$N_{-1} - N_0 + N_1 - N_2 \pm \ldots (-1)^{n-1} N_n = 0 \tag{8}$$

where $N_{-1} = N_n = 1$. Note that then with all even dimensions $N_{-1} + (-1)^{n-1} N_n = 0$, and hence ($n$ even):

$$N_0 - N_1 + N_2 - N_3 + \ldots + N_{n-1} = 0 \tag{9}$$

---

*Cf. Ludwig Schläfli, *Theorie der vielfachen Kontinuität*, Bern, 1901.

whereas in odd dimensions we have $N_{-1} + (-1)^{n-1} N_n = 2$ and hence (n odd):

$$N_0 - N_1 + N_2 - N_3 \pm \ldots - N_{n-1} = 2 \tag{10}$$

Thus every closed figure in n-dimensional space ($n \geq 0$) has exactly one n-dimensional element, namely itself; and also just one negadimensional element ($N_{-1} = 1$), which might be considered the "program," connectivity plan or "DNA" for the figure, expressing its *intension* whereas the positive dimensional elements express its *extension*.

4.1. Odd and Even Dimensions Have Characteristic Topologies

We see that odd and even dimensions have different topologies in the physics of those dimensions. Thus light in even-dimensional spaces would behave very differently than in odd-dimensional spaces. Crystal structure would also differ markedly between spaces of odd and even dimension. So also would vector fields on (hyper) spheres of odd dimension differ characteristically from those in even dimension.

As we saw in discussing Eq. (1), the oddness or evenness of dimension also affects sphere packing. It is also reflected in the nature of prime numbers since there is only a single even prime, namely 2. In 1982 (unpublished) we showed that if p is odd, then only if p is prime is it true that

$$(p^7 - p^3) \left[\sum_{n=1}^{\frac{1}{2}(p-1)} \left(\sum_{(p-n)} d^3 \cdot \sum_{(n)} d^3 \right) \right]^{-1} = 240 \tag{11}$$

whereas if p is even, it is prime only if the constant in Eq. (11) is $\frac{1}{2}(240) = 120$ and the upper index of the first summation sign is $(p-1)$. If p is not prime the right-hand side of Eq. (11) is less than 240. It is interesting that 240 is also the maximal number of equal hyperspheres that fit around a like central hypersphere in 8-space and hence with a 7-dimensional hypersurface. It can be shown by implications from E. Hecke's work on elliptic modular functions [10] that 240 is a minimal constant and cannot be improved. In Eq. (11), $\sum_{(r)} d^k$ denotes the sum of the kth powers of the divisors of r from 1 through r.

4.2. Algebraic Spaces, Fermat Primes, and Class Numbers

There are deep connections between higher dimensional algebras (see Sec. 3.1 above) and higher dimensions D of the form $D = 2^n$. If $n = 0$ or $D = 1$ we have the ordinary (real) algebra of the number line or axis of reals. If $n = 1$ or $D = 1$ we have the ordinary algebra of the number line or axis of reals. If $n = 1$ or $D = 2$ we have the algebra of complex numbers requiring a plane (two dimensions) for its representation. If $n = 2$ ($D = 4$) we have the 4-dimensional algebraic space of quaternions as well as the 4-D algebras containing zero divisors: those whose basis is $i_n, \varepsilon_n, i_0, \varepsilon_0$, or $\varepsilon_a, \varepsilon_b, \varepsilon_c$ where $\varepsilon_a \varepsilon_b = -i_c, \varepsilon_b i_c = \varepsilon_a$ and $i_c \varepsilon_a = \varepsilon_b$.

If $n = 3$ ($D = 8$) we have the 8-dimensional space of octonions or Cayley-Graves-Dickson numbers of the related 8-D algebras containing zero divisors, with bases $\varepsilon_a, \varepsilon_b, \varepsilon_c, i_a, i_b, i_c, \varepsilon_0$ and i_0 where any two factors of subscripts a and b yield a product of subscript c; b and c, one of subscript a; and c and a, one of subscript b. Finally, if $n = 4$ ($D = 16$) we have our 16-dimensional algebra already discussed.

A table of these algebras and their corresponding dimensions will be found in the author's reference 6 of this chapter. Beyond 16 dimensions there are no more circular algebras or those whose power orbits are circles. The power orbit of an algebra is the locus traced out by the real powers of its base elements in its algebraic space, and all the hypernumbers beyond ε have nonciruclar power orbits, just as all the algebras containing ε and/or i have circular power orbits.

The dimension 16 as a limit also appears importantly in the theory of Fermat primes or prime numbers of the form $2^n + 1$ when n is a real non-negative integer; namely $2, 3, 5, 17, 257$ and 65537 for $n = 0, 1, 2, 4, 8$ and 16 respectively. A remarkable theorem stemming from our own research, is that there exists no Fermat prime F, where $F > 2^{16} + 1$.

The number 16 as a dimensional limit appears strikingly also in the theory of class numbers of algebraic fields based in the roots of the mth roots of unity or the nth roots of negative unity where $n = m - 1$ or $2^n = \frac{1}{2}(2^m)$. Let us consider then the number fields $K(2^n)$, and specifically their class numbers h_n, recalling that only if $h = 1$ does the field retain the important property of unique factorization, whereby any composite number in the field can have only a single set of prime factors.

It was conjectured by Heinrich Weber and amplifiedly shown by Helmut Hasse (*Über de Klassenzahl Abelscher Zahlenkörper*, Berlin, 1952, pp. 105–106) that $K(2^4) = K(16)$ is the highest such field with unique factorization. Hasse also showed that for all $K(2^n)$, whether they have unique factorization or not, their relative class number is of the form 1 modulo 16.

5.0. True or Topological Fractional Dimensions

As mentioned in Sec. 2.1, there is little or nothing on the nature and mathematics of truly fractional dimensions in the literature. The only approach to such a comprehension was begun by Oliver Heaviside in his concepts of fractional differentiation and integration and his brilliant use of those concepts in physical applications. Actually, however, such comprehension had been implied since Leonhard Euler's tremendous breakthrough in conceiving the gamma function, which by the same token implied that there were such things as fractional combinations and permutations; and the writer has explored these in as yet unpublished investigations which connect them with fractional dimensions.

5.1. The Geometry of True Noninteger Dimensions

Our preliminary investigation of true fractional dimensions appeared in 1972 [11]. Both summarizing some salient points and also continuing that investigation, we find that fractional dimensions between 0 and 1 are, geometrically, the nodes of certain wave forms governed by the basic equation, where $0 \leq D \leq 1$ is the dimension:

$$D = \sin^2 \theta \qquad (12)$$

where θ is the slope of the dimension-generating wave function at the origin, i.e., the arc whose tangent is the value of the function first derivative at the origin. Thus if the wave function is y then $\theta = \tan^{-1} y'$ at $x = 0$ where y' means dy/dx.

Although geometers have easily generated lines by moving points; surfaces, by moving lines; and (hyper) volumes by moving (hyper) surfaces, the process of *half*-generated lines, surfaces, et al. has not been considered; although, algebraically, a half-integrated (or a half-differentiated) function can be precisely expressed, notably first by Oliver Heaviside who removed such expressions from the status of mere mathematical curiosity to that of profoundly useful and elegant physical application.

But though fractional dimensions have thus appeared and been used in algebra and in function theory, the *geometry* of true fractional dimensions (which, as we have seen, goes far deeper than the mere kinkiness of so-called "fractals") – is still *terra incognita*.

We know we can build any solid from a series of thin cross sections, and a surface from a series of thin strips. In fact, this yields the process of integration as the thickness of the slices or the width of the strips diminishes without limit.

Now proceeding dimensionally downward, we see that a line can also be generated from a series of short segments, which approach points as the length diminishes and their numbers increase. It is thus natural to conceive of a line as generated by a colinear set of points of increasing density.

For simplicity let us conceive these points as a set of nodes generated by two sine waves $180°$ out of phase and both waves intersecting at the origin. Let $0 \leq D \leq 1$ be a true or topological dimension. If the wavelength were infinitely long the set would be reduced to only one point $(D = 0)$ in the finite realm, i.e., the origin itself, and the density of the set would be 0. If the wavelength and amplitude approached zero, the nodal set would become infinitely dense and a line $(D = 1)$ would be generated. It is thus quite natural that fractional dimensions between the point $(D = 0)$ and the line $(D = 1)$ should be considered as nodal-point sets of *finite* density.

The simplest form of such a wave function generating these fractional dimensions can be expressed as a quasi-differential equation:

$$yy_0' = \pm \sin \left(xy_0'^2 \right) \tag{13}$$

where y_0' is defined as dy/dx at $x = 0$. The simplest explicit solution of Eq. (13) is

$$y = \pm(\lambda/2\pi)^{1/2} \sin \left(2\pi x/\lambda \right). \tag{14}$$

Hence

$$y_0' = (\lambda/2\pi)^{1/2} \left(2\pi/\lambda \right) = (2\pi/\lambda)^{1/2} = \tan \theta \tag{15}$$

is the slope at the origin of the positive wave in Eq. (13).

Now we can define $0 \leq D \leq 1$, more explicitly than in Eq. (12), as

$$D = \sin^2 \theta = \tan^2 \theta/(1 + \tan^2 \theta) = \left(1 + \frac{\lambda}{2\pi} \right)^{-1}. \tag{16}$$

Thus if $D = 1/4, \lambda = 6\pi$ and $\theta = \pi/6$; if $D = 1/2$, then $\lambda = 2\pi$ and $\theta = \pi/4$; if $D = 0, \lambda = \infty$ and $\theta = 0$; and if $D = 1, \lambda = 0$ and $\theta = \pi/2$. So our initial conditions are fulfilled.

The density of the nodes of Eq. (13) measures the (fractional) dimension. Since these nodes form a point set, ordinary measure theory, which makes all such sets to have measure zero, is as invalid here as if we were to equate to zero all fractions between 0 and 1. What we now need is a measure of density that will serve to distinguish between such sets. Now the nodal density is measured most simply by the frequency f of the wave function, and f in turn is measured most simply as the reciprocal of the wavelength λ. Hence

$$f = \frac{D}{2\pi(1 - D)} \tag{17}$$

for any fractional dimension $0 \leq D \leq 1$, the wavelength being given by

$$\lambda = [2\pi(1 - D)]/D \tag{18}$$

and the square of the amplitude by

$$A^2 = (1/D) - 1 \,. \tag{19}$$

Thus fractional dimensions between 0 and 1 are point sets, which in conventional dimension theory have measure zero and are indistinguishable from dimension zero. So we need a considerably deepened theory of dimension to handle either nega- or fractional dimensions; and now we have it.

5.2. Applications to Cosmology and Philosophy

In this deepened theory, integer dimensions, as we have seen, are associated with waves of infinite frequency and hence with nodal point-sets of infinite density. But in *physics* there is no such thing as infinitesimal or infinite: there is only the very minute or the very large. Hence in applying this deepened dimensional concept physically, it shows us that the space of our universe may be nearly integer but cannot be exactly so.

Actual space must thus involved a very small fractional dimension and so be quantized in terms of the very small lattice constant that would emerge out of the nodal spacing of the very high-frequency waves that would be necessary to generate a near-integer dimensional space – a constant on the order of 10^{-13} cm. or less. Thus quantum physics and quantized time as well (our continuous time can be shown to be intimately related to the first negadimension) follow at once from the nature of fractional dimension theory.

To conclude we turn to that passage in Oswald Spengler's remarkable essay on the meaning of numbers [12] where he notes that "the sense of form of the sculptor, the painter, the composer is essentially mathematical in its nature.... Mathematics goes beyond observation and dissection, and in its highest moments find the way by vision, not abstraction.... Thus the word 'creative' means much more in the mathematical sphere than it does in the pure sciences – Newton, Gauss and Riemann were artists, and we know with what suddennes their great conceptions came to them. 'A mathematician', said old Karl Weierstrass, 'who is not at the same time something of a poet will never be a full mathematician'."

References

1. B. Mandelbrot, *Fractals: Form, Chance and Dimension*, W. H. Freeman, San Fracisco, 1977.

2. *Science News*, 20 August 1977, 122.

3. C. Musès, *Explorations in Mathematics*, p. 67f. in UNESCO's Impact of Science, vol. 27, no. 1, 1977, reference 12, p. 315f. of cited reference.

4. H. S. M. Coxeter, *Regular Polytopes*, Macmillan, New York, 2nd ed., 1963, p. 122.

5. H. S. M. Coxeter, in *Lectures on Modern Mathematics*, ed. T. L. Saaty, Academic Press, New York, 1966.

6. C. Musès, in J. Appl. Math. Comp. **6** (1980), 63–94. See also K. Demys, J. Math. Phys. (2) **28** (1987) p. 339.

7. K. Carmody, *Circular and Hyperbolic Quaternions, Octonians, and Sedenions, and Finite Fourier Series*, J. Appl. Math. Comp. **28** (1988), 47–72 and 89–96.

8. Journal for Mathmatical Physics (2) **28** (1987) 339.

9. C. Musès, Book Review of *Sphere Packings, Lattices and Groups* (Conway and Sloane) in Kybernetes (3) **18** (1989), 65–66.

10. E. Hecke, Abh. Math. Sem. Univ. Hamburg **5** (1927), 199–224; also *Mathematische Werke*, Vandenhoeck & Ruprecht, Göttingen, 1959, pp. 641–486, as well as Hecke's Lecture Notes (modular and quadratic forms), Institute for Advanced Study, Princeton, 1938.

11. C. Musès, *Fractional Dimensions and Their Experiential Meaning*, J. Study of Consc. **5** (1972), 315–318.

12. Oswald Spengler, *Der Untergang des Abendlandes: Gestalt und Wirklichkeit*, ch. 2. The citation is based on pages 61 and 62 of C. F. Atkinson's English translation published by Alfred A. Knopf at New York in 1945.

Supplementary Note

If the theory of authentic or topological fractional dimensions of space here outlined is to be applied physically, it will be found necessary to employ not only the ordinary quantum physics but a quantum theory that includes the *phase* of the guiding probability wave function. In the papers and texts of quantum physics we repeatedly find the caveat "accurate only to the phase" or some such remark, which means in effect that the phase is not being taken into account.

In such applications to omit the phase, however, makes as much sense as not taking it into account in, say, the polarization or interference effects of light or any other wave phenomena, including even sound. Interactions of waves cannot be validly understood without considering phase and its attendant phenomena.

It is little known, and must be more investigated in view of the fact that the physical space of our universe is fractional dimensional — that Erwin Schrödinger himself pointed the way of the solution of this problem in his original paper on wave mechanics [*Annalen der Physik*, 79 (1926)]. That discussion was unfortunately omitted in his later papers. In view of the physical importance of fractional dimensional waves interacting with Schrödinger/de Broglie waves, that omission in quantum field theory should now be rectified.

From Schrödinger's pioneer work it can be shown that $\nabla^2 \tau = \nabla(1/w)$ where w is the phase velocity of the wave in question. More explicitly, the coefficient of the phase term of $-i\psi$ is

$$2\pi E \sqrt{2m/h}[\partial/\partial x(\sqrt{E_x - V_x}/E_x)$$
$$+ \partial/\partial y(\sqrt{E_y - V_y}/E_y) + \partial/\partial z(\sqrt{E_z - V_z}/E_z)],$$

where E is the total energy, V is potential energy, m the electron mass, and h Planck's constant. It is thus apparent that variations in phase could occur through variations in the potential energy V alone: $V \to V + \delta V$, since the energy E of the stable state, as well as π, m and h, are constant.

Such variations would take time to be equilibrized by counterchanges in the kinetic portion of E, and meanwhile the phase of the wave would be altered, causing consequent and appropriate changes in the probabilities of the corresponding events by (anti)resonance effects among the other probability waves involved. Thus, the phase term allows us to understand how a shift (say, in awareness of bomb, for example) that effects potential energy only, can alter event-patterns and change physical history by probability-phase triggering.

A new science of qualitative time emerges: the value of t, the fundamental independent variable for all phenomena, thus contributes ineluctably to the energy structure of any situation; and interactions among probability waves guide outcomes on all levels. Quantum physics so leads into chronotopology (see, for instance, chapters 1 and 3 of the present author's *Destiny and Control in Human Systems*, in the series Frontiers in System Science, Kluwer-Nijhoff, Dordrecht, 1985).

In conclusion to this supplementary note, I first announced some years ago the existence of *hypernumbers*, in the sense of numbers with arithmetics that were perfectly consistent but still lay beyond the "hypercomplex" numbers, i.e. those that are based on quaternions and octonions. At that time it became almost simultaneously evident that two (out of the seven elemental species of hypernumbers I had found) would cause the least trauma to logical familiarity, and would hence offer the greatest probability of earlier incorporation into application. It will be remembered that even Gauss, the actual first discoverer of the applications of i, sensed despite his great reputation, the strong rejection due to unfamiliarity and hence did not publish his findings before Bolyai and Lobatchevsky.

Those two hypernumbers with the lowest probability of rejection as too "alien" are those elemental forms of unit for which (like 1 and i) $u^0 = 1$ is true. We noted at the time that that did not necessarily imply that $x^{-1} = 1/x$ where x is a number in the u system, e.g. $1 \pm u$. Indeed, we showed that if $(1 + u)(1 - u) = 0$, then $x^0 = 1$ could *not* be true. We also pointed out that $1/(1 \pm \varepsilon)$ are divisors of infinity, just as $(1 \pm \varepsilon)$ are divisors of zero. For any such divisor x, x^{-1} does not equal $1/x$.

The two elemental units, other than 1 or i, for which $x^0 = 1$ holds, we have termed ε and w. The first of these important hyperunits is defined by $\varepsilon^2 = 1, \varepsilon \neq \pm 1$; and the second, by $(\pm w)^2 = -1 \pm w$, the signs being taken in order and separately. The ε-type numbers we termed countercomplex since they follow a hyperbolic counter theorem to de Moivre's circular theorem for complex numbers. De Moivre's theorem reads $\exp \theta i = \cos \theta + i \sin \theta$, whereas our new theorem for countercomplex numbers is $\exp \theta \varepsilon = \cosh \theta +$

$\varepsilon \sinh \theta$, where θ is any number that is real, complex or countercomplex.

The arithmetic of w is more unfamiliar than that of ε, since the *power* orbits of both i and ε are still circles even though *exponentially* ε is hyperbolic and i circular as we have just seen. The power orbits of $(+w)$ and $(-w)$, however, are orthogonal ellipses which bespeak greater unfamiliarity.

Already there are deep applications of ε to quadratic number fields of positive discriminant and their ideals. The applications of w are understandably more recondite and seem to be suggested as playing a fundamental rôle in the avoidance of unwarranted infinities in quantum field theory that arise from neglecting to include w in the basic assumptions of the still inadequate field theory, now able to be bettered.

C. Musès
Mathematics & Morphology Research Centre
Editorial & Research Offices
45911 Silver Avenue
Sardis, B.C.
Canada V2R 1Y8

SECOND COHOMOLOGY SPACES AND FLEXIBLE LIE-ADMISSIBLE ALGEBRAS

Hyo Chul Myung and Arthur A. Sagle

Following the associative and Lie algebra theory of Hochschild and Chevalley-Eilenberg, the second cohomology group is introduced for a flexible Lie-admissible algebra and is suggested by algebraic deformations of a Lie-admissible algebra. A brief application of this is made to Lie-admissible deformations of a commutative associative algebra in particular reference to deformations with a Lie-admissible Poisson bracket as infinitesimal.

1. Introduction

Since the brief introduction of Lie-admissible algebras by Albert [1] in 1948, very little about the structure of these algebras has been known until the subject recently came to the attention of physicists. During the last few years the theory of Lie-admissible algebras has seen a considerable growth both in theoretical mathematics and its applications to physics and mechanics [12, 14]. Many of the known Lie-admissible algebras are obtained from deformations or cohomology extensions of given algebras [3, 10, 11]. Following Hochschild [8] for associative algebras and Chevalley and Eilenberg [4] for Lie algebras, we briefly discussed in [11] the second cohomology space of a Lie-admissible algebra which was also suggested by the theory of algebraic deformations [11]. The present paper concerns second cohomology spaces of flexible Lie-admissible algebras. Our discussion is based on 2-cocycles of an algebra with values in its bimodule. We make a brief application of this to algebraic deformations of a Lie-admissible algebra.

Throughout, let A denote a nonassociative algebra with multiplication denoted by xy over a field F of characteristic $\neq 2$. We define an anti-commutative algebra A^- with multiplication $[x, y] = xy - yx$ defined on the vector space A. The Jacobian J is a trilinear map $A \times A \times A \to A$ given by

$$J(x, y, z) = [[x, y], z] + [[y, z], x] + [[z, x], y] \tag{1}$$

for $x, y, z \in A$. In case A^- is a Lie algebra, that is, the Jacobi identity $J(A, A, A) = 0$ holds, A is called a *Lie-admissible algebra*. Associative and Lie algebras are basic examples of Lie-admissible algebras, but there are many more. In fact, it is readily seen that the assumption of Lie-admissibility alone is too broad a condition to obtain a fruitful structure theory [12].

There is a best understood class of Lie-admissible algebras which has been actively studied using representation theory of Lie algebras. For this, let H denote a trilinear map $A \times A \times A \to A$ defined by

$$H(x, y, z) = y[x, z] + [x, y]z - [x, yz] \tag{2}$$

for all $x, y, z \in A$. It can be shown [12] that the identity $H(A, A, A) = 0$ holds, or equivalently the adjoint map $\mathrm{ad}_x : A \to A$ defined by $\mathrm{ad}_x(y) = [x, y]$ is a derivation of A for all $x \in A$, if and only if A satisfies the flexible identity $(xy)x = x(yx)$ and A is Lie-admissible. Thus, A is called *flexible Lie-admissible*, originally introduced by Albert [1], if $H(A, A, A) = 0$ holds. Associative and Lie algebras are clearly flexible Lie-admissible, but there are many dispersed sources of these algebras which are not associative or Lie [12]. An important example is obtained from the simple Lie algebras of type A_n ($n \geq 2$). Thus, let $\mathrm{sl}(n + 1, F)$ be the Lie algebra of $(n + 1) \times (n + 1)$ trace zero matrices over F of characteristic 0. For a fixed scalar $\alpha \neq 0$, define a product "$*$" on the vector space $\mathrm{sl}(n + 1, F)$ by

$$x * y = \frac{1}{2}[x, y] + \alpha \left(xy + yx - \frac{2}{n+1}(\mathrm{tr}\, xy)I \right) \tag{3}$$

for $x, y \in \mathrm{sl}(n + 1, F)$, where $[x, y]$ is the Lie algebra product, xy the matrix product, I the identity matrix, and tr denotes the trace. Then $(\mathrm{sl}(n + 1, F), *)$ is flexible Lie-admissible, but not associative or Lie. In fact, this is the only non-Lie simple flexible Lie-admissible algebra A over an algebraically closed field of characteristic 0 with A^- simple [12].

2. Bimodules for Lie-Admissible Algebras

Following the standard theory of bimodules, we briefly introduce bimodules for Lie-admissible algebras A over a field F (of characteristic $\neq 2$). Thus, let M be a vector space over F and suppose that there exist two bilinear maps $M \times A \to M$ given by $(u, a) \to au$ and $(u, a) \to ua, u \in M, a \in A$. Let $E = M \oplus A$ be the vector space direct sum and define a product in E by

$$(u + a)(v + b) = ub + av + ab \tag{4}$$

for $u, v \in M$ and $a, b \in A$. The algebra E defined by (4) is called the split null extension of A by M. If E is Lie-admissible, then M is called a *Lie-admissible bimodule* for A or simply an A-bimodule.

Lemma 1. Let A be a Lie-admissible algebra over F and E be the split null extension of A by M. Then, M is an A-bimodule if and only if

$$J(A, A, M) = 0, \tag{5}$$

where $J(\quad, \quad, \quad)$ is the Jacobian in E given by (1).

Proof. Let $x = u + a, y = v + b, z = w + c$ be elements in E with $u, v, w \in M, a, b, c \in A$. By direct computation we obtain $J(x, y, z) = J(a, b, w) + J(b, c, u) + J(c, a, v) + J(a, b, c) = J(a, b, w) + J(b, c, u) + J(c, a, v)$, since $MM = 0$ and A is Lie-admissible. Since $J(\quad, \quad, \quad)$ is skew in x, y, z, we have that $J(E, E, E) = 0$ if and only if $J(A, A, M) = 0$. $\square$

Thus, by (5) any A-bimodule M is regarded as a Lie module for A^- under the composition $a \cdot u = -u \cdot a = [a, u]$ for $u \in M, a \in A$. However, every Lie module for A^- is not necessarily converted into an A-bimodule (see [11]). Any Lie-admissible algebra A becomes an A-bimodule under the left and right multiplications in A. Such a bimodule is called a regular A-bimodule.

Let A be a flexible Lie-admissible algebra over F, so that the identity $H(a, b, c) = 0$ of (2) holds for all $a, b, c \in A$. A vector space M over F with compositions $(u, a) \to ua, (u, a) \to au$ for $u \in M, a \in A$ is called a *flexible Lie-admissible bimodule* for A or an A-bimodule if the split null extension

$E = M \oplus A$ of A by M is flexible Lie-admissible.

Lemma 2. Let A be a flexible Lie-admissible algebra over F and let $E = M \oplus A$ be the split null extension of A by M. Then, M is an A-bimodule if and only if

$$H(A, A, M) = H(A, M, A) = H(M, A, A) = 0.\qquad(6)$$

Proof. For $u, v, w \in M$ and $a, b, c \in A$, using $H(a, b, c) = 0$ and $MM = 0$, we compute $H(u + a, v + b, w + c) = H(a, b, w) + H(a, v, c) + H(u, b, c)$. Thus (6) implies $H(E, E, E) = 0$, so that E is flexible Lie-admissible. If $H(E, E, E) = 0$ then the last identity for $c = 0, b = 0$ or $c = 0$ gives relation (6). $\square$

3. Second Cohomology Spaces

Throughout we let M denote an A-bimodule for a Lie-admissible algebra over the field F with compositions $(u, a) \to ua, (u, a) \to au$ for $u \in M, a \in A$. Denote by $C^2(A, M)$ the F-space of bilinear maps: $A \times A \to M$, and each element in $C^2(A, M)$ is called a 2-cochain of A in M. For $f \in C^2(A, M)$, define the skew symmetric 2-cochain f^- by

$$f^-(a, b) = f(a, b) - f(b, a)$$

for $a, b \in A$. Let $E = M \oplus A$ be the vector space direct sum. For a 2-cochain h of A in M, define a product in E by

$$(u + a)(v + b) = ub + av + h(a, b) + ab\qquad(7)$$

for $u, v \in M$ and $a, b \in A$. The algebra E defined by (7) is called the *null extension* of A determined by M and h and is denoted by (E, h). The split null extension is a special case of this with $h = 0$. If (E, h) is a Lie-admissible algebra, then (E, h) is called the *null Lie-admissible extension* or simply the null LA-extension of A by M and h. Similarly, for a flexible Lie-admissible algebra A and an A-bimodule M, if (E, h) is flexible Lie-admissible under product (7), then (E, h) is called the *null FLA-extension* of A by M and h.

A null extension (E, h) is said to be *equivalent* to the null extension (E, h') if there is an algebra isomorphism ω of (E, h) to (E, h') such that $\omega|_M$ is the identity map on M and $\omega(a) - a \in M$ for all $a \in A$. If (E, h) is equivalent to (E, h'), then we also say that h is equivalent to h' and write $h \sim h'$. We first give a condition that (E, h) be a null LA-extension or a null FLA-extension of A by M and h.

Lemma 3. (i) For a Lie-admissible algebra A and an A-bimodule M, the extension (E, h) defined by (7) is a null LA-extension if and only if the 2-cochain $h \in C^2(A, M)$ satisfies the identity

$$[h^-(a,b), c] + [h^-(b, c), a] + [h^-(c, a), b] \\ + h^-([a, b], c) + h^-([b, c], a) + h^-([c, a], b) = 0 \tag{8}$$

for $a, b, c \in A$.

(ii) If A is flexible Lie-admissible, then (E, h) is a null FLA-extension of A by M and h if and only if h satisfies the identity

$$h(b, [a, c]) + h([a, b], c) - [a, h(b, c)] \\ + bh^-(a, c) + h^-(a, b)c - h^-(a, bc) = 0 \tag{9}$$

for $a, b, c \in A$.

Proof. Let $u + a, v + b, w + c$ be elements in $E = M \oplus A$ for $u, v, w \in M, a, b, c \in A$. For part (i), we compute the Jacobian J in E as

$$J(u + a, v + b, w + c) = J(a, v, c) + J(u, b, c) + J(a, b, w) + J(a, b, c) \\ + [h^-(a, b), c] + [h^-(b, c), a] + [h^-(c, a), b] \\ + h^-([a, b], c) + h^-([b, c], a) + h^-([c, a], b) ,$$

noting $MM = 0$. Since A is Lie-admissible and M is an A-bimodule, the first four terms vanish, so that (E, h) is Lie-admissible if and only if identity (8) holds. For part (ii), we compute the trilinear map H of (2) in (E, h) and obtain $H(u + a, v + b, w + c) = H(a, b, c) + H(a, b, w) + H(u, b, c) + H(a, v, c) + h(b, [a, c]) + h([a, b], c) - [a, h(b, c)] + bh^-(a, c) + h^-(a, b)c - h^-(a, bc)$. Since A is flexible Lie-admissible and M is an A-bimodule, by Lemma 2 the first four terms vanish again. This shows that (E, h) is a null FLA-extension if and only if h satisfies identity (9). $\square$

In view of Lemma 3, we define a 2-cochain h of A in M to be a 2-*cocycle* of A in M if h satisfies identity (8) or (9), according as A is Lie-admissible or flexible Lie-admissible. Denote by $Z^2(A, M)$ the set of all 2-cocycles of A in M, which forms an F-subspace of $C^2(A, M)$. For a Lie algebra L and an L-module, recall that a skew symmetric cochain f of L in M (so $f^- = 2f$) is a 2-cocycle if f satisfies identity (8). Thus, if A is Lie-admissible then

$$Z^2(A, M) = \{h \in C^2(A, M) | h^- \in Z^2(A^-, M)\}$$

with M viewed as a Lie module for A^- under the composition $a \cdot u = [a, u] = au - ua, a \in A, u \in M$.

For an associative algebra A and an A-bimodule M (i.e., $(au)b = a(ub)$ for $a, b \in A, u \in M$), as is well known, an $h \in C^2(A, M)$ is a 2-cocycle if h satisfies

$$ah(b, c) - h(ab, c) + h(a, bc) - h(a, b)c = 0 \tag{10}$$

for $a, b, \in A$, which is equivalent to the associativity of the null extension (E, h) defined by (7). Hence Lemma 3 shows that the definition of a 2-cocycle for a Lie-admissible or flexible Lie-admissible algebra generalizes that for a Lie or associative algebra. Two-cocycles given by (8) were extensively used for the construction of various Lie-admissible algebras [3, 11]. Identity (9) first appeared in [11] as an identity satisfied by infinitesimals of algebraic deformations of flexible Lie-admissible algebras, although it was not referred to as a 2-cocycle. The following result gives the definition of the second cohomology space based on 2-cocycles.

Theorem 4. Let A be a Lie-admissible or flexible Lie-admissible algebra and let M be an A-bimodule. For 2-cocycles $h, h' \in Z^2(A, M), h \sim h'$ holds if and only if there is a linear map $\sigma \in \text{Hom}(A, M)$ such that

$$h'(a, b) = h(a, b) - \sigma(ab) + a\sigma(b) + \sigma(a)b \tag{11}$$

for all $a, b \in A$. The relation "$\sim$" defines an equivalence relation in $Z^2(A, M)$. If $H^2(A, M)$ denotes the set of equivalence classes $\bar{h}$ determined by $h \in Z^2(A, M)$, then $H^2(A, M)$ becomes an F-space under the operations $\bar{h} + \bar{h'} = \overline{h + h'}$ and $\alpha \bar{h} = \overline{\alpha h}, \alpha \in F$. The F-space $H^2(A, M)$ so defined is called the *second cohomology space* of A in M.

Proof. For $h, h' \in Z^2(A, M)$, let (E, h) and (E, h') be the null extensions of A by M defined by (7). Assume $h \sim h'$, so that (E, h) is equivalent to (E, h'). Hence there is an algebra isomorphism $\omega : (E, h) \to (E, h')$ such that $\omega|_M = \mathrm{id}$ and $\omega(a) - a \in M$ for all $a \in A$. Letting $\sigma(a) = a - \omega(a)$ for $a \in A$, we find that $\sigma \in \mathrm{Hom}\,(A, M)$. Denoting by "·" the product in (E, h') for $u, v \in M, a, b \in A$, we obtain

$$
\begin{aligned}
\omega[(u + a)(v + b)] &= \omega(ub + av + h(a, b) + ab) \\
&= ub + av + h(a, b) + ab - \sigma(ab) \\
&= \omega(u + a) \cdot \omega(v + b) \\
&= (u - \sigma(a) + a) \cdot (v - \sigma(b) + b) \\
&= ub + av - \sigma(a)b - a\sigma(b) + h'(a, b) + ab\,,
\end{aligned}
$$

which gives identity (11).

Suppose that identity (11) holds. Define a linear map $\omega : (E, h) \to (E, h')$ by $\omega(u + a) = u - \sigma(a) + a$ for $u \in M, a \in A$. We have $\omega[(u + a)(v + b)] = \omega(ub + av + h(a, b) + ab) = ub + av + h'(a, b) - a\sigma(b) - \sigma(a)b + ab$ (by (11)) $= (u - \sigma(a) + a) \cdot (v - \sigma(b) + b) = \omega(u + a) \cdot \omega(v + b)$. Thus, ω is the desired isomorphism of (E, h) to (E, h'). The equivalence of the relation "$\sim$" now follows immediately from identity (11) which also shows that the operations in $H^2(A, M)$ are well defined. $\square$

The introduction of $H^2(A, M)$ by relation (11) is parallel to that given by [9] for certain variety of algebras, but the present one is more direct approach. In the associative and Lie algebra theory, $H^2(A, M)$ and more generally $H^n(A, M)$ for $n = 1, 2, \ldots$ are definable in terms of n-cocycles and n-coboundaries of A in M. It is not known whether n-cocycles and n-coboundaries for $n \geq 1$ are definable in a Lie-admissible algebra and its bimodule. However, it is possible to define 2- and 3-coboundaries for a Lie-admissible algebra. This is suggested by identities (8), (9) and leads to $H^2(A, M)$ in a way parallel to the associative and Lie theory.

For $\sigma \in \mathrm{Hom}\,(A, M) = C^1(A, M)$, define the 2-coboundary operator δ by

$$
(\delta\sigma)(a, b) = \sigma(a)b + a\sigma(b) - \sigma(ab) \tag{12}
$$

for $a, b \in A$. Thus, δ is a linear map: $C^1(A, M) \to C^2(A, M)$. For $f \in C^2(A, M)$, define δf to be a 3-cochain in $C^3(A, M)$ given by

$$
\begin{aligned}
(\delta f)(a, b, c) = &[f^-(a, b)c] + [f^-(b, c), a] + [f^-(c, a), b] \\
&+ f^-([a, b], c) + f^-([b, c], a) + f^-([c, a], b)
\end{aligned} \tag{13}
$$

if A is Lie-admissible, and for flexible Lie-admissible A, δf is defined as

$$
\begin{aligned}
(\delta f)(a,b,c) =&f(b,[a,c]) + f([a,b],c) - [a,f(b,c)] \\
&+ bf^-(a,c) + f^-(a,b)c - f^-(a,bc)
\end{aligned}
\tag{14}
$$

for $a,b,c \in A$. In both cases, we note that the kernel of δ in $C^2(A,M)$ is $Z^2(A,M)$. Let $B^2(A,M)$ denote the F-space of 2-coboundaries, i.e., $B^2(A,M) = \delta(C^1(A,M))$. We show $\delta^2 = 0$ and in particular $B^2(A,M) \subset Z^2(A,M)$.

Lemma 5. For $C^1(A,M) \xrightarrow{\delta} C^2(A,M) \xrightarrow{\delta} C^3(A,M)$ defined by (12)–(14), we have $\delta^2 = 0$.

Proof. If A is Lie-admissible, then for $\sigma \in C^1(A,M)$, by (12) $(\delta\sigma)^-$ is a 2-coboundary of the Lie algebra A^- in M viewed as a Lie module under the adjoint action and $(\delta\sigma)^- = \delta\sigma^- \in B^2(A^-,M)$. Hence the result follows from the Lie algebra case. Assume that A is flexible Lie-admissible. For $\sigma \in C^1(A,M)$ and $a,b,c \in A$, we obtain by (12) and (14)

$$
\begin{aligned}
(\delta^2\sigma)(a,b,c) =&(\delta\sigma)(b,[a,c]) + (\delta\sigma)([a,b],c) - [a,(\delta\sigma)(b,c)] \\
&+ b(\delta\sigma)^-(a,c) + (\delta\sigma)^-(a,b)c - (\delta\sigma)^-(a,bc) \\
=&b\sigma([a,c]) + \sigma(b)[a,c] - \sigma(b[a,c]) + [a,b]\sigma(c) \\
&+ \sigma([a,b])c - \sigma([a,b]c) - [a,b\sigma(c)] - [a,\sigma(b)c] \\
&+ [a,\sigma(bc)] + b[a,\sigma(c)] + b[\sigma(a),c] - b\sigma([a,c]) \\
&+ [a,\sigma(b)]c + [\sigma(a),b]c - \sigma([a,b])c - [\sigma(a),bc] \\
&- [a,\sigma(bc)] + \sigma([a,bc]) \\
=& -\sigma(H(a,b,c)) + H(\sigma(a),b,c) + H(a,\sigma(b),c) + H(a,b,\sigma(c)) \\
=&0
\end{aligned}
$$

by Lemma 2, since A is flexible Lie-admissible and M is an A-bimodule. Hence $\delta^2 = 0$. $\square$

Theorem 6. $H^2(A,M) \simeq Z^2(A,M)/B^2(A,M)$.

Proof. By Lemma 5 we have $B^2(A,M) \subset Z^2(A,M)$. Define a map

$H^2(A, M) \to Z^2(A, M)/B^2(A, M)$ by $\bar{h} \to h + B^2(A, M)$ for $h \in Z^2(A, M)$. By (11) and (12) the map is well defined and is a linear isomorphism. $\square$

As noted in [11], by (12) the kernel $Z^1(A, M)$ of δ in $C^1(A, M)$ is the set of derivations of A in M. A derivation of A in M is inner if it can be extended to an inner derivation of the split null extension of A by M. Letting $B^1(A, M)$ denote the F-space of inner derivations of A in M, as in the associative or Lie case we define the *first cohomology space $H^1(A, M)$* of A in M by $H^1(A, M) = Z^1(A, M)/B^1(A, M)$.

4. Applications to Deformations

As discussed in [11], the basic definitions and development of algebraic deformations of Lie-admissible algebras are patterned after those of associative and Lie algebras [8, 4] and are suggested by the second cohomology space of a Lie-admissible algebra. Bimodules for A considered here are regular bimodules for A and thus a 2-cocycle of A with values in itself is a bilinear map: $A \times A \to A$ which satisfies identity (8) or (9), where the bimodule operation is the algebra multiplication in A. Two-cochains are assumed to be skew symmetric and symmetric in the Lie and commutative associative theories of Chevalley-Eilenberg [4] and Harrison [7]. We do not make this assumption for the present discussion. This enables us to deform Lie algebras and commutative algebras into Lie-admissible algebras which are not necessarily Lie. Thus if A is a commutative algebra and so is trivially flexible Lie-admissible, then a 2-cochain $h \in C^2(A, A)$ is a 2-cocycle if h satisfies the identity

$$h^-(a, bc) = bh^-(a, c) + h^-(a, b)c \tag{15}$$

for $a, b, c \in A$, using (9). The classical Poisson bracket is a well-known example of identity (15) and thus a 2-cocycle of a commutative associative algebra of C^∞-functions defined on a C^∞-manifold.

We briefly recall some basic facts about deformations [11]. Let A be a Lie-admissible algebra over the field F. Let $F[[t]]$ denote the formal power series ring in indeterminate t over F and let $F((t)) = K$ be the field of quotients of $F[[t]]$. Any 2-cochain $f \in C^2(A, A)$ can be extended to a unique 2-cochain, also denoted by f, in $C^2(A_K, A_K)$, where $A_K = K \otimes_F A$

is the scalar extension of A to K. A 2-cochain $f \in C^2(A_K, A_K)$ is said to be *defined over* F if f is an extension of a 2-cochain in $C^2(A, A)$.

Suppose that there is given a bilinear map $\pi = \pi_t : A_K \times A_K \to A_K$ that is expressed in the form

$$\pi(x, y) = \pi_t(x, y) = \sum_{i=0}^{\infty} t^i f_i(x, y) \tag{16}$$

for $x, y \in A_K$ where each f_i is a 2-cochain in $C^2(A_K, A_K)$ defined over F and $f_0(x, y) = xy$ is the product in A_K. Denote by A_t the algebra A_K with multiplication π_t. For convenience, we also write $\pi_t(x, y) = \pi(x, y) = x * y$ for $x, y \in A_K$, so that

$$[x, y]^* = x * y - y * x = \pi_t^-(x, y).$$

If A_t is Lie-admissible or flexible Lie-admissible, then A_t or π_t is called a (one-parameter family of) *Lie-admissible* (LA) or *flexible Lie-admissible* (FLA) *deformation* of A. In this case, the 2-cochain f_1 is called the *infinitesimal* or *differential* of the deformation. For $f, g \in C^2(A, A)$ or $C^2(A_K, A_K)$, define $f \circ g$ and $f \bigtriangleup g$ to be 3-cochains given by

$$(f \circ g)(x, y, z) = f(g(x, y), z) + f(g(y, z), x) + f(g(z, x), y), \tag{17}$$
$$(f \bigtriangleup g)(x, y, z) = g(y, f(x, z)) + g(f(x, y), z) - f(x, g(y, z)) \tag{18}$$

for $x, y, z \in A$.

Theorem 7. Let A be a Lie-admissible or flexible Lie-admissible algebra over F. The algebra A_t defined by (16) is an LA or FLA deformation of A if and only if the infinitesimal f_1 is a 2-cocycle in $Z^2(A, A)$ and the f_i satisfies respectively the (obstruction) equations

$$\sum_{\substack{i+j>0 \\ i+j=k}} f_i^- \circ f_j^- = -\delta f_k^-, \tag{19}$$

$$\sum_{\substack{i+j>0 \\ i+j=k}} f_i^- \bigtriangleup f_j = -\delta f_k \tag{20}$$

for $k = 2, 3, \ldots$ where δ is the coboundary operator defined by (13) and (14).

Proof. We note that Lie-admissibility and flexible Lie-admissibility of A_t are equivalent to the equations $\pi_t^- \circ \pi_t^- = 0$ and $\pi_t^- \bigtriangleup \pi_t = 0$. By bilinearity of the maps $(f, g) \to f \circ g$ and $(f, g) \to f \bigtriangleup g$, from these equations and (16) we obtain

$$\sum_{\substack{i,j \geq 0 \\ i+j=k}} f_i^- \circ f_j^- = 0 , \tag{21}$$

$$\sum_{\substack{i,j \geq 0 \\ i+j=k}} f_i^- \bigtriangleup f_j = 0 \tag{22}$$

for $k = 0, 1, \ldots$. When $k = 0$, (21) and (22) give the Jacobi identity and the identity H for A given by (1) and (2). For $k = 1$, (21) and (22) imply the equations

$$f_0^- \circ f_1^- + f_1^- \circ f_0^- = 0 , \tag{23}$$

$$f_0^- \bigtriangleup f_1 + f_1^- \bigtriangleup f_0 = 0 . \tag{24}$$

Since $f_0(x, y) = xy$ is the product in A, (23) and (24) show by (8) and (9) that f_1 is a 2-cocycle in $Z^2(A, A)$. For $k \geq 2$, (21) gives $\sum_{i,j>0, i+j=k} f_i^- \circ f_j^-$ $= -f_0^- \circ f_k^- - f_k^- \circ f_0^- = -\delta f_k^-$ by (13). Similarly, from (14) and (22) we obtain (20). Conversely, if f_1 is a 2-cocycle in $Z^2(A, A)$ and if (19) or (20) holds, then using (13) or (14) relation (21) or (22) follows from (23) or (24). Thus A_t is an LA or FLA deformation of A. $\square$

An interesting special case of Theorem 7 is

Corollary 8. Let A be a flexible Lie-admissible algebra over F and let $f_2, f_3, \ldots$ be symmetric 2-cochains of A. Then the algebra A_t defined by (16) is an FLA deformation of A if and only if f_1 is a 2-cocycle of A and is flexible Lie-admissible, i.e., $f_1^- \bigtriangleup f_1 = 0$, and f_i for $i \geq 2$ satisfies the identity

$$[x, f_i(y, z)] = f_i(y, [x, z]) + f_i([x, y], z) \tag{25}$$

for $x, y, z \in A$. In this case, $\pi_t^- = f_0^- + t f_1^-$.

Proof. Since $f_i^- = 0$ for $i \geq 2$ and $f_0^- \vartriangle f_0 = 0$, Eq. (22) is equivalent to the relations $f_1^- \vartriangle f_0 + f_0^- \vartriangle f_1 = 0, f_1^- \vartriangle f_1 = 0$ and $f_0^- \vartriangle f_i = 0$ for $i \geq 2$. The second equation means that the algebra (A, f_1) with multiplication f_1 is flexible Lie-admissible while the last is equivalent to (25). $\square$

If, in particular, A is commutative, then (25) trivially holds, so that A_t is an FLA deformation if and only if the infinitesimal f_1 is a 2-cocycle, i.e., f_1 satisfies (15) and is flexible Lie-admissible. As an application of Corollary 8, we have

Theorem 9. Let A be a finite-dimensional flexible Lie-admissible algebra over an algebraically closed field F of characteristic 0 such that A^- is simple. Then any FLA deformation A_t of A with f_i symmetric for $i \geq 1$ is given by

$$x * y = \frac{1}{2}[x, y] + \left(\sum_{i=0}^{\infty} \alpha_i t^i \right) x \# y, \alpha_i \in F, i \geq 0 \qquad (26)$$

for $x, y \in A_K$, where $x \# y = 0$ for all $x, y \in A_K$ if A^- is not of type $A_n (n \geq 2)$ and $x \# y = xy + yx - (2/(n+1))(\operatorname{tr} xy)I$ if A^- is of type $A_n (n \geq 2)$. In the latter case, we identify $A^- = \operatorname{sl}(n+1, F), xy$ denotes the matrix product, tr the trace and I is the identity matrix.

Proof. For $f \in C^2(A, A)$, denote by (A, f) the algebra with product $f(x, y)$ defined on the F-space A. We first note that the deformation A_t of A given by (26) is clearly an FLA deformation of A with all f_i symmetric for $i \geq 1$, since it is easy to see that $f_i(x, y) \equiv \alpha_i(x \# y)$ for $i \geq 1$ is a symmetric 2-cochain satisfying (25) [12]. Assume that A_t defined by (16) is an FLA deformation of A and the f_i are symmetric for $i \geq 1$. By Corollary 8 the f_i $(i \geq 1)$ satisfy (25). Thus, by the known classification [12], each $f_i (i \geq 1)$ must be a constant multiple of "#", i.e., $f_i(x, y) = \alpha_i x \# y$ for $\alpha_i \in F$ and $i \geq 1$, where $\alpha_i = 0$ for all $i \geq 1$ if A^- is not of type A_n $(n \geq 2)$. Let $x \cdot y$ denote the product in A. It is well known [12] that the product $x \cdot y$ is given by $x \cdot y = \frac{1}{2}[x, y] + \alpha_0 x \# y$ for some $\alpha_0 \in F$. Substituting these in (16) gives (26), as desired. $\square$

In the remainder of this section, we focus on some examples of LA deformations of a commutative associative algebra. Let A be a commutative associative algebra over F with identity element 1. The associative algebra $C^1(A, A) = \mathrm{Hom}_F(A, A)$ over F is made into a unital A-module under the operation $(a\sigma)(x) = a\sigma(x)$ for $a, x \in A$ and $\sigma \in C^1(A, A)$. For $\sigma, \tau \in C^1(A, A)$, define the *cup product* $\sigma \smile \tau$ to be a 2-cochain that belongs to $C^2(A, A)$, given by

$$(\sigma \smile \tau)(x, y) = \sigma(x)\sigma(y)$$

for $x, y \in A$. Let $C^1 = C^1(A, A)$ and let $C^1 \smile C^1$ denote the (unital) A-module spanned by $\sigma \smile \tau, \sigma, \tau \in C^1$. Then, $C^1 \smile C^1$ becomes an associaticve F-algebra under the product

$$(\sigma \smile \tau)(\sigma' \smile \tau') = \sigma\sigma' \smile \tau\tau'.$$

We are particularly interested in an A-submodule D of C^1 which is spanned by a commuting set $\{\sigma_i \,|\, i \in I\}$ of derivations of A, i.e., $\sigma_i\sigma_j = \sigma_j\sigma_i$ for all $i, j \in I$. Thus, every element $\sigma \in D$ is expressed as $\sigma = \sum a_i\sigma_i$ for $a_i \in A, i \in I$. Note that each element in D is a derivation of A but D is not necessarily a commuting set. In fact, for $\sigma, \tau \in D$ and $a, b \in A$ we have

$$[a\sigma, b\tau] = ab[\sigma, \tau] + a\sigma(b)\tau - b\tau(a)\sigma.$$

Let $D \smile D$ denote the A-submodule spanned by $\sigma_i \smile \sigma_j, i, j \in I$. An element P in $D \smile D$ is called a *Lie-admissible* (LA) *Poisson bracket* (relative to D) if the algebra (A, P) with multiplication $P(x, y)$ defined on the F-space A is Lie-admissible, i.e., $P^- \circ P^- = 0$.

Since A is commutative, for $\sigma, \tau \in D$ we have $(\sigma \smile \tau)(x, y) = \sigma(x)\tau(y)$ $= \tau(y)\sigma(x) = (\tau \smile \sigma)(y, x)$ for $x, y \in A$ and hence $(\sigma \smile \tau)^- = \sigma \smile \tau$ $-\tau \smile \sigma$. Any element $P \in D \smile D$ is expressed as

$$P = \sum_{i,j=1}^{n} a_{ij}\sigma_i \smile \sigma_j \,, a_{ij} \in A \,, i, j \in I \,. \tag{27}$$

Then we have

$$P^- = \sum a_{ij}\sigma_i \smile \sigma_j - \sum a_{ij}\sigma_j \smile \sigma_i = \sum (a_{ij} - a_{ji})\sigma_i \smile \sigma_j$$
$$= \sum a_{ij}^-\sigma_i \smile \sigma_j \,,$$

where we set $a_{ij}^- = a_{ij} - a_{ji}$ for $i, j = 1, \dots, n$. It is easy to see that P given by (27) satisfies the relation $P(a, bc) = bP(a, c) + P(a, b)c$ for $a, b, c \in A$, so that P is a 2-cocycle of A by (15).

Lemma 10. Let $P \in D \smile D$ be expressed by (27). If the equations

$$\sum_{k=1}^{n} [a_{kq}^- \sigma_k(a_{ij}^-) + a_{ki}^- \sigma_k(a_{jq}^-) + a_{kj}^- \sigma_k(a_{qi}^-)] = 0 \tag{28}$$

hold for $i, j, q = 1, \dots, n$, then P is an LA Poisson bracket. If, in particular, the a_{ij} are scalars in $F1$, then P is an LA Poisson bracket.

Proof. For $x, y, z \in A$, we compute the Jacobian $(P^- \circ P^-)(x, y, z)$ in (A, P) as

$$
\begin{aligned}
&P^-(P^-(x, y), z) + P^-(P^-(y, z), x) + P^-(P^-(z, x), y) \\
&= \sum a_{kq}^- \sigma_k(a_{ij}^-)\sigma_i(x)\sigma_j(y)\sigma_q(z) + \sum a_{kq}^- a_{ij}^-(\sigma_k\sigma_i)(x)\sigma_j(y)\sigma_q(z) \\
&\quad + \sum a_{kq}^- a_{ij}^- \sigma_i(x)(\sigma_k\sigma_j)(y)\sigma_q(z) + \sum a_{ki}^- \sigma_k(a_{jq}^-)\sigma_i(x)\sigma_j(y)\sigma_q(z) \\
&\quad + \sum a_{ki}^- a_{jq}^- \sigma_i(x)(\sigma_k\sigma_j)(y)\sigma_q(z) + \sum a_{ki}^- a_{jq}^- \sigma_i(x)\sigma_j(y)(\sigma_k\sigma_q)(z) \\
&\quad + \sum a_{kj}^- \sigma_k(a_{qi}^-)\sigma_i(x)\sigma_j(y)\sigma_q(z) + \sum a_{kj}^- a_{qi}^- \sigma_i(x)\sigma_j(y)(\sigma_k\sigma_q)(z) \\
&\quad + \sum a_{kj}^- a_{qi}^-(\sigma_k\sigma_i)(x)\sigma_j(y)\sigma_q(z),
\end{aligned}
$$

where each sum is taken over $i, j, k, q = 1, \dots, n$. Since $\sigma_1, \dots, \sigma_n$ commute, the two sums involving factor "$\sigma_k\sigma_\ell$" on the right side cancel each other, e.g., for the two sums involving $\sigma_k\sigma_i$ we have

$$
\begin{aligned}
&\sum a_{kq}^- a_{ij}^-(\sigma_k\sigma_i)(x)\sigma_j(y)\sigma_q(z) + \sum a_{kj}^- a_{qi}^-(\sigma_k\sigma_i)(x)\sigma_j(y)\sigma_q(z) \\
&= \sum (a_{kq}^- + a_{qk}^-)a_{ij}^-(\sigma_k\sigma_i)(x)\sigma_j(y)\sigma_q(z) = 0,
\end{aligned}
$$

since $\sigma_k\sigma_i = \sigma_i\sigma_k$ and $a_{kq}^- + a_{qk}^- = 0$. Thus, the right side of the above computation reduces to

$$\sum_{i,j,q}\sum_k [a_{kq}^- \sigma_k(a_{ij}^-) + a_{ki}^- \sigma_k(a_{jq}^-) + a_{kj}^- \sigma_k(a_{qi}^-)]\sigma_i(x)\sigma_j(y)\sigma_q(z),$$

which vanishes, if Eq. (28) holds for $i, j, q = 1, \ldots, n$. Hence (28) implies the Jacobi identity $P^- \circ P^- = 0$. $\square$

If there exist elements $x_1, \ldots, x_n$ in A such that $\sigma_i(x_j) = \delta_{ij}$ for $i, j = 1, \ldots, n$, then from the last equation in the proof of Lemma 10 it follows that (28) is equivalent to the Jacobi identity $P^- \circ P^- = 0$. Important examples of such an algebra A arise from the polynomial algebra $F[x_1, \ldots, x_n]$, the formal power series algebra $F[[x_1, \ldots, x_n]]$, or the commutative associative algebra of C^∞-functions on a C^∞-manifold with a local chart $(x_1, \ldots, x_n)$. In the first two cases, if F is of characteristic $p > 0$, A is truncated to the quotient algebra of A by the ideal $(x_1^p, \ldots, x_n^p)$ generated by $x_1^p, \ldots, x_n^p$. This algebra has been used for the study of nodal noncommutative Jordan algebras and of simple Lie algebras of characteristic $p > 0$ [14, 15]. For these algebras A, any derivation of A has the form

$$a_1(x)\frac{\partial}{\partial x_1} + \ldots + a_n(x)\frac{\partial}{\partial x_n}$$

where the $\partial/\partial x_i$ are the usual partial differential operators. Let P be a 2-cochain given by

$$P = \sum_{i,j=1}^{n} a_{ij}(x)\frac{\partial}{\partial x_i} \smile \frac{\partial}{\partial x_j}. \tag{29}$$

By Lemma 10, P is an LA Poisson bracket if and only if the following differential equations hold:

$$\sum_{k=1}^{n}\left(a_{kq}^- \frac{\partial a_{ij}^-}{\partial x_k} + a_{ki}^- \frac{\partial a_{jq}^-}{\partial x_k} + a_{kj}^- \frac{\partial a_{qi}^-}{\partial x_k} \right) = 0, \, i, j, q = 1, \ldots, n. \tag{30}$$

If the $a_{ij}(x)$ are skew symmetric, i.e., $a_{ij}(x) = -a_{ji}(x)$, then $P^- = 2P$ and this is the case of classical Poisson bracket. Associative deformations with classical Poisson bracket as infinitesimal have been studied in the analytic theory of deformations [2, 5]. When A is a commutative associative algebra of C^∞-functions defined on a C^∞-manifold (of dimension $2n$), an LA Poisson bracket P given by (29) has been called a general Poisson bracket by Santilli and the study of this is a central subject of the generalization of classical mechanics initiated by Santilli [13]. In this case, one introduces a new product $x * y$ in A by

$$x * y = xy + P(x, y), \, x, y \in A \tag{31}$$

where xy denotes the product in A. This algebra is regarded as a specialization of the deformation given by (16) with $f_1 = P$, and $f_i = 0$ for $i \geq 2$ and $t = 1$. In general, the algebra defined by (31) is not flexible [11].

Let P be a 2-cochain of A given by (27). For a more specific example, in the rest of our discussion we assume that the a_{ij} are scalars in $F1$ and F is of characteristic 0, so that P is necessarily Lie-admissible. Viewing $(P^-)^m$ for $m \geq 0$ as an element in $C^1 \smile C^1 \subset C^2(A, A)$, it is easy to see [11] that

$$(P^-)^m = \begin{cases} \text{symmetric for } m \text{ even} \\ \text{skew symmetric for } m \text{ odd}, \end{cases} \tag{32}$$

where we put $(P^-)^0 = 1 \smile 1$. From a theorem of Gerstenhaber [6] we have that the deformation of A given by

$$\pi_t = \exp tP^- = \sum_{r=0}^{\infty} \frac{t^r}{r!} (P^-)^r \tag{33}$$

is associative, i.e., $\pi_t(\pi_t(x, y), z) = \pi_t(x, \pi_t(y, z))$ for $x, y, z \in A_K$. Hence π_t is Lie-admissible. Define

$$\sinh tP^- = \frac{1}{2}[\exp tP^- - \exp(-tP^-)],$$

$$\cosh tP^- = \frac{1}{2}[\exp tP^- + \exp(-tP^-)],$$

so that

$$\sinh tP^- = \sum_{r=0}^{\infty} \frac{t^{2r+1}}{(2r+1)!} (P^-)^{2r+1},$$

$$\cosh tP^- = \sum_{r=0}^{\infty} \frac{t^{2r}}{(2r)!} (P^-)^{2r}.$$

Denoting $\pi_t^+(x, y) = \pi_t(x, y) + \pi_t(y, x)$ for $x, y \in A$, we have

$$\pi_t^{\pm}(x, y) = \pi_t(x, y) \pm \pi_t(y, x) = \sum_{r=0}^{\infty} \frac{t^r}{r!}[(P^-)^r(x, y) \pm (P^-)^r(y, x)]$$

$$= \sum \frac{t^r}{r!}[(P^-)^r \pm (-1)^r (P^-)^r](x, y)$$

using (32). This gives

$$\pi_t^- = 2 \sinh tP^-, \quad \pi_t^+ = 2 \cosh tP^-.$$

Since π_t is associative, π_t^- and π_t^+ respectively define a Lie and Jordan algebra products on A_K, i.e., letting $\pi_t^+(x,y) = x \cdot y$, then $x \cdot y = y \cdot x$ and $(x \cdot x) \cdot (y \cdot x) = [(x \cdot x) \cdot y] \cdot x$ for $x, y \in A_K$.

Using this, we consider two deformations of A with P as infinitesimal given by

$$
\begin{aligned}
\zeta_t &= 1 \smile 1 + 2tP + \sum_{r=1}^{\infty} \frac{t^{2r+1}}{(2r+1)!}(P^-)^{2r+1} \\
&= 1 \smile 1 + t(2P - P^-) + \sinh tP^- \\
\eta_t &= tP + \cosh tP^- .
\end{aligned}
$$

Both ζ_t and η_t are Lie-admissible, since $\zeta_t^- = 2\sinh tP^-$ and $\eta_t = tP^-$. It is easily seen that ζ_t and η_t are not in general flexible [11].

As a final application, for $\lambda, \mu \in F$ with $\lambda \neq \pm\mu$, let $\pi_t^{(\lambda,\mu)}$ be defined by

$$
\pi_t^{(\lambda,\mu)}(x,y) = \lambda \pi_t(x,y) + \mu \pi_t(y,x)
$$

for $x, y \in A_K$, called the (λ, μ)-*mutation* of π_t. Since π_t is associative, it is well known [11] that $\pi_t^{(\lambda,\mu)}$ is (not associative) flexible Lie-admissible. We can also express $\pi_t^{(\lambda,\mu)}$ as

$$
\pi_t^{(\lambda,\mu)} = (\lambda + \mu)\cosh tP^- + (\lambda - \mu)\sinh tP^-
$$

[11]. Thus, $\pi_t^{(\lambda,\mu)}$ is an FLA deformation of the commutative associative algebra $A(\lambda,\mu)$ with product $(\lambda + \mu)xy$, which is isomorphic to A.

References

1. A. A. Albert, *Power associative rings*, Trans. Amer. Math. Soc. **64** (1948) 552–593.
2. F. Bayen, M. Flato, C. Fronsdal, A. Lichnerowicz, and D. Sternheimer, *Deformation theory and quantization*, I. *Deformations of symplectic structures*, Ann. of Phys. 111 (1978), 61–110; II. *Physical applications*, Ann. of Phys. *111* (1978) 111–151.
3. G. M. Benkart, *The construction of examples of Lie-admissible algebras*, Hadronic J. **5** (1982) 431–493.
4. C. Chevalley and S. Eilenberg, *Cohomology theory of Lie groups and Lie algebras*, Trans. Amer. Math. Soc. **63** (1948) 85–124.

5. M. Flato, A. Lichnerowicz and D. Sternheimer, *Deformations of Poisson brackets, Dirac brackets and applications*, J. Math. Phys. **17** (1976) 1754–1762.

6. M. Gerstenhaber, *On the deformation of rings and algebras*, Ann. Math. **79** (1964) 59–103; *On the deformation of rings and algebras: II.* Ann. Math. **84** (1966) 1–19; *On the deformation of rings and algebras: III*, Ann. Math. **88** (1968) 1–88.

7. D. K. Harrison, *Commutative algebras and cohomology*, Trans. Amer. Math. Soc. **104** (1962) 191–204.

8. G. Hochschild, *On the cohomology groups of an associative algebra*, Ann. Math. **46** (1945) 58–67.

9. N. Jacobson, *Structure and Representations of Jordan Algebras*, Amer. Math. Soc. Colloq. Publ., Vol. 39, Amer. Math. Soc. Providence, RI, 1968.

10. M. Kôiv and J. Lôhmus, *Generalized deformations of nonassociative algebras. Definitions and some simple examples*, Hadronic J. **3** (1979) 53–78.

11. H. C. Myung, *The exponentiation and deformation of Lie-admissible algebras*, Hadronic J. **5** (1982) 771–903.

12. H. C. Myung, *Malcev-Admissible Algebras*, Birkhäuser, Boston-Basel-Stuttgart, 1986.

13. R. M. Santilli, *Lie-Admissible Approach to the Hadronic Structure*, Vols. I and II, Hadronic Press, Nonantum, MA, 1978 and 1982.

14. R. D. Schafer, *Nodal noncommutative Jordan algebras and simple Lie algebras of characteristic p*, Trans. Amer. Math. Soc. **94** (1960) 310–326.

15. D. R. Scribner, *Lie-admissible, nodal, noncommutative Jordan algebras*, Trans. Amer. Math. Soc. **154** (1971) 105–111.

Hyo Chul Myung
Department of Mathematics
University of Northern Iowa
Cedar Falls, Iowa 50614
USA

Arthur A. Sagle
Department of Mathematics
University of Hawaii – Hilo
Hilo, Hawaii 96720
USA

THE MATH. HERITAGE OF C.F. GAUSS (pp. 562-572)
edited by George M. Rassias
©1991 World Scientific Publ. Co. Singapore

NEW FIBONACCI AND LUCAS IDENTITIES

S. A. Obaid

We obtain some new trigonometric identities, Chebyshev identities, Fibonacci and Lucas identities. Some of these identities may be used in solving certain boundary value problems.

1. Introduction

Recently Bassali, Rung and the author [1], [8], [9], [11] developed some new trigonometric identities. Some of these identities are essential for solving the flexure and torsion of certain beams (see [11], [12]) whose cross sections are

$$r = a \cos^n \left(\frac{\theta}{n} \right), \, a > 0, \, (-\pi < \theta \leq \pi), \, n = 2, 3, 4, \ldots \qquad (1.1)$$

It is interesting that these new trigonometric identities can be used to derive new Chebyshev, Fibonacci and Lucas identities. In this paper we introduce additional related trigonometric identities and derive the corresponding new Fibonacci and Lucas identities. These identities can be used to solve boundary value problems with Dirichlet and Neumann type boundary conditions (see [7], [9]).

Our proofs use some combinatorial identites of Knuth [4] and Riordan [10].

2. Trigonometric Identities

In this section we list six trigonometric identities but delay their proofs to the next section. These identities yield new Fibonacci and Lucas identities.

Identity 1. For $n \geq 1$

$$2^{2n-1}(\sin \phi)^{2n} = (-1)^n \cos 2n\phi$$
$$+ \sum_{m=1}^{n} (-1)^{m-1} 2^{2n-2m} (\sin \phi)^{2n-2m} \cos(2m-2)\phi$$
$$+ \sum_{m=1}^{n-1} (-1)^{m-1} 2^{2n-2m-1} (\sin \phi)^{2n-2m-1} \sin(2m-1)\phi .$$

$$(2.1)$$

Identity 2. For $n \geq 1$

$$2^{2n}(\sin \phi)^{2n+1} = (-1)^n \sin(2n+1)\phi$$
$$+ \sum_{m=1}^{n} (-1)^{m-1} 2^{2n-2m+1} (\sin \phi)^{2n-2m+1} \cos(2m-2)\phi$$
$$+ \sum_{m=1}^{n} (-1)^{m-1} 2^{2n-2m} (\sin \phi)^{2n-2m} \sin(2m-1)\phi .$$

$$(2.2)$$

Identity 3. For $n \geq 1$

$$2^{2n}(\sin \phi)^{2n} \cos \phi = (-1)^n \cos(2n+1)\phi$$
$$+ \sum_{m=1}^{n} (-1)^{m-1} 2^{2n-2m} (\sin \phi)^{2n-2m} \cos(2m-1)\phi$$
$$\sum_{m=1}^{n-1} (-1)^{m-1} 2^{2n-m-1} (\sin \phi)^{2n-2m-1} \sin 2m\phi .$$

$$(2.3)$$

Identity 4. For $n \geq 1$

$$
2^{2n+1}(\sin \phi)^{2n+1} \cos \phi = (-1)^n \sin(2n + 2)\phi
$$
$$
+ \sum_{m=1}^{n} (-1)^{m-1} 2^{2n-2m+1}(\sin \phi)^{2n-2m+1} \cos(2m - 1)\phi
$$
$$
\sum_{m=1}^{n} (-1)^{m-1} 2^{2n-2m}(\sin \phi)^{2n-2m} \sin 2m\phi .
$$

$$(2.4)$$

Identity 5. For $n \geq 1$

$$
2^n(\cos \phi)^{2n} = \sum_{m=0}^{n} \binom{n}{m} (\cos \phi)^n \cos(n - 2m)\phi
$$
$$
= \sum_{m=1}^{n} 2^{m-n+1} \binom{2n - m - 1}{n - 1} (\cos \phi)^m \cos m\phi .
$$

$$(2.5)$$

Identity 6. For ≥ 1 and $k \geq 1$

$$
(-1)^{n-k}(\csc \phi)^{2n+2k} \cos(2n - 2k)\phi
$$
$$
= \sum_{m=0}^{n+k} (-1)^m 2^{2n+2k-2m} E_{2n+2k-2m}(\csc \phi)^{2m} \cos 2m\phi ,
$$

$$(2.6)$$

where $E_0 = 1$,

$$
E_{2s} = \frac{1}{2}\left[\binom{2n}{s} + \binom{2k}{s} - \sum_{l=0}^{s-1} E_{2l} \binom{2n + 2k - 2l}{2n + 2k - 2s} \right] , \qquad (2.7)
$$

$s = 1, 2, 3, \ldots , n + k$.

3. Proofs of Identities 1–6

We convert to a complex polynomial identity by letting $z = e^{i\phi} = \cos \phi + i \sin \phi$. Then $\cos m\phi = (z^m + z^{-m})/2, \sin m\phi = (z^m - z^{-m})/2i$.

Using these relations, identity (2.1) becomes

$$(1 - z^2)^{2n} = 1 + z^{2n+2} - \sum_{m=1}^{n} (1 - z^2)^{2n-2m} (z^{4m-2} + z^2)$$
$$+ \sum_{m=1}^{n-1} (1 - z^2)^{2n-2m-1} (z^{4m} - z^2).$$

Replacing z^2 by t gives

$$(1 - t)^{2n} = 1 + t^{n+1} - \sum_{m=1}^{n} (1 - t)^{2n-2m} (t^{2m-1} + t)$$
$$+ \sum_{m=1}^{n-1} (1 - t)^{2n-2m-1} (t^{2m} - t). \qquad (3.1)$$

Thus, it is sufficient to prove (3.1). We proceed to show that the coefficient, A, of $t^l, 0 < l < 2n$ on the left side of (3.1) is equal to A^*, the coefficient of $t^l, 0 < l < 2n$ on the right side of the identity. Elementary computations yield

$$A = (-1)^l \binom{2n}{l}.$$

The right-hand side of (3.1) takes the form

$$-\sum_{m=1}^{n} \left[\sum_{j=0}^{2n-2m} (-1)^j \binom{2n - 2m}{j} \{t^{j+2m-1} + t^{j+1}\} \right]$$
$$+ \sum_{m=1}^{n-1} \left[\sum_{j=0}^{2n-2m-1} (-1)^j \binom{2n - 2m - 1}{j} \{t^{j+2m} - t^{j+1}\} \right].$$

Thus

$$A^* = -\sum_{m=1}^{n} (-1)^{l-1} \binom{2n - 2m}{l - 2m + 1} - \sum_{m=1}^{n} (-1)^{l-1} \binom{2n - 2m}{l - 1}$$
$$+ \sum_{m=1}^{n-1} (-1)^l \binom{2n - 2m - 1}{l - 2m} - \sum_{m=1}^{n-1} (-1)^{l-1} \binom{2n - 2m - 1}{l - 1}$$
$$= (-1)^l \left[\sum_{m=0}^{2n-1} \binom{2n - m - 2}{2n - l - 1} + \sum_{m=0}^{2n-1} \binom{2n - m - 2}{l - 1} \right].$$

Using the identity (see [10], p. 7), $\sum_{i=0}^{s-k} \binom{s-i-1}{k-1} = \binom{s}{k}$, $s > k$, with $s = 2n-1, k = 2n-l, i = m$ in the first sum and with $s = 2n-1, k = l, i = m$ in the second sum we have

$$A^* = (-1)^l \left[\binom{2n-1}{2n-l} + \binom{2n-1}{l} \right] = (-1)^l \left[\binom{2n-1}{l-1} + \binom{2n-1}{l} \right]$$
$$= (-1)^l \binom{2n}{l},$$

which completes the proof.

The proof of identity 2 is similar and thus is omitted. Since the proofs of identities 3 and 4 are similar, we will give only the proof to the latter identity.

Proof of Identity 4

Following the same steps used in proving identity 1 we can rewrite identity 4 in the form

$$(1-t)^{2n+1}(1+t) = 1 - t^{2n+2} - \sum_{m=1}^{n}(1-t)^{2n-2m}(t - t^{2m+1})$$
$$- \sum_{m=1}^{n}(1-t)^{2n-2m+1}(t + t^{2m}). \tag{3.2}$$

Consequently, it is sufficient to prove (3.2). The coefficient B of $t^l, 0 < l < 2n + 2$ on the left-hand side of (3.2) is given by $B = (-1)^l\left[\binom{2n+1}{l} - \binom{2n+1}{l-1}\right]$, and the coefficient B^* of $t^l, 0 < l < 2n + 2$ on the right-hand side of (3.2) takes the form

$$B^* = (-1)^l \sum_{m=1}^{n} \binom{2n-2m}{l-1} - (-1)^l \sum_{m=1}^{n} \binom{2n-2m}{l-2m-1}$$
$$+ (-1)^l \sum_{m=1}^{n} \binom{2n-2m+1}{l-1} - (-1)^l \sum_{m=1}^{n} \binom{2n-2m+1}{l-2m}$$

or,

$$B^* = (-1)^l \left[\sum_{m=1}^{2n-l+1} \binom{2n-m}{l-1} - \sum_{m=1}^{l-1} \binom{2n-m}{2n-l+1} \right]$$

$$= (-1)^l \left[\binom{2n}{l} - \binom{2n}{2n-l+2} \right]$$

$$= (-1)^l \left[\binom{2n}{l} - \binom{2n}{l-2} \right]$$

$$= (-1)^l \left[\binom{2n}{l} + \binom{2n}{l-1} \right] - (-1)^l \left[\binom{2n}{l-1} + \binom{2n}{l-2} \right]$$

$$= (-1)^l \left[\binom{2n+1}{l} - \binom{2n+1}{l-1} \right],$$

where we used the identity 9 on p. 54 of Knuth [4]. This completes the proof.

Identity 5 is a combination of two identities. The one on the left is known and the identity of the right has been proved lately (see references [6], [9]). Here we outline its proof.

Let $\zeta = \xi + i\eta = z^{-1/n}$, where $z = re^{i\theta}$. The family of curves K_n given by $\xi = 1$ takes the form

$$r = \cos^n \left(\frac{\theta}{n} \right), \tag{3.3}$$

which can be rewritten in the complex form

$$z^{-1/n} + z^{*1/n} = 2, \tag{3.4}$$

where z^* is the complex conjugate of z. Multiplying both sides by $(zz^*)^{1/n}$ we have

$$2(zz^*)^{1/n} = z^{1/n} + z^{*1/n}.$$

The above relation yields

$$2^{2n} zz^* = 2^n (z^{1/n} + z^{*1/n})^n. \tag{3.5}$$

But we know that (see [11], p. 445)

$$2^{2n} zz^* = \sum_{m=1}^{n} 2^m \binom{2n-m-1}{n-1} (z^{m/n} + z^{*m/n}). \tag{3.6}$$

Using the binomial theorem, the right-hand side of (3.5) reduces to

$$2^n (z^{1/n} + z^{*1/n})^n = 2^n \sum_{m=0}^{n} \binom{n}{m} z^{m/n} z^{*(n-m)/n}. \tag{3.7}$$

Equating the right-hand side of Eqs. (3.6) and (3.7), replacing z by $\cos^n\left(\frac{\theta}{n}\right)$ $e^{i\theta/n}$, taking the real part of both sides, and replacing θ/n by ϕ yield identity 5.

Proof of Identity 6

As before we convert identity 6 to a complex polynomial identity. It is straight forward to change (2.6) into

$$z^{4n} + z^{4k} = \sum_{m=0}^{n+k} E_{2n+2k-2m}(1 - z^2)^{2n-2m+2}(1 + z^{4m}).$$

Replacing z^2 by t and setting $l = n + k - m$ we obtain

$$t^{2n} + t^{2k} = \sum_{l=0}^{n+k} E_{2l}(1 - t)^{2l}(1 + t^{2n+2k-2l}). \tag{3.8}$$

Our objective is to find E_{2l} so that both sides of (3.8) are equal. Differentiating both sides of (3.8) $2s$ times $(0 \leq s \leq n + k)$ we have

$$\binom{2n}{2s}t^{2n-s} + \binom{2k}{2s}t^{2k-s}$$
$$= \sum_{l=0}^{n+k}\sum_{j=0}^{s} E_{2l}\binom{2l}{j}\binom{2n+2k-2l}{2s-j}(t-1)^{2l-j}[t^{2n+2k-2l-2s+j} + \delta_{0(2s-j)}]$$

where δ_{lj} is the Kronecker delta. Setting $t = 1$ on both sides we arrive at

$$\binom{2n}{2s} + \binom{2k}{2s} = \sum_{l=0}^{n+k} E_{2l}\binom{2n+2k-2l}{2s-2l}[1 + \delta_{0(2s-sl)}],$$

or $E_0 = 1$, and

$$E_{2s} = \frac{1}{2}\left[\binom{2n}{2s} + \binom{2k}{2s} - \sum_{l=0}^{s-1} E_{2l}\binom{2n+2k-2l}{2s-2l}\right], \quad s = 1, 2, \ldots, n + k,$$

which is the required result.

Example. If we set $n = 2, k = 1$ in (2.6), (2.7) we obtain the identity

$$\csc^6\phi\cos 2\phi = \csc^6\phi\cos 6\phi + 16\csc^4\phi\cos 4\phi + 80\csc^2\phi\cos 2\phi + 64.$$

4. Chebyshev Identities

It is well known that the Chebyshev polynomials of the first and second kind $T_n(x), U_n(x)$ are related to trigonometric functions according to the following formulas (see [5])

$$T_n(x) = \frac{n}{2} \sum_{m=0}^{[n/2]} (-1)^m \frac{(n-m-1)!}{m!(n-2m)!} (2x)^{n-2m},$$

$$U_n(x) = \sum_{m=0}^{[n/2]} (-1)^m \binom{n-m}{m} (2x)^{n-2m},$$

$$T_n(\cos\phi) = \cos n\phi, \quad U_n(\cos\phi) = \sin(n+1)\phi/\sin\phi.$$

In view of the above relations, identities 1–6 may be rewritten in terms of the Chebyshev polynomials as follows:

$$2^{2n-1}(1-x^2)^n =$$

$$(-1)^n T_{2n}(x) + \sum_{m=1}^{n} (-1)^{m-1} 2^{2n-2m}(1-x^2)^{n-m} T_{2m-2}(x)$$

$$+ \sum_{m=1}^{n-1} (-1)^{m-1} 2^{2n-2m-1}(1-x^2)^{n-m} U_{2m}(x). \tag{4.1}$$

$$2^{2n}(1-x^2)^n =$$

$$(-1)^n U_{2n}(x) + \sum_{m=1}^{n} (-1)^{m-1} 2^{2n-2m+1}(1-x^2)^{n-m} T_{2m-2}(x)$$

$$+ \sum_{m=1}^{n-1} (-1)^{m-1} 2^{2n-2m}(1-x^2)^{n-m} U_{2m}(x). \tag{4.2}$$

$$2^{2n}x(1-x^2)^n =$$

$$(-1)^n T_{2n+1}(x) + \sum_{m=1}^{n} (-1)^{m-1} 2^{2n-2m}(1-x^2)^{n-m} T_{2m-1}(x)$$

$$+ \sum_{m=1}^{n-1} (-1)^{m-1} 2^{2n-2m-1}(1-x^2)^{n-m} U_{2m-1}(x). \tag{4.3}$$

$$2^{2n+1}x(1-x^2)^n =$$

$$(-1)^n U_{2n+1} + \sum_{m=1}^{n}(-1)^{m-1}2^{2n-2m+1}(1-x^2)^{n-m}T_{2m-1}(x)$$

$$+ \sum_{m=1}^{n}(-1)^{m-1}2^{2n-2m}(1-x^2)^{n-m}U_{2m-1}(x). \tag{4.4}$$

$$2^{2n}x^{4n} =$$

$$\binom{2n}{n}x^{2n} + 2\sum_{m=0}^{n-1}\binom{2n}{m}x^{2n}T_{2n-2m}(x)$$

$$= \sum_{m=1}^{2n}2^{m+1-2n}\binom{4n-m-1}{2n-1}x^m T_m(x). \tag{4.5}$$

$$(-1)^{n-k}T_{2n-2k}(x) =$$

$$\sum_{m=0}^{n+k}(-1)^m 2^{2n+2k-2m}E_{2n+2k-2m}(1-x^2)^{n+k-m}T_{2m}(x), \tag{4.6}$$

where E_{2s} is given by relation (2.7).

Also we may rewrite several new Chebyshev identities from Obaid and Rung [9] in the form

$$2^n x^{n+1} = T_{n+1}(x) + \sum_{m=1}^{n}2^{n-m}T_{m-1}(x). \tag{4.7}$$

$$n2^{2n}x^{2n-1} = \sum_{m=1}^{n}m2^m\binom{2n-m-1}{n-1}x^{m-1}U_m(x). \tag{4.8}$$

For $n \geq k \geq 1$,

$$2^{n+k}x^{n+k-1}[nU_{n-k}(x) - kU_{n-k-2}(x)]$$

$$= \sum_{m=1}^{n}m2^m\left[\binom{n+k-m-1}{k-1} + \binom{n+k-m-1}{n-1}\right]x^{m-1}U_m(x). \tag{4.9}$$

For $n \geq k \geq 1$,

$$2^{n+k}x^{n+k-1}[nT_{n-k+1}(x) - kT_{n-k-1}(x)]$$

$$= \sum_{m=1}^{n}m2^m\left[\binom{n+k-m-1}{k-1} - \binom{n+k-m-1}{n-1}\right]x^{m-1}T_{m+1}(x). \tag{4.10}$$

5. Fibonacci and Lucas Identities

The Fibonacci sequence is defined by the recurrence relation $F_{m+2} = F_{m+1} + F_m, m \geq 0$, with $F_0 = 0, F_1 = 1$ and the Lucas sequence is defined by $L_{m+2} = L_{m+1} + L_m, m \geq 0$, with $L_0 = 2, L_1 = 1$. It is well known (see Byrd [2], [3]) that these sequences are related to the Chebyshev polynomials according to the two formulas

$$T_m\left(\frac{i}{2}\right) = \frac{(i)^m}{2} L_m \,, \; (i = \sqrt{-1}) \,, \tag{5.1}$$

$$U_m\left(\frac{i}{2}\right) = (i)^m F_{m+1} \,. \tag{5.2}$$

It is straight forward that when we use (5.1) and (5.2) in the identities (4.1)–(4.10) we obtain ten new Fibonacci and Lucas identities. Here we introduce only three of them as an illustration. Setting $x = i/2$ in (4.1) and using (5.1) and (5.2) we obtain

$$5^n = L_{2n} + \sum_{m=1}^{n} 5^{n-m} L_{2m-2} + \sum_{m=1}^{n-1} 5^{n-m} F_{2m-1} \,. \tag{5.3}$$

Similarly, identities (4.8) and (4.10) reduce to

$$n = \sum_{m=1}^{n} (-1)^n m \binom{2n-m-1}{n-1} F_{m+1} \,, \tag{5.4}$$

for $n > k \geq 1$,

$$(-1)^n[nL_{n-k+1} + kL_{n-k-1}] = \sum_{m=1}^{n} (-1)^m m \left[\binom{n+k-m-1}{k-1} - \binom{n+k-m-1}{n-1} \right] L_{m+1} \,, \tag{5.5}$$

respectively.

Acknowledgement

The author wishes to express his gratitude to Professor Evelyn Obaid for her valuable comments.

References

1. W. A. Bassali and S. A. Obaid, *On the torsion of elastic bars*, J. Appl. Math Mech. **61** (1981) 639–650.
2. P. F. Byrd, *Expansion of analytic functions in polynomials associated with Fibonacci numbers*, Fibonacci Quarterly **1** (1963) 16–28.
3. P. F. Byrd, *Expansion of analytic functions in terms involving Lucas numbers and similar number sequences*, Fibonacci Quarterly **3** (1965) 101–114.
4. D. E. Knuth, *Fundamental Algorithms* I, Wiley, New York, 1968.
5. W. Magnus, F. Oberhettinger, and R. P. Soni, *Formulas and Theorems for the Special Functions of Mathematical Physics*, Springer-Verlag, New York, 1966.
6. V. Mangulis, *Handbook of Series for Scientists and Engineers*, Academic Press, New York, 1965.
7. S. A. Obaid, *Flexure of beams with certain curvilinear cross sections*, J. Appl. Math. Phys. **34** (1983) 439–449.
8. S. A. Obaid, *Two new trigonometric formulas with appliations*, Appl. Math. Letters (4) **2** (1989) 1–5.
9. S. A. Obaid and D. C. Rung, *New mathematical identities with applications to flexure and torsion*, J. Appl. Math. Phys. **40** (1989) 93–110.
10. J. Riordan, *Combinatorial Identities*, Wiley, New York, 1968.
11. D. C. Rung and S. A. Obaid, *Combinatorics and flexure*, J. Appl. Math. Phys. **36** (1985) 443–459.
12. I. S. Sokolinkiff, *Mathematical Theory of Elasticity*, 2nd Edition, McGraw-Hill, New York, 1956.

S. A. Obaid
Department of Mathematics and
Computer Science
San Jose State University
San Jose, California 95192
USA

THE MATH. HERITAGE OF C.F. GAUSS (pp. 573-584)
edited by George M. Rassias
©1991 World Scientific Publ. Co. Singapore

HYPER-KÄHLER METRICS AND MONOPOLES

Henrik Pedersen

We consider a general hyper-Kähler metric in dimension 4 with a S^1-action compatible with the hyper-Kähler structure. We prove that such a metric can be described in terms of the S^1-monopole coming from the twistor space of the metric.

Keywords: Hyper-Kähler Geometry, Einstein Metrics, Monopoles, Twistor Space, Holomorphic Geometry.

1. Introduction

Let (M, g) be a $4n$-dimensional Riemannian manifold with three almost complex structures I, J and K satisfying the quaternion algebra identities

$$I^2 = J^2 = K^2 = -1, \quad IJ = -JI = K \tag{1}$$

etc. Assume that g is Hermitian with respect to I, J and K, i.e.,

$$g(IX, IY) = g(X, Y), \quad X, Y \in TM \tag{2}$$

etc. Then (M, g) is called a *hyper-Kähler* manifold iff the complex structures are covariant constant or equivalently iff I, J and K are integrable and the Kähler forms $\omega_1, \omega_2, \omega_3$ are closed, where

$$\omega_1(X, Y) = g(IX, Y), \quad X, Y \in TM \tag{3}$$

Math. Class. No.: 32L25; 53C25.

etc. [1,8]. The twistor space Z of such a hyper-Kähler metric is the complex $2n + 1$-dimensional manifold consisting of the compatible complex structures on M [6,14]. It is a generalization of Penrose's non-linear graviton construction [13].

Recently, [6,9], Hitchin et al. described the general hyper-Kähler metric — and its twistor space — in dimension $4n$ with n commuting Killing fields which preserve I, J and K. From their description of the metric in the case $n = 1$ it is easily seen that

$$g = V d\bar{x} \cdot d\bar{x} + V^{-1}(d\tau + A)^2 \tag{4}$$

where $d\bar{x} \cdot d\bar{x}$ is the canonical metric on $\mathbb{R}^3, V$ is a function on $\mathbb{R}^3$ and A a 1-form on $\mathbb{R}^3$. Moreover (V, A) satisfy the *monopole* equations on $\mathbb{R}^3$

$$*dV = -dA . \tag{5}$$

The twistor space of the metric is a line bundle over the complex surface $T\mathbb{CP}^1$ trivial on holomorphic sections of $T\mathbb{CP}^1 \to \mathbb{CP}^1$. Such a line bundle corresponds to a S^1-monopole on $\mathbb{R}^3$ [5,7,15] and we prove that this monopole coincides with the monopole appearing in the metric. Finally, we have some remarks on the sheaf cohomological aspects of the computations.

2. The Metric and Its Twistor Space

We shall review briefly the work of Hitchin et al. [6], but only in the 4-dimensional case: Let M be a 4-dimensional hyper-Kähler manifold with a free action of $\mathbb{R}$ on it. We assume that this action extends to a free holomorphic action of $\mathbb{C}$ on the twistor space Z. Then Z becomes a principal $\mathbb{C}$ bundle over $T\mathbb{CP}^1$.

Remarks. Strictly, Z is a $\mathbb{C}$ bundle over some open subset of $T\mathbb{CP}^1$: Since the solution $dA = - * dV$ is contained in the metric (4) we need V to be positive (or negative). But if V is positive and harmonic then V must be constant. This is easily seen using Poisson's integral formula and Gauss' law of arithmetic mean. Thus, global non-trivial solutions do not exist.

The twistor correspondence between M and Z is the usual one: Points of M parametrizes twistor lines in Z. The projection of the twistor lines in Z to $T\mathbb{CP}^1$ gives the 3-parameter family of sections of $T \to \mathbb{CP}^1$: If ζ denotes the affine coordinate on the complex line $\mathbb{CP}^1$ and η is a coordinate along the fibre of $T\mathbb{CP}^1$, then a real holomorphic section of $T\mathbb{CP}^1$ can be written as

$$\eta = z - x\zeta - \bar{z}\zeta^2 \tag{6}$$

where $z \in \mathbb{C}$ and $x \in \mathbb{R}$. To find the full 4-parameter family we describe Z in terms of a transition function: Let $U, \widetilde{U}$ be the usual cover of $T\mathbb{CP}^1$. Then we have coordinates (ξ, η, ζ) on $\mathbb{C} \times U$ and $(\widetilde{\xi}, \widetilde{\eta}, \widetilde{\zeta})$ on $\mathbb{C} \times \widetilde{U}$ related by

$$\begin{aligned}
\widetilde{\xi} &= \xi + \frac{\partial H}{\partial \eta}(\eta, \zeta) \\
\widetilde{\eta} &= \zeta^{-2}\eta \\
\widetilde{\zeta} &= \zeta^{-1}
\end{aligned} \tag{7}$$

on $\mathbb{C} \times U \cap \widetilde{U}$. Here H is a holomorphic function defined on $\mathbb{C} \times U \cap \widetilde{U} = \mathbb{C}^2 \times \mathbb{C}^*$. It is the Hamiltonian function for a symplectic vector field with respect to a sympletic form ω on the fibres of $Z \to T\mathbb{CP}^1 \to \mathbb{CP}^1$. Now, we seek a holomorphic function $\widetilde{\xi}$ of $\widetilde{\zeta}$ and a function ξ of ζ which satisfy

$$\widetilde{\xi}(\zeta^{-1}) = \xi(\zeta) + \frac{\partial H}{\partial \eta}(\eta(\zeta), \zeta) \tag{8}$$

where $\eta(\zeta)$ is given in (6) i.e., we seek curves in Z which project to a fixed section of $T\mathbb{CP}^1$ under the projection $Z \to T\mathbb{CP}^1$. Expanding in power series

$$\begin{aligned}
\widetilde{\xi} &= \sum_{n=0}^{\infty} a_n \zeta^{-n} \\
\xi &= \sum_{n=0}^{\infty} b_n \zeta^n \\
\frac{\partial H}{\partial \eta} &= \sum_{n=-\infty}^{\infty} c_n \zeta^n
\end{aligned} \tag{9}$$

we obtain from (8) and the residue theorem

$$\begin{aligned}
b_n &= -c_n \\
&= -\frac{1}{2\pi i} \int_\Gamma \zeta^{-(n+1)} \frac{\partial H}{\partial \eta} d\zeta; \quad n = 1, 2, \ldots
\end{aligned} \tag{10}$$

(and similarly with a_n)

$$a_0 - b_0 = \frac{1}{2\pi i} \int_\Gamma \zeta^{-1} \frac{\partial H}{\partial \eta} d\zeta \ . \tag{11}$$

Thus, from (11) we see that the coefficients a_0 and b_0 are not uniquely determined and this 1-dimensional ambiguity gives the full 4-parameter family of twistor lines (in order for the lines to be real we must demand that $a_0 = -\bar{b}_0$).

The manifold M which parametrizes the twistor lines is diffeomorphic to the fibres of $Z \to \mathbb{CP}^1$ and the twistor lines intersect the fibre at $\zeta = 0$ at a point with I-complex coordinates

$$\eta(0) = z, \quad \xi(0) = b_0 \equiv u \ . \tag{12}$$

To find x as a function of u and z we consider the solution to the Laplace equation in $\mathbb{R}^3$

$$F_{xx} + F_{z\bar{z}} = 0 \tag{13}$$

defined by

$$F(x, z, \bar{z}) = \frac{1}{2\pi i} \int_\Gamma \zeta^{-2} H(z - x\zeta - \bar{z}\zeta, \zeta) d\zeta \ . \tag{14}$$

Then it follows from (11), (12), (14) and the reality condition, that

$$F_x = u + \bar{u} \tag{15}$$

which determines x implicitly as a function of u and z.

To detemine the metric we use the fact that the symplectic form ω on the fibres of Z is given in terms of the symplectic forms $\omega_1, \omega_2, \omega_3$:

$$\omega = (\omega_2 + i\omega_3) + 2\omega_1 \zeta - (\omega_2 - i\omega_3)\zeta^2 \ . \tag{16}$$

From this it is shown that

$$\omega_1 = i(\bar{\partial}(F_z) \wedge dz + du \wedge \bar{\partial}x) \ . \tag{17}$$

By implicit differentiation of (15) we find that

$$F_{xx}x_{\bar{z}} + F_{x\bar{z}} = 0$$
$$F_{xx}x_{\bar{u}} = 1 \ . \tag{18}$$

Then from (13), (17) and (18) we get the metric

$$g = 2\mathrm{Re}[(F_{xx} + F_{zx}F_{xx}^{-1}F_{x\bar{z}})dz \otimes d\bar{z}$$
$$- F_{xx}^{-1}F_{zx}dz \otimes d\bar{u} - F_{xx}^{-1}F_{x\bar{z}}du \otimes d\bar{z} + F_{xx}^{-1}du \otimes d\bar{u}] . \qquad (19)$$

This ends the summary of [6,9].

Now, let us prove that the metric has the form in (4): Introduce a real coordinate

$$y = i(\bar{u} - u) \qquad (20)$$

then from (15) we obtain

$$du = \frac{1}{2}(d\,F_x + i\,dy) . \qquad (21)$$

Let V be the function on $\mathbb{R}^3$ given by

$$V^{-1} = g\left(\frac{\partial}{\partial y}, \frac{\partial}{\partial y}\right) \qquad (22)$$

and let θ be the connection in the principle $\mathbb{R}$-bundle M given by

$$\theta = V g\left(\frac{\partial}{\partial y}, \cdot\right) = dy + A \qquad (23)$$

where A is a 1-form on $\mathbb{R}^3$. Then we easily get

$$(V, A) = (2F_{xx}, i(F_{xz}dz - F_{x\bar{z}}d\bar{z})) . \qquad (24)$$

Furthermore, we compute the quotient metric

$$g - V^{-1}\theta^2 = V \mathrm{Re}(dz \otimes d\bar{z}) + \frac{1}{4}V\,dx^2$$
$$= V\,d\bar{x} \cdot d\bar{x} .$$

Remark. $T\mathbb{CP}^1$ is the mini-twistor space of $\mathbb{R}^3$ [5]: Points ($z = x_2 + i\,x_3, x = 2x_1$) of $\mathbb{R}^3$ parametrizes the real sections (6) of $T\mathbb{CP}^1$ and the canonical metric $d\bar{x} \cdot d\bar{x}$ on $\mathbb{R}^3$ is induced by the discriminant of (6).

Hence, the metric has the form given in (4). Moreover (V, A) is a monopole: We have the $*$ operator defined by

$$\langle \alpha, \beta \rangle \,\mathrm{vol} = \alpha \wedge *\bar{\beta}$$

and

$$\text{vol} = \frac{i}{4} dz \wedge d\bar{z} \wedge dx$$

so

$$*dz = -\frac{i}{2} dz \wedge dx$$
$$*dz = \frac{i}{2} d\bar{z} \wedge dx$$
$$*dx = i\, dz \wedge d\bar{z}$$

and we easily get

$$dA = - * dV .$$

Remark. The metrics of Gibbons and Hawking [3] are all of the form in (4). Indeed, in dimension 4, a hyper-Kähler metric is just a self-dual Einstein metric with vanishing cosmological constant.

3. The Monopole Induced by the Twistor Space

We have described a hyper-Kähler manifold with $\mathbb{R}$ acting on it. If we exponentiate our description from before we get the set-up with a S^1-action. Thus the twistor space becomes a $\mathbb{C}^*$ bundle and we let L be the associated vector bundle on $T\mathbb{CP}^1$ trivial on sections and with transition function

$$g_{01} = \exp \frac{\partial H}{\partial \eta} . \tag{25}$$

Now, such a line bundle determines a S^1-monopole [5]: Since L is trivial on lines we obtain a line bundle $\tilde{L}$ on $\mathbb{R}^3$ by

$$\tilde{L}_x = H^0(\mathbb{CP}_x, \mathcal{O}(L))$$

where $\mathbb{CP}_x$ is the line in (6) corresponding to $(z, x) \in \mathbb{R}^3$. If we fix a point (ζ_0, η_0) in $T\mathbb{CP}^1$, Eq. (6) describes a null plane $\pi(\zeta_0, \eta_0)$ and we obtain a flat connection ∇_π on π by trivializing $\tilde{L}$:

$$\psi \cdot \tilde{L}\,|_\pi \xrightarrow{\ \sim\ } L_{(\zeta_0, \eta_0)} \tag{26}$$

where ψ evaluates a section on $\mathbb{CP}_x$ in the point (ζ_0, η_0). It follows from the holomorphic descriptions that the connections ∇_π gives us a connection ∇ in L which coincides with the connections ∇_π along null lines. To describe ∇_π we seek a non-vanishing holomorphic section of L on $\mathbb{CP}_x$: From (8) and (25) we see that $(\exp \xi, \exp \tilde{\xi})$ is such a section. Also, we trivialize $\tilde{L}$ on $\mathbb{R}^3$ by

$$s : \mathbb{R}^3 \to \tilde{L} \qquad (27)$$

where s at x is the section of L on $\mathbb{CP}_x$ given by $(\exp \xi, \exp \tilde{\xi})$. Now, suppose we had a function f on $\mathbb{R}^3$ such that fs satisfied $\psi(fs) = 1$,

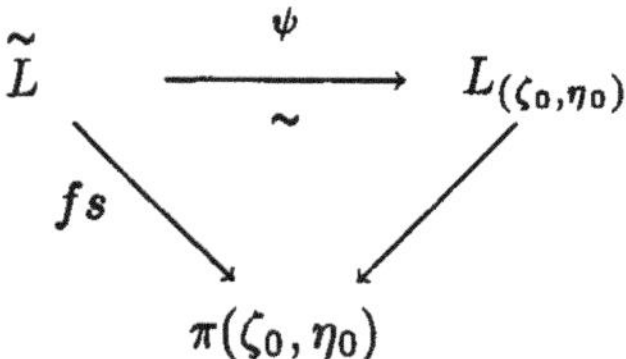

then $\nabla_\pi(fs) = 0$. We have

$$\psi(fs) = f(x) \exp \xi(\zeta_0) \qquad (28)$$

so

$$f = \exp(-\xi(\zeta_0)) \ .$$

Then, from (9) we get

$$
\begin{aligned}
0 &= \nabla_\pi(\exp(-\xi(\zeta_0))s) \\
&= -d\xi(\zeta_0) \cdot \exp(-\xi) \cdot s + \exp(-\xi)\theta_\pi s
\end{aligned}
$$

where θ_π is the connection form of ∇_π in the frame s. Thus

$$
\begin{aligned}
\theta_\pi &= d\xi(\zeta_0) \\
&= \sum_{n=0} db_n \zeta_0^n \ .
\end{aligned} \qquad (29)
$$

Now, on a null plane we have

$$dx = \zeta_0^{-1}dz - \zeta_0 d\bar{z} \qquad (30)$$

so since the metric is given by

$$dx^2 + 4dzd\bar{z} \qquad (31)$$

we see that the null lines are given by (6) together with

$$\zeta_0^{-1}dz + \zeta_0 d\bar{z} = 0 \ . \tag{32}$$

Consider the connection form of ∇ with respect to the frame s

$$A = fdz + gdx + hd\bar{z} \ . \tag{33}$$

On a null line we have

$$A = (h - 2g\zeta_0 - f\zeta_0^2)d\bar{z} \tag{34}$$

which coincide with

$$\begin{aligned}
\theta_\pi = \{&(b_0)_{\bar{z}} + (-2(b_0)_x + (b_1)_{\bar{z}})\zeta_0 \\
&+ (-2(b_1)_x - (b_0)_z + (b_2)_{\bar{z}})\zeta_0^2\}d\bar{z} \ .
\end{aligned} \tag{35}$$

Thus, we get

$$\begin{aligned}
h &= (b_0)_{\bar{z}} \\
2g &= 2(b_0)_x - (b_1)_{\bar{z}} \\
f &= (b_0)_z + 2(b_1)_x - (b_2)_{\bar{z}}
\end{aligned} \tag{36}$$

and from (10), (12), (14) and (21) it follows that

$$\begin{aligned}
(b_0)_x &= \frac{1}{2}F_{xx} \\
(b_0)_z &= \frac{1}{2}F_{xz} \\
(b_0)_{\bar{z}} &= \frac{1}{2}F_{x\bar{z}} \\
(b_1)_x &= -F_{xz} \\
(b_1)_{\bar{z}} &= -F_{z\bar{z}} \\
(b_2)_{\bar{z}} &= -F_{xz} \ .
\end{aligned} \tag{37}$$

(For example: $b_2 = -\frac{1}{2\pi i}\int_\Gamma \zeta^{-3}\frac{\partial H}{\partial \eta}d\zeta$ so $(b_2)_{\bar{z}} = \frac{1}{2\pi i}\int_\Gamma \zeta^{-1}\frac{\partial^2 H}{\partial \eta^2}d\zeta = -F_{xz}$.)
This gives

$$A = -\frac{1}{2}\{F_{xz}dz - F_{x\bar{z}}d\bar{z}\} \ . \tag{38}$$

Next, to find the potential V we note that on the null planes

$$\nabla - \nabla_\pi = \frac{i}{2}Vdt \tag{39}$$

where the null lines in π are given by $t = $ constant. This follows from the fact that ∇ and ∇_π agree along null lines and from the holomorphic description it is clear that V is independent of the plane π. Thus from (32) we get

$$A - \theta_\pi = \frac{i}{2}V(\zeta_0^{-1}dz + \zeta_0 d\bar{z}) \tag{40}$$

and this gives

$$V = iF_{xx} . \tag{41}$$

Compare (38) and (41) with (24) and we have proved that the monopole comming from the twistor space is equal to the monopole contained in the metric (up to a scalar multiple).

Remark. This situation of having an Einstein metric given in terms of a monople has been considered earlier in a different setting [7,11].

4. Sheaf Theoretical Considerations

There is a bijective correspondence between the sheaf cohomology group $H^1(T\mathbb{CP}^1, \mathcal{O})$ and monopoles in $\mathbb{R}^3$. Also, it is known that we have a bijective correspondence between $H^1(T\mathbb{CP}^1, \mathcal{O}(-2))$ and solutions to the Laplace equations in $\mathbb{R}^3$ [2, 4, 7, 10]: If $[f(\zeta, \eta)d\zeta] \in H^1(T\mathbb{CP}^1, \mathcal{O}(-2))$, then the function

$$F(x, z, \bar{z}) = \frac{1}{2\pi i} \int_\Gamma f(\zeta, z - x\zeta - \bar{z}\zeta^2)d\zeta$$

satisfy the Laplace equation: $F_{xx} + F_{z\bar{z}} = 0$. Furthermore the group $H^1(T\mathbb{CP}^1, \mathcal{O})$ corresponds to the holomorphic line bundles trivial on sections of $T\mathbb{CP}^1$: We have the short exact sequence

$$0 \rightarrow \mathbb{Z} \rightarrow \mathcal{O} \xrightarrow{\exp} \mathcal{O}^* \rightarrow 0$$

and we know that $H^1(T\mathbb{CP}^1, \mathcal{O}^*)$ consists of the isomorphism classes of holomorphic line bundles on $T\mathbb{CP}^1$. Also, $H^2(T\mathbb{CP}^1, \mathbb{Z}) \cong H^2(\mathbb{CP}^1, \mathbb{Z}) \cong \mathbb{Z}$. Then, from the long exact sequence on cohomology we get

$$0 \rightarrow H^1(T\mathbb{CP}^1, \mathcal{O}) \xrightarrow{\exp} H^1(T\mathbb{CP}^1, \mathcal{O}^*) \xrightarrow{\delta} \mathbb{Z} \rightarrow .$$

Since the coboundary map δ is the Chern class, we see that the image

$$\exp(H^1(T\mathbb{CP}^1, \mathcal{O})) \subseteq H^1(T\mathbb{CP}^1, \mathcal{O}^*)$$

consists of the line bundles with vanishing Chern class. Thus if $L \in \exp(H^1(T\mathbb{CP}^1, \mathcal{O}))$ and $\mathbb{CP}^1$ is a section of $T\mathbb{CP}^1$ we get

$$\text{degree}(L|_{\mathbb{CP}^1}) = \int_{\mathbb{CP}^1} c_1(L) = 0 \ .$$

Hence from a class $\left[\frac{\partial H}{\partial \eta}(\zeta, \eta)\right]$ in $H^1(T\mathbb{CP}^1, \mathcal{O})$ we obtain the line bundle with transition function $\exp\left[\frac{\partial H}{\partial \eta}\right]$ — trivial on plane sections — and this bundle gives the monopole as described above.

Next, we shall describe an isomorphism

$$H^1(T\mathbb{CP}^1, \mathcal{O}) \xrightarrow{\sim} H^1(T\mathbb{CP}^1, \mathcal{O}(-2)) \ .$$

Consider "differentiation along fibres", d_F, defined by the composite map

$$\mathcal{O}_{T\mathbb{CP}^1} \xrightarrow{d} \Omega^1_{T\mathbb{CP}^1} \xrightarrow{\pi} \Omega^1_F$$

where Ω^1_X is the sheaf of germs of holomorphic 1-forms on X and F is the fibre of the projection

$$T\mathbb{CP}^1 \xrightarrow{p} \mathbb{CP}^1 \ .$$

Now, it is obvious that $\Omega^1_F \cong p^*\mathcal{O}(-2)$. We then have the short exact sequence

$$0 \to \mathcal{O}_{\mathbb{CP}^1} \to \mathcal{O}_{T\mathbb{CP}^1} \xrightarrow{d_F} p^*\mathcal{O}(-2) \to 0$$

and from the long exact sequence on cohomology we get

$$\to H^1(\mathbb{CP}^1, \mathcal{O}) \to H^1(T\mathbb{CP}^1, \mathcal{O}) \xrightarrow{d_F} H^1(T\mathbb{CP}^1, \mathcal{O}(-2))$$
$$\to H^2(\mathbb{CP}^1, \mathcal{O}) \to \ .$$

Hence, since $H^1(\mathbb{CP}^1, \mathcal{O}) = 0 = H^2(\mathbb{CP}^1, \mathcal{O})$ we obtain the isomorphism

$$H^1(T\mathbb{CP}^1, \mathcal{O}) \underset{\sim}{\xrightarrow{d_F}} H^1(T\mathbb{CP}^1, \mathcal{O}(-2))$$

which sends the monopole represented by $\frac{\partial H}{\partial \eta}$ into the solution to the Laplacian represented by $\frac{\partial^2 H}{\partial \eta^2}$.

Now, the hyper-Kähler metric is given in terms of the solution to the Laplacian

$$F = \int_\Gamma H\zeta^{-2}d\zeta$$

represented by $H\zeta^{-2}d\zeta \in H^1(T\mathbb{CP}^1, \mathcal{O}(-2))$. But the monopole (V, A) represented by $\frac{\partial H}{\partial \eta} \in H^1(T\mathbb{CP}^1, \mathcal{O})$ also gives a solution V to the Laplacian and we have seen that

$$V = F_{xx} = \int_\Gamma \frac{\partial^2 H}{\partial \eta^2}d\zeta \ .$$

Hence V is represented by $d_F\left[\frac{\partial H}{\partial \eta}\right] \in H^1(T\mathbb{CP}^1, \mathcal{O}(-2))$.

Remark. Most of the results above can be generalized to hyper-Kähler geometry in dimension $4n$ [12]. This involves a generalization of the Bogomolny equations whose possible physical implications remains to be seen.

Acknowledgement

The ideas discussed here on the twistorial aspects of $\mathbb{R}^3$ are heavily dependent on lectures given by N. J. Hitchin at the Mathematical Institute, Oxford.

References

1. E. Calabi, *Métriques Kählériennes et Fibrés Holomorphes*, Ann. Scient. Ec. Norm. Sup. 4^{E} série, t.12 (1979) 269–294.
2. M. G. Eastwood, *The generalized Penrose-Ward transform*, Math. Proc. Camb. Phil. Soc. **97** (1985) 165–187.
3. G. W. Gibbons, S. W. Hawking, *Gravitational multi-instantons*, Phys. Lett. **78B** (1978) 430–432.
4. N. J. Hitchin, *Linear field equations on self-dual spaces*, Proc. Roy. Soc. Lond. **A370** (1980) 173–191.
5. N. J. Hitchin, *Monopoles and geodesics*, Commun. Math. Phys. **83** (1982) 579–602.
6. N. J. Hitchin, A. Karlhede, U. Lindström, M. Rocek, *Hyper-Kähler metrics and supersymmetry*, Commun. Math. Phys. **108** (1987) 535–589.

7. P. E. Jones, *Minitwistors*, D. Phil. Thesis, Oxford, 1984.

8. S. Kobayashi, K. Nomizu, *Foundation of Differential Geometry*, vol. II, Wiley, New York, 1969 p. 148.

9. U. Lindström, M. Rocek, *Scalar tensor duality and $N = 1, 2$ nonlinear σ-models*, Nucl. Phys. **B222** (1983) 285–309.

10. M. K. Murray, *A Penrose transform for the twistor space of an even dimensional conformally flat Riemannian manifold*, Ann. Global Anal. Geom. **4**, No. 1 (1986) 71–88.

11. H. Pedersen, *Einstein metrics, spinning top motions and monopoles*, Math. Ann. **274** (1986) 35–59.

12. H. Pedersen, Y. S. Poon, *Hyper-Kähler metrics and a generalization of the bogomolny equations*, Commun. Math. Phys. **117** (1988) 569–580.

13. R. Penrose, *Nonlinear gravitons and curved twistor theory*, Gen. Relativity and Gravition **7** (1976) 31–52.

14. S. Salamon, *Quaternionic Kähler manifolds*, Invent. Math. **67** (1982) 143–171.

15. R. S. Ward, *On self-dual gauge fields*, Phys. Lett. **61A** (1977) 81–82.

Henrik Pedersen
Department of Mathematics
Odense University
Campusvej 55
DK-5230 Odense M
Denmark

THE MATH. HERITAGE OF C.F. GAUSS (pp. 585-604)
edited by George M. Rassias
©1991 World Scientific Publ. Co. Singapore

ON CERTAIN MATHEMATICAL PROBLEMS CONNECTED WITH THE USE OF THE COMPLEX VARIABLE BOUNDARY ELEMENT METHOD TO THE PROBLEMS OF PLANE HYDRODYNAMICS. GAUSS' VARIANT OF THE PROCEDURE

Titus Petrila

This work is devoted to some aspects of the approximation of complex functions on boundary of different domains in view of constructing of a complex variable boundary element method (CVBEM) for boundary problems joined to Laplace operator.

The first two sections deal with the approximation by means of interpolating functions on Fejer nodes or by "piecewise" Lagrangian interpolating polynomials. In both cases the convergence of the approximation as well as the convergence of the whole CVBEM procedure is established.

The third section contains the case when the domain is the exterior of a Jordan curve C with an angular point. Some appropriate approximations for this situation based on the results of the first sections are developed.

In the next section, a generalization in the complex plane of the Gauss interpolating polynomial for the case of the contours with some angular points is envisaged. This "generalized" Gauss interpolating function is very convenient to be used to applications in plane hydrodynamics which is the purpose of the final section.

The fifth and final section gives a general method of approaching the main problems of plane hydrodynamics. This method also leads to a joined CVBEM based upon the approximations considered before. As an example the case of the fluid flow induced by a mobile profile with a sharp trailing edge performing a rototranslation in the mass of the ideal incompressible potential flow is analysed.

The complex variable boundary element method (CVBEM) and its application to the problems of the plane hydrodynamics already constituted the object of several investigations [2–4]. Essentially, this variant of BEM is applied to those problems where the boundary problem is formulated in terms of an unknown holomorphic function whose real or imaginary part (or a combination of these ones) is given on the boundary. Thus, using Cauchy's formula for this function, one obtains immediately the integral representation associated to the given boundary problem, which constitutes in fact the working instrument of BEM. In addition, using a system of interpolating functions for the unknown function in the boundary points, one can obtain directly, by the same Cauchy's formula, the solution of our problem in any point of the considered domain. In this way CVBEM avoids the construction and solution of the integral equation on the boundary (which were singular in our case, too), and implicitly, does not need the use of quadrature formulae (sometimes, not even the approximation of the boundary [3, 4]).

The purpose of this paper is to survey some aspects of the problems concerning the approximation of the unknown holomorphic function by means of interpolating functions. We shall differentiate some working directions, all developed towards the convergency of the respective approximation process. Additionally we try to use these procedures to some practical problems of the profiles theory.

1. Let $f(z)$ be a holomorphic function in a domain D bounded by the closed, simple, rectifiable (Jordan) curve C. Suppose meanwhile that $f(z)$ is holomorphic in $D \cup C \equiv K$, which is compact from $\mathbb{C}$, and whose complement $K^C = \mathbb{C} \backslash K$ is a simply connected domain. Then there will exist the conformal mappings normed in the neighbourhood of the infinite point

$$z = \psi(w) = cw + c_0 + c_1/w + \dots \quad (c > 0) \, ,$$

defined on $\{w \mid |w| > 1\}$ and valued into K^C.

Our purpose is to interpolate the function $f : K \to \mathbb{C}$. Denote by $z_k^{(n)}(k, n = \overline{0, n})$ the interpolating nodes from K; we have therefore the "nodal matrix"

$$z_0^{(0)}$$

$$z_0^{(1)} \quad z_1^{(1)}$$

$$\cdots \cdots \cdots \cdots \cdots \cdots$$

$$z_0^{(n)} \quad z_1^{(n)} \quad \ldots \quad z_n^{(n)} \; ,$$

and, accordingly, the polynomials

$$\omega_n(z) = \prod_{k=0}^{n} \left(z - z_k^{(n)} \right), \quad n = 0, 1, \ldots \; .$$

If $L_n(z)$ are the Lagrange interpolating polynomials, then one knows that the "error" is

$$f(z) - L_n(z) = (2\pi i)^{-1} \int_C (\omega_n(z)/\omega_n(s)) f(s)(s-z)^{-1} ds \quad (s \in C) \; .$$

Definition 1.1. The nodes $z_n^{(n)}$ are called to be uniformly distributed on K if $(M_n)^{1/(n+1)} \to c$, where

$$M_n = \max\{|\omega_n(z)| \, | \, z \in K\} \; ,$$

and

$$c = (d\psi/dw)_{w=\infty} \; .$$

Thus the following theorem (Kalmar, Walsh) [5] holds:

Theorem. $L_n(z) \to f(z)$ $(n \to \infty, z \in K)$ for every f holomorphic on K if and only if the interpolating nodes $z_k^{(n)} \in K$ are uniformly distributed on K.

We shall deal further with the construction of a system of nodes, uniformly distributed on K, system which will totally belong to the boundary C. Implicitly, the restriction of the approximation $L_n(z)|_C$ will converge uniformly to $f(t), t \in C$, and this is essential for the development of CVBEM.

Since the boundary C is a Jordan curve, it is obvious that the function $\psi : \{w|\,|w| > 1\} \to K^C$ will admit a continous extension on $\{w|\,|w| \geq 1\}$. Then we may introduce:

Definition 1.2. The points $z_k^{(n)} = \psi(e^{2\pi ik/(n+1)})$, $k = \overline{0,n}$, are called the Fejer nodes of n-th order on K.

Thus the following theorem (Fejer) [4] holds:

Theorem. The Fejer nodes are uniformly distributed on K.

It is obvious that these Fejer nodes, which constitute the images of the $(n + 1)$-th order roots of the unity through ψ, will all lie in the points of the contour C; since we need only the approximation of $f(t)$ in the points $t \in C$, this choice of the interpolating nodes will be extremely suitable.

Unfortunately, the hypothesis that the function f is holomorphic on $D \cup C$ is a "too strong" hypothesis for practical problems as, for instance, those belonging to profile theory. Generally, only the continuity of $f(t)$ on C will be ensured, but not the continuity and so much the holomorphicity on $D \cup C$ (Fatou's result). That is why Cauchy's integral which we shall use will be in fact only a Cauchy-type integral for which the Plemelj-Privaloff jump relations hold. Of course, this deadlock could be broken if the data on boundary $f(t)$ satisfy, in the points of the contour C, the following singular integral equation

$$f(z_0) = (\pi i)^{-1} \oint_C f(t)(t - z_0)^{-1}dt \ ,$$

which also represents the necessary and sufficient condition for the validity of Cauchy's formula

$$f(z) = (\pi i)^{-1} \int_C f(t)(t - z)^{-1}dt$$

in this case, too [6].

Suppose that Cauchy's integral (see above) is used to express our function $f(z)$. The previously established convergency of the Lagrange interpolating system on the Fejer nodes will ensure the convergency of the integrals

$$(2\pi i)^{-1} \int_C L_n(t)(t - z)^{-1}dt$$

towards $f(z)$ in any point inside the domain D.

In this way, the calculation of the Cauchy-type integrals of the form $(2\pi)^{-1}\int_C t^k(t-z)^{-1}dt, k = \overline{0, n(n+1)/2}$, involved by the approximation process, will not be difficult; they are respectively equal to $z^k, k = \overline{0, n(n+1)/2}$. This is also in agreement with the fact that the error of the procedure, estimated on the boundary, will be "conserved" in every point inside our domain.

2. Suppose now that the unknown function $f(z)$ is holomorphic only in the domain D, while its restriction $f(t)$ in the points of the boundary C is continuous for $t \in C$. Obviously, Cauchy's integral, which will be now only a Cauchy-type integral, will exist in this case, too, allowing us to write that

$$f(z) = (2\pi i)^{-1}\int_C f(t)(t-z)^{-1}dt$$

holds for every point $z \in D$.

Since our purpose is a convergent approximation of the function f just in the points of the contour C — which will then allow, via Cauchy's integral, the construction of an approximation for the unknown function f in every point inside D — consider some set of nodal points (nodes), $z_0, z_1, \ldots, z_n(z_n \equiv z_0)$ lying on C. These nodes, distributed counterclockwise, will divide the boundary C into the boundary elements $C_j(j = \overline{1, n})$, where C_j is the simple arc linking the points z_{j-1} and z_j.

Let now $\widetilde{f}(t)$ be the following approximation of $f(t)$:

$$\widetilde{f}(t) = \sum_{j=1}^{n} f_j L_j(t)\,,$$

where $f_j = f(z_j)$, while $L_j(t)$ are the Lagrange interpolating polynomials constructed respectively on each arc, namely

$$L_j(t) = \begin{cases} (t - z_{j-1})/(z_j - z_{j-1}), & \text{for } t \in C_j\ ; \\ (t - z_{j+1})/(z_j - z_{j+1}), & \text{for } t \in C_{j+1}\ ; \\ 0, & \text{otherwise}\ , \end{cases}$$

and then

$$\widetilde{f}(t)|_{C_j} = f_{j-1}(t - z_j)/(z_{j-1} - z_j)$$
$$+ f_j(t - z_{j-1})/(z_j - z_{j-1})\,.$$

Therefore the following approximation for Cauchy's intergral will be obtained:

$$f^*(z) = \sum_{j=1}^{n} f_j \widetilde{L}_j(z) \, ,$$

in which

$$\widetilde{L}_j(z) = (2\pi i)^{-1} \int_C L_j(t)(t-z)^{-1} dt$$
$$= (2\pi i)^{-1}(((z - z_{j-1})/(z_j - z_{j-1})) \cdot \ln((z - z_j)/(z - z_{j-1}))$$
$$+ ((z - z_{j+1})/(z_j - z_{j+1})) \cdot \ln((z - z_{j+1})/(z - z_j))) \, ,$$

and the principal branch for the complex logarithm was chosen.

Obviously, the function $f(z)$ will be holomorphic in $D \cup C$ and, implicitly, continuous on $D \cup C$, such that its values f_k in the nodes z_k are obtained by simple substitution. More concretely, we shall have

$$f_k \equiv f^*(z_k) = \sum_{j=1}^{n} f_j \widetilde{L}_j(z_k), \quad k = \overline{1,n} \, .$$

Denoting then $u_k + i v_k \equiv f_k$ and $L_{jk} \equiv M_{kj} + i N_{kj} = \widetilde{L}_j(z_k)$, we are guided to the solution of the linear algebraic system of $2n$ equations with $2n$ unknowns

$$u_k = \sum_{j=1}^{n} M_{kj} u_j - \sum_{j=1}^{n} N_{kj} v_j \, ,$$
$$v_k = \sum_{j=1}^{n} M_{kj} v_j + \sum_{j=1}^{n} N_{kj} u_j \, .$$

Solving this system — using, of course, the boundary problem data, too — one will obtain the approximation $\widetilde{f}(t)$ of the function $f(t)$, and then, via Cauchy's formula, the solution of the proposed boundary problem in all points of the domain D [2, 3].

Let us study now the convergency of this approximation procedure; more precisely, we shall try to see in what conditions $\widetilde{f}(t)$ tends to $f(t)$, and, implicitly, $f^*(z) \to f(z)$.

Defintion 2.1. A division $d := z_0, z_1, \ldots, z_n (z_0 \equiv z_n)$ of the curve C will be called "acceptable" if for each $t \in C_j$ $(j = \overline{1,n})$ the following condition is fulfilled

$$\max\{|t - z_j|, |t - z_{j-1}|\} < |z_j - z_{j-1}| \ .$$

Let now d be an "acceptable" division of the boundary C, and let $\delta = \max_j |z_j - z_{j-1}|$ be the norm of this division. Then the following theorem holds:

Theorem. If $\widetilde{f}(t) = \sum_{j=1}^{n} f_j L_j(t)$ is the Lagrange linear "piecewise" approximation (that is, constructed on each arc C_j of the contour C) of the function $f(t)$ corresponding to the "acceptable" division d of the norm δ, then

$$\lim_{\delta \to 0} \widetilde{f}(t) = f(t), \quad t \in C \ .$$

Proof. Since the function $f(t)$ is continuous on the closed contour C of finite length, it will be uniformly continuous on C, too. But then, for every $\varepsilon > 0$ there exists $\eta > 0$ such that the inequality $|f(t_1) - f(t_2)| < \varepsilon/2$ holds for every couple of points $t_1, t_2 \in C$ such that $|t_1 - t_2| < \eta$. A similar relation can be written for $\widetilde{f}(t)$ (which also is uniformly continuous on C), too. Considering then an "acceptable" division of the contour C such that $\delta < \eta$ in every point $t \in C$ which belongs to a C_k, too, we have

$$|f(t) - \widetilde{f}(t)| = |f(t) - f(z_k) + \widetilde{f}(z_k) - \widetilde{f}(t)|$$
$$\leq |f(t) - f(z_k)| + |\widetilde{f}(z_k) - \widetilde{f}(t)| < \varepsilon/2 + \varepsilon/2 = \varepsilon \ .$$

Since t is an arbitrary point of C, it will result that

$$\lim_{\delta \to 0} |f(t) - \widetilde{f}(t)| = 0 \ .$$

Remark. The convergency of the above Lagrange interpolating system towards the function $f(t)$, whatever the system of "acceptable" nodes with $\delta \to 0$ is, is not surprising since this interpolating system consists of "piecewise" Lagrange functions which essentially are splines of the first order. The results keep their validity even if the above "segmentary" Lagrange approximation is replaced on certain arcs C_j by arbitrary powers of this one. This generalization will be important while considering contours C with angular points, as for instance in the case of the profiles with sharp trailing edges.

The above theorem allows lastly to establish the convergency of the whole procedure. More concretely, the following final result holds, too:

Theorem. If $f^*(z) = \sum_{j=1}^{n} f_j \widetilde{L}_j(z)$ is the linear Lagrange "piecewise" approximation of the function $f(z)$, corresponding to the "acceptable" division d of norm δ, then

$$\lim_{\delta \to 0} f^*(z) = f(z)$$

holds in every point $z \in D$.

Proof. From the previous theorem, we have immmediately

$$\begin{aligned}
\lim_{\delta \to 0} f^*(z) &= \lim_{\delta \to 0} (2\pi i)^{-1} \int_C \widetilde{f}(t)(t-z)^{-1} dt \\
&= (2\pi i)^{-1} \int_C (\lim_{\delta \to 0} \widetilde{f}(t))(t-z)^{-1} dt \\
&= (2\pi i)^{-1} \int_C f(t)(t-z)^{-1} dt = f(z)
\end{aligned}$$

and the theorm is proved.

As to the delimitation of the procedure error, observe that:

$$|f^*(z) - f(z)| \leq (2\pi)^{-1} \sum_{j=1}^{n} \int_{C_j} (|f(t) - \widetilde{f}(t)|/|t-z|)|dt| \; ;$$

then, denoting $\Delta = \min_{t \in C} |t - z|$, and choosing an "acceptable" division d of the contour with the norm $\delta \in (\eta/k, \eta)$, where η was constructed such that $|f(t) - \widetilde{f}(t)| < \varepsilon$, while k was chosen such that $n\eta/k \leq l(C)$, we have

$$|f^*(z) - f(z)| \leq n\eta\varepsilon/(2\pi R) \leq kl(C)\varepsilon/(2\pi R) \; .$$

3. Suppose this time that our domain D represents the exterior of a simple, closed, rectifiable (Jordan) contour C, which has in the point $z_F \in C$

an angular point in which the angle of the semitangents (counterclockwise oriented) is $\pi - \mu\pi$, with $-1 \leq \mu < 0$. The unknown function $f(z)$ will be holomorphic in this domain D, in the infinite point, too, whose neighbourhood it admits the expansion:

$$f(z) = a_0 + a_1/z + \ldots + a_n/z^n + \ldots \, .$$

Assume that $f(z)$ is continous on the boundary C where its real or imaginary part (or a combination of these ones) is known.

Cauchy's formula for the function $f(z)$ and domain $D \cup C$ will then allow us to write

$$f(z) = (2\pi i)^{-1} \int_{\overrightarrow{C}} f(t)(t - z)^{-1} dt + a_0, \quad z \in D \, .$$

The behaviour of this integral representation in the neighbourhood of the angular point z_F will be of the type

$$f(z) - f(z_F) = (z - z_F)^{1/(1-\mu)} h(z), \quad h(z) \neq 0 \, .$$

Such a behaviour entails for the derivative a behaviour of the form $0((z - z_F)^{\mu/(1-\mu)})$, that is, the derivative becomes unbounded in z_F for $-1 \leq \mu < 0$.

In order to use CVBEM in this case, too, by means of the same Cauchy's formula, we could work according to the manner described in Sec. 2. More exactly, the linear interpolation must also correspond now to the behaviour of $f(z)$ in $V(z_F)$; in other words, in this neighbourhood and in the points $t \in C$ (assuming that z_F is a node of the "acceptable" division d), the function $f(z)$ must be interpolated through

$$\widetilde{f}(t) = \begin{cases} f_F + (f_{F-1} - f_F)((t - z_F)/(z_{F-1} - z_F))^{1/(1-\mu)}, & \text{for } t \in C_F \, ; \\ f_F + (f_{F+1} - f_F)((t - z_F)/(z_{F+1} - z_F))^{1/(1-\mu)}, & \text{for } t \in C_{F+1} \, . \end{cases}$$

Accordingly, in the approximation on the whole boundary C:

$$\widetilde{f}(t) = \sum_{j=1}^{n} f_j L_j(t) \, ,$$

for $j \neq F - 1, F, F + 1$ the expressions of the polynomials $L_j(t)$ are identical to those given in Sec. 2, while for $j = F - 1, F, F + 1$ we have now

$$L_{F-1}(t) = \begin{cases} (t - z_{F-2})/(z_{F-1} - z_{F-2}), & \text{for } t \in C_{F-1} \, ; \\ (t - z_F)/(z_{F-1} - z_F)^{1/(1-\mu)}, & \text{for } t \in C_F \, ; \\ 0, & \text{otherwise} \, ; \end{cases}$$

$$L_F(t) = \begin{cases} 1 - ((t - z_F)/(z_{F-1} - z_F))^{1/(1-\mu)}, & \text{for } t \in C_F \ ; \\ 1 - ((t - z_F)/(z_{F+1} - z_F))^{1/(1-\mu)}, & \text{for } t \in C_{F+1} \ ; \\ 0, & \text{otherwise} \ ; \end{cases}$$

$$L_{F+1}(t) = \begin{cases} (t - z_{F+2})/(z_{F+1} - z_{F+2}), & \text{for } t \in C_{F+2} \ ; \\ ((t - z_F)/(z_{F+1} - z_F))^{1/(1-\mu)}, & \text{for } t \in C_{F+1} \ ; \\ 0, & \text{otherwise} \ . \end{cases}$$

By the theorems of convergence and the remark connected with them from Sec. 2, the convergence of the procedure is ensured. In addition, it is possible to calculate directly all coefficients of the algebraic system obtained, without boundary approximations or use of quadrature formulae [3, 4].

We propose now to imagine a more rapidly convergent approximation procedure, with a "densering" of the nodes in the neighbourhood of the angular point. Such an approaching manner would be extremely useful if we have to study, for instance, the plane, incompressible, and potential fluid flow generated by rototranslation into the ideal fluid mass having a uniform (complex) speed W_∞, of a profile with sharp trailing edge [7].

The Lagrange interpolating system constructed on the Fejer nodes which was developed in Sec. 1 would provide a more rapidly convergent approximation, with the possibility to describe more correctly the fluid flow in the neighbourhood of the trailing edge, where there are high velocity gradients. Unfortunately, with our actual hypotheses, the function $f(z)$ will not be holomorphic in $D \cup C$. Moreover, although $D \cup C$ is a compact and simply connected set in the "chord" metric (on the Riemann sphere), it no longer has the same properties in the Euclidean metric, in which we are working.

Suppose now that we "compactify" the domain D, considering its "approximation" D^* as bounded on one hand by a circumference $|z| = R$ (of a sufficiently large radius R in order to have $f(z)\big|_{|z|=R} \cong a_0$), and on the other hand by the "rounded" contour $C^* \subset D$, between whose points z_{C^*} and the points z_C of C a bijection which observes the condition $|z_C - z_{C^*}| = \varepsilon$, with a suitably given (sufficently small) $\varepsilon > 0$, can be established. This bijection could practically be performed by associating the points z_C and z_{C^*} lying on the chord $z_0 z_C$, where z_0 is the centre of the profile bounded by C.

It is obvious that the compactness of D^* and the holomorphicity of $f(z)$ in the points of C^*, too, would allow the use of the procedure described in

Sec. 1. Thus, choosing the system of Fejer nodes on the contour C^*, we shall have $L_n(z_{C^*}) \rightrightarrows f(z_{C^*})$ in the points of this contour. With this, we have to calculate only

$$\left[\lim_{z_{C^*} \to z_C} L_n(z_{C^*}) \right]_{\text{col}}$$

(where the index "col" signifies the restriction according to which z_0, z_C, z_{C^*} are collinear), the result representing a system of interpolating functions for $f(t), t \in C$.

Let us now assume that our contour (profile) C has the sharp trailing edge $z_F(A_0)$ belonging to a "dihedral" zone $A^- A_0 A^+$. Obviously, the trailing edge $A_0(z_F)$ could be a Fejer node, for instance that corresponding to $e^0 \equiv 1$. But, beginning with a certain rank N, the Fejer nodes bordering "symmetrically" the trailing edge will lie in the dihedral zone $A^- A_0 A^+$. Since, moreover, for $p > N$ we also have

$$|\psi(e^0) - \psi(e^{2\pi i/(p+1)})| \cong |\psi(e^{2\pi i p/(p+1)}) - \psi(e^0)| = h \ ,$$

we may consider in this zone, at a given instant, the "equidistant" nodes (on the directions $A^- A_0$ and $A_0 A^+$, respectively), and use automatically the Lagrange interpolating formula in Gauss' variant (G_n). Of course, this one could in its turn be modified, imposing in addition a behaviour of the type $0((z - z_F)^{1/(1-\mu)})$ in the neighbourhood of z_F.

4. Consider again a bounded domain D, delimited by the simple, closed, rectifiable (Jordan) contour C which has an angular point in the point z_F. Suppose that the acute angle of the semitangents in this point, oriented from the angular point, is $\alpha = \pi + \mu\pi$, where $-1 \leq \mu < 0$. It is obvious that this choice corresponds to a jump of the semitangents, by exterior conturning of C, by $\mu\pi$, as in Sec. 3.

Also consider again a function $f(z)$, holomorphic in D and continuous on C. In what follows we propose to construct a new interpolating function L_n^g, more general than the above constructed "piecewise" Lagrange functions, which also satifies, besides the identity conditions of a set of nodes, the behaviour of $f(z)$ in the neighbourhood of z_F (behaviour generated by the integral representation of $f(z)$ by means of Cauchy's formula), namely

$$L_n^g(z) - L_n^g(z_F) = 0((z - z_F)^q) \ ,$$

where hereafter (but only in this section) $q \equiv 1/(1 - \mu)$.

Let again d be an "acceptable" division of $n + 1$ nodes, $z_0, z_1, \ldots, z_n$, of the contour C, division which contains the point z_F, too. It is clear that the function

$$\omega_g(z) \equiv (z - z_0)(z - z_1) \ldots (z - z_{F-1})(z - z_F)^q (z - z_F) \ldots (z - z_n)$$

can be associated to this division, with the specification of a corresponding branch. Then the following theorem holds:

Theorem. The function

$$L_n^g(z) = (2\pi i)^{-1} \int_C (f(t)(\omega_g(t) - \omega_g(z))/((t - z)\omega_g(t)))dt \ ,$$

"generalized" polynomial of the $(n + q)$-th degree, will fulfill both the conditions

$$L_n^g(t_k) = f(t_k), \quad k = \overline{0, n}$$

and the behaviour condition

$$L_n^g(z) - L_n^g(z_F) = 0((z - z_F)^q) \ .$$

In addition, the remainder associated to this formula will be

$$R_n(z) \equiv f(z) - L_n^g(z) = (2\pi i)^{-1} \int_C (f(t)/(t - z))(\omega_g(z)/\omega_g(t))dt \ .$$

The proof of this result is immediate. The fact that the integrand contains a quantity of the degree $n + q - 1 = n + \mu q$ in z, while the integration is performed with respect to t, will determine for the function L_n^g the required degree in z. Additionally, since $\omega_g(z_k) = 0, k = \overline{0, n}$, the application of Cauchy's formula will give at once $L_n^g(z_k) = f(z_k), k = \overline{0, n}$. Also, the behaviour of $\omega_g(z)$ into $V(z_F)$ leads automatically to the same behaviour for $L_n^g(z)$.

Lastly, the expression of the remainder associated to the interpolating function $L_n^g(z)$ is an immediate consequence of the same Cauchy's formula. Obviously, this expression of the remainder will allow the delimitation of the procedure error in any point of D.

Suppose now that the angular point $z_F \equiv a_0(A_0)$ is the vertex of an angle $A^- A_0 A^+$ whose sides $A^- A_0$ and $A_0 A^+$ belong to our contour C, and whose measure is still $\alpha = \pi + \mu\pi$. If we denote $\theta_0 = \arg \overrightarrow{A^- A_0}$, it is obvious that $\arg \overrightarrow{A_0 A^+} = \theta_0 + \mu\pi$.

Consider now on the side $A_0 A^-$ the equidistant nodes of step h

$$a_k = a_0 + kh \exp(i(\theta_0 + \pi)), \quad k = -1, -2, \ldots, -n,$$

and on the side $A_0 A^+$ the equidistant nodes of the same step h

$$a_k = a_0 + kh \exp(i(\theta_0 + 2\pi + \mu\pi)), \quad k = 1, 2, \ldots, n.$$

By choosing this system of nodes, the angular point $z_F \equiv a_0$ was "framed" symmetrically on each direction by n nodes. Our purpose is to construct now a generalized interpolating function on this system of $2n + 1$ nodes. With this end in view we shall use the technique of obtaining Gauss interpolating functions on equidistant nodes.

Let us consider at first the Lagrange interpolating polynomial constructed on the previously introduced system of nodes, that is

$$
\begin{aligned}
L_{2n+1}(z) = {}& f(a_0) + (z - a_0)[a_0, a_1; f] \\
& + (z - a_0)(z - a_1)[a_0, a_1, a_{-1}; f] \\
& + (z - a_0)(z - a_1)(z - a_{-1})[a_0, a_1, a_{-1}, a_2; f] + \ldots \\
& + (z - a_0)(z - a_1)\ldots(z - a_n)[a_0, a_1, a_{-1}, \ldots, a_n, a_{-n}; f].
\end{aligned}
$$

Denoting then $t = \exp(i(\theta_0 + \pi))$ and $s = \exp(i(\pi + \mu\pi))$, if one performs the change of variable $z = a_0 + uht$, which will introduce the dimensionless variable u, we shall respectively have for the divided differences $[a_0, a_1, a_{-1}, \ldots, a_k, a_{-k}; f]$ and $[a_0, a_1, a_{-1}, \ldots, a_{k-1}, a_{-k+1}, a_k; f]$ the expressions:

$$
\begin{aligned}
[a_0, &a_1, a_{-1}, \ldots, a_k, a_{-k}; f] \\
&= (ht)^{-2k}(f(a_k)/((-1)^{2k} k!(k + s) \cdot (k + 2s) \ldots (k + ks)) \\
&\quad + \ldots + f(a_0)/((-1)^k k! k! s^k) + \ldots \\
&\quad + f(a_k)/((ks + k)(ks + k - 1) \ldots (ks + 1)k! s^k)),
\end{aligned}
$$

and

$$
\begin{aligned}
[a_0, &a_1, a_{-1}, \ldots, a_{k-1}, a_{-k+1}, a_k; f] \\
&= (ht)^{-2k+1} \cdot (f(a_{-k+1})/((-1)^{2k-1} 1!(k - 1)!(k - 1 + s)(k - 1 + 2s) \ldots \\
&\quad \ldots (k - 1 + ks)) + \ldots + f(a_0)/((-1)^k (k - 1)! k! s^k) + \ldots \\
&\quad + f(a_k)/((ks + k - 1) \cdot (ks + k - 2) \ldots (ks + 1)k! s^k)).
\end{aligned}
$$

But then the Lagrange interpolating polynomial in Gauss' variant (corresponding to the equidistant nodes $a_{\pm k}, k = \overline{0,n}$) will be written in the dimensionless variable u as follows

$$
\begin{aligned}
G_{2n+1}(u) = f_0 + \sum_{k=1}&((u+k-1)(u+k-2)\ldots\\
&\ldots u(u-s)\ldots(u-(k-1)s)\cdot\\
&\cdot(f(a_{-k+1})/((-1)^{2k-1}1!(k-1)!(k-1+s)(k-1+2s)\ldots\\
&\ldots(k-1+ks)) + \ldots + f(a_0)/((-1)^k(k-1)!k!s^k) + \ldots\\
&\ldots + f(a_k)/((ks+k-1)\ldots(ks+1)k!s^k)) + (u+k-1)\ldots\\
&\ldots u(u-s)\ldots(u-ks)(f(a_{-k})/((-1)^{2k}k!(k+s)(k+2s)\ldots\\
&\ldots(k+ks)) + \ldots + f(a_0)/((-1)^k k!k!s^k) + \ldots\\
&\ldots + f(a_k)/((ks+k)\cdot(ks+k-1)\ldots s^k k!)))\;.
\end{aligned}
$$

In the peculiar case of 3 or 5 nodes — cases which will especially be used in the hydrodynamics applications — we shall have accordingly

$$
G_3(u) = f_0 + u(f_1 - f_0)/s + u(u-s)(-f_0/s + f_1/(s(s+1)) + f_{-1}/(s+1))
$$

and

$$
\begin{aligned}
G_5(u) = f_0 &+ u(f_1 - f_0)/s + u(u-s)(-f_0/s + f_1/(s(s+1))\\
&+ f_{-1}/(s+1)) + u(u-s)(u+1)(f_0/(2!s^2)\\
&- f_1/(s^2(s+1)) - f_{-1}/((1+s)(1+2s))\\
&+ f_2/(2s^2(2s+1)) + u(u-s)(u+1)(u-2s)\\
&\cdot(f_0/(2!2!s^2) - f_{-1}/((1+s)(1+2s))\\
&- f_1/((s+2)(s+1)s^2) + f_{-2}/(2!(2+s)(2+2s))\\
&+ f_2/((2s+2)(2s+1)2!s^2))\;.
\end{aligned}
$$

Suppose now that, instead of Gauss interpolating polynomials, we want to construct "generalized" Gauss interpolating polynomials, $G^g_{2n+1}(u)$. These interpolating functions $G^g_{2n+1}(u)$ will still observe the identity condition on the set of the nodes, namely

$$
G^g_{2n+1}(-ks) = f_{-k}, \quad k = \overline{0,n}; \quad G^g_{2n+1}(ks) = f_k, \quad k = \overline{1,n}\;,
$$

and will also fulfill the behaviour condition in the neighbourhood of the point $z_F = a_0(A_0)$, more exactly

$$G^g_{2n+1}(u) - G^g_{2n+1}(0) = 0(u^q) \; .$$

Restricting ourselves to the case of 3 or 5 nodes, we shall obtain the following "generalized" Gauss interpolating polynomial of the $(1+q)$-th, respectively $(3+q)$-th degree, that is

$$\begin{aligned}
G^g_3(u) = {} & f_0 + u^q(f_1 - f_0)/s^q + u^q(u-s)((f_{-1} - f_0)/((-1)^{q+1} \\
& \cdot (1+s)) + (f_1 + f_0)/(s^q(1+s)))
\end{aligned}$$

and

$$\begin{aligned}
G^g_5(u) = {} & f_0 + u^q(f_1 - f_0)/s^q + u^q(u-s)((f_{-1} - f_0)/((-1)^{q+1} \\
& \cdot (1+s)) + (f_1 - f_0)/(s^q(1+s))) \\
& + u^q(u-s)(u+1)((f_2 - f_0)/(2^q \cdot s^{q+1}(s+1)) \\
& - (f_1 - f_0)/(s^{q+1}(s+1)) - (f_{-1} - f_0)/((-1)^{q+1} \cdot (1+s)^2) \\
& - (f_1 - f_0)/((1+s)^2 s^q)) + u^q(u-s)(u+1)(u-2s) \\
& \cdot (-(f_2 - f_0)/((-2)^q(2+s)(2+2s)) \\
& + (f_2 - f_0)/((2+2s)(s+1)2^q s^{q+1}) \\
& - (f_1 - f_0)/(s^{q+1}(s+1)(2+2s)) \\
& - (f_{-1} - f_0)/((-1)^{q+1}(1+s)^2(2+2s)) \\
& - (f_1 - f_0)/(s^q(1+s)^2(2+2s))) \; .
\end{aligned}$$

Of course, these interpolating functions G^g_{2n+1} can also be expressed under the integral form given by means of the above theorem. In addition, the remainder of the interpolation and, implicitly, the error of the CVBEM approximation process which uses these functions will be delimited by the remainder estimating formula given by the same theorem.

5. Let us now consider a plane, incompressible, inviscid fluid flow generated by a displacement (rototranslation) of a profile (C) in the mass of the fluid, the fluid having already a given basic flow of complex velocity $w_B(z)$, and which superposes over the first flow. We shall suppose that the contour C of the profile (C) has an angular point z_F where the angle of the semitangents is equal to $\pi + \mu\pi(-1 \le \mu < 0)$. Precisely, the profile (C) presents a dihedral point in $z_F(A_0)$ "framed" by the segments $A^- A_0$ and $A_0 A^+$ which belong to C and determine the angle $\alpha = \pi + \mu\pi$. Also suppose that the basic flow presents some given singularities (vortices, sources, etc.). We denote by $f(z;t)$ the complex potential joined to the flow and by $w(z;t) = df/dz$ its complex velocity, t being the time which could explicitly appear, the flow being a nonstationary one.

We remark that the function $w(z;t)$ is a holomorphic function in the whole outside of the profile (C) which also includes the infinite point. In the neighbourhood of this point the function $w(z;t)$ has an expansion of the type

$$w(z;t) = w_\infty + (\Gamma/(2\pi i))/z + b_2/z^2 + b_3/z^3 + \ldots$$

where $w_\infty = \lim_{|z| \to \infty} (df/dz)$ is the complex velocity of the fluid at great distances, and Γ is a real function of time (which could be a constant or even zero) called the circulation of the flow and which represents the multiformity period of the real part of the complex potential f.

It is just this regularity of the complex speed $w(z;t)$ that makes us use in the sequel this function and not the complex potential $f(z;t)$ as we would have been tempted to (in fact, $f(z;t)$ presents a logarithmic singularity at infinity). Of course, the behaviour of the complex velocity in the neighbourhood of the angular point (the sharp trailing edge) z_F, i.e., $w(z;t) = 0((z - z_F)^{\mu/(1-\mu)})$ requires us — to avoid the unboundedness of w in z_F — to choose the circulation such that [8]

$$\Gamma = L \cdot l + M \cdot m + N \cdot \omega \, ,$$

where the uniquely determined coefficients L, M, N depend on the considered profile (C). Here $l(t), m(t)$ are the components of the translation speed in a point $z_A \in (C)$ — evaluated in a mobile system of coordinates $0xy$ centered in $z_A \equiv 0$ — and $\omega(t)$ is the instantaneous rotation of the profile.

Let now the function $w_B(z)$ be the complex velocity of the basic flow. This function (which is given) belongs to a class (a) of functions having the properties:

(1a) They are holomorphic functions in the domain D_1 (the whole $0xy$-plane, the infinite point being included), except a number Q of points z_r

placed at a finite distance and which represent singular points for these functions; let D_1^* be the domain D_1 from which one has taken off the singular points $\{z_r\}_{r=\overline{1,Q}}$ and let $w_B(\infty)$ be the value of the limit $\lim\limits_{|z|\to\infty} w_B(z)$ which obviously exists and is finite.

(2a) If $\Gamma_B = \int_\Delta w_B(z)dz$ (where Δ is a simple, rectifiable curve surrounding all singularities z_r) is the circulation of the basic flow, this is equal to $\sum\limits_{r=1}^{Q} \Gamma_r$, i.e., with the sum of the circulations of all given singularities of the flow.

As to the unknown function $w(z)$ — the complex velocity of the resultant flow obtained by the above mentioned superposition — it will be looked for in a class of functions (b) satisfying the properties:

(1b) They are holomorphic functions in the domain $D = D_1\backslash(\overline{C})$ (where one supposes that during the displacement of (C) we have $(\overline{C}) \subset D_1$, i.e., the profile (C) does not cross the points $\{z_r\}_{r=\overline{1,Q}}$ which always belong to the outside of (C)) except the same points $\{z_r\}_{r=\overline{1,Q}}$ which are the singular points of the same nature as for $w_B(z)$; at infinity, their behaviour is identical with that of $w_B(z)$, namely $\lim\limits_{|z|\to\infty} w(z) = w(\infty) = w_B(\infty)$.

(2b) Let $s = s(\theta)$ be the parametrical equation of the Jordan curve C. We suppose that s is a 2π-periodic function, bounded and derivable in $[0, 2\pi)\backslash\{\theta_0\}$, so that $\dot{s}(\theta) \neq 0$ and $|\dot{s}(\theta)| < M, M$ being a finite constant. In the neighbourhood of the trailing edge $z_F = s(\theta_0) \in C$, where the angle of the semitangents is $\pi - \mu\pi$, we have

$$w(z) = (z - z_F)^{\mu/(1-\mu)}g(z), \quad g(z_f) \neq 0 .$$

(3b) In the points of the curve C, the functions $w(s(\theta))$ belong to the class H^*, that is, they are Hölderian functions on C except the angular point $z_F = s(\theta_0)$, in whose neighbourhood one has

$$w(s(\theta)) = w^*(s(\theta))/(s(\theta) - s(\theta_0))^{\mu/(\mu-1)} ,$$

where $w^* \in H_0$ in the same neighbourhood, which means that $W^*(s(\theta))$ is separately Hölderian on the upper side and on the lower side of the profile in the neighbourhood of $z_F = s(\theta_0)$.

(4b) In the points of the curve C they satisfy, except the angular point, the following boundary condition:

There is a real continuous function $V(\theta)$ such that for every $\theta \in [0, 2\pi)\backslash\{\theta_0\}$ one has

$$w(s(\theta)) = V(\theta)(\dot{s}(\theta)/|\dot{s}(\theta)|) + l + im + i\omega(s(\theta) - z_A) ,$$

where $z_A \in (C)$ and the significances of $l(t), m(t), \omega(t)$ were already precised.

(5b) They fulfill the equality $\int_C w(z)dz = \Gamma$ (the singularity in z_F being weak $(0 < \mu/(\mu - 1) < 1)$, the integral is convergent), the circulation of the flow Γ being chosen such that one has the boundedness of the velocity in z_F, i.e., [8] $\Gamma = L \cdot l + M \cdot m + N \cdot n$, where the coefficients L, M, N are given with the obstacle (C).

Let us now consider the function $w(z) - w_B(z)$. This function known together with $w(z)$ being holomorphic in the outside of (C), the Cauchy formula is valid in D and we immediately have

$$w(\xi) = w_B(\xi) - (2\pi i)^{-1} \int_C w(z)(z - \xi)^{-1}dz$$
$$+ (2\pi i)^{-1} \int_C w_B(z)(z - \xi)^{-1}dz$$

for $\xi \in D$. Observe that, since $w(s(\theta)) - w_B(s(\theta)) \in H^*$, while C is a sectionally smooth curve, the Cauchy-type integral exists.

The existence and uniqueness of the solution of the proposed problem (of the function $w(z)$, looked for under the above representation) have been studied earlier [7].

Considering then a system of nodal points, $z_0, z_1, \ldots, z_{F-1}, z_F, z_{F+1}, \ldots, z_n \equiv z_0$, on the curve C, a manner for the effective construction of an approximate solution by CVBEM would be to use the piecewise Lagrange interpolating functions on each arc C_j, taking into account the behaviour into $V(z_F)$, namely to use the approximation

$$\tilde{w}(s(\theta)) - w_B(s(\theta)) = \sum_{j=1}^{n}(w_j - w_{B_j})L_j(s(\theta)) ,$$

where L_j could be obtained from those previously written in Sec. 3, by replacing $1/(1 - \mu)$ by $\mu/(1 - \mu)$.

In what follows we want to introduce the other manner of approaching the approximation, based on the choice of Fejer nodes. Precisely, having a profile (C) with a "dihedral" zone $A^- A_0 A^+$, there exist a rank N so that for $n > N$ the counterimages of these Fejer nodes frame "symmetrically" the trailing edge A_0 and lie on this dihedral zone. Moreover, since for $p > N$ one has

$$|\psi(e^0) - \psi(e^{2\pi i/(p+1)})| \cong |\psi(e^{2\pi ip/(p+1)}) - \psi(e^0)| = h ,$$

(what we have already stated in Sec. 3), a Gauss interpolation function for at least 3 nodes lying on the dihedral zone and which also takes into account the jump of the semitangents is envisage (Sec. 4). Of course, the equidistant nodes of this zone will be completed by the other non-equidistant ones, the counterimages of the whole system of Fejer nodes, so that we have the approximation

$$\widetilde{w}(s(\theta)) - w_B(s(\theta)) = \sum_{\nu=-m}^{m} (w_\nu - w_{B_\nu})G_\nu + \sum_{j=2m+2}^{n} (w_j - w_{B_j})L_j ,$$

where L_j are the corresponding fundamental Lagrange polynomials.

Once this Gauss interpolation is written, we should use it firstly in the neighbourhood of $z_f(A_0)$, which confers to the procedure a higher degree of approximation than the piecewise Lagrange interpolation. On the rest of the contour C — the nondihedral zone — for the sake of simplicity a piecewise Lagrange interpolation could be introduced. This mixed procedure of interpolation leads to a good degree of approximation of the CVBEM, which is also a convergent one. Additionally, the general calculus required by CVBEM could be performed without any difficulty, the real algebraic homogeneous system to which we are led again have a unique solution which fulfills the condition of an "a priori" given circulation.

As a final remark, we state that the same algorithm for CVBEM could be used for an arbitrary system of profiles performing independent displacements in the mass of the fluid, in the possible presence of some walls, namely practically for the majority of the models of plane hydrodynamics.

References

1. C. F. Gauss, *Principia generalia theoriae figurae fluidorum in statu equilibrii*, Göttingae Sumtibus Dieterichianis, 1830.

2. T. V. Hromadka II, *The Complex Variable Boundary Element Method*, Springer-Verlag, Berlin, 1984.
3. D. Homentcovschi, D. Cocora and R. Magureanu, *Some Developments of the CVBEM. Application to the Mixed Boundary-Value Problem for the Laplace Equation*, Increst-Bucharest, Preprint ser. Math., No. 13/1981.
4. T. Petrila, *An improved CVBEM for plane hydrodynamics*, Anal. Numér. Théor. Approx. **16**, No. 2 (1987) 149–157.
5. D. Gaier, *Vorlesungen über Approximationen im Komplexen*, Biekhäuser Verlag, Basel, Boston, Stuttgart, 1980.
6. C. Iacob, *Introduction mathématique à la mécanique des fluides*, Ed. Acad. R. P. R., Bucureşti, Gauthier-Villars, Paris, 1959.
7. T. Petrila, *Mathematical Models in Plane Hydrodynamics*, Publ. House Romanian Acad., Bucharest, 1981 (in Romanian).
8. G. Couchet, *La condition de Joukowsky en mouvements non-stationnaires*, Faculté Sci. Montpellier, Secrétariat des Mathématiques, Publ. No. 74/1969–70.
9. C. A. Brebbia, J. C. F. Telles and L. C. Wrobel, *Boundary Element Techniques. Theory and Application in Engineering*, Springer-Verlag, Berlin, Heidelberg, New York, Tokyo, 1984.
10. T. Petrila and C. Gheorghiu, *Finite Element Methods and Applications*, Publ. House Romanian Acad., Bucharest, 1987 (in Romanian).

Titus Petrila
University of Cluj-Napoca
Romania

THE MATH. HERITAGE OF C.F. GAUSS (pp. 605-620)
edited by George M. Rassias
©1991 World Scientific Publ. Co. Singapore

A SURVEY ON THE POINCARÉ CONJECTURE
OF THE TOPOLOGY OF 3-MANIFOLDS

George M. Rassias

In this paper, I will give an account of the main developments on the approach followed by C. DnPapakyriakopoulos in his effort to prove the (as yet unanswered) Poincaré conjecture of the Topology of 3-Manifolds.

Introduction

As it is well-known the Poincaré conjecture asserts that any simply-connected, closed (i.e. compact without boundary) 3-manifold is a 3-sphere, i.e., any homotopy 3-sphere is a 3-sphere.

C. D. Papakyriakopoulos [11, 12] reduced the Poincaré conjecture to two conjectures that will be mentioned below.

1. Let $M = H \cup H'$, where M is a closed orientable 3-manifold, H and H' are handlebodies of the same genus $p(\geq 0)$ and $N = bdH = bdH' = H \cap H'$ is a closed orientable surface of genus p. Then $M = H \cup H'$ is said to be a *Heegaard Splitting of genus p* of M. Let $A_i, B_i, i = 1, \ldots, p$ be a fundamental system of N based at the point o, i.e., the A, B's are simple (without multiple points) loops on N based at o, having only the point o in common, and such that if we cut N along the A, B's we obtain a 2-cell with boundary p

$$\prod_{i=1}^{p} A_i B_i A_i^{-1} B_i^{-1} \, .$$

From now on, unless otherwise stated we suppose that $p \geq 2$.

Let A_i, B_i be a system as above such that $A_i \simeq 0$ in H (i.e. A_i are homotopic to zero, in H).

Let $\Phi : N \to N$ be a homeomorphism such that $\Phi(A_i) \simeq 0$ in H', and let $\psi : F \to F$ be the automorphism induced by Φ, where

$$\Pi_1(N, o) \approx F = \left(a_1, b_1, \ldots, a_p, b_p : \prod_{i=1}^{p} [a_i, b_i] \right), \quad p \geq 2 \, .$$

Denote by $A = \langle a_1, \ldots, a_p \rangle, \tilde{\Delta} = A \cap \psi(A), \hat{\Delta} = \langle A, \psi(A) \rangle$.

Theorem 1.1 ([11]). $\Pi_1(M, o) \approx F | \hat{\Delta}$. Therefore M is simply-connected if and only if $\hat{\Delta} = F$.

Lemma 1.2. The groups $A \big|_{\tilde{\Delta}}, \psi(A) \big|_{\tilde{\Delta}}, \hat{\Delta} \big|_{\tilde{\Delta}}$, and $F \big|_{\tilde{\Delta}}$ have no elements of finite order. $\hat{\Delta}, \tilde{\Delta}$ are normal subgroups of F, such that $\tilde{\Delta} \subset \hat{\Delta}$ and $1 \neq \tilde{\Delta} \neq F$.

Theorem 1.3 ([11]). There exists on N a loop L which is $\not\simeq 0$ on N, but it is $\simeq 0$ both in H and H'. However, L is not simple, generally.

Let $\tau : \tilde{H} \to H$ or $\tau' : \tilde{H}' \to H'$ be the universal coverings of H or H' respectively, and let $\tilde{N} = bd\tilde{H}, \tilde{N}' = bd\tilde{H}'$.

Lemma 1.4 ([11]). $\tilde{N}$ and $\tilde{N}'$ are planar surfaces i.e., homeomorphic to open connected subsets of a 2-sphere. Let us consider the diagram where $\gamma\beta : \tilde{D} \to N$ and $\gamma : \hat{D} \to N$ are the regular coverings corresponding to the normal subgroups $\tilde{\Delta}$ and $\hat{\Delta}$ of F respectively. Moreover $\gamma\beta\alpha : \bar{D} \to N$ is the regular covering corresponding to a normal subgroup $\bar{\Delta}$ of F, where $1 \neq \bar{\Delta} \subset \tilde{\Delta}$ and which will be defined more precisely later on. Note that D is the Poincaré model of the hyperbolic plane. The problems related with $\bar{\Delta}$ and $\bar{D}$ form the main difficulties of the method. We emphasize that, $\bar{\Delta}$ may be equal or different to $\tilde{\Delta}$.

Consider the sequence

$$D \xrightarrow{\zeta} \dot{D} \xrightarrow{\eta} N$$

where $\eta : \dot{D} \to N$ is the *regular* covering corresponding to the (not necessarily proper) *normal* subgroup $\dot{\Delta} \neq 1$ of F, and $\zeta : D \to \dot{D}$ is the universal covering of D.

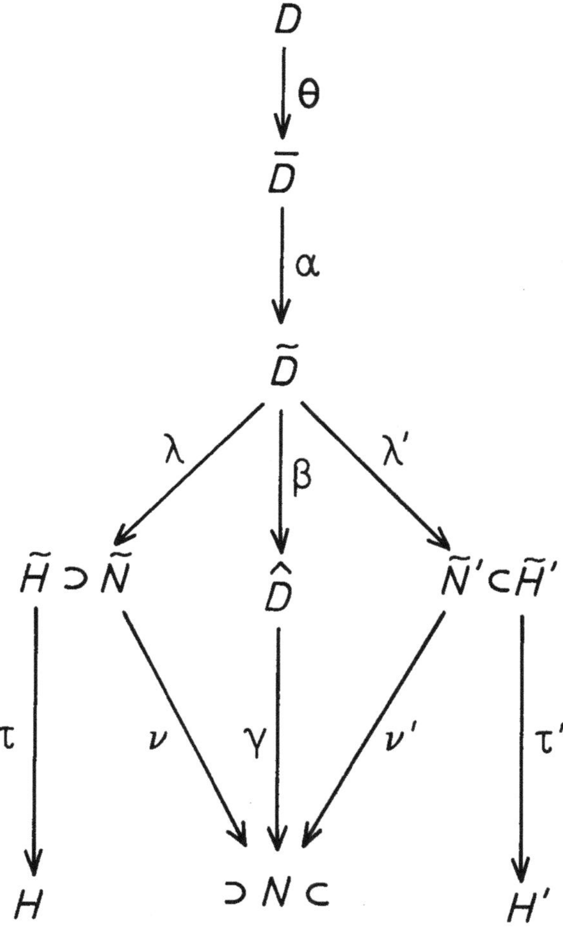

2. Definition. An oriented closed curve $\dot{G}$ on $\dot{D}$, which has at most "double" points and which is $\not\simeq 0$ on $\dot{D}$, is called a *geodesoid* if $\zeta^{-1}(\dot{G})$ consists of simple arcs, such that any two different ones have at most one point in common, but no endpoint in common.

Lemma 2.1 ([11]). Let $\dot{G}$ be a geodesoid on $\dot{D}$ and let $\dot{J}$ be an oriented closed curve on $\dot{D}$, which has at most "double" points, such that $\dot{J} \simeq \dot{G}$ on $\dot{D}$. Then the number of double points of $\dot{G}$ is $\leq$ the number of double points of $\dot{J}$, i.e., $c(\dot{G}) \leq c(\dot{J})$, and if equality holds then $\dot{J}$ is a geodesoid.

Now consider the sequence

$$D \xrightarrow{\theta} \bar{D} \xrightarrow{\xi} \dot{D} \xrightarrow{\eta} N$$

where $\eta : \dot{D} \to N$ and $\eta\xi : \bar{D} \to N$ are *regular* coverings, corresponding to the *normal* subgroups $\dot{\Delta}$ and $\bar{\Delta}$ of F, and having the following properties

(2.2) $\bar{\Delta}$ is a nontrivial proper subgroup of $\dot{\Delta}$, but $\dot{\Delta}$ may not be a proper subgroup of F.

(2.3) $\dot{\Delta}\big|_{\bar{\Delta}}$ has no elements of finite order

(2.4) $\bar{D}$ is planar

Let $(\dot{D}, \bar{D})$ be the set of all geodesoids of $\dot{D}$, which are covered just *once* by *simple* geodesoids of $\bar{D}$. It is proved that $(\dot{D}, \bar{D}) \neq \emptyset$.

For any $\dot{G} \in (\dot{D}, \bar{D})$, Papakyriakopoulos uses the notions of *complexity* $c(\dot{G})$ and *complementary complexity* $C(\dot{G})$ of $\dot{G}$, in the same way he defined them in [13] for circuits. He calls $\dot{G} \in (\dot{D}, \bar{D})$ an *extremal* element of this set, if there is no $\dot{J} \in (\dot{D}, \bar{D})$ such that, either $c(\dot{J}) < c(\dot{G})$, or $c(\dot{J}) = c(\dot{G})$ and $C(\dot{J}) < C(\dot{G})$.

Theorem 2.5 ([11]). Under the conditions (2.2)–(2.4), any extremal element of $(\dot{D}, \bar{D})$ is simple.

3. The following theorem provides necessary and sufficient conditions for the decomposition of 3-manifolds. The notations and conventions here are those of Sec. 1.

Theorem 3.1 ([12]). There exists a simple loop L on N such that $L \not\simeq 0$ on N, and $L \simeq 0$ both in H and H' if and only if, there exists a normal subgroup $\bar{\Delta}$ of F, such that $1 \neq \bar{\Delta} \subset \tilde{\Delta}, F\big|_{\bar{\Delta}}$ has no elements of finite order and $\bar{D}$ is planar.

Proof ([12]). Consider the diagram of Sec. 1. Here we use the fact that $\bar{\Delta} \subset \tilde{\Delta}$. Suppose now that $\dot{\Delta} = F$, whence $\dot{D} = N$. Let L be an extremal element of $(N, \bar{D})$. By theorem 2.5, L is a simple loop on N. Let $\bar{L}$ be a loop on $\bar{D}$ covering L. Then

$$L = \gamma\beta\alpha(\bar{L}) = \nu\lambda\alpha(\bar{L}) \simeq 0 \quad \text{in} \quad H,$$
$$= \nu'\lambda'\alpha(\bar{L}) \simeq 0 \quad \text{in} \quad H'$$

because $\lambda\alpha(\bar{L}) \simeq 0$ in $\tilde{H}$, $\lambda'\alpha(\bar{L}) \simeq 0$ in $\tilde{H}'$, because $\tilde{H}$ and $\tilde{H}'$ are universal coverings of H and H' respectively. Moreover $L \not\simeq 0$ on N, because L is a geodesoid on N. So sufficiency was proved.

The necessity is proved as follows. Let us suppose that there exists a simple loop L on N, such that $L \not\simeq 0$ on N, and $L \simeq 0$ both in H and H'. It is no loss of generality to suppose that, the base point of L is 0, i.e. that $\pi_1(N,o)$ is based at o. It can be shown that we can suppose that L separates N. Let l be the element of $\pi_1(N,o)$ corresponding to L. Let $\bar{\Delta}$ be the smallest normal subgroup of F containing l, and let $\xi : \bar{D} \to N$ be the regular covering corresponding to $\bar{\Delta}$.

Let N' be the 2-complex obtained by attaching a 2-cell to N along L the obvious way. By Seifert and Threlfall [27] we have $\pi_1(N') \approx F\big|_{\bar{\Delta}}$.

By [33] it is easily seen that N' is aspherical. Since the fundamental group of an aspherical polyhedron of finite dimension has no element of finite order, $F\big|_{\bar{\Delta}}$ has no elements of finite order, then it can be shown that $\bar{D}$ is planar.

Up to now we did not need to introduce "coordinates" in M. But the condition $1 \neq \bar{\Delta} \subset \tilde{\Delta}$ does depend on them, because $\tilde{\Delta}$ does. So let A_i, B_i and $A_i', B_i'', i = 1, \ldots, p$, be two fundamental systems of N with the same base point 0, such that the A's or A''s form a canonical system of H or H' respectively, with the property: A_j, B_j and $A_j', B_j'', j = 1, \ldots, d < p$ lie on the closure of the one component of $N - L$, and A_k, B_k and $A_k', B_k'', k = d + 1, \ldots, p$, lie on the other. Thus

$$l = \prod_{j=1}^{d} [a_j, b_j] \quad \text{or} \quad \prod_{j=1}^{d} [a_j', b_j'']$$

with respect to the "coordinates" a_i, b_i or a_i', b_i'' of $\pi_1(N, 0)$. Hence $1 \neq \bar{\Delta} \subset \tilde{\Delta}$, where $\tilde{\Delta}$ is defined through the "coordinates" A_i, B_i and $A_i', i = 1, \ldots, p$.

The above theorem 3.1 helps us explain the cause of the difficulty of the classification problem for orientable closed 3-manifolds.

3.2. In Papakyriakopoulos's [11] words, "the complexity of the classification problem for orientable closed 3-manifolds is due to the fact that: there exist covering spaces of planar surfaces which are not planar". Let us suppose that any regular covering space of any planar surface is planar. Then $\tilde{D}$ is planar, because so is $\tilde{N}$, see Lemma 1.4. By Lemma 1.2, $F\big|_{\tilde{\Delta}}$ has no elements of finite order and $1 \neq \tilde{\Delta}$. Then by Theorem 3.1, Dehn's lemma [15], and induction on p would follow that any orientable closed 3-manifold would be the composition of a 3-sphere with handles and a finite number of lens spaces. However this is not true, because there are orientable closed 3-manifolds with finite not cyclic fundamental group [27, pp. 216–218]. Hence the cause of our difficulties is the following.

3.3 *Generally, $\tilde{D}$ is not planar*

The following remarks of R. H. Fox indicate that Proposition 3.2 is not as awkward as it may look: let $\tilde{S} \to S$ be a finite sheeted branched covering over the 2-sphere S. Deleting the branch points of $\tilde{S}$ and their projections on S, we obtain an (unbranched) covering $\tilde{S}' \to S'$. It is well-known that among all possible $\tilde{S}$ there are surfaces of genus > 0. Therefore among all possible $\tilde{S}'$ there are nonplanar surfaces, while all possible S' are planar.

4. Homology Spheres and Homotopy Spheres

Suppose M is a *homology* 3-sphere, and let T be a handlebody of genus $p \geq 1$, semi-linearly imbedded in M, and let $N = bdT$.

Theorem 4.1 ([11]). If M is a homology 3-sphere, then there exists a system of oriented simple closed curves $X_j, Y_j, j = 1, \ldots, p$ on N such that: (i) the X- or Y-curves are disjoint; (ii) X_j meets only Y_j and only at one point; (iii) if we cut N along X_j and Y_j we obtain a 2-sphere with p holes, having boundaries $X_j Y_j X_j^{-1} Y_j^{-1}$; (iv) $X_j \simeq 0$ in T, and $Y_j \sim 0$ in $\mathrm{cl}(M - T)$.

The case $p = 1$ has been proved by M. Dehn ([1]). The case $p \geq 2$ is proved by induction and the proof is rather geometric. C. D. Papakyriakopoulos calls $X_j, Y_j, j = 1, \ldots, p$ *a meridian longitude* system of $N \subset M$ and he calls the above theorem the "meridian-longitude theorem".

Suppose that M is a *homotopy* 3-sphere and let $M = T \cup T'$ be a Heegaard splitting of it. By Theorem 4.1, there is a meridian-longitude system X_j, Y_j or X_j', Y_j' of $T \subset M$ or $T' \subset M$ respectively. Theorem 1.1, in conjunction with an algebraic lemma enable us to perform two *elementary transformations* of the system X_j, Y_j by means of which, C. D. Papakyriakopoulos obtains the following

Theorem 4.2 ([11]). If M is a homotopy 3-sphere, and if $X'_j, Y'_j, j = 1, \ldots, p(\geq 1)$, is a meridian-longitude system of $T' \subset M$, then there exists a meridian-longitude system X_j, Y_j of $T \subset M$, such that $Y_j \sim X'_j$ on N.

By an isotopic deformation of each one of the pairs X_j, Y_j and X'_j, Y'_j on N, Papakyriakopoulos obtains new systems A_j, B_j, and A'_j, B'_j based at 0, such that both A_j, B_j and $A'_j, B'_j, j = 1, \ldots, p$, are fundamental systems of N. Denoting by a_j, b_j, a'_j the elements of $\pi_1(N, 0)$ corresponding to A_j, B_j, A'_j respectively and by using Theorem 4.2, he obtains $a'_j = b_j \tau_j$, where $\tau_j \in [\Phi, \Phi]$, and Φ is the free group freely generated by $a_1, b_1, \ldots, a_p, b_p$.

5. Reduction of the Poincaré Conjecture

Consider the special case where one of the τ's, say τ_1, is 1. Define $\bar{\Delta} = \langle [a_1, a'_1] \rangle$. Then using the Poincaré model of the hyperbolic plane, Papakyriakopoulos proves that

$$1 \neq \bar{\Delta} \subset \langle a_1, \ldots, a_p \rangle \cap \psi(\langle a_1, \ldots, a_p \rangle)$$

$$F \big| \bar{\Delta} \approx \left(a_1, b_1, \ldots, a_p, b_p : \prod_{i=1}^{p} [a_i, b_i], [a_1, b_1] \right)$$

$$\approx (a_1, b_1 : [a_1, b_1]) * \left(a_2, b_2, \ldots, a_p, b_p : \prod_{i=2}^{p} [a_i, b_i] \right)$$

and therefore it has no elements of finite order. Finally the regular covering $\bar{D}$ of N corresponding to $\bar{\Delta}$ is planar, because there exists a simple loop homotopic on N to the loop representing $[a_1, b_1]$. Thus by Theorem 3.1 there is on N a *simple* loop L, which is $\not\simeq 0$ on N, but $\simeq 0$ in both the handlebodies T and T'.

The above suggests that something similar may happen in the general case, where all τ's are $\neq 1$. This leads us to the formulation of the following two conjectures.

5.1. For some subset $(m, \ldots, n)$ of $(1, \ldots, p)$ the group $(a_1, b_1, \ldots, a_p, b_p :$

$\prod\limits_{i=1}^{p}[a_i, b_i], [a_m, b_m\tau_m], \ldots , [a_n, b_n\tau_n])$ has no elements of finite order, where $\tau_m, \ldots , \tau_n \in [\Phi, \Phi]$.

5.2. The regular covering space $\bar{D}$ of N, corresponding to $\bar{\Delta} = \langle[a_m, b_m\tau_m], \ldots , [a_n, b_n\tau_n]\rangle$ is planar.

In our problem $b_m\tau_m = a'_m, \ldots , b_n\tau_n = a'_n$ are represented by simple loops on N. This condition may turn out to be very important for the validity of the conjectures 5.1 and 5.2. It can be seen that the group of covering translations $F|_{\bar{\Delta}}$ of the regular covering $\bar{D} \to N$ is the one appearing in 5.1.

C. D. Papakyriakopoulos ([11]) states "Conjectures 5.1 and 5.2 imply the Poincaré conjecture, however they seem to be very hard problems indeed."

6. Improvement of the Conjectures

G. A. Swarup [(31)] examines Papayriakopoulos' reduction and studies two reductions of the Poincaré conjecture. The first is group theoretic and is an improvement over Papakyriakopoulos' reduction [(12)]. The second reduces the conjecture to a special case of it. His method also gives a new proof of a crucial theorem in his reduction.

P.1 Conjecture. Let $G_p : \{a_1, b_1, , \ldots , a_p, b_p ; \prod\limits_{i=1}^{p}[a_i, b_i], p > 1\}$ and let $Q_p = \{a_1, b_1, \ldots , a_p, b_p; \prod\limits_{i=1}^{p}[a_i, b_i], [a_1, b_1\tau]\}$, where $\tau \in [\Phi_p, \Phi_p]$, Φ_p being the free group generated by $a_1, b_1, \ldots , a_p, b_p$. Let T_p be an orientable surface of genus p and identify $\pi_1(T_p)$ with G_p. Then

(a) Q_p is torsion-free

(b) The cover of T_p corresponding to the kernel of the natural map $\varphi_p : G_p \to Q_p$ is planar.

E. S. Rapaport ([17]) proved P.1.(a) and Papakyriakopoulos showed that P.1 implies the Poinaré conjecture ([12]). Papakyriakopoulos ([12, 14]) also considered the question whether P.1(b) is group theoretic.

G. A. Swarup ([32]) considers the following

P.2 Conjecture. The group Q_p defined above is a nontrivial free-product.

He shows

A. Theorem. P.1 $\Longrightarrow$ P.2 $\Longrightarrow$ Poincaré conjecture. Moreover, P.1 is group-theoretic.

Swarup ([32]) states that the crucial step in Papakyriakopoulos ([12]) reduction of Poincaré conjecture to P.1 is a theorem which connects the problem of finding nontrivial simple loops in a certain normal subgroup with regular planar covers subordinate to it. This result was strengthened by Maskit ([8]); Lemmas 1 and 2 below imply his theorem. These lemmas connect the approach of Papakyriakopoulos with that of Stallings ([30]).

Let

$$\{e\} \to L \xrightarrow{i} G \xrightarrow{\varphi} H \to \{e\}$$

be an exact sequence of groups, where G is the fundamental group of a closed surface T. We are interested in finding a criterion which asserts — the existence of a simple loop on T which represents a nontrivial element of L.

1. **Lemma** ([32]). (a) With the above notation there is a nontrivial simple loop on T whose nth power for some n represents an element of L if there is a factorization $\varphi = qp : G \xrightarrow{q} Q \xrightarrow{p} H$ such that $H^1(Q; Z_2Q) \neq 0$.

(b) If H is torsion-free, there is a nontrivial simple loop on T which represents an element of L if and only if φ factors through a nontrivial free-product.

2. **Lemma** ([32]). With the notation of Lemma 1, there is a nontrivial simple loop on T whose nth power for some n represents an element of L, if there is a normal subgroup $K(\neq \{e\})$ of G with $K \subset L$ and such that the cover of T corresponding to K is planar. The above condition is necessary [8].

Proof ([32]). Let Q be the quotient of G by K and let $\tilde{T}$ be the cover of T corresponding to K. Consider the following commutative diagram

$$0 \to H^1(Q\,;\,\mathbb{Z}_2Q) \to H^1(G\,;\,\mathbb{Z}_2Q) \to H^1(K\,;\,\mathbb{Z}_2) \to$$

$$\downarrow$$

$$H_c^1(\tilde{T}\,;\,\mathbb{Z}_2) \to H^1(\tilde{T}\,;\,\mathbb{Z}_2)$$

$$\downarrow$$

$$H_1(\tilde{T}\,;\,\mathbb{Z}_2) \to H_1^\infty(\tilde{T}\,;\,\mathbb{Z}_2)$$

The first row is exact ([6, p. 354]). It is well-known that the map in the second row is zero if and only if $\tilde{T}$ is planar. Since $K \neq \{e\}, H_1(\tilde{T}\,;\,\mathbb{Z}_2) \neq 0$. Thus, if $\tilde{T}$ is planar we have $H^1(Q;\,\mathbb{Z}_2Q) \approx H^1(G;\,\mathbb{Z}_2Q) \neq 0$. This completes the proof using Lemma 1.

Next, let T be a two-sided orientable surface in a 3-manifold M. Let M_1, M_2 be the two parts into which T divides M; if $M - T$ is connected we write $M_2 = M_1$. We have two injections $i_1 : T \to M_1$ and $i_2 : T \to M_2$ and let φ_1, φ_2 be the induced maps in the fundamental groups. By the loop theorem, $\varphi_i(T)$, is a free product of orientable surface groups and free cyclic groups and therefore $\varphi_i(T)$ is torsion free. Consider

$$\varphi_1 \times \varphi_2 : \pi_1(T) \to \pi_1(M_1) \times \pi_1(M_2)$$

and let φ denote the induced map of $\pi_1(T)$ onto the image H of $\varphi_1 \times \varphi_2$ and let L be the kernel of φ. Then Lemma 2 implies the following.

3. **Theorem** (Papakyriakopoulos, Maskit). With the above notation, there is a simple loop on T which represents a nontrivial element of L if and only if there is a normal subgroup $K \neq \{e\}$ of $\pi_1(T)$ with $K \subset L$ and such that the cover of T corresponding to K is planar.

That P.1 $\implies$ Poincaré conjecture is proved using the above theorem when $(T; M_1, M_2)$ is a Heegaard decomposition of a homotopy 3-sphere M. In this case $\varphi_1 \times \varphi_2$ is surjective and the normal subgroup K_p generated by the image of $[a_1, b_1\tau]$ is in the kernel of $\varphi_1 \times \varphi_2$ for some $\tau \in [\Phi_p, \Phi_p]$ (see [12]). Let φ_p, Q_p be as in P.1. Let $\tilde{T}$ be the cover of T corresponding to K_p. As in Lemma 2, $\tilde{T}$ is planar if and only if $H^1(Q_p;\,\mathbb{Z}_2Q_p) \to H^1(G_p;\,\mathbb{Z}_2Q_p)$ is an isomorphism. This shows that the conjecture P.1 is group-theoretic. If P.1 is true, then

$$H^1(Q_p;\,\mathbb{Z}_2Q_p) \approx H^1(G_p;\,\mathbb{Z}_2Q_p) \approx H_1(K_p;\,\mathbb{Z}_2) \neq 0\,.$$

Since Q_p is torsion-free, by [29], Q_p is a free product. Thus P.1 $\implies$ P.2. If Q_p is a free product, then by Lemma 1 we can find a simple loop on T which represents nontrivial element in $K_p \subset$ kernel of $\varphi_1 \times \varphi_2$ and Poincaré conjecture follows by induction. This proves theorem A.

Remark ([32]). The above proof shows that Papakyriakopoulos's approach (via Theorem 3 above) really involves factoring $\varphi_1 \times \varphi_2$ through a free product and hence it is contained in Stallings approach [30]. However, Papakyriakopoulos has given possible candidates (Kp above) for finding simple loops and what we have done above seems to make his conjectures more accessible. The lemmas above can be generalized and can be used to prove theorem 3 in Maskit's paper [8].

7. Conjecture P.2 and so P.1 is Not in General True

S. Moran ([10]) disproves the Conjecture P.2 of Swarup in the following theorem which in turn disproves the conjecture P.1 of Papakyriakopoulos

Theorem ([10]). The conjecture P.2 is not in general true. Hence the conjecture P.1 is not in general true.

Proof ([10]). Let $G_1 = \langle a_1, b_1; (a_1, b_1 c) \rangle$, where c is any fixed element of the commutator subgroup $F(\{a_1, b_1\})'$ of the free group $F(\{a_1, b_1\})$ so that $(a_1, b_1 c)$ is not conjugate to $(a_1, b_1)^{\pm 1}$ in $F(\{a_1, b_1\})$. For example one could take

$$c = (a_1, b_1).$$

Take $G_2 = \langle a_2, b_2, \ldots, a_n, b_n; - \rangle$ and $G = G_1 \underset{H}{*} G_2$, where $H = \langle h; - \rangle$ and the amalgamating isomorphisms are given by

$$\varphi_1(h) = (a_1, b_1) \quad \text{and} \quad \varphi_2(h) = (a_n, b_n)^{-1} \ldots (a_2, b_2)^{-1}.$$

Now G is an example of a group of the type denoted by

$$Q_n = \langle a_1, b_1, \ldots, a_n, b_n \; ; \prod_{i=1}^{n} (a_i, b_i), (a_1, b_1 \tau) \rangle$$

above. This is so since G_1 is torsion free by Magnus, Karrass and Solitar [7, §4.4, Theorem. 4.12].

Also $(a_1, b_1) \neq e$ in G_1, because of the assumption on c and Magnus, Karrass and Solitar [7, §4.4, Theorem 4.12]. We show below that G cannot be decomposed into a proper free product. Suppose that contrary to the above assertion we have that $G = X * Y$, where X and Y are nontrivial groups. The rank of G is $2n$, since G/G' is a free abelian group of rank $2n$. The rank of G_1 is 2 and the group G_1 is torsion free by Magnus, Karrass and Solitar [7, §4.1, p. 192]. Hence if G_1 is a proper free product, then G_1 is a free group of rank 2, by theorem of B. H. Neumann on the rank of a free product. However the group G_1 is clearly not a free group. So G_1 cannot be decomposed into a proper free product. Hence by the Kurosh subgroup theorem for a free product, it follows that

$$(*) \qquad G_1 \subseteq g^{-1} X g \quad \text{for some element} \quad g \quad \text{of} \quad G.$$

Hence

$$G\big|_{\bar{G}_1 G} \cong X/(\overline{gG_1g^{-1}}^X) * Y.$$

Also

$$G\big|_{\bar{G}_1 G} \cong G_2/\overline{\langle \varphi_2(h) \rangle}^{G_2} \quad \text{is a surface group},$$

since G has generators $a_1, b_1, a_2, b_2, \ldots, a_n, b_n$ and defining relations

$$\prod_{i=1}^{n} (a_i, b_i) = e \quad \text{and} \quad (a_1, b_1 c) = e$$

$G\big|_{\bar{G}_1 G}$ cannot be both a surface group and a proper free product because of a result of A. Shenitzer [Prop. 5.14 of [5]]. Hence

$$X = \overline{gG_1g^{-1}}^X \quad \text{and} \quad Y \cong G_2/\overline{\langle \varphi_2(h) \rangle}^{G_2}.$$

Thus the rank of Y is $2n - 2 \geq 2$.

All conjugates of Y intersect the subgroup H (of G) trivially. For

$$X \cap \overline{Y}^G = e \quad \text{and} \quad X = \overline{gG_1g^{-1}}^X \supseteq gHg^{-1}.$$

Hence, by the subgroup theorem of H. Neumann for free products with amalgamation [5] the group Y is either a proper free product or is contained in some conjugate of one of the groups G_1 and G_2. However,

(i) Y cannot be a proper free product, by the above mentioned result of A. Shenitzer, since it is a surface group.

$$
\text{On the Poincaré Conjecture}
$$

(ii) Y cannot be contained in a conjugate of G_1, since this would imply by (*) that Y is conjugate to a subgroup of X.

(iii) Y cannot be contained in a conjugate of G_2, since if it were Y would be a free group (as G_2 is a free group) which is false (a surface group is not a free group).

Hence none of these possibilities can occur.

Remark. As it was previously mentioned E. S. Rapaport [17] proved (affirmatively) Conjecture P.1.(a). Therefore the above theorem shows that conjecture P.1.(b). does not in general hold.

8. McCool's Counterexample to Conjecture P.2

J. McCool ([9]) gives the following counterexample to Swarup's conjecture P2 which we recall P.2 Conjecture ([32])

Let $Q_p = \{a_1, b_1, \ldots, a_p, b_p;\ \prod_{i=1}^{p}[a_i, b_i], [a_1, b_1\tau]\}$, where $p > 1$ and $\tau \in [\Phi_p, \Phi_p]$, Φ_p being the free group on $a_1, b_1, \ldots, a_p, b_p$. Then Q_p is a nontrivial free product.

Counterexample ([9]). Take $p = 2, \tau = [a_1, b_1]$ and suppose that $Q_2 = A * B$, where A and B are nontrivial. Now we observe that, with a slight change of notation

$$
\begin{aligned}
Q_2 &= \{a,b,c,d\,; [a,b],[c,d],[a,bab\bar{a}\bar{b}]\} \\
&= \{a,b,c,d,x,y;\ bab\bar{x},\ bx\bar{b}\bar{y},\ ab\bar{a}\bar{b}[c,d],\ abab\bar{a}\bar{b}\bar{a}bab\bar{a}\bar{b}\} \\
&= \{a,b,c,d,x,y;\ bab\bar{x},\ bx\bar{b}\bar{y},\ a\bar{x}[c,d],\ ax\bar{y}\bar{x}y\bar{x}\} \\
&= \{b,c,d,x,y;\ bx\bar{y}xy\bar{x}\bar{b}\bar{x},\ bx\bar{b}\bar{y},\ x\bar{y}xy\bar{x}^2[c,d]\} \\
&= \{b,c,d,x,y,z;\ \bar{z}x\bar{y},\ bx\bar{y}xy\bar{x}\bar{b}\bar{x},\ bx\bar{b}\bar{y},\ x\bar{y}xy\bar{x}^2[c,d]\} \\
&= \{b,c,d,x,z;\ bzx\bar{z}\bar{b}\bar{x},\ bx\bar{b}\bar{x}z,\ zx\bar{z}\bar{x}[c,d]\}\,.
\end{aligned}
$$

Let $K = \{z, x, c, d;\ [z,x][c,d]\}$. Then using the Freiheitssatz ([7] or [5]), we see that the subgroup of K generated by x, z is free of rank 2, and so therefore are the subgroups generated by $zx\bar{z}, x$ and $x, \bar{z}x$, respectively. It follows that Q_2 is an HNN-extension ([5]) of K by the stable letter b, and hence that the subgroup of Q_2 generated by z, x, c, d is naturally isomorphic to K. Now K is not cyclic, and from the results of ([28]) K is indecomposable with respect to free products. It then follows from the

Kurosh subgroup theorem that K is contained in a conjugate of A or B, say B. By conjugating the original A and B, we may asume that K is contained in B.

Since b and K generate Q_2, it follows that $b \notin B$. Now the second relator of the last presentation of Q_2 above gives $bx\bar{b} = \bar{z}x$, i.e., conjugation of the element x of B by b gives the element $\bar{z}x$ of B. Consideration of the normal form of b shows immediately that this implies $b \in B$, which is a contradiction. Thus Q_2 must be indecomposable with respect to free products.

Remark. Because of the previously mentioned counterexamples of S. Moran and J. McCool to Swarup's conjecture and therefore to Papakyriakopoulos's conjecture it is of great interest in finding out weaker versions that still imply the Poincaré conjecture.

References

1. M. Dehn, *Über die Topologie des dreidimensionalen Raumes*, Math. Ann. **69** (1910) 137–168.
2. R. H. Fox, *On the asphericity of regions in 3-space*, Bull. Amer. Math. Soc. **55** (1949) 521.
3. R. H. Fox, *Free differential Calculus I*, Ann. of Math. **57** (1953) 547–560.
4. B. V. Kerékjártó, *Vorlesungen über Topologie*, Springer, Berlin, 1923.
5. R. C. Lyndon and P. E. Schupp, *Combinatorial Group Theory*, Springer, Berlin, 1977.
6. S. MacLane, *Homology*, Springer-Verlag, Berlin and New York, 1963.
7. W. Magnus, A. Karrass and D. Solitar, *Combinatorial Group Theory*, Interscience, New York, 1966.
8. B. Maskit, *A theorem on planar covering surfaces, with applications to 3-manifolds*, Ann. of Math. (2) **81** (1965) 341–355.
9. J. McCool, *A counterexample to conjectures of Papakyriakopoulos and Swarup*, Proc. Amer. Math. Soc. **81** (1981) 193–194.
10. S. Moran, *On a conjecture of Papakyriakopoulos*, Bull. Amer. Math. Soc. **4** (1981) 189–191.
11. C. D. Papakyriakopoulos, *A reduction of the Poincaré conjecture to other conjectures*, Bull. Amer. Math. Soc. **68** (1962) 360–366.
12. C. D. Papakyriakopoulos, *A reduction of the Poincaré conjecture to group theoretic conjectures*, Annals of Math. (2) **77** (1963) 250–305.
13. C. D. Papakyriakopoulos, *On solid tori*, Proc. London Math. Soc. (3) **7** (1957) 281–299.

14. C. D. Papakyriakopoulos, *Attaching 2-dimensional cells to a complex*, Ann. of Math. (2) **78** (1963) 205–222.

15. C. D. Papakyriakopoulos, *On Dehn's lemma and the asphericity of knots*, Ann. of Math. (2) **66** (1957) 1–26.

16. C. D. Papakyriakopoulos, *Planar regular coverings of orientable closed surfaces*, in Knots, Groups and 3-Manifolds, ed. L. P. Neuwirth, papers dedicated to the memory of R. H. Fox, Ann. of Math. Studies **84**, Princeton Univ. Press, (1975) pp. 261–292.

17. E. S. Rapaport, *Proof of a conjecture of Papakyriakopoulos*, Ann. of Math. (2) **79** (1964) 506–513.

18. G. M. Rassias, *On the genus problem of 3-dimensional manifolds and the Poincaré conjecture*, Bull. de la Soc. Royale des Sci. de Liège, 50e année, no. **3–4** (1981) 125–128.

19. G. M. Rassias, *Covering of differentiable manifolds with open disks*, Archiv der Math. (4) **38** (1982) 366–367.

20. G. M. Rassias, *On the Morse-Smale characteristic of a differentiable manifold*, Bull. Australian Math. Soc. (2) **20** (1979) 281–283.

21. G. M. Rassias, *New results and research problems in Morse-Smale and Stallings theories*, Teubner-Texte zur Math., Band **70**, Teubner Verlag, Leipzig (1984) 7–16.

22. G. M. Rassias, *Counterexamples to a conjecture of René Thom*, Bull. de L'Acad. Polonaise des Sci., vol. XXIX, no. 5–6 (1981) 311–316.

23. G. M. Rassias, *Homeomorphisms of the sphere and a generalization of the Whyburn conjecture for compact connected manifolds with boundary*, vol. L11, Fasc. 2 (1987) 199–203.

24. G. M. Rassias, *On the monotone union and monotone intersection properties of topological manifolds*, Comptes Rendus Math. de L'Acad. des Sci., Canada, (6) **4** (1982) 327–330.

25. G. M. Rassias, *Poincaré's conjecture in topology*, Teubner-Texte zur Math., Band **76**, Teubner Verlag, Leipzig (1985) 8–25.

26. K. Reidemeister, *Kommutative Fundamentalgruppen*, Monatsh. Math. Phys. **43** (1936) 20–28.

27. H. Seifert and W. Threlfall, *Lehrbuch der Topologie*, Teubner, Leipzig, 1934.

28. A. Shenitzer, *Decomposition of a group with a single defining relation into a free product*, Proc. Amer. Math. Soc. **6** (1955) 273–279.

29. J. R. Stallings, *Group Theory and 3-Dimensional Manifolds*, Yale Univ. Press, 1971.

30. J. R. Stallings, *How not to prove the Poincaré conjecture*, Ann. of Math. Studies **60**, Princeton Univ. Press, Princeton, N.J. (1966) 83–88.

31. G. Stratopoulos and G. M. Rassias (ed.), *Geometry and Topology*, World Scientific Publ. Co., Singapore, 1989.

32. G. A. Swarup, *Two reductions of the Poincaré conjecture*, Bull. Amer. Math. Soc. (N.S) **1** (1979) 774–777.

33. J. H. C. Whitehead, *On the asphericity of regions in a 3-sphere*, Fund. Math. **32** (1939) 149–166.

George M. Rassias
Department of Mathematics
The American College of Greece
Aghia Paraskevi GR 15310
Athens
Greece

THE MATH. HERITAGE OF C.F. GAUSS (pp. 621-644)
edited by George M. Rassias
©1991 World Scientific Publ. Co. Singapore

ADJOINT CONNECTIONS ON GROUP MANIFOLDS AND GAUGE TRANSFORMATIONS

Hanno Rund

With any set of r independent 1-forms $\{\pi^\alpha : \alpha = 1, \ldots, r\}$ on an r-parameter Lie group G there is associated a set of 2-forms $\{\Pi^\alpha : \alpha = 1, \ldots, r\}$, the so-called Maurer-Cartan forms. The latter are nonlinear in the former: this implies that any substitution applied to $\{\pi^\alpha\}$ that preserves the structure of $\{\Pi^\alpha\}$ must be non-homogeneous with components in distinct cotangent spaces of G. By means of such pre-gauge transformations the exterior derivatives of the matrices of the adjoint representation of G can be decomposed to yield a theory of connections and curvature on G that is distinct from any theory of affine connections. It is subsequently assumed that the group parameters are functions of the coordinates of a given base manifold M. This gives rise to a bundle structure and a process by which s-forms can be projected from G to M. The projection to M of the pre-gauge transformations and the connection and curvature forms on G yields the standard gauge transformations on M, together with the associated theory of gauge connections and curvature.

Introduction

In much of the literature on non-Abelian gauge fields the underlying gauge transformations are introduced with little or no intrinsic motivation. From a practical point of view this is entirely justifiable, since any application of these transformations immediately demonstrates that they are entirely appropriate for the purpose for which they are designed. Moreover, it is certainly possible to trace the genesis of gauge transformations within the more abstract setting of the theory of Lie algebra-valued connection 1-forms on a principal fibre bundle ([2]). However, since the structural group is a fundamental ingredient of any gauge theory, one may speculate as to whether the peculiar non-homogeneous structure of gauge transformations may have its origin in the group *per se* without reference to the base manifold on which the transformation is actually performed.

In the present note it will be shown that certain nonhomogeneous structures do in fact exist on the group manifold. To this end we shall consider a r-dimensional Lie group G, and with a given set of r independent 1-forms $\{\pi^\alpha : \alpha = 1, \dots, r\}$ on G we shall associate a system of 2-forms $\{\Pi^\alpha : \alpha = 1, \dots, r\}$ whose structure is suggested by the well-known Maurer-Cartan equations. Since the latter are nonlinear in $\{\pi^\alpha\}$ the application of a linear transformation to $\{\pi^\alpha\}$ cannot transform Π^α into a similar Maurer-Cartan system. Accordingly a class of nonhomogeneous substitutions is constructed which is such as to preserve the Maurer-Cartan structure. Because of the formal similarity between these substitutions and standard gauge transformations they are called *pre-gauge transformations.* (This nomenclature was suggested by D. G. B. Edelen.) By means of the latter the exterior derivatives of the matrices of the adjoint representation of G can be decomposed to yield a system of equations whose formal structure is identical with that of the equations that characterize affine connections on manifolds. This gives rise to an analogue of Cartan's equations of structure and therefore suggests the definition of a so-called adjoint connection on G (although this is not a connection in the usual sense); from this a theory of adjoint curvature on G is developed in accordance with the standard pattern.

A remarkable feature of this analysis is the fact that the pre-gauge transformations inevitably involve 1-forms that have components in the cotangent space $\Lambda^1(T_u)$ at some $u \in G$ as well as components in the cotangent space $\Lambda^1(T_e)$ at the identity element e of G. Moreover, the resulting Maurer-Cartan forms are always confined to $\Lambda^2(T_e)$. This state of affairs suggests a generalization of the aforementioned theory of adjoint connections that encompasses connection 1-forms that admit *ab initio* a direct sum decomposition with respect to the cotangent spaces of distinct elements u and v of G. The resulting theory of curvature has a rich structure, being once more closely associated with systems of Maurer-Cartan forms. General pre-gauge transformations are then defined by the requirement that these Maurer-Cartan forms be confined to $\Lambda^2(T_v)$, and the existence of such transformations is demonstrated by the construction of explicit solutions of systems of partial differential equations that must be satisfied if such confinements are to be realized.

In the subsequent analysis it is assumed that a base manifold M is given (the latter usually being identified with space-time), it being supposed that the group parameters depend differentiably on the local coordinates of points of M. This gives rise to the construction of a local bundle space B over M, this space being endowed with a class of sections, each section σ_u being defined by a unique element $u \in G$. A mapping of s-forms from $\Lambda^s(T_u)$ to σ_u is defined, and the composition of this with the pull-back σ_u^* of the maps $M \to B$ defined by the sections gives rise to a projection of s-forms on G to s-forms on M. The images under such projections of the general pre-gauge transformations on G are found to have a structure that is identical with that of finite gauge transformations on M. The resulting standard theory of curvature on M then emerges effortlessly by direct projection from G to M of the adjoint curvature theory on G.

In order to render this article self-contained, the remainder of this introduction is devoted to a brief description of some formal properties of Lie groups that will be used below. The group composition on G, by which the product $w = uv$ of any pair $u \in G, v \in G$ is defined, is a map $\phi : G \times G \to G$. The local coordinates of u, v will be denoted by $\{u^\alpha : \alpha = 1, \ldots, r\}$, $\{v^\alpha, \alpha = 1, \ldots, r\}$ respectively, and the coordinate presentation of ϕ is given by the composition functions ϕ^α on $G \times G$, the values of which prescribe the coordinates w^α of $w \in G : w^\alpha = \phi^\alpha(u, v)$. Since $ue = eu = u$ for all $u \in G$, where e denotes the identity element whose coordinates are always taken to be $(0, \ldots, 0)$, the expansion of ϕ^α must be of the form

$$w^\alpha = \phi^\alpha(u, v) = u^\alpha + v^\alpha + a_\beta{}^\alpha{}_\gamma u^\beta v^\gamma + \ldots, \tag{1.1}$$

where $a_\beta{}^\alpha{}_\gamma$ are constants that depend on the choice of the local coordinates on G. (Here, and in the sequel, Greek indices range from 1 to $r = \dim G$; the summation convention is operative throughout in respect of these indices.) The structure constants of G are defined as

$$C_\beta{}^\alpha{}_\gamma = a_\beta{}^\alpha{}_\gamma - a_\gamma{}^\alpha{}_\beta, \tag{1.2}$$

and the associativity of G entails that these satisfy the Jacobi identities ([3], p. 380)

$$C_\sigma{}^\alpha{}_\beta C_\epsilon{}^\sigma{}_\gamma + C_\sigma{}^\alpha{}_\epsilon C_\gamma{}^\sigma{}_\beta + C_\sigma{}^\alpha{}_\gamma C_\beta{}^\sigma{}_\epsilon = 0. \tag{1.3}$$

In terms of the notation

$$\Phi^\alpha_\beta(u, v) = \frac{\partial \phi^\alpha(u, v)}{\partial u^\beta}, \quad \Psi^\alpha_\beta(u, v) = \frac{\partial \phi^\alpha(u, v)}{\partial v^\beta}, \tag{1.4}$$

the following matrix elements are introduced:

$$\lambda_\beta^\alpha(u) = \Phi_\beta^\alpha(u, u^{-1}), \quad \chi_\beta^\alpha(u) = \Psi_\beta^\alpha(u, 0). \tag{1.5}$$

From (1.1) it is immediately evident that

$$\Phi_\beta^\alpha(u, 0) = \delta_\beta^\alpha, \quad \Psi_\beta^\alpha(0, v) = \delta_\beta^\alpha, \quad \lambda_\beta^\alpha(0) = \delta_\beta^\alpha, \quad \chi_\beta^\alpha(0) = \delta_\beta^\alpha. \tag{1.6}$$

The first set of entries in (1.5) defines a system of 1-form fields $\{\lambda^\alpha : \alpha = 1, \dots, r\}$ on G:

$$\lambda_u^\alpha = \lambda_\beta^\alpha(u) du^\beta \tag{1.7}$$

in which the subscript u on the left-hand side refers to the representative of these fields in the cotangent space $\Lambda^1(T_u)$ of G at u. The relations

$$\frac{\partial \lambda_\beta^\alpha}{\partial u^\gamma} - \frac{\partial \lambda_\gamma^\alpha}{\partial u^\beta} = C_\nu{}^\alpha{}_\nu \lambda_\gamma^\nu \lambda_\beta^\mu \tag{1.8}$$

([4], p. 134) imply that

$$d\lambda_u^\alpha = \frac{1}{2} C_\beta{}^\alpha{}_\gamma \lambda_u^\beta \wedge \lambda_u^\gamma. \tag{1.9}$$

The combined sets of entries in (1.5) define the matrices of the adjoint representation of G ([1], p. 9, [4], p. 131), namely

$$G_\beta^\alpha(u) = \lambda_\epsilon^\alpha(u) \chi_\beta^\epsilon(u). \tag{1.10}$$

These satisfy the partial differential equations

$$\frac{\partial G_\beta^\alpha(u)}{\partial u^\gamma} = C_\nu{}^\alpha{}_\mu \lambda_\gamma^\nu(u) G_\beta^\mu(u), \tag{1.11}$$

while the algebraic relations

$$G_\epsilon^\alpha(u) C_\beta{}^\epsilon{}_\gamma = C_\nu{}^\alpha{}_\mu G_\beta^\nu(u) G_\gamma^\mu(u) \tag{1.12}$$

are valid identically. It may also be shown ([5]) that, if $w = uv$,

$$\lambda_\epsilon^\alpha(w) \Phi_\beta^\epsilon(u, v) = \lambda_\beta^\alpha(u), \quad \Psi_\epsilon^\alpha(u, v) \chi_\beta^\epsilon(v) = \chi_\beta^\alpha(w), \tag{1.13}$$

which may be used to show that

$$\Psi_\beta^\alpha(u, v) = \Phi_\epsilon^\alpha(u, v) \chi_\gamma^\epsilon(u) \lambda_\beta^\gamma(v), \tag{1.14}$$

and

$$G_\beta^\alpha(w) = G_\epsilon^\alpha(u) G_\beta^\epsilon(v). \tag{1.15}$$

2. Pre-gauge Transformations and Adjoint Connections

To a given system of 1-forms $\{\pi^\alpha : \alpha = 1,\dots,r\}$ on G we assign a set of 2-forms

$$\Pi^\alpha = d\pi^\alpha + \frac{1}{2}C_\beta{}^\alpha{}_\gamma \pi^\beta \wedge \pi^\gamma. \tag{2.1}$$

We shall call these the Maurer-Cartan (MC) forms associated with the given system in view of their structural similarity with the well-known Maurer-Cartan equations that are associated with the Lie algebra of G. Because of the nonlinearity inherent in (2.1) it is evident that the effect of a linear transformation $\pi^\alpha \to \pi'^\alpha = A_\beta^\alpha \pi^\beta$ on (2.1) is not in general represented by 2-forms that are linear combinations of the MC forms associated with π'^α. In fact, the overall situation is considerably more complicated, as is shown by the following

Lemma. Let $\{\overset{\circ}{\pi}{}^\alpha : \alpha = 1,\dots,r\}$ denote a set of r independent but otherwise arbitrary 1-forms in $\Lambda^1(T_e)$. Then the 1-forms defined by

$$\pi_u^\alpha = G_\beta^\alpha(u)\overset{\circ}{\pi}{}^\beta - \lambda_u^\alpha, \tag{2.2}$$

are such that the relationship between the respective MC forms is given by

$$\Pi_u^\alpha = G_\beta^\alpha(u)\overset{\circ}{\Pi}{}^\beta. \tag{2.3}$$

Proof. Because of (1.11) and (1.9) the exterior derivative of (2.2) can be written as

$$d\pi_u^\alpha = C_\beta{}^\alpha{}_\gamma G_\sigma^\gamma \lambda_u^\beta \wedge \overset{\circ}{\pi}{}^\sigma + G_\beta^\alpha d\overset{\circ}{\pi}{}^\beta - \frac{1}{2}C_\beta{}^\alpha{}_\gamma \lambda_u^\beta \wedge \lambda_u^\gamma,$$

while (2.2) also entails that

$$\frac{1}{2}C_\beta{}^\alpha{}_\gamma \pi_u^\beta \wedge \pi_u^\gamma = \frac{1}{2}C_\beta{}^\alpha{}_\gamma G_\sigma^\beta G_\tau^\gamma \overset{\circ}{\pi}{}^\sigma \wedge \overset{\circ}{\pi}{}^\tau - C_\beta{}^\alpha{}_\gamma G_\sigma^\gamma \lambda_u^\beta \wedge \overset{\circ}{\pi}{}^\sigma$$

$$+ \frac{1}{2}C_\beta{}^\alpha{}_\gamma \lambda_u^\beta \wedge \lambda_u^\gamma.$$

When these relations are added, two pairs of terms cancel, leaving us with

$$d\pi_u^\alpha + \frac{1}{2}C_\beta{}^\alpha{}_\gamma \pi_u^\beta \wedge \pi_u^\gamma = G_\beta^\alpha d\overset{\circ}{\pi}{}^\beta + \frac{1}{2}C_\beta{}^\alpha{}_\gamma G_\sigma^\beta G_\tau^\gamma \overset{\circ}{\pi}{}^\sigma \wedge \overset{\circ}{\pi}{}^\tau.$$

The identity (1.12) is now applied to the second term on the right-hand side to yield

$$d\pi_u^\alpha + \frac{1}{2}C_\beta{}^\alpha{}_\gamma \pi_u^\beta \wedge \pi_u^\gamma = G_\beta^\alpha d\overset{o}{\pi}{}^\beta + \frac{1}{2}G_\epsilon^\alpha C_\beta{}^\epsilon{}_\gamma \overset{o}{\pi}{}^\beta \wedge \overset{o}{\pi}{}^\gamma ,$$

which implies the assertion (2.3) in view of the definition (2.1).

The relations (2.2) represent a special case of the pre-gauge transformations referred to in the introduction. Their significance is not solely due to the structural simplicity of (2.3). Indeed, the most striking feature of (2.2) is revealed by the following observation. First, since $\overset{o}{\pi}{}^\alpha \in \Lambda^1(T_e)$, while $\lambda_u^\alpha \in \Lambda^1(T_u)$, it is evident from (2.2) that $\pi_u^\alpha \in \Lambda^1(T_e) \oplus \Lambda^1(T_u)$. Second, this direct sum decomposition is such that the resulting MC forms Π_u^α are elements of $\Lambda^2(T_e)$, the space of 2-forms on G at the identity e, this conclusion being implied by (2.3) because $\overset{o}{\Pi}{}^\alpha \in \Lambda^2(T_e)$. However, by way of contrast we note that, for arbitrary $\mu^\alpha \in \Lambda^1(T_e) \oplus \Lambda^1(T_u)$, the corresponding MC forms are elements of the space $\Lambda^2(T_e \oplus T_v)$, of which $\Lambda^2(T_e)$ is a proper subspace, since $\dim \Lambda^2(T_e \oplus T_v) = r(2r - 1)$, while $\dim \Lambda^2(T_e) = \frac{1}{2}r(r - 1)$. We therefore recognize that *the special transition* $\overset{o}{\pi}{}^\alpha \rightarrow \pi_u^\alpha$ *as exemplified by the pre-gauge transformation* (2.2) *has the effect of confining the resulting MC forms to the space* $\Lambda^2(T_e)$. This phenomenon will provide the motivation for a subsequent construction of a more general nature.

Another immediate consequence of the special structure of (2.2) emerges as a result of the elimination of the 1-forms λ_u^ν from (1.11) by means of (2.2):

$$dG_\beta^\alpha = C_\nu{}^\alpha{}_\mu G_\beta^\mu(u)(G_\sigma^\nu(u)\overset{o}{\pi}{}^\sigma - \pi_u^\nu) ,$$

which, with the aid of (1.12), can be reduced to

$$dG_\beta^\alpha = G_\epsilon^\alpha(u)C_\gamma{}^\epsilon{}_\beta \overset{o}{\pi}{}^\gamma - G_\beta^\epsilon(u)C_\gamma{}^\alpha{}_\epsilon \pi_u^\gamma . \tag{2.4}$$

This clearly suggests the construction of the following 1-forms:

$$\overset{o}{\tilde{\omega}}_\beta{}^\epsilon = C_\gamma{}^\epsilon{}_\beta \overset{o}{\pi}{}^\gamma , \quad \tilde{\omega}_\beta{}^\epsilon(u) = C_\gamma{}^\epsilon{}_\beta \pi_u^\gamma , \tag{2.5}$$

for this allows us to express (2.4) as

$$dG_\beta^\alpha = G_\epsilon^\alpha(u)\overset{o}{\tilde{\omega}}_\beta{}^\epsilon - G_\beta^\epsilon(u)\tilde{\omega}_\epsilon{}^\alpha(u) , \tag{2.6}$$

which has a structure that is identical with that of the well-known equations that display the relationship between two sets of affine connection 1-forms on a neighborhood of a manifold referred to two distinct local coordinate systems. Although (2.6) does not not refer in any way to an affine connection on G, being a direct result solely of the given group structure, it nevertheless leads directly as follows to equations of structure (in the sense of E. Cartan).

Exterior differentiation of (2.6) yields

$$0 = dG_\epsilon^\alpha \wedge \overset{o}{\widetilde{\omega}}_\beta{}^\epsilon + G_\epsilon^\alpha d\overset{o}{\widetilde{\omega}}_\beta{}^\epsilon - dG_\beta^\epsilon \wedge \widetilde{\omega}_\epsilon{}^\alpha - G_\beta^\epsilon \wedge d\widetilde{\omega}_\epsilon{}^\alpha \,,$$

from which we eliminate dG_β^α by means of (2.6). After some simplification it is thus found that

$$G_\epsilon^\alpha(u)\overset{o}{\widetilde{\Omega}}_\beta{}^\epsilon = G_\beta^\epsilon(u)\widetilde{\Omega}_\epsilon{}^\alpha(u)\,, \tag{2.7}$$

where we have put

$$\overset{o}{\widetilde{\Omega}}_\beta{}^\alpha = d\overset{o}{\widetilde{\omega}}_\beta{}^\alpha + \overset{o}{\widetilde{\omega}}_\epsilon{}^\alpha \wedge \overset{o}{\widetilde{\omega}}_\beta{}^\epsilon \,, \quad \widetilde{\Omega}_\beta^\alpha(u) = d\widetilde{\omega}_\beta{}^\alpha(u) + \widetilde{\omega}_\epsilon{}^\alpha(u) \wedge \widetilde{\omega}_\beta{}^\epsilon(u)\,, \tag{2.8}$$

these being the required equations of structure. Consequently, by analogy with the standard theory of affine connections, we shall refer to the 1-forms (2.5) and the 2-forms (2.8) as the *adjoint connection* and the *adjoint curvature forms* respectively.

The relationship between the latter and the MC forms (2.1) is readily established as follows. Let us substitute from the second member of (2.5) in the second member of (2.8) to obtain

$$\widetilde{\Omega}_\beta{}^\alpha(u) = C_\epsilon{}^\alpha{}_\beta d\pi^\epsilon + C_\sigma{}^\alpha{}_\epsilon C_\tau{}^\epsilon{}_\beta \pi^\sigma \wedge \pi^\tau \,. \tag{2.9}$$

By means of the Jacobi identities (1.3) the quadratic term can be expressed as

$$\begin{aligned} C_\sigma{}^\alpha{}_\epsilon C_\tau{}^\epsilon{}_\beta \pi^\sigma \wedge \pi^\tau &= \frac{1}{2}(C_\sigma{}^\alpha{}_\epsilon C_\tau{}^\epsilon{}_\beta - C_\tau{}^\alpha{}_\epsilon C_\sigma{}^\epsilon{}_\beta)\pi^\sigma \wedge \pi^\tau \\ &= -\frac{1}{2}C_\beta{}^\alpha{}_\epsilon C_\sigma{}^\epsilon{}_\tau \pi^\sigma \wedge \pi^\tau \,. \end{aligned} \tag{2.10}$$

When this is substituted in (2.9), the definition (2.1) being taken into account, it is found that

$$\widetilde{\Omega}_\beta{}^\alpha(u) = C_\epsilon{}^\alpha{}_\beta \Pi^\epsilon \,. \tag{2.11}$$

This is the relation that we have been seeking: it shows that *the adjoint curvature 2-forms are elements of the space* $\Lambda^2(T_e)$ since this is the case for the MC forms. Again, it should be emphasized that the latter are defined as in (2.1) in terms of the 1-forms π_u^α prescribed by the pre-gauge transformation (2.2).

3. General Adjoint Connection and Curvature Forms

The definition (2.2) of the 1-forms π_u^α involves a set of 1-forms $\overset{\circ}{\pi}{}^\alpha \in \Lambda^1(T_e)$ and another set of 1-forms $\lambda_u \in \Lambda^1(T_u)$. We shall now consider a rather more general situation, namely one in which the two given sets of 1-forms are respectively elements of $\Lambda^1(T_u)$ and $\Lambda^1(T_v)$, where u and v are arbitrary elements of the group G. Accordingly we now introduce a set of 1-forms

$$\pi^\alpha_{(u,v)} = H_\beta^\alpha(u,v)du^\beta + K_\beta^\alpha(u,v)dv^\beta \,, \tag{3.1}$$

in which the coefficients H_β^α and K_β^α are initially arbitrary except for the requirement that they are class C^2 functions of the coordinates of u and v. (Whenever there is no danger of confusion, the subscripts (u,v) on $\pi^\alpha_{(u,v)}$ and similar objects will be omitted.) As before the MC forms associated with the system (1.1) are defined as in (2.1), and it is readily verified that these forms can be represented as

$$\begin{aligned}
\Pi^\alpha_{(u,v)} &= d\pi^\alpha + \frac{1}{2}C_\epsilon{}^\alpha{}_\beta \pi^\epsilon \wedge \pi^\beta \\
&= -\frac{1}{2}H_\beta{}^\alpha{}_\gamma du^\beta \wedge du^\gamma - L_\beta{}^\alpha{}_\gamma du^\beta \wedge dv^\gamma - \frac{1}{2}K_\beta{}^\alpha{}_\gamma dv^\beta \wedge dv^\gamma \,,
\end{aligned} \tag{3.2}$$

where

$$H_\beta{}^\alpha{}_\gamma = \frac{\partial H_\beta^\alpha}{\partial u^\gamma} - \frac{\partial H_\gamma^\alpha}{\partial u^\beta} - C_\nu{}^\alpha{}_\mu H_\beta^\nu H_\gamma^\mu \,, \tag{3.3}$$

$$L_\beta{}^\alpha{}_\gamma = \frac{\partial H_\beta^\alpha}{\partial v^\gamma} - \frac{\partial K_\gamma^\alpha}{\partial u^\beta} - C_\nu{}^\alpha{}_\mu H_\beta^\nu K_\gamma^\mu \,, \tag{3.4}$$

$$K_\beta{}^\alpha{}_\gamma = \frac{\partial K_\beta^\alpha}{\partial v^\gamma} - \frac{\partial K_\gamma^\alpha}{\partial v^\beta} - C_\nu{}^\alpha{}_\mu K_\beta^\nu K_\gamma^\mu \,. \tag{3.5}$$

The relation (3.2) shows that $\Pi^\alpha_{(u,v)} \in \Lambda^2(T_u \oplus T_v)$: it is an exemplification of the direct sum decomposition

$$\Lambda^2(T_u \oplus T_v) = \Lambda^2(T_u) \oplus \Lambda^2(T_u \times T_v) \oplus \Lambda^2(T_v) \,. \tag{3.6}$$

Guided by the structure of (2.5) we now introduce the *general adjoint connection* 1-forms as

$$\omega_\beta{}^\alpha(u,v) = C_\epsilon{}^\alpha{}_\beta \pi^\epsilon_{(u,v)} \,, \tag{3.7}$$

together with the *general adjoint curvature* 2-forms

$$\Omega_\beta{}^\alpha(u,v) = d\omega_\beta{}^\alpha(u,v) + \omega_\epsilon{}^\alpha(u,v)\omega_\beta{}^\epsilon(u,v) \,. \tag{3.8}$$

The analysis outlined in (2.9) and (2.10) may be applied directly to these definitions to yield the relations

$$\Omega_\beta{}^\alpha(u,v) = C_\epsilon{}^\alpha{}_\beta \Pi^\epsilon_{(u,v)} \,. \tag{3.9}$$

The 1-forms (3.7) are now used in accordance with the usual procedure to define an operation of adjoint exterior differentiation: given a set of differentiable s-forms $\{\mu^\alpha : \alpha = 1, \dots, r\}$ in $\Lambda^s(T_u \oplus T_v)$, where $0 \le s \le r$, we define

$$D\mu^\alpha = d\mu^\alpha + \omega_\beta{}^\alpha \wedge \mu^\beta = d\mu^\alpha + C_\epsilon{}^\alpha{}_\beta \pi^\epsilon \wedge \mu^\beta \,. \tag{3.10}$$

For the $(s+1)$-forms $D\mu^\alpha$ we then have

$$D(D\mu^\alpha) = d(d\mu^\alpha + \omega_\beta{}^\alpha \wedge \mu^\beta) + \omega_\beta{}^\alpha \wedge (d\mu^\beta + \omega_\gamma{}^\beta \wedge \mu^\gamma) \,,$$

which is readily simplified by means of (3.8) to

$$D(D\mu^\alpha) = \Omega_\beta{}^\alpha(u,v)\mu^\beta \,. \tag{3.11}$$

Similarly, when the definition (3.10) is applied to the MC forms (2.1) it is found that

$$\begin{aligned}
D\Pi^\alpha &= d(d\pi^\alpha + \frac{1}{2}C_\epsilon{}^\alpha{}_\beta \pi^\epsilon \wedge \pi^\beta) + C_\epsilon{}^\alpha{}_\beta \pi^\epsilon \wedge (d\pi^\beta + \frac{1}{2}C_\gamma{}^\beta{}_\nu \pi^\gamma \wedge \pi^\nu) \\
&= \frac{1}{2}C_\epsilon{}^\alpha{}_\beta C_\gamma{}^\beta{}_\nu \pi^\epsilon \wedge \pi^\gamma \wedge \pi^\nu \,.
\end{aligned}$$

Since the cubic term on the right-hand side vanishes identically by virtue of the Jacobi identity (1.3), it is thus inferred that

$$D\Pi^\alpha = 0 \,, \tag{3.12}$$

which represents *the Bianchi indentities associated with the general adjoint connection.*

For future reference we should briefly glance at the coordinate presentation of derivatives such as (3.10). For a system of 0-forms X^α, that is, a set of r differentiable functions of the coordinates of the group elements u and v, we have by (3.10) and (3.1)

$$DX^\alpha = dX^\alpha + C_\epsilon{}^\alpha{}_\beta X^\beta \pi^\epsilon = \frac{\partial X^\alpha}{\partial u^\gamma} du^\gamma + \frac{\partial X^\alpha}{\partial v^\gamma} dv^\gamma$$
$$+ C_\epsilon{}^\alpha{}_\beta X^\beta (H^\epsilon_\gamma du^\gamma + K^\epsilon_\gamma dv^\gamma),$$

which we shall write as

$$DX^\alpha = X^\alpha_{|\gamma} du^\gamma + X^\alpha_{\|\gamma} dv^\gamma, \tag{3.13}$$

where we have put

$$X^\alpha_{|\gamma} = \frac{\partial X^\alpha}{\partial u^\gamma} + C_\epsilon{}^\alpha{}_\beta H^\epsilon_\gamma X^\beta, \quad X^\alpha_{\|\gamma} = \frac{\partial X^\alpha}{\partial v^\gamma} + C_\epsilon{}^\alpha{}_\beta K^\epsilon_\gamma X^\beta. \tag{3.14}$$

For a set of 1-forms

$$\mu^\alpha = \mu^\alpha_\beta du^\beta + \nu^\alpha_\beta dv^\beta, \tag{3.15}$$

it is found similarly that

$$D\mu^\alpha = \mu^\alpha_{\gamma|\beta} du^\beta \wedge du^\gamma + \left(\nu^\alpha_{\gamma|\beta} - \mu^\alpha_{\beta\|\gamma}\right) du^\beta \wedge dv^\gamma + \nu^\alpha_{\gamma\|\beta} dv^\beta \wedge dv^\gamma, \tag{3.16}$$

where we have written, as in (3.14),

$$\mu^\alpha_{\gamma|\beta} = \frac{\partial \mu^\alpha_\gamma}{\partial u^\beta} + C_\nu{}^\alpha{}_\sigma H^\nu_\beta \mu^\sigma_\gamma, \quad \mu^\alpha_{\beta\|\gamma} = \frac{\partial \mu^\alpha_\beta}{\partial v^\gamma} + C_\nu{}^\alpha{}_\sigma K^\nu_\gamma \mu^\sigma_\beta, \tag{3.17}$$

with corresponding derivatives for ν^α_β. The expression (3.16) is equivalent to

$$D\mu^\alpha = D\mu^\alpha_\beta \wedge du^\beta + D\nu^\alpha_\beta \wedge dv^\beta, \tag{3.18}$$

in which the effect of D as applied to the 0-forms $\mu^\alpha_\beta, \nu^\alpha_\beta$ is prescribed by (3.13).

The two cases $s = 0$ and $s = 1$, described above, can be combined if it is assumed that $\mu^\alpha = DX^\alpha$, in which case one would have $\mu^\alpha_\beta = X^\alpha_{|\beta}, \nu^\alpha_\beta = X^\alpha_{\|\beta}$ in consequence of (3.13) and (3.15). With the aid of (3.16) it is then found that, on the one hand,

$$D(DX^\alpha) = -\frac{1}{2}(X^\alpha_{|\beta|\gamma} - X^\alpha_{|\gamma|\beta}) du^\beta \wedge du^\gamma - (X^\alpha_{|\beta\|\gamma} - X^\alpha_{\|\gamma|\beta}) du^\beta \wedge dv^\gamma$$
$$- \frac{1}{2}(X^\alpha_{\|\beta\|\gamma} - X^\alpha_{\|\gamma\|\beta}) dv^\beta \wedge dv^\gamma, \tag{3.19}$$

while, on the other hand, by virtue of (3.11),

$$D(DX^{\alpha}) = \Omega_{\mu}{}^{\alpha}(u,v)X^{\mu}\,. \tag{3.20}$$

But when (3.2) is substituted in (3.9) it is found that

$$\Omega_{\mu}{}^{\alpha} = -\frac{1}{2}C_{\epsilon}{}^{\alpha}{}_{\mu}H_{\beta}{}^{\epsilon}{}_{\gamma}du^{\beta}\wedge du^{\gamma} - C_{\epsilon}{}^{\alpha}{}_{\mu}L_{\beta}{}^{\epsilon}{}_{\gamma}du^{\beta}\wedge dv^{\gamma} - \frac{1}{2}C_{\epsilon}{}^{\alpha}{}_{\mu}K_{\beta}{}^{\epsilon}{}_{\gamma}dv^{\beta}\wedge dv^{\gamma}\,. \tag{3.21}$$

Thus a companion of (3.19) with (3.20) yields the following important relations:

$$X^{\alpha}_{|\beta|\gamma} - X^{\alpha}_{|\gamma|\beta} = C_{\epsilon}{}^{\alpha}{}_{\mu}H_{\beta}{}^{\epsilon}{}_{\gamma}X^{\mu}\,, \tag{3.22}$$

$$X^{\alpha}_{|\beta\|\gamma} - X^{\alpha}_{\|\gamma|\beta} = C_{\epsilon}{}^{\alpha}{}_{\mu}L_{\beta}{}^{\epsilon}{}_{\gamma}X^{\mu}\,, \tag{3.23}$$

$$X^{\alpha}_{\|\beta\|\gamma} - X^{\alpha}_{\|\gamma\|\beta} = C_{\epsilon}{}^{\alpha}{}_{\mu}K_{\beta}{}^{\epsilon}{}_{\gamma}X^{\mu}\,. \tag{3.24}$$

In conclusion let us consider the identity (3.12). By means of a direct calculation based on (3.1) and (3.2) it may be shown that

$$\begin{aligned}
D\Pi^{\alpha} = {} & -\frac{1}{2}H_{\beta}{}^{\alpha}{}_{\gamma|\epsilon}du^{\beta}\wedge du^{\gamma}\wedge du^{\epsilon} - \left(L_{\gamma}{}^{\alpha}{}_{\epsilon|\beta} + \frac{1}{2}H_{\beta}{}^{\alpha}{}_{\gamma\|\epsilon}\right)du^{\beta}\wedge du^{\gamma}\wedge dv^{\epsilon} \\
& - \left(\frac{1}{2}K_{\gamma}{}^{\alpha}{}_{\epsilon|\beta} + L_{\beta}{}^{\alpha}{}_{\gamma\|\epsilon}\right)du^{\beta}\wedge dv^{\gamma}\wedge dv^{\epsilon} \\
& - \frac{1}{2}K_{\beta}{}^{\alpha}{}_{\gamma\|\epsilon}dv^{\beta}\wedge dv^{\gamma}\wedge dv^{\epsilon}\,,
\end{aligned} \tag{3.25}$$

where, as in (3.17),

$$H_{\beta}{}^{\alpha}{}_{\gamma\|\epsilon} = \frac{\partial H_{\beta}{}^{\alpha}{}_{\gamma}}{\partial v^{\epsilon}} + C_{\nu}{}^{\alpha}{}_{\mu}K_{\epsilon}{}^{\nu}H_{\beta}{}^{\mu}{}_{\gamma}\,, \quad K_{\gamma}{}^{\alpha}{}_{\epsilon|\beta} = \frac{\partial K_{\gamma}{}^{\alpha}{}_{\epsilon}}{\partial u^{\beta}} + C_{\nu}{}^{\alpha}{}_{\mu}H_{\beta}{}^{\nu}K_{\gamma}{}^{\mu}{}_{\epsilon}\,. \tag{3.26}$$

According to (3.3) and (3.5) the coefficients $H_{\beta}{}^{\alpha}{}_{\gamma}$ and $K_{\beta}{}^{\alpha}{}_{\gamma}$ that appear in (3.2) are skew-symmetric in their subscripts. Thus the substitution of (3.25) in (3.12) yields the following system:

$$H_{\beta}{}^{\alpha}{}_{\gamma|\epsilon} + H_{\gamma}{}^{\alpha}{}_{\epsilon|\beta} + H_{\epsilon}{}^{\alpha}{}_{\beta|\gamma} = 0\,, \tag{3.27}$$

$$H_{\beta}{}^{\alpha}{}_{\gamma\|\epsilon} + L_{\gamma}{}^{\alpha}{}_{\epsilon|\beta} - L_{\beta}{}^{\alpha}{}_{\epsilon|\gamma} = 0\,, \tag{3.28}$$

$$K_{\gamma}{}^{\alpha}{}_{\epsilon|\beta} + L_{\beta}{}^{\alpha}{}_{\gamma\|\epsilon} - L_{\beta}{}^{\alpha}{}_{\epsilon\|\gamma} = 0\,, \tag{3.29}$$

$$K_{\beta}{}^{\alpha}{}_{\gamma\|\epsilon} + L_{\gamma}{}^{\alpha}{}_{\epsilon\|\beta} - K_{\epsilon}{}^{\alpha}{}_{\beta\|\gamma} = 0\,. \tag{3.30}$$

Clearly these relations represent *the coordinate presentation of the Bianchi identities* (3.12).

4. Confinement of Maurer-Cartan Forms: General Pre-gauge Transformations

Guided by the example of the pre-gauge transformations of Sec. 2, we now require that the 1-forms (3.1) be such that the associated MC forms do not have components in $\Lambda^2(T_u \times T_v)$. From (3.2) it is evident that this is tantamount to the stipulation that

$$L_\beta{}^\alpha{}_\gamma = 0. \tag{4.1}$$

Because of (3.28) and (3.29) this condition imposes restrictions on $H_\beta{}^\alpha{}_\gamma$ and $K_\beta{}^\alpha{}_\gamma$ also, namely

$$H_\beta{}^\alpha{}_{\gamma\|\epsilon} = 0, \tag{4.2}$$

together with

$$K_\gamma{}^\alpha{}_{\epsilon|\beta} = 0. \tag{4.3}$$

The explicit representation of these expressions is given in (3.26), which indicates that the equations (4.2) and (4.3) in turn entail integrability conditions. The commutation relations (3.22) and (3.24) suggest that these conditions are satisfied for *both* systems whenever

$$C_\nu{}^\alpha{}_\mu H_\beta{}^\mu{}_\gamma K_\sigma{}^\mu{}_\tau = 0, \tag{4.4}$$

as may also be verified by direct calculation. This motivates the imposition of an additional condition, namely

$$H_\beta{}^\alpha{}_\gamma = 0, \tag{4.5}$$

in which case (4.4) is certainly satisfied. According to (3.2) we then have

$$\Pi^\alpha_{(u,v)} = -\frac{1}{2}K_\beta{}^\alpha{}_\gamma dv^\beta \wedge dv^\gamma. \tag{4.6}$$

This shows that the MC forms (3.2) associated with the 1-forms (3.1) are confined to $\Lambda^2(T_v)$ whenever

(i) the coefficients H^α_β are solutions of the system (4.5), and

(ii) the coefficients K_β^α are such that the system (4.1) is satisfied, the integrability conditions of the latter being guaranteed by (4.5).

A comparison of (3.3) with (1.8) shows that a solution of the Eqs. (4.5) is given by

$$H_\beta^\alpha(u, v) = -\lambda_\beta^\alpha(u) \,. \tag{4.7}$$

In this case H_β^α is independent of the coordinates of the element $v \in G$, and the resulting expression (3.4) for $L_\beta{}^\alpha{}_\gamma$ shows that (4.1) may be reduced to

$$\frac{\partial K_\gamma^\alpha}{\partial u^\beta} = -C_\nu{}^\alpha{}_\mu H_\beta^\nu K_\gamma^\mu = C_\nu{}^\alpha{}_\mu \lambda_\beta^\nu(u) K_\gamma^\beta \,, \tag{4.8}$$

where, in the second step, we have used (4.7). This is a system of partial differential equations for K_γ^β whose structure is identical with that of the system (1.11), the latter being satisfied by the matrices $(G_\beta^\alpha(u))$ of the adjoint representation. Since these represent unique solutions of (1.11), with $G_\beta^\alpha(0) = \delta_\beta^\alpha$ by virtue of (1.10) and (1.7), it follows that the most general solution of (4.8) is given by

$$K_\beta^\alpha(u, v) = G_\epsilon^\alpha(u) M_\beta^\epsilon(v) \,, \tag{4.9}$$

in whch the factors M_β^ϵ are *arbitrary* functions of the coordinates of v. The resulting 1-forms (3.1) are thus given by

$$\pi_{(u,v)}^\alpha = -\lambda_\beta^\alpha(u) du^\beta + G_\epsilon^\alpha(u) M_\beta^\epsilon(v) dv^\beta \,. \tag{4.10}$$

If we write

$$\lambda_u^\alpha = \lambda_\beta^\alpha(u) du^\beta \,, \quad M_v^\alpha = M_\beta^\alpha(v) dv^\beta \,, \tag{4.11}$$

this can also be expressed as

$$\pi_{(u,v)}^\alpha = G_\beta^\alpha(u) M_v^\beta - \lambda_u^\alpha \,. \tag{4.12}$$

This relation prescribes a procedure that may be applied to a set of arbitrary 1-forms $M_v^\alpha \in \Lambda^1(T_v)$ in order to generate a set of 1-forms $\pi_{(u,v)}^\alpha \in \Lambda^1(T_u) \oplus \Lambda^1(T_v)$ which are such that their associated MC forms are contained entirely in $\Lambda^2(T_v)$. In particular, when $v = e$, we can write $M_v^\beta = M_e^\beta = \overset{\circ}{\pi}{}^\beta \in \Lambda^1(T_e)$, in which case (4.12) reduces to the pre-gauge transformation (2.2). We shall therefore call (4.12) — with arbitrary $M_v^\beta \in \Lambda^1(T_v)$ — a *general pre-gauge transformation*.

634   *H. Rund*

When the solution (4.9) is substituted in (3.5) it is found that

$$
\begin{aligned}
K_\beta{}^\alpha{}_\gamma(u,v) = &\, G_\epsilon^\alpha(u)\left(\frac{\partial M_\beta^\epsilon(v)}{\partial v^\gamma} - \frac{\partial M_\gamma^\epsilon(v)}{\partial v^\beta}\right) \\
&- C_\nu{}^\alpha{}_\mu G_\sigma^\nu(u)G_\tau^\mu(u)M_\beta^\sigma(v)M_\gamma^\tau(v)\,.
\end{aligned}
$$

By means of (1.12) this can be reduced to

$$
K_\beta{}^\alpha{}_\gamma(u,v) = G_\epsilon^\alpha(u)M_\beta{}^\epsilon{}_\gamma(v)\,,
\tag{4.13}
$$

where we have put, by analogy with (3.5),

$$
M_\beta{}^\alpha{}_\gamma(v) = \frac{\partial M_\beta^\alpha(v)}{\partial v^\gamma} - \frac{\partial M_\gamma^\alpha(v)}{\partial v^\beta} - C_\nu{}^\alpha{}_\mu M_\beta^\nu(v)M_\gamma^\mu(v)\,.
\tag{4.14}
$$

In terms of this notation we can also express the MC forms associated with the 1-forms M_v^α as

$$
dM_v^\alpha + \frac{1}{2}C_\nu{}^\alpha{}_\mu M_v^\nu \wedge M_v^\mu = -\frac{1}{2}M_\beta{}^\alpha{}_\gamma(v)dv^\beta \wedge dv^\gamma\,.
\tag{4.15}
$$

Thus, in view of (4.13), it is now evident that the MC forms (4.6) admit the representation

$$
\begin{aligned}
\Pi_{(u,v)}^\alpha &= -\frac{1}{2}G_\epsilon^\alpha(u)M_\beta{}^\epsilon{}_\gamma(v)dv^\beta \wedge dv^\gamma \\
&= G_\beta^\alpha(u)\big(dM_v^\alpha + \frac{1}{2}C_\nu{}^\alpha{}_\mu M_v^\nu \wedge M_v^\mu\big)\,.
\end{aligned}
\tag{4.16}
$$

This shows that the MC forms generated by the 1-forms (4.12) are related in a linear manner to the MC forms associated with the 1-forms M_v^α.

Remark. The above analysis concerning the general pre-gauge transformation (4.12) is entirely dependent on the special solution (4.7) of this system (4.5). However, this solution is by no means unique. In fact, an equally acceptable solution is given by

$$
H_\beta^\alpha(u,v) = \hat{\chi}_\beta^\alpha(u)
\tag{4.17}
$$

in which $\hat{\chi}_\beta^\alpha$ denotes the inverse of the elements χ_α^β defined in (1.5). The corresponding solution of (4.1) is then given by

$$
K_\beta^\alpha(u,v) = \hat{G}_\epsilon^\alpha(u)N_\beta^\epsilon(v)\,,
\tag{4.18}
$$

where the entries N_β^ϵ are arbitrary functions of the coordinates of v, and $\hat{G}_\epsilon^\alpha(u) = G_\epsilon^\alpha(u^{-1})$. This is readily verified by means of the equations

$$\frac{\partial \hat{G}_\beta^\alpha(u)}{\partial u^\gamma} = -C_\nu{}^\alpha{}_\mu \hat{\chi}_\gamma^\nu(u)\hat{G}_\beta^\mu(u)\,, \tag{4.19}$$

which are a direct consequence of (1.11) and (1.12). The solutions (4.17) and (4.18) then give rise to the following counterpart of (4.10), namely

$$\pi_{(u,v)}^{\prime\alpha} = \hat{\chi}_\beta^\alpha(u)du^\beta + \hat{G}_\epsilon^\alpha(u)N_\beta^\epsilon(v)dv^\beta\,, \tag{4.20}$$

which is also a general pre-gauge transformation. Again, the substitution of (4.18) in (3.5) yields

$$K_\beta^{\prime\,\alpha}{}_\gamma = \hat{G}_\epsilon^\alpha(u)N_\beta^\epsilon{}_\gamma(v)\,, \tag{4.21}$$

where

$$N_\beta^\alpha{}_\gamma(v) = \frac{\partial N_\beta^\alpha(v)}{\partial v^\gamma} - \frac{\partial N_\gamma^\alpha(v)}{\partial v^\beta} - C_\nu{}^\alpha{}_\mu N_\beta^\nu(v)N_\gamma^\mu(v)\,. \tag{4.22}$$

Accordingly the MC forms (4.6) generated by 1-forms (4.20) can be expressed as

$$\Pi_{(u,v)}^{\prime\alpha} = -\frac{1}{2}\hat{G}_\epsilon^\alpha(u)N_\beta^\epsilon{}_\gamma(v)dv^\beta \wedge dv^\gamma = \hat{G}_\beta^\alpha(u)\left(dN_v^\beta + \frac{1}{2}C_\nu{}^\beta{}_\mu N_v^\nu \wedge N_v^\mu\right)\,, \tag{4.23}$$

in which

$$N_v^\alpha = N_\beta^\alpha(v)dv^\beta \tag{4.24}$$

denotes set of essentially arbitrary 1-forms in $\Lambda^1(T_v)$.

5. Position-dependence of the Group Parameters and Finite Gauge Transformations

Thus far our analysis has been concerned exclusively with a given Lie group G. It is now supposed that in addition an n-dimensional differentiable manifold M is given, whose local coordinates are denoted by $\{x^j : j = 1,\ldots,n\}$. (In physical applications M is usually identified with the space-time manifold). Moreover, it is assumed that the parameters $\{u^\alpha : \alpha = $

$1, \ldots, r\}$ of each $u \in G$ are dependent on the coordinates of points of M in a manner to be described as follows.

For the purposes of this construction we shall restrict ourselves to a coordinate neighborhood U of the identity $e \in G$, and to a coordinate neighborhood V on M. The position-dependence of the group parameters entails the assignment of an r-tuple $\{u^\alpha(x^j) : \alpha = 1, \ldots, r\} \in \mathbb{R}^r$ to each pair $u \in U, x \in V$, in which u^α, x^j refer to the parameters of u and the coordinates of x respectively. For each $x \in V$ we construct the set

$$Y_x = \{u^\alpha(x^j) : u \in U\} \subset \mathbb{R}^r \,,$$

it being assumed that this assignment is such that $u^\alpha(x^j) \neq v^\alpha(x^j)$ on Y_x whenever $u \neq v$ on G. This clearly suggests a local fibre bundle structure for which M is the base space, and

$$B = \bigcup_{x \in V} Y_x$$

is the bundle space, the natural projection $p : B \to M$ being such that $p(z) = x$ for all $z \in Y_x$. In accordance with the standard pattern it is supposed that a map $\phi : V \times U \to p^{-1}(V)$ is given, which is such that, for any $x \in V$, and any $u \in U$, one has $p \circ \phi(x, u) = x$. Thus $\phi(x, u) \in Y_x$, and for each $x \in V$ this defines a map $\phi_x : U \to Y_x$, with $z = \phi_x(u) = \phi(x, u)$. It is assumed that every ϕ_x is a homeomorphism that depends differentiably on x. A further restriction on ϕ_x is the requirement that $\phi_x(e) = \{0, \ldots, 0\} \in Y_x \subset \mathbb{R}^r$ for all $x \in V$. In terms of a given chart (U, h) on G, for which $h(u) = \{u^\alpha : \alpha = 1, \ldots, r\}$, with $h^{-1}\{u^\alpha\} = u$, we construct the homeomorphism $h_x : h(U) \to Y_x$ by putting $h_x = \phi_x \circ h^{-1}$. Thus for any $\{u^\alpha\} \in h(U)$ we have

$$h_x(u^\alpha) = \phi_x \circ h^{-1}(u^\alpha) = \phi_x(u) = z \,.$$

In particular, since any chart on G must be such that $h(e) = \{0, \ldots, 0\}$, it is evident that

$$h_x(0, \ldots, 0) = \phi_x \circ h^{-1}(0, \ldots, 0) = \phi_x(e) = \{0, \ldots, 0\} \,.$$

By construction, each $z \in Y_x$ determines a unique set of group parameters $\{u^\alpha : \alpha = 1, \ldots, r\} = h_x^{-1}(z)$, and the coordinates z^α of z are defined as $z^\alpha = u^\alpha(x^j)$ in accordance with the original definition of Y_x. This naturally gives rise to coordinates on the bundle space B: if $z \in B$ is on the fibre Y_x, its $(n + r)$ coordinates are given by $z^j = x^j, (j = 1, \ldots, n)$, and $z^\alpha = u^\alpha(x^j), (\alpha = 1, \ldots, r)$. According to this scheme, the point $x \in V$ locates the fibre in which z is contained, and $u \in U$ determines the position of z on that fibre. (In the sequel all lower case Latin indices $j, h, \ldots$ range from 1 to n, and the summation convention is operative in respect of these indices also.)

With any given $u \in U$ one can associate a unique n-dimensional subspace of B whose parametric representation is given by

$$z^j = x^j, \quad z^\alpha = u^\alpha(x^j), \tag{5.1}$$

in which $u^\alpha = h(u)$, there being one such subspace for every $u \in U$. Since each h_x is a homeomorphism the set of all such subspaces constitutes an n-dimensional foliation of B, each leaf of which is a section $\sigma_u : M \to B$, with $\sigma_u(x) = z \in Y_x$ as prescribed by (5.1). As usual, this section induces the map $(\sigma_u)_* : T_x(M) \to T_z(B)$, for which

$$(\sigma_u)_* \frac{\partial}{\partial x^j} = \frac{\partial z^h}{\partial x^j} \frac{\partial}{\partial z^h} + \frac{\partial z^\alpha}{\partial x^j} \frac{\partial}{\partial z^\alpha} = \frac{\partial}{\partial z^j} + \frac{\partial u^\alpha(x^h)}{\partial x^j} \frac{\partial}{\partial z^\alpha}, \tag{5.2}$$

together with the pull-back $\sigma_u^* : T_z^*(B) \to T_x^*(M)$, for which

$$(\sigma_u^*)dz^j = dx^j, \quad (\sigma_u^*)dz^\alpha = \frac{\partial u^\alpha(x^h)}{\partial x^j} dx^j. \tag{5.3}$$

Thus, for any 1-form ω on B, given by

$$\omega = \omega_j dz^j + \omega_\alpha dz^\alpha, \tag{5.4}$$

we have

$$(\sigma_u^*)\omega = \left(\omega_j + \omega_\alpha \frac{\partial u^\alpha(x^h)}{\partial x^j} \right) dx^j. \tag{5.5}$$

Let $\nu = \nu_{\alpha_1 \ldots \alpha_s}(u) du^{\alpha_1} \wedge \ldots \wedge du^{\alpha_s}$ be an s-form in $\Lambda^s(T_u)$ for some given $u \in U$, it being assumed that the coefficients of ν are completely skew-symmetric in all subscripts. This single s-form generates an s-form field on σ_u, namely

$$\psi_u(\nu) = \frac{1}{s!} \nu_{\alpha_1 \ldots \alpha_s}(u(x^l)) \frac{\partial(u^{\alpha_1}, \ldots, u^{\alpha_s})}{\partial(x^{j_1}, \ldots, x^{j_s})} dz^{j_1} \wedge \ldots \wedge dz^{j_s}. \tag{5.6}$$

In particular,

$$\psi_u(du^\alpha) = \frac{\partial u^\alpha(x^l)}{\partial x^j} dz^j \,. \tag{5.7}$$

Because of the identity

$$\frac{\partial(u^{\alpha_1}, \ldots, u^{\alpha_s})}{\partial(x^{j_1}, \ldots, x^{j_s})} = \delta^{\alpha_1 \ldots \alpha_s}_{\beta_1 \ldots \beta_s} \frac{\partial u^{\beta_1}}{\partial x^{j_1}} \cdots \frac{\partial u^{\beta_s}}{\partial x^{j_s}} \,, \tag{5.8}$$

we can express (5.6) as

$$\psi_u(\nu) = \nu_{\alpha_1 \ldots \alpha_s}\big(u(x^l)\big) \frac{\partial u^{\alpha_1}}{\partial x^{j_1}} \cdots \frac{\partial u^{\alpha_s}}{\partial x^{j_s}} dz^{j_1} \wedge \ldots \wedge dz^{j_s} \,. \tag{5.9}$$

Consequently

$$d(\psi_u(\nu)) = (-1)^s \frac{\partial \nu_{\alpha_1 \ldots \alpha_s}\big(u(x^l)\big)}{\partial u^{\alpha_{s+1}}} \frac{\partial u^{\alpha_1}}{\partial x^{j_1}} \cdots \frac{\partial u^{\alpha_s}}{\partial x^{j_s}} \frac{\partial u^{\alpha_{s+1}}}{\partial x^{j_{s+1}}} dz^{j_1}$$
$$\wedge \ldots \wedge dz^{j_s} \wedge dz^{j_{s+1}} \,.$$

However, since

$$d\nu = (-1)^s \frac{\partial \nu_{\alpha_1 \ldots \alpha_s}(u)}{\partial u^{\alpha_{s+1}}} du^{\alpha_1} \wedge \ldots \wedge du^{\alpha_s} \wedge du^{\alpha_{s+1}} \,,$$

it follows directly that

$$\psi_u(d\nu) = d(\psi_u(\nu)) \,. \tag{5.10}$$

Similarly, it is evident that for $\nu \in \Lambda^s(T_u)$, and $\mu \in \Lambda^t(T_u)$, with $s + t \leq \min(r, n)$, one has

$$\psi_u(\nu \wedge \mu) = \psi_u(\nu) \wedge \psi_u(\mu) \,. \tag{5.11}$$

If $f \in \Lambda^0(T_u)$, that is, if f is a real-valued function of the parameters $(u^1, \ldots, u^r)$ of u, the above definition of ψ_u entails that $\psi_u(f) = f_u \in \Lambda^0(\sigma_u)$, whose values on σ_u are given by

$$f_u(x^l) = \psi_u(f(u^\alpha)) = f(u^\alpha(x^l)) \,. \tag{5.12}$$

A special case of (5.1) occurs when $u = e$, which gives rise to the identity section σ_e with parametric representation $z^j = x^j, z^\alpha = 0$, since $\phi_x(e) = 0$. Consequently $\partial z^\alpha/\partial x^j = \partial u^\alpha/\partial x^j = 0$ on σ_e. Equivalently, since $u^\alpha(x^j) = 0$ for all $x \in V$ whenever $u = e$, we have

$$\left(\frac{\partial u^\alpha}{\partial x^j}\right)_{u=e} = 0 \,. \tag{5.13}$$

Thus (5.6) implies that

$$\psi_u(\nu) = 0 \quad \text{for any} \quad \nu \in \Lambda^s(T_e) , \tag{5.14}$$

while it is evident from (5.12) that for any $f \in \Lambda^0(T_e)$ we must write $f_e(x^l) = \psi_e(f(u^\alpha)0 = f(0,\dots,0)$, which is constant on σ_e.

The domain of the maps $\psi_u : \Lambda^1(T_u) \to \Lambda^1(\sigma_u)$ can be extended to $\Lambda^1(T_u \oplus T_v)$ in an obvious manner. If $\nu = \nu_u + \nu_v \in \Lambda^1(T_u \oplus T_v)$, we define

$$\psi_u(\nu) = \psi_u(\nu_u) , \quad \psi_v(\nu) = \psi_v(\nu_v) , \tag{5.15}$$

where, for any $v \in U$, the counterpart of (5.6) is given by

$$\psi_v(dv^\alpha) = \frac{\partial v^\alpha(x^l)}{\partial x^j} dz^j . \tag{5.16}$$

Thus, for the 1-forms (3.1) we have

$$\psi_u\big(\pi^\alpha_{(u,v)}\big) = \widetilde{H}^\alpha_\beta(x^l)\frac{\partial u^\beta(x^l)}{\partial x^j} dz^i , \tag{5.17}$$

and

$$\psi_v\big(\pi^\alpha_{(u,v)}\big) = \widetilde{K}^\alpha_\beta(x^l)\frac{\partial v^\beta(x^l)}{\partial x^j} dz^i , \tag{5.18}$$

where

$$\widetilde{H}^\alpha_\beta(x^l) = H^\alpha_\beta\big(u^\lambda(x^l), v^\lambda(x^l)\big) , \quad \widetilde{K}^\alpha_\beta(x^l) = K^\alpha_\beta\big(u^\lambda(x^l), v^\lambda(x^l)\big) . \tag{5.19}$$

We now observe that (5.17) and (5.18) represent 1-form fields on the sections σ_u and σ_v respectively, which allows for the application of the maps σ_u^* and σ_v^*. The composition of the latter with ψ_u and ψ_v gives rise to a map $\theta : \Lambda^1(T_u \oplus T_v) \to \Lambda^1(M)$ in the following manner:

$$\theta\big(\pi^\alpha_{(u,v)}\big) = \sigma_u^* \circ \psi_u\big(\pi^\alpha_{(u,v)}\big) + \sigma_v^* \circ \psi_v\big(\pi^\alpha_{(u,v)}\big) . \tag{5.20}$$

For the sake of brevity we shall denote these 1-forms by $\widetilde{\pi}^\alpha$. Because of (5.17), (5.18), and (4.3) their coordinate presentation is given by

$$\widetilde{\pi}^\alpha = \left(\widetilde{H}^\alpha_\beta(x^l)\frac{\partial u^\beta(x^l)}{\partial x^j} + \widetilde{K}^\alpha_\beta(x^l)\frac{\partial v^\beta(x^l)}{\partial x^j} \right) dx^j . \tag{5.21}$$

Let us now turn to the general pre-gauge transformation (4.10). If we write

$$\widetilde{G}^\alpha_\beta(x^l) = G^\alpha_\beta\big(u^\lambda(x^l)\big) , \quad \widetilde{\lambda}^\alpha_\beta(x^l) = \lambda^\alpha_\beta\big(u^\lambda(x^l)\big) , \tag{5.22}$$

we have as a special case of (5.21)

$$\widetilde{\pi}^\alpha = \left(\widetilde{G}^\alpha_\epsilon(x^l) M^\epsilon_\beta(v^\lambda(x^l)) \frac{\partial v^\beta(x^l)}{\partial x^j} - \widetilde{\lambda}^\alpha_\beta(x^l) \frac{\partial u^\beta(x^l)}{\partial x^j} \right) dx^j \, . \quad (5.23)$$

In this particular context it is appropriate to adopt the following notation:

$$\widetilde{\pi}^\alpha = A^\alpha_u = A^\alpha_j(x^l) dx^j \, , \tag{5.24}$$

together with

$$\overset{\circ}{A}{}^\alpha = \overset{\circ}{A}{}^\alpha_j(x^l) dx^j = M^\alpha_\beta\big(v^\lambda(x^l)\big) \frac{\partial v^\beta(x^l)}{\partial x^j} dx^j \, , \tag{5.25}$$

in terms of which (5.23) can be expressed as

$$A^\alpha_j(x^l) = \widetilde{G}^\alpha_\beta(x^l) \overset{\circ}{A}{}^\beta_j(x^l) - \widetilde{\lambda}^\alpha_\beta(x^l) \frac{\partial u^\beta(x^l)}{\partial x^j} \, , \tag{5.26}$$

which has the structure of a finite gauge transformation on the base manifold M.

The notation (5.25) is justified in view of the following circumstance. If we identify $u \in G$ with identity e in (5.26), the second term on the right-hand side vanishes identically by virtue of (5.13), while by (5.22) and (1.6) we have $\widetilde{G}^\alpha_\beta(x^l) = G^\alpha_\beta(0) = \delta^\alpha_\beta$. Thus it follows that

$$A^\alpha_j(x^l)\big|_{u=e} = \overset{\circ}{A}{}^\alpha_j(x^l) \, . \tag{5.27}$$

The gauge transformation (5.26) *describes the effect of the action of the element* $u \in G$ *on the functions* $\overset{\circ}{A}{}^\alpha_j$ (the latter often being referred to as "gauge potentials"). In passing we observe that the gauge transformations of classical eletromagnetic theory emerge from (5.26) as special cases when the group G is abelian. For such groups the matrices of the adjoint representation are given by $G^\alpha_\beta(u) = \delta^\alpha_\beta$ for all $u \in G$ in consequence of (1.11) and (1.6) since the structure constants vanish, while (1.8) shows that $d\lambda^\alpha_u = 0$. This implies the local existence of functions $f^\alpha \in \Lambda^0(T_u)$ such that $\lambda^\alpha_u = df^\alpha$. With $f^\alpha_u \in \Lambda^0(M)$ given by (5.12), it follows with the aid of (5.10) that the relation (5.26) reduces to the system

$$A^\alpha_u = \overset{\circ}{A}{}^\alpha - df^\alpha_u \, ,$$

each member of which has the structure of a "classical" gauge transformation.

We shall conclude this section by establishing an important property of the gauge transformation (5.26). In terms of the notation (5.22) we shall write

$$\widetilde{\lambda}_u^\alpha = \widetilde{\lambda}_\beta^\alpha(x^l)\frac{\partial u^\beta(x^l)}{\partial x^j}dx^j\,, \tag{5.28}$$

so that (5.26) can be expressed as

$$A_u^\alpha = G_\beta^\alpha\big(u^\lambda(x^l)\big)\overset{\circ}{A}{}^\beta - \widetilde{\lambda}_u^\alpha\,. \tag{5.29}$$

Let us keep the element $v \in G$ fixed (so that, by (5.25), the functions $\overset{\circ}{A}{}_j^\alpha$ are fixed), while varying the element $u \in G$. For any other $u' \in G$ the counterpart of (5.29) is

$$A_{u'}^\alpha = G_\beta^\alpha\big(u'^\lambda(x^l)\big)\overset{\circ}{A}{}^\beta - \widetilde{\lambda}_u^\alpha\,, \tag{5.30}$$

and similarly, with $w = uu'$,

$$A_w^\alpha = G_\beta^\alpha\big(w^\lambda(x^l)\big)\overset{\circ}{A}{}^\beta - \widetilde{\lambda}_w^\alpha\,. \tag{5.31}$$

Since

$$w^\beta(x^l) = \phi^\beta\big(u^\epsilon(x^l), u'^\epsilon(x^l)\big)\,,$$

we have, in terms of (1.4),

$$\frac{\partial w^\beta}{\partial x^j} = \Phi_\epsilon^\beta(u, u')\frac{\partial u^\epsilon}{\partial x^j} + \Psi_\epsilon^\beta(u, u')\frac{\partial u'^\epsilon}{\partial x^j}\,,$$

to which we apply (1.14) to obtain

$$\frac{\partial w^\beta}{\partial x^j} = \Phi_\epsilon^\beta(u, u')\left[\frac{\partial u^\epsilon}{\partial x^j} + \chi_\tau^\epsilon(u)\lambda_\sigma^\tau(u')\frac{\partial u'^\sigma}{\partial x^j}\right]\,.$$

Because of (1.13) and (1.10) this yields

$$\lambda_\beta^\alpha(w)\frac{\partial w^\beta}{\partial x^j} = \lambda_\beta^\alpha(u)\frac{\partial u^\beta}{\partial x^j} + G_\epsilon^\alpha(u)\lambda_\beta^\epsilon(u')\frac{\partial u'^\beta}{\partial x^j}\,,$$

that is, in terms of (5.28)

$$\widetilde{\lambda}_w^\alpha = \widetilde{\lambda}_u^\alpha + G_\beta^\alpha\big(u^\lambda(x^l)\big)\widetilde{\lambda}_{u'}^\alpha\,, \quad (w = uu')\,. \tag{5.32}$$

When (5.32) is substituted in (5.31) it is found with the aid of (1.15) that

$$A_w^\alpha = G_\epsilon^\alpha\big(u^\lambda(x^l)\big)G_\beta^\epsilon\big(u'^\lambda(x^l)\big)\overset{\circ}{A}{}^\beta - \widetilde{\lambda}_u^\alpha - G_\beta^\alpha\big(u^\lambda(x^l)\big)\widetilde{\lambda}_{u'}^\alpha\,,$$

so that, by (5.30)

$$A_w^\alpha = G_\beta^\alpha\big(u^\lambda(x^l)\big)A_{u'}^\alpha - \widetilde{\lambda}_u^\alpha\,. \tag{5.33}$$

This is the result that we have been seeking: it displays the effect of the action of $u \in G$ on the 1-form fields $A^\alpha_{u'}$ on M by which $A^\alpha_{u'} \to A^\alpha_w$, where $w = uu'$.

6. Adjoint Connection and Curvature Forms on the Base Manifold

The 1-forms fields A^α on the base manifold M as constructed in the previous section give rise to adjoint connection and curvature forms on M whose properties are formally identical with those of the corresponding forms on the Lie group G as described in Secs. 2 and 3. Because of these similarities the formalism on M can be transcribed from G without further calculation.

Exterior differentiation of the first member of (5.22) and subsequent substitution from (1.11) yields

$$d\widetilde{G}^\alpha_\beta = C_\epsilon{}^\alpha{}_\sigma \widetilde{G}^\sigma_\beta \widetilde{\lambda}^\epsilon_u \, . \tag{6.1}$$

The 1-forms $\widetilde{\lambda}^\epsilon_u$ that appear in this relation are eliminated by means of (5.29) to give

$$d\widetilde{G}^\alpha_\beta = C_\epsilon{}^\alpha{}_\sigma \widetilde{G}^\sigma_\beta \big(\widetilde{G}^\epsilon_\gamma \overset{\circ}{A}{}^\gamma - A^\epsilon_u\big) = \widetilde{G}^\alpha_\epsilon C_\gamma{}^\epsilon{}_\beta \overset{\circ}{A}{}^\gamma - \widetilde{G}^\epsilon_\beta C_\gamma{}^\alpha{}_\epsilon A^\gamma_u \, , \tag{6.2}$$

where, in the second step, we have used (1.12). Again, the structure of the right-hand side suggests the introduction of the following 1-forms on M:

$$\pi_\beta{}^\alpha = C_\epsilon{}^\alpha{}_\beta A^\epsilon_u \, , \qquad \overset{\circ}{\pi}{}_\beta{}^\alpha = C_\epsilon{}^\alpha{}_\beta \overset{\circ}{A}{}^\epsilon \, , \tag{6.3}$$

in terms of which (6.2) can be expressed as

$$d\widetilde{G}^\alpha_\beta = \widetilde{G}^\alpha_\epsilon \overset{\circ}{\pi}{}_\beta{}^\epsilon - \widetilde{G}^\epsilon_\beta \pi_\epsilon{}^\alpha \, . \tag{6.4}$$

This is clearly the analogue of the relation (2.6); it provides a similar motivation for the interpretation of *the 1-forms* (6.3) *as adjoint connection 1-forms on M*. Since the 1-forms A^α_u have been recognized as representing the result of the action $u \in G$ on $\overset{\circ}{A}{}^\alpha$, one can consider the 1-forms $\pi_\beta{}^\alpha$ as resulting from $\overset{\circ}{\pi}{}_\beta{}^\alpha$ by the action of u.

The concomitant *adjoint curvature 2-forms on M* are defined as in (2.8):

$$\Pi_\beta{}^\alpha = d\pi^\alpha_\beta + \pi^\alpha_\epsilon \wedge \pi_\beta{}^\epsilon \, , \qquad \overset{\circ}{\Pi}{}_\beta{}^\alpha = d\overset{\circ}{\pi}{}_\beta{}^\alpha + \overset{\circ}{\pi}{}_\epsilon{}^\alpha \wedge \overset{\circ}{\pi}{}_\beta{}^\epsilon \, , \tag{6.5}$$

the effect of the action of $u \in G$ on $\overset{\circ}{\Pi}{}_\beta{}^\alpha$ being given by

$$\Pi_\beta{}^\alpha = \tilde{G}^\alpha_\epsilon \overset{o}{\Pi}{}_\gamma{}^\epsilon \widehat{\tilde{G}}_\beta{}^\gamma \,, \tag{6.6}$$

which is the counterpart of (2.7).

Now let us write

$$F_u^\alpha = \theta\big(\Pi^\alpha_{(u,v)}\big) \,, \tag{6.7}$$

in which the argument refers to the MC forms (4.16) for fixed $v \in G$. From the definition of θ it follows that

$$F_u^\alpha = -\frac{1}{2} F_j{}^\alpha{}_h \, dx^j \wedge dx^h \tag{6.8}$$

where

$$F_j{}^\alpha{}_h = G^\alpha_\epsilon\big(u^\lambda(x^l)\big) M_\beta{}^\epsilon{}_\gamma\big(v^\lambda(x^l)\big) \frac{\partial v^\beta}{\partial x^j} \frac{\partial v^\gamma}{\partial x^h} \cdot \tag{6.9}$$

If $\overset{o}{F}{}^\alpha$ denotes the MC forms associated with the 1-forms $\overset{o}{A}{}^\alpha$, that is, if

$$\overset{o}{F}{}^\alpha = d\overset{o}{A}{}^\alpha + \frac{1}{2} C_\nu{}^\alpha{}_\mu \overset{o}{A}{}^\nu \wedge \overset{o}{A}{}^\mu \,, \tag{6.10}$$

we have

$$\overset{o}{F}{}^\alpha = -\frac{1}{2} \overset{o}{F}{}_j{}^\alpha{}_h \, dx^j \wedge dx^h \,, \tag{6.11}$$

where

$$\overset{o}{F}{}_j{}^\alpha{}_h = \frac{\partial \overset{o}{A}{}^\alpha_j}{\partial x^h} - \frac{\partial \overset{o}{A}{}^\alpha_h}{\partial x^j} - C_\nu{}^\alpha{}_\mu \overset{o}{A}{}^\nu_j \overset{o}{A}{}^\mu_h \,. \tag{6.12}$$

Because of (5.25) and (4.14) this is equivalent to

$$\overset{o}{F}{}_j{}^\alpha{}_h = M_\beta{}^\alpha{}_\gamma\big(v^\lambda(x^l)\big) \frac{\partial v^\beta}{\partial x^j} \frac{\partial v^\gamma}{\partial x^h} \cdot \tag{6.13}$$

A comparison of this with (6.9) yields the required relation

$$F_j{}^\alpha{}_h = G^\alpha_\beta\big(u^\lambda(x^l)\big) \overset{o}{F}{}_j{}^\beta{}_h \,, \tag{6.14}$$

which displays the effect of the action of $u \in G$ on the MC forms (6.10).

In conclusion we note that the adjoint connection 1-forms (6.3) give rise to an operator D of adjoint exterior differentiation on M in a manner that is directly analogous to the operation on G as described by (3.10). The counterpart on M of (3.12) on G is the well-known Bianchi identity

$$DF^\alpha = 0 \,. \tag{6.15}$$

In short, the adjoint curvature theory projected from G to M is nothing other than the standard gauge formalism.

References

1. A. O. Barut and R. Raczka, *Theory of Group Representations and Applications*, Polish Scientific Publishers, Warsaw, 1977.
2. W. Drechsler and M. E. Mayer, *Fiber Bundle Tecniques in Gauge Theories*, Lecture Notes in Physics **67**, Springer-Verlag, New York, 1977.
3. L. S. Pontrjagin, *Topological Groups*, 2nd Edition, Gordon and Breach, New York, 1966.
4. H. Rund, *Differential-geometric and variational background of classical gauge field theories*, Aeq. Math. **24** (1982), 121–174.
5. H. Rund, *Differential-geometric structures associated with families of r-parameter transformation groups*, Geometry and Topology (ed. G. Rassias), World Scientific Publ., Singapore (1989), pp. 259–280.

Hanno Rund
Program in Applied Mathematics
University of Arizona
Tucson, Arizona 85721
USA

THE MATH. HERITAGE OF C.F. GAUSS (pp. 645-651)
edited by George M. Rassias
©1991 World Scientific Publ. Co. Singapore

AN ABSTRACT FIXED-POINT THEOREM OF VANDERBAUWHEDE-VANGILS TYPE

Krzysztof P. Rybakowski

We state a general fixed-point theorem for a finite scale of Banach spaces which implies and unifies many results on the existence of C^m-center and center-like manifolds, and, in addition, yields precise and explicit formulas for all their Fréchet derivatives of order $\leq m$.

1. Introduction

In their work [7] Vanderbauwhede and VanGils present an interesting approach to the existence proof of C^m-manifolds for ODEs, which was previously used by Diekmann and VanGils [1] for Volterra equations.

In both cases the center manifold φ is a solution of a fixed-point equation

$$\varphi(x) = f(x, \varphi(x)) \, .$$

Here, $f : Z \times BC^\eta \to BC^\eta$ is a uniform contraction in its second argument. Z is some Banach space and the space $BC^\eta = BC^\eta(\mathbb{R}, \mathbb{R}^n), \eta > 0$, is the Banach space of all continuous maps $y : \mathbb{R} \to \mathbb{R}^n$ such that the norm

$$\|y\|_\eta = \sup\{e^{-\eta|s|}\|y(s)\| \, | \, s \in \mathbb{R}\}$$

is finite.

The main difficultly in proving that $\varphi \in C^m$ is the fact that f is of class C^m only if regarded as a map from $Z \times BC^\mu$ to BC^η where $m\mu < \eta$.

The proof given in [7] is a technical induction argument on p for $1 \leq p \leq m$, which, when repeated for Volterra equations, yields the result in [1]. (cf. a slightly different proof in [5]).

In this paper we present a very general fixed-point theorem for a finite scale of Banach spaces which immediately implies the above-mentioned results in [1], [7] and analogous results of [6] for difference equations.

This theorem is not only an abstract formulation of what is really needed to make the arguments in [6] and [7] work, but it also yields a precise formula for all the derivatives $D^p\varphi, 1 \leq p \leq m$.

This latter formula is important in actual computations of higher-order expansions of, say, bifurcating periodic solutions etc.

In order to state the fixed-point theorem we need some notation which is introduced in the next section.

The fixed-point theorem is stated and commented upon in Sec. 3.

For the proofs we refer the reader to [4].

2. Notation and Terminology

If $X_1, \ldots, X_n$ and Y are Banach spaces then, as usual, $\mathcal{L}^n(X_1 \times \ldots \times X_n, Y)$ is the Banach space of all bounded n-linear maps $B : X_1 \times \ldots \times X_n \to Y$.

Given any set A and $j \in \mathbb{N}$ we denote by A^j the set of all j-tuples of elements of A, or, equivalently, of all maps $f : \{1, \ldots, j\} \to A$.

In particular, set $A := \{1, 2\} \times \mathbb{N}$. Then every $\mu \in A^j$ can be written as $\mu = \lambda \times \nu$ where $\lambda \in \{1, 2\}^j$ and $\nu \in \mathbb{N}^j$. In this case, set

$$\bar{\mu} := \max\{1, \sum{}' \nu(s)\},$$

where the sum $\sum'$ is extended over all $s, 1 \leq s \leq j$, with $\lambda(s) = 2$. Furthermore, for l with $1 \leq l \leq j$, set μ_{l+} to be the j-tuple defined as

$$\mu_{l+}(l) = (\lambda(l), \nu(l) + 1) \quad \text{and}$$
$$\mu_{l+}(s) = \mu(s) \qquad \text{for } s \neq l.$$

Finally, for $k \in \{1, 2\}$, set μ^k to be the $(j + 1)$-tuple defined as

$$\mu^k(1) = (k, 1) \quad \text{and}$$
$$\mu^k(s + 1) = \mu(s) \quad \text{for } 1 \leq s \leq j.$$

Now let M be any finite set of positive integers. For the definition of an *ordered partition* β of M and the notations $\mathcal{PO}(M), \mathcal{P}_n$ and $c(\beta)$ we refer the reader to [2].

Let $\beta \in \mathcal{PO}(M)$ be any ordered partition of M with

$$c(\beta) =: (j, n_1, \ldots, n_j).$$

For $\lambda \in \{1,2\}^j$ arbitrary define $\mu = \mu(\lambda, \beta) \in A^j$ as

$$\mu(s) = (\lambda(s), n_s)$$

for $1 \leq s \leq j$.

Finally, set

$$W(\beta) := \{\lambda \in \{1,2\}^j \,|\, n_s = 1 \quad \text{whenever} \quad \lambda(s) = 1\}.$$

3. Statement of the Fixed-Point Theorem

In this section we shall consider the following hypotheses (H1)–(H6):

[H1] m is a positive integer and $b : \{1, \ldots, m\} \to \mathbb{R}$ is a nondecreasing map with $b(n) \geq n$ for $1 \leq n \leq m$.

[H2] X is a Banach space.
For all $s \in S := \{0, 1, \ldots, m\} \cup b(\{1, \ldots, m\})$, E_s is a Banach space such that for $s, t \in S$ with $s < t$, $E_s \subset E_t$ and the inclusion induced embedding $J_{s,t} : E_s \to E_t$ is continuous.

We write $J_s := J_{s,b(s)}$ and $\tilde{E}_s := E_{b(s)}$ for $1 \leq s \leq m$. For j with $1 \leq j \leq m$ let W_j be the set of all

$$\mu = \lambda \times \nu \in (\{1,2\} \times \mathbb{N})^j \text{ with } \nu(s) = 1 \text{ whenever } \lambda(s) = 1.$$

Set $W^* = \{\mu = \lambda \times \nu \in W_1 | \lambda(1) = 2\}$. Fix j with $1 \leq j \leq m$ and $\mu = \lambda \times \nu$ with $\nu(s) \leq m$ for all s.

Define

$$Z_\mu = \prod_{s=1}^{j} Z_\mu(s) \quad (\text{resp. } \tilde{Z}_\mu = \prod_{s=1}^{j} \tilde{Z}_\mu(s))$$

where $Z_{\mu(s)} = \tilde{Z}_{\mu(s)} = X$ if $\lambda(s) = 1$ and $Z_{\mu(s)} = E_{\nu(s)}(\text{resp. } \tilde{Z}_{\mu(s)} = \tilde{E}_{\nu(s)})$ if $\lambda(s) = 2$.

By e_μ (resp. i_μ) we denote the canonical embedding from $\mathcal{L}^j(Z_\mu, E_{\bar\mu})$ to $\mathcal{L}^j(Z_\mu, \tilde E_{\bar\mu})$ (resp. to $\mathcal{L}^j(Z_\mu, \tilde E_{\bar\varrho})$) where $\varrho = \mu^2$).

[H3] $U \subset X$ and $V \subset E_1$ are open sets. $f : U \times V \to E_1, \varphi_0 : U \to E_0$ and $\varphi : U \to V$ are given maps such that φ_0 is continuous,

$$C := \varphi_0(U) \subset V, \quad \varphi = J_{0,1} \circ \varphi_0 \quad \text{and} \tag{1}$$
$$\varphi(x) = f(x, \varphi(x))$$

for $x \in U$.

By $a : U \times C \to U \times V$ we denote the inclusion induced embedding.

[H4] For all j with $1 \leq j \leq m$ and $\mu = \lambda \times \nu \in W_j$ with $|\nu| := \nu(1) + \ldots + \nu(j) \leq m$

$$f_\mu : U \times V \to \mathcal{L}^j(Z_\mu, E_{\bar\mu})$$

and

$$\tilde f_\mu : U \times V \to \mathcal{L}^j(\tilde Z_\mu, \tilde E_{\bar\mu})$$

are given maps such that for all $(x, y) \in U \times V$ and all l with $1 \leq l \leq j$,

$$\tilde f_\mu(x, y)|Z_\mu = J_{\bar\mu} \circ (f_\mu(x, y))$$

and, if $|\nu| < m$ and $\lambda(l) = 2$,

$$\tilde f_\varrho(x, y)|\tilde Z_\mu = J_{b(\bar\mu), b(\bar\varrho)} \circ (\tilde f_\mu(x, y))$$

where $\varrho = \mu_{l+}$.

Moreover, if $\mu \notin W^*$ then the map $\tilde f_\mu$ is continuous. On the other hand, if $\mu \in W^*$ then the map $e_\mu \circ f_\mu$ is continuous. Finally, there is a $\kappa, 0 < \kappa < 1$, such that for all $\mu \in W^*$ with $\mu(1) =: (1, n)$ and all $x \in U, y, z \in V$

$$\|f(x, y) - f(x, z)\|_{E_1} \leq \kappa \|y - z\|_{E_1} \tag{2}$$
$$\|f_\mu(x, \varphi(x))\|_{L^1(E_n, E_n)} \leq \kappa \tag{3}$$
$$\|\tilde f_\mu(x, \varphi(x))\|_{L^1(\tilde E_n, \tilde E_n)} \leq \kappa.$$

[H5] The partial derivatives

$$D_1(f \circ a) : U \times C \to \mathcal{L}^1(X, E_1)$$
$$D_2(J_1 \circ f) : U \times V \to \mathcal{L}^1(E_1, \tilde{E}_1)$$

exist and are continuous.

Moreover,

$$D_1(f \circ a) = f_{(1,1)} \circ a$$

and

$$D_2(J_1 \circ f)(x, y) = J_1 \circ (f_{(2,1)}(x, y))$$

for $(x, y) \in U \times V$.

Here we use the obvious identifications $(\{1, 2\} \times \mathbb{N})^1 \cong \{1, 2\} \times \mathbb{N}$.

[H6] For all $1 \leq j \leq m$, and $\mu = \lambda \times \nu \in W_j$ with $|\nu| < m$ the partial derivatives

$$D_1(f_\mu \circ a) : U \times C \to \mathcal{L}^1(X, \mathcal{L}^j(Z_\mu, E_{\bar{\mu}}))$$

and

$$D_2(i_\mu \circ f_\mu) : U \times V \to \mathcal{L}^1(E_1, \mathcal{L}^j(Z_\mu, E_{\bar{\varrho}})),$$

where $\varrho = \mu^2$, exist and are continuous.

Moreover, for all $(z_1, \ldots, z_j) \in Z_\mu$

$$D_1(f_\mu \circ a)(x, y)(h)(z_1, \ldots, z_j) = (f_\gamma \circ a)(x, y)(h, z_1, \ldots, z_j)$$

for all $(x, y) \in U \times C, h \in X$ and

$$D_2(i_\mu \circ f_\mu)(x, y)(w)(z_1, \ldots, z_j) = J_{\bar{\varrho}}(f_\varrho(x, y)(w, z_1, \ldots, z_j))$$

for all $(x, y) \in U \times V, w \in E_1$, where $\gamma := \mu^1$ and $\varrho := \mu^2$.

We can now state the following

Theorem A (see [4]). Assume hypotheses [H1]–[H6]. Then the map $\varphi : U \to V$ is locally Lipschitz continuous. Moreover, for all n with $1 \leq$

$n \leq m$, the map $J_{1,b(n)} \circ \varphi : U \to \tilde{E}_n$ is n-times continuously differentiable. There exists a map. $\varphi_{(n)} : U \to \mathcal{L}^n(X^n, E_n)$ such that

$$D^n(J_{1,b(n)} \circ \varphi)(x)(h_1, \ldots, h_n) = J_n(\varphi_{(n)}(x)(h_1, \ldots, h_n))$$

for all $x \in U, (h_1, \ldots, h_n) \in X^n$.

The maps $\varphi_{(n)}, 1 \leq n \leq m$, satisfy the following formula (for all $x \in U, (h_1, \ldots, h_n) \in X^n$):

$$\varphi_{(n)}(x)(h_1, \ldots, h_n) = \sum_{\beta \in \mathcal{P}_n} \Lambda_\beta \, . \tag{4}$$

Here, for $\beta \in \mathcal{P}_n$ with $c(\beta) =: (j, n_1, \ldots, n_j)$

$$\Lambda_\beta := \sum_{\lambda \in W_\beta} f_{\mu(\lambda, \beta)}(x, \varphi(x))(\alpha_1(\lambda, \beta), \ldots, \alpha_j(\lambda, \beta))$$

with $\alpha_s(\lambda, \beta) := h_{\beta(s)(1)}$ if $\lambda(s) = 1$, and

$$\alpha_2(\lambda, \beta) := \varphi_{(n_s)}(x)(h_{\beta(s)(1)}, \ldots, h_{\beta(s)(n_s)}) \text{ if } \lambda(s) = 2 \, .$$

Remarks

1. By hypotheses (H1) and (H2), the spaces $E_s, s \in S$, form *a scale of* Banach spaces.
2. Formula (2) shows that φ is uniquely determined by equation (1).
3. For $\mu = \lambda \times \nu \in W_j$ the map f_μ is a substitute for the partial Fréchet derivative $D_{\lambda(1), \ldots, \lambda(j)} f$, since the latter, in general, does not exist in the present situation.
4. For $m = 1$, the hypotheses of Theorem A are only insignificantly more general than those of Theorem 3 in [7] while the assertions and the proofs of these theorems are the same in this case.
5. In view of (3), relation (4) uniquely determines $\varphi_{(n)}$ by recursion. One can actually give a more explicit formula for $\varphi_{(n)}$ (using trees) which does not contain any terms $\varphi_{(p)}, p < n$, but is made up of terms containing exclusively combinations of the maps f_μ. (cf. [3] for related results.)

References

1. O. Diekmann and S. A. VanGils, *Invariant manifolds for Volterra integral equations of convolution type*, J. D. E. **54** (1984) 139–180.

2. K. P. Rybakowski, *Formulas for higher-order finite expansions of composite maps*, to appear in The Mathematical Heritage of C. F. Gauss, Geometry, Topology and Number Theory, G. M. Rassias, ed., World Scientific Publ. Co.

3. K. P. Rybakowski, *Formulas for higher-order Fréchet derivatives of composite maps, implicitly defined maps and solutions of differential equations*, to appear in J. Nonl. Andl., TMA.

4. K. P. Rybakowski, *An implicit function theorem for a scale of Banach spaces and smoothness of invariant manifolds*, to appear.

5. A. Vanderbauwhede, *Center manifolds, normal forms and elementary bifurcations*, Dynamics Reported, **2** (1989) 89–170.

6. A. Vanderbauwhede, *Invariant manifolds in infinite dimensions*, in Dynamics of Infinite Dimensional Systems, S. N. Chow and J. Hale (eds.), NATO ASI series, Vol. 37 (1987), pp. 409–420.

7. A. Vanderbauwhede and S. A. VanGils, *Center manifolds and contractions on a scale of Banach spaces*, J. Funct. Anal. **72** (1987) 209–224.

Krzysztof P. Rybakowski
Universität Freiburg
Institut für Angewandte Mathematik
Hermann-Herder-Str. 10
7800 Freiburg i. Br.
West Germany

THE MATH. HERITAGE OF C.F. GAUSS (pp. 652-669)
edited by George M. Rassias
©1991 World Scientific Publ. Co. Singapore

FORMULAS FOR HIGHER-ORDER FINITE
EXPANSIONS OF COMPOSITE MAPS

Krzysztof P. Rybakowski

We survey a number of explicit formulas for higher-order Fréchet derivatives (resp. higher-order finite expansions) of the composite of two or more mappings.

1. Introduction

In this paper we survey a number of higher-order chain rules, i.e., explicit formulas which express a higher-order Fréchet derivative $D^m(g \circ f)(x_0)$ of the composite $g \circ f$ of two maps f and g at x_0 in terms of the derivatives $D^k g(y_0)$ and $D^k f(x_0)$ where $y_0 = f(x_0)$ and k runs from 1 to m. We also discuss a recent extension to the case of the composite of an arbitrary number of mappings.

Despite their elementary character these formulas appear rather infrequently in the literature and, even then, they are not always stated correctly (see the remarks following Theorem 2).

On the other hand, explicit formulas for higher order derivatives of composite maps are important in some applications (see e.g., [2] and the remarks made in [6]). In many cases, they can simplify awkward inductive arguments (which, because of their very awkwardness, are often left to the reader).

There are two different ways to prove these formulas. The first is by induction on the derivative order m (see e.g., Abraham and Robbin [2]). The second method is explicitly used by Fraenkel [5]. It is simpler and more general than the inductive approach, since it can also be used with maps which, rather than being m-times differentiable, merely have a finite expansion at the given points. Actually the second approach is little more than the familiar expand-in-power-series-and-equate-coefficients method used, say, in perturbation theory.

In this paper we shall use Fraenkel's method together with some elementary combinatorial analysis to prove a few formulas for finite expansions of $g \circ f$. These formulas will in turn yield a number of known higher-order chain rules for the two-mappings case. Finally we state a recent extension of one of the formulas to the many-mapping case.

2. Notation and Terminology

The letters E, F and G (possibly with indices, like E_i) denote arbitrary normed linear spaces over the same real or complex field. All norms are denoted by $\| \cdot \|$. $\mathbf{N}$ is the set of all positive integers and $\mathbf{N}_0$ is the set of all nonnegative integers. For $m \in \mathbf{N}, \sigma(m)$ is the set of all permutations of $\{1, \dots, m\}$. For every set M and every $p \in \mathbf{N}_0$

$$M^p = \underbrace{M \times \dots \times M}_{p \text{ times}} \quad \text{if } p \neq 0, \quad M^0 = \{\emptyset\} .$$

If $p \geq 1$ and $\alpha = (\alpha(1), \dots, \alpha(p)) = (\alpha_1, \dots, \alpha_p) \in \mathbf{N}_0^p$ is a multiindex and $x_1, \dots, x_p$ are arbitrary symbols, then for $x := (x_1, \dots, x_p)$ we define

$$\begin{aligned}
x^\alpha :=\ & (x_1)^{\alpha(1)} \dots (x_s)^{\alpha(s)} \dots (x_p)^{\alpha(p)} \\
:=\ & (\underbrace{x_1, \dots, x_1}_{\alpha(1) \text{ times}}, \dots, \underbrace{x_s, \dots, x_s}_{\alpha(s) \text{ times}}, \dots, \underbrace{x_p, \dots, x_p}_{\alpha(p) \text{ times}}) ,
\end{aligned}$$

if α is not the zero multiindex and $x^0 := (x_1)^0 \dots (x_s)^0 \dots (x_p)^0 := \emptyset$.

Example. Let $\alpha = (1, 0, 3)$, then $(x_1)^{\alpha(1)}(x_2)^{\alpha(2)}(x_3)^{\alpha(3)} = (x_1, x_3, x_3, x_3)$.

Let M be any finite set. We define the sets $\mathcal{Z}^p(M)$ recursively as follows: $\mathcal{Z}^0(M) = M$ and $\mathcal{Z}^{p+1}(M)$ is the set of all nonempty finite sequences of elements of $\mathcal{Z}^p(M)$.

Thus $\beta \in \mathcal{Z}^{p+1}(M)$ if and only if there is a number $n \in \mathbf{N}$, called the *length of β* and denoted by $n = n(\beta)$, such that β is a map from $\{1, \dots, n\}$

to $\mathcal{Z}^p(M)$. Set $\mathcal{Z}^\infty(M) := \bigcup\limits_{p=0}^{\infty} \mathcal{Z}^p(M)$. If $M = \{1, \ldots, m\}, m \in \mathbf{N}$, then we write $\mathcal{Z}_m^p = \mathcal{Z}^p(M), p \in \mathbf{N}_0 \cup \{\infty\}$.

We identify $\beta \in \mathcal{Z}^p(M), p \geq 1$, with the n-tuple $(\beta(1), \ldots, \beta(n))$, where $n = n(\beta)$. But note that we distinguish between, say, 5, (5) and ((5)). Thus, if $5 \in M$, then $(5) \in \mathcal{Z}^1(M), ((5)) \in \mathcal{Z}^2(M)$ and so on.

Example. Let $M = \{0, 5, 7\}$. Then $(0, 5) \in \mathcal{Z}^1(M)$ and has length 2, $((0, 5), (3), (3)) \in \mathcal{Z}^2(M)$ and has length 3, $(((0, 5))) \in \mathcal{Z}^3(M)$ and has length 1.

If $M \subset \mathbf{Z}$, then we define the map $q : \mathcal{Z}^\infty(M) \to \mathbf{Z}$ recursively as follows: $q[\beta] = \beta$ if $\beta = m \in \mathcal{Z}^0(M)$ and $q[\beta] = \sum\limits_{s=1}^{n(\beta)} q[\beta(s)]$ if $\beta \in \mathcal{Z}^{p+1}(M)$. Intuitively, $q[\beta]$ is the sum of all the "lowest order elements" of β.

Example. $q[(1, 1)] = 2, q[((1, 3), (2))] = 6$.

If M is ordered by a strict order relation $<$, then we assume on $\mathcal{Z}^p(M)$ the *lexicographic order with respect to last element*. This is defined recursively by stipulating that whenever $\beta_i \in \mathcal{Z}^{p+1}(M)$ has length n_i for $i = 1, 2$, then $\beta_1 < \beta_2$ if and only if $\beta_1(n_1) < \beta_2(n_2)$.

An *ordered partition of M* is an element β of $\mathcal{Z}^2(M)$ such that 1–3 below hold:

1. β is increasing, i.e., $\beta(1) < \beta(2) < \ldots < \beta(n)$ on $\mathcal{Z}^1(M)$ where $n = n(\beta)$,
2. for every $s, 1 \leq s \leq n, \beta(s)$ is increasing, i.e., $\beta(s)(1) < \beta(s)(2) < \ldots < \beta(s)(r_s)$ on $\mathcal{Z}^0(M) = M$ where $r_s = n(\beta(s))$,
3. the sets $M_s := \{\beta(s)(t) | 1 \leq t \leq r_s\}, s = 1, \ldots, n$, are pairwise disjoint and $\bigcup\limits_{s=1}^{n} M_s = M$.

The set of all ordered partitions of M is denoted by $\mathcal{PO}(M)$. If $m \in \mathbf{N}$ and $M = \{1, \ldots, m\}$ then we set $\mathcal{P}_m = \mathcal{PO}(M)$.

For $\beta \in \mathcal{PO}(M)$ we set $c(\beta) = (n, r_1, \ldots, r_n)$ where $n = n(\beta)$ and $r_s = n(\beta(s))$ for $1 \leq s \leq n$.

Example. Let $\beta = ((3), (2, 4), (1, 5))$. Then $\beta \in \mathcal{P}_5$ and $c(\beta) = (3, 1, 2, 2)$.

For the proofs of the following results the reader is referred e.g. to [4].

If $m \in \mathbf{N}$ and $A : E^m \to F$ is m-linear, then Sym A is the symmetriza-
tion of A, i.e.,

$$(\mathrm{Sym}\, A)(h_1, \ldots, h_m) = \frac{1}{m!} \sum_{\pi \in \sigma(m)} A(h_{\pi(1)}, \ldots, h_{\pi(m)})$$

for $(h_1, \ldots, h_m) \in E^m$.

A map $P : E \to F$ is called a *homogeneous polynomial (of degree m)* if
there exists an m-linear map $A : E^m \to F$ such that $Ah^m = A(\underbrace{h, \ldots, h}_{m \text{ times}}) =$
$P(h)$ for all $h \in E$. We say that A *generates* P. This definition also applies
to the case $m = 0$ if we agree that *any* map A from $E^0 = \{\emptyset\}$ to F be
called 0-*linear* and *symmetric*. Thus we have $Ah^0 = A(\emptyset) = P(h)$ for
every $h \in E$, and so a homogeneous polynomial of degree 0 is any constant
map $P : E \to F$.

Note that, given P, the map A, in general, is not uniquely determined.
However, for every homogeneous polynomial P of degree $m \geq 0$ there exists
a unique symmetric m-linear map $\hat{A}$ which generates P. If p is continuous
then so is $\hat{A}$.

We use the common Landau o-notation: Let $U \subset E$ be open, $x_0 \in U$,
$f, g : U \to E$ and $\gamma : U \to \mathbf{R}$, be arbitrary.

We write

$$f(x) \equiv g(x) \quad \mathrm{mod}\ o(\gamma(x)) \quad \text{as} \quad x \to x_0$$

or, less formally,

$$f(x) = g(x) + o(\gamma(x)) \quad \text{as} \quad x \to x_0$$

if for every $\epsilon > 0$ there is a $\delta > 0$ such that $\|x - x_0\| < \delta$ implies $x \in U$ and
$\|f(x) - g(x)\| \leq \epsilon \gamma(x)$.

If γ has the form

$$\gamma(x) = \|x - x_0\|^\alpha$$

then we drop the phrase 'as $x \to x_0$'.

A map $\Psi : E \to F$ is called an *m-expansion of f at x_0* if there are
homogeneous polynomials $P_k : E \to F$ of degree $k, k = 0, \ldots, m$, such
that:

1. $\Psi(x) = \sum\limits_{k=0}^{m} P_k(x - x_0)$ for all $x \in E$,
2. Ψ is continuous at x_0,
3. $f(x) \equiv \Psi(x) \mod o(\|x - x_0\|^m)$.

If an m-expansion Ψ of f at x_0 exists then the polynomials $P_k, k = 0, \ldots, m$, are uniquely determined and continuous. The map $x \mapsto P_k(x - x_0)$ is called the *total k-th order term of f at x_0*.

If f is m-times (Fréchet) differentiable at x_0, then the unique m-expansion Ψ of f at x_0 exists and is given by the m-th order Taylor polynomial $T_{x_0}^m f$ of f at x_0:

$$\Psi(x) = \left(T_{x_0}^m f\right)(x) = \sum_{k=0}^{m} \frac{1}{k!} D^k f(x_0)(x - x_0)^k \ .$$

3. The Main Results

In this section, we first derive a few formulas which express the m-expansion of the compsite map $g \circ f$ at a point x_0 in terms of the m-expansion of f at x_0 and the m-expansion of g at $f(x_0)$. (Theorem 1 and Corollaries 1–3.) This will imply the corresponding formulas for $D^m(g \circ f)(x_0)$. (Theorem 2) At the end of this section we state a generalization of Theorem 2 to the composition of an arbitrary number of mappings. (Theorem 3)

We begin by a simple, yet basic result.

Proposition 1. Let U (resp. V) be open in E (resp. F), $x_0 \in U$ (resp. $y_0 \in V$) be arbitrary and $\alpha \geq 1$ be an arbitrary real number.

Let $f, \tilde{f} : U \to V$ and $g, \tilde{g} : V \to G$ be maps satisfying the following assumptions:

1. $\tilde{f}(x_0) = y_0$,
2. $f(x) \equiv \tilde{f}(x) \mod o(\|x - x_0\|^\alpha)$,
3. $g(y) \equiv \tilde{g}(y) \mod o(\|y - y_0\|^\alpha)$,
4. $\tilde{f}$ is Lipschitz continuous at x_0, i.e., there is a neighborhood $U_1 \subset U$ of x_0 and a constant $L > 0$ such that $\|\tilde{f}(x) - \tilde{f}(x_0)\| \leq L\|x - x_0\|$ for $x \in U_1$,

5. $\tilde{g}$ is Lipschitz continuous in a neighborhood $V_1 \subset V$ of y_0, i.e., there is an $M > 0$ such that $\|\tilde{g}(y_1) - \tilde{g}(y_2)\| \le M\|y_1 - y_2\|$ for $y_1, y_2 \in V_1$.

Under these hypotheses

$$(g \circ f)(x) \equiv (\tilde{g} \circ \tilde{f})(x) \quad \mod o(\|x - x_0\|^\alpha) \, .$$

Proof. Our hypotheses imply that f and $\tilde{f}$ are continuous at x_0 and $f(x_0) = \tilde{f}(x_0) = y_0$. Therefore we can assume that $U_1 = U$ and $V_1 = V$.

Define $\beta := f - \tilde{f}, \gamma := g - \tilde{g}$ and let $\epsilon, 0 < \epsilon < 1$, be arbitrary. There is a $\delta_1 > 0$ such that $\|y - y_0\| < \delta_1$ implies $y \epsilon V$ and

$$\|\gamma(y)\| \le \epsilon\|y - y_0\|^\alpha \, .$$

Moreover, there is a $\delta_2 > 0$ such that $\|x - x_0\| < \delta_2$ implies $x \in U$ and

$$\|f(x) - y_0\| < \delta_1 \, .$$

Finally, there is a $\delta > 0, \delta < \min\{1, \delta_2\}$ such that $\|x - x_0\| < \delta$ implies

$$\|\beta(x)\| \le \epsilon\|x - x_0\|^\alpha \, .$$

Putting everything together we get for every $\|x - x_0\| < \delta$

$$
\begin{aligned}
\|(g \circ f)(x) - (\tilde{g} \circ \tilde{f})(x)\| &= \|\gamma(\tilde{f}(x) + \beta(x)) + \tilde{g}(f(x)) - \tilde{g}(\tilde{f}(x))\| \\
&\le \|\gamma(\tilde{f}(x) + \beta(x))\| + \|\tilde{g}(f(x)) - \tilde{g}(\tilde{f}(x))\| \\
&\le \epsilon\|\tilde{f}(x) + \beta(x) - y_0\|^\alpha + M\|f(x) - \tilde{f}(x)\| \\
&\le \epsilon \left(L\|x - x_0\| + \epsilon\|x - x_0\|^{\alpha-1}\|x - x_0\|\right)^\alpha \\
&\quad + M\,\epsilon\|x - x_0\|^\alpha \\
&\le \epsilon((L + 1)^\alpha + M)\|x - x_0\|^\alpha \, .
\end{aligned}
$$

The proposition is proved.

We can now state the following

Theorem 1. Let $m \in \mathbf{N}$ be arbitrary and assume the following hypotheses:

1. U (resp. V) is open in E (resp. F) and $x_0 \in U$ (resp. $y_0 \in V$) is arbitrary.
2. $f : U \to V$ and $g : V \to G$ are given mappings with $f(x_0) = y_0$.
3. $P : E \to F$ and $Q : F \to G$ are maps (polynomials) of the form

$$P(x) = \sum_{k=0}^{m} A_k(x - x_0)^k, \quad Q(y) = \sum_{k=0}^{m} B_k(y - y_0)^k$$

where for every $k = 0, \dots, m, A_k : E^k \to F$ and $B_k : F^k \to G$ are k-linear and continuous maps.

4. $f(x) \equiv P(x) \mod o(\|x - x_0\|^m)$
   $g(y) \equiv Q(y) \mod o(\|y - y_0\|^m)$.

Under these assumptions

$$(g \circ f)(x) \equiv \sum_{k=0}^{m} R_k(x - x_0) \bmod o(\|x - x_0\|^m) \tag{1}$$

where $R_0(h) \equiv (g \circ f)(x_0)$ and $R_k, k = 1, \dots, m$, is given by

$$R_k(h) = \sum_{j=1}^{m} \sum_{\substack{r_1, \dots, r_j = 1 \\ r_1 + \dots + r_j = k}}^{k} B_j\left(A_{r_1} h^{r_1}, \dots, A_{r_s} h^{r_s}, \dots, A_{r_j} h^{r_j}\right) \tag{2}$$

for $h \in E$. R_k is a homogeneous polynomial generated by the k-linear continuous map $C_k : E^k \to G$ given, for every $(h_1, \dots, h_k) \in E^k$, by

$$C_k(h_1, \dots, h_k) = \sum_{j=1}^{k} \sum_{\substack{r_1, \dots, r_j = 1 \\ r_1 + \dots + r_j = k}}^{k} C_{k,j,r_1,\dots,r_j}(h_1, \dots, h_k) \tag{3}$$

where

$$
\begin{aligned}
C_{k,j,r_1,\dots,r_j}(h_1, \dots, h_k) = &B_j(A_{r_1}(h_1, \dots, h_{r_1}), \dots \\
&\dots, A_{r_s}(h_{r_1 + \dots + r_{s-1} + 1}, \dots, h_{r_1 + \dots + r_s}), \dots \\
&\dots, A_{r_j}(h_{k - r_j + 1}, \dots, h_k)) .
\end{aligned}
\tag{4}
$$

Remarks. Thus, for $k = 0, \dots, m$, the total k-th order term of f at x_0 is given by the map $x \mapsto R_k(x - x_0)^k$ where R_k is defined in (2).

Proof. It is clear that (4) defines a k-linear continuous map $C_{k,j,r_1,\dots,r_j}$. Consequently C_k is k-linear, continuous and generates R_k. Thus we only have to prove formula (1).

Our hypotheses imply that $f - P$ is continuous at x_0 and $f(x_0) - P(x_0) = 0$. Thus $P(x_0) = y_0$. Moreover, P and Q are locally Lipschitz continuous on E. Thus all hypotheses of Proposition 1 are satisfied and so

$$(g \circ f)(x) \equiv (Q \circ P)(x) \mod o(\|x - x_0\|^m) . \tag{5}$$

(5) implies that $(Q \circ P)(x_0) \doteq (g \circ f)(x_0) = g(y_0)$. Now, writing $h := x - x_0$ for short, we obtain

$$
\begin{aligned}
(Q \circ P)(x) &= g(y_0) + \sum_{j=1}^{m} B_j (P(x) - y_0)^j \\
&= g(y_0) + \sum_{j=1}^{m} B_j \left(\sum_{j=1}^{m} A_r h^r \right)^j \\
&= g(x_0) + \sum_{j=1}^{m} \sum_{r_1,\ldots,r_j=1}^{m} B_j(A_{r_1} h^{r_1}, \ldots, A_{r_\bullet} h^{r_\bullet}, \ldots, A_{r_j} h^{r_j}) .
\end{aligned}
\tag{6}
$$

Now it is obvious that

$$B_j(A_{r_1} h^{r_1}, \ldots, A_{r_\bullet} h^{r_\bullet}, \ldots, A_{r_j} h^{r_j}) = o\left(\|h\|^{r_1 + \cdots + r_j - 1}\right) . \tag{7}$$

Rearranging terms in (6) and using (7) we easily obtain

$$
\begin{aligned}
(Q \circ P)(x) &= g(y_0) + \sum_{k=1}^{m} \sum_{j=1}^{m} \sum_{\substack{r_1,\ldots,r_j=1 \\ r_1 + \cdots + r_j = k}}^{m} B_j(A_{r_1} h^{r_1}, \ldots \\
&\qquad \ldots, A_{r_\bullet} h^{r_\bullet}, \ldots, A_{r_j} h^{r_j}) + o\left(\|h\|^m\right) \\
&= g(y_0) + \sum_{k=1}^{m} \sum_{j=1}^{k} \sum_{\substack{r_1,\ldots,r_j=1 \\ r_1 + \cdots + r_j = k}}^{m} B_j(A_{r_1} h^{r_1}, \ldots \\
&\qquad \ldots, A_{r_\bullet} h^{r_\bullet}, \ldots, A_{r_j} h^{r_j}) + o\left(\|h\|^m\right) \\
&= \sum_{k=0}^{m} R_k(x - x_0) + o\left(\|x - x_0\|^m\right) .
\end{aligned}
\tag{8}
$$

This proves (1).

Note that in Theorem 1 we do not assume A_k or B_k to be symmetric, nor does it follow that C_k is symmetric. If the maps A_k or B_k (or both) are symmetric, then certain summands in (2) occur repeatedly. Collecting such terms leads to formulas with fewer summands but more complicated coefficients. This is the contents of the following Corollaries.

Corollary 1. Assume all hypotheses of Theorem 1. In addition, suppose that the maps $B_k, k = 1, \ldots, m$ are symmetric.

For $k = 1, \ldots, m$, define

$$J_k = \left\{ \alpha \in \mathbf{N}_0^k \,\middle|\, \sum_{i=1}^k i\alpha_i = k \right\} .$$

Then for every $k = 1, \ldots, m$ and every $h \in E$

$$C_k h^k = \sum_{\alpha \in J_k} \frac{|\alpha|}{\alpha_1! \ldots \alpha_k!} B_{|\alpha|}((A_1 h^1)^{\alpha_1}, \ldots, (A_s h^s)^{\alpha_s}, \ldots, (A_k h^k)^{\alpha_k}) . \tag{9}$$

Proof. Define

$$L_k = \big\{ (j, r_1, \ldots, r_j) \,|\, 1 \le j \le k, \quad (r_1, \ldots, r_j) \in \mathbf{N}^j, \quad r_1 + \ldots + r_j = k \big\},$$
$$1 \le k \le m .$$

Define the map $\phi : L_k \to \mathbf{N}_0^k$ by $\phi(j, r_1, \ldots, r_j) := \alpha := (\alpha_1, \ldots, \alpha_k)$ where $\alpha_i = \sharp\{n \,|\, 1 \le n \le k, r_n = i\}$.

If $(j, r_1, \ldots, r_j) \in L_k$ and $\alpha = \phi(j, r_1, \ldots, r_j)$ then it is clear that

$$|\alpha| = \alpha_1 + \ldots + \alpha_k = j \tag{10}$$

and

$$\sum_{j=1}^k i\alpha_i = r_1 + \ldots + r_j = k . \tag{11}$$

(11) implies that ϕ maps L_k into J_k. For $\alpha \in J_k$ let

$$(w_1^{(\alpha)}, \ldots, w_j^{(\alpha)}) = (\underbrace{1, \ldots, 1}_{\alpha_1 \text{ times}}, \ldots, \underbrace{s, \ldots, s}_{\alpha_s \text{ times}}, \ldots, \underbrace{k, \ldots, k}_{\alpha_k \text{ times}}) .$$

Then it is immediate that $(j, r_1, \ldots, r_j) \in \phi^{-1}\{\alpha\}$ if and only if $j = |\alpha|$ and there is a permutation π of $\{1, \ldots, j\}$ such that $r_i = w_{\pi(i)}^{(\alpha)}$ for all $i = 1, \ldots, j$. In particular, we obtain

$$\sharp\phi^{-1}\{\alpha\} = \frac{j!}{\alpha_1! \ldots \alpha_k!} = \frac{|\alpha|}{\alpha_1! \ldots \alpha_k!} . \tag{12}$$

Now Theorem 1, (10), (12) and the symmetry of the maps B_k imply

$$
\begin{aligned}
C_k h^k &= \sum_{j=1}^{k} \sum_{\substack{r_1,\ldots,r_j=1 \\ r_1+\ldots+r_j=k}}^{k} B_j(A_{r_1} h^{r_1}, \ldots, A_{r_s} h^{r_s}, \ldots, A_{r_j} h^{r_j}) \\
&= \sum_{(j,r_1,\ldots,r_j)\in L_k} B_j(A_{r_1} h^{r_1}, \ldots, A_{r_s} h^{r_s}, \ldots, A_{r_j} h^{r_j}) \\
&= \sum_{\alpha\in J_k} \left(\sum_{(j,r_1,\ldots,r_j)\in\phi^{-1}\{\alpha\}} B_j(A_{r_1} h^{r_1}, \ldots, A_{r_s} h^{r_s}, \ldots, A_{r_j} h^{r_j}) \right) \\
&= \sum_{\alpha\in J_k} \left(\sum_{(j,r_1,\ldots,r_j)\in\phi^{-1}\{\alpha\}} 1 \right) B_{|\alpha|}((A_1 h^1)^{\alpha_1}, \ldots \\
&\qquad \ldots, (A_s h^s)^{\alpha_s}, \ldots, (A_k h^k)^{\alpha_k}) \\
&= \sum_{\alpha\in J_k} \frac{|\alpha|}{\alpha_1!\ldots\alpha_k!} B_{|\alpha|}((A_1 h^1)^{\alpha_1}, \ldots, (A_s h^s)^{\alpha_s}, \ldots, (A_k h^k)^{\alpha_k}) .
\end{aligned}
\tag{13}
$$

The proof is complete. $\quad$.

Corollary 2. Assume all hypotheses of Theorem 1. In addition, suppose that all the maps $A_k, B_k, k = 1,\ldots,m$, are symmetric.

For $k = 1,\ldots,m, \beta \in \mathcal{P}_k$ with $c(\beta) = (j,r_1,\ldots,r_j)$ and $(h_1,\ldots,h_k) \in E^k$ let $d_s := d_s(j,r_1,\ldots,r_j) := r_1 + \ldots + r_s$ for $s = 1,\ldots,j, d_0 := 0$ and define

$$
\begin{aligned}
\Gamma(\beta, h_1,\ldots,h_k) &= \frac{j!\,r_1!\ldots r_j!}{k!} B_j(A_{r_1}(h_{\beta(1)(1)},\ldots,h_{\beta(1)(r_1)}),\ldots \\
&\quad \ldots, A_{r_s}(h_{\beta(s)(d_{s-1}+1)},\ldots,h_{\beta(s)(d_s)}),\ldots, A_{r_j}(h_{\beta(j)(d_{j-1}+1)},\ldots \\
&\quad \ldots, h_{\beta(j)(d_j)})) .
\end{aligned}
\tag{14}
$$

Then

$$
(\operatorname{Sym} C_k)(h_1,\ldots,h_k) = \sum_{\beta\in\mathcal{P}_k} \Gamma(\beta, h_1,\ldots,h_k) .
\tag{15}
$$

Proof. Let L_k be defined as in the proof of Corollary 1. For $(j,r_1,\ldots,r_j) \in L_k$ and $1 \le s \le j$ set

$$
K_s = K_s(j,r_1,\ldots,r_j) = \left\{ r \in \mathbf{N} \,|\, d_{s-1} + 1 \le r \le d_s \right\} .
$$

Let $\text{Sh} = \text{Sh}(j, r_1, \ldots, r_j)$ be the set of all permutations $\pi \in \sigma(k)$ such that π is increasing on K_s for $1 \le s \le j$.

For $\pi \in \text{Sh}(j, r_1, \ldots, r_j)$ let $\Delta_\pi(j, r_1, ldots, r_j)$ be the set of all permutations $\pi' \in \sigma(k)$ which can be obtained from π by arbitrarily permuting the elements *within* the sets $\pi(K_s)$ for $1 \le s \le j$.

It is easily shown that

$$\Delta_{\pi_1}(j, r_1, \ldots, r_j) \cap \Delta_{\pi_2}(j, r_1, \ldots, r_j) = \emptyset \tag{16}$$

for $\pi_1, \pi_2 \in \text{Sh}(j, r_1 \ldots, r_j), \pi_1 \ne \pi_2$,

$$\sigma(k) = \bigcup_{\pi \in \text{Sh}(j, r_1, \ldots, r_j)} \Delta_\pi(j, r_1, \ldots, r_j) \,, \tag{17}$$

$$\sharp \Delta_\pi(j, r_1, \ldots, r_j) = r_1! \ldots r_j! \,, \tag{18}$$

$$C_{(k, j, r_1, \ldots, r_j)}(h_{\pi'(1)}, \ldots, h_{\pi'(k)}) = C_{(k, j, r_1, \ldots, r_j)}(h_{\pi(1)}, \ldots, h_{\pi(k)}) \tag{19}$$

for $\pi' \in \Delta_\pi(j, r_1, \ldots, r_j)$. (The last property follows from our symmetry assumption of the A_l.)

Properties (16)–(19) and Theorem 1 imply

$$(\text{Sym}\, C_k)(h_1, \ldots, h_k)$$
$$= \frac{1}{k!} \sum_{\pi \in \sigma(k)} \sum_{(j, r_1, \ldots, r_j) \in L_k} C_{(k, j, r_1, \ldots, r_j)}(h_{\pi(1)}, \ldots, h_{\pi(k)})$$
$$= \sum_{(j, r_1, \ldots, r_j) \in L_k} \frac{1}{k!} \sum_{\pi \in \text{Sh}(j, r_1, \ldots, r_j)} \sum_{\pi' \in \Delta_\pi(j, r_1, \ldots, r_j)}$$

$$C_{(k, j, r_1, \ldots, r_j)}(h_{\pi'(1)}, \ldots, h_{\pi'(k)})$$
$$= \sum_{(j, r_1, \ldots, r_j) \in L_k} \sum_{\pi \in \text{Sh}(j, r_1, \ldots, r_j)} \frac{r_1! \ldots r_j!}{k!} B_j(A_{r_1}(h_{\pi(1)}, \ldots$$
$$\ldots, h_{\pi(r_1)}), \ldots, A_{r_s}(h_{\pi(r_1 + \ldots + r_{s-1} + 1)}, \ldots$$
$$\ldots, h_{\pi(r_1 + \ldots + r_s)}), \ldots, A_{r_j}(h_{\pi(k - r_j + 1)}, \ldots$$
$$\ldots, h_{\pi(k)})) \,. \tag{20}$$

Define

$$W = \{(j, r_1, \ldots, r_j, \pi) | (j, r_1, \ldots, r_j) \in L_k \text{ and } \pi \in \text{Sh}(j, r_1, \ldots, r_j)\} \,.$$

For $(j, r_1, \ldots, r_j, \pi) \in W$ set

$$\gamma(s)(t) = \pi(r_1, + \ldots + r_{s-1} + t)$$

for $1 \le t \le r_s$ and $1 \le s \le j$. Then there is a unique permutation $\rho = \rho(j, r_1, \ldots, r_j, \pi) \in \sigma(j)$ such that $\beta = \beta(j, r_1, \ldots, r_j, \pi) \in \mathcal{Z}_k^2$ defined by $\beta(s) = \gamma(\rho(s))$ for $1 \le s \le j$ is in $\mathcal{P}_k$.

Define the map $\phi : W \to \mathcal{P}_k$ by

$$\phi(j, r_1, \ldots, r_j, \pi) = \beta(j, r_1, \ldots, r_j, \pi) \ .$$

Using the definition of ϕ, the symmetry of the B_l together with (14) and (20), it is now clear how to complete the proof of the corollary.

Remarks. Permutations in $\mathrm{Sh}(j, r_1, \ldots, r_j)$ may be called $(r_1, \ldots, r_j)$-*shuffles* in obvious extension of the definition given in [1]. Hence the symbol Sh.

The first part of the above argument gives the following

Corollary 3. Assume all hypotheses of Theorem 1 and, in addition, suppose that the maps $A_k, k = 1, \ldots, m$ are symmetric.

Then for every $k = 1 \ldots, m$

$$(\mathrm{Sym}\, C_k(h_1, \ldots, h_k)$$

$$= \sum_{j=1}^{k} \sum_{\substack{r_1, \ldots, r_j = 1 \\ r_1 + \ldots + r_j = k}}^{k} \sum_{\pi \in \mathrm{Sh}(j, r_1, \ldots, r_j)} \frac{r_1! \ldots r_j!}{k!} C_{k, j, r_1, \ldots, r_j}(h_{\pi(1)}, \ldots, h_{\pi(k)})$$

$$\tag{21}$$

where $\mathrm{Sh}(j, r_1, \ldots, r_j)$ is defined as above.

Using the preceding results we can now easily prove a number of formulas expressing the higher order derivatives $D^m(g \circ f)(x_0)$ in terms of $D^k g(f(x_0))$ and $D^k f(x_0)$ for $k = 1, \ldots, m$.

Theorem 2. Let U (resp. V) be open in E (resp. F) and $x_0 \in U$ (resp. $y_0 \in V$) be arbitrary.

Let $f : U \to V$ and $g : V \to G$ be mappings with $f(x_0) = y_0$.

Suppose that $m \in \mathbf{N}$ and f is m-times differentiable at x_0 while g is m-times differentiable at y_0.

For $k = 1, \ldots, m, \beta \in \mathcal{P}_k$ with $c(\beta) = (j, r_1, \ldots, r_j)$ and $(h_1, \ldots, h_k) \in E^k$ let $d_s := d_s(j, r_1, \ldots, r_j) := r_1 + \ldots + r_s$ for $s = 1, \ldots, j, d_0 := 0$ and define

$$\Lambda(\beta, h_1, \ldots, h_k) = D^j g(y_0)(D^{r_1} f(x_0)(h_{\beta(1)(1)}, \ldots, h_{\beta(1)(r_1)}), \ldots$$
$$\ldots, D^{r_s} f(x_0)(h_{\beta(s)(d_{s-1}+1)}, \ldots, h_{\beta(s)(d_s)}), \ldots$$
$$\ldots, D^{r_j} f(x_0)(h_{\beta(j)(d_{j-1}+1)}, \ldots, h_{\beta(j)(d_j)})) \; . \tag{22}$$

Then the following formulas hold for all $h \in E$ (resp. for all $(h_1, \ldots, h_k) \in E^k$):

$$D^k(g \circ f)(x_0)h^k = \sum_{j=1}^{k} \sum_{\substack{r_1, \ldots, r_j = 1 \\ r_1 + \ldots + r_j = k}}^{k} \frac{k!}{j! r_1! \ldots r_j!}$$
$$\times D^j g(y_0)(D^{r_1} f(x_0)h^{r_1}, \ldots, D^{r_s} f(x_0)h^{r_s}, \ldots, D^{r_j} f(x_0)h^{r_j}) \tag{23}$$

$$D^k(g \circ f)(x_0)h^k = \sum_{\alpha \in J_k} \frac{k!}{\alpha_1! \ldots \alpha_k! \cdot (1!)^{\alpha_1}(2!)^{\alpha_2} \ldots (k!)^{\alpha_k}}$$
$$\times D^{|\alpha|} g(y_0)((D^1 f(x_0)h^1)^{\alpha_1}, \ldots$$
$$\ldots, (D^s f(x_0)h^s)^{\alpha_s}, \ldots, (D^k f(x_0)h^k)^{\alpha_k}) \; . \tag{24}$$

Here J_k is defined as

$$J_k = \left\{ \alpha \in \mathbf{N}_0^k \;\middle|\; \sum_{i=1}^{k} i\alpha_i = k \right\} \; .$$

$$D^k(g \circ f)(x_0)(h_1, \ldots, h_k)$$
$$= \sum_{\pi \in \sigma(k)} \sum_{j=1}^{k} \sum_{\substack{r_1, \ldots, r_j = 1 \\ r_1 + \ldots + r_j = k}}^{k} \frac{1}{j! r_1! \ldots r_j!}$$
$$\times D^j g(y_0)(D^{r_1} f(x_0)(h_{\pi(1)}, \ldots, h_{\pi(r_1)}), \ldots$$
$$\ldots, D^{r_s} f(x_0)(h_{\pi(r_1+\ldots+r_{s-1}+1)}, \ldots, h_{\pi(r_1+\ldots+r_s)}), \ldots$$
$$\ldots, D^{r_j} f(x_0)(h_{\pi(k-r_j+1)}, \ldots, h_{\pi(k)})) \tag{25}$$

$$D^k(g \circ f)(x_0)(h_1, \ldots, h_k)$$

$$= \sum_{\substack{j=1 \\ }}^{k} \sum_{\substack{r_1,\ldots,r_j=1 \\ r_1+\ldots+r_j=k}}^{k} \sum_{\pi \in \mathrm{Sh}(k,j,r_1,\ldots,r_j)} \frac{1}{j!}$$

$$\times D^j g(y_0)(D^{r_1}f(x_0)(h_{\pi(1)}, \ldots, h_{\pi(r_1)}), \ldots$$

$$\ldots, D^{r_s}f(x_0)(h_{\pi(r_1+\ldots+r_{s-1}+1)}, \ldots, h_{\pi(r_1+\ldots+r_s)}), \ldots$$

$$\ldots, D^{r_j}f(x_0)(h_{\pi(k-r_j+1)}, \ldots, h_{\pi(k)})) \tag{26}$$

$$D^k(g \circ f)(x_0)(h_1, \ldots, h_k) = \sum_{\beta \in \mathcal{P}_k} \Lambda(\beta, h_1, \ldots, h_k) . \tag{27}$$

Proof. All assumptions of Theorem 1 and Corollaries 1–3 hold with

$$A_k = \frac{1}{k!}D^k f(x_0), \quad B_k = \frac{1}{k!}D^k g(y_0), \quad k = 1, \ldots, m .$$

From the uniqueness of m-expansions it follows that

$$C_k h^k = \frac{1}{k!}D^k(g \circ f)(x_0)h^k$$

for all $h \in E$. This implies that

$$\mathrm{Sym}\, C_k = \frac{1}{k!}D^k(g \circ f)(x_0) .$$

Now an application of Theorem 1 yields formulas (23) and (25), Corollary 1 implies (24), Corollary 2 implies (27) and Corollary 3 yields (26). The proof is complete.

Remarks. Formula (24) is due to de Bruno [5] for the case $E = F = G = \mathbf{R}$ and it appears in [8] in the general version stated here.

Abraham and Robbin [1] recognized the importance of having an explicit higher-order chain rule. They also prove (by induction) a predecessor of formula (25) (stating it somewhat incorrectly and with less explicit coefficients). Formulas (23) and (25) are given by Fraenkel [6], with a proof similar to the one above. Formula (25) is also stated in [9, p.7] and in [1, p.91] with a misprint: the factor $j!$ is missing in the coefficient $1/(j!r_1!\ldots r_j!)$ (in our notation). Formula (26) is stated in [1,p.92] with the coefficient

$1/j!$ missing. Formula (27) is stated in Nelson [9,p.6]. It is also stated and proved (by induction) in [7] and, for a special case, in [3].

We shall now generalize Theorem 1 to the case of the composite of an arbitrary number of mappings.

Theorem 3. Assume the following hypotheses:

1. $p \in \mathbf{N}_0$ and $m \in \mathbf{N}$ are arbitrary.
2. For $i = 0, \ldots, p, U_i$ is open in E_i and $a_i \in E_i$ is arbitrary.
3. For $i = 0, \ldots, p - 1, f_i : U_i \to U_{i+1}$ and $f_p : U_p \to E_{p+1}$ are given mappings with $f_i(a_i) = a_{i+1}$ for $i = 0, \ldots, p - 1$.
4. For $i = 0, \ldots, p, P_i : E_i \to E_{i+1}$ are maps (polynomials) of the form
$$P_i(x_i) = \sum_{k=0}^{m} A_{i,k}(x_i - a_i)^k$$
   for $x_i \in E_i$, where $A_{i,k} : E_i^k \to E_{i+1}, k = 0, \ldots, m$, is a continuous k-linear map.
5.
$$f_i(x_i) \equiv P_i(x_i) \mod o(\|x_i - a_i\|^m)$$
   for $i = 0, \ldots, p$.

Set
$$E := E_0 \quad \text{and} \quad f = f_p \circ f_{p-1} \circ \ldots \circ f_0 .$$

For $k = 1, \ldots, m, l = 0, \ldots, p$ and $\beta \in \mathcal{Z}_k^l$ with $q(\beta) = k$ define the maps
$$C_k^l(\beta) : E^k \to E_{l+1} \tag{28}$$
recursively as follows:

- if $\beta \in \mathcal{Z}_k^0$ with $q(\beta) = k$ (i.e., $\beta = k$) then
$$C_k^0(\beta)(h_1, \ldots, h_k) = A_{0,k}(h_1, \ldots, h_k) \tag{29}$$
   for all $(h_1, \ldots, h_k) \in E^k$,
- if $0 \le l \le p - 1$ and $\beta \in \mathcal{Z}_k^{k+1}$ with $q(\beta) = k, n(\beta) =: j$ and $q(\beta(s)) =: r_s$ for $s = 1, \ldots, j$ then
$$\begin{aligned}
C_k^{l+1}(\beta)(h_1, \ldots, h_k) = A_{l+1,j}(C_{r_1}^l(\beta(1))(h_1, \ldots, h_{r_1}), \ldots \\
\ldots, C_{r_s}^l(\beta(s))(h_{r_1 + \ldots + r_{s-1} + 1}, \ldots \\
\ldots, h_{r_1 + \ldots + r_s}), \ldots, C_{r_j}^l(\beta(j)) \\
(h_{k - r_j + 1}, \ldots, h_k))
\end{aligned} \tag{30}$$

for all $(h_1, \ldots, h_k) \in E^k$.

Write

$$C_k(\beta) = C_k^p(\beta) \ .$$

Under these assumptions

$$f(x) \equiv \sum_{k=0}^{m} R_k(x - a_0) \ \mathrm{mod} \, o(\|x - a_0\|^m) \tag{31}$$

where $R_0(h) \equiv f(a_0)$ and $R_k, k = 1, \ldots, m$, is the homogeneous polynomial of degree k given by

$$R_k(h) = \sum_{\substack{\beta \in \mathcal{Z}_k^p \\ q(\beta) = k}} C_k(\beta) h^k \tag{32}$$

for all $h \in E$.

The map $x \mapsto R_k(x - a_0)$ is the total k-th order term of the composite map f at a_0.

Remarks. This result generalizes Theorem 1. It leads immediately to a corresponding extension of formulas (23) and (25).

There also exists an extension of Corollary 2 which implies a 'coefficient-free' higher-order chain rule for the many-mapping case, corresponding to formula (27).

For details, the reader is referred to [10].

We end this paper by explaining how to calculate, iteratively, the terms $C_k(\beta) h^k$ occurring in the definition of R_k.

Let $\beta \in \mathcal{Z}^l(M)$ be arbitrary.

Step 1. If $l = 0$ then $\beta = s$ for some $s \in M$. Replace the symbol β by the string of symbols $0 : s$.

If $l \geq 1$ then $\beta = (\beta(1), \ldots, \beta(n))$ with $n = n(\beta)$. Replace the string $(\beta(1), \ldots, \beta(n))$ by the string $l : n(\beta(1), \ldots, \beta(n))$. Do the same for all $\beta(r)$, all $\beta(r)(s)$ and so on.

Step 2. If $\beta \in \mathcal{Z}_k^p$ with $q(\beta) = k$ is given then perform Step 1. Then replace all occurences of $l : n$ for $l \geq 1$ by $A_{l,n}$ and all occurences of $0 : n$ by $A_{0,n} h^n$. The resulting string of symbols gives $C_k(\beta) h^k$.

Example 1. Let $\beta = (1,3,2)$. Then $k = 6, p = 1$. We obtain

$$(1,3,2) \to 1 : 3(1,3,2) \to 1 : 3(0 : 1, 0 : 3, 0 : 2)$$

and so

$$C_k(\beta)h^k = A_{1,3}(A_{0,1}h^1, A_{0,3}h^3, A_{0,2}h^2) \, .$$

Example 2. Let $\beta = ((1,3,2),(5,5))$. Then $k = 16, p = 2$. We obtain

$$((1,3,2),(5,5)) \to 2 : 2((1,3,2),(5,5)) \to 2 : 2(1 : 3(1,3,2), 1 : 2(5,5))$$
$$\to 2 : 2(1 : 3(0 : 1, 0 : 3, 0 : 2), 1 : 2(0 : 5, 0 : 5))$$

and so

$$C_k(\beta)h^k = A_{2,2}(A_{1,3}(A_{0,1}h^1, A_{0,3}h^3, A_{0,2}h^2), A_{1,2}(A_{0,5}h^5, A_{0,5}h^5)) \, .$$

References

1. R. Abraham, J. E. Marsden and T. Ratiu, *Manifolds, Tensor Analysis, and Applications*, Addison-Wesley Publishing Comp., Reading Mass., 1983.
2. R. Abraham and J. Robbin, *Transversal Mappings and Flows*, Addison-Wesley Publishing Comp., Reading, Mass., 1967.
3. J. C. Butcher, *The Numerical Analysis of Ordinary Differential Equations*, John Wiley and Sons Ltd, Chichester-New York-Brisbane-Toronto-Singapore, 1987.
4. S. B. Chae, *Holomorphy and Calculus in Normed Spaces*, Marcel Dekker, Inc., New York and Basel, 1985.
5. F. de Bruno, *Note sur une nouvelle formule du calcul différentiel*, Quart. J. Math. 1 (1856) 359–360.
6. L. E. Fraenkel, *Formulae for high derivatives of composite functions*, Math. Proc. Cambr. Philos. Soc. 83 (1978) 159–165.
7. W. Gähler, *Grundstrukturen der Analysis*, Vol. 2, Birkhäuser Verlag, Basel, 1978.
8. M. A. Krasnoselski et al., *Approximate Solutions of Operator Equations*, Wolters-Noordhoff Publishers, Groningen, 1972.
9. E. Nelson, *Topics in Dynamics, I: Flows*, Princeton Univ. Press, Princeton, N. J., 1969.

10. K. P. Rybakowski, *Formulas for higher-order Fréchet derivatives of composite maps, implicitly defined maps and solutions of differential equations*, J. Nonl. Anal., TMA, to appear.

Krzysztof P. Rybakowski
Universität Freiburg
Institut für Angewandte Mathematik
Hermann-Herder-Str. 10
7800 Freiburg i. Br.
West Germany

THE MATH. HERITAGE OF C.F. GAUSS (pp. 670-684)
edited by George M. Rassias
©1991 World Scientific Publ. Co. Singapore

INFINITE ORDER DIFFERENTIAL OPERATORS IN GENERALIZED FOCK SPACES

John Schmeelk

Laurent Schwartz, the principal architect of distribution theory, presented the impossibility of extending a form of multiplication to distribution theory. There have been many varieties of partial solutions to this problem. Some of the solutions contain heuristic computations done by physicists in quantum field theory. A recent profound technique developed by J. Colombeau culminates with multiplication and integration theory for distributions. We will develop this theory in the spirit of a sequence approach much like fundamental sequences are to distributions. However, in the new tempered distribution theory the indexing can be noncountable. Different types of differentiation are then developed and termed mixed differentiation. The time variable will contain differentiation in the Colombeau sense and the space variable will contain differentiation in a specialized functional sense. Classical distributional derivatives will also be contained in the development. The three forms of differentiation will then be studied in scales of Fréchet spaces. This process culminates with a new state space vector representation which will enjoy the property of multiplying tempered distributions.

1. Introduction

Modern physics requires much from the mathematical performance of tempered distributions. Recent applications such as renormalization [45, 68] suspension of equi-sized spheres in a heterogeneous media [46] and shock waves [12] all manifest physical properties that can be modeled implementing special classes of generalized functions. The impossibility of multiplication [65] has in many cases been circumvented by a mathematical concept of a direct or tensor product. However, a new generalized function theory developed by J. Colombeau [7–12] performs distributional multiplicating in a milieu that has a rich physical collection of applications. It is this entirely new approach developed by J. Colombeau which will provide us with the mathematical machinery to multiply the δ and have examples such as δ^2.

We will also implement a functional derivative developed in the sense of pseudo topologies [33]. This will enable us to have a derivative in an infinite number of truly "distinct directions" within an infinite dimensional space. The spaces we develop will satisfy the inclusion,

$$\mathcal{H}_0(\mathbb{R}^k \times \mathbb{S}'; \mathbb{R}) \subset \mathcal{H}_m(\mathbb{R}^k \times \mathbb{S}'; \mathbb{R}) \subset \mathcal{H}(\mathbb{R}^k \times \mathbb{S}'; \mathbb{R}) \qquad (1.1)$$

where $\mathcal{H}_0$ is an algebraic ideal in the space $\mathcal{H}_m(\mathbb{R}^k \times \mathbb{S}'; \mathbb{R})$. This will enable us to have a state vector which will enjoy distributional multiplication and functional differentiation.

Our strategy is to review the J. Colombeau algebra, $\mathcal{G}(\Omega)$ as developed by T. D. Todorov [67]. We will include some very fundamental results for this algebra. We will then introduce the space, $\Gamma^{p,sB}$, having the requisite functional differention properties. We then combine $\Gamma^{p,sB}$ spaces with its Colombeau-Todorov inclusions,

$$\mathcal{G}_0^{\mathcal{D}+} \subset \mathcal{G}_m^{\mathcal{D}+} \subset \mathcal{G}^{\mathcal{D}+}$$

to formulate our space, $\mathcal{H}(\mathbb{R}^k \times \mathbb{S}'; \mathbb{R})$ having appropriate subspaces satisfying expression (1.1). We conclude with an infinite order functional derivative operator which is controlable in our scale of spaces.

2. The Colombeau-Todorov Algebra

The space of rapid descent test functions, $\mathbb{S}(\mathbb{R}^k)$, are $C^\infty(\mathbb{R}^k)$ functions which together with all of their derivatives tend to zero faster than any power of $1/\|t\|$ as $\|t\| \to \infty$. Herein the notation, $\|t\|$, denotes the standard k-dimensional Euclidean norm. The Todorov development implements the functions, $\mathcal{D}(\mathbb{R}^k)$, which are $C^\infty(\mathbb{R}^k)$ functions having compact support. We replace these test functions with rapid descent test functions so as to accommodate a Fourier transform so widely used in applications.

The growth conditions required of rapid descent test functions are captured and better expressed by an increasing sequence of norms,

$$\|\phi\|_m = \sup_{\substack{|\alpha| \leq m \\ t \in \mathbb{R}^k}} [(1+t_1^2)\dots(1+t_k^2)]^m |D^{(\alpha_1,\dots,\alpha_k)}\phi(t_1,\dots,t_k)|$$

$$\stackrel{\triangle}{=} \sup_{\substack{|\alpha| \leq m \\ t \in \mathbb{R}^k}} [(1+t^2)]^m |D^\alpha \phi(t)|. \qquad (2.1)$$

It is easily shown that the growth conditions and sequential uniform convergence are satisfied if and only if the norms in expression (2.1), $m = 0, 1 \ldots,$ and all $|\alpha| = \sum_{i=1}^{k} \alpha_i \leq m$ and all $D^\alpha = \frac{\partial^{\alpha_1 + \cdots \alpha_k}}{\partial t_1^{\alpha_1} \ldots \partial t_k^{\alpha_k}}$ are finite.

We refine the space, $\mathbb{S}(\mathbb{R}^k)$, into a descreasing sequence of subspaces following the Colombeau construction for the $\mathcal{A}_q$ spaces in $\mathcal{D}(\mathbb{R}^k)$. These subspaces are defined as

$$\mathbb{S}_q(\mathbb{R}^k) = \{\phi \in \mathbb{S}(\mathbb{R}^k) : \int_{\mathbb{R}^k} \phi(\tau)d\tau = 1, \int_{\mathbb{R}^k} \tau^\alpha \phi(\tau)d\tau = 0, 1 \leq |\alpha| \leq q\}.$$
$$(2.2)$$

Clearly the sets, $\mathbb{S}_q(\mathbb{R}^k)$, satisfy for $q = 1, \ldots,$ the inclusions, $\mathbb{S}(\mathbb{R}^k) \supset \mathbb{S}_1(\mathbb{R}^k) \supset \ldots \supset \mathbb{S}_q(\mathbb{R}^k) \supset \ldots.$

Lemma 2.3. If $\phi(t)\epsilon \, \mathbb{S}_q(\mathbb{R}^1)$ for $q = 1, 2, \ldots,$ then $\int_{\mathbb{R}^1} \tau^m \phi^{(m)}(\tau)d\tau = m!(-1)^m$ for $m = 1, 2, \ldots$ and all $\phi(\tau)\epsilon \, \mathbb{S}_q(\mathbb{R}^1)$.

Proof. A traditional induction on m will prove the lemma.

Theorem 2.4. If $\phi(t)\epsilon \, \mathbb{S}(\mathbb{R}^1)$, then $\phi(t)\epsilon \, \mathbb{S}_q(\mathbb{R}^1)$ iff $(-1)^m / m! t^m \phi^{(m)}(t)\epsilon \, \mathbb{S}_q(\mathbb{R}^1)$.

Proof. Again a traditional induction on m will prove the theorem.

Similar results hold for $\mathbb{S}_q(\mathbb{R}^k), k > 1$. We indicate the Lemma 2.3 and Theorem 2.4 prove a uniform result on growth of derivatives in some sense when the rapid descent test functions are members of $\mathbb{S}_q(\mathbb{R}^k), q = 1, 2, \ldots.$

The set, $\mathscr{G}^{\mathbb{S}_+}$ is in spirit of ultrapowers developed by T. D. Todorov [67]. The set, $\mathbb{S}_+ = \mathbb{S} \times \mathbb{R}_+$, where $\mathbb{S}$ are rapid descent test functions and $\mathbb{R}_+$ are real positive numbers. The set, $\mathscr{G}^{\mathbb{S}_+}$, will denote all countable or uncountable sequences having $\mathbb{S}_+$ as the domain, and $C^\infty(\mathbb{R}^k)$ as the range. An example of a nontrivial function belonging to $\mathscr{G}^{\mathbb{S}_+}$ is as follows. Select a tempered distribution, $t(\xi)\epsilon \, \mathbb{S}'(\mathbb{R}^k)$, and define

$$T : \mathbb{S}(\mathbb{R}^k) \times \mathbb{R}_+ \to C^\infty(\mathbb{R}^k)$$
$$T : (\phi, \epsilon) \sim \to T_{\phi\otimes\epsilon}(t) \stackrel{\triangle}{=} \, < t(\xi), \frac{1}{\epsilon^k}\phi\left(\frac{\xi - t}{\epsilon}\right).$$
$$(2.5)$$

It is clear that $T_{\phi \otimes \epsilon}(t) \in C^{\infty}(\mathbb{R}^k)$ for each $\phi \otimes \epsilon \in \mathcal{S}_+$.

Definition 2.6. $F \epsilon \;\; \mathcal{G}^{\mathcal{S}+}$ is moderate if and only if for every compact subset $\mathbb{K} \subset \mathbb{R}^k$ and every partial derivative operator,

$$D_t^\alpha = \frac{\partial^{\alpha_1 + \ldots + \alpha_k}}{\partial t_1^{\alpha_1} \ldots \partial t_k^{\alpha_k}},$$

there is a $q_0 \in N$ such that if $\phi(t) \epsilon \; \mathbb{S}_{q_0}(\mathbb{R}^k)$, then there is an $\epsilon_0 > 0$ and positive constant, $C > 0$, such that

$$|D^\alpha F_{\phi \otimes \epsilon}(t)| < C \epsilon^{-q_0} \tag{2.7}$$

for every $t \in K$ and $0 < \epsilon < \epsilon_0$. The test of all moderate "sequences" in $\mathcal{G}^{\mathcal{S}+}$ will be denoted by $\mathcal{G}_m^{\mathcal{S}+}$.

Theorem 2.8. Every tempered distribution can be associated to a moderate functional via the construction given in expression (2.5).

Proof. See references [10, 67].

Definition 2.9. A sequence, $F \in \;\; \mathcal{G}^{\mathcal{S}+}$ is null if and only if for every compact subset, $K \subset \mathbb{R}^k$ and every increasing real valued function, $\gamma(q)$ such that $\lim_{q \to \infty} \gamma(q) = \infty$ and every partial derivative operator,

$$D_t^\alpha = \frac{\partial^{\alpha_1 + \ldots + \alpha_k}}{\partial t_1^{\alpha_1} \ldots \partial t_k^{\alpha_k}},$$

there is a $q_0 \in N$ such that if $\phi(t) \in \mathbb{S}_q(\mathbb{R}^k)$ for $q \geq q_0$, then there exists an $\epsilon_0 > 0$ and positive constant, C, such that

$$|D_t^\alpha F_{\phi \otimes \epsilon}(t)| \leq C \epsilon^{\gamma(q) - q_0}$$

$\forall t \in K$ and $0 < \epsilon < \epsilon_0$. The set of all null sequences in $\mathcal{G}^{\mathcal{S}+}$ will be denoted by $\mathcal{G}_0^{\mathcal{S}+}$. It is immediate that $\mathcal{G}_0^{\mathcal{S}+} \subset \mathcal{G}_m^{\mathcal{S}+} \subset \mathcal{G}^{\mathcal{S}+}$. However, implementing the Leibnitz formula for product differentiation, it becomes evident that $\mathcal{G}_0^{\mathcal{S}+}$ is an algebraic ideal under pointwise multiplication. Thus we are able to "measure" two functions in $\mathcal{G}_m^{\mathcal{S}+}$ to be "close", iff the difference, $F - G$, is a member of $\mathscr{G}_0^{\mathbf{S}+}$. This relation defines an equivalence relation on $\mathscr{G}_m^{\mathbf{S}+}$. Therefore the coset classes, $\mathscr{G}_m^{\mathbf{S}+} / \mathscr{G}_0^{\mathbf{S}+}$, can be multiplied and termed the new tempered distributions.

We close this section with the example δ^2. Denote the Colombeau regularization of δ to be Δ defined by

$$\Delta_{\phi \otimes \epsilon}(t) = \left\langle \delta(\xi), \frac{1}{\epsilon^k} \phi\left(\frac{\xi - t}{\epsilon}\right) \right\rangle = \frac{1}{\epsilon^k} \phi\left(\frac{-t}{\epsilon}\right).$$

Clearly $\Delta_{\phi \otimes \epsilon}(t)$ is a member of $\mathscr{G}_m^{\mathbf{S}+}$ by Theorem 2.4. The zero functional $\vec{0}$ is a member of $\mathscr{G}_0^{\mathbf{S}+}$. Thus the coset class containing $\Delta_{\phi \otimes \epsilon}(t)$ has a representative,

$$\Delta_{\phi \otimes \epsilon}(t) + \vec{0}.$$

When we multiply the coset class with itself recalling that $\mathscr{G}_0^{\mathbf{S}+}$ is an ideal in $\mathscr{G}_0^{\mathbf{S}+}$, we have a representation of δ^2 to be

$$\delta^2 \leftrightarrow \frac{1}{\epsilon^{2k}} \phi^2\left(\frac{-t}{\epsilon}\right).$$

3. The Space, $\Gamma^p = \bigcup_{s \geq 1} \Gamma^{p, sB}$.

For each $s \geq 1$ the space, $\Gamma^{p, sB}(p > 1, B = \{B_i\}_{i=0}^{\infty}, B_i > B_j, j > i)$, is called an infinite dimensional Fock space. The p and $B_i, i \geq 0$ are all real numbers. These spaces are topological spaces of real-valued functionals on the space of real valued tempered distributions, $\mathbf{S}'(\mathbb{R}^k)$. The set of functionals which are members of $\Gamma^{p, sB}$ are all $C^\infty(\mathbf{S}'(\mathbb{R}^k); \mathbb{R})$ in the sense of functional differentiation [6]. We also require if $\Phi \in \Gamma^{p, SB}$, then

$$\Phi(x) = \sum_{q=0}^{\infty} a_q x^q = \sum_{q=0}^{\infty} a_q[x, \dots, x] \tag{3.1}$$

where $a_0 \in \mathbb{R}$ and a_q are multilinear continuous functionals on $\mathbf{S}'(\mathbb{R}^k) \times \dots \times \mathbf{S}'(\mathbb{R}^k)$, ($q$ copies, $q \geq 1$) to $\mathbb{R}$.

We control the growth of these functionals by equipping this vector space with the following increasing sequence of norms.

$$|||\Phi|||_{sB_m} = \sup_q \frac{\|a_q\|_m q!^{1/p}}{(sB_m)^q} < \infty, m = 0, 1, \dots \tag{3.2}$$

where

$$\|a_q\|_m = \sup_{\|x\|_{-m} \le 1} |a_q x^q| < \infty \quad m = 0, 1, \dots ,$$

$$\|x\|_{-m} = \sup_{\|\phi\|_m \le 1} |\langle x, \phi \rangle|, \, m = 0, 1, \dots , \phi \in \mathbb{S}(\mathbb{R}^k), x \in \mathbb{S}'(\mathbb{R}^k) .$$

The natural topology induced by the increasing sequence of norms given in expression (3.2) is easily seen to be a Frèchet space [31, 70]. It is also clear for $1 \le s \le s'$ implies $\Gamma^{p,sB} \subset \Gamma^{p,s'B}$ and the cannonical injection,

$$J_{s's} : \Gamma^{p,sB} \mapsto \Gamma^{p,s'B} ,$$

is continuous. We take the union over all the sets $\Gamma^{p,sB}, s \ge 1$, and form a directed scale of Frèchet spaces,

$$\Gamma^{p,sB} = \bigcup_{s \ge 1} \Gamma^{p,sB} .$$

Properties regarding this scale of Frèchet spaces are developed in references [60–63].

4. The Space, $\mathcal{H}(\mathbb{R}^k \times \mathbb{S}' ; \mathbb{R})^{\mathbb{S}+}$.

The space of new tempered Fock functionals, $\mathcal{H}(\mathbb{R}^k \times \mathbb{S}' ; \mathbb{R})^{\mathbb{S}+}$, is in the spirit of the space, $\mathscr{Y}^{\mathbb{S}+}$. The Fock functional, $G \in \mathcal{H}(\mathbb{R}^k \times \mathbb{S}' ; \mathbb{R})^{\mathbb{S}+}$, is defined as

$$\begin{aligned} G &: \mathbb{S}(\mathbb{R}^k) \times \mathbb{R}_+ \to C^\infty(\mathbb{R}^k \times \mathbb{S}' ; \mathbb{R}) \\ G &: \phi \times \epsilon \rightsquigarrow G_{\phi \otimes \epsilon}(t, x) \end{aligned} \tag{4.1}$$

where $G_{\phi \otimes \epsilon}(t, x)$ is $C^\infty(\mathbb{R}^k)$ in $t \in \mathbb{R}^k$ and $C^\infty(\mathbb{S}')$ in x in the functional derivative sense.

Example 4.2. Select a tempered distribution, $f(\xi) \, \mathscr{Y} \mathbb{S}'(\mathbb{R}^k)$ and $\epsilon > 0, \phi \in \mathbb{S}(\mathbb{R}^k)$ and $\Phi \in \Gamma^{p,sB}$. Then consider the functional,

$$F \cdot \Phi_{\phi \otimes \epsilon}|t_1, \dots, t_k, x) \triangleq \left\langle f(\xi), \frac{1}{\epsilon^k} \phi\left(\frac{\xi - t}{\epsilon} \right) \right\rangle \cdot \Phi(x) .$$

It is immediate that $F \cdot \Phi_{\phi \otimes \epsilon} \in \mathcal{H}(\mathbb{R}^k \times \mathbb{S}' ; \mathbb{R})^{\mathbb{S}+}$.

Definition 4.3. $G \in \mathcal{H}(\mathbb{R}^k \times \mathbb{S}' ; \mathbb{R})^{\mathbb{S}+}$ is moderate if and only if for every compact subset, $K \in \mathbb{R}^k$, and every mixed partial differential operator,

$$D_t^\alpha D_h^n \triangleq \frac{\partial^{\alpha_1 + \ldots + \alpha_k}}{\partial t_1^{\alpha_1} \ldots \partial t_k^{\alpha_s}} \cdot D_h^n \, ,$$

and every $x \in \mathbf{S}'(\mathbb{R}^k)$, there is an $q_0 \in N$ such that if $\phi(t) \in \mathbf{S}_{q_0}(\mathbb{R}^k)$, then there is a positive constant, C and $\epsilon_0 > 0$ such that $|D_t^\alpha D_h^n G_{\phi \times \epsilon}(t_1, \ldots, t_k, x)| < C\epsilon^{-q_0}$ for every $t \in K$ and $0 < \epsilon < \epsilon_0$. The set of all moderate "sequences" in $\mathcal{H}(\mathbb{R}^k \times \mathbf{S}'; \mathbb{R})^{\mathbf{S}+}$ will be denoted by $\mathcal{H}_m(\mathbb{R}^k \times \mathbf{S}'; \mathbb{R})^{\mathbf{S}+}$.

Theorem 4.4. The regularization of a tempered distribution, $t(\xi)$, and a Fock functional, $\Phi(x) \in \Gamma^{p,s B}$, is moderate.

Proof. The regularization of a tempered distribution, $t(\xi)$, and a Fock functional $\Phi(x)$, is given by the formula,

$$T \cdot \Phi_{\phi \otimes \epsilon}(t_1, \ldots, t_k, x) \triangleq \left\langle t(\xi), \frac{1}{\epsilon^k} \phi \left(\frac{\xi - t}{\epsilon} \right) \right\rangle \cdot \Phi(x) \, .$$

Theorem 2.8 proves that $\langle t(\xi), \frac{1}{\epsilon^k} \phi(\frac{\xi - t}{\epsilon}) \rangle$ is moderate. Now we also must apply the functional derivative, D_h^n, to $\Phi(x)$.

Since $D_h^n \Phi(x) = \sum_{q=0}^{\infty} \frac{q!}{(q-n)!} a_{q+n} x^q h^n$, we consider partial sums,

$$\left| \sum_{q=0}^{\infty} \frac{q!}{(q-n)!} a_{q+n} x^q h^n \right|$$

$$\leq \sum_{q=0}^{N_0} \frac{q!}{(q-n)!} \|a_{q+n}\|_m \|x\|_{-m}^q \|h\|_{-m}^n$$

$$= \sum_{q=0}^{N_0} \frac{q!}{(q-n)!} \|a_{q+n}\|_m \|x\|_{-m}^q \|h\|_{-m}^n \frac{(q+n)!^{1/p}}{(s'B_m)^{q+n}} \frac{(s'B_m)^{q+n}}{(q+n)!^{1/p}} \qquad (4.5)$$

$$\leq \|\|\Phi\|\|_{s'B_m} \bullet \sum_{q=0}^{N_0} \frac{q!}{(q-n)!^{1/p}} \|x\|_{-m}^q \|h\|_{-m}^n$$

$$\leq K \|\|\Phi\|\|_{s'B_m} \, .$$

Since the bound given in expression (4.5) is true for any N_0, we have

$$|D_t^\alpha \cdot D_h^n T \cdot \Phi_{\phi \otimes \epsilon}(t_1, \ldots, t_k, x)| \leq C\epsilon^{-q_0}$$

provided we keep q_0 sufficiently large and $\phi \in \mathbf{S}_q(\mathbb{R}^k)$.

Definition 4.6. $G \in \mathcal{H}(\mathbb{R}^k \times \mathbf{S}';\mathbb{R})^{\mathbf{S}+}$ is null if and only if for every compact subset, $K \subset \mathbb{R}^k$, and every increasing positive real valued function, $\gamma(q)$, where $\lim_{q\to\infty} \gamma(q) = \infty$, and every mixed partial differential operator,

$$D_t^\alpha \cdot D_h^n \triangleq \frac{a^{\alpha_1 + \ldots + \alpha_k}}{at_1^{\alpha_1} \ldots at_k^{\alpha_k}} \cdot D_h^n ,$$

and every $x \epsilon \, \mathbf{S}'(\mathbb{R}^k)$. Under these conditions there exists a $q_0 \in N$ such that if $\phi(t) \in \mathbf{S}_q(\mathbb{R}^k)$ for $q \geq q_0$, then there is an $\epsilon_0 > 0$ and $C > 0$ such that

$$|D_t^\alpha \cdot D_h^n G_{\phi \otimes \epsilon}(t_1, \ldots, t_k, x)| < C\epsilon^{\gamma(q)-q_0}$$

for every $t \in K$ and $0 < \epsilon < \epsilon_0$.

The set of null sequences in $\mathcal{H}(\mathbb{R}^k \times \mathbf{S}';\mathbb{R})^{\mathbf{S}+}$ will be denoted by $\mathcal{H}_0(\mathbb{R}^k \times \mathbf{S}';\mathbb{R})^{\mathbf{S}+}$. We observe that

$$\mathcal{H}_0(\mathbb{R}^k \times \mathbf{S}';\mathbb{R})^{\mathbf{S}+} \subset \mathcal{H}_m(\mathbb{R}^k \times \mathbf{S}';\mathbb{R})^{\mathbf{S}+} .$$

The pointwise product of a $G \in H_0(\mathbb{R}^k \times \mathbf{S}';\mathbb{R})^{\mathbf{S}+}$ and a $W \in \mathcal{H}_m(\mathbb{R}^k \times \mathbf{S}';\mathbb{R})^{\mathbf{S}+}$ is well defined and by the Leibnitz formula for higher order mixed derivatives implies the product, $W \cdot G_{\phi \otimes \epsilon}(t_1, \ldots, t_k, x)$, is a member of $\mathcal{H}_0(\mathbb{R}^k \times \mathbf{S}';\mathbb{R})^{\mathbf{S}+}$. We conclude that $\mathcal{H}_0(\mathbb{R}^k \times \mathbf{S}';\mathbb{R})^{\mathbf{S}+}$ is an ideal in $\mathcal{H}_m(\mathbb{R}^k \times \mathbf{S}';\mathbb{R})^{\mathbf{S}+}$.

Definition 4.7. Two sequences, $G, H \in \mathcal{H}_m(\mathbb{R}^k \times \mathbf{S}';\mathbb{R})^{\mathbf{S}+}$ are related if and only if $G - H \in \mathcal{H}_0(\mathbb{R}^k \times \mathbf{S}';\mathbb{R})^{\mathbf{S}+}$. Again we have an equivalence relation on $\mathcal{H}_m(\mathbb{R}^k \times \mathbf{S}';\mathbb{R})^{\mathbf{S}+}$ and we can multiply the coset classes in the traditional sense.

5. Infinite Order Differentiation

Infinite order partial differentiation, functional differentiation or distributional differentiation can be studied in scales of Frechet space. We will examine infinite order functional differentiation in the space, $\mathcal{H}(\mathbb{R}^k \times$

$\mathbb{S}';\mathbb{R})$ $^{\mathcal{S}+}$. A few scalar valued function growth conditions must first be reviewed.

Definition 5.1. A scalar valued function, $g(t)$, has strict order of growth $\leq \lambda$ iff $\exists K > 0$ such that $|g(t)| < K \exp \epsilon |t|^{\lambda}$.

Lemma 5.2. If an entire scalar valued function, $g(t) = \sum\limits_{\nu=0}^{\infty} c_{\nu} t^{\nu}$, is of minimal type and strict order of growth $\leq \lambda$, then the coefficients, $\{c_{\nu}\}_{\nu=0}^{\infty}$ satisfy

$$|c_{\nu}| \leq C_{\epsilon}\left(\frac{\epsilon e \lambda}{\nu}\right)^{\nu/\lambda} \leq \frac{(\epsilon e \lambda)^{\nu/\lambda}}{\nu!^{1/\lambda}}.$$

Proof. [34, pg. 194]. We select a functional, $F \cdot \Phi_{\phi \otimes \epsilon}(t_1 \ldots t_r, x) \in \mathcal{H}_m(\mathbb{R}^k \times \mathbb{S}'; \mathbb{R})$ $^{\mathcal{S}+}$, where $\Phi \in \Gamma^{pB}$. We will apply the functional derivative, D_h, to our functional $F \cdot \Phi_{\phi \otimes \epsilon}(t_1 \ldots t_k, x)$. Since F is a constant for any $(t_1 \ldots t_k) \in \mathbb{R}^k$, our functional derivative, D_h, will only be applied to the $\Phi \in \Gamma^{sB}$. Under these conditions, we can prove the following theorem.

Theorem 5.3. If $\Phi \in \Gamma^{pB}$ and $g(t)$ is an entire function of minimal type and strict order of growth, q, where $\frac{1}{p} + \frac{1}{q} = 1$, then the infinite order functional differential operator,

$$g(D_h) = \sum_{\nu=0}^{\infty} c_{\nu}(D_h)^{\nu}, \tag{5.4}$$

defined by

$$g(D_h)\Phi = \sum_{\nu=0}^{\infty} c_{\nu} D_h^{\nu} \Phi$$

is a linear continuous transformation on Γ^{pB} to Γ^{pB}.

Proof. We select an arbitrary space, $\Gamma^{p,sB}$, in the scale of Frèchet spaces, Γ^{pB}. Then select a space, $\Gamma^{p,s'B}$, with $s' \geq 2s$. Now examine the operator defined in expression (5.4) having $\Gamma^{p,sB}$ as its domain and $\Gamma^{p,s'B}$

as its range. The claim is that

$$g(D_h)\Phi(\cdot) = \sum_{q=0}^{\infty}\left(\sum_{q=0}^{\infty} c_\nu \frac{(q+\nu)!}{q!} a_{q+\nu} h^\nu \; ; \right)$$

is a member of $\Gamma^{p,s'B}$. Since $h \in S'$, $\exists$ smallest integer, $m_0, h \in E_{-m_0}$. Consequently, one must examine the following two cases:

$$\text{Case } 1\,;\, m \geq m_0 \quad , \qquad \text{Case } 2\,:\, m < m_0 \,.$$

Case 1. After one selects any norm, $\||\cdot\||_{s'B_m}$ in the space, $\Gamma^{p,s'B}$, the following computations will be studied.

$$\||g(D_h)\Phi\||_{s'B_m} = \sup_q \frac{\|\sum\limits_{\nu=0}^{\infty} c_\nu \frac{1}{q!}(q+\nu)! a_{q+\nu}\|_m q!^{1/p}}{(s'B_m)^q} \tag{5.5}$$

$$\leq \sup_q \frac{q!^{1/p}}{(s'B_m)} \sum_{\nu=0}^{\infty} \frac{|c_\nu|(q+\nu)!\|a_{q+\nu}\|_m (\|h\|_{-m})^\nu}{q!}\,.$$

Since $\Phi \in \Gamma^{p,sB}$, there holds that

$$\|a_{q+\nu}\|_m \leq M_m (sB_m)^{q+\nu}(q+\nu)!^{1/p} \tag{5.6}$$

and the Taylor coefficients of $g(t)$ satisfy

$$|c_\nu| \leq C_\epsilon \frac{(\epsilon e\lambda)^{\nu/\lambda}}{\nu!^{1/p}} \quad \text{where } \lambda = \frac{p}{p-1}\,. \tag{5.7}$$

Using the estimates given in expression (5.6) and (5.7) in expression (5.5), gives us

$$\leq \sup_q \frac{M_m C_\epsilon}{(s'B_m)^q} \sum_{\nu=0}^{\infty} \frac{(\epsilon e p/p - 1)^{\nu[1-1/p]}(q+\nu)!^{1-1/p}(sB_m)^{q+\nu}(\|h\|_{-m})^\nu}{\nu!^{1/p}q!^{1-1/p}}$$

$$\leq \sup_q \frac{M_m C_\epsilon}{(s'B_m)^q} \sum_{\nu=0}^{\infty} \frac{(\epsilon e p/p - 1)^{\nu[1-1/p]}}{\nu!^{1/p}q!^{1-1/p}}(q+\nu)!^{1-1/p}[sB_m\|h\|_{-m}]^\nu$$

$$\leq \sup_q C\left(\frac{sB_m}{(s'B_m)}\right)^q \sum_{\nu=0}^{\infty}(\epsilon e p/p - 1)^{\nu[1-1/p]}p[2^{\nu+q}]^{1-1/p}[sB_m\|h\|_{-m}]^\nu$$

$$\leq \sup_q C\left(\frac{2s}{s'}\right)^q \sum_{\nu=0}^{\infty}(\epsilon e p/p - 1)^{\nu[1-1/p]}[2^\nu]^{1-1/p}[sB_m\|h\|_m]^\nu\,.$$

$$\tag{5.8}$$

For $\epsilon > 0$ sufficiently small, the series in expression (5.7) converges and sup q exists if $2s < s'$.

Case 2. We indicate for $m < m_0$ that the following estimate,

$$|||g(D_n)\Phi|||_{s'B_m} \leq |||g(D_h)\Phi|||_{s'B_{m_0}}$$

and computations proceed as in Case 1 with m_0.

Partial differentiation and infinite order partial differentiation operators can also be treated in a scale of Frèchet spaces. There are some adjustments that must be made in controlling the growth of all of the derivatives within the given scale of Frèchet spaces. Several wonderful results exhibiting these properties are developed in references [51–53]. Moreover distributional derivatives and infinite order distributional derivative operators can likewise be treated in a similar milieu. All of these results then allow for a very general setting to construct difficult physics models.

Bibliography

1. V. Bargmann, *Remarks on a Hilbert space of analytic functions*, Proceedings N.A.S., 48, (1962) 199–204.
2. V. Bargmann, *On a Hilbert space of analytic functions and an associated integral transform*, Communications on Pure and Applied Mathematics 19 (1961) 187–214.
3. V. Bargmann, *A family of related function spaces application to distribution theory Part II*, Communications on Pure and Applied Mathematics 20 (1967) 1–101.
4. Victor M. Bogdan, *Existence of solutions to differential equations of relativistic mechanics involving Lorentzian time delays*, to appear in Journal of Math. Anal. Appl.
5. N. N. Bogoliubov, D. V. Shirkov, *Introduction to the Theory of Quantized Fields*, John Wiley & Sons, New York, 1976.
6. Frank Cholewinski, *Generalized fock spaces and associated operators*, Siam Journal of Mathematical Analysis (1) 15 (1984) 177–202.
7. J. F. Colombeau, *Some aspects of infinite-dimensional holomorphy in mathematical physics*, Aspects of Mathematics and its Applications, ed. J. A. Barroso, Elsevier, 1986, pp. 253–263.
8. J. F. Colombeau, *Differential Calculus and Holomorphy*, North Holland Mathematical Studies No. 64, North Holland Pub. Co., New York, 1982.

9. J. F. Colombeau, *New Generalized Functions and Multiplication of Distributions*, North Holland Mathematical Studies No. 84, North Holland Pub. Co., New York, 1984.

10. J. F. Colombeau, *Elementary Introduction to New Generalized Functions*, North Holland Mathematical Studies No. 113, North Holland Pub. Co., New York, 1985.

11. J. F. Colombeau, *A multiplication of distributions*, J. of Math. Ann. Appl. (1) **94** (1983) 96–115.

12. J. F. Colombeau and A. Y. Le Roux, *Generalized functions and products appearing in equations of physics*, preprint.

13. F. Constantinescu, *Distributions and their Applications in Physics*, Pergammon Press, New York, 1980.

14. K. Cooke and J. Wiener, *Distributional and analytic solutions of functional differential equations*, J. of Math. Anal. Appl. (1) **98** (1984) 111–129.

15. Nikolic Takaci Despotiv, *On the distributional Stieltjes transformation*, International J. Math. and Math. Sci. (2) **9** (1986) 313–317.

16. P. A. M. Dirac, *Principles of Quantum Mechanics*, Oxford University Press, England, 1967.

17. P. Duchateau, *New proofs and generalization of theorems of existence and uniqueness for the Goursat problem*, Applicable Analysis **2** (1972) 61–78.

18. P. Duchateau, *The Cauch-Goursat Problem*, Mem. Amer. Math Soc. No. 118, 1972.

19. P. Duchateau, *A Holmgren type theorem for pseudo differential operators in Gevrey classes*, J. Diff. Equ. **13** (1973) 319–328.

20. T. Dwyer, *Partial differential equations in Fischer-Fock spaces for the Hilbert-Schmidt holomorphy type*, Bull. Amer. Math. Soc. **77** (1971) 725–730.

21. T. Dwyer, *Holomoprhic Fock representations and partial differential equations on countably Hilbert spaces.* Bull. Amer. Math. Soc. **79** (1973) 1045–1050.

22. T. Dwyer, *Partial differential equations in holomorphic Fock spaces*, Functional Analysis and Applications, ed. L. Nachbin, Lecture Notes in Math. No. 384, Springer-Verlag, Berlin-Heidelberg-New York, 1974, pp. 252–259.

23. T. Dwyer, *Homomorphic representation of tempered distributions and weighted Fock spaces*, Analyse Fonctionnelle et Applications, ed. L. Nashbin, Actuallite's Sci. et Industr. No. 1367, Hermann, Paris, 1975, pp. 95–118.

24. T. Dwyer, *Differential equations of infinite order in vector-valued homomorphic Fock spaces*, Infinite-Dimensional Holomorphy and Applications, ed. M. Matos, North-Holland Mathematics Studies, North-Holland Publ. Co., Amsterdam, 1977, pp. 167–200.

25. T. Dwyer, *Equations differentielles d'Ordre infini dans des espaces localement covexes*, C. R. Acad. Sci. Paris, Ser A, **280** (1975) 1439–1442.

26. T. Dwyer, *Duallite' des espaces de fonctions entie' en dimension infinie*, Ann. Inst. Fourier (Grenoble) **26** (1976) 151–195.

27. T. Dwyer, *Differential operator of infinite order in locally covex spaces, I*, Rendiconti di matematica **10**, Serie VI (1977) 149–179.

28. T. Dwyer, *Differential operator of infinite order in locally covex spaces, II*, Rendiconti di matematica (2) **10** (1977) 278–293.

29. T. Dwyer, *Analytic evolution equations in Banach spaces*, Proc. 1977 Dublin Conference on Vector Space Measures, ed. S. Cineen, Lecture Notes in Math., Springer-Verlag, Berlin-Heidelberg-New York, 1978, pp. 48–61.

30. T. Dwyer, *Infinite dimensional analytic systems*, Proc. 1977 Conference on Decision and Control, IEEE, pp. 285–290.

31. Avner Fridman, *Generalized Functions and Partial Differential Equations*, Prentice Hall, Inc., New Jersey, 1963.

32. K. O. Friedrichs and H. N. Shapiro, *Integration of Functionals*, New York University Notes, 1957.

33. A. Frolicher and W. Bucher, *Calculus in Vector Spaces without Norm*, Lecture Notes in Mathematics, 30, Springer-Verlag, New York, 1966.

34. I. M. Gelfand and G. E. Shilov, *Generalized Functions Volume II*, translated by Morris D. Friedman, Amiel Feinstein, Christian P. Peltzer, Academic Press, Inc., New York, 1968.

35. Daniel Kastler, *Introduction a L'Electrodynamique Quantique*, Dunod, Paris, 1961.

36. G. Köthe, *Dualitat in der Funktionentheorie*, J. Reine Angew. Math. **191** (1953) 30–49.

37. G. Köthe and O. Toeplitz, *Lineare Raume mit Unendlichvielen Koordinaten und Ringe unendlicher Matrizen*, J. Reine and Angew. Math. **171** (1934) 193–226.

38. G. Köthe, *Topological Vector Spaces I*, Springer-Verlag Inc., New York, 1969.

39. P. Kristensen, I. Mejlbo, and E. T. Poulsen, *Tempered distribution in infinitely many dimensions I, canonical field operators*, Communications Mathematical Physics, **I** (1965) 175–214.

40. P. Kristensen, I. Mejlbo, and E. T. Poulsen, *Tempered distribution in infinitely many dimensions II, displacement operators*, Math. Scand. **14** (1964) 129–150.

41. P. Kristensen, I. Mejlbo, and E. T. Poulsen, *Tempered distribution in infinitely many dimensions III, linear transformations of field operators*, Commun. Math. Physics **6** (1967) 29–48.

42. M. J. Lighthill, *Fourier Analysis and Generalized Functions*, Cambridge University Press, England, 1964.

43. T. P. G. Liverman, *Generalized Functions and Direct Operational Methods*, Prentice-Hall, Englewood Cliff, New Jersey, 1964.

44. G. Marinescu, *Espaces Vectorials Pseudo-topolgiques et theorie des Distributions*, Berlin, VEB deutscher Verlag der Wissenshaften, 1963.

45. E. G. Manovkian, *Renormalization*, Academic Press, New York, 1983.

46. K. Markov, *Application of Volterra-Weiner series for bounding the overall conductivity of heterogeneous media I. General procedure*, Siam J. Appl.

Math. **47** (1987) 836–870.

47. M. Oberguggenberger, *Generalized solutions to semilinear hyperbolic systems*, Mh. Math. **103** (1987) 133–144.

48. M. Oberguggenberger, *Weak limits of solutions to semilinear hyperbolic systems*, Math. Ann. **274** (1986) 599–607.

49. M. Oberguggenberger, *Products of distributions*, J. Reine Angew. Math. **365** (1986) 1–11.

50. J. Persson, *Invariance of the Cauchy Problem for Distributional Differential Equations*, Proceedings of the 1987 Generalized Function Conference at Dubrovnik, Yugoslavia, Dellen Pub., Co., 1988.

51. Jan Persson, *Fundamental theorems for linear measure differential equations*, Math. Scand. **62** (1988) 19–43.

52. Stevan Pilipovic, *Structural Theorems for Periodic Ultradistributions*, Proceedings of the A.M.S. (2) **98** (1986) 261–266.

53. Stevan Pilipovic, *On the quasiasymptotic behavior of the Stieltjes transformation of distributions*, Publication de L'Institut Mathematique **40** (1986) 143–152.

54. C. K. Raju, *Products and composition with the Dirac delta function*, J. Phys A: Math. Gen. **15** (1982) 381–396.

55. Charles M Roumieu, *Sur quelques extensions de la notion de distributions*, Ann. Scient. E. Norm. Sup. **77** (1960) 41–121.

56. Jan Rzewuski, *On a triplet including the Hilbert space of entire functionals*, Bull. de L' Academic Polonaise des Sciences, Ser. des Sciences Math, Astro and Physics, (7) **17** (1969) 459–466.

57. Jan Rzewuski, *On a Hilbert space of functional power series*, Bull. de L' Academic Polonaise des Sciences, Ser. des Sciences Math, Astro and Physics, (11) **18** (1970) 677–685.

58. Jan Rzewuski, *On entire functionals in quantum field theory*, Reports on Mathematical Physics (1) **1** (1971) 1–27.

59. Jan Rzewuski, *Some estimates for generating functionals with an application to quantum field theory*, Bull. de L'academie Ponaise des Sciences, Ser. des Sciences Math., Astro and Physics (3) **19** (1971) 235–249.

60. John Schmeelk, *An infinite dimensional Laplacian operator*, Journal of Differential Equation

66. A. Tacaki, *A note on the distributional Stieltjes transforms*, Math. Proc. Comb. Phil. Soc. **94** (1983) 523–527.
67. T. D. Todorov, *Sequential approach to Colombeaus theory of generalized functions*, International Center for Theoretical Physics, IC 126 (1987), Trieste, Italy.
68. G. Velo and A. S. Wightman, eds., *Renormalizatin Theory*, Proceedings of the NATO Advaced Study Institue, International School of Mathematical Physics in Sicily, Italy, August 1975., D. Reidel Pub. Co., Boston, 1976.
69. U. S. Vladimirov, Y. N. Drozzinov, and B. I. Zavialow, *Tauberian Theorems for Generalized Functions*, Kluwer Academic Pub., Boston, Mass., 1988.
70. Armen Zemanian, *Distribution Theory and Transform Analysis*, McGraw Hill Book Co., New York, 1965.

John Schmeelk
Department of Mathematical Sciences
Virginia Commonwealth University
Richmond, Virginia 23284-2014
USA

THE MATH. HERITAGE OF C.F. GAUSS (pp. 685-711)
edited by George M. Rassias
©1991 World Scientific Publ. Co. Singapore

GAUSS AND THE ELECTRODYNAMIC FORCE

Domina Eberle Spencer and Shama Y. Uma

The paper traces the development of the electrodynamic force equation from the pioneer suggestion of Gauss to the present. It shows that the two basic ideas suggested by Gauss, that the electrodynamic force equation should be a function of the relative velocity of source and receiver and that it should be retarded, are once again incorporated in the most recent proposals. Such formulations are called Gaussian Electrodynamics.

The idea that all electromagnetic phenomena can be explained by the application of a single equation for the electrodynamic force originated in the fertile brain of Karl Friedrich Gauss in July 1835. He wrote [1] that "Two elements of electricity in a state of relative motion attract or repel one another, but not in the same way as if they are in a state of relative rest". Gauss derived an equation for the force between two moving charges, but he was sufficiently dissatisfied with this equation to never publish it.

Even so Gauss was the originator of the basic idea of searching for an electrodynamic force equation. In the nineteenth century, his younger colleagues at Göttingen, Weber [2] and Riemann [3], proposed other equations. In the twentieth century further work on this subject was done by Ritz [4], Bush [5], Warburton [6] and Moon and Spencer [7]. Recently, Mirchandaney, Spencer, Uma and Mann [8] have proposed a postulational formulation which yields 120 possible equations.

The unpublished seed planted by Gauss may yet bear the fruit of leading to a single electrodynamic equation from which all eletromagnetic phenomena can be derived, which we propose to call *Gaussian Electrodynamics*.

1. Pre-Gaussian Fields

Before the time of Gauss, three fields were well known: the gravitational field, the electrostatic field and the magnetostatic field. In the year of Gauss' birth (April 30, 1777), Lagrange [9] first showed that the gravitational field could be derived from a scalar potential φ. Using vector analysis, we would today write,

$$\mathcal{F} = -\text{grad } \varphi \tag{1}$$

where $\mathcal{F}$ is the force per unit mass in the gravitational field (the acceleration of gravity). In 1785, Laplace showed [10] that this scalar potential satisfied the equation which we now call Laplace's equation

$$\nabla^2\varphi = 0 \tag{2}$$

in any region of space which contained no masses. Thus, when the fourteen year old Gauss entered the Collegium Carolinum in Brunswick in February 1792, the scalar potential was already well established. From 1795 to 1798, Gauss studied at the University of Göttingen, then at the University of Helmstedt in 1798. His thesis was a masterpiece in which he proved the fundamental theorem of algebra for the first time.

The young Gauss must also have studied the gravitational equations of Newton and the mathematically identical equation for the electrostatic force which had been proposed [11] by Coulomb in 1785. For Gauss the future was shaped by the discovery of Ceres by Giuseppe Piazzi of Palermo on January 1, 1800. Gauss' work on calculating the orbit of Ceres led to his appointment in 1807 as director of the Göttingen Observatory.

It was not until 1813 that Poisson [12] showed that within a medium containing a distributed mass density δ, the differential equation satisfied by the gravitational potential was

$$\nabla^2\varphi = 4\pi G\delta, \tag{3}$$

which we now call Poisson's equation. Poisson also pointed out that the scalar potential φ could be employed in electrostatics and in 1824 extended [13] his work to magnetostatics (in a current free region).

The three vector fields, the gravitational, electrostatic and magnetostatic (in free space) could all be summarized by the same form of mathematical equation. The field vector $\mathcal{F}$ was in the radial direction and varied inversely as the square of the radius

$$\mathcal{F} = \mathbf{a}_r \frac{K}{r^2} \, . \tag{4}$$

Only the constant K must be defined differently for each field. For all three fields

$$\mathcal{F} = -\mathrm{grad} \ \varphi$$
$$\mathrm{curl} \ \mathcal{F} = 0 \tag{5}$$
$$\mathrm{div}\mathcal{F} = -\nabla^2 \varphi \, .$$

These equations could not yet be written in the concise vector notation which had yet to be invented, but their content was well known to Gauss. Since the curl of the field vectors vanished, the fields were conservative: the line integral

$$\oint_C \mathcal{F} \cdot \mathbf{ds} = 0 \tag{6}$$

about any closed path. And any of these three vector fields could be expressed as a gradient of a scalar potential.

2. The Electrodynamic Equation of Gauss

In 1820 the neat world in which the three known physical fields could be expressed by identical mathematics was shattered by Oersted's discovery [14] of an interreaction of electric and magnetic fields. That same year Ampère demonstrated that there was a force between two current carrying conductors. By 1823, Ampère had completed [15] his experimental work. Maxwell says [16],

"The experimental investigation by which Ampère established the laws of the mechanical action between electric currents is one of the most brilliant achievements in science.

The whole, theory and experiment, seems as if it had leaped full grown and full armed, from the brain of the 'Newton of electricity'. It is perfect in form, and unassailable in accuracy, and it is summed up in a formula from which all the phenomena may be deduced, and which must always remain the cardinal formula of electro-dynamics."

In vector notation, Ampère's equation for the force on current element $I_2\mathbf{ds}_2$ produced by current element $I_1\mathbf{ds}_1$, Fig. 1, can be written

$$d^2\mathbf{F}_A = -\frac{I_1 I_2}{4\pi\epsilon_0 c^2 r^2}\mathbf{a}_r \left[2(\mathbf{ds}_1 \cdot \mathbf{ds}_2) - 3(\mathbf{ds}_1 \cdot \mathbf{a}_r)(\mathbf{ds}_2 \cdot \mathbf{a}_r)\right] \tag{7}$$

where $\mathbf{ds}_1$ and $\mathbf{ds}_2$ are elements of distance in the direction tangential to the current carrying conductors and $\mathbf{a}_r$ is a unit vector directed from $\mathbf{ds}_1$ to $\mathbf{ds}_2$.

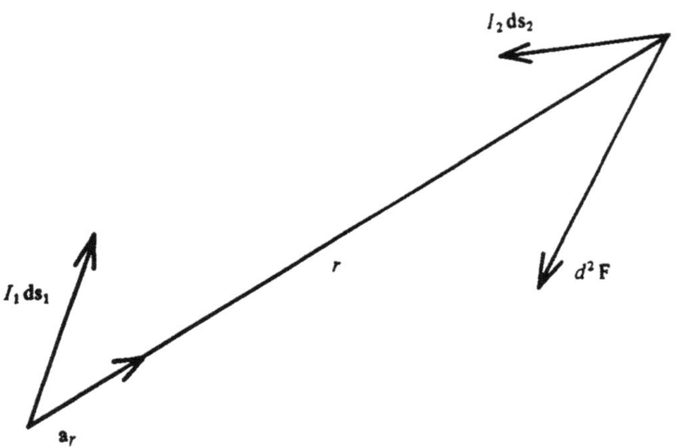

Fig. 1. The force between current elements.

When Gauss turned his attention to the study of Ampère's equation it was 1835. Twelve years had passed since the publication of Ampère's equation. Gauss was the first to set himself the task of deriving this equation from a fundamental equation for the force between moving charges. In this historical setting it was natural for Gauss to assume that the electrodynamic force, like the Coulomb force, must be a radial force, must vary inversely as the square of the distance and must be expressed as the gradient of a scalar potential.

If we assume that a vector field is expressed in terms of a scalar potential φ, Mirchandaney, Spencer, Uma and Mann [17] have shown that if

$$\varphi(x(t), y(t), z(t), t) = \int \frac{u(\xi(\tau), \eta(\tau), \zeta(\tau), \tau)dV}{(r)_R^n} \tag{8}$$

and

$$\mathcal{F} = -\text{grad } \varphi \tag{9}$$

then

$$\begin{cases} \text{curl } \mathcal{F} = -\text{curl grad } \varphi \equiv 0 \\ \text{div } \mathcal{F} = -\nabla^2\varphi, \end{cases} \tag{10}$$

$$\mathcal{F} = \int \left[\frac{nu(\tau)}{(r)_R^{n+1}} + \frac{1}{c(r)_R^n} \frac{du(\tau)}{d\tau} \right] \mathbf{a}_{r_R} dV , \tag{11}$$

$$\operatorname{div} \mathcal{F} = - n(n-1) \int \frac{u(\tau) dV}{(r)_R^{n+2}} - \frac{2(n-1)}{c} \int \frac{1}{(r)_R^{n+1}} \frac{du(\tau)}{d\tau} dV$$

$$- \frac{1}{c^2} \int \frac{1}{(r)_R^n} \frac{d^2 u(\tau)}{d\tau^2} dV , \tag{12}$$

where the force field $\mathcal{F}$ and the scalar potential φ are defined at time t but are produced by a potential which depends on a function u which may be defined at an earlier time $\tau, t > \tau$.

Gauss had no basis for suspecting that the effect might be retarded so for him, in 1835, $\tau = t$. He assumed that the force must vary inversely as the square of the distance and considered only the case of steady currents or constant velocities. So in Eq. (11), $n = 1$ and $\frac{du}{dt} = 0$. But Gauss had the great foresight to say that the scalar potential must be a function of the *relative velocity*.

Suppose that in the non-rotating coordinate system in which $\mathcal{F}$ is to be calculated, the charge Q moves with velocity $\mathbf{v}$, Fig. 2. The test charge is at point P and moves with velocity $\mathbf{u}$. Then the numerator of the expression for the potential must, according to Gauss be a function of the relative velocity $(\mathbf{v} - \mathbf{u})$. The scalar potential which Gauss used was in effect

$$\varphi_G = \frac{Q}{4\pi\epsilon_0 r} \left[1 + \frac{(\mathbf{v} - \mathbf{u}) \cdot (\mathbf{v} - \mathbf{u})}{c^2} - \frac{3}{2c^2} \{ (\mathbf{v} - \mathbf{u}) \cdot \mathbf{a}_r \}^2 \right] \tag{13}$$

and his equation for the force per unit charge was (in the mks system)

$$\mathcal{F}_G = \mathbf{a}_r \frac{Q}{4\pi\epsilon_0 r^2} \left[1 + \frac{(\mathbf{v} - \mathbf{u}) \cdot (\mathbf{v} - \mathbf{u})}{c^2} - \frac{3}{2c^2} \{ (\mathbf{v} - \mathbf{u}) \cdot \mathbf{a}_r \}^2 \right] . \tag{14}$$

Gauss did not publish his equation because he was not satisfied with it and he gave no details of how he arrived at its form.

However, we can apply Gauss' electrodynamic equation, to the derivation of Ampère's equation. In Gauss' day, it was known that the net charge on a current-carrying conductor was zero. But it was not known whether the positive or negative charge was in motion or whether both were in motion.

In order to derive Ampère's equation for the force between current elements, Eq. (7), from Gauss' equation, Eq. (14), let us proceed as he could have. Not knowing whether positive or negative charge or both are

                    D. E. Spencer & S. Y. Uma

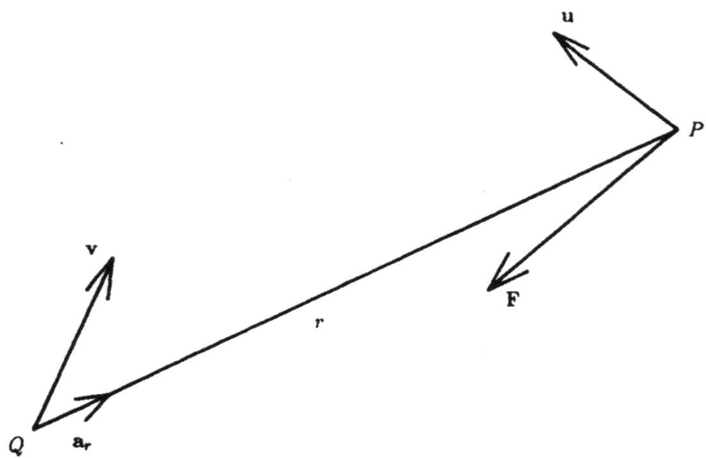

Fig. 2. Force $\mathcal{F}$ on a test charge at point P.

in motion, he could have assigned arbitrary velocities to both positive and negative charges. In $\mathbf{ds}_1$ assume that the velocity of the positive charge is $\mathbf{v}_+$ and the velocity of the negative charge is $\mathbf{v}_-$. In $\mathbf{ds}_2$, the corresponding velocities are $\mathbf{u}_+$ and $\mathbf{u}_-$. Then, for the force between positive charges in Eq. (14), $\mathbf{v} = \mathbf{v}_+$, and $\mathbf{u} = \mathbf{u}_+$ and

$$\mathbf{F}_{++} = \mathbf{a}_r \frac{|Q_1 Q_2|}{4\pi\epsilon_0 r^2}\left[1 + \frac{(\mathbf{v}_+ - \mathbf{u}_+)\cdot(\mathbf{v}_+ - \mathbf{u}_+)}{c^2} - \frac{3}{2c^2}\{(\mathbf{v}_+ - \mathbf{u}_+)\cdot\mathbf{a}_r\}^2\right].$$

$$(15_{++})$$

Similarly, for the positive charge in $\mathbf{ds}_1$ and the negative charge in $\mathbf{ds}_2$, $\mathbf{v} = \mathbf{v}_+$, and $\mathbf{u} = \mathbf{u}_-$ and

$$\mathbf{F}_{+-} = -\mathbf{a}_r \frac{|Q_1 Q_2|}{4\pi\epsilon_0 r^2}\left[1 + \frac{(\mathbf{v}_+ - \mathbf{u}_-)\cdot(\mathbf{v}_+ - \mathbf{u}_-)}{c^2} - \frac{3}{2c^2}\{(\mathbf{v}_+ - \mathbf{u}_-)\cdot\mathbf{a}_r\}^2\right].$$

$$(15_{+-})$$

Also for the negative charge in $\mathbf{ds}_1$ and the positive charge in $\mathbf{ds}_2$, $\mathbf{v} = \mathbf{v}_-$, and $\mathbf{u} = \mathbf{u}_+$ and

$$\mathbf{F}_{-+} = -\mathbf{a}_r \frac{|Q_1 Q_2|}{4\pi\epsilon_0 r^2}\left[1 + \frac{(\mathbf{v}_- - \mathbf{u}_+)\cdot(\mathbf{v}_- - \mathbf{u}_+)}{c^2} - \frac{3}{2c^2}\{(\mathbf{v}_- - \mathbf{u}_+)\cdot\mathbf{a}_r\}^2\right]$$

$$(15_{-+})$$

and for the force between the negative charge in $\mathbf{ds}_1$ and the negative charge in $\mathbf{ds}_2, \mathbf{v} = \mathbf{v}_-$, and $\mathbf{u} = \mathbf{u}_-$ and

$$\mathbf{F}_{--} = \mathbf{a}_r \frac{|Q_1 Q_2|}{4\pi\epsilon_0 r^2} \left[1 + \frac{(\mathbf{v}_- - \mathbf{u}_-) \cdot (\mathbf{v}_- - \mathbf{u}_-)}{c^2} - \frac{3}{2c^2} \{(\mathbf{v}_- - \mathbf{u}_-) \cdot \mathbf{a}_r\}^2 \right].$$

$$(15_{--})$$

The total force between current elements derived from Gauss equation is

$$\mathbf{F}_G = \mathbf{F}_{++} + \mathbf{F}_{+-} + \mathbf{F}_{-+} + \mathbf{F}_{--}. \tag{16}$$

Substitution of Eq. (15) into Eq. (16) gives

$$\mathbf{F}_G = \mathbf{a}_r \frac{|Q_1 Q_2|}{4\pi\epsilon_0 c^2 r^2} \left[-2(\mathbf{v}_- - \mathbf{v}_+) \cdot (\mathbf{u}_- - \mathbf{u}_+) \right.$$
$$\left. + 3((\mathbf{v}_- - \mathbf{v}_+) \cdot \mathbf{a}_r)((\mathbf{u}_- - \mathbf{u}_+) \cdot \mathbf{a}_r) \right]. \tag{17}$$

Let Q_e be the electronic charge. Since the conductor is uncharged, the number of positive and negative charges in each current element must be equal. Suppose that N_1 and N_2 are the number of free electrons per unit volume in current elements $\mathbf{ds}_1$ and $\mathbf{ds}_2$ and cross sectional areas of the conductors are A_1 and A_2. Then

$$|Q_1| = N_1 A_1 |Q_e| |\mathbf{ds}_1|, \quad |Q_2| = N_2 A_2 |Q_e| |\mathbf{ds}_2|. \tag{18}$$

It is customary to express the Ampère equation in terms of $I_1, I_2, \mathbf{ds}_1$ and $\mathbf{ds}_2$. The current density in each conductor is

$$\mathbf{J}_1 = \rho_{1+}\mathbf{v}_+ - \rho_{1-}\mathbf{v}_-, \quad \mathbf{J}_2 = \rho_{2+}\mathbf{u}_+ - \rho_{2-}\mathbf{u}_-. \tag{19}$$

But the charge densities are

$$|\rho_{1+}| = |\rho_{1-}| = N_1|Q_e|, \quad |\rho_{2+}| = |\rho_{2-}| = N_2|Q_e|. \tag{20}$$

Substitution of Eq. (20) into Eq. (19) gives

$$\mathbf{J}_1 = N_1|Q_e|(\mathbf{v}_+ - \mathbf{v}_-), \quad \mathbf{J}_2 = N_2|Q_e|(\mathbf{u}_+ - \mathbf{u}_-). \tag{21}$$

Therefore, the current in each conductor is

$$I_1 = |\mathbf{J}_1|A_1 = N_1|Q_e|A_1|\mathbf{v}_+ - \mathbf{v}_-|,$$
$$I_2 = |\mathbf{J}_2|A_2 = N_2|Q_e|A_2|\mathbf{u}_+ - \mathbf{u}_-|. \tag{22}$$

Multiplication of Eq. (22) by the appropriate element of distance gives, by Eq. (18)

$$I_1|\mathbf{ds_1}| = N_1 A_1 |Q_e| \, |\mathbf{ds_1}| \, |\mathbf{v_+} - \mathbf{v_-}| = |Q_e| \, |\mathbf{v_+} - \mathbf{v_-}|,$$
$$I_2|\mathbf{ds_2}| = N_2 A_2 |Q_e| \, |\mathbf{ds_2}| \, |\mathbf{u_+} - \mathbf{u_-}| = |Q_e| \, |\mathbf{u_+} - \mathbf{u_-}|.$$

$$(23)$$

Therefore, Eq. (17) becomes identical to Eq. (7)

$$d^2\mathbf{F}_G = \mathbf{a}_r \frac{I_1 I_2}{4\pi\epsilon_0 c^2 r^2}\left[-2\mathbf{ds_1}\cdot\mathbf{ds_2} + 3(\mathbf{ds_1}\cdot\mathbf{a}_r)(\mathbf{ds_2}\cdot\mathbf{a}_r)\right]$$
$$= d^2\mathbf{F}_A.$$

$$(24)$$

Therefore, the Ampère equation for the force between current elements can be derived from the Gauss equation for all possible theories of electronic conduction: for the modern theory in which $\mathbf{v_+} = \mathbf{u_+} = 0$ and for the Fechner theory, widely believed in the nineteenth century, in which $\mathbf{v_+} = -\mathbf{v_-}$ and $\mathbf{u_+} = -\mathbf{u_-}$.

Thus, we can draw two important conclusions about the Gauss equation:

1. The Gauss electrodynamic equation can be derived from a scalar potential φ, if the relative velocity is constant.
2. The Gauss electrodynamic equation permits the derivation of the Ampère equation for the force between current elements.

3. The Electrodynamic Equation of Weber

Even though Gauss did not publish his electrodynamic equation in 1835, it is certain that Wilhelm Weber knew this result and had discussed it with Gauss. For in 1833 Gauss and Weber collaborated on the invention of the electric telegraph at Göttingen. It was not until 1837 that Weber left Göttingen for Leipzig. On March 19, 1845 Gauss [18] wrote to Weber and referred to the electrodynamic speculations with which he had been occupied long before. He said that he would have published his electrodynamic equation if he could have established what he considered the keystone of electrodynamics: the deduction of the force acting between electric particles in motion from the consideration of an action between them, not instantaneous, but propagated in time, in a similar manner to that of light. Thus, it appears that Gauss was the first to suspect that the potential should be retarded.

However, the honor of publishing the first electrodynamic equation must be attributed to Wilhelm Weber [2]. Like the Gauss equation, the Weber equation assumes a force directed along the radial line joining source and receiver. Weber also includes an acceleration term which varies inversely with the distance r. In the mks system, the Weber equation for the force per unit charge can be written as

$$\mathcal{F}_W = \mathbf{a}_r \frac{Q}{4\pi\epsilon_0 r^2}\left[1 - \frac{1}{2c^2}\left(\frac{dr}{dt}\right)^2 + \frac{r}{c^2}\frac{d^2r}{dt^2}\right]. \tag{25}$$

In order to write this equation in a form similar to that given for the Gauss equation, we must express the derivatives of r in terms of $\mathbf{u}, \mathbf{v}$ and $\mathbf{a}_r$, Fig. 3. The source charge Q is at point (ξ, η, ζ, t) and the receiver at $P(x, y, z, t)$. The velocities of source and receiver are

$$\mathbf{v} = \frac{d\xi}{dt}\mathbf{a}_x + \frac{d\eta}{dt}\mathbf{a}_y + \frac{d\zeta}{dt}\mathbf{a}_z \,,$$

$$\mathbf{u} = \frac{dx}{dt}\mathbf{a}_x + \frac{dy}{dt}\mathbf{a}_y + \frac{dz}{dt}\mathbf{a}_z \,. \tag{26}$$

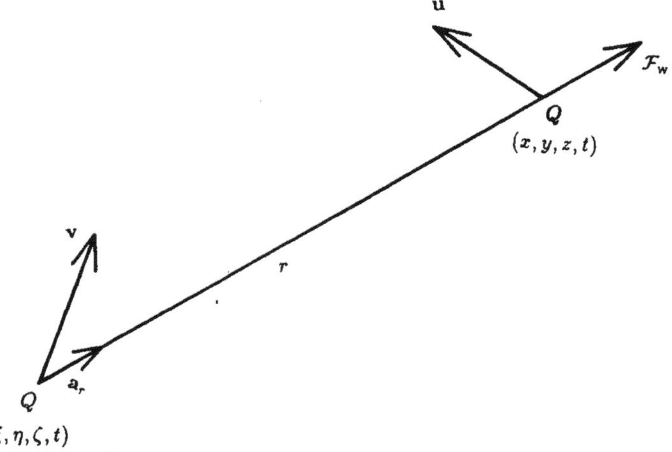

Fig. 3. The Weber force $\mathcal{F}_W$ on the test charge at point P produced by charge Q.

The distance r from source to receiver is

$$r = \left[(\xi - x)^2 + (\eta - y)^2 + (\zeta - z)^2\right]^{1/2}. \tag{27}$$

Differentiating

$$\frac{dr}{dt} = \frac{\xi - x}{r}\left[\frac{d\xi}{dt} - \frac{dx}{dt}\right] + \frac{\eta - y}{r}\left[\frac{d\eta}{dt} - \frac{dy}{dt}\right] + \frac{\zeta - z}{r}\left[\frac{d\zeta}{dt} - \frac{dz}{dt}\right]. \tag{28}$$

But the unit vector $\mathbf{a}_r$ is

$$\mathbf{a}_r = \frac{x - \xi}{r}\mathbf{a}_x + \frac{y - \eta}{r}\mathbf{a}_y + \frac{z - \zeta}{r}\mathbf{a}_z. \tag{29}$$

Substituting Eqs. (26) and (29) into Eq. (28),

$$\frac{dr}{dt} = -(\mathbf{v} - \mathbf{u}) \cdot \mathbf{a}_r. \tag{30}$$

Differentiation of Eq. (28) gives

$$\begin{aligned}
\frac{d^2r}{dt^2} =& \frac{1}{r}\left[\left[\frac{d\xi}{dt} - \frac{dx}{dt}\right]^2 + \left[\frac{d\eta}{dt} - \frac{dy}{dt}\right]^2 + \left[\frac{d\zeta}{dt} - \frac{dz}{dt}\right]^2\right] \\
&+ \frac{1}{r}\left[(\xi - x)\left[\frac{d^2\xi}{dt^2} - \frac{d^2x}{dt^2}\right] + (\eta - y)\left[\frac{d^2\eta}{dt^2} - \frac{d^2y}{dt^2}\right]\right. \\
&+ \left.(\zeta - z)\left[\frac{d^2\zeta}{dt^2} - \frac{d^2z}{dt^2}\right]\right] - \frac{1}{r^2}\left[(\xi - x)\left[\frac{d\xi}{dt} - \frac{dx}{dt}\right]\right. \\
&+ (\eta - y)\left[\frac{d\eta}{dt} - \frac{dy}{dt}\right] + \left.(\zeta - z)\left[\frac{d\zeta}{dt} - \frac{dz}{dt}\right]\right]\frac{dr}{dt}.
\end{aligned} \tag{31}$$

Substitution of Eqs. (26), (29) and (30) into Eq. (31) gives

$$\frac{d^2r}{dt^2} = \frac{(\mathbf{v} - \mathbf{u}) \cdot (\mathbf{v} - \mathbf{u})}{r} - \left[\frac{d\mathbf{v}}{dt} - \frac{d\mathbf{u}}{dt}\right] \cdot \mathbf{a}_r - \frac{\{(\mathbf{v} - \mathbf{u}) \cdot \mathbf{a}_r\}^2}{r}. \tag{32}$$

Substitution of Eqs. (30) and (32) into Eq. (25) gives

$$\begin{aligned}
\mathcal{F}_W =& \mathbf{a}_r \frac{Q}{4\pi\epsilon_0 r^2}\left[1 + \frac{(\mathbf{v} - \mathbf{u}) \cdot (\mathbf{v} - \mathbf{u})}{c^2} - \frac{3\{(\mathbf{v} - \mathbf{u}) \cdot \mathbf{a}_r\}^2}{2c^2}\right] \\
&- \mathbf{a}_r \frac{Q}{4\pi\epsilon_0 c^2 r}\left[\frac{d\mathbf{v}}{dt} - \frac{d\mathbf{u}}{dt}\right] \cdot \mathbf{a}_r.
\end{aligned} \tag{33}$$

If the current is constant, this equation reduces to the Gauss equation, Eq. (14).

The Weber equation can be expressed as

$$\mathcal{F}_W = -\text{grad } \varphi_W$$

if

$$\varphi_W = \frac{Q}{4\pi\epsilon_0 r}\left[1 + \frac{1}{2c^2}[(\mathbf{v} - \mathbf{u}) \cdot \mathbf{a}_r]^2\right]. \tag{34}$$

Thus, the Weber equation like the Gauss equation can be derived from a scalar potential and permits the derivation of the Ampère equation for the force between current elements. In fact the Weber equation is merely a generalization of the Gauss equation to accelerated charges. However, Weber was willing to publish this equation without waiting to solve the problem of retardation.

The Weber equation is still regarded by a number of scientists as a viable description of electrodynamics. Recent papers based on this equation have been published by Wesley [19], Pappas [20], Phipps [21] and Graneau [22]. An important paper extending the use of this form of equation to gravitation has been published by Assis [23].

Another Göttingen professor, Bernard Riemann, suggested a modified form of the electrodynamic equation in a course taught in the summer of 1861. The Riemann electrodynamic equation was notable because it is the first equation in which the force is not in the radial direction. Riemann [3] has components in the radial direction, in the direction of the relative velocity and in the direction of the acceleration. Riemann was also the first to attempt to include retardation. However, the Riemann equation was not published during his lifetime. As no one attempts to use the Riemann electrodynamic equation today, we will not consider it further in this paper.

4. The Vector Potential

The necessity for describing the electromagnetic field in terms of both a scalar and a vector potential does not follow from the experiments of Ampère. The experimental basis for the introduction of the vector potential lies in the work of Faraday.

In 1791, Michael Faraday was born a blacksmith's son in a family of ten children. At fourteen he was apprenticed to a bookbinder. His only education came from reading the books which he bound, particularly the *Encyclopedia Britannica*. At twenty one, he had attended public lectures given by Sir Humphrey Davy. Finally he dared to apply for a position as an assistant in Davy's laboratory in the Royal Institution. Thus began a career marked by brilliant and thorough experiments but totally innocent of mathematics.

During the years from 1813 to 1821, Faraday worked with Davy in chemistry. News of Ampère's experiments in Paris reached Farady in Lon-

don and aroused Faraday's interest in electromagnetism. His first paper in
this field [24] appeared in 1821. It was called a "Historical sketch of electromagnetism". Faraday not only studied the history, but painstakingly
repeated every experiment. When Davy died in 1829, the 38 year old Faraday succeeded him as director of the Royal Institution. Ten years after his
first electromagnetic paper, in 1831, at the age of 40 Faraday was ready to
publish his monumental paper [25] on induction. He set himself the task of
determining the laws of induction of currents. For this purpose he developed the concept of lines of magnetic force filling all space. He proved that
currents could be induced in three ways: by varying the strength of current in a nearby conductor, by bringing a magnet near the circuit, and by
moving a ciruit in the presence of another current or a magnet. In 1832 he
proved that induction depends on relative motion. In 1837 he studied the
difference between insulators and conductors and in 1838 he extended his
theory of induction to dielectrics. From 1846 to 1851 Faraday studied the
interaction of magnetism and light and speculated on the electromagnetic
theory of light. His last paper was published in 1855 at the age of 64.

In 1845 Franz Neumann [26] was the first to introduce the vector potential $\mathbf{A}$. In his memoir Neumann attempted a mathematical deduction of
the laws of induction by utilizing Ampère's forces between current elements.
Today we would write his conclusion by saying that

$$\mathbf{B} = \text{curl } \mathbf{A}$$
$$\mathbf{E} = -\text{grad } \varphi - \frac{\partial \mathbf{A}}{\partial t}. \tag{35}$$

In 1847, Thomson [27] independently came to the conclusion that the
magnetic field must be expressed as the curl of a vector potential $\mathbf{A}$.

Maxwell was born in Scotland, in 1831, in the year in which Faraday's
great paper on induction was published. Unlike Faraday, Maxwell received
a thorough education in both mathematics and physics at Edinburgh and
at Cambridge. At the age of 24, in 1855, in the year in which Faraday published his last paper, Maxwell published his first paper [28]. He attempted
to find a mathematical description of Faraday's great experimental work
on induction using Thomson's mathematics. In modern notation his 1855
paper utilized the definition of $\mathbf{B}$ and $\mathbf{E}$ in terms of scalar and vector potentials as in Eq. (35) and concluded that

$$\text{div } \mathbf{A} = 0, \quad \text{div } \mathbf{B} = 0, \quad \text{curl } \mathbf{H} = \mathbf{J}.$$

From 1856 to 1860, Maxwell was Professor of Natural Philosophy at Marischal College in Aberdeen, Scotland. During the years 1860–1865 when Maxwell was teaching at King's College in London, Faraday was living in retirement at Hampton Court and there was the opportunity for personal contact between the grand experimentalist and the gifted interpreter. By 1858, Maxwell wrote that Faraday had "the nucleus of everything electric since 1830". In 1861 Maxwell devised a mechanical conception of the electromagnetic field involving interlocking vortices and had concluded [29] that

$$\text{curl } \mathbf{E} = -\frac{\partial \mathbf{B}}{\partial t}.$$

By 1864 in a paper [30] "A dynamical theory of the electromagnetic field" read to the Royal Society, Maxwell had concluded that

$$\text{div } \mathbf{D} = \rho, \quad \text{div } \mathbf{J} = -\frac{\partial \rho}{\partial t}, \quad \text{curl } \mathbf{H} = \mathbf{J} + \frac{\partial \mathbf{D}}{\partial t}.$$

By 1868, Maxwell was proposing [31] that the potentials φ and $\mathbf{A}$ had no physical significance and could be abandoned. Only the equations relating the electric and magnetic fields pictured by Faraday as lines filling all of space were needed:

$$
\begin{aligned}
\text{div } \mathbf{B} &= 0 & \text{curl } \mathbf{E} &= -\frac{\partial \mathbf{B}}{\partial t} \\
\text{div } \mathbf{D} &= \rho & \text{curl } \mathbf{H} &= \mathbf{J} + \frac{\partial \mathbf{D}}{\partial t}.
\end{aligned}
\tag{36}
$$

In 1865 at the age of 34, Maxwell resigned from King's College and retired to his family estate, Glenlair, to write a connected account of his electromagnetic theory. In 1871, he was invited to Cambridge as Professor of Experimental Physics. His great *Treatise of Electricity and Magnetism* was published [32] in 1873. While working on the revision for the second edition, Maxwell died in 1879.

The Maxwell equations were developed as a description of Faraday's experiments. Yet Maxwell was not unaware of the work of Gauss and Weber, for he devoted the last chapter of his book to a beautiful exposition of both theories. He clearly states that the Ampère force is produced by the "relative velocity of the electricity in the two elements". Had Maxwell lived longer he might well have found an equation for the force between moving charges which could be reconciled with Faraday's experiments.

5. Retarded Scalar and Vector Potentials

The propagation of electrical effects through space occurs at a finite velocity. This was first taken into account by Ludwig Lorenz [33] of Copenhagen in 1867. He followed the qualitative ideas of Gauss and Riemann, that the electric and magnetic fields were delayed effects of the source. Lorenz was the first to formulate the idea of retardation in a mathematical equation. He proposed that electromagnetic fields could be derived from scalar and vector potentials $\varphi(t)$ and $\mathbf{A}(t)$ which were defined at time t in terms of the charge density $\rho(\tau)$ and $\mathbf{J}(\tau)$ at an earlier time τ:

$$\varphi(t) = \frac{1}{4\pi\epsilon_0} \int \frac{\rho(\tau)}{r} d\mathcal{V}$$
$$\mathbf{A}(t) = \frac{\mu_0}{4\pi} \int \frac{\mathbf{J}(\tau)}{r} d\mathcal{V} \tag{37}$$

where $\tau = t - r/c$. Using these retarded potentials and the definitions

$$\mathbf{B} = \text{curl } \mathbf{A}$$
$$\mathbf{E} = -\text{grad } \varphi - \frac{\partial \mathbf{A}}{\partial t} \tag{35}$$

Lorenz independently derived the major equations

$$\text{div } \mathbf{A} = -\frac{1}{c^2} \frac{\partial \varphi}{\partial t}$$

$$\text{div } \mathbf{B} = 0 \qquad \text{curl } \mathbf{E} = -\frac{\partial \mathbf{B}}{\partial t} \tag{38}$$

$$\text{div } \mathbf{D} = \rho \qquad \text{curl } \mathbf{H} = \mathbf{J} + \frac{\partial \mathbf{D}}{\partial t}$$

from Eqs. (35) and (37). A short paragraph [32] is devoted to Lorenz's 1867 paper at the end of Maxwell's Chapter XX, but the importance of the retarded potentials is not fully recognized.

6. The Force on a Moving Charge

It is remarkable that all of the beautiful work of Maxwell and Lorenz was done without any attempt to relate the electric and magnetic fields to the force on a moving charge. In 1889, Heaviside [34] suggested in effect that the force per unit charge $\mathcal{F}$

$$\mathcal{F} = \mathbf{E} + \mathbf{v} \times \mathbf{B} \tag{39}$$

where $\mathbf{v}$ is the velocity of the test charge in the coordinate system in which $\mathcal{F}$ is defined.

In 1892 Hendrik Antoon Lorenz [35] in Leiden returned in part to the fundamental idea of Gauss by attempting to describe all electromagnetic theory in terms of the motion of electrons. Lorenz employed retarded potentials

$$\varphi = \frac{Q}{4\pi\epsilon_0 r} \quad \text{and} \quad \mathbf{A}(t) = \frac{\mu_0}{4\pi}\frac{Q\mathbf{v}(\tau)}{r} \tag{40}$$

but did not use the relative velocity recommended by Gauss. From these potentials, he was able to derive Maxwell's equations for the field of any moving electron. Lorenz also derived the Heaviside equation for the force on a moving charge by means of the Lagrangian.

In a recent paper [17], Mirchandaney, Spencer, Uma and Mann have attempted to place the subject of electrodynamics on a firm postulational basis. This work is an extension of the postulational development suggested by Moon and Spencer [36].

The electric and magnetic field vectors $\mathbf{E}$ and $\mathbf{B}$ are *defined* in terms of scalar and vector potentials by

$$\mathbf{B} = \operatorname{curl} \mathbf{A}$$
$$\mathbf{E} = -\operatorname{grad} \varphi - \frac{\partial \mathbf{A}}{\partial t}. \tag{35}$$

Moon and Spencer [36] have shown that two of the Maxwell equations follow from these definitions as a direct consequence of vector identities:

$$\operatorname{div} \mathbf{B} = 0$$
$$\operatorname{curl} \mathbf{E} = -\frac{\partial \mathbf{B}}{\partial t}. \tag{36}$$

The force equation of Heaviside is employed to define the force per unit charge $\mathcal{F}$,

$$\mathcal{F} = \mathbf{E} + \mathbf{w} \times \mathbf{B} \tag{41}$$

but with the generalization that $\mathbf{w}$ may be either an absolute or a relative velocity. Note that the field vectors depend on the definition of three quantities $\varphi, \mathbf{A}$ and $\mathbf{w}$. The remaining Maxwell equations can always [36] be written as

$$\operatorname{div} \mathbf{E} = -\left[\nabla^2\varphi - \frac{1}{c^2}\frac{\partial^2\varphi}{\partial t^2}\right] - \frac{\partial}{\partial t}\left[\operatorname{div} \mathbf{A} + \frac{1}{c^2}\frac{\partial\varphi}{\partial t}\right]$$
$$\operatorname{curl} \mathbf{H} - \frac{\partial \mathbf{D}}{\partial t} = -\frac{1}{\mu}\left[\mathbf{A} - \frac{1}{c^2}\frac{\partial^2\mathbf{A}}{\partial t^2}\right] + \frac{1}{\mu}\operatorname{grad}\left[\operatorname{div} \mathbf{A} + \frac{1}{c^2}\frac{\partial\varphi}{\partial t}\right]. \tag{42}$$

The scalar and vector potentials are defined as

$$\varphi(t) = \int \frac{u(\tau)}{r} d\mathcal{V} ,$$
$$\mathbf{A}(t) = \int \frac{\mathbf{U}(\tau,t)}{r} d\mathcal{V} . \tag{43}$$

Thus, the entire set of Maxwell equations and the force equation depend on four postulates:

    (1) the scalar potential φ,

    (2) the vector potential $\mathbf{A}$,

    (3) the velocity $\mathbf{w}$,

    (4) the relation between τ and t (or the postulate on the velocity of light).

Two postulates for the scalar potential are

Postulate φ_1
$$\varphi(t) = \frac{Q}{4\pi\epsilon_0 (r)_R} \tag{44_{φ_1}}$$

where $(r)_R$ is a radius defined differently for each postulate on the velocity of light.

Postulate φ_2
$$\varphi(t) = \frac{Q}{4\pi\epsilon_0 (r)_R} \frac{1}{\left[1 - \frac{\mathbf{v}(\tau) \cdot \mathbf{a}_{r_R}}{c}\right]} \tag{44_{φ_2}}$$

where $\mathbf{v}(\tau)$ is the velocity of the source at time τ and $\mathbf{a}_{r_R}$ is a vector pointing from source to receiver along the radius $(r)_R$.

The first of these is the classical scalar potential used by Maxwell, Lorenz and Lorenz. The second is a postulate suggested by Liènard [37] and Wiechert [38].

Five possible definitions of the retarded vector potential $\mathbf{A}$ have been considered:

Postulate A_1
$$\mathbf{A}(t) = \frac{Q\mathbf{v}(\tau)}{4\pi\epsilon_0 c^2 (r)_R} \tag{$45\mathrm{A}_1$}$$

where $\mathbf{v}$ is the velocity of the source at time τ in the coordinate system in which $\mathbf{A}$ is defined at a later time t.

$$\textit{Postulate A}_2 \qquad \mathbf{A}(t) = \frac{Q\mathbf{v}(\tau)}{4\pi\epsilon_0 c^2(r)_R\left[1 - \frac{\mathbf{v}(\tau)\cdot\mathbf{a}_{r_R}}{c}\right]} \qquad (45\text{A}_2)$$

suggested by Liènard [37] and Wiechert [38] and recently studied by Whitney [39].

In an attempt to incorporate the basic idea that the vector potential should depend on relative velocity rather than absolute velocity, as suggested by Gauss, three additional postulates for the vector potential [8] have been considered:

$$\textit{Postulate A}_3 \qquad \mathbf{A}(t) = \frac{Q[\mathbf{v}(\tau) - \mathbf{u}(t)]}{8\pi\epsilon_0 c^2(r)_R}, \qquad (45\text{A}_3)$$

where $\mathbf{u}$ is the velocity of the receptor.

$$\textit{Postulate A}_4 \qquad \mathbf{A}(t) = \frac{Q[\mathbf{v}(t) - \mathbf{u}(t)]}{8\pi\epsilon_0 c^2(r)_R}, \qquad (45\text{A}_4)$$

$$\textit{Postulate A}_5 \qquad \mathbf{A}(t) = \frac{Q[\mathbf{v}(\tau) - \mathbf{u}(\tau)]}{8\pi\epsilon_0 c^2(r)_R}. \qquad (45\text{A}_5)$$

The postulate on the velocity $\mathbf{w}$ which was first suggested by Heaviside [34] in 1889 and was subsequently derived independently by Lorenz [35] in 1892 is

$$\textit{Postulate w}_1 \qquad \mathbf{w} = \mathbf{u}(t). \qquad (46\text{w}_1)$$

But if we are to follow Gauss in always employing relative velocity, it is necessary to consider three other possibilities:

$$\textit{Postulate w}_2 \qquad \mathbf{w} = \mathbf{u}(t) - \mathbf{v}(\tau), \qquad (46\text{w}_2)$$

$$\textit{Postulate w}_3 \qquad \mathbf{w} = \mathbf{u}(t) - \mathbf{v}(t), \qquad (46\text{w}_3)$$

$$\textit{Postulate w}_4 \qquad \mathbf{w} = \mathbf{u}(\tau) - \mathbf{v}(\tau). \qquad (46\text{w}_4)$$

Three postulates on the velocity of light have been considered. Moon, Spencer and Moon [40] have shown that all of the postulates on the velocity of light can be expressed in the same mathematical form

$$(r)_R = c(t - \tau) \tag{47}$$

if $(r)_R$ is suitably defined, Fig. 4.

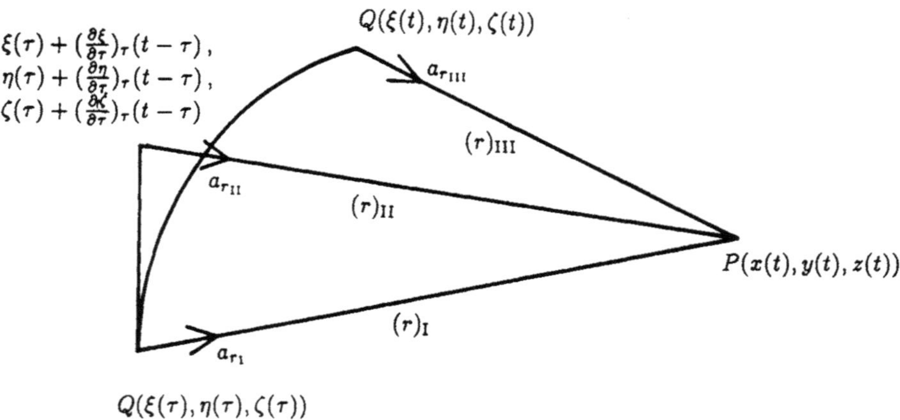

Fig. 4. The distances $(r)_I$, $(r)_{II}$ and $(r)_{III}$ which occur in three postulates on the velocity of light.

The oldest postulate on the velocity of light was suggested by Einstein [41] in 1905. It has recently been shown by Moon, Spencer and Moon [42] that in order to predict the Michelson-Gale [43] fringe shift, the 1905 postulate of Einstein must be modified to the form suggested by Einstein [44] in 1907:

Postulate I*. The velocity of light in free space is always a constant c irrespective of the motion of source or receiver in any coordinate system which is not in rotation.

$$(r)_I = c(t - \tau) \tag{48I}$$

where

$$(r)_I = \left[[x(t) - \xi(\tau)]^2 + [y(t) - \eta(\tau)]^2 + [z(t) - \zeta(\tau)]^2 \right]^{1/2}. \tag{49I}$$

Here as shown in Fig. 4, $(r)_\mathrm{I}$ is the distance from where the source was at time τ to where the receiver is at time t.

The next postulate on the velocity of light was suggested [4] by the Swiss physicist Walter Ritz working at Göttingen in 1908.

Postulate II. The velocity of light is a constant c with respect to the point to which the source would have moved at the instant of reception t if it had continued to move at the velocity of emission $\frac{d\xi(\tau)}{d\tau}, \frac{d\eta(\tau)}{d\tau}, \frac{d\zeta(\tau)}{d\tau}$,

$$(r)_\mathrm{II} = c(t - \tau) \tag{48II}$$

where

$$\begin{aligned}
(r)_\mathrm{II} = \Bigg[&\left[x(t) - \xi(\tau) - \frac{d\xi(\tau)}{d\tau}(t - \tau) \right]^2 \\
&+ \left[y(t) - \eta(\tau) - \frac{d\eta(\tau)}{d\tau}(t - \tau) \right]^2 \\
&+ \left[z(t) - \zeta(\tau) - \frac{d\zeta(\tau)}{d\tau}(t - \tau) \right]^2 \Bigg]^{1/2} .
\end{aligned} \tag{(49II)}$$

The third postulate was suggested by Moon and Spencer [45] in 1956 in order to retain the concept of universal time. It has recently been modified by Moon, Spencer and Moon [42] in order to predict the Michelson-Gale [43] fringe shift:

Postulate III*. In any coordinate system which is not moving with respect to the source and is not in rotation, the velocity of light in free space is a constant c:

$$(r)_\mathrm{III} = c(t - \tau) \tag{48III}$$

where

$$(r)_\mathrm{III} = \left[[x(t) - \xi(t)]^2 + [y(t) - \eta(t)]^2 + [z(t) - \zeta(t)]^2 \right]^{1/2} . \tag{49III}$$

Here $(r)_\mathrm{III}$ is the distance between source and receiver at the instant of reception, Fig. 4.

Note that there are coordinate systems in which the velocity of light is not a constant according to all three postulates. In a coordinate system attached to the laboratory, the velocity of light is not quite constant according to both Postulates I* and III*.

If we consider all possible combinations of these postulates, we have 30 possible equations for $\mathbf{E}$, 15 equations for $\mathbf{B}$ and 120 possible equations for $\mathcal{F}$.

However, in the very important special case in which velocities are low and retardation has a negligible effect,

$$\frac{u}{c} \ll 1, \quad \frac{v}{c} \ll 1 \quad \tau \approx t, \quad \frac{\partial \tau}{\partial t} \approx 1, \tag{50}$$

all of the 120 equations for $\mathcal{F}$ reduce to two. The first is the classical equation of Lorenz in which

$$\varphi = \frac{Q}{4\pi\epsilon_0 r}, \quad \mathbf{A}_C(t) = \frac{Q\mathbf{v}(t)}{4\pi\epsilon_0 c^2 r}, \quad \mathbf{w}_C = \mathbf{u}(t). \tag{51C}$$

Here the velocities $\mathbf{v}(t)$ and $\mathbf{u}(t)$ are velocities of source and receiver relative to the coordinate system in which the force $\mathcal{F}$ is defined. Introduction of Eq. (51C) into Eqs. (35) and (41) gives the classical expression

$$\begin{aligned}
\mathcal{F}_C =& \frac{Q}{4\pi\epsilon_0 (r)^2}\left[\mathbf{a}_r + \frac{\mathbf{u}(t) \times (\mathbf{v}(t) \times \mathbf{a}_r)}{c^2}\right] \\
&- \frac{Q}{4\pi\epsilon_0 c^2 (r)}\left[\frac{d\mathbf{v}(t)}{dt} - \frac{\mathbf{u}(t)}{c} \times \left[\frac{d\mathbf{v}(t)}{dt} \times \mathbf{a}_r\right]\right].
\end{aligned} \tag{52C}$$

However, if we follow Gauss, Weber and Riemann in employing relative velocities the expressions for $\varphi, \mathbf{A}$ and $\mathbf{w}$ become,

$$\varphi = \frac{Q}{4\pi\epsilon_0 r}, \quad \mathbf{A}_R(t) = \frac{Q(\mathbf{v}(t) - \mathbf{u}(t))}{8\pi\epsilon_0 c^2 r}, \quad \mathbf{w}_R = \mathbf{u}(t) - \mathbf{v}(t). \tag{51R}$$

The relative velocity form of the electrodynamic force equation is

$$\begin{aligned}
\mathcal{F}_R =& \frac{Q}{4\pi\epsilon_0 (r)^2}\left[\mathbf{a}_r - \frac{(\mathbf{v}(t) - \mathbf{u}(t)) \times ((\mathbf{v}(t) - \mathbf{u}(t)) \times \mathbf{a}_r)}{2c^2}\right] \\
&- \frac{Q}{8\pi\epsilon_0 c^2 (r)}\left[\frac{d\mathbf{v}(t)}{dt} - \frac{d\mathbf{u}(t)}{dt}\right. \\
&\left. + \frac{(\mathbf{v}(t) - \mathbf{u}(t))}{c} \times \left[\left[\frac{d\mathbf{v}(t)}{dt} - \frac{d\mathbf{u}(t)}{dt}\right] \times \mathbf{a}_r\right]\right].
\end{aligned} \tag{52R}$$

7. The Force Between Current Elements

Now let us apply the classical and relative velocity forms of the electrodynamic force equations to the calculation of the force between current elements. Proceeding as in Sec. 2, for the force between positive charges, $\mathbf{v} = \mathbf{v}_+, \mathbf{u} = \mathbf{u}_+$ and

$$\mathbf{F}_{C++} = \frac{|Q_1 Q_2|}{4\pi\epsilon_0 (r)^2}\left[\mathbf{a}_r + \frac{\mathbf{u}_+ \times (\mathbf{v}_+ \times \mathbf{a}_r)}{c^2}\right]$$
$$- \frac{|Q_1 Q_2|}{4\pi\epsilon_0 c^2 (r)}\left[\frac{d\mathbf{v}_+}{dt} - \frac{\mathbf{u}_+}{c} \times \left[\frac{d\mathbf{v}_+}{dt} \times \mathbf{a}_r\right]\right]. \qquad (53\text{C}_{++})$$

and

$$\mathbf{F}_{R++} = \frac{|Q_1 Q_2|}{4\pi\epsilon_0 (r)^2}\left[\mathbf{a}_r - \frac{(\mathbf{v}_+ - \mathbf{u}_+) \times ((\mathbf{v}_+ - \mathbf{u}_+) \times \mathbf{a}_r)}{2c^2}\right]$$
$$- \frac{|Q_1 Q_2|}{8\pi\epsilon_0 c^2 (r)}\left[\frac{d\mathbf{v}_+}{dt} - \frac{d\mathbf{u}_+}{dt}\right.$$
$$\left. + \frac{(\mathbf{v}_+ - \mathbf{u}_+)}{c} \times \left[\left[\frac{d\mathbf{v}_+}{dt} - \frac{d\mathbf{u}_+}{dt}\right] \times \mathbf{a}_r\right]\right]. \qquad (53\text{R}_{++})$$

Similarly for positive charge in $\mathbf{ds}_1$ and negative charge in $\mathbf{ds}_2, \mathbf{v} = \mathbf{v}_+, \mathbf{u} = \mathbf{u}_-$ and

$$\mathbf{F}_{C+-} = \frac{-|Q_1 Q_2|}{4\pi\epsilon_0 (r)^2}\left[\mathbf{a}_r + \frac{\mathbf{u}_- \times (\mathbf{v}_+ \times \mathbf{a}_r)}{c^2}\right]$$
$$+ \frac{|Q_1 Q_2|}{4\pi\epsilon_0 c^2 (r)}\left[\frac{d\mathbf{v}_+}{dt} - \frac{\mathbf{u}_-}{c} \times \left[\frac{d\mathbf{v}_+}{dt} \times \mathbf{a}_r\right]\right] \qquad (53\text{C}_{+-})$$

and

$$\mathbf{F}_{R+-} = \frac{-|Q_1 Q_2|}{4\pi\epsilon_0 (r)^2}\left[\mathbf{a}_r - \frac{(\mathbf{v}_+ - \mathbf{u}_-) \times ((\mathbf{v}_+ - \mathbf{u}_-) \times \mathbf{a}_r)}{2c^2}\right]$$
$$+ \frac{|Q_1 Q_2|}{8\pi\epsilon_0 c^2 (r)}\left[\frac{d\mathbf{v}_+}{dt} - \frac{d\mathbf{u}_-}{dt}\right.$$
$$\left. + \frac{(\mathbf{v}_+ - \mathbf{u}_-)}{c} \times \left[\left[\frac{d\mathbf{v}_+}{dt} - \frac{d\mathbf{u}_-}{dt}\right] \times \mathbf{a}_r\right]\right]. \qquad (53\text{R}_{+-})$$

Similarly for negative charge in $\mathbf{ds}_1$ and positive charge in $\mathbf{ds}_2, \mathbf{v} = \mathbf{v}_-, \mathbf{u} = \mathbf{u}_+$ and

$$\mathbf{F}_{C-+} = \frac{-|Q_1 Q_2|}{4\pi\epsilon_0 (r)^2}\left[\mathbf{a}_r + \frac{\mathbf{u}_+ \times (\mathbf{v}_- \times \mathbf{a}_r)}{c^2}\right]$$
$$+ \frac{|Q_1 Q_2|}{4\pi\epsilon_0 c^2 (r)}\left[\frac{d\mathbf{v}_-}{dt} - \frac{\mathbf{u}_+}{c} \times \left[\frac{d\mathbf{v}_-}{dt} \times \mathbf{a}_r\right]\right] \qquad (53\text{C}_{-+})$$

and

$$\mathbf{F}_{R-+} = \frac{-|Q_1 Q_2|}{4\pi\epsilon_0 (r)^2}\left[\mathbf{a}_r - \frac{(\mathbf{v}_- - \mathbf{u}_+) \times ((\mathbf{v}_- - \mathbf{u}_+) \times \mathbf{a}_r)}{2c^2}\right]$$
$$+ \frac{|Q_1 Q_2|}{8\pi\epsilon_0 c^2 (r)}\left[\frac{d\mathbf{v}_-}{dt} - \frac{d\mathbf{u}_+}{dt}\right.$$
$$\left. + \frac{(\mathbf{v}_- - \mathbf{u}_+)}{c} \times \left[\left[\frac{d\mathbf{v}_-}{dt} - \frac{d\mathbf{u}_+}{dt}\right] \times \mathbf{a}_r\right]\right]. \qquad (53\text{R}_{-+})$$

Finally for negative charge in each current element, $\mathbf{v} = \mathbf{v}_-, \mathbf{u} = \mathbf{u}_-$ so

$$\mathbf{F}_{C--} = \frac{|Q_1 Q_2|}{4\pi\epsilon_0 (r)^2}\left[\mathbf{a}_r + \frac{\mathbf{u}_- \times (\mathbf{v}_- \times \mathbf{a}_r)}{c^2}\right]$$
$$- \frac{|Q_1 Q_2|}{4\pi\epsilon_0 c^2 (r)}\left[\frac{d\mathbf{v}_-}{dt} - \frac{\mathbf{u}_-}{c} \times \left[\frac{d\mathbf{v}_-}{dt} \times \mathbf{a}_r\right]\right]. \qquad (53\text{C}_{--})$$

and

$$\mathbf{F}_{R--} = \frac{|Q_1 Q_2|}{4\pi\epsilon_0 (r)^2}\left[\mathbf{a}_r - \frac{(\mathbf{v}_- - \mathbf{u}_-) \times ((\mathbf{v}_- - \mathbf{u}_-) \times \mathbf{a}_r)}{2c^2}\right]$$
$$- \frac{|Q_1 Q_2|}{8\pi\epsilon_0 c^2 (r)}\left[\frac{d\mathbf{v}_-}{dt} - \frac{d\mathbf{u}_-}{dt}\right.$$
$$\left. + \frac{(\mathbf{v}_- - \mathbf{u}_-)}{c} \times \left[\left[\frac{d\mathbf{v}_-}{dt} - \frac{d\mathbf{u}_-}{dt}\right] \times \mathbf{a}_r\right]\right]. \qquad (53\text{R}_{--})$$

The total force between current elements is

$$\mathbf{F} = \mathbf{F}_{++} + \mathbf{F}_{+-} + \mathbf{F}_{-+} + \mathbf{F}_{--}. \qquad (54)$$

Substitution of Eq. (37) into Eq. (38) gives

$$\mathbf{F}_C = \frac{|Q_1 Q_2|}{4\pi\epsilon_0 c^2 r^2}(\mathbf{u}_- - \mathbf{u}_+) \times \{(\mathbf{v}_- - \mathbf{v}_+) \times \mathbf{a}_r\}$$
$$+ \frac{|Q_1 Q_2|}{4\pi\epsilon_0 c^3 r}(\mathbf{u}_- - \mathbf{u}_+) \times \left[\left[\frac{d\mathbf{v}_-}{dt} - \frac{d\mathbf{v}_+}{dt}\right] \times \mathbf{a}_r\right]. \qquad (55\text{C})$$

Similarly,

$$\mathbf{F}_R = \frac{|Q_1 Q_2|}{8\pi\epsilon_0 c^2 r^2}\left[(\mathbf{u}_- - \mathbf{u}_+) \times \{(\mathbf{v}_- - \mathbf{v}_+) \times \mathbf{a}_r\}\right.$$
$$\left. + (\mathbf{v}_- - \mathbf{v}_+) \times \{(\mathbf{u}_- - \mathbf{u}_+) \times \mathbf{a}_r\}\right] \qquad (55\text{R})$$
$$+ \frac{|Q_1 Q_2|}{8\pi\epsilon_0 c^3 r}\left[(\mathbf{u}_- - \mathbf{u}_+) \times \left[\left[\frac{d\mathbf{v}_-}{dt} - \frac{d\mathbf{v}_+}{dt}\right] \times \mathbf{a}_r\right]\right.$$
$$\left. + (\mathbf{v}_- - \mathbf{v}_+) \times \left[\left[\frac{d\mathbf{u}_-}{dt} - \frac{d\mathbf{u}_+}{dt}\right] \times \mathbf{a}_r\right]\right].$$

Substitution of Eq. (23) into Eq. (55) gives

$$d^2\mathbf{F}_C = \frac{1}{4\pi\epsilon_0 c^2}\left[\frac{I_1(t)I_2(t)}{r^2} + \frac{I_2(t)\frac{dI_1(t)}{dt}}{cr}\right]\{\mathbf{ds}_2 \times (\mathbf{ds}_1 \times \mathbf{a}_r)\} \qquad (56C)$$

and

$$d^2\mathbf{F}_R = \frac{I_1(t)I_2(t)}{8\pi\epsilon_0 c^2 r^2}\left[\mathbf{ds}_1 \times (\mathbf{ds}_2 \times \mathbf{a}_r) + \mathbf{ds}_2 \times (\mathbf{ds}_1 \times \mathbf{a}_r)\right]$$
$$+ \frac{1}{8\pi\epsilon_0 c^3 r}\left[I_1(t)\frac{dI_2(t)}{dt}\mathbf{ds}_1 \times (\mathbf{ds}_2 \times \mathbf{a}_r)\right.$$
$$\left. + I_2(t)\frac{dI_1(t)}{dt}\mathbf{ds}_2 \times (\mathbf{ds}_1 \times \mathbf{a}_r)\right]. \qquad (56R)$$

In the steady state, Eq. (56C) reduces to

$$d^2\mathbf{F}_C = d^2\mathbf{F}_G = \frac{I_1 I_2}{4\pi\epsilon_0 c^2 r^2}\mathbf{ds}_2 \times (\mathbf{ds}_1 \times \mathbf{a}_r) \qquad (57C)$$

which was first suggested by Grassmann [46] in 1845. Similarly [47]

$$d^2\mathbf{F}_R = \frac{I_1 I_2}{8\pi\epsilon_0 c^2 r^2}\left[\mathbf{ds}_1 \times (\mathbf{ds}_2 \times \mathbf{a}_r) + \mathbf{ds}_2 \times (\mathbf{ds}_1 \times \mathbf{a}_r)\right]. \qquad (57R)$$

The force $d^2\mathbf{F}_C$ does not satisfy Newton's third law and is always perpendicular to the current element on which it acts. However, the expression based on the Gauss-Weber-Riemann idea of relative velocities satisfies Newton's law and predicts components of $d^2\mathbf{F}_R$ which may be tangential to $\mathbf{ds}_2$.

Conclusions

The paper traces the effect of Gauss' basic idea that all electromagnetic phenomena should be derivable from a single equation for the force between moving charges, which should be a retarded function of the relative velocity of source and receiver. The masterful formulation of Faraday's experiments in terms of electric and magnetic lines of force has been so successful that the idea of an electrodynamic force equation has been nearly forgotten by most physicists. However, recent research has suggested a large family of 120 possible electrodynamic force equations. Twenty-seven of these possible equations belong to the Gaussian family. At low velocities, there are only two possible equations. These yield two possible expressions for the force between current elements. In the steady state, one of these was suggested by Grassmann [46] in 1845. It predicts that tangential forces are impossible. The relative velocity formulation was introduced in 1987 [47] and was called the Riemann force in honor of Bernard Riemann who was closest to formulating such an equation.

The determination of whether any electrodynamic equation of the Gaussian form is valid will, of course, depend on experiment. All previous experiments must be analyzed to see if it is possible to determine which of the proposed postulates is valid. A few papers have analyzed some key experiments. Analysis of the data on binary stars [48] shows that Postulate II must be discarded unless we wish to employ Riemannian space for all large distances. If we wish to keep the simplicity of Euclidean space, it is necessary to employ either Postulate I* or Postulate III*. Thus, one third of the possible electrodynamic equations are eliminated at one fell stroke. The analysis of the Edwards [49] effect appears to eliminate the non-Gaussian equations. At low velocities, the question is whether the tangential forces between current elements are real or not. Perhaps it will be possible to prove that there is one and only one possible equation for the force between moving charges. Perhaps this equation belongs to the family envisioned by Gauss.

References

1. K. F. Gauss, *Zur mathematischen Theorie der elektrodynamischen Wirkung*, Werke, Göttingen, 1867, Vol. V, p. 602.

2. W. Weber, *Elektrodynamische Maassbestimmungen*, Ann. der Physik and Chemie (2) **73** (1848), p. 193.

3. B. Riemann, *Schwere, Elektricität und Magnetismus*, ed. K. Hattendorf, C. Rümpler, Hannover, 1875, p. 326.

4. W. Ritz, *Recherches critiques sur l'électrodynamique générale*, Ann. Chim. Phys. **13** (1908), p. 1451.

5. V. Bush, *The force between moving charges*, J. Math. Phys. (3) **5** (1926), p. 129.

6. F. W. Warburton, *Reciprocal electric force*, Phys. Rev. **69** (1946), p. 40.

7. P. Moon and D. E. Spencer, *Electromagnetism without magnetism: An historical sketch*, Amer. J. Phys. **22** (1954), p. 210; *Interpretation of the Ampère force*, J. Franklin Inst. **257** (1954), p. 203; *The Coulomb force and the Ampère force*, J. Franklin Inst. **257** (1954), p. 305; *A new electrodynamics*, J. Franklin Inst. **257** (1954), p. 369.

8. A. S. Mirchandaney, D. E. Spencer, S. Y. Uma and P. J. Mann, *The electromagnetic fields of a moving charge*, to be published in Hadronic J. Supplement.

9. J. L. Lagrange, Mém. de Berlin, 1777.

10. P. S. Laplace, Mém. de l'Acad, 1785, p. 113.

11. C. A. Coulomb, Mém. de l'Acad, 1785.

12. S. D. Poisson, Bull. de la Soc. Philomathique **3** (1813), p. 388.

13. S. D. Poisson, *Mémoire sur la théorie du magnétisme*, Mém. de l'Acad **5** (1824), p. 247.

14. H. C. Oersted, *Experimenta circa Conflictus Electrici in Acum Magneticam*, Copenhagen, 1820.

15. A. M. Ampère, *Mémoire sur la théorie mathématique des phénomènes électrodynamiques, uniquement déduite de l'expérience*, Mém. de l'Acad. des Sci. **VI** (1823), p. 175.

16. J. C. Maxwell, *A Treatise on Electricity and Magnetism*, Vol. II, Third Ed., Clarendon Press, Oxford, 1904, p. 175.

17. A. S. Mirchandaney, D. E. Spencer, S. Y. Uma and P. J. Mann, *The theory of retarded potentials*, Hadronic J. **11** (1988), p. 231.

18. K. F. Gauss, Werke, p. 627.

19. J. P. Wesley, *Weber electrodynamics extended to include radiation*, Speculations in Science and Technology (1) **10** (1986), p. 47.

20. P. T. Pappas, *The original Ampère force and Biot-Savart and Lorentz forces*, Nuovo Cimento **76B** (1983), p. 189.

21. T. E. Phipps, Jr., *Heretical Verities: Mathematical Themes in Physical Description*, Classic Non-fiction Library, Urbana, Ill., 1986; T. E. Phipps and T. E. Phipps, Jr., *Observation of Ampère forces in Mercury*, to be published.

22. P. Graneau, *Ampère-Neumann Electrodynamics of Metals*, Hadronic Press, Nonantum, Mass., 1985.

23. A. K. T. Assis, *On Mach's principle*, Found. Phys. Lett. **2** (1989), p. 301.

24. M. Faraday, *Historical sketch of electromagnetism*, Ann. Phil. **2** (1821), p. 195.

25. M. Faraday, *Experimental Researches in Electricity*, London, 1844, 1849, 1855.

26. F. E. Neumann, *Die mathematischen Gesetze der inductirten elektrischen Stroeme*, Akademie der Wissenschaften, Berlin, 1845, p. 1.

27. W. Thomson, Camb. and Dub. Math. Jour. **2** (1847), p. 61.

28. J. C. Maxwell, *Maxwell's Scientific Papers*, Vol. I, Cambridge, 1890, p. 155.

29. J. C. Maxwell, *Maxwell's Scientific Papers*, Vol. I, Cambridge, 1890, p. 451.

30. J. C. Maxwell, *A dynamical theory of the electromagnetic field*, Proc. Royal Soc. **13** (1864), p. 531.

31. J. C. Maxwell, *Maxwell's Scientific Papers*, Vol. II, Cambridge, 1890, p. 125.

32. J. C. Maxwell, *Treatise on Electricity and Magnetism*, Vols. I and II, Oxford, 1873.

33. L. Lorenz, Oversigt over det. K. danske Vid. Selskaps Forhandlinger, 1867, p. 26; Ann. d. Phys. **81** (1867), p. 243; Phil. Mag. **34** (1867), p. 287.

34. E. Whittaker, *A History of the Theories of Aether and Electricity*, Philosophical Library, New York, 1951, p. 310.

35. H. A. Lorentz, *The Theory of Electrons*, G. E. Stechert & Co., New York, N. Y., 1906.

36. P. Moon and D. E. Spencer, *A postulational approach to electromagnetism*, J. Franklin Inst. **259** (1955), p. 293; *Foundations of Electrodynamics*, D. Van Nostrand, Princeton, N. J., 1960.

37. A. Liénard, *Champ électrique et magnétique produit par une charge électrique concentrée en un point et animée d'un movement quelconque*, l'Eclairage Electrique **16** (1898), p. 5, 53, 106.

38. E. Wiechert, *Elektrodynamische Elementargesetze*, Ann. d. Phys. **4** (1901), p. 680.

39. C. K. Whitney, *On the Liénard-Wiechert potentials*, Hadronic J. **11** (1988), p. 257.

40. P. Moon, D. E. Spencer and E. E. Moon, *Universal time and the velocity of light*, Physics Essays (2) **4** (1989), p. 368.

41. A. Einstein, *Zur Electrodynamik bewegter Körper*, Ann. d. Phys. **17** (1905), p. 891.

42. P. Moon, D. E. Spencer and E. E. Moon, *The Michelson-Gale experiment and its effect on the postulates on the velocity of light*, to be published in Physics Essays.

43. A. A. Michelson, *The effect of the earth's rotation on the velocity of light*, Part I, Astrop. J. **61** (1925), p. 137; A. A. Michelson and H. G. Gale, *The effect of the earth's rotation on the velocity of light*, Part II, Astrop. J. **61** (1925), p. 140.

44. A. Einstein, *Uber das Relativitatsprinzip und die aus demeselben gezogenen Folgerungen*, Jahrbuch der Radioaktivitat, IV, p. 422–462; V, p. 98–99 (Berichtigungen), 1907.

45. P. Moon and D. E. Spencer, *On the establishment of universal time*, Phil. of Sci. **23** (1956), p. 216.

46. H. Grassmann, *Neue Theorie der Elektrodynamik*, Ann. d. Phys. **64** (1845), p. 1.

47. P. Moon, D. E. Spencer, S. Y. Uma and P. J. Mann, *The Riemann force*, to be published in the Proceedings of the International Congress on Relativity and Gravitation, Munich, West Germany, April, 1988.

48. P. Moon, D. E. Spencer and E. E. Moon, *Binary stars from three viewpoints*, Physics Essays **2** (1989), p. 275.

49. D. E. Spencer, S. Y. Uma and P. J. Mann, *The force between moving charges and the Edwards effect*, The Thorny Way of Truth, International Publishers, East and West, Part IV, Graz, Austria, 1989.

Domina Eberle Spencer
Department of Mathematics
University of Connecticut
Storrs, CT 06268
USA

Shama Y. Uma
Department of Mathematics and
Computer Science
Bridgewater State College
Bridgewater, MA 02325
USA

SOME OPERATIONAL TECHNIQUES IN THE THEORY OF GENERALIZED GAUSSIAN AND CLAUSENIAN FUNCTIONS

H. M. Srivastava

In this paper we aim at presenting several applications of operational techniques in the theory of special functions including, for example, certain classes of generalized hypergeometric functions of one and more variables. The Laplace transform operator $\mathcal{L}$ and its inverse $\mathcal{L}^{-1}$, and the Riemann-Liouville fractional derivative operator D_z^μ, are among the various linear operators which are shown to be capable of yielding fruitful results. Relevant connections of the results considered here with those available in the literature are also provided.

1. Introduction, Definitions, and Preliminaries

Some of the most important classes of functions of mathematical analysis, which are used rather frequently in engineering and physical sciences, are special or limiting cases of the familiar hypergeometric function defined by (cf. [17])

$$F(a, b; c; z) = 1 + \frac{a \cdot b}{1 \cdot c} z + \frac{a(a + 1) \cdot b(b + 1)}{1 \cdot 2 \cdot c(c + 1)} z^2 + \dots , \qquad (1.1)$$

which was introduced, in the year 1812, by Carl Friedrich Gauss (1777–1855).

As long ago as 1828, Thomas Clausen (1801–1885) proved the only case in which the square of a certain *Gaussian hypergeometric function* of argument z can be expressed in terms of a higher-order hypergeometric function of the same argument. This higher-order hypergeometric function, occurring in Clausen's identity, is usually referred to in the literature as the *Clausenian hypergeometric function*.

With a view to recalling Clausen's identity, we begin by introducing the generalized Gaussian and Clausenian hypergeometric function defined, in the notations of Leo Pochhammer (1841–1920) and Ernest William Barnes (1874–1954), by

$$
{}_pF_q(\alpha_1,\ldots,\alpha_p\,;\,\beta_1,\ldots,\beta_q\,;z)
$$

$$
\equiv {}_pF_q
\begin{bmatrix}
\alpha_1,\ldots,\alpha_p\,; \\[4pt]
\qquad\qquad\quad z \\[4pt]
\beta_1,\ldots,\beta_q\,;
\end{bmatrix}
$$

$$
= \sum_{n=0}^{\infty} \frac{(\alpha_1)_n \ldots (\alpha_p)_n}{(\beta_1)_n \ldots (\beta_q)_n} \frac{z^n}{n!}
\tag{1.2}
$$

$$
(p \leqq q+1\,; p < q+1 \text{ and } |z| < \infty\,; p = q+1 \text{ and } z \in \mathcal{U} = \{z : |z| < 1\}\,;
$$
$$
p = q+1,\, z \in \partial\mathcal{U},\text{ and } \mathrm{Re}\,(\omega) > 0),
$$

where

$$
\omega = \sum_{j=1}^{q} \beta_j - \sum_{j=1}^{p} \alpha_j\,,
\tag{1.3}
$$

and $(\lambda)_n$ denotes the Pochhammer symbol given by

$$
(\lambda)_n = \frac{\Gamma(\lambda+n)}{\Gamma(\lambda)} =
\begin{cases}
1, & \text{if } n = 0, \\
\lambda(\lambda+1)\ldots(\lambda+n-1), & \text{if } n \in \mathbb{N} = \{1,2,3,\ldots\},
\end{cases}
\tag{1.4}
$$

it being understood, as also in the definition (1.1), that no zeros appear in the denominators involved.

The Gaussian hypergeometric function (1.1) is the special case $p = 2$ and $q = 1$ of the function defined by (1.2). On the other hand, a special case of (1.2) when $p = 3$ and $q = 2$ is the aforementioned Clausenian hypergeometric function. Indeed, as already pointed out by Barnes [5], the generalized hypergeometric function ${}_pF_q$ originated with Clausen [9] and was studied, among others, by Johannes Karl Thomae (1840–1921), Édouard Jean–Baptiste Goursat (1858–1936), and Pochhammer whose voluminous work on the subject provides a detailed development of the theory.

We now state Clausen's identity in the form[†]:

$$
\{{}_2F_1(a,b;a+b+\tfrac{1}{2};z)\}^2 = {}_3F_2
\begin{bmatrix}
2a, 2b, a+b\,; \\[4pt]
\qquad\qquad\qquad z \\[4pt]
a+b+\tfrac{1}{2}, 2a+2b\,;
\end{bmatrix},
\tag{1.5}
$$

[†]The *second* denominator parameter on the right-hand side of Clausen's identity (1.5) appears *erroneously* in Erdélyi *et al.* [12, p. 185, Equation 4.3(1)].

where we have used the notation (1.2) for both sides. More generally, Goursat [18] proved that the product

$$_2F_1(a, b; c; z)\,_2F_1(a', b'; c'; z)$$

is a generalized hypergeometric $_pF_q$ function of argument z only if *either*

$$\begin{cases} a' = 1 + a - c, b' = 1 + b - c, c' = 2 - c, \\ c - a - b = n + \frac{1}{2} \ (n \in \mathbb{Z} = \{0, \pm 1, \pm 2, \dots\}) \end{cases} \tag{1.6}$$

or

$$\begin{cases} a - a' \in \mathbb{Z}, b - b' \in \mathbb{Z}, c - c' \in \mathbb{Z}, \\ c + c' - a - a' - b - b' = n \ (n \in \mathbb{N}_0 = \mathbb{N} \cup \{0\}). \end{cases} \tag{1.7}$$

The usefulness of Clausen's identity (1.5), and indeed also of numerous other hypergeoemtric identities of this type, cannot be overemphasized. For example, it was used by Askey and Gasper [2] (and, subsequently, by Gasper [14]) to prove the important hypergeometric inequality:

$$\frac{(a+2)_n}{n!} {}_3F_2 \left[\begin{array}{c} -n, a + n + 2, \frac{1}{2}(a + 1); \\ \\ a + 1, \frac{1}{2}(a + 3); \end{array} x \right] \geqq 0 \tag{1.8}$$
$$(0 < x < 1; a \geqq -2; n \in \mathbb{N}_0),$$

whose special case was required by de Branges [11] in order to complete his proof of the Milin conjecture of 1971 for univalent functions, containing (as a special case) the Robertson conjecture of 1936, which (in turn) implies the celebrated Bieberbach conjecture of 1916. (See, for details, [38, Sec. 25].) More recently, Clausen's identity (1.5) was used in expressing a certain definite integral (involving the Gaussian hypergeometric function) in terms of the square of a hypergeometric function (cf. [10]). For various interesting further extensions of Clausen's identity (1.5), involving generalized *basic* hypergeometric functions, one may see the works of Jackson [22], Singh [30], Srivastava and Jain [35], Gasper [15], and Gasper and Rahman [16].

The main object of the present paper is to show how readily certain operational techniques would apply to yield generalizations of various hypergeometric identities like (1.5) involving a double hypergeometric function defined by (cf. [23], [1, p. 150], and [37, p. 63])

$$
F^{p:r;u}_{q:s;v}
\left[
\begin{array}{c}
\alpha_1,\ldots,\alpha_p : \gamma_1,\ldots,\gamma_r ; \rho_1,\ldots,\rho_u ; \\[2mm]
\beta_1,\ldots,\beta_q : \delta_1,\ldots,\delta_s ; \sigma_1,\ldots,\sigma_v ;
\end{array}
\quad x,y
\right]
$$

$$
= \sum_{m,n=0}^{\infty}
\frac{(\alpha_1)_{m+n}\cdots(\alpha_p)_{m+n}(\gamma_1)_m\cdots(\gamma_r)_m(\rho_1)_n\cdots(\rho_u)_n}{(\beta_1)_{m+n}\cdots(\beta_q)_{m+n}(\delta_1)_m\cdots(\delta_s)_m(\sigma_1)_n\cdots(\sigma_v)_n}
\frac{x^m}{m!}\frac{y^n}{n!},
\tag{1.9}
$$

which was introduced in 1931 by Joseph Kampé de Fériet (1893–1982). The operational techniques, which we shall illustrate in this paper, are based upon the classical Laplace transform:

$$
\mathcal{L}\{f(t):s\} = \int_0^{\infty} e^{-st} f(t)dt = F(s),
\tag{1.10}
$$

its inverse

$$
\mathcal{L}^{-1}\{F(s):t\} = \frac{1}{2\pi i}\int_{\sigma-i\infty}^{\sigma+i\infty} e^{st} F(s)ds = f(t),
\tag{1.11}
$$

and the Riemann-Liouville fractional derivative operator D_z^{μ} defined by (cf. [13, Vol. II, Chapter 13] and [37, Chapter 5])

$$
D_z^{\mu}\{f(z)\} =
\begin{cases}
\frac{1}{\Gamma(-\mu)} \int_0^z (z-\zeta)^{-\mu-1} f(\zeta)d\zeta, & \operatorname{Re}(\mu) < 0, \\[2mm]
\frac{d^m}{dz^m} D_z^{\mu-m}\{f(z)\}, & m-1 \leq \operatorname{Re}(\mu) < m \ (m \in \mathbb{N}).
\end{cases}
\tag{1.12}
$$

In our investigation we shall require a number of operational formulas involving the operators $\mathcal{L}, \mathcal{L}^{-1}$, and D_z^{μ}. Operational images (or operational representations) of many classes of special functions in the Laplace transform (1.10) can be obtained by appealing to Euler's integral:

$$
\int_0^{\infty} e^{-st} t^{\lambda-1} dt = \frac{\Gamma(\lambda)}{s^{\lambda}}
\tag{1.13}
$$
$$
(\min\{\operatorname{Re}(s),\ \operatorname{Re}(\lambda)\} > 0).
$$

On the other hand, computation of the inverse Laplace transform (1.11) is usually based upon Hankel's contour integral in the *equivalent* form (see, e.g., [39, p. 245, Example 1] and [25, p. 17, Eq. 2.7(5)])

$$
\frac{1}{2\pi i}\int_{\sigma-i\infty}^{\sigma+i\infty} e^s s^{-z} ds = \frac{1}{\Gamma(z)}
\tag{1.14}
$$
$$
(\sigma > 0;\ \operatorname{Re}(z) > 0).
$$

By applying the integral formulas (1.13) and (1.14) to the generalized hypergeometric $_pF_q$ function defined by (1.2), it is readily seen that (cf. [13, Vol. I, p. 219, Eq. 4.23(17)])

$$\mathcal{L}\left\{ t^{\lambda-1}\,_pF_q \left[\begin{array}{c} \alpha_1,\ldots,\alpha_p\,; \\ \\ \beta_1,\ldots,\beta_q\,; \end{array} \; zt \; \right] : s \right\}$$
$$= \frac{\Gamma(\lambda)}{s^\lambda}\,_{p+1}F_q \left[\begin{array}{c} \lambda,\alpha_1,\ldots,\alpha_p\,; \\ \\ \beta_1,\ldots,\beta_q\,; \end{array} \; \frac{z}{s} \; \right] \tag{1.15}$$

$$(\mathrm{Re}(\lambda) > 0\,;\; \mathrm{Re}\,(s) > 0 \text{ if } p < q\,;\; \mathrm{Re}\,(s) > \mathrm{Re}\,(z) \text{ if } p = q)$$

and [op. cit., p. 297, Eq. 5.21 (1)]

$$\mathcal{L}^{-1}\left\{ s^{-\mu}\,_pF_q \left[\begin{array}{c} \alpha_1,\ldots,\alpha_p\,; \\ \\ \beta_1,\ldots,\beta_q\,; \end{array} \; \frac{z}{s} \; \right] : t \right\}$$
$$= \frac{t^{\mu-1}}{\Gamma(\mu)}\,_pF_{q+1} \left[\begin{array}{c} \alpha_1,\ldots,\alpha_p\,; \\ \\ \mu,\beta_1,\ldots,\beta_q\,; \end{array} \; zt \; \right] \tag{1.16}$$
$$(\mathrm{Re}\,(\mu) > 0\,;\; p \leq q + 1)\,,$$

which incidentally follows easily from (1.15).

Finally, it is not difficult to observe from (1.12) that

$$D_z^\mu\{z^\lambda\} = \frac{\Gamma(\lambda+1)}{\Gamma(\lambda-\mu+1)} z^{\lambda-\mu} \;(\mathrm{Re}\,(\lambda) > -1)\,, \tag{1.17}$$

which, in view of the definition (1.2), yields the operational formula:

$$D_z^{\lambda-\mu}\left\{ z^{\lambda-1}\,_pF_q \left[\begin{array}{c} \alpha_1,\ldots,\alpha_p\,; \\ \\ \beta_1,\ldots,\beta_q\,; \end{array} \; z \; \right] \right\}$$
$$= \frac{\Gamma(\lambda)}{\Gamma(\mu)} z^{\mu-1}\,_{p+1}F_{q+1} \left[\begin{array}{c} \lambda,\alpha_1,\ldots,\alpha_p\,; \\ \\ \mu,\beta_1,\ldots,\beta_q\,; \end{array} \; z \; \right] \tag{1.18}$$
$$(\mathrm{Re}\,(\lambda) > 0\,;\; |z| < \infty \text{ when } p \leq q\,;\; z \in \mathcal{U} \text{ when } p = q + 1)\,.$$

2. An Illustrative Example

In the theory of special functions one finds numerous applications of operational techniques involving various integral transforms. The Laplace transform operator $\mathcal{L}$ and its inverse $\mathcal{L}^{-1}$, which have indeed been exploited a great deal, are capable of yielding fruitful results (see, e.g., [37, Sec. 4.1]). However, in the problem of augmentation of parameters of a generalized hypergeometric function (1.2), these operators obviously lack in the sense that any single application of the Laplace transform operator $\mathcal{L}$ (or its inverse $\mathcal{L}$) can augment only one numerator (or denominator) parameter (as already exhibited by the operational formulas (1.15) and (1.16) above). On the other hand, the operational formula (1.18) provides another process of augmentation of parameters of a hypergeometric function in which a given $_pF_q$ transforms into $_{p+1}F_{q+1}$. In fact, by comparing (1.18) with (1.15) and (1.16), the (Riemann-Liouville) fractional derivative operator D_z^μ may be looked upon as being the "union" of the Laplace transform operator $\mathcal{L}$ and its inverse $\mathcal{L}^{-1}$.

For a detailed survey of the various other processes of augmentation of parameters of the hypergeometric $_pF_q$ function (by means, for example, of the Euler or Beta transformation), see the work of Rainville [29, Sec. 56], and of Srivastava and Manocha [37, Sec. 4.3].

Our first example illustrating the usefulness of the operators $\mathcal{L}, \mathcal{L}^{-1}$, and D_z^μ stems from Clausen's identity (1.5). If we multiply each member of (1.5) by $z^{\lambda-1}$, and then operate upon both sides of the resulting equation by $D_z^{\lambda-\mu}$, we obtain

$$
\begin{aligned}
& F_{1:1;1}^{1:2;2}\left[\begin{array}{ccc} \lambda : a,b & ; a,b & ; \\ & & z,z \\ \mu : & a+b+\tfrac{1}{2}; a+b+\tfrac{1}{2}; & \end{array} \right] \\
& = {}_4F_3\left[\begin{array}{cc} \lambda, 2a, 2b, a+b; & \\ & z \\ \mu, a+b+\tfrac{1}{2}, 2a+2b; & \end{array} \right],
\end{aligned}
\tag{2.1}
$$

where we have made use of the operational formula (1.18) as well as the following two-variable consequence of (1.17):

$$D_z^{\lambda-\mu} \left\{ z^{\lambda-1} F_{q:s;v}^{p:r;u} \left[\begin{array}{c} (\alpha_p) : (\gamma_r); (\rho_u); \\ \\ (\beta_q) : (\delta_s); (\sigma_v); \end{array} \; z, z \right] \right\}$$

$$= \frac{\Gamma(\lambda)}{\Gamma(\mu)} z^{\mu-1} F_{q+1:s;v}^{p+1:r;u} \left[\begin{array}{c} \lambda, (\alpha_p) : (\gamma_r); (\rho_u); \\ \\ \mu, (\beta_q) : (\delta_s); (\sigma_v); \end{array} \; z \right], \tag{2.2}$$

where, for convenience, (α_p) abbreviates the array of p parameters

$$\alpha_1, \dots, \alpha_p,$$

with similar interpretations for (β_q), *et cetera*.

Alternatively, in view of the integral formulas (1.13) and (1.14), and their such consequences as (1.15) and (1.16), we can derive (2.1) by suitably applying the operators $\mathcal{L}$ and $\mathcal{L}^{-1}$.

The form of the hypergeometric identity (2.1) suggests the existence of a generalization which we first state as follows (cf. [36, p. 29, Eq. 1.3 (36)]):

$$F_{q:1;1}^{p:2;2} \left[\begin{array}{c} \lambda_1, \dots, \lambda_p : a, b \quad\quad ; a, b \quad\quad ; \\ \\ \mu_1, \dots, \mu_q : a+b+\frac{1}{2} ; a+b+\frac{1}{2} ; \end{array} \; z, z \right]$$

$$= {}_{p+3}F_{q+2} \left[\begin{array}{c} \lambda_1, \dots, \lambda_p, 2a, 2b, a+b; \\ \\ \mu_1, \dots, \mu_q, a+b+\frac{1}{2}, 2a+2b; \end{array} \; z \right] \tag{2.3}$$

$$(p, q \in \mathbb{N}_0 ; p \leq q ; p < q \text{ and } |z| < \infty ; p = q \text{ and } z \in \mathcal{U} ;$$

$$p = q, z \in \partial\mathcal{U}, \text{ and } \mathrm{Re}(\kappa) > -\frac{1}{2}),$$

where

$$\kappa = \sum_{j=1}^{q} \mu_j - \sum_{j=1}^{p} \lambda_j. \tag{2.4}$$

The hypergeometric identity (2.1) is simply the case $p = q = 1$ of the general result (2.3). Thus, in order to complete our operational derivation of Formula (2.3), we can prove (2.3) by using the principle of mathematical induction on p and q in one of the following two ways:

Method 1. Apply the inverse Laplace transform operator $\mathcal{L}^{-1}$ to show that Formula (2.3) holds true also when q is replaced by $q+1$. Then apply the Laplace transform operator $\mathcal{L}$ to show that Formula (2.3) is valid also when p is replaced by $p + 1$.

Method 2. Apply the fractional derivative operator D_z^μ to show, in just *one* step, that Formula (2.3) holds true also when p and q are replaced by $p + 1$ and $q + 1$, respectively.

Clearly, Method 1 would make use of the integral formulas (1.13) and (1.14), and their such consequences as the operational formulas (1.15) and (1.16). Method 2, on the other hand, is based upon the fractional derivative formulas (1.18) and (2.2).

Letting $b \to \infty$ in (2.3), we obtain the hypergeometric identity:

$$
\begin{aligned}
F_{q:0;0}^{p:1;1} & \left[\begin{array}{c} \lambda_1, \ldots, \lambda_p : a ; a ; \\ \\ \mu_1, \ldots, \mu_q : - ; - ; \end{array} \quad z, z \right] \\
&= {}_{p+1}F_q \left[\begin{array}{c} 2a, \lambda_1, \ldots, \lambda_p; \\ \\ \mu_1, \ldots, \mu_q; \end{array} \quad z \right],
\end{aligned}
\tag{2.5}
$$

which incidentally follows also from the known result (cf. [36, p. 28, Eq. 1.3 (32)]):

$$
\begin{aligned}
F_{q:0;0}^{p:1;1} & \left[\begin{array}{c} \lambda_1, \ldots, \lambda_p : a ; b ; \\ \\ \mu_1, \ldots, \mu_q : - ; - ; \end{array} \quad z, z \right] \\
&= {}_{p+1}F_q \left[\begin{array}{c} a + b, \lambda_1, \ldots, \lambda_p ; \\ \\ \mu_1, \ldots, \mu_q ; \end{array} \quad z \right]
\end{aligned}
\tag{2.6}
$$

upon setting $b = a$.

3. Further Examples and Generalizations

The hypergeometric identity (2.3), and each of its numerous companions presented systematically by Srivastava and Karlsson [36, pp. 28–32], would obviously qualify as a reduction formula for a double hypergeometric function. Some of these hypergeometric reduction formulas were shown recently by Miller ([26], [27]) to be intimately connected with representations for incomplete Lipschitz-Hankel integrals of cylindrical functions including, for example, the Bessel functions $J_\nu(z), I_\nu(z), K_\nu(z), Y_\nu(z)$, *et cetera*.

Moreover, following Srivastava ([32], [33], [34]), Buschman and Srivastava [7], and Karlsson [24], each of these reduction formulas can be extended to hold true for series with essentially arbitrary terms. For example, if $\{\Omega_n\}_{n=0}^{\infty}$ is a suitably bounded sequence of complex numbers, it is not difficult to prove an extension of the hypergeometric identity (2.3) in the form:

$$\sum_{m,n=0}^{\infty} \Omega_{m+n} \frac{(a)_m (a)_n (b)_m (b)_n}{(a+b+\frac{1}{2})_m (a+b+\frac{1}{2})_n} \frac{z^{m+n}}{m!n!}$$
$$= \sum_{n=0}^{\infty} \Omega_n \frac{(2a)_n (2b)_n (a+b)_n}{(a+b+\frac{1}{2})_n (2a+2b)_n} \frac{z^n}{n!} \, , \tag{3.1}$$

provided that the series involved converge absolutely.

The hypergeometric reduction formula (2.3) follows from the series identity (3.1) when

$$\Omega_n = \frac{(\lambda_1)_n \ldots (\lambda_p)_n}{(\mu_1)_n \ldots (\mu_q)_n} \quad (n \in \mathbb{N}_0) . \tag{3.2}$$

There is yet another direction in which a hypergeometric identity like (2.3) can be extended further. While discussing certain relations in planetary theory, Cayley [8] discovered the interesting theorem:

If

$$(1-z)^{a+b-c} {}_2F_1(2a, 2b; 2c; z) = \sum_{n=0}^{\infty} \Delta_n z^n \, , \tag{3.3}$$

then

$${}_2F_1(a, b; c+\frac{1}{2}; z) {}_2F_1(c-a, c-b; c+\frac{1}{2}; z) = \sum_{n=0}^{\infty} \frac{(c)_n}{(c+\frac{1}{2})_n} \Delta_n z^n . \tag{3.4}$$

Cayley had stated his theorem *without* proof. It was not until forty years later that a proof of Cayley's theorem was published by Orr [28], who discussed the differential equations satisfied by the product of two Gaussian hypergeometric series, and obtained several additional results. For numerous further theorems of Cayley and Orr type, the interested reader may see the works of Bailey ([3], [4]), Burchnall and Chaundy [6], Henrici ([19], [20], [21]), and Slater [31].

For $c = a + b$, Cayley's theorem reduces at once to Clausen's identity (1.5). Consequently, by applying the operational techniques of the preceding section *mutatis mutandis*, Cayley's theorem would lead us to an interesting generalization of the hypergeometric identity (2.3), given by

Theorem 1. If

$$
F_{q:1;0}^{p:2;1}
\begin{bmatrix}
\lambda_1,\ldots,\lambda_p : 2a,2b;c-a-b; \\[4pt]
\mu_1,\ldots,\mu_q : \quad 2c; \qquad\quad -;
\end{bmatrix} \quad z,z
= \sum_{n=0}^{\infty} \xi_n z^n ,
\tag{3.5}
$$

then

$$
F_{q:1;1}^{p:2;2}
\begin{bmatrix}
\lambda_1,\ldots,\lambda_p : \quad a,b;c-a,c-b; \\[4pt]
\mu_1,\ldots,\mu_q : c+\tfrac{1}{2}; \qquad c+\tfrac{1}{2};
\end{bmatrix} \quad z,z
$$
$$
= \sum_{n=0}^{\infty} \frac{(c)_n}{(c+\tfrac{1}{2})_n} \xi_n z^n ,
\tag{3.6}
$$

provided that the series involved converge absolutely.

For $p = q = 0$, Theorem 1 immediately yields Cayley's result. On the other hand, the hypergeometric identity (2.3) is a special case of Theorem 1 when $c = a + b$.

Each of the aforementioned results of Cayley and Orr type can also be generalized to a form analogous to that of Theorem 1. For the convenience of the interested reader, we present some of these generalizations contained in Theorems 2 to 6 below.

Theorem 2. If

$$
F_{q:1;0}^{p:2;1}
\begin{bmatrix}
\lambda_1,\ldots,\lambda_p : 2a,2b;c-a-b+\tfrac{1}{2}; \\[4pt]
\mu_1,\ldots,\mu_q : \quad 2c ; \qquad\qquad -;
\end{bmatrix} \quad z,z
= \sum_{n=0}^{\infty} \eta_n z^n ,
\tag{3.7}
$$

then

$$
F_{q:1;1}^{p:2;2}
\begin{bmatrix}
\lambda_1,\ldots,\lambda_p : a,b;c-a+\tfrac{1}{2},c-b+\tfrac{1}{2}; \\[4pt]
\mu_1,\ldots,\mu_q : \quad c ; \qquad\qquad c+1;
\end{bmatrix} \quad z,z
$$
$$
= \sum_{n=0}^{\infty} \frac{(c+\tfrac{1}{2})_n}{(c+1)_n} \eta_n z^n ,
\tag{3.8}
$$

provided that the series involved converge absolutely.

Theorem 3. If

$$F_{q:1;0}^{p:2;1}\left[\begin{array}{c}\lambda_1,\ldots,\lambda_p\ :\ 2a-1,2b\ ;\ c-a-b+\frac{1}{2};\\[2mm]\mu_1,\ldots,\mu_q\ :\quad 2c-1\ ;\qquad\qquad -;\end{array}\ z,z\right]=\sum_{n=0}^{\infty}\zeta_n z^n,\quad(3.9)$$

then

$$F_{q:1;1}^{p:2;2}\left[\begin{array}{c}\lambda_1,\ldots,\lambda_p\ :\ a,b\,;\,c-a+\frac{1}{2},c-b-\frac{1}{2};\\[2mm]\mu_1,\ldots,\mu_q\ :\quad c\ ;\qquad\qquad c;\end{array}\ z,z\right]$$

$$=\sum_{n=0}^{\infty}\frac{(c-\frac{1}{2})_n}{(c)_n}\zeta_n z^n,\qquad(3.10)$$

provided that the series involved converge absolutely.

Theorem 4. If

$$F_{q:2;0}^{p:3;1}\left[\begin{array}{c}\lambda_1,\ldots,\lambda_p\ :\qquad 2a,2b,c\,;\,c-a-b+\frac{1}{2}\ ;\\[2mm]\mu_1,\ldots,\mu_q\ :2c,a+b+\frac{1}{2}\ ;\qquad\qquad -\ ;\end{array}\ z,z\right]=\sum_{n=0}^{\infty}\theta_n z^n,$$

$$(3.11)$$

then

$$F_{q:1;1}^{p:2;2}\left[\begin{array}{c}\lambda_1,\ldots,\lambda_p\ :\qquad a,b\,;\,c-a+\frac{1}{2},c-b+\frac{1}{2};\\[2mm]\mu_1,\ldots,\mu_q\ :a+b+\frac{1}{2}\ ;\qquad 2c-a-b+\frac{1}{2};\end{array}\ z,z\right]$$

$$=\sum_{n=0}^{\infty}\frac{(c+\frac{1}{2})_n}{(2c-a-b+\frac{1}{2})_n}\theta_n z^n,\qquad(3.12)$$

provided that the series involved converge absolutely.

Theorem 5. If

$$F_{q:2;0}^{p:3;1}\left[\begin{array}{c}\lambda_1,\ldots,\lambda_p\ :\qquad 2a,2b,c\ ;\,c-a-b+\frac{1}{2};\\[2mm]\mu_1,\ldots,\mu_q\ :2c,a+b+\frac{1}{2}\ ;\qquad\qquad -;\end{array}\ z,z\right]$$

$$(3.13)$$

$$=\sum_{n=0}^{\infty}\varphi_n z^n,$$

then

$$
F_{q:1;1}^{p:2;2}
\left[
\begin{array}{l}
\lambda_1,\ldots,\lambda_p : a+\tfrac{1}{2}, c-a+\tfrac{1}{2}; b, c-b; \\[2mm]
\mu_1,\ldots,\mu_q : \qquad\quad c+\tfrac{1}{2};\quad c+\tfrac{1}{2};
\end{array}
\; z,z
\right]
\tag{3.14}
$$
$$
= \sum_{n=0}^{\infty} \frac{(a+b+\tfrac{1}{2})_n}{(c+\tfrac{1}{2})_n}\varphi_n z^n\,,
$$

provided that the series involved converge absolutely.

Theorem 6. If

$$
F_{q:2;0}^{p:3;1}
\left[
\begin{array}{l}
\lambda_1,\ldots,\lambda_p : 2a,2b,c+\tfrac{1}{2}; c-a-b; \\[2mm]
\mu_1,\ldots,\mu_q : 2c, a+b+\tfrac{1}{2}; \qquad -;
\end{array}
\; z,z
\right]
= \sum_{n=0}^{\infty} \psi_n z^n\,,
\tag{3.15}
$$

then

$$
F_{q:1;1}^{p:2;2}
\left[
\begin{array}{l}
\lambda_1,\ldots,\lambda_p : a,b \qquad\quad ; c-a,c-b \quad ; \\[2mm]
\mu_1,\ldots,\mu_q : a+b+\tfrac{1}{2}; 2c-a-b+\tfrac{1}{2};
\end{array}
\; z,z
\right]
\tag{3.16}
$$
$$
= \sum_{n=0}^{\infty} \frac{(c)_n}{(2c-a-b+\tfrac{1}{2})_n}\psi_n z^n\,,
$$

provided that the series involved converge absolutely.

By assigning suitable special values to $p,q,a,b,$ and c, we can deduce a number of interesting consequences of each of the above theorems. In particular, Theorem 4 with

$$
p = q = 0 \quad \text{and} \quad c = a+b
$$

yields the hypergeometric identity:

$$
{}_2F_1\left(a,b; a+b+\frac{1}{2}; z\right){}_2F_1\left(a+\frac{1}{2}, b+\frac{1}{2}; a+b+\frac{1}{2}; z\right)
$$
$$
= (1-z)^{-\frac{1}{2}}{}_3F_2\left[
\begin{array}{c}
2a,2b,a+b; \\[2mm]
a+b+\tfrac{1}{2}, 2a+2b;
\end{array}
\; z
\right],
\tag{3.17}
$$

which, in view of Euler's transformation:

$$_2F_1(a,b;c;z) = (1-z)^{c-a-b}\,_2F_1(c-a,c-b;c;z)\,, \qquad (3.18)$$

follows also as an immediate consequence of Clausen's identity (1.5).

The assertion made in each of our results in this section (Theorems 1 to 6) can be restated as a linear transformation between two double hypergeometric functions of different orders. Thus, by means of a fairly simple analysis, Theorem 1 yields the double hypergeometric transformation:

$$F^{p+1:2;2}_{q+1:1;1}\left[\begin{array}{c} c+\frac{1}{2},\lambda_1,\ldots,\lambda_p: \quad a,b;c-a,c-b\,; \\[2mm] c,\mu_1,\ldots,\mu_q\,:\,c+\frac{1}{2}; \qquad c+\frac{1}{2}\,; \end{array}\ z,z\right]$$
$$= F^{p:2;1}_{q:1;0}\left[\begin{array}{c} \lambda_1,\ldots,\lambda_p:2a,2b;\ c-a-b\,; \\[2mm] \mu_1,\ldots,\mu_q\,: \quad 2c; \qquad\qquad -; \end{array}\ z,z\right] \qquad (3.19)$$

or, equivalently,

$$F^{p:2;2}_{q:1;1}\left[\begin{array}{c} \lambda_1,\ldots,\lambda_p:a,b \quad;c-a,c-b; \\[2mm] \mu_1,\ldots,\mu_q:c+\frac{1}{2}; \qquad c+\frac{1}{2}; \end{array}\ z,z\right]$$
$$= F^{p+1:2;1}_{q+1:1;0}\left[\begin{array}{c} c,\lambda_1,\ldots,\lambda_p:2a,2b;c-a-b; \\[2mm] c+\frac{1}{2},\mu_1,\ldots,\mu_q: \quad 2c; \qquad -; \end{array}\ z,z\right], \qquad (3.20)$$

which, for $c = a + b$, reduces immediately to the hypergeometric reduction formula (2.3).

In a similar manner, Theorems 2 to 6 lead us to the following double hypergeometric transformations:

$$F^{p:2;2}_{q:1;1}\left[\begin{array}{c} \lambda_1,\ldots,\lambda_p:a,b;c-a+\frac{1}{2},c-b+\frac{1}{2}; \\[2mm] \mu_1,\ldots,\mu_q: \quad c; \qquad\qquad c+1\,; \end{array}\ z,z\right]$$
$$= F^{p+1:2;1}_{q+1:1;0}\left[\begin{array}{c} c+\frac{1}{2},\lambda_1,\ldots,\lambda_p:2a,2b;c-a-b+\frac{1}{2}; \\[2mm] c+1,\mu_1,\ldots,\mu_q: \quad 2c; \qquad\qquad -\ ; \end{array}\ z,z\right], \qquad (3.21)$$

which, for $c = a + b - \frac{1}{2}$, yields a known hypergeometric reduction formula

[36, p. 29, Eq. 1.3 (37)];

$$F_{q:1;1}^{p:2;2}\left[\begin{matrix}\lambda_1,\ldots,\lambda_p:a,b;c-a+\frac{1}{2},c-b-\frac{1}{2};\\[2mm]\mu_1,\ldots,\mu_q:\quad c;\qquad\qquad c;\end{matrix}\quad z,z\right]$$

$$=F_{p+1:1;0}^{p+1:2;1}\left[\begin{matrix}c-\frac{1}{2},\lambda_1,\ldots,\lambda_p:2a-1,2b;c-a-b+\frac{1}{2};\\[2mm]c,\mu_1,\ldots,\mu_q:\quad 2c-1;\qquad\qquad -;\end{matrix}\quad z,z\right],$$

(3.22)

which, for $c=a+b-\frac{1}{2}$, yields another known hypergeometric reduction formula [36, p. 29, Eq. 1.3 (38)];

$$F_{q:1;1}^{p:2;2}\left[\begin{matrix}\lambda_1,\ldots,\lambda_p:\qquad a,b;c-a+\frac{1}{2},c-b+\frac{1}{2};\\[2mm]\mu_1,\ldots,\mu_q:a+b+\frac{1}{2};\qquad 2c-a-b+\frac{1}{2};\end{matrix}\quad z,z\right]$$

$$=F_{q+1:2;0}^{p+1:3;1}\left[\begin{matrix}c+\frac{1}{2},\lambda_1,\ldots,\lambda_p:2a,2b,c\;;c-a-b+\frac{1}{2};\\[2mm]2c-a-b+\frac{1}{2},\mu_1,\ldots,\mu_q:2c,a+b+\frac{1}{2};\qquad -;\end{matrix}\quad z,z\right],$$

(3.23)

which, for $c=a+b-\frac{1}{2}$, yields one of the aforementioned known hypergeometric reduction formulas [36, p. 29, Eq. 1.3 (37)];

$$F_{q:1;1}^{p:2;2}\left[\begin{matrix}\lambda_1,\ldots,\lambda_p:a+\frac{1}{2},c-a+\frac{1}{2};b,c-b;\\[2mm]\mu_1,\ldots,\mu_q:\qquad c+\frac{1}{2};\;c+\frac{1}{2};\end{matrix}\quad z,z\right]$$

$$=F_{q+1:2;0}^{p+1:3;1}\left[\begin{matrix}a+b+\frac{1}{2},\lambda_1,\ldots,\lambda_p:\quad 2a,2b,c;c-a-b+\frac{1}{2};\\[2mm]c+\frac{1}{2},\mu_1,\ldots,\mu_q:2c,a+b+\frac{1}{2};\qquad -;\end{matrix}\quad z,z\right],$$

(3.24)

which, for $c=a+b-\frac{1}{2}$, yields one of the aforementioned known hypergeometric reduction formulas [36, p. 29, Eq. 1.3 (38)];

$$F_{q:1;1}^{p:2;2}\left[\begin{matrix}\lambda_1,\ldots,\lambda_p:\qquad a,b;\quad c-a,c-b;\\[2mm]\mu_1,\ldots,\mu_q:a+b+\frac{1}{2};2c-a-b+\frac{1}{2};\end{matrix}\quad z,z\right]$$

$$=F_{q+1:2;0}^{p+1:3;1}\left[\begin{matrix}c,\lambda_1,\ldots,\lambda_p:2a,2b,c+\frac{1}{2}\;;c-a-b;\\[2mm]2c-a-b+\frac{1}{2},\mu_1,\ldots,\mu_q:2c,a+b+\frac{1}{2};\qquad -\;;\end{matrix}\quad z,z\right],$$

(3.25)

which, for $c=a+b$, also yields the hypergeometric reduction formula (2.3).

Now we turn to certain interesting further generalizations of some of the above theorems. In fact, if we make use of the Cayley-Orr type identities given by Burchnall and Chaundy [6], we shall obtain several additional assertions contained in the following theorems.

Theorem 7. If

$$F^{p:3;1}_{q:2;0}\left[\begin{array}{l}\lambda_1,\ldots,\lambda_p : 2k+a-b-c-\frac{1}{2}, 2k+b-c-a-\frac{1}{2}, k-\frac{1}{2}; c-k+\frac{1}{2};\\[2mm]\mu_1,\ldots,\mu_q : \qquad\qquad\qquad\qquad 2k-1, 2k-c; \qquad -;\end{array}\ z,z\right]$$
$$=\sum_{n=0}^{\infty}\Theta_n z^n,\tag{3.26}$$

then

$$F^{p:2;2}_{q:1;1}\left[\begin{array}{l}\lambda_1,\ldots,\lambda_p : a,b; k-a, k-b;\\[2mm]\mu_1,\ldots,\mu_q : \quad c; \qquad 2k-c;\end{array}\ z,z\right]\tag{3.27}$$
$$=\sum_{n=0}^{\infty}\frac{(k)_n}{(c)_n}\Theta_n z^n \quad \left(c=a+b\pm\frac{1}{2}\right),$$

provided that the series involved converge absolutely.

Theorem 8. If

$$F^{p:3;1}_{q:2;0}\left[\begin{array}{l}\lambda_1,\ldots,\lambda_p : a-b+\frac{1}{2}, b-a+\frac{1}{2}, k-\frac{1}{2}; k-\frac{1}{2};\\[2mm]\mu_1,\ldots,\mu_q : \qquad\qquad\qquad c, 2k-c; \qquad -;\end{array}\ z,z\right]=\sum_{n=0}^{\infty}\Phi_n z^n,\tag{3.28}$$

then

$$F^{p:2;2}_{q:1;1}\left[\begin{array}{l}\lambda_1,\ldots,\lambda_p : a,b; k-a, k-b;\\[2mm]\mu_1,\ldots,\mu_q : \quad c; \qquad 2k-c;\end{array}\ z,z\right]\tag{3.29}$$
$$=\sum_{n=0}^{\infty}\frac{(k)_n}{(2k-1)_n}\Phi_n z^n \quad \left(c=a+b\pm\frac{1}{2}\right),$$

provided that the series involved converge absolutely.

Theorem 9. If

$$
F_{q:2;0}^{p:3;1}
\begin{bmatrix}
\lambda_1,\ldots,\lambda_p : & a-b+\tfrac{1}{2}, b-a+\tfrac{1}{2}, c;\ c; \\[2mm]
\mu_1,\ldots,\mu_q : 2c-a-b+\tfrac{1}{2}, a+b+\tfrac{1}{2};\ -;
\end{bmatrix} z, z
\tag{3.30}
$$
$$
= \sum_{n=0}^{\infty} \Psi_n z^n
$$

then

$$
F_{q:1;1}^{p:2;2}
\begin{bmatrix}
\lambda_1,\ldots,\lambda_p : & a,b; c-a+\tfrac{1}{2}, c-b+\tfrac{1}{2}; \\[2mm]
\mu_1,\ldots,\mu_q : a+b+\tfrac{1}{2}; & 2c-a-b+\tfrac{1}{2};
\end{bmatrix} z, z
\tag{3.31}
$$
$$
= \sum_{n=0}^{\infty} \frac{(c+\tfrac{1}{2})_n}{(2c)_n} \Psi_n z^n \,,
$$

provided that the series involved converge absolutely.

For $2k = c+1$, Theorem 7 and Theorem 8 would obviously become identical. If, in Theorem 7, we choose the value $c = a+b-\tfrac{1}{2}$, and replace a, b, and k by $c-a+\tfrac{1}{2}, c-b+\tfrac{1}{2}$, and $c+\tfrac{1}{2}$, respectively, so that (as a consequence) c is to be replaced by $2c-a-b+\tfrac{1}{2}$, we shall obtain Theorem 4. These same choices in Theorem 8 instead will yield Theorem 9. It should be remarked in passing that the alternative choice $c = a+b+\tfrac{1}{2}$ in Theorem 7 and Theorem 8 would lead us to essentially the same results.

Just as Theorems 1 to 6, our last assertions (Theorems 7, 8, and 9) can be restated as the following double hypergeometric transformations:

$$
F_{q:1;1}^{p:2;2}
\begin{bmatrix}
\lambda_1,\ldots,\lambda_p : a,b; k-a, k-b; \\[2mm]
\mu_1,\ldots,\mu_q :\ c;\quad 2k-c;
\end{bmatrix} z, z
$$
$$
= F_{q+1:2;0}^{p+1:3;1}
\begin{bmatrix}
k,\lambda_1,\ldots,\lambda_p : 2k+a-b-c-\tfrac{1}{2}, 2k+b-c-a-\tfrac{1}{2},\ k-\tfrac{1}{2};\ c-k+\tfrac{1}{2}; \\[2mm]
c,\mu_1,\ldots,\mu_q :\quad\quad 2k-1, 2k-c;\quad -;
\end{bmatrix} z, z
$$
$$
\left(c = a+b\pm\tfrac{1}{2} \right),
\tag{3.32}
$$

which provides an interesting generalization of the transformation formula
(3.23);

$$
F_{q:1;1}^{p:2;2}\left[\begin{array}{c}\lambda_1,\ldots,\lambda_p : a,b\,;\,k-a,k-b\,; \\ \\ \mu_1,\ldots,\mu_q : \quad c\,; \qquad 2k-c\,;\end{array}\; z,z\right]
$$

$$
= F_{q+1:2;0}^{p+1:3;1}\left[\begin{array}{c}k,\lambda_1,\ldots,\lambda_p : a-b+\frac{1}{2},b-a+\frac{1}{2},k-\frac{1}{2}\,;\,k-\frac{1}{2}\,; \\ \\ 2k-1,\mu_1,\ldots,\mu_q : \qquad\qquad c,2k-c\,; \qquad -\,;\end{array}\; z,z\right]
$$

$$
\left(c=a+b\pm\frac{1}{2}\right);
$$

$$\tag{3.33}$$

$$
F_{q:1;1}^{p:2;2}\left[\begin{array}{c}\lambda_1,\ldots,\lambda_p : \qquad a,b\,;\,c-a+\frac{1}{2},c-b+\frac{1}{2}\,; \\ \\ \mu_1,\ldots,\mu_q : a+b+\frac{1}{2}\,; \qquad 2c-a-b+\frac{1}{2}\,;\end{array}\; z,z\right]
$$

$$
= F_{q+1:2;0}^{p+1:3;1}\left[\begin{array}{c}c+\frac{1}{2},\lambda_1,\ldots,\lambda_p : \qquad a-b+\frac{1}{2},b-a+\frac{1}{2},c\,;\,c\,; \\ \\ 2c,\mu_1,\ldots,\mu_q : 2c-a-b+\frac{1}{2},a+b+\frac{1}{2}\,;\,-\,;\end{array}\; z,z\right],
$$

$$\tag{3.34}$$

which, as we observed above, corresponds to a special case of (3.33) when
$k = c + \frac{1}{2}$.

Finally, each of the double hypergeometric transformations recorded
above as corollaries of Theorems 1 to 9 can be extended as a series identity
analogous to (3.1). Thus, under the usual convergence requirements, it can
be shown, for every bounded sequence $\{\Omega_n\}_{n=0}^{\infty}$ of complex numbers, that

$$
\sum_{m,n=0}^{\infty}\Omega_{m+n}\frac{(c+\frac{1}{2})_{m+n}(a)_m(b)_m(c-a)_n(c-b)_n}{(c)_{m+n}(c+\frac{1}{2})_m(c+\frac{1}{2})_n}\frac{z^{m+n}}{m!n!}
$$

$$
= \sum_{m,n=0}^{\infty}\Omega_{m+n}\frac{(2a)_m(2b)_m(c-a-b)_n}{(2c)_m}\frac{z^{m+n}}{m!n!}
$$

$$\tag{3.35}$$

or, equivalently,

$$
\sum_{m,n=0}^{\infty}\Omega_{m+n}\frac{(a)_m(b)_m(c-a)_n(c-b)_n}{(c+\frac{1}{2})_m(c+\frac{1}{2})_n}\frac{z^{m+n}}{m!n!}
$$

$$
= \sum_{m,n=0}^{\infty}\Omega_{m+n}\frac{(c)_{m+n}(2a)_m(2b)_m(c-a-b)_n}{(c+\frac{1}{2})_{m+n}(2c)_m}\frac{z^{m+n}}{m!n!},
$$

$$\tag{3.36}$$

which, for $c = a + b$, readily yields the series identity (3.1);

$$\sum_{m,n=0}^{\infty} \Omega_{m+n} \frac{(a)_m (b)_m (c-a+\frac{1}{2})_n (c-b+\frac{1}{2})_n}{(c)_m (c+1)_n} \frac{z^{m+n}}{m!n!} \tag{3.37}$$

$$= \sum_{m,n=0}^{\infty} \Omega_{m+n} \frac{(c+\frac{1}{2})_{m+n} (2a)_m (2b)_m (c-a-b+\frac{1}{2})_n}{(c+1)_{m+n} (2c)_m} \frac{z^{m+n}}{m!n!} ;$$

$$\sum_{m,n=0}^{\infty} \Omega_{m+n} \frac{(a)_m (b)_m (c-a+\frac{1}{2})_n (c-b-\frac{1}{2})_n}{(c)_m (c)_n} \frac{z^{m+n}}{m!n!} \tag{3.38}$$

$$= \sum_{m,n=0}^{\infty} \Omega_{m+n} \frac{(c-\frac{1}{2})_{m+n} (2a-1)_m (2b)_m (c-a-b+\frac{1}{2})_n}{(c)_{m+n} (2c-1)_m} \frac{z^{m+n}}{m!n!} ;$$

$$\sum_{m,n=0}^{\infty} \Omega_{m+n} \frac{(a)_m (b)_m (c-a+\frac{1}{2})_n (c-b+\frac{1}{2})_n}{(a+b+\frac{1}{2})_m (2c-a-b+\frac{1}{2})_n} \frac{z^{m+n}}{m!n!}$$

$$= \sum_{m,n=0}^{\infty} \Omega_{m+n} \frac{(c+\frac{1}{2})_{m+n} (2a)_m (2b)_m (c)_m (c-a-b+\frac{1}{2})_n}{(2c-a-b+\frac{1}{2})_{m+n} (2c)_m (a+b+\frac{1}{2})_m} \frac{z^{m+n}}{m!n!} ; \tag{3.39}$$

$$\sum_{m,n=0}^{\infty} \Omega_{m+n} \frac{(a+\frac{1}{2})_m (c-a+\frac{1}{2})_m (b)_n (c-b)_n}{(c+\frac{1}{2})_m (c+\frac{1}{2})_n} \frac{z^{m+n}}{m!n!}$$

$$= \sum_{m,n=0}^{\infty} \Omega_{m+n} \frac{(a+b+\frac{1}{2})_{m+n} (2a)_m (2b)_m (c)_m (c-a-b+\frac{1}{2})_n}{(c+\frac{1}{2})_{m+n} (2c)_m (a+b+\frac{1}{2})_m} \frac{z^{m+n}}{m!n!} ; \tag{3.40}$$

$$\sum_{m,n=0}^{\infty} \Omega_{m+n} \frac{(a)_m (b)_m (c-a)_n (c-b)_n}{(a+b+\frac{1}{2})_m (2c-a-b+\frac{1}{2})_n} \frac{z^{m+n}}{m!n!}$$

$$= \sum_{m,n=0}^{\infty} \Omega_{m+n} \frac{(c)_{m+n} (2a)_m (2b)_m (c+\frac{1}{2})_m (c-a-b)_n}{(2c-a-b+\frac{1}{2})_{m+n} (2c)_m (a+b+\frac{1}{2})_m} \frac{z^{m+n}}{m!n!} ; \tag{3.41}$$

$$\sum_{m,n=0}^{\infty} \Omega_{m+n} \frac{(a)_m (b)_m (k-a)_n (k-b)_n}{(c)_m (2k-c)_n} \frac{z^{m+n}}{m!n!}$$

$$= \sum_{m,n=0}^{\infty} \Omega_{m+n} \frac{(k)_{m+n} (2k+a-b-c-\frac{1}{2})_m (2k+b-c-a-\frac{1}{2})_m (k-\frac{1}{2})_m (c-k+\frac{1}{2})_n}{(c)_{m+n} (2k-1)_m (2k-c)_m} \frac{z^{m+n}}{m!n!}$$

$$(c = a+b \pm \tfrac{1}{2}); \tag{3.42}$$

$$\sum_{m,n=0}^{\infty} \Omega_{m+n} \frac{(a)_m (b)_m (k-a)_n (k-b)_n}{(c)_m (2k-c)_n} \frac{z^{m+n}}{m!n!}$$

$$= \sum_{m,n=0}^{\infty} \Omega_{m+n} \frac{(k)_{m+n} (a-b+\frac{1}{2})_m (b-a+\frac{1}{2})_m (k-\frac{1}{2})_m (k-\frac{1}{2})_n}{(2k-1)_{m+n} (c)_m (2k-c)_m} \frac{z^{m+n}}{m!n!}$$

$$(c = a+b \pm \tfrac{1}{2}); \tag{3.43}$$

$$\sum_{m,n=0}^{\infty} \Omega_{m+n} \frac{(a)_m (b)_m (c-a+\frac{1}{2})_n (c-b+\frac{1}{2})_n}{(a+b+\frac{1}{2})_m (2c-a-b+\frac{1}{2})_n} \frac{z^{m+n}}{m!n!}$$

$$= \sum_{m,n=0}^{\infty} \Omega_{m+n} \frac{(c+\frac{1}{2})_{m+n} (a-b+\frac{1}{2})_m (b-a+\frac{1}{2})_m (c)_m (c)_n}{(2c)_{m+n} (2c-a-b+\frac{1}{2})_m (a+b+\frac{1}{2})_m} \frac{z^{m+n}}{m!n!} , \tag{3.44}$$

which is, of course, contained in the series identity (3.43).

For various special values of the parameters c and k, many of these series identities would simplify considerably. As an illustration, by setting $c = a + b - \frac{1}{2}$ in (3.37) or (3.39), we obtain the series identity:

$$\sum_{m,n=0}^{\infty} \Omega_{m+n} \frac{(a)_m(a)_n(b)_m(b)_n}{(a+b-\frac{1}{2})_m(a+b+\frac{1}{2})_n} \frac{z^{m+n}}{m!n!}$$
$$= \sum_{n=0}^{\infty} \Omega_n \frac{(2a)_n(2b)_n(a+b)_n}{(2a+2b-1)_n(a+b+\frac{1}{2})_n} \frac{z^n}{n!} , \tag{3.45}$$

which is a particularly remarkable companion of (3.1).

Acknowledgements

The present investigation was supported, in part, by the Natural Sciences and Engineering Research Council of Canada under Grant OGP0007353.

References

1. P. Appell and J. Kampé de Fériet, *Fonctions Hypergéométriques et Hypersphériques; Polynômes d' Hermite*, Gauthier-Villars, Paris, 1926.

2. R. Askey and G. Gasper, *Positive Jacobi polynomial sums*. II, Amer. J. Math. **98** (1976) 441–484.

3. W. N. Bailey, *Some theorems concerning products of hypergeometric series*, Proc. London Math. Soc. (2) **38** (1935) 377–384.

4. W. N. Bailey, *Generalized Hypergeometric Series*, Cambridge Tracts in Mathematics and Mathematical Physics, No. **32**, Cambridge University Press, Cambridge, London, and New York, 1935; Reprinted by Stechert-Hafner Service Agency, New York and London, 1964.

5. E. W. Barnes, *The asymptotic expansion of integral functions defined by generalized hypergeometric series*, Proc. London Math. Soc. (2) **5** (1907) 59–166.

6. J. L. Burchnall and T. W. Chaundy, *The hypergeometric identities of Cayley, Orr, and Bailey*, Proc. London Math. Soc. (2) **50** (1949) 56–74.

7. R. G. Buschman and H. M. Srivastava, *Series identities and reducibility of Kampé de Fériet functions*, Math. Proc. Cambridge Philos. Soc. **91** (1982) 435–440.

8. A. Cayley, *On a theorem relating to hypergeometric series*, Philos. Mag. (4) **16** (1858) 356–357.

9. T. Clausen, *Ueber die Fälle, wenn die Reihe von der Form*
$$y = 1 + \frac{\alpha \cdot \beta}{1 \cdot \gamma} x + \frac{\alpha(\alpha+1) \cdot \beta(\beta+1)}{1 \cdot 2 \cdot \gamma(\gamma+1)} x^2 + etc.$$
ein Quadrat von der Form
$$z = 1 + \frac{\alpha' \cdot \beta' \cdot \delta'}{1 \cdot \gamma' \cdot \epsilon'} x + \frac{\alpha'(\alpha'+1) \cdot \beta'(\beta'+1) \cdot \delta'(\delta'+1)}{1 \cdot 2 \cdot \gamma'(\gamma'+1) \cdot \epsilon'(\epsilon'+1)} x^2 + etc.$$
hat, J. Reine Angew. Math. **3** (1828) 89–91.

10. S. Lj. Damjanović and M. L. Glasser, *A hypergeometric function integral* (Problem 85-19), SIAM Rev. **28** (1986) 572–573.

11. L. de Branges, *A proof of the Bieberbach conjecture*, Acta Math. **154** (1985) 137–152.

12. A. Erdélyi, W. Magnus, F. Oberhettinger, and F. G. Tricomi, *Higher Transcendental Functions*, Vol. I, McGraw-Hill, New York, Toronto, and London, 1953.

13. A. Erdélyi, W. Magnus, F. Oberhettinger, and F. G. Tricomi, *Tables of Integral Transforms*, Vols. I and II, McGraw-Hill, New York, Toronto, and London, 1954.

14. G. Gasper, *A short proof of an inequality used by de Branges in his proof of the Bieberbach, Robertson and Milin conjectures*, Complex Variables Theory Appl. **7** (1986) 45–50.

15. G. Gasper, *q-Extensions of Clausen's formula and of the inequalities used by de Branges in his proof of the Bieberbach, Robertson, and Milin conjectures*, SIAM J. Math. Anal. **20** (1989) 1019–1034.

16. G. Gasper and M. Rahman, *A nonterminating q-Clausen formula and some related product formulas*, SIAM J. Math. Anal. **20** (1989) 1270–1282.

17. C. F. Gauss, *Disquisitiones Generales Circa Seriem Infinitam*
$$1 + \frac{\alpha \cdot \beta}{1 \cdot \gamma} x + \ldots,$$
Göttingen thesis, 1812; Comment. Soc. Reg. Sci. Göttingenis Recent. **2** (1813). [Reprinted in *Carl Friedrich Gauss Werke* (12 vols.), Vol. 3, pp. 123–162 (see also pp. 207–230), Göttingen, 1870–1933.]

18. É. Goursat, *Sur les Fonctions Hypergéométriques d' Ordre Supérieur*, Ann. Sci. École Norm. Sup. (2) **12** (1883) 261–285 and 395–430.

19. P. Henrici, *Über die Funktionen von Gegenbauer*, Arch. Math. (Basel) **5** (1954) 92–98.

20. P. Henrici, *On generating functions of the Jacobi polynomials*, Pacific J. Math. **5** (1955) 923–931.

21. P. Henrici, *On the product of two Kummer series*, Canad. J. Math. **10** (1958) 463–467.

22. F. H. Jackson, *The q^θ equations whose solutions are products of solutions of q^θ equations of lower order*, Quart. J. Math. (Oxford) **11** (1940) 1–17.

23. J. Kampé de Fériet, *Les Fonctions Hypergéométriques d' Ordre Supérieur à Deux Variables*, C. R. Acad. Sci. Paris **173** (1921) 401–404.

24. P. W. Karlsson, *Some reduction formulae for double power series and Kampé de Fériet functions*, Nederl. Akad. Wetensch. Indag. Math. **46** (1984) 31–36.

25. Y. L. Luke, *The Special Functions and Their Approximations*, Vol. I, Academic Press, New York, San Francisco, and London, 1969.

26. A. R. Miller, *Incomplete Lipschitz-Hankel integrals of Bessel functions*, J. Math. Anal. Appl. **140** (1989) 476–484.

27. A. R. Miller, *Reduction formulae for Kampé de Fériet functions* $F_{q:1;0}^{p:2;1}$, J. Math. Anal. Appl. **151** (1990), 428–437.

28. W. M. Orr, *Theorems relating to the product of two hypergeometric series*, Trans. Cambridge Philos. Soc. **17** (1899) 1–15.

29. E. D. Rainville, *Special Functions*, Macmillan, New York, 1960; Reprinted by Chelsea Publishing Company, Bronx, New York, 1971.

30. V. N. Singh, *The basic analogues of identities of the Cayley-Orr type*, J. London Math. Soc. **34** (1959) 15–22.

31. L. J. Slater, *Generalized Hypergeometric Functions*, Cambridge University Press, Cambridge, London, and New York, 1966.

32. H. M. Srivastava, *On the reducibility of Appell's function F_4*, Canad. Math. Bull. **16** (1973) 295–298.

33. H. M. Srivastava, *Some generalizations of Carlson's identity*, Boll. Un. Mat. Ital. A (5) **18** (1981) 138–143.

34. H. M. Srivastava, *Reduction and summation formulae for certain classes of generalised multiple hypergeometric series arising in physical and quantum chemical applications*, J. Phys. A: Math. Gen. **18** (1985) 3079–3085.

35. H. M. Srivastava and V. K. Jain, *q-Series identities and reducibility of basic double hypergeometric functions*, Canad. J. Math. **38** (1986) 215–231.

36. H. M. Srivastava and P. W. Karlsson, *Multiple Gaussian Hypergeometric Series*, Halsted Press (Ellis Horwood Limited, Chichester), John Wiley and Sons, New York, Chichester, Brisbane, and Toronto, 1985.

37. H. M. Srivastava and H. L. Manocha, *A Treatise on Generating Functions*, Halsted Press (Ellis Horwood Limited, Chichester), John Wiley and Sons, New York, Chichester, Brisbane, and Toronto, 1984.

38. H. M. Srivastava and S. Owa (Editors), *Univalent Functions, Fractional Calculus, and Their Applications*, Halsted Press (Ellis Horwood Limited, Chichester), John Wiley and Sons, New York, Chichester, Brisbane, and Toronto, 1989.

39. E. T. Whittaker and G. N. Watson, *A Course of Modern Analysis*, Fourth Edition, Cambridge University Press, Cambridge, London, and New York, 1927.

H. M. Srivastava
Department of Mathematics and Statistics
University of Victoria
Victoria, British Columbia V8W 3P4
Canada

GAUSS' GAMMA MULTIPLICATION THEOREM: ANALOGUES AND EXTENSIONS

Kenneth B. Stolarsky

In "Circa Seriem Infinitam $1 + \frac{\alpha\beta}{1\cdot\gamma}x + ...$" (Werke, Königlichen Gesellschaft der Wissenschaften zu Göttingen, 1876, Band 3, pp. 125–162) Gauss develops the theory of the $F(\alpha,\beta,\gamma;x)$ hypergeometric series, and its summation for $x=1$ in terms of gamma functions. In connection with this he also develops, on the basis of its functional equation (pp. 144–150), the theory of the gamma function itself, with special attention to the "theorema elegans" now called the "Gauss multiplication theorem." Here we follow a similar path towards various such multipication theorems for certain analytic functions that satisfy *second order difference equations*

$$F(n+1)=nF(n)^{\alpha}F(n+1)^{\beta}$$

where $\beta \geq 0$ and $\alpha+\beta=1$. For $\beta=0$ this reduces to the functional equation of the usual gamma function.

1. Introduction

In "Circa Seriem Infinitam $1 + \frac{\alpha\beta}{1\cdot\gamma}x + ...$" (Werke, Königlichen Gesellschaft der Wissenschaften zu Göttingen, 1876, Band 3, pp. 125–162) Gauss develops the theory of the $F(\alpha,\beta,\gamma;x)$ hypergeometric series, and its summation for $x=1$ in terms of gamma functions. In connection with this he also develops, on the basis of its functional equation (pp. 144–150), the theory of the gamma function itself, with special attention to the "theorema elegans" now called the "Gauss multiplication theorem." Here we follow a similar path towards various such multiplication theorems for certain analytic functions that satisfy *second* order difference equations

$$F(n + 1) = nF(n)^{\alpha}F(n + 1)^{\beta} \qquad (1.1)$$

where $\beta \geq 0$ and $\alpha + \beta = 1$. For $\beta = 0$ this reduces to the functional equation of the usual gamma function. Our results are related to a formal method (described herein) for reducing the order of certain functional difference equations. We also examine some possible convexity characterizations of our functions.

We remark that Gauss' notation differs slightly from ours, especially in that he uses Πs for what we called $\Gamma(s + 1)$. In his own words ([12], pp. 145–146]), "Definimus ... functionem Πz per valorem producti

$$\frac{1 \cdot 2 \cdot 3 \ldots k \cdot k^{z}}{(z + 1)(z + 2)(z + 3) \ldots (z + k)}$$

pro $k = \infty \ldots$. Pro valore integro negativo ipsius z erit valor functionis Πz infinite magnus; pro valoribus integris non negativus habemus $\Pi 0 = 1$, $\Pi 1 = 1$, $\Pi 2 = 2, \ldots$ etc".

Every solution of the functional equation (1.1) satisfying the natural growth condition (5.8) has a duplication formula. The following solution seems to be the most natural and interesting.

For $\beta > 0$ and s complex (but neither zero nor a negative real number) let

$$q(s) = (1 + \beta)^{-1}\left\{\frac{\Gamma'(s)}{\Gamma(s)} + \beta \int_{0}^{1} t^{s-1}(\beta + 1)^{-1}dt\right\} \qquad (1.2)$$

and

$$F(s) = F(s, \beta) = A \exp \int_{1}^{s} q(t)dt \qquad (1.3)$$

where $A = F(1)$ is a constant and the path of integration is a line segment. Then $F(s, \beta)$ satisfies (1.1) and

$$F(s + 1)F(s)^{\beta} = F(2)F(1)^{\beta}\Gamma(s + 1) ; \qquad (1.4)$$

for $\beta = 1$ (see (7.14)) this reduces to the usual Legendre duplication formula. These facts are proved (after some basic estimates are obtained in Secs. 3 and 4) in Secs. 5, 6 and 7. Theorem 4 in Sec. 7 then gives a generalization of the Gauss multiplication formula.

In Sec. 9 we characterize (Theorem 6) the above $F(s)$ from the collection of all solutions of (1.4) as the "most natural" for $\beta \geq 1$ (and hence for all positive β if the solution is to be analytic in β). As in [1, pp. 14–15], the characterization is by a convexity condition. A "less natural" solution to (1.4) is considered in Sec. 8.

The secondary object of this paper is the introduction of a purely formal method for reducing the order of certain functional difference equations (Sec. 2). Theorem 2 is in fact a proof of the validity of the reduction method in a particular cae. These equations are exponentiated versions of

$$u(m) = t(m) + [a_0 u(m) + a_1 u(m-1) + \ldots + a_m u(0)] \qquad (1.5)$$

where $a_0 = 0, a_n = 0$ for all sufficiently large n, and

$$\sum_{m=0}^{\infty} a_m = 1 . \qquad (1.6)$$

We remark that equations of this type can be considered as somewhat degenerate cases of the renewal equation [11, pp. 278–308].

2. A Formal Reduction Process

Consider the difference equation

$$F(s+m) = \phi(s)F(s+m-1)^{a_1}F(s+m-2)^{a_2}\ldots F(s)^{a_m} \qquad (2.1)$$

where $s > 0, m$ is a positive integer, $a_i \geq 0$, and

$$\sum_{i=1}^{m} a_i = 1 . \qquad (2.2)$$

Say there is an $f(s)$ such that

$$f(s+1) = \phi(s)f(s) . \qquad (2.3)$$

Then if G satisfies the equation

$$G(s+m-1)G(s+m-2)^{b_2}G(s+m-3)^{b_3}\ldots G(s)^{b_m}$$
$$= f(s), \qquad m \geq 2 , \qquad (2.4)$$

it follows that

$$G(s+m)G(s+m-1)^{b_2}\ldots G(s+1)^{b_m} = f(s+1) = \phi(s)f(s)$$
$$= \phi(s)G(s+m-1)G(s+m-2)^{b_2}\ldots G(s)^{b_m} \tag{2.5}$$

or

$$G(s+m) = \phi(s)G(s+m-1)^{1-b_2}G(s+m-2)^{b_2-b_3}\ldots$$
$$\ldots G(s+1)^{b_{m-1}-b_m}G(s)^{b_m} . \tag{2.6}$$

Thus if we set

$$b_k = a_k + a_{k+1} + \ldots + a_m \tag{2.7}$$

Eq. (2.6) becomes (2.1) with G in place of F. Thus we have "reduced" Eq. (2.1) to the difference equation (2.4) of lower order. This is, of course, purely formal; in the absence of a uniqueness theorem there is no reason why G should coincide with the original function F.

We now examine (2.3). If $\phi(s) = 1$ we take $f(s) = 1$. If $f_n(s)$ is defined by

$$f_n(s)^{(-1)^n} = \prod_k \left\{ (s+2k)^{\binom{n}{2k}}(s+2k-1)^{-\binom{n}{2k-1}} \right\} \tag{2.8}$$

for $n > 0$, then

$$f_n(s+1) = f_{n+1}(s)f_n(s) , \tag{2.9}$$

so if $\phi(s) = f_{n+1}(s)$ for $n \geq 0$ we can take $f(s) = f_n(s)$. Here

$$f_1(s) = 1 + s^{-1}, \quad f_0(s) = s . \tag{2.10}$$

If $\phi(s) = s$ the above pattern breaks down, but we can take $f(s) = \Gamma(s)$. For $\phi(s) = \Gamma(s)$ we can use the function $G(s)$ studied by Barnes [6] that satisfies

$$G(s+1) = \Gamma(s)G(s) ,$$

and so forth. Thus the gamma function is a member of a hierarchy of functions, with the identically one function at one extreme, and functions of very rapid growth at the other.

We shall see in Sec 6 that the functional equation (2.9) provides the key to producing a continuum of duplication formulae.

3. Lemmas on $f_n(s)$

Set

$$y_n(s) = f_n(s)^{(-1)^n} . \qquad (3.1)$$

We shall establish various facts about the behavior of $y_n(s)$ for $s > 0$. Since

$$\sum_k \binom{n}{2k} = \sum_k \binom{n}{2k-1} \qquad (3.2)$$

we have from (2.8) that

$$\lim_{s \to \infty} y_n(s) = 1 . \qquad (3.3)$$

Thus

$$y_n(s) = 1 + a_1 s^{-1} + a_2 s^{-2} + \dots . \qquad (3.4)$$

Next, an easy induction argument shows that

$$y_n'(s)/y(s) = \frac{d}{ds} \log y_n(s)$$

$$= \frac{d}{ds} \sum_{k=0}^{n} \left\{ \binom{n}{2k} \log(s+2k) - \binom{n}{2k-1} \log(s+2k-1) \right\}$$

$$= \frac{n!}{s(s+1)\dots(s+n)} . \qquad (3.5)$$

This shows that $y_n(s)$ is an increasing function, and also that it is logarithmically concave. Equation (3.5) also shows that $y_n'(s) = 0(s^{-n-1})$, and hence

$$y_n(s) = 1 + a_n s^{-n} + a_{n+1} s^{-n-1} + \dots . \qquad (3.6)$$

It is also easy to see that $a_n = -(n-1)!$ Equation (3.5) moreover yields the representation

$$y_n(s) = \exp\left(-\int_s^\infty [n!/t(t+1)\dots(t+n)]\, dt \right) . \qquad (3.7)$$

From (3.7) we can obtain a bound on $\log y_n(s)$ that is independent of n for $n \geq 2$. Since

$$\int_s^\infty \frac{n!}{t(t+1)\dots(t+n)} dt \leq \frac{1}{s} \int_0^\infty \frac{dt}{(t+1)\left(\frac{t}{2}+1\right)\dots\left(\frac{t}{n}+1\right)}$$

$$\leq \frac{2}{s} \int_0^\infty \frac{dt}{(t+1)^2} = \frac{2}{s} \qquad (3.8)$$

we have

$$-2/s \leq \log y_n(s) \leq 0 \; ; \tag{3.9}$$

the upper bound is immediate from the fact that $y_n(s)$ is increasing.

4. Asymptotic Behavior of $F(s)$

Let $\beta > 0, \alpha = 1 - \beta$, and let $F(s) = F(s; \beta)$ be a function defined and positive for $s > 0$ such that

$$F(s + 1) = sF(s)^{1-\beta} F(s - 1)^{\beta} \; . \tag{4.1}$$

Let

$$E(k) = \left(1 + (-1)^{k+1}\beta^k\right)/(1 + \beta), \quad k = 0, 1, 2, \dots \; . \tag{4.2}$$

It is easy to see that $E(0) = 0, E(1) = 1, E(2) = \alpha$,

$$E(k + 2) = \alpha E(k + 1) + \beta E(k), \quad k > 1 \; , \tag{4.3}$$

and

$$E(k + 1) = 1 - \beta E(k) \; . \tag{4.4}$$

Lemma. If $u > m + 1$, then

$$F(u) = F(u - m)^{E(m+1)} F(u - m - 1)^{1-E(m+1)} \prod_{k=1}^{m} (u - k)^{E(k)} \; . \tag{4.5}$$

Proof. For $m = 1$ this is simply (4.1). From

$$\begin{aligned}
F(u &- m)^{E(m+1)} F(u - m - 1)^{1-E(m+1)} \\
&= (u - m - 1)^{E(m+1)} F(u - m - 1)^{\alpha E(m+1)+1-E(m+1)} \\
&\quad \times F(u - m - 2)^{\beta E(m+1)} \\
&= (u - m - 1)^{E(m+1)} F(u - m - 1)^{E(m+2)} \\
&\quad \times F(u - m - 2)^{1-E(m+2)}
\end{aligned} \tag{4.6}$$

we see that its truth for m_1 implies its truth for $m_1 + 1$, provided $u > (m_1 + 1) + 1$.

Now let $u = x + n$ and $m = n$ in (4.5). Thus

$$F(n + x) = F(x)^{E(n+1)} F(x - 1)^{1-E(n+1)} \prod_{k=1}^{n} (n + x - k)^{E(k)} , \qquad (4.7)$$

so

$$\log F(n + x) = E(n + 1) \log F(x) + [1 - E(n + 1)] \log F(x - 1)$$
$$+ \sum_{k=1}^{n} E(n + 1 - k) \log(x + k - 1) . \qquad (4.8)$$

The last sum on the right is

$$\frac{1}{1 + \beta} \left\{ \sum_{k=1}^{n} \log(x + k - 1) + (-1)^n \beta^{n+1} \sum_{k=1}^{n} (-\beta)^{-k} \log(x + k - 1) \right\} . \qquad (4.9)$$

By Stirling's formula [11, p. 64], the first sum in (4.9) is

$$\log \Gamma(n + x) - \log \Gamma(x) = n \log n - n + \left(x - \frac{1}{2} \right) \log n$$
$$+ \frac{1}{2} \log 2\pi - \log \Gamma(x) + x(x - 1)/2n + 1/12n + 0(n^{-3/2}) . \qquad (4.10)$$

We shall apply summation by parts to the second sum in (4.9). Define

$$\sigma_m^{(1)} = \sum_{j=1}^{m} (-\beta)^{-j} = (1 + \beta)^{-1} (-\beta)^{-m} - (1 + \beta)^{-1} \qquad (4.11)$$

and

$$\sigma_m^{(d+1)} = \sum_{j=1}^{m} \sigma_j^{(d)} = (1 + \beta)^{-d-1} (-\beta)^{-m} + 0(m^d) . \qquad (4.12)$$

Thus (recall the notation (3.1))

$$\sum_{k=1}^{n} (-\beta)^{-k} \log(x + k - 1)$$
$$= \sigma_n^{(1)} \log y_0(x + n - 1) + \sigma_{n-1}^{(2)} \log y_1(x + n - 2)$$
$$+ \sigma_{n-2}^{(3)} \log y_2(x + n - 3) + \sigma_{n-3}^{(4)} \log y_3(x + n - 4)$$
$$+ \sigma_{n-4}^{(5)} \log y_4(x + n - 5) + \ldots + \sigma_{n-q}^{(q+1)} \log y_q(x + n - q - 1)$$
$$+ \sum_{k=1}^{n-q-1} \sigma_k^{(q+1)} \log y_{q+1}(x + k - 1) . \qquad (4.13)$$

Let q be fixed and n large; in particular, assume $n > q + 1$. The last sum on the right of (4.13) is

$$(1 + \beta)^{-q-1} \sum_{k=1}^{n-q-1} (-\beta)^{-k} \log y_{q+1}(x + k - 1)$$

$$+ 0(n^q) \sum_{k=1}^{n-q-1} \left| \log y_{q+1}(x + k - 1) \right| . \tag{4.14}$$

By (3.9), the error term in (4.14) is

$$0(x^{-1} n^{q+1}) . \tag{4.15}$$

Now break the first sum in (4.14) into two parts, the first for $1 < k \le n/2$ and the second for $n/2 < k \le n - q - 1$. From (3.9) and (3.6) these sums are respectively

$$0(\beta^{-1-n/2} x^{-1}), \quad 0(\beta^{-n+q+1} q! \, [n/2 + x - 1]^{-q-1}) . \tag{4.16}$$

Next, the first $(q + 1)$ terms on the right of (4.13) are (recall (3.9))

$$\sum_{j=0}^{q} \sigma_{n-j}^{(j+1)} \log y_j (x + n - j - 1)$$

$$= \sum_{j=0}^{q} (1 + \beta)^{-j-1} (-\beta)^{-n+j} \log y_j (x + n - j - 1)$$

$$+ 0 \left[n^q \log \frac{x + 1 + n}{x + 1} \right] . \tag{4.17}$$

Thus (4.15), (4.16), (4.17) yield, for $0 < \beta < 1$,

$$(-1)^n \beta^{n+1} \sum_{k=1}^{n} (-\beta)^{-k} \log(x + k - 1)$$

$$= \frac{\beta}{1 + \beta} \log(x + n - 1) - \frac{\beta^2}{(1 + \beta)^2} \log y_1 (x + n - 2)$$

$$+ \frac{\beta^3}{(1 + \beta)^3} \log y_2 (x + n - 3) + \ldots$$

$$+ \frac{\beta(-\beta)^q}{(1 + \beta)^{q+1}} \log y_q (x + n - q - 1) + 0(n^{-q-1}) . \tag{4.18}$$

Hence from (4.10) and (4.18) we have

$$\sum_{k=1}^{n} E(n+1-k)\log(x+k-1)$$

$$= \frac{1}{1+\beta}\Big\{ n\log n - n + \Big(x - \frac{1}{2}\Big)\log n + \frac{1}{2}\log 2\pi - \log\Gamma(x)$$

$$+ \frac{\beta}{1+\beta}\log(x+n-1) - \frac{\beta^2}{(1+\beta)^2}\log y_1(x+n-2) + \dots \Big\}.$$

$$(4.19)$$

This and (4.8) now lead us to the following fundamental formula for F, namely

$$\log F(x+n)^{1+\beta} = \log F(x) + \beta\log F(x-1) + n\log n - n$$

$$+ \Big(x - \frac{1}{2} + \frac{\beta}{1+\beta}\Big)\log n + \frac{1}{2}\log 2\pi - \log\Gamma(x)$$

$$+ \Big[\frac{1}{12} + \frac{x(x-1)}{2} + \frac{\beta}{1+\beta}(x-1) + \frac{\beta^2}{(1+\beta)^2}\Big]\frac{1}{n}$$

$$+ 0(n^{-3/2}).$$

$$(4.20)$$

5. Duplication Formulae

Let $F(s) = F(s;\beta)$ be the function studied in Sec. 4. If (4.20) is written first with x replaced by $x+1$, and then by 2, and these equations are subtracted, we obtain

$$\log F(n+x+1) - \log F(n+2) - (1+\beta)^{-1}(x-1)\log n$$

$$= (1+\beta)^{-1}\Big[\beta\log F(x) + \log F(x+1) - \beta\log F(1)$$

$$- \log F(2) - \log\Gamma(x+1) + 0\Big(\frac{1}{n}\Big)\Big],$$

$$(5.1)$$

so

$$\lim_{n\to\infty} \frac{F(n+x+1)}{n^{(x-1)/(1+\beta)}F(n+2)} = \Big\{ \frac{F(x+1)F(x)^\beta}{F(2)F(1)^\beta\Gamma(x+1)} \Big\}^{\frac{1}{1+\beta}}.$$

$$(5.2)$$

An expository note of B. Berndt [9, esp. pp. 6–7] shows how the theory of the gamma function can be developed from the following simple result.

Theorem 1. If s is not a non-positive integer, then there exists a unique function $F(s)$ satisfying

$$F(1) = 1 , \tag{5.3}$$

$$F(s + 1) = sF(s) , \tag{5.4}$$

and

$$\lim_{n \to \infty} \frac{F(s+n)}{n^s F(n)} = 1 . \tag{5.5}$$

This approach is close to that of Gauss [12, pp. 144–147] and is probably quite similar to the original approach of Euler. We now obtain an analogous result.

Theorem 2. If $\alpha + \beta = 1$ where $0 \le \beta < 1$ and the function $F(s)$ satisfies

$$F(1) > 0, \quad F(2) > 0 , \tag{5.6}$$

$$F(s + 1) = sF(s)^\alpha F(s - 1)^\beta, \quad s > 1 , \tag{5.7}$$

and

$$\lim_{n \to \infty} \frac{F(n+s)}{n^{s/(1+\beta)} F(n)} = c(\beta) > 0 \tag{5.8}$$

where $c(\beta)$ is independent of s, then

$$F(s + 1)F(s)^\beta = F(2)F(1)^\beta \Gamma(s + 1) . \tag{5.9}$$

Proof. From (5.2), with $s = x - 1$, we obtain (5.9) with an extra factor of $c(\beta)^{1+\beta}$ on the right. By setting $x = 1$, we find that this factor is 1.

Theorem 3. Let $0 \le \beta < 1$. Let $F(s)$ be a function defined for $s > 0$ such that $F(1) > 0, F(2) > 0$, and (5.9) is valid. Then (5.7) and (5.8) are valid with $c(\beta) = 1$.

Proof. Since

$$\Gamma(s+1) = s\Gamma(s) = sF(s)F(s-1)^\beta/F(2)F(1)^\beta , \qquad (5.10)$$

we immediately deduce (5.7) from (5.9). Thus (5.2) is valid, so (5.8) follows with $c(\beta) = 1$.

6. The "Natural" Solution

Let A be a positive constant and $0 \le \beta < 1$. In this section we study the function $F(s) = F(s; \beta)$ satisfying (5.6)–(5.8) that is defined by

$$F(s) = A\Gamma(s)^{\frac{1}{1+\beta}} \prod_{n=0}^{\infty} f_n(s)^{(-1)^n \beta/(1+\beta)^{n+2}} . \qquad (6.1)$$

By (3.8) we have

$$\left| \log f_n(s)^{(-1)^n} \right| \le 2/s, \quad s > 0 , \qquad (6.2)$$

so the product converges uniformly and absolutely for $s > \delta > 0$. It is now easy to show that

$$F(s+1)F(s)^\beta = A^{1+\beta}\Gamma(s+1) \qquad (6.3)$$

by exploiting (2.9). As in the proof of Theorem 3, it follows from (6.3) that $F(s)$ satisfies (5.6)–(5.8). From (3.5) we see, moreover, that $F(s)/\Gamma(s)^{\frac{1}{1+\beta}}$ is logarithmically concave.

Greater insight into the analytic nature of $F(s)$ can be obtained by starting from the logarithmic derivative of (6.1). As usual, we define the psi function by

$$\psi(s) = \Gamma'(s)/\Gamma(s) . \qquad (6.4)$$

Let s be a complex number not on the closure of the negative real axis, and set

$$q(s) = (1+\beta)^{-1}\psi(s) + \sum_{n=0}^{\infty} \beta n!/(1+\beta)^{n+2} s(s+1)\ldots(s+n) \qquad (6.5)$$

and

$$F(s) = A \exp \int_1^s q(t)\, dt \tag{6.6}$$

where $A = F(1) > 0$ and where the path of integration is a straight line. Then $F(s)$ satisfies the differential equation

$$F'(s)/F(s) = q(s) \ . \tag{6.7}$$

Lemma. Let

$$\begin{aligned}
Q(k) = {} & \sum_{n=k+1}^{\infty} \frac{(n-1)!\beta}{(1+\beta)^{n+1}(s+1)\ldots(s+n)} \\
& + \sum_{n=k}^{\infty} \frac{n!\beta^2}{(1+\beta)^{n+2}(s+1)\ldots(s+n)} \\
& - k!\beta/(1+\beta)^{k+1}s(s+1)\ldots(s+k) \ .
\end{aligned} \tag{6.8}$$

Then $Q(k) = 0$ for $k = 0, 1, 2, \ldots$.

Proof. It is easy to show that $Q(k) = Q(k+1)$; the result follows by letting $k \to \infty$.

It is now straightforward to deduce

$$\frac{F'(s+1)}{F(s+1)} + \beta\frac{F'(s)}{F(s)} = \psi(s+1) \tag{6.9}$$

from (6.7), (6.8) with $k = 0$, and the identity

$$\psi(s+1) = \frac{1}{s} + \psi(s) \ . \tag{6.10}$$

Upon integrating both sides of (6.9) we obtain

$$F(s+1)F(s)^{\beta} = F(2)F(1)^{\beta}\Gamma(s+1) \ . \tag{6.11}$$

By means of the identity

$$\int_0^1 (1-t)^n t^{s-1} dt = \frac{n!}{s(s+1)\ldots(s+n)} \tag{6.12}$$

we can transform (6.5) into the simpler form

$$q(s) = \frac{1}{1+\beta}\left\{\psi(s) + \beta \int_0^1 \frac{t^{s-1}}{\beta+t}\, dt\right\}. \tag{6.13}$$

From (6.10) and (6.13) it is easy to verify, *independently of the preceding material in this chapter*, that (6.9) is satisfied with $q(s)$ in place of the logarithmic derivative of $F(s)$. Also, by rewriting the second term in (6.13) as

$$\int_0^1 t^{s-1}dt - \int_0^1 \frac{t^s dt}{\beta+t} \tag{6.14}$$

we see, for $s > 0$, that

$$q(s) = q(s,\beta) \to \psi(s) \tag{6.15}$$

as $\beta \to 0$. Thus for $s > 0$ we have

$$F(s) = F(s,\beta) \to A\Gamma(s) \tag{6.16}$$

as $\beta \to 0$. The fact that

$$F(s) = F(s,\beta) \to A2^{(s-1)/2}\Gamma[(s+1)/2] \tag{6.17}$$

as $\beta \to 1$ is less immediate. We defer the demonstration to the next section, where a more general result is obtained.

From (6.5) and (6.6) we see that for $\beta > 0$ the function $q(s)$ is analytic except at the *strictly* negative integers; a simple calculation shows that the singularity at $s = 0$ is removable. In fact, for s near $-k$ the two terms of (6.5) behave like

$$-(1+\beta)^{-1}(s+k)^{-1}, \quad (1+\beta)^{-1}(-1)^k \beta^{-k}(s+k)^{-1} \tag{6.18}$$

respectively (these terms recall $E(k)$; see (4.2)). Since the sum of the terms in (6.18) is nonzero for $k > 0$, the function $F(s)$ is the exponential of a function with branch points at the strictly negative integers.

It is in fact the case that any $F(s)$ satisfying (5.9) for negative s must be singular at the strictly negative integers. A simple induction shows that for $\beta > 0$ the relationship

$$|F(s+1)||F(s)|^\beta = |\Gamma(s+1)| \tag{6.19}$$

implies that

$$|F(s)| \sim (s+k)^{\sum_{m=1}^{k}(-\beta)^{-m}} \prod_{j=1}^{k-1}(j!)^{(-\beta)^{j-k}} \tag{6.20}$$

as $s \to -k$, where k is a positive integer.

Rademacher's excellent exposition of the theory of the classical gamma function [15, pp. 30–46] shows that the "real reason" why $\Gamma(s)$ has integral representations involving e^{-s} is that $\Gamma(s)$ has a simple pole with residue $(-1)^n/n!$ at $-n$, and is otherwise well behaved on the negative real axis. Formula (6.20) seems to rule out such phenomena for $0 < \beta < 1$ and $1 < \beta$. Also note that, in spite of the continuity result (6.16), singularities of "arbitrarily high order" are present for *any* positive β with $0 < \beta < 1$, no matter how small β is.

The logarithmic concavity of $F(s)\Gamma(s)^{-\frac{1}{1+\beta}}$ is an immediate consequence of (6.5) and (6.7). It can also be seen immediately by differentiating the second term of (6.13).

Note that (6.6) only provides us with a one parameter family of solutions to (5.9), since

$$F(2)/F(1) = \exp \int_1^2 q(t) \, dt \tag{6.21}$$

is a well defined function of β.

We now give a very simple proof of a very general duplication formula.

Proposition. If explicit solutions $q_1(s), q_2(s)$ can be found to the difference equation

$$f(s+1) + \beta f(s) = y(s) \tag{6.22}$$

for two distinct values of β, say β_1 and β_2 respectively, then the difference equation

$$f(s+1) + \beta_2 f(s) = q_1(s+1) \tag{6.23}$$

can be solved explicitly. In fact,

$$f(s) = \frac{\beta_2}{\beta_2 - \beta_1} q_2(s) - \frac{\beta_1}{\beta_2 - \beta_1} q_1(s) \tag{6.24}$$

is a solution.

Proof. This follows by substituting (6.24) into (6.23).

All the functions appearing in (6.23) satisfy the same sort of difference equation, so it has a right to be called a "duplication formula", especially for $\beta_2 = 1$. For $\beta_2 = \beta, \beta_1 = -1, y(s) = s^{-1}$, and

$$q_1(s) = \psi(s), \quad q_2(s) = \int_0^1 t^{s-1}(\beta+1)^{-1}dt , \tag{6.25}$$

Eqs. (6.23) and (6.24) assert that $f(s) = q(s)$, where $q(s)$ is given by (6.13), and that

$$q(s+1) + \beta q(s) = \Gamma'(s+1)/\Gamma(s+1) . \tag{6.26}$$

But (6.26) is equivalent to (6.9), and we have already shown that (6.9) yields our original generalized duplication formula.

7. The Gauss Multiplication Theorem

By analogy with the $q(s)$ of Sec. 6, define

$$q(s) = q(s;m) = q(s;m;\alpha_1,\dots,\alpha_n)$$
$$= \frac{1}{1+\beta_1}\left\{\psi(s) + \int_0^1 \frac{t^{s-1}(\beta_1 t^{m-1} + \beta_2 t^{m-2} + \dots + \beta_m)}{t^m + \alpha_1 t^{m-1} + \dots + \alpha_m} \, dt\right\}$$

$$\tag{7.1}$$

where $s > 0$,

$$\beta_j = \alpha_j + \dots + \alpha_m , \tag{7.2}$$

and $\alpha_i \geq 0$ for $i = 1,\dots,m$. Also, set $\alpha_0 = 1$. Note that formula (7.2) recalls (2.7). We shall establish an extension of the Gauss multiplication theorem

$$\Gamma(s) = (m+1)^{s-\frac{1}{2}}(2\pi)^{-m/2}\Gamma\left(\frac{s}{m+1}\right)\Gamma\left(\frac{s+1}{m+1}\right)\dots\Gamma\left(\frac{s+m}{m+1}\right) , \tag{7.3}$$

or more precisely of its logarithmic derivative

$$\psi(s) = \log(m+1) + (m+1)^{-1}\sum_{j=0}^{m}\psi\left(\frac{s+j}{m+1}\right) , \tag{7.4}$$

for the function $q(s)$. We use the well-known fact that

$$\psi(s) = -s^{-1} - \gamma - \sum_{k=1}^{\infty}\left(\frac{1}{s+k} - \frac{1}{k}\right) \tag{7.5}$$

where γ denotes Euler's constant.

Proposition. For $s > 0$ we have

$$q(s+m) + \alpha_1 q(s+m-1) + \ldots + \alpha_m q(s) = \psi(s+m) . \tag{7.6}$$

Proof. Since

$$\int_0^1 t^{s-1} dt = s^{-1} , \tag{7.7}$$

it follows that

$$\sum_{i=0}^m \alpha_i q(s+m-i)$$

$$= (1+\beta_1)^{-1} \sum_{i=0}^m \alpha_i \psi(s+m-i) + (1+\beta_1)^{-1} \sum_{k=1}^m \frac{\beta_k}{s+m-k} \tag{7.8}$$

upon evaluating the integral. We now make double use of

$$\psi(s+n) = \sum_{j=1}^n \frac{1}{s+n-j} + \psi(s) . \tag{7.9}$$

By (7.9) the first term on the right of (7.8) is

$$(1+\beta_1)^{-1} \sum_{i=0}^m \sum_{j=1}^{m-i} \frac{\alpha_i}{s+m-i-j} + \psi(s)$$

$$= (1+\beta_1)^{-1} \sum_{k=1}^m \sum_{i+j=k} \frac{\alpha_i}{s+m-(i+j)} + \psi(s)$$

$$= (1+\beta_1)^{-1} \sum_{k=1}^m \frac{1+\beta_1-\beta_k}{s+m-k} + \psi(s) . \tag{7.10}$$

The results follows upon substituting (7.10) back into (7.8) and making a second application of (7.9).

Proposition. For $s > 0$ we have

$$\int_0^1 t^{s-1} \frac{mt^{m-1} + (m-1)t^{m-2} + \ldots + 1}{t^m + t^{m-1} + \ldots 1} \, dt$$

$$= \frac{1}{m+1} \left[(m+1)\psi\left(\frac{s+m}{m+1}\right) - \sum_{j=0}^m \psi\left(\frac{s+j}{m+1}\right) \right] . \tag{7.11}$$

Proof. The left hand side is

$$\int_0^1 t^{s-1}\frac{(1+t+\ldots+t^{m-1})-mt^m}{1-t^{m+1}}\,dt$$

$$=\sum_{k=0}^{\infty}\left(\sum_{j=0}^{m-1}[(s+(m+1)k+j)]^{-1}-m[s+k(m+1)+m]^{-1}\right)$$

$$=\sum_{j=0}^{m}[(s+j)^{-1}-(s+m)^{-1}]+(m+1)^{-1}\sum_{j=0}^{m}\sum_{k=1}^{\infty}\{[(s+j)$$

$$\times(m+1)^{-1}+k]^{-1}-k^{-1}-[(s+m)(m+1)^{-1}+k]^{-1}+k^{-1}\}\,,$$

$$(7.12)$$

and by (7.5) this is the right hand side.

Theorem 4. If $\alpha_1 = \ldots = \alpha_m = 1$, then (7.6) is the logarithmic derivative form of the Gauss multiplication formula.

Proof. From (7.4) and the previous Proposition,

$$q(s) = (m+1)^{-1}\left[\log(m+1)+\psi\left(\frac{s+m}{m+1}\right)\right]\,.\qquad(7.13)$$

Thus, upon summing the left side of (7.6), we see that (7.6) is simply (7.4) with s replaced by $s+m$.

Corollary. If $F(s)$ is defined by (6.6), then

$$\lim_{\beta\to 1}F(s) = A2^{(s-1)/2}\Gamma[(s+1)/2]\,.\qquad(7.14)$$

Proof. By (6.7) and (7.13) with $m=1$ we have

$$F'(s)/F(s) = \frac{1}{2}\{\log 2 + \psi[(s+1)/2]\}\,.\qquad(7.15)$$

Upon integrating (7.15) we obtain

$$F(s) = K2^{s/2}\Gamma[(s+1)/2]\,.\qquad(7.16)$$

The constant K is determined by the fact that $F(1) = A$.

In general, we note that if $A > 0$, then

$$F(s) = A \exp \int_1^s q(t;m)dt \tag{7.17}$$

defines a function that satisfies

$$F(s+m)F(s+m-1)^{\alpha_1}\ldots F(s)^{\alpha_m}$$
$$= [F(m+1)F(m)^{\alpha_1}\ldots F(1)^{\alpha_m}/m!]\Gamma(s+m) \tag{7.18}$$

in the complement of the closure of the negative real axis for any set of α_i such that $t^m + \alpha_1 t^{m-1} + \ldots + \alpha_m$ has no zero in $[0, 1]$. For $\alpha_1 > 0$ and $\sum \alpha_i = 1$, it is likely that (7.18) implies a growth condition on $F(s)$ analogous to those derived for the $m = 1$ case.

We can also establish, as in Sec. 6, a very general form of the multiplication theorem, this time in the context of mth order difference equations.

Proposition. If $\theta \neq 1$ and

$$q_1(s+m) + \beta_{11}q_1(s+m-1) + \ldots + \beta_{1m}q_1(s) = y(s)$$
$$q_2(s+m) + \beta_{21}q_2(s+m-1) + \ldots + \beta_{2m}q_2(s) = y(s)$$
$$\tag{7.19}$$

where
$$\beta_{21}/\beta_{11} = \beta_{22}/\beta_{12} = \ldots = \beta_{2m}/\beta_{1m} = \theta , \tag{7.20}$$

then

$$f(s+m) + \beta_{21}f(s+m-1) + \ldots + \beta_{2m}f(s) = q_1(s+m) \tag{7.21}$$

is satisfied by

$$f(s) = (1-\theta)^{-1}q_1(s) - \theta(1-\theta)^{-1}q_2(s) . \tag{7.22}$$

Proof. This is a straightforward calculation.

8. Other Solutions of (5.9)

By inspection of (5.9) it seems reasonable that for $0 < \beta < 1$ and $F(2)F(1)^\beta \neq 0$ the function $F(s)$ (if defined for s negative) will grow in modulus something like $\operatorname{expexp} |s_n|$ for sequences $s_n \to -\infty$, even if s_n stays away from the negative integers. We now prove that an $F(s)$ with a somewhat slower rate of growth cannot assume a finite nonzero value at any positive integer.

Theorem 5. If $F(s)$ is a solution of (5.9) such that

$$\lim_{k \to \infty} \beta^k \log |F(s - k)| = 0 \tag{8.1}$$

for any fixed s, where s is not an integer, then

$$|F(s)| = K^{\frac{1}{1+\beta}} |\Gamma(s)|^{\frac{1}{1+\beta}} \prod_{j=1}^{\infty} |s - j|^{(-1)^{j-1}\beta^j/(1+\beta)} . \tag{8.2}$$

Proof. Set $K = |F(2)F(1)^\beta|$. From (5.9) we have

$$\log |F(s + 1)| + \beta \log |F(s)| = \log |\Gamma(s + 1)| + \log K \tag{8.3}$$

which implies

$$\log |F(s + 1)| = \sum_{k=0}^{m} (-1)^k \beta^k \log |\Gamma(s + 1 - k)|$$
$$+ (-\beta)^{m+1} \log |F(s - m)| + \frac{\log K}{1 + \beta}[1 - (-\beta)^{m+1}] . \tag{8.4}$$

It now follows from (8.1) that

$$\log |F(s + 1)| = \sum_{k=0}^{\infty} (-\beta)^k \log |\Gamma(s + 1 - k)| + (1 + \beta)^{-1} \log K$$
$$= \log |\Gamma(s + 1)| - \beta \log |\Gamma(s)| + (1 + \beta)^{-1} \log k$$
$$+ \sum_{k=2}^{\infty} (-\beta)^k \Big[\log |\Gamma(s)| - \sum_{j=1}^{k-1} \log |s - j|\Big] . \tag{8.5}$$

We next make use of the operational identity

$$\sum_{k=2}^{\infty}\sum_{j=1}^{k-1} = \sum_{j=1}^{\infty}\sum_{k=j+1}^{\infty} . \tag{8.6}$$

The interchange of summation is easily justifed in this case, and we obtain

$$\log|F(s+1)| = \log|\Gamma(s+1)| - \frac{\beta}{1+\beta}\log|\Gamma(s)| + (1+\beta)^{-1}\log K$$
$$- \sum_{j=1}^{\infty}\frac{(-\beta)^{j+1}}{1+\beta}\log|s-j| \tag{8.7}$$

or

$$|F(s+1)| = K^{\frac{1}{1+\beta}}|\Gamma(s+1)||\Gamma(s)|^{-\frac{\beta}{1+\beta}}\prod_{j=1}^{\infty}|s-j|^{-(-\beta)^{j+1}/(1+\beta)} . \tag{8.8}$$

From

$$|\Gamma(s+1)| = K^{-1}|F(s+1)||F(s)|^{\beta} \tag{8.9}$$

we obtain (8.2).

It is easy to verify directly that the right side of (8.2) satisfies (5.9). Also, by multiplying both sides of (8.2) by

$$\exp[(\sin\pi s)^{-2}] \tag{8.10}$$

we obtain for $s > 0$ a C^{∞} solution of (5.9) that is zero at every integer.

Nörlund [14. pp. 37–44] has studied the problem of finding smooth solutions to a given first order linear difference equation. There is no uniqueness here, but Nörlund defines "principal solutions" and shows that in many respects they are the "most natural" solutions. Here, however, the $|F(s)|$ given by (8.2) is the "principal solution" of (5.9) for $s > 0$, yet the $F(s)$ of Sec. 6 seems far more natural. Nörlund also gives references to various papers in which "non-principal solutions" are studied.

9. Logarithmic Concavity

We now determine to what extend the function $F(s,\beta)$ defined in Sec. 6 is characterized by the logarithmic concavity of

$$u(s) = F(s,\beta)\Gamma(s)^{-1/(1+\beta)} . \tag{9.1}$$

Since

$$u(s+1)u(s)^\beta = Ks^{\beta/(1+\beta)} \tag{9.2}$$

by (5.9), where $K = F(2)F(1)^\beta$, our problem is to characterize the solutions of Eq. (9.6).

Lemma. If $f(s)$ is concave, $f''(s) \to 0$ as $s \to \infty$, and $p(s)$ is a nonconstant periodic function, then the sum

$$a(s) = f(s) + p(s) \tag{9.3}$$

is not concave.

Proof. Without loss of generality, assume $p(s)$ has period 1, that $p(0) = p(1) = 0$, and that $p(\theta) < 0$ for some fixed $0 < \theta < 1$. If the line segment joining $(n, a(n))$ and $(n+1, a(n+1))$ lies below $(n+t, a(n+t))$ for any $0 < t < 1$, we have

$$\theta f(n+1) + (1-\theta)f(n) - f(n+\theta) < p(n+\theta) = p(\theta) < 0 \; . \tag{9.4}$$

By Taylor's theorem with remainder, the left side of (9.4) equals

$$\frac{1}{2}\theta f''(s_1) - \frac{1}{2}\theta^2 f''(s_2) \tag{9.5}$$

for certain points s_1, s_2 in the interval $(n, n+1)$. But this yields a contradiction for n sufficiently large.

Theorem 6. Let $\beta \geq 1$. Then for any constant c there is a unique concave function $f(s)$ such that

$$f(s+1) + \beta f(s) = \frac{\beta}{1+\beta}\log s + c \; . \tag{9.6}$$

Proof. We first treat the case $\beta = 1$. We easily see that (9.6) has a solution

$$f(s) = \frac{1}{2}\log\left[\Gamma\left(\frac{s+1}{2}\right)\Big/\Gamma\left(\frac{s}{2}\right)\right] + c_1 \tag{9.7}$$

for some constant c_1. From

$$\frac{1}{2}\left[\psi\left(\frac{s+1}{2}\right)\psi\left(\frac{s}{2}\right)\right] = s^{-1} - (s+1)^{-1} + (s+2)^{-1} - \ldots \tag{9.8}$$

we deduce

$$f''(s) = -s^{-2} + (s+1)^{-2} - (s+2)^{-2} + \ldots \tag{9.9}$$

so $f''(s) \to 0$ as $s \to \infty$. If $h(s)$ is any other solution of (9.6), then the difference $d(s) = h(s) - f(s)$ satisfies

$$d(s+1) + d(s) = 0 \tag{9.10}$$

and hence is a periodic function. Hence, by the above Lemma, $h(s)$ cannot be concave.

For $\beta > 1$ we use a different argument. The concavity of f implies, for $0 < s < 1$, that

$$f(n+1) - f(n) \le s^{-1}[f(n+s) - f(n)] \le f(n) - f(n-1) \tag{9.11}$$

and hence

$$sf(n+1) - sf(n) + f(n) \le f(n+s) \le sf(n) - sf(n-1) + f(n) . \tag{9.12}$$

From (9.6) we obtain

$$f(s+n) = \sum_{m=0}^{n-1}(-\beta)^m \frac{\beta}{1+\beta}\log(s+n-m-1)$$
$$+ c\sum_{m=0}^{n-1}(-\beta)^m + (-\beta)^n f(s) . \tag{9.13}$$

Thus, for n even, (9.12) and (9.13) yield

$$-\sigma\beta^{-n} + \beta^{-n}[sf(n+1) - sf(n) + f(n)]$$
$$\le f(s) \le -\sigma\beta^{-n} + \beta^{-n}[sf(n) - sf(n-1) + f(n)] \tag{9.14}$$

where σ is the sum of the two sums on the right of (9.13). From (9.11) we deduce

$$f(1) - f(2) + f(n+1) \le f(n) \tag{9.15}$$

and from this and (9.6) with $s = n$, we obtain

$$f(n+1) \ll \log n \; . \tag{9.16}$$

By concavity, $f(n)$ and $f(n+1)$ will have the same sign for n sufficiently large. Thus by (9.6) we see that $f(n)$ is non-negative for all sufficiently large n, so (9.16) provides a bound on the modulus of $f(n)$. Hence, by letting $n \to \infty$ in (9.14), we have

$$f(s) = \lim_{n \to \infty} [-\sigma \beta^{-2n}], \quad 0 < s < 1 \; . \tag{9.17}$$

This establishes uniqueness for $0 < s \leq 1$; the uniqueness for all $s > 0$ now follows from (9.13). On the other hand a solution exists, namely $f(s) = \log u(s)$ where $u(s)$ is given by (9.1) and $F(s, \beta)$ by (6.5) and (6.6).

We remark that the above proof, as well as the standard characterization of $\Gamma(u)$ by means of logarithmic convexity, only uses the convexity hypothesis for u large. We now show, however, that there can be no such result for $0 < \beta < 1$.

Proposition. Let $0 < \beta < 1$. Then there is a constant $c(\beta)$ and infinitely many functions $f(s)$, concave for s sufficiently large, such that

$$f(s+1) + \beta f(s) = \frac{\beta}{1+\beta} \log s + c(\beta) \; . \tag{9.18}$$

Proof. Let

$$f(s) = \frac{\beta}{1+\beta} \int_0^1 \frac{t^{s-1} - 1}{(\beta + t) \log t} \, dt \; . \tag{9.19}$$

Then a simple calculations shows that

$$f'(s+1) + \beta f'(s) = \frac{\beta}{1+\beta} \frac{1}{s} \; , \tag{9.20}$$

so (9.18) follows. Next, for $\beta < 1 - \epsilon < 1$ and s large,

$$\begin{aligned} f''(s) &= \frac{\beta}{1+\beta} \int_0^1 \frac{t^{s-1} \log t}{\beta + t} \, dt \\ &\leq \frac{\beta}{1+\beta} \int_\beta^{1-\epsilon} \leq \frac{\beta}{1+\beta} \cdot \frac{\log(1-\epsilon)}{\beta+1-\epsilon} \cdot \frac{(1-\epsilon)^s}{2s} < 0 \; ; \end{aligned} \tag{9.21}$$

in particular, $f(s)$ is concave. We now observe that for large s the second derivative of

$$h(s) = f(s) + K\beta^s \sin \pi s \qquad (9.22)$$

is negative, whatever the value of the constant K. But $h(s)$ satisfies (9.18), so the result is established.

Thus the real variable characterization problem for $0 < \beta < 1$ remains open. Perhaps what we need is the function that is concave on the largest possible subinterval of $(0, \infty)$, or that is best behaved for $|\tau| < \infty$ (where $s = \sigma + i\tau$).

10. Remarks

For a reasonably complete survey of work on the gamma function prior to 1905 see [13]. There are a number of excellent short introductions to the gamma function, each from a somewhat different point of view. We mention [1, 4, 9, 10], [16, pp. 30–46] and [17, pp. 235–264]. More recently R. Askey [2] has found q-analogues of many of the most important properties. An excellent source for characterizations of the gamma function and the literature pertaining to this is [13]. The deepest investigations into gamma functions and their generalizations seem to be those of Barnes [4–8]; of particular interest in connection with our reduction method is [6].

The Gauss multiplication theorem has been generalized in different ways by Schobloch [14, p. 198] and Askey [2, p. 131]. More closely related to the present paper are the "Multiplikationstheoreme der Hauptlösungen" (that is, of principle solutions to linear difference equations) in Nörlund [15, p. 4] that lead to the orginal Gauss multiplication theorem (essentially formula (14), p. 103 of [15]). His results, however, do not contain those of the present paper. The integral on the right of (6.13), with $\beta = 1$, was introduced into the study of the gamma function by Legendre (see 14, p. 169]). Integrals resembling the one on the right of (7.1) occur in the study of difference equations with constant coefficients [15, pp. 295–298]).

References

1. E. Artin, *The Gamma Function*, Holt, Rinehart, and Winston, New York, 1964.

2. R. Askey, *The q-gamma and q-beta functions*, J. Applicable Analysis **8** (1978) 125–141.

3. S. B. Bank and R. P. Kaufman, *A note a Hölder's theorem concerning the Gamma function*, Math. Ann. **232** (1978) 115–120.

4. E. W. Barners, *The theory of the Gamma function*, Mess. Math. (2) **29** (1900) 64–128.

5. E. W. Barnes, *The genesis of the double Gamma functions*, Proc. London Math. Soc. (1) **31** (1900) 358–381.

6. E. W. Barnes, *The theory of the G-function*, Quart. J. Pure Appl. Math. **31** (1900) 264–314.

7. E. W. Barnes, *The theory of the double Gamma function*, Phil. Trans. London **196A** (1901) 265–387.

8. E. W. Barnes, *On the theory of the multiple Gamma function*, Trans. Camb. Phil. Soc. **19** (1904) 374–425.

9. B. Berndt, *Rudiments of the theory of the gamma function*, unpublished notes.

10. P. J. Davis, *Leonard Euler's integral: A historical profile of the gamma function*, Amer. Math. Monthly **66** (1959) 849–869.

11. W. Feller, *An Introduction to Probability Theory and Its Applications*, vol. I, 2nd ed., John Wiley, New York, 1960.

12. C. F. Gauss, *Werke, Königlichen Gesellschaft der Wissenschaften zu Göttingen*, 1876, Band 3, pp. 144–150.

13. M. Kuczma, *Functional Equations in a Single Variable*, Monografie Matematyczne vol. 46, Polish Scientific Publishers, Warsaw, 1968.

14. N. Nielsen, *Die Gammafunktion*, Chelsea, New York, 1965 (reprint of 1906 edition).

15. N. Nörlund, *Vorlesungen über Differenzenrechnung*, Chelsea, New York, 1954.

16. H. Rademacher, *Topics in Analytic Number Theory*, Springer-Verlag, New York, 1973.

17. E. T. Whittaker and G. N. Watson, *A Course of Modern Analysis*, 4th ed., Cambridge Univ. Press, Cambridge, 1962 (reprint of 1927 edition).

Kenneth B. Stolarsky
Department of Mathematics
University of Illinois
1409 West Green Street
Urbana, Illinois 61801
USA

THE MATH. HERITAGE OF C.F. GAUSS (pp. 758-768)
edited by George M. Rassias
©1991 World Scientific Publ. Co. Singapore

SOME PROPERTIES OF HEREDITARILY LOCALLY CONNECTED CONTINUA RELATED TO THE HANN-MAZURKIEWICZ THEOREM

L. B. Treybig

It is apparently a still open question as to whether or not each hereditarily locally connected Hausdorff continuum M is the continuous image of an arc (ordered continuum). In this paper such an M is shown to have several properties that it must have if it is such an image. Some additional questions about similar properties are also raised.

In what follows let K' denote the class of compact hausdorff spaces that are the continuous image of a compact ordered space, and let C denote the class of continua that are the continuous image of an arc (ordered continuum).

In [6] Pearson gave an argument to show that there is an example of a hereditarily locally connected continuum which is not in C. In [5] Nikiel points out that Pearson's example is not compact, and raises

Problem 1. Is each hereditarily locally connected continuum in C?

In the interval between the submission of this manuscript and its publication J. Nikiel has solved Problem 1 in the affirmative.

In [7] Pearson shows that if X is a hereditarily indecomposible continuum and Y is an upper semi-continuous decompositiion of X into Peano continua such that the decomposition space Y is rim finite, then X is in C. Also in [7] Pearson gives several equivalent conditions that a hereditarily locally connected Suslinian continuum be in C. In [18] Whyburn develops a number of properties and characterizations of hereditarily locally connected metric continua, including the fact that such continuum is a rational curve. In [8] Simone extends a number of Wilder's results to non-metric continua.

It is the purpose of this paper to give some results about hereditarily locally connected continua which may prove useful in the further study of Problem 1. In particular

Theorem 1. If $x_1, x_2, x_3, \ldots$ is a sequence of points in a hereditarily locally connected continuum M, then some subsequence of $x_1, x_2, x_3, \ldots$ converges.

is a theorem which must hold in each member of C (see [10]). Also

Theorem 2. If H and K are mutually separated closed subsets of the hereditarily locally connected continuum M, then there is a countable set S such that $\bar{S}$ separates H from K in M.

is a theorem which must be true for an element of C since Mardešić [3] has shown that elements of K' are locally peripherally metric. It is also a theorem which would seem to be needed in trying to extend the techniques used in the proof of the characterization given by Theorem 3 of [13]. In addition

Theorem 3. If X is a countable subset of the hereditarily locally connected continuum M, then there is a countable subset Y of M such that (1) $X \subset Y$, and (2) if Q is a component of $M - \overline{Y}$, there do not exist disjoint arcs $a_1 b_1$ and $a_2 b_2$ such that seg $a_i b_i \subset Q$ for $i = 1, 2$ and $\{a_1, b_1, a_2, b_2\} \subset \mathrm{Bd}\,(Q)$.

is likewise a theorem which must hold for a member of C (see [11]).

In this paper a continuum is a compact connected Hausdorff space and an arc is an ordered continuum. A continuum is locally connected provided it has a basis of connected open sets, and is hereditarily locally connected provided every subcontinuum of it is locally connected. For most of the other definitions of this paper, including basic theory of upper semi-continuous collections, see Whyburn [18].

Some other papers in this area are given by Cornette [1], Mardešić [2], Nikiel [4], Treybig and Ward [9], Treybig [11], Ward [16], [17], Tuncali [14], Tymchatyn [15], and Young [19].

Proof of Theorem 1. If some sequence of indices $n_1, n_2, n_3, \ldots$ satisfy $n_1 < n_2 < n_3 < \ldots$ and $x_{n_1} = x_{n_2} = x_{n_3} = \ldots$, then $x_{n_1}, x_{n_2}, x_{n_3}, \ldots$ converges.

Thus we may suppose without loss of generality that $x_i = x_j$ implies $i = j$. There is a point x of M such that x is a limit point of $\{x_1, x_2, x_3, \ldots\}$ and we suppose without loss of generality that no $x_i = x$. We also note that if a, b are two points of M, then any subcontinuum of M which is irreducible from a and b is an arc from a to b.

Case 1. Suppose there is an arc $\alpha = x_j x$ from x_j to x which contains infinitely many x_i. Then, some subsequence $x_{n_1}, x_{n_2}, x_{n_3}, \ldots$ satisfies $n_1 < n_2 < n_3 < \ldots$ and is monotone relative to the order from x_j to x on α. Clearly $x_{n_1}, x_{n_2}, x_{n_3}, \ldots$ converges.

Case 2. Suppose case 1 does not occur, but for some arc $x_j x$, if U is an open set containing x, then the component Q of $U - x$ containing a segment (open interval) yx of $x_j x$ contains infinitely many x_i.

In this case we may construct (1) sequences of distinct points $x_{n_1}, x_{n_2}, x_{n_3}, \ldots ; z_{n_1}, z_{n_2}, z_{n_3}, \ldots$ where $n_1 < n_2 < n_3 < \ldots$, (2) a sequence of arcs $\alpha_1, \alpha_2, \alpha_3, \ldots$, and (3) a sequence of open sets $Q_1, Q_2, Q_3, \ldots$ such that for each positive i (1) $x_{n_i} \in Q_i - x_j x$, (2) α_i is an arc from x_{n_i} to x which meets $x_j x$ in an arc $z_{n_i} x$, and (3) $\alpha_{i+1} \subset Q_{i+1} \subset \overline{Q}_{i+1} \subset Q_i - \mathrm{Cl}(\alpha_i - z_{n_i} x)$. Since the sequence $z_{n_1}, z_{n_2}, z_{n_3}, \ldots$ clearly converges to a point z, if $x_{n_1}, x_{n_2}, x_{n_3}, \ldots$ does not converge to z, then the continuum $\mathrm{Cl}(\alpha_1 \cup \alpha_2 \cup \alpha_3 \cup \ldots)$ is not locally connected, a contradiction. See Theorem 4 of [8].

Case 3. Suppose neither Case 1 nor Case 2 holds. In this case we may construct (1) a sequence of distinct points $x_{n_1}, x_{n_2}, x_{n_3}, \ldots$, where $n_1 < n_2 < n_3 < \ldots$, (2) a sequence $Q_1, Q_2, Q_3, \ldots$ of connected open sets containing x, and (3) a sequence of arcs $\alpha_1, \alpha_2, \alpha_3, \ldots$ such that for each i (1) α_i is an arc from x_{n_i} to x lying in Q_i, (2) $\overline{Q}_{i+1} \subset Q_i$, and (3) the component R_{i+1} of $Q_{i+1} - x$ which contains a segment $z_{n_i} x$ of α_i contains no x_j. If $x_{n_1}, x_{n_2}, x_{n_3}, \ldots$ does not converge to x, then $\mathrm{Cl}\,(\alpha_1 \cup \alpha_2 \cup \alpha_3 \cup \ldots)$ is a subcontinuum of M which is not locally connected, a contradiction.

Before we proceed with the proof of Theorem 2 we need the following

Lemma 1. Let M be a hereditarily locally connected continuum and let G be an upper semi-continuous decomposition of M into continua so that M/G is a Peano continuum. Let $f : [0,1] \to M/G$ be a continuous onto map. Then if (1) $0 \leq x < y \leq 1$ and (2) $(x,y) \cap f^{-1}(f(x))$ and $(x,y) \cap f^{-1}(f(y))$ are both void, there exists a unique point $f(x^+)(f(y^-))$ of the continuum $f(x)(f(y))$ such that if U is an open set containing $f(x^+)(f(y^-))$ there is a point z of (x,y) such that $\cup\{f(t) : x < t < z\} \subset U (\cup\{f(t) : z < t < y\} \subset U)$.

Proof for $f(x^+)$. If the Lemma does not hold there exist two points p, q of $f(x)$ so that if U, V are open sets containing p, q, respectively, and $z \in (x,y)$, then $\cup\{f(t) : x < t < z\}$ intersects both U and V. Let U, V be disjoint connected open sets containing p, q respectively, so that $\overline{U} \subset M - \overline{V}$.

Since M/G is a Peano continuum there exist (1) sequences of points $a_1, a_2, a_3, \ldots ; b_1, b_2, b_3, \ldots$ converging to x and (2) a sequence $W_1, W_2, W_3, \ldots$ of open sets in M/G such that (1) for each $i, x < a_{i+1} < b_{i+1} < a_i < b_i < y$, (2) $f(x) = \cap_{i=1}^{\infty} W_i$, (3) for each $i, f([a_i, b_i]) \subset W_i$ and $\cup\{f(t) : a_i \leq t \leq b_i\}$ intersects both U and V and (4) for each $i, \overline{W}_{i+1} \subset W_i - f([a_i, b_i])$. Then, the continuum $f(x) \cup \overline{U} \cup (\cup\{f(t) : t \in \cup_{p=1}^{\infty}[a_i, b_i]\})$ fails to be locally connected at a point of $\overline{V}$, a contradiction.

Proof of Theorem 2. There exist open sets U_1, V_1, U_2, V_2 such that $H \subset U_1 \subset \overline{U}_1 \subset U_2, K \subset V_1 \subset \overline{V}_1 \subset V_2$ and $\overline{U}_2 \subset M - \overline{V}_2$. We let $H_1, H_2, H_3, \ldots$ denote a sequence of fintie covers of M by connected open sets such that for each n (1) H_{n+1} is a star refinement of H_n, and (2) no element of H_n intersects two of $H, K, \mathrm{Bd}(U_1), \mathrm{Bd}(U_2), \mathrm{Bd}(V_1)$, and $\mathrm{Bd}\,(V_2)$.

Define a relation T on M such that xTy if and only if $x \in \cap_{n=1}^{\infty} st(y, H_n)$. With the aid of Lemma 4 of [12] we find that T is an equivalence relation on M such that each element of the set G of equivalence classes modulo T is a locally connected continuum. It is also shown in [12] that G is an upper semi-continuous decomposition of M so that M/G is a Peano continuum, which by [18] is also hereditarily locally connected.

There is an onto continuous map $f : [0,1] \to M/G$ such that (1) if

$g \in G$, then $f^{-1}(g)$ is totally disconnected, and (2) $f(0) \cap H$ and $f(1) \cap K$ are non-void. There is a sequence $F_1, F_2, F_3, \ldots$ of finite subsets of $[0,1]$ such that (1) $0, 1 \in F_1$, (2) for each n $(i) F_n \subset F_{n+1}$ and (ii) if the points of F_n are listed in increasing order as $x_1, x_2, \ldots, x_j$, then $x_{p+1} - x_p < \frac{1}{n}$ for $n = 1, \ldots, j-1$, and (3) if $x, y \in \cup_{n=1}^{\infty} F_n$ and $x \neq y$, then $f(x) \neq f(y)$.

Let $\phi_1 : M \to M/G$ be the natural map, and note that $f^{-1}(\phi_1(H))$, $f^{-1}(\phi_1(K))$, and $f^{-1}(\phi_1(M - (U_2 \cup V_2)))$ are disjoint closed subsets of $[0,1]$. Since the set of all components $[0, 1] - (f^{-1}(\phi_1(H)) \cup f^{-1}(\phi_1(K)))$ is a cover of $f^{-1}(\phi_1(M - (U_2 \cup V_2)))$ by open intervals, it has a finite subcover $(r_1, s_1), \ldots, (r_m, s_m)$. There is a positive integer k such that the distance from any point of $f^{-1}(\phi_1(M - (U_2 \cup V_2)))$ to any point of $\{r_1, \ldots, r_m, s_1, \ldots, s_m\}$ is greater than $\frac{3}{k}$.

For each $n \geq k$ let $G_n = f^{-1}(f(F_n))$ and let $y \in Y_n$ if and only if for some component (r, s) of $[0, 1] - G_n$ such that $[r, s] \subset$ some (r_i, s_i), the point $y = f(r^+)$ or $f(s^-)$. We note that each Y_n is countable and let $S = \cup_{n=k}^{\infty} Y_n$.

Suppose $\overline{S}$ does not separate H from K in M. There is an arc $\alpha = hk$ from a point h of H to a point k of K where $\alpha \subset M - \overline{S}$. Let cd be a subarc of hk so that $c \in \text{Bd}(U_2), d \in \text{Bd}(V_2)$ and seg $cd \subset M - (U_2 \cup V_2)$.

Since no element of G intersects both Bd (U_2) and Bd (V_2) and since $\cup\{\phi_1(t) : t \in cd\}$ is a continuum, there is a point x of seg cd so that $x \in g_x \in G$ and $g_x \subset M - (\overline{U}_2 \cup \overline{V}_2)$. Let $g_x \subset R$ open. There is an open set R' so that $g_x \subset R' \subset (M - (\overline{U}_2 \cup \overline{V}_2)) \cap R$ and R' is the union of elements of G. Let (r, s) be a component of $f^{-1}(\phi_1(R'))$. Since $(r, s) \subset$ some (r_p, s_p), there exists an integer $n \geq k$ and points x, x' of X_n such that $r < x < x' < s$. There is a component (c, d) of $[0, 1] - G_n$ so that $x \leq c < d \leq x'$, so therefore $f(c^+) \in S$. It is now an easy argument to show g_x contains a point of $\overline{S}$, and so each element g of G lying in $M - (\overline{U}_2 \cup \overline{V}_2)$ contains a point $y(g)$ of $\overline{S}$.

We now let $x_1, x_2, x_3, \ldots$ be an increasing sequence of points in the order from c to d on arc cd such that (1) if $x_i \in g_{x_i} \in G$, then $g_{x_i} \subset M - (\overline{U}_2 \cup \overline{V}_2)$ and (2) if $i < j$ then $g_{x_i} \neq g_{x_j}$. Since the sequence $x_1, x_2, x_3, \ldots$ converges to a point x of cd, if $y(g_{x_1}), y(g_{x_2}), y(g_{x_3}), \ldots$ does not converge to x, we obtain a contradiction as above. Therefore $x \in \overline{S}$, a contradiction, so $\overline{S}$ separates H from K in M.

Corollary 1. With M as above, if P is an element of the open set U, there is a countable set S and an open set V such that $P \in V \subset \overline{V} \subset U$ and Bd $(V) \subset \overline{S} \subset U - V$. Since M is locally connected, V may also be chosen to be connected.

Proof of Theorem 3. Let the points of X be labeled $P_1, P_2, P_3, \ldots$ so that if $i \neq j$ then $P_i \neq P_j$.

Let $H_1, H_2, H_3, \ldots$ be a sequence of finite covers of M by connected open sets such that if n is a positive integer then (1) H_{n+1} is a star refinement of H_n and there exist elements $h_1, h_2, \ldots, h_n$ of H_n such that $P_i \in h_i, i = 1, \ldots, n$, and $\bar{h}_1, \bar{h}_2, \ldots, \bar{h}_n$ are disjoint, (2) if $V \in H_n$ there is a countable set S_V such that Bd $(V) \subset \overline{S}_V \subset M - V$, and (3) if h_1, h_2, h_3, h_4 are elements of $\cup_{i=1}^n H_i$ and h is a connected union of elements of $\cup_{i=1}^n H_i$ such that

(i) h intersects h_j for $j = 1, \ldots, 4$ but $\bar{h}_i$ and $\bar{h}_j$ are disjoint if $i \neq j$, and

(ii) there exist disjoint arcs $\alpha_i = a_i b_i, i = 1, 2$, such that each α_i has its end points a_i, b_i in different ones, say h_{i_1}, h_{i_2}, respectively, of h_1, h_2, h_3, h_4, but does not intersect the closures of the other two, and each $\alpha_i - \cup_{p=1}^4 h_p$ is a segment lying in h, then for some such pair $\alpha_i = a_i b_i, i = 1, 2$ there exist finite chains $C^i = (c_1^i, c_2^i, \ldots, c_{j_i}^i)$ of connected open sets such that

(i) each $\alpha_i \subset \cup_{p=1}^{j_i} c_p^i \subset h \cup h_{i_1} \cup h_{i_2} - \cup\{\bar{h}_p : p \in \{1, \ldots, 4\} - \{i_1, i_2\}\}$,

(ii) $a_i \in c_1^i \subset h_{i_1}$ and $b_i \in c_{j_i}^i \subset h_{i_2}$ for each i,

(iii) $\cup_{p=1}^{i_1} c_p^1 \subset M - \cup_{p=1}^{j_2} c_p^2$,

(iv) there exists k_i $(i = 1, 2)$ such that Cl $(c_{k_i}^i) \subset h - \cup_{p=1}^4 \bar{h}_p$, and

(v) if $i = 1$ or 2 and if U, V are elements of H_{n+1} such that U intersects both α_i and V, then $U \cup V \subset$ some c_p^i.

To see how to construct the above, suppose $H_1, \ldots, H_n$ have been constructed.

Let $U_1, \ldots, U_{n+1}$ be open sets containing $P_1, \ldots, P_{n+1}$, respectively, such that $\overline{U}_1, \ldots, \overline{U}_{n+1}$ are disjoint. Let G denote the cover $\{U_1, \ldots, U_{n+1}, M - \{P_1, \ldots, P_{n+1}\}\}$ of M.

Let $(\alpha_{a_i b_i}, C_{a_i b_i}), i = 1, 2, \ldots, m$, be a representative set of arcs and finite chains chosen for the various collections (h_1, h_2, h_3, h_4, h) formed from $\cup_{i=1}^n H_i$ as described in (3) above. For each $i = 1, \ldots, m$ let K_i be the cover $C_{a_i b_i} \cup \{M - \alpha_{a_i b_i}\}$ of M.

Let $H'_n, G', K'_1, \ldots, K'_m$ be finite open covers of M by connected open sets such that (1) $H'_n, G', K'_1, \ldots, K'_m$ are star refinements of $H_n, G, K_1,$

..., K_m, respectively, and (2) if $h \in H'_n \cup G' \cup K'_1 \cup \ldots \cup K'_m$ then h satisfies Corollary 1 (i.e., there is a countable set S_h such that $\mathrm{Bd}\,(h) \subset \overline{S}_h \subset M - h$).

We define T' such that $h \in T'$ if and only if for some point P of M, h is the component containing P of the intersection of all the elements of $H'_n \cup G' \cup K'_1 \cup \ldots \cup K'_m$ containing P. Note that if the intersection above is written as $\cap^z_{i=1} U_i$ and h is the component containing P, then $\mathrm{Bd}\,(h) \subset S_h = \mathrm{Cl}\,(\cup^z_{i=1} S_{U_i}) \subset M - h$. We let H_{n+1} be a finite cover of M by elements of T'.

We let $Y = X \cup (\cup\{S_V : V \in \cup^\infty_{n=1} H_n\})$. Also, we define a relation T on M such that xTy if and only if $x \in \cap^\infty_{n=1} st(y, H_n)$. As noted in the proof of Theorem 2, T is an equivalence relation on M, the set G of equivalence classes modulo T is an upper semi-continuous decomposition of M into continua, and M/G is a Peano continuum.

Now suppose Q is a component of $M - \overline{Y}$ and there exist disjoint arcs $x_1 x_2$ and $x_3 x_4$ such that seg $x_1 x_2 \cup$ seg $x_3 x_4 \subset Q$ and $x_i \in \mathrm{Bd}\,(Q)$ for $i = 1, 2, 3, 4$. We show that $Q \subset g \in G$. For if not, then Q intersects distinct elements g_1, g_2 of G, so let $x \in g_1 \cap Q$ and $y \in g_2 \cap Q$. Since $y \notin g_1$, there is a positive integer n such that $y \notin st(x, H_n)$. If $x \in U \in H_n$ then $\overline{S}_U \cap Q$ is not void, a contradiction. Thus $Q \subset g \in G$ and $\mathrm{Bd}\,(Q) \subset g$.

For each positive integer n let $W_n = st(g, H_n) = \cup\{w : w \in H_n$ and w intersects $g\}$. We note that (1) for each positive integer $n, g \subset W_{n+1} \subset \overline{W}_{n+1} \subset W_n$, and (2) $\cap^\infty_{n=1} W_n = g$. To see that (1) holds let $P \in \mathrm{Bd}\,(W_{n+1})$ and let $P \in h \in H_{n+1}$. There exist $x \in g$ and $k \in H_{n+1}$ such that $x \in k$ and k intersects h. But there eixsts $U \in H_n$ such that $P \in h \cup k \subset U \subset W_n$. To see that (2) holds suppose $P \in \cap^\infty_{n=1} W_n - g$. For each positive integer n there exists $x_n \in g$ such that $x_n \in st(P, H_n)$. There is a limit point x of $\{x_1, x_2, \ldots\}$ in g, and a positive integer N such that $x \notin st(P, H_N)$. Let $h, k \in H_{N+1}$, where $x \in h$ and $P \in k$. Then h, k are disjoint, for if not, there exists $U \in st(P, H_N)$ with $h \cup k \subset U$, a contradiction. Thus $h \subset M - st(P, H_{N+1})$. There exist $i \geq N+1$ such that $x_i \in h$. We thus find, by induction, that $x_i \notin st(P, H_i) \subset st(P, H_{N+1})$. Therefore $P \in g$.

Returning now to the main argument, let $U_i, i = 1, \ldots, 4$, be connected open sets such that (1) $x_i \in U_i, i = 1, \ldots, 4$ and (2) $\overline{U}_i$ and $\overline{U}_j$ are disjoint if $i \neq j$.

There is a positive integer N such that if $n \geq N$, then $W_n - g$ contains no continuum β_n which intersects two of the $U_i, i = 1, 2, 3, 4$. For if there is no such N a typical case is as follows: There is an increasing sequence

$n_1, n_2, n_3, \ldots$ of positive integers and a sequence of continua $\beta_1, \beta_2, \beta_3, \ldots$ such that for each i

(1) $\beta_i \subset W_{n_i} - \overline{W}_{n_{i+1}}$, and

(2) β_i intersects both U_1 and U_2.

The continuum $\overline{U}_1 \cup g \cup (\cup_{n=1}^\infty \beta_n)$ fails to be locally connected at a point of $\overline{U}_2$, a contradiction. Thus there is such an N as above.

Let $k_i \in H_N, i = 1, 2, 3, 4$, such that $x_i \in k_i$. We let $h = \cup_{p=1}^4 k_i$ and note that $Q \subset h$.

Since G is upper semi-continuous the set $W'_N = \cup \{w : w \in G \text{ and } w \subset W_N\}$ is an open set in M containing g. There exists a point $y_i, i = 1, \ldots, 4$, of $M - g$ contained in the component of $h \cap U_i \cap W'_N$ which contains x_i. We let $y_i \in g_i \in G$ for $i = 1, 2, 3, 4$, and note that g_1, g_2, g_3, g_4, g are distinct elements of G and are subsets of W_N.

There is a positive integer $L \geq N$ so that if $n \geq L$, then Cl $(st(y_i, H_n))$ $\subset W'_N \cap (M - \text{Cl}(st(y_j, H_n)))$ for $i, j \in \{1, 2, 3, 4, \}$ and $i \neq j$. Let $y_i \in h_i \in H_L, i = 1, \ldots, 4$. Using the arcs $x_1 x_2$ and $x_3 x_4$ and the component of $h \cap U_i \cap W'_N$ containing $x_i, i = 1, \ldots, 4$, we may construct disjoint arcs $y_1 y_2$ and $y_3 y_4$ lying in $h \cup (\cup_{p=1}^4 h_p)$. The arcs $y_1 y_2$ and $y_3 y_4$ may now be used to help construct arcs $\alpha_i = a_i b_i$ $(i = 1, 2)$ as in part (3) of the definition of H_{n+1}, where $n = L$. Thus there exist such arcs $\alpha_i = a_i b_i$ and finite chains C^i for $i = 1, 2$, as in part (3) of the definition of H_{L+1} where no pair of intersecting elements of H_{L+1} contain points of α_1 and α_2, respectively. Since no element of G intersects both α_1 and α_2, then one of α_1, α_2 is a continuum which lies is $W_N - g$ and intersects two of U_1, U_2, U_3 and U_4. This involves a contradiction.

Theorem 4. If $M, X, Y,$ and Q are as in Theorem 3, then Bd (Q) is finite.

Proof. Suppose Bd (Q) is infinite, and let L be a limit point of Bd (Q). Let $x \in Q$ and let $x x_1$ ad $x x_2$ be arcs such that $x x_i - x_i \subset Q, x_i \in \text{Bd}(Q)$, and $x_i \neq L$ (for $i = 1, 2$). Then $x x_1 \cup x x_2$ contains an arc $x_1 x_2$ from x_1 to x_2. There is an open set R containing L such that $\overline{R} \subset M - x_1 x_2$.

There exist arcs $z_i y_i, i = 1, 2, \ldots$, such that each $y_i \in \text{Bd}(Q) \cap R$ and $z_i y_i - y_i \subset$ a component Q_i of $Q - x_1 x_2$ and where $y_i = y_j$ implies $i = j$. If some $Q_i = Q_j$, where $i \neq j$, then there is an arc $y_i y_j$ from y_i to y_j in $Q_i \cup \{y_i, y_j\}$. The arcs $x_1 x_2$ and $y_i y_j$ contradict the conclusion of Theorem

3. Therefore we assume that $i \neq j$ implies $Q_i \neq Q_j$.

For each $i = 1, 2, 3, \ldots$ there is an arc $z_i b_i$ such that $b_i \in \operatorname{seg} x_1 x_2$ and $z_i b_i - b_i \subset Q_i$. Suppose for example that b_1, b_2, b_3, b_4 are distinct and occur in the order from x_1 to x_2 on $x_1 x_2$. Then if $b_1 b_2$ and $b_3 b_4$ denote arcs on $x_1 x_2$, then there is an arc $y_1 y_2$ from y_1 to y_2 in $z_1 y_1 \cup z_1 b_1 \cup b_1 b_2 \cup z_2 b_2 \cup z_2 y_2$ and an arc $y_3 y_4$ from y_3 to y_4 in $z_3 y_3 \cup z_3 b_3 \cup b_3 b_4 \cup z_4 b_4 \cup z_4 y_4$. The arcs $y_1 y_2$ and $y_3 y_4$ contradict the conclusion of Theorem 3. Therefore we may assume without loss of generality that $b_i = b_j$ for all i, j, and that there exists points t_1, t_2 of $x_1 x_2$ such that x_1, t_1, b_1, t_2, x_2 occur in that order on $x_1 x_2$, and where $x_1 t_1 - x_1$ and $t_2 x_2 - x_2$ contain boundary points of Q_i for at most three integers i.

By Urysohn's Lemma there is a continuous map $f : M \to [0, 1]$ such that $f(t_1 t_2) = 0$ and $f(M - Q) = 1$, where $t_1 t_2$ is an arc on $x_1 x_2$. For each i let $s_i \in Q_i \cap f^{-1}(\frac{1}{2})$. Some point s is a limit point of $\{s_1, s_2, s_3, \ldots\}$. Since $s \in Q$ and M is locally connected there is a connected open set V containing s such that $V \subset Q - t_1 t_2$ if $s \in x_1 x_2$, and where $V \subset Q - x_1 x_2$ if $s \notin x_1 x_2$. If $s \notin x_1 x_2$ then V intersects two components Q_i and Q_j, so $Q_i = Q_j$, a contradiction. If $s \in x_1 x_2$, then $s \in \operatorname{seg} x_1 t_1 \cup \operatorname{seg} t_2 x_2$, so infinitely many components Q_i have a boundary point in $\operatorname{seg} x_1 t_1 \cup \operatorname{seg} t_2 x_2$, a contradiction.

Corollary 2. If M is a hereditarily locally connected continuum and no finite set separates M, then M is separable.

Question 1. If M is a hereditarily locally connected continuum, is $s(M)$ (the least cardinal of a dense subset of M) $= w(M)$ (the least cardinal of a basis for M)?

Question 2. If M is a hereditarily locally connected continnum and C is a countable subset of M, does the exist a countable subset D of M such that $C \subset D$ and each component of $M - \overline{D}$ has a boundary of cardinality ≤ 2?

If such an M as above is the continuous image of an arc, then by [11] the questions above must have affirmative answers.

References

1. J. L. Cornettte, *Image of a Hausdorff arc is cyclically extensible and reducible*, Trans. Amer. Math. Soc. **199** (1974) 225–267.

2. S. Mardešić, *On the Hahn-Mazurkiewicz theorem in non-metric spaces*, Proc. Amer. Math. Soc. **11** (1960) 919–927.

3. S. Mardešić, *Images of ordered compacta are locally peripherally metric*, Pacific J. Math. **23** (1967) 557–568.

4. J. Nikiel, *Concerning continuous images of arcs*, Mathematical Institute, University of Warsaw, Preprint 67 (1986).

5. J. Nikiel, *Some problems on continuous images of compact ordered spaces*, Mathematical Institute, University of Warsaw, Preprint 72 (1986).

6. B. J. Pearson, *A hereditarily locally connected continnum which is not the continuous image of an arc*, Glasnik Mat. **18** (1983) 335–341.

7. B. J. Pearson, *A Hahn-Mazurkiewicz theorem for hereditarily locally connected continua*, Glasnik Mat. **19** (1984) 323–333.

8. J. N. Simone, *Concerning hereditarily locally connected continua*, Coll. Math. **39** (1978) 243–251.

9. L. B. Treybig and L. E. Ward, Jr., *The Hahn-Mazurkiewicz problem*, Topology and Order Structures, Part I, eds. H. R. Bennett and D. I. Lutzer, Stichting Mathematisch Centrum, 1981, pp. 95–105.

10. L. B. Treybig, *Concerning continuous images of compact ordered spaces*, Proc. Amer. Math. Soc. **15** (1964) 866–871.

11. L. B. Treybig, *Concerning continua which are continuous images of compact ordered spaces*, Duke Math. J. **32** (1965) 417–422.

12. L. B. Treybig, *"Extending" maps of arcs to maps of ordered continua*, Topology and Order Structures, Part I, eds. H. R. Bennett and D. J. Lutzer, Stichting Mathematisch Centrum, 1981, pp. 83–93.

13. L. B. Treybig, *A characterization of spaces that are the continuous image of an arc*, Topology Appl. **24** (1986) 229–239.

14. H. M. Tuncali, *Analogues of Treybig's product theorem*, submitted for publication.

15. E. D. Tymchatyn, *The Hahn-Mazurkiewicz theorem for finitely Suslinian continua*, Gen. Top. and its Appl. **7** (1977) 123–127.

16. L. E. Ward, Jr., *The Hahn-Mazurkiewicz theorem for rim-finite continua*, Gen. Top. and its Appl. **6** (1976) 183–190.

17. L. E. Ward, Jr., *A generalization of the Hahn-Mazurkiewicz theorem*, Proc. Amer. Math. Soc. **58** (1976) 369–374.

18. G. T. Whyburn, *Analytic Topology*, Amer. Math. Soc. Coll. Publ. **28** (1942).
19. G. S. Young, *Representations of Banach spaces*, Proc. Amer. Math. Soc. **13** (1962) 667–668.

L. B. Treybig
Texas A&M University
College Station, Texas 77843
USA

THE MATH. HERITAGE OF C.F. GAUSS (pp. 769-784)
edited by George M. Rassias
©1991 World Scientific Publ. Co. Singapore

CONVEX NONHOLONOMIC HYPERSURFACES

Constantin Udrişte and Oltin Dogaru

Keywords: Nonholonomic hypersurfaces, extrema with nonholonomic constraints, convexity.

Let D be an open set in $R^n, n \geq 3$. Let Σ_{x_0} be the family of all images of integral manifolds through $x_0 \in D$ of a Pfaff equation on D which is not completely integrable. The pair $\left(\bigcup_{x_0 \in D} \Sigma_{x_0}, D \right)$ is called a nonholonomic hypersurface.

Section 1 analyses the properties of the Weingarten map and of the second fundamental form for a nonholonomic hypersurface. Section 2 refers to extrema with a nonholonomic constraint, supplementing our results in the papers [2], [3], [4], and to connection between this theory and the second fundamental form of the nonholonomic hypersurface. Using the theory of extrema with a nonholonomic constraint, Sec. 3 gives a possible meaning for the convexity of nonholonomic hypersurfaces and describes the interdependence between convexity and the second fundamental form.

We dedicate this paper to the memory of C. F. Gauss (1777-1855), the founder of differential geometry of surfaces and to Gh. Vrănceanu (1900-1979), the founder of the differential geometry of nonholonomic manifolds.

AMS Subject Classification: 52A40, 53A07, 49B99.

1. Nonholonomic Hypersurfaces

Let D be an open set in $R^n, n \geq 3$, and

$$\omega(x) = \sum_{i=1}^{n} \omega_i(x)dx^i = 0, \quad x = (x^1, \ldots, x^n) \tag{1}$$

a Pfaff equation, where $\omega_i : D \to R, i = 1, \ldots, n$, are functions of class C^1 such that

$$\text{rank}\,[\omega_i(x)] = 1, \quad \forall\, x \in D . \tag{2}$$

Let $m \in \{1, \ldots, n-1\}$ and I a compact interval in R^m. An injective and regular function

$$r : I \to D, \quad r = (x^1, \ldots, x^n),$$
$$x^1 = x^1(u^1, \ldots, u^m), \ldots, x^n = x^n(u^1, \ldots, u^m) ,$$

of class C^2, which satisfies

$$\sum_{i=1}^{n} \omega_i(r(u))\frac{\partial x^i}{\partial u^k} = 0, \quad \forall\, u = (u^1, \ldots, u^m) \in I, \quad k = 1, \ldots, m ,$$

is called an m-dimensional integral manifold of Eq. (1).

Let x_0 be a fixed point in D. We say that the integral manifold $r : I \to D$ of (1) passes through x_0 if $\exists\, u_0 \in I$ such that $r(u_0) = x_0$. We observe that there is always an infinity of integral curves (case $m = 1$) passing through x_0.

The Pfaff equation (1) is called completely integrable if, for every point $x_0 \in D$, there exists an integral hypersurface of Eq. (1) passing through x_0 (case $m = n - 1$). In this case the family of integral hypersurfaces can be written in the form $F(x) = c$ with $dF(x) = \omega(x)$ and the integral hypersurface through x_0 contains any other integral manifold through x_0.

If the Pfaff equation (1) is not completely integrable then through x_0 there can pass integral manifolds which have in common only the point x_0.

Suppose that the Pfaff equation (1) is not completely integrable. Denote by M_{x_0} the image of an integral manifold through x_0 and by Σ_{x_0} the family of all images of the integral manifolds through x_0. Let $\Sigma = \bigcup_{x_0 \in D} \Sigma_{x_0}$. Then the pair (Σ, D) is called a nonholonomic hypersurface on D attached to Eq. (1).

To the Pfaff equation (1) and to the point $x_0 \in D$ we can attach the hyperplane

$$H_{x_0} = \left\{ x \in R^n \;\middle|\; \sum_{i=1}^{n} \omega_i(x_0)(x^i - x_0^i) = 0 \right\}.$$

Since the linear manifolds tangent to integral manifolds which pass through x_0 are included in H_{x_0}, the hyperplane H_{x_0} is called the tangent hyperplane at x_0 of (Σ, D).

Remark. The correspondence $x_0 \to H_{x_0} \subset T_{x_0}D$ defines a function H on D which is an $(n-1)$-dimensional distribution. A vector field Y belongs to the distribution H if $Y(x_0) \in H_{x_0}, \forall x_0 \in D$ or equivalently $\omega(Y) = 0$ on D. Obviously, for every $x_0 \in D$ there exists a neighbourhood U of x_0 and $n-1$ vector fields $Y_1, \ldots, Y_{n-1}$ of class C^1 on U such that $\{Y_1(x_0), \ldots, Y_{n-1}(x_0)\}$ is a basis of H_{x_0}. The set $\{Y_1, \ldots, Y_{n-1}\}$ is called a local basis of the distribution H. The integral manifolds of the Pfaff equation are called integral manifolds of the distribution H.

The distribution H is not involutive (i.e., $Y, Z \in H$ and Y, Z of class C^1 do not imply $[Y, Z] \in H$) since the Pfaff equation is not completely integrable.

The quadratic form associated to the identity on H_{x_0}, i.e., the real function $v \to (v, v), v \in H_{x_0}$, is called the first fundamental form of the nonholonomic hypersurface (Σ, D) at the point x_0 and is denoted by g_{x_0}. It is remarked that this function is a scalar product (the restriction to H_{x_0} of the scalar product on R^n). The function $x_0 \to g_{x_0}, x_0 \in D$ is called the first fundamental form of (Σ, D) and is denoted by g.

Let ξ be the unit vector field on D given by $\frac{1}{\|\omega\|}(\omega_1, \ldots, \omega_n)$, where $\|\omega\| = \sqrt{\omega_1^2 + \ldots + \omega_n^2}$. Let M_{x_0} be the image of an integral manifold through x_0 which admits the orthonormal vector fields $\xi_1, \ldots, \xi_{n-m}$. One of these fields coincides with the restriction of ξ to M_{x_0} if and only if $M_{x_0} \subset H_{x_0}$ ($\Rightarrow M_{x_0}$ is an hyperplane manifold).

We substitute the condition (2) with the strong condition $\sum_{i=1}^{n} \omega_i^2(x) = 1$. Then $\xi = (\omega_1, \ldots, \omega_n)$. Let $v \in H_{x_0}$. By differentiating the identity $(\xi, \xi) = 1$ with respect to v we find $(D_v\xi, \xi)(x_0) = 0$ and hence $D_v\xi \in H_{x_0}$. The linear map

$$S_{x_0} : H_{x_0} \to H_{x_0}, \quad S_{x_0}(v) = -D_v\xi$$

is called the Weingarten map of (Σ, D) at x_0. The geometric meaning of S_{x_0} can be deduced from the formula

$$D_v \xi = \frac{d}{dt} \xi(\alpha(t)) \Big|_{t=t_0} \, ,$$

where $\alpha : I \to D$ is an integral curve passing through x_0 at the moment t_0 with the velocity vector $\alpha'(t_0) = v$. Therefore S_{x_0} measures (up to sign) the rate of change of the direction of ξ as one passes through x_0 along $\alpha(I)$. Since H_{x_0} is identified to $\xi(\alpha(t_0))^{\perp}$, H_{x_0} turns as the normal ξ turns and so $S_{x_0}(v)$ can be interpreted as a measure of the turning of H_{x_0} as one passes through x_0 along $\alpha(I)$. Thus S_{x_0} contains information about the shape of (Σ, D) at x_0.

The most important properties of the Weingarten map of a nonholonomic hypersurface are included in the following two theorems.

Theorem. Let (Σ, D) be the nonholonomic hypersurface attached to the Pfaff equation (1) with $\sum_{i=1}^{n} \omega_i^2(x) = 1$. Let $x_0 \in D$ and $v \in H_{x_0}$. Then for every integral curve $\alpha : I \to D$, with $\alpha(t_0) = x_0, \alpha'(t_0) = v$, the relation

$$(\alpha''(t_0), \xi(x_0)) = (S_{x_0}(v), v)$$

holds good.

Proof. By differentiating the identity

$$\sum_{i=1}^{n} \omega_i(x(t)) \frac{dx^i}{dt} = 0$$

with respect to t we find

$$\sum_{i,j=1}^{n} \frac{\partial \omega_i}{\partial x^j}(x(t)) \frac{dx^i}{dt} \frac{dx^j}{dt} + \sum_{i=1}^{n} \omega_i(x(t)) \frac{d^2 x^i}{dt^2} = 0 \, .$$

For $t = t_0$, we obtain the relation from the statement.

Consequently the normal component $(\alpha''(t_0), \xi(x_0))$ of the acceleration is the same for all integral curves passing through x_0 with velocity v. Particularly, if the normal component of acceleration is non-zero for some integral curve α with $\alpha'(t_0) = v$, then it is non-zero for all integral curves passing through x_0 with the same velocity. The normal component of acceleration is forced by the shape of (Σ, D) at x_0 and it can be computed directly from the value of S_{x_0} on v.

Theorem. The Weingarten map S_{x_0} is not self-adjoint, that is there exist $v, w \in H_{x_0}$ such that

$$(S_{x_0}(v), w) \neq (v, S_{x_0}(w)) \ .$$

Proof. The computation shows that

$$(S_{x_0}(v), w) = - \sum_{i,j=1}^{n} \frac{\partial \omega_i}{\partial x^j}(x_0) v^i w^j$$

and the matrix $\left[\dfrac{\partial \omega_i}{\partial x^j}\right]$ is not symmetric.

Let $S_{x_0}^* : H_{x_0} \to H_{x_0}$ be the adjoint of S_{x_0}, that is the linear map (unique) defined by $(S_{x_0}^*(v), w) = (v, S_{x_0}(w)), \forall v, w \in H_{x_0}$. The matrix of $S_{x_0}^*$ with respect to an orthonormal basis of H_{x_0} is the transpose of the matrix of S_{x_0} relative to that basis.

The linear map $\frac{1}{2}(S_{x_0} + S_{x_0}^*)$ is self-adjoint. The quadratic form associated to this map, i.e., the function

$$\Omega_{x_0} : H_{x_0} \to R, \quad \Omega_{x_0}(v) = \frac{1}{2}(S_{x_0}(v) + S_{x_0}^*(v), v)$$

is called the second fundamental form of the nonholonomic hypersurface (Σ, D) at x_0. We observe that

$$\Omega_{x_0}(v) = (\alpha''(t_0), \xi(x_0)) \ ,$$

where $\alpha : I \to D$ is an integral curve with $\alpha(t_0) = x_0$ and $\alpha'(t_0) = v$. The function $x_0 \to \Omega_{x_0}$ is called the second fundamental form of (Σ, D) and is denoted by Ω. This function can be obtained by usually differentiating the Pfaff form

$$-\sum_{i=1}^{n} \omega_i(x) dx^i \left(\sum_{i=1}^{n} \omega_i^2(x) = 1 \right)$$

and then taking the symmetric part, i.e.,

$$\Omega = -\frac{1}{2} \sum_{i,i=1}^{n} \left(\frac{\partial \omega_i}{\partial x^j} + \frac{\partial \omega_j}{\partial x^i} \right)(x) dx^i dx^j \ .$$

Particularly, if $v \in H_{x_0}, \|v\| = 1$, then $k(v) = \Omega_{x_0}(v)$ is called the normal curvature of (Σ, D) at x_0 in the direction v. For $k(v) > 0$, every M_{x_0} bends toward ξ in the direction v; for $k(v) < 0$, every M_{x_0} bends aways from ξ in the direction v.

The extrema of normal curvature are proper values of the linear map $\frac{1}{2}(S_{s_0} + S_{x_0}^*)$. This, being a self-adjoint linear map, admits real proper values $k_1(x_0), \ldots, k_{n-1}(x_0)$ and the orthonormal proper vectors $e_1, \ldots, e_{n-1}$ which constitute a basis in H_{x_0}. The proper values $k_1(x_0), \ldots, k_{n-1}(x_0)$ are called principal curvatures of (Σ, D) at x_0, and proper vectors $e_1, \ldots, e_{n-1}$ are called principal directions. If the principal curvatures are ordered so that $k_1(x_0) \leq \ldots \leq k_{n-1}(x_0)$, then $k_{n-1}(x_0) = \max\limits_{\|v\|=1} k(v), k_{n-2}(x_0) = \max\limits_{\|v\|=1, v \perp e_{n-1}} k(v), k_{n-3}(x_0) = \max\limits_{\|v\|=1, v \perp \{e_{n-2}, e_{n-1}\}} k(v), \ldots, k_1(x_0) = \min\limits_{\|v\|=1} k(v).$

Since $v = \sum\limits_{i=1}^{n-1} (v, e_i)e_i = \sum\limits_{i=1}^{n-1} (\cos \theta_i)e_i$, we find the Euler formula

$$k(v) = \sum_{i=1}^{n-1} k_i(x_0) \cos^2 \theta_i \ .$$

The number

$$K(x_0) = \det \frac{1}{2}(S_{x_0} + S_{x_0}^*) = k_1(x_0) \ldots k_{n-1}(x_0)$$

is called the Gauss-Kronecker curvature of (Σ, D) at x_0. The function $x_0 \to K(x_0), x_0 \in D$ is called the Gauss-Kronecker curvature of (Σ, D).

The number

$$H(x_0) = \frac{1}{n-1} \mathrm{tr} \ \frac{1}{2}(S_{x_0} + S_{x_0}^*) = \frac{1}{n-1} \sum_{i=1}^{n-1} k_i(x_0)$$

is called the mean curvature of (Σ, D) at x_0. The function $x_0 \to H(x_0)$, $x_0 \in D$ is called the mean curvature of (Σ, D).

Remarks. (1) If (Σ, D) is described by (1) and (2), then

$$\Omega = -\frac{1}{2\|\omega\|} \sum_{i,j=1}^{n} \left(\frac{\partial \omega_i}{\partial x^j} + \frac{\partial \omega_j}{\partial x^i} \right)(x) dx^i dx^j \ .$$

(2) If $\dfrac{\partial \omega_i}{\partial x^j} + \dfrac{\partial \omega_j}{\partial x^i} = 0, i,j = 1,\ldots,n$, i.e., ξ is a Killing vector field on D, then (Σ, D) is called a nonholonomic hyperplane. The nonholonomic hyperplanes of R^n are characterized by the Pfaff equation $\displaystyle\sum_{i=1}^{n}\Big(\sum_{j=1}^{n} a_{ij}x_j + a_i\Big)dx^i = 0$ with the conditions

$$a_{ij} = -a_{ji},$$
$$a_{il}(a_{kj} - a_{jk}) + a_{jl}(a_{ik} - a_{ki}) + a_{kl}(a_{ji} - a_{ij}) \neq 0, \quad i,j,k,l = 1,\ldots,n.$$

(3) If

$$\frac{\partial \omega_i}{\partial x^j} + \frac{\partial \omega_j}{\partial x^i} = \psi\delta_{ij}, \quad i,j = 1,\ldots,n,$$

i.e., ξ is a conformal vector field on D, then (Σ, D) is called a nonholonomic hypersphere. The nonholonomic hyperspheres of R^n are characterized by the Pfaff equation

$$\sum_{i=1}^{n}\Big(\frac{1}{2}x_i\sum_{j=1}^{n} c_j x_j - \frac{1}{4}c_i\sum_{j=1}^{n} x_j^2 + \sum_{j=1}^{n} c_{ij}x_j + d_i\Big)dx^i = 0, \quad c_{ij} + c_{ji} = \frac{c}{2}\delta_{ij},$$

with the conditions which ensure that this Pfaff equation is not completely integrable. For example, the nonholonomic surface defined by the Pfaff equation

$$(x + y)dx + (y - x)dy + zdz = 0, \quad (x,y,z) \neq (0,0,0)$$

is a nonholonomic sphere.

Open problem. What properties of (Σ, D) can be derived from the skew-symmetric linear map $\frac{1}{2}(S_{x_0} - S_{x_0}^*)$, i.e., from the skew part of S_{x_0}?

2. Extrema with a Nonholonomic Constraint

In Mechanics, a completely integrable Pfaff equation is called a holonomic constraint, and a Pfaff equation which is not completely integrable is called a nonholonomic constraint.

We will present a theory of extrema with constraint (1), i.e., a theory in which the classical constraints are the integral manifolds of the Pfaff equation (1). If the Pfaff equation (1) is completely integrable, then the theory developed by us is reduced to the method of Lagrange multiplier. If the Pfaff equation (1) is not completely integrable, then our theory is new.

Definition. Let $f : D \to R$ be a real function of class C^1. We say that $x_0 \in D$ is a minimum (maximum) point of f constrained by (1) if for every integral manifold $r : I \to D$ of Eq. (1) which passes through $x_0 (\exists\, u_0 \in I$ with $r(u_0) = x_0)$ there exists a neighborhood $V_{x_0} \subset D$ of x_0 such that $f(x) \geq (\leq)f(x_0), \forall\, x \in V_{x_0} \cap r(I)$.

If in this definition we consider only integral curves we obtain the notion of constrained extremum in the sense of [2], [3], [4]. To make the distinction we will utilize the namings "constrained extremum on integral manifolds", "constrained extremum on integral curves" respectively. Every point of constrained extremum on integral manifolds is a point of constrained extremum on integral curves.

Theorem. $x_0 \in D$ is an extremum point of $f : D \to R$ constrained by (1) iff for every integral manifold $r : I \to D, r(u_0) = x_0$ there exists a neighborhood $I_{u_0} \subset I$ of u_0 such that $f(r(u)) - f(x_0)$ has constant sign on I_{u_0}.

Proof. ($\Rightarrow$) Let $r : I \to D$ be an integral manifold through x_0 and $V_{x_0} \subset D$ a neighborhood of x_0 for which $f(x) - f(x_0)$ has a constant sign on $r(I) \cap V_{x_0}$. Then I_{u_0} is the connected component of $r^{-1}(V_{x_0})$ which contains u_0.

($\Leftarrow$) The set $r(I_{u_0})$ is open in $r(I)$ because r is a homeomorphism $I \to r(I)$. Consequently $\exists\, V_{x_0} \subset D$ with $V_{x_0} \cap r(I) = r(I_{u_0})$ and $f(x) - f(x_0)$ has constant sign on $V_{x_0} \cap r(I)$.

Theorem [2], [3]. If $x_0 \in D$ is a constrained extremum point on integral curves of $f : D \to R$ constrained by (1), then there exists $\lambda \in R$ such that

$$\frac{\partial f}{\partial x^i}(x_0) + \lambda \omega_i(x_0) = 0, \quad i = 1, \dots, n \ .$$

Geometric interpretation. Let $f : D \to R$ be a real function of class C^1 and $x_0 \in D$. Let Σ_{x_0} be the family of all integral manifolds through x_0 of the Pfaff equation (1). One observes that, for every $M_{x_0} \in \Sigma_{x_0}$, the relation

$$\text{grad}_D f(x_0) = \text{grad}_{M_{x_0}} f_{M_{x_0}}(x_0) + a(x_0)\xi(x_0)$$

is satisfied, where $f_{M_{x_0}}$ is the restriction of f to $M_{x_0}, a(x_0) = (\text{grad}_D f(x_0),$ $\xi(x_0)) = df(x_0)(\xi(x_0))$, and $\xi = \frac{1}{\|\omega\|}(\omega_1, \ldots, \omega_n)$. Hence $\text{grad}_{M_{x_0}} f_{M_{x_0}}$ does not depend on M_{x_0} but only on f, on x_0 and on Σ_{x_0}. Therefore we will denote this vector by $\text{grad}_{\Sigma_{x_0}} f(x_0)$. From the above relation it follows that $\text{grad}_{\Sigma_{x_0}} f(x_0)$ is the horizontal part of $\text{grad}_D f(x_0)$. Consequently it becomes obvious that if $x_0 \in D$ is a critical point of f, then x_0 is a critical point of $f_{M_{x_0}}, \forall M_{x_0} \in \Sigma_{x_0}$. Also one observes that x_0 is a critical point of $f_{M_{x_0}}$ iff

$$\text{grad}_D f(x_0) = a(x_0)\xi(x_0) \in H_{x_0}^{\perp} .$$

When x_0 is a critical point of $f_{M_{x_0}}$, but is not a critical point of f, the hyperplane H_{x_0} coincides with the tangent hyperplane at x_0 to the level hypersurface $f(x) = f(x_0)$.

Definition. The points $x \in D$ which are solutions of the system

$$\frac{\partial f}{\partial x^i}(x) + \lambda\omega_i(x) = 0, \quad i = 1, \ldots, n \tag{4}$$

are called critical points of f constrained by the Pfaff equation (1).

One remarks that any critical point of f is also a constrained critical point.

Theorem. Let Ω_{x_0} be the second fundamental form at x_0 of the nonholonomic hypersurface (Σ, D) described by the Pfaff equation (1). If x_0 is a constraint critical point of the function $f : D \to R$ of class C^2 with constraint (1) and if the quadratic form

$$d^2 f(x_0) - \lambda\|\omega(x_0)\|\Omega_{x_0} \tag{5}$$

constrained by

$$\sum_{i=1}^{n} \omega_i(x_0)dx^i = 0 \tag{6}$$

is positive (negative) definite, then x_0 is a point of a strict minimum (maximum) constrained by (1).

Proof. Let $r : I \to D, r = (x^1, \ldots, x^n), x^i = x^i(u^1, \ldots, u^m), i = 1, \ldots, n$, an integral manifold of the equation (1) with $r(u_0) = x_0$ and $\varphi(u) = f(r(u))$. Then

$$\varphi(u) - \varphi(u_0) = d\varphi(u_0)(u - u_0) + \frac{1}{2}d^2\varphi(u_0)(u - u_0, u - u_0)$$
$$+ \mathcal{O}(\|u - u_0\|^2) ,$$
$$\frac{\partial \varphi}{\partial u^k}(u) = \sum_{i=1}^{n} \frac{\partial f}{\partial x^i}(r(u))\frac{\partial x^i}{\partial u^k}(u),$$
$$d\varphi(u) = \sum_{i=1}^{n} \frac{\partial f}{\partial x^i}(r(u)) \sum_{k=1}^{m} \frac{\partial x^i}{\partial u^k}(u)du^k ,$$
$$d^2\varphi(u) = \sum_{i,j=1}^{n} \frac{\partial^2 f}{\partial x^i \partial x^j}(r(u)) \sum_{k,l=1}^{m} \frac{\partial x^i}{\partial u^k}(u)\frac{\partial x^j}{\partial u^l}(u)du^k du^l$$
$$+ \sum_{i=1}^{n} \frac{\partial f}{\partial x^i}(r(u)) \sum_{k,l=1}^{m} \frac{\partial^2 x^i}{\partial u^k \partial u^l}(u)du^k du^l .$$

Since r is an integral manifold we have

$$\sum_{i=1}^{n} \omega_i(r(u))\frac{\partial x_i}{\partial u^k}(u) = 0, \quad \forall u \in I .$$

By the chain rule for derivatives we obtain

$$\frac{1}{2}\sum_{i,j=1}^{n} \left(\frac{\partial \omega_i}{\partial x^j} + \frac{\partial \omega_j}{\partial x^i}\right)(r(u))\frac{\partial x^i}{\partial u^k}(u)\frac{\partial x^j}{\partial u^l}(u) + \sum_{i=1}^{n} \omega_i(r(u))\frac{\partial^2 x^i}{\partial u^k \partial u^l}(u) = 0 .$$

$$(7)$$

Let $x_0 = r(u_0)$ be a constrained critical point. Then

$$0 = \frac{\partial \varphi}{\partial u^k}(u_0) = \sum_{i=1}^{n} \frac{\partial f}{\partial x^i}(r(u_0))\frac{\partial x^i}{\partial u^k}(u_0) = -\lambda \sum_{i=1}^{n} \omega_i(r(u_0))\frac{\partial x^i}{\partial u^k}(u_0) ,$$

i.e., $d\varphi(u_0) = 0$. In addition, by (7), we find

$$
\begin{aligned}
d^2\varphi(u_0) &= \sum_{i,j=1}^{n} \frac{\partial^2 f}{\partial x^i \partial x^j}(r(u_0)) \sum_{k,l=1}^{m} \frac{\partial x^i}{\partial u^k}(u_0)\frac{\partial x^j}{\partial u^l}(u_0)du^k du^l \\
&\quad - \lambda \sum_{i=1}^{n} \omega_i(r(u_0)) \sum_{k,l=1}^{m} \frac{\partial^2 x^i}{\partial u^k \partial u^l}(u_0)du^k du^l \\
&= \sum_{i,j=1}^{n} \frac{\partial^2 f}{\partial x^i \partial x^j}(x_0)dx^i dx^j + \frac{\lambda}{2} \sum_{i,j=1}^{n} \left(\frac{\partial \omega_i}{\partial x^j} + \frac{\partial \omega_j}{\partial x^i}\right)(x_0)dx^i dx^j \\
&= d^2 f(x_0) - \lambda\|\omega(x_0)\|\Omega_{x_0} \ .
\end{aligned}
$$

Since (5) constrained by (6) is positive (negative) definite it follows that $d^2\varphi(u_0) > 0 \, (< 0)$, i.e., $\varphi(u) - \varphi(u_0) > 0 \, (< 0), \forall u \in I_{u_0} \subset I, u \neq u_0$. Consequently there exists V_{x_0} such that $r(I_{u_0}) = V_{x_0} \cap r(I)$ and $f(x) > (<)f(x_0), \forall x \in V_{x_0} \cap r(I) - \{x_0\}$.

Geometric interpretation. If the Pfaff equation (1) is not completely integrable, then the quadratic form in the previous theorem is the horizontal part of $d^2 f(x_0)$, i.e., the second order differential at $r(u_0) = x_0$ of the restriction of f to the nonholonomic hypersurface.

Theorem. If x_0 is a point of constrained extremum on integral curves and the quadratic form (5) with the constraint (6) is (either positive or negative) definite, then x_0 is a point of constrained extremum on integral manifolds.

Proof. If x_0 is a point of constrained extremum on integral curves, then x_0 is a constrained critical point. Adding the hypothesis that the quadratic form (5) with the constraint (6) is (either positive or negative) definite, it follows that $x_0 = r(u_0)$ is a point of extremum for $\varphi(u) = f(r(u))$.

Theorem [2], [3]. If $x_0 \in D$ is a point of constrained extremum on integral curves of the function $f : D \to R$ of class C^2, then the quadratic form (5) constrained by (6) is either positive or negative semidefinite.

From these theorems and from the fact that every point of constrained extremum on integral manifolds is a point of constrained extremum on integral curves it follows that for such points at which the quadratic form (5) constrained by (6) is either positive or negative definite the two notions coincide. For "doubtful" such points, i.e., such points at which the quadratic form (5) constrained by (6) is semidefinite and degenerate, we do not have a definite answer yet.

3. Convex Nonholonomic Hypersurfaces

Let (Σ, D) be the nonholonomic hypersurface described by the Pfaff equation (1) with the condition (2). The hyperplane

$$H_{x_0} = \left\{ x \in R^n \ \middle| \ \sum_{i=1}^{n} \omega_i(x_0)(x^i - x_0^i) = 0 \right\}$$

determines the semispaces

$$H_{x_0}^- = \left\{ x \in R^n \ \middle| \ \sum_{i=1}^{n} \omega_i(x_0)(x^i - x_0^i) \leq 0 \right\},$$

$$H_{x_0}^+ = \left\{ x \in R^n \ \middle| \ \sum_{i=1}^{n} \omega_i(x_0)(x^i - x_0^i) \geq 0 \right\}.$$

Definition. The nonholonomic hypersurface (Σ, D) is called convex at $x_0 \in D$ if for every integral manifold $M_{x_0} \in \Sigma_{x_0}$ there exists an open neighborhood $V_{x_0} \subset D$ of x_0 such that all the sets $V_{x_0} \cap M_{x_0}$ are contained either in $H_{x_0}^-$ or in $H_{x_0}^+$.

Let the function

$$F_{\omega(x_0)}(x) = \sum_{i=1}^{n} \omega_i(x_0)(x^i - x_0^i).$$

It is obvious that (Σ, D) is convex at x_0 iff x_0 is an extremum point of $F_{\omega(x_0)}$ constrained by the Pfaff equation (1).

A nonholonomic hypersurface (Σ, D) convex at x_0 is said to be strictly convex at x_0 if $V_{x_0} \cap M_{x_0} \cap H_{x_0} = \{x_0\}$.

Theorem. If (Σ, D) is a nonholonomic hypersurface convex at $x_0 \in D$, then the second fundamental form Ω_{x_0} of (Σ, D) at x_0 is (either negative or postive) semidefinite.

Proof. For $v \in H_{x_0}$, let $\alpha : I \to D, \alpha(t) = (x^1(t), \ldots, x^n(t))$ be an integral curve with $\alpha(t_0) = x_0, \alpha'(t_0) = v$. Suppose, for example, that there exists V_{x_0} with $\alpha(I) \cap V_{x_0} \subset H_{x_0}^+$. Then the function

$$f : I \to R, f(t) = \sum_{i=1}^{n} \omega_i(x_0)(x^i(t) - x_0^i)$$

has the properties $f(t) \geq 0, \forall t \in I$ and $f(t_0) = 0$. In this way $t_0 \in I$ is a global minimum point and hence

$$0 = f'(t_0) = \sum_{i=1}^{n} \omega_i(x_0)\frac{dx^i}{dt}(t_0) = \sum_{i=1}^{n} \omega_i(x_0)v^i \ ,$$

$$0 \leq f''(t_0) = \sum_{i=1}^{n} \omega_i(x_0)\frac{d^2 x^i}{dt^2}(t_0) = \Omega_{x_0}(v), \quad \forall\, v \in H_{x_0} \ .$$

In the hypothesis $\alpha(I) \cap V_{x_0} \subset H_{x_0}^-$ one obtains the inequality $\Omega_{x_0}(v) \leq 0, \forall\, v \in H_{x_0}$.

Variant of proof. The point x_0 is an extremum point of the function $F_{\omega(x_0)}$ with the nonholonomic constraint (1).

The converse of the previous theorem is not true. To show this, we consider the Pfaff equation

$$\eta = dz - 2xdx - 3y^2xdy = 0, \quad (x, y, z) \in R^3 \ .$$

As $d\eta = -3y^2 dx \wedge dy$, the Pfaff equation $\eta = 0$ is not completely integrable. It defines a nonholonomic surface (Σ, R^3) which is constituted only from integral curves. Let the point be $O(0, 0, 0)$. Since

$$\xi(x, y, z) = \left(\frac{-2x}{\sqrt{1 + 4x^2 + 9x^2y^4}}, \frac{-3y^2x}{\sqrt{\cdots}}, \frac{1}{\sqrt{\cdots}} \right) \ ,$$

we find $\xi(0, 0, 0) = (0, 0, 1)$ and hence $H_0 : z = 0$ is the tangent plane to (Σ, R^3) at O. On the other hand, if $\alpha : I \to R^3, \alpha(t) = (x(t), y(t), z(t))$,

$\alpha(0) = (0,0,0), z'(t) - 2x(t)x'(t) - 3y^2(t)x(t)y'(t) = 0$, i.e., $\alpha(I) \in \Sigma_O$ then $z''(0) = 2x'^2(0) \geq 0$ and hence $f(t) = (\alpha(t) - 0, \xi(0,0,0)) = z(t)$ satisfies $f''(0) = z''(0) \geq 0$. Consequently the second fundamental form Ω_0 of (Σ, R^3) at O is positive semidefinite. However (Σ, R^3) is not convex at O since there exist integral curves which traverse the plane $H_0 : z = 0$ at the origin. Such a curve is $x = 1, z = y^3$. Hence in any neighborhood of the origin there exist points of the curves in Σ_O placed below the tangent plane and above the tangent plane H_0.

Remark. The curve $\alpha : R \to R^3, \alpha = (x,y,z), x = -t, y = t, z = t^2(1 - \frac{3}{4}t^2)$ belongs to Σ_0. The variation table

t	$-\infty$		$-2/\sqrt{3}$		0		$2/\sqrt{3}$		∞
z	∞	$-$	0	$+$ $+$ $+$			0	$-$	∞

shows that $\alpha(R)$ traverses H_0 at the points $\left(\frac{2}{\sqrt{3}}, \frac{-2}{\sqrt{3}}, 0\right)$ and $\left(-\frac{2}{\sqrt{3}}, \frac{2}{\sqrt{3}}, 0\right)$.

Theorem. Let (Σ, D) be the nonholonomic hypersurface defined by (1) and (2). If the second fundamental form Ω_{x_0} is either positive or negative definite, then (Σ, D) is strictly convex at x_0.

Proof. Without loss of generality, we suppose $\sum_{i=1}^{n} \omega_i^2(x) = 1$. By computation one proves that any point $x_0 \in D$ is a constrained critical point of $F_{\omega(x_0)}(x) = \sum_{i=1}^{n} \omega_i(x_0)(x^i - x_0^i)$, with $\lambda = -1$. On the other hand, for $F_{\omega(x_0)}$, the quadratic form (5) reduces to Ω_{x_0}. Consequently x_0 is a strict local extremum point of $F_{\omega(x_0)}$ constrained by (1), i.e., the nonholonomic hypersurface (Σ, D) is strictly convex at x_0.

Remark. The second fundamental form of the nonholonomic hypersurface (Σ, D) depends only on $\omega_i(x)$. It can be positive or negative definite (semidefinite) on an open ball, without necessarily considering its restriction to the integral manifolds of (1); see the next examples.

Examples. (1) The Pfaff equation $-(x+z)dx + dy + xdz = 0$ is not completely integrable. The second fundamental form of the nonholonomic surface defined by this equation is dx^2. It is positive semidefinite.

(2) Let $\eta = (3x^2 + 4y)dx + 2ydy + 2zdz = 0$ and $D = R^3 - \{(0,0,0)\}$. It follows that $d\eta = 4dy \wedge dx$ and hence $\eta \wedge d\eta \neq 0$, i.e., the Pfaff equation $\eta = 0$ is not completely integrable. The second fundamental form can be obtained by usually differentiating the form $-\eta$. Hence $\Omega = -2(3xdx^2 + dy^2 + dz^2 + 2dxdy)$. For $x > \frac{1}{3}$ this quadratic form is negative definite. Consequently the nonholonomic surface described by $\eta = 0$ is strictly convex at any point in the region $x > \frac{1}{3}$.

Theorem. Let $f : D \to R$ be of class C^2 and Ω_{x_0} be the second fundamental form at x_0 of (Σ, D). If $d^2 f(x_0)$ is positive (negative) definite and if there exists $\lambda \neq 0$ such that the quadratic form $d^2 f(x_0) - \lambda \|\omega(x_0)\| \Omega_{x_0}$ constrained by $\sum_{i=1}^{n} \omega_i(x_0)dx^i = 0$ is negative (positive) semidefinite, then Ω_{x_0} is either positive or negative definite.

Proof. The relations $d^2 f(x_0) > 0, d^2 f(x_0) - \lambda \|\omega(x_0)\| \Omega_{x_0} \leq 0$ imply $0 < d^2 f(x_0) \leq \lambda \|\omega(x_0)\| \Omega_{x_0}$ and hence $\Omega_{x_0} > 0$.

Corollaries. If there exists $\lambda \neq 0$ such that the quadratic form $\sum_{i=1}^{n} dx_i^2 - \lambda \|\omega(x_0)\| \Omega_{x_0}$ constrained by $\sum_{i=1}^{n} \omega_i(x_0)dx^i = 0$ is negative (positive) semidefinite, then

(1) the nonholonomic hypersurface (Σ, D) is convex at x_0;

(2) the trace of Ω_{x_0} cannot be zero (the nonholonomic hypersurface cannot be minimal at x_0);

(3) if the Gauss-Kronecker curvature is nowhere zero, then the nonholonomic hypersurface (Σ, D) is strictly convex at each point $x \in D$.

Proof. (3) Ω_{x_0} is (either positive or negative) definite. As Ω is nondegenerate everywhere, it follows that Ω is definite everywhere.

References

1. R. Miron, *Geometrizarea sistemelor Pfaff în spaţii riemanniene cu metrică nedefinită*, St. Cerc. Şt. Acad. RPR. Iaşi, Matematică, 1 (1958).
2. C. Udrişte and O. Dogaru, *Extrema with nonholonomic constraints*, Buletinul Institutului Politehnic Bucureşti, t.L (1988), 1–8.
3. C. Udrişte and O. Dogaru, *Mathematical programming problems with nonholonomic constraints*, Seminarul de mecanică, Universitatea din Timişoara 14 (1988).
4. C. Udrişte and O. Dogaru, *Extreme condiţionate pe orbite*, Buletinul Institutului Politehnic Bucureşti, sub tipar.
5. Gh. Vrănceanu, Opera matematică, Editura Academiei RSR, Bucureşti, 1969, 1971,...

Constantin Udrişte and Oltin Dogaru
Dept. of Mathematics I
Polytechnic Institute
Splaiul Independenţei 313
Bucharest
Romania

THE MATH. HERITAGE OF C.F. GAUSS (pp. 785-809)
edited by George M. Rassias
©1991 World Scientific Publ. Co. Singapore

THE MEASURE OF COVERING THE EUCLIDEAN SPACE BY GROUP-TRANSLATES OF A SET

B. Uhrin

Let $L \subset R^n$ be a point lattice (integer combination of r linearly independent vectors) and $S \subset R^n$ be a linear subspace such that $\mathrm{lin}(L) \cap S = \{\theta\}$, where $\mathrm{lin}(L)$ is the linear subspace spanned by L. Denote $M := L \oplus S$ (the direct sum). Let $A \subset R^n$ be a bounded measurable set. The paper deals with the question: what portion of R^n is covered by the family $\{(A+u)\}_{u \in M}$? A number $\alpha(M,A)$ (called the measure of covering) that measures the portion covered, is introduced. Four descriptions of $\alpha(M,A)$ are given. Some basic properties of $\alpha(M,A)$ are investigated, say, how $\alpha(M_1,A)$ and $\alpha(M_2,A)$ are related to each other if $M_1 \subset M_2$, or, how $\alpha(M,t_1A)$ is related to $\alpha(M,t_2A)$, where $t_1 > t_2 \geq 0$? The main result of the paper is: the function $\alpha(M,tK).t^{-n+k}$ is a non-decreasing function of $t \geq 0$, where $K \subset R^n$ is a convex body and k is the dimension of M. This theorem has many interesting corollaries especially for discrete M (i.e., $S=\{\theta\}$), among others an improvement of the successive minima theorem in the geometry of numbers. Extensions of the results for discrete M to more general M are also studied.

1. Introduction

Let $R^k, k \geq 0$, denote the k-dimensional Euclidean space, i.e., the linear space of ordered k-tuples of real numbers (by definition $R^0 = \{\theta\}$, the zero element). Let μ_k mean the Lebesque measure in R^k (by definition $\mu_0 = 1$). The cardinality of the finite set H is denoted by $|H|$, the empty set $\emptyset$ has the cardinality 0. For non-empty sets $A, B \subseteq R^k$, the algebraic sum $A + B$ is defined as the set $\{a + b : a \in A, b \in B\}$. If for any $z \in A + B$ there are unique $x \in A, y \in B$, such that $z = x + y$, then $A + B$ will be denoted by $A \oplus B$ (the direct sum). For $x, y \in R^k$ the scalar (or inner) product $\langle x, y \rangle$ is defined as $\sum_{i=1}^{k} x_i y_i$. The $x, y \in R^k$ are orthogonal, in notation $x \perp y$, if $\langle x, y \rangle = 0$. For $A, B \subseteq R^k$, $A \perp B$ means that $a \perp b$ for all $a \in A, b \in B$. The vectors $b_1, b_2, \ldots, b_k \in R^n$ are linearly independent if $\sum_{i=1}^{k} x_i b_i = \theta$ implies $x_1 = \ldots = x_k = 0$. Given such vectors, the set $\mathrm{lin}(b_1, \ldots, b_k) := \left\{ \sum_{i=1}^{k} x_i b_i : x_i \text{ real numbers} \right\}$ is a k-dimensional linear subspace of R^n and the set $\mathrm{int}\,(b_1, \ldots, b_k) := \left\{ \sum_{i=1}^{k} u_i b_i : u_i \text{ integers} \right\}$ is a k-dimensional point lattice in R^n. The (b_i) is a basis of the subspace and the lattice, respectively. In general, given any set $A \subset R^n$, the linear hull $\mathrm{lin}(A)$ is defined as the linear subspace of minimal dimension containing the A. Given a point lattice $L \subset R^n$, the linear subspace $\mathrm{lin}(L)$ is called the carrying subspace of L. The dimension of any linear subspace $F \subset R^n$ will be denoted by $\dim\,(F)$. For any set $A \subseteq R^n$, the $\dim\,(A)$ is the dimension of the linear hull of A. The basis $b_1, \ldots, b_k \in R^n$ of the linear subspace $\mathrm{lin}\,(b_1, \ldots, b_k)$ is called orthonormal if $\langle b_i, b_i \rangle = 1, i = 1, 2, \ldots, k$, and $\langle b_i, b_j \rangle = 0$ for $1 \leq i < j \leq k$. Let $E \subseteq R^n$ be a k-dimensional linear subspace and (b_i) be its orthonormal basis. The set $A \subseteq E$ is called measurable if the set $\tilde{A} := \left\{ (x_1, x_2, \ldots, x_k) : \sum_{i=1}^{k} x_i b_i \in A \right\}$ is Lebesque measurable in R^k and the measure in E is defined as $\mu(A) := \mu_k(\tilde{A})$. We note that if the basis (b_i) used for the definition of μ is not orthonormal, then μ is not quite consistent with the k-dimensional "volume" in E (for volumes and measures see, e.g., [7]).

If for linear subspaces $E, F \subseteq R^n, E \cap F = \{\theta\}$, then $E + F = E \oplus F$ and $E \oplus F$ is a linear subspace of dimension $\dim\,(E) + \dim(F)$. Let μ' and μ'' be the measures in E and F, respectively. Then we can define the measure in $E \oplus F$ in two different ways. First, $E \oplus F$ can be considered as a Cartesian "product" of the two spaces. In this case we can define measurable sets and the product measure $\mu' \times \mu''$ in $E \oplus F$ in the usual way starting from measurable parallelotopes $A \oplus B$ and defining $\mu' \times \mu''(A \oplus B) := \mu'(A) \cdot \mu''(B)$ (for details see, e.g., [2], [4], [8]). Denote by μ the product measure $\mu' \times \mu''$. Let $A \subset E \oplus F$ be a bounded measurable set. Then the set $(A - x) \cap E$ is measurable in E for all $x \in F$, the function $\mu'((A - x) \cap E)$ is a measurable function of x on F and

$$\mu(A) = \int_F \mu'((A - x) \cap E)d\mu''(x). \tag{1.1}$$

This is a special case of the Fubini theorem (see [8]). The second way of defining measurability in $E \oplus F$ is to consider the $E \oplus F$ as a linear subspace, to take an orthonormal basis in it and to define the measurability analogously as in E. The measure defined in this way does not coincide in general with $\mu' \times \mu''$ but if $E \perp F$ then the two approaches yield the same result. So if $E \perp F$ then (1.1) also holds for μ defined in the linear subspace $E \oplus F$.

Let us recall another identity. Let $L \subset R^n$ be a k-dimensional point lattice generated by the basis $(b_1, \ldots, b_k)$ (for $k = 0$ we take $L = \{\theta\}$). Let $B \subset \text{lin}\,(L)$ be a measurable bounded set and $\mu'(B)$ be its measure. Then

$$\mu'(B) = \int_Q |(B - x) \cap L|d\mu'(x), \tag{1.2}$$

where $Q := \left\{ \sum_{i=1}^k \lambda_i b_i : 0 \leq \lambda_i < 1, i = 1, 2, \ldots, k \right\}$ is the basic cell of L in $\text{lin}\,(L)$ belonging to the basis (b_i). This identity is a special case of the identity on p. 36 in [13].

The identities (1.1) and (1.2) play important roles in our investigations.

The paper is organized as follows: Sec. 2 contains the definition and different forms of the measure of covering, together with a basic identity (a consequence of (1.1)) and some other general properties. In Sec. 3 we prove the main result of the paper that concerns the behaviour of the measure of covering R^n by the sets $tK, t \geq 0$, where $K \subset R^n$ is a convex body. The main result have many interesting corollaries for discrete subgroups of R^n. These are proved in Sec. 4. Sec. 5 investigates, to what extent can the discrete results be generalized to arbitrary closed subgroups. Finally, Sec. 6 contains some concluding remarks.

At the end of the Introduction, let us say a few words on the preliminaries of this paper. Let $\Lambda \subset R^n$ be an n-dimensional point lattice and P be a basic cell of Λ. Let $A \subset R^n$ be a bounded measurable set. In [10] we proved that

$$|(A - A) \cap \Lambda| \geq \frac{2\mu_n(A)}{\mu_n(\varphi(A))} - 1, \tag{1.3}$$

where

$$\varphi(A) := \{x \in P : (A - x) \cap \Lambda \neq \emptyset\}. \tag{1.4}$$

This is a sharpening of the first main theorem in the geometry of numbers, of the Minkowski-Blichfeldt theorem. It took some time until we discovered that $\mu_n(\varphi(A))$ is equal, as concerns its numerical value, to the quantity appearing in a proof of the second main theorem in the geometry of numbers, of the successive minima theorem. Namely let $\Lambda_0 \subset R^n$ be the lattice of integer points of R^n.

Then the result in [3] p. 215 implies that for a symmetric convex body $K \subset R^n$ (i.e., $K = -K$), we have

$$\mu_n(\varphi_0(tK)) \begin{cases} = t^n \mu_n(K) & \text{if} \quad 0 \leq t \leq \lambda_1/2, \\[2em] \geq t^{n-k} \mu_n(K) \cdot \prod_{i=1}^{k} (\lambda_i/2) & \text{if} \quad \lambda_k/2 \leq t \leq \lambda_{k+1}/2, \end{cases}$$
$$k = 1, 2, \ldots, n, \tag{1.5}$$

where λ_i are the successive minima of K (with respect to Λ_0). For arbitrary (not necessarily symmetric) convex bodies K a result in [5], p. 61, yields

$$\mu_n(\varphi_0(\nu_n K)) \geq \mu_n(K) \cdot \prod_{i=1}^{n} \nu_i, \tag{1.6}$$

where ν_i are the successive minima of $K - K$ (with respect to Λ_0).

They were definitely the results (1.3), (1.5) and (1.6) that turned our attention to the quantity $\mu_n(\varphi(A))$. Our last inspiration came from the paper [1], especially the decomposition $A = \bigcup_{i=1} A(i)$ and the quantity

$$\sum_{i \geq 1} \frac{\mu_n(A(i))}{i} \tag{1.7}$$

introduced in [1] was interesting to us.

We shall see that the $\mu_n(\varphi(A))$ is equal to (1.7) (the "fourth form" of the measure of covering, see Sec. 2, here we need (1.2)).

These were the starting points of our research and in the subsequent sections a report on some results can be found.

2. The Measure of Covering: The Definition and Basic Properties

In what follows we shall write (1.1), (1.2) and all similar expressions in a simplified way denoting by μ all measures and by $\int dx$ all integrals, i.e., when writing $\mu(H)$ and $\int f(x)dx$ it will always be clear from the context in what linear subspaces μ and $\int dx$ are to be understood.

(1.1) and (1.2) make the following definitions meaningful. Let A, F and E be as in (1.1). We call the set

$$\pi(A) := \{x \in F \: : \: \mu((A - x) \cap E) > 0\} \tag{2.1}$$

the measurable projection of A into F along E.

Let B, L, Q be as in (1.2). We call the set

$$\psi(B) := \{x \in Q \: : \: |(B - x) \cap L| > 0\} \tag{2.2}$$

the canonical projection of B into Q.

Let $V \subseteq R^n$ be a linear subspace, $L \subset V$ be an r-dimensional point lattice, Q a basic cell of L in $\operatorname{lin}(L)$, let $S \subset V$ be an s-dimensional linear subspace, such that $\operatorname{lin}(L) \cap S = \{\theta\}$, $W := \left\{ \sum_1^s \lambda_i b_i : 0 \leq \lambda_i \leq 1, i = 1, \ldots, s \right\}$ be the basic cell of an orthonormal basis $(b_1, \ldots, b_s)$ of S and $T \subseteq V$ be a p-dimensional linear subspace such that $(\operatorname{lin}(L) \oplus S) \perp T$ and $\operatorname{lin}(L) \oplus S \oplus T = V$ (any of r, s, p can be zero). Denote $M := L \oplus S$.

Let $C \subset \operatorname{lin}(L) \oplus T$ be a bounded measurable set and denote $P := Q \oplus T$. Then we call the set

$$\varphi(C) := \{z \in P : \: |(C - z) \cap L| > 0\} \tag{2.3}$$

the canonical projection of C into P.

Let $A \subset V$ be a bounded measurable set. We call the number

$$\alpha(M, A) := \mu(\varphi(\pi(A)) \oplus W) \tag{2.4}$$

the measure of covering V by M-translates of A, where $\pi(A)$ is the measurable projection of A into $\operatorname{lin}(L) \oplus T$ along S and $\varphi(\pi(A))$ is the cannonical projection of $\pi(A)$ into P. One can easily see that

$$\alpha(M, A) = \alpha(M, A + x), \quad x \in V \tag{2.5}$$

and

$$\alpha(M,A) \le \alpha(M,B) \quad \text{if} \quad A \subseteq B \subset V. \tag{2.6}$$

$\alpha(M,A)$ can be given two more equivalent forms. We call the function

$$\varphi(z) := \sum_{1}^{r} \{\xi_i\} b_i + y, \quad z = \sum_{1}^{r} \xi_i b_i + y \in \operatorname{lin}(L) \oplus T, \ y \in T, \tag{2.7}$$

the canonical projection of lin $(L) \oplus T$ *into* P, where $0 \le \{\xi\} < 1$ is the number such that $\xi - \{\xi\}$ is integer (the fractional part of ξ). One can easily see that P is in a one-to-one correspondence with the quotient space $(\operatorname{lin}(L) \oplus T)/L$, hence φ is a representation of the usual canonical quotient map $\operatorname{lin}(L) \oplus T \to (\operatorname{lin}(L) \oplus T)/L$. Now we can easily see that

$$\varphi(C) = \bigcup_{z \in C} \varphi(z) = \bigcup_{u \in L} ((C + u) \cap P), \quad \text{for} \quad C \subset \operatorname{lin}(L) \oplus T, \tag{2.8}$$

where $\varphi(C)$ is the set (2.3) and $\varphi(z)$ is the map (2.7).

This implies

$$\alpha(M,A) = \mu\left(\left(\bigcup_{z \in \pi(A)} \varphi(z)\right) \oplus W\right) = \mu\left(\left(\bigcup_{u \in L} ((\pi(A) + u) \cap P)\right) \oplus W\right). \tag{2.9}$$

It is important to note that the $\alpha(M,A)$ defined by (2.4) does not depend on W and T, i.e., it has the same value for any W and T satisfying the given conditions. This is a consequence of the fact that given a $W \subset S$ the number

$$\mu(\bar{\pi}(A) \oplus W), \tag{2.10}$$

does not depend on H, where $H \subset V$ is any linear subspace such that $H \cap S = \{\theta\}, H \oplus S = V$ and $\bar{\pi}(A)$ is the measurable projection of A into H along S. The latter fact can easily be seen if the set A is of the form

$$\left\{ \sum_{i=1}^{s} x_i \bar{b}_i + \sum_{i=1}^{q} y_i c_i \ : \ 0 \le x_i \le v_i, \quad 0 \le y_i \le w_i \right\},$$

where $s + q = \dim(V)$ and $\bar{b}_i \in S, c_i \perp S$ are mutually orthogonal vectors in V (the "rectangular parallelotope"). After that usual density consideration, prove the statement for any bounded measurable $A \subset V$. It is clear that the $\varphi(\pi(A))$ occuring in (2.4) can be written in the form $\pi(\overline{A})$, where $\overline{A}$ is a bounded measurable set of V.

We call the $M := L \oplus S \subset V$ *of regular form (or being in regular form)* if $\lin(L) \perp S$.

One can easily see that to any $M := L \oplus S$ there is a point lattice $\hat{L}$ such that $\dim(\hat{L}) = \dim(L), M = \hat{L} \oplus S$ and $\lin(\hat{L}) \perp S$, i.e., any M can be written in regular form. Of course, any discrete M (i.e., when $S = \{\theta\}$) is of regular form. To find $\hat{L}$, it is enough to take the orthogonal complement subspace $\hat{H}$ of S in V and after that take $\hat{L} := \mathrm{int}\,(\hat{b}_1, \ldots, \hat{b}_r)$, where $\hat{b}_i = \hat{H} \cap (S + b_i)$ and (b_i) is a basis of L.

Let M and T be such that

$$(\lin(L) \oplus T) \perp S. \tag{2.11}$$

Then

$$\alpha(M, A) = \mu(\varphi(\pi(A))). \tag{2.12}$$

Below, if we speak of regular form of $M = L \oplus S$, we always think of the triple L, S, T satisfying also (2.11).

$\alpha(M, A)$ can be considered as an extension of the measure $\mu(A)$, because if $M = \{\theta\}$, i.e., $L = S = \{\theta\}$ then automatically $\alpha(M, A) = \mu(A)$ where μ is the measure in V. In this case $\alpha(M, A)$ is by (1.1) an integral of the "measures of coverings" of its lower dimensional sections. The following lemma shows that this is also true for general M.

Lemma 2.1. Let $A \subset V$ be a bounded measurable set, $M := L \oplus S \subset V' \subset V \subseteq R^n$, where V', V are linear subspaces and let $F \subset V$ be a linear subspace such that $V' \perp F$ and $V' \oplus F = V$. Let α' and α mean the measures of covering V' and V by M-translates of sets, respectively. Then we have

$$\alpha(M, A) = \int_F \alpha'(M, (A - x) \cap V')dx. \tag{2.13}$$

Proof. It is clear that the subspaces T' and T occuring in the definitions of α' and α, respectively, can be choosen so that $T = T' \oplus F$. Let π', φ' and π, φ be the projections occuring in the definitions of α' and α, respectively. Then (2.13) is a consequence of (1.1) and the identity

$$\mu(((\varphi(\pi(A)) \oplus W) - x) \cap V') = \mu(\varphi'(\pi'((A - x) \cap V')) \oplus W), \; x \in F. \tag{2.14}$$

Let $x \in F$ and denote $B = A - x$.

(2.14) is a consequence of the following relations.

$$\pi'(B \cap V') = \pi(B) \cap V', \tag{2.15}$$

$$\varphi'(\pi'(B \cap V')) = \varphi(\pi'(B \cap V')), \tag{2.16}$$

$$\varphi(\pi(B) \cap V') = \varphi(\pi(B)) \cap V', \tag{2.17}$$

$$(\varphi(\pi(B)) \cap V') \oplus W = (\varphi(\pi(B)) \oplus W) \cap V', \tag{2.18}$$

$$(\varphi(\pi(A)) \oplus W) \cap (V' + x) = ((\varphi(\pi(B)) \oplus W) \cap V') + x. \tag{2.19}$$

Since T' occurs both in the definitions of π, φ, and π', φ', one can check (2.15)–(2.19) using a "three-dimensional picture" $\mathrm{lin}\,(L) \oplus S \oplus F$.

Let us explain shortly what "group-translates" in the title of the paper means. Considering R^n as a locally compact topological Abelian group, where the addition of vectors is the group operation and the Lebesgue measure is the Haar measure, it is well known that any closed subgroup M of R^n can be written in the form $L \oplus S$, where L is a point lattice and S a linear subspace (see, e.g., [9]). So our M-translates are in fact "closed group"-translates. Now, given a bounded measurable set $A \subset V \subseteq R^n$ containing θ and a closed subgroup $M \subset V$, one can ask what portion of V is covered by the family $\mathcal{A}(M) := \{(A + u)\}_{u \in M}$, i.e., we are interested in evaluating $\mu((A + M) \cap G(\rho))$, where $G(\rho)$ is a Euclidean ball in V centred at θ with a sufficiently large radius when compared with the "size" of A.

We claim that if $(A - x) \cap S \neq \emptyset \Rightarrow \mu((A - x) \cap S) > 0$, then

$$\mu((A + M) \cap G(\rho)) \sim \alpha(M, A) C(\rho) \tag{2.20}$$

where $C(\rho)$ is a constant depending on ρ only and "$\sim$" means approximately, i.e., when $\rho \to \infty$.

(2.20) can be explained in at least two ways.

Let M and T satisfy (2.11). In this case we have, by (2.9) and (2.12)

$$\alpha(M, A) = \mu\left(\bigcup_{u \in L}((\pi(A) + \mu) \cap P)\right). \tag{2.21}$$

If now the family $\{\pi(A) + u)\}_{u \in L}$ covers the space $\mathrm{lin}\,(L) \oplus T$ then

$$P = \bigcup_{u \in L}((\pi(A) + u) \cap P) \tag{2.22}$$

(so if $T \neq \{\theta\}$, then $\alpha(M, A) = +\infty$). Hence the right-hand side of (2.22) is really a set that indicates to what extent the space $\text{lin}\,(L) \oplus T$, hence the space $\text{lin}\,(L) \oplus T \oplus S$, is covered by M-translates of A.

(2.20) is more transparent from a following fourth form of $\alpha(M, A)$.

To write it we need the decomposition of a set proposed in [1].

Let $L \subset \overline{V} \subseteq R^n$ be a point lattice and $\overline{V}$ a linear subspace. Let $C \subset \overline{V}$ be a bounded set. We call the set

$$D(L, C) := \{u \in L \,:\, u \neq \theta,\, C \cap (C + u) \neq \emptyset\} \tag{2.23}$$

the dispersion of C with respect to L.
Clearly

$$D(L, C) = D(L, C + x), \quad x \in \overline{V}. \tag{2.24}$$

Denote

$$L(x) := \{u \in D(L, C) \,:\, x \in C + u\}, \quad x \in C \tag{2.25}$$

and

$$C(i) := \{x \in C \,:\, |L(x)| = i - 1\}, \quad i = 1, 2, \dots . \tag{2.26}$$

Obviously

$$C = \bigcup_{i \geq 1} C(i), \quad \mu(C) = \sum_{i \geq 1} \mu(C(i)). \tag{2.27}$$

Assertion 2.2. Let $M = L \oplus S \subset V$ be of regular form. Then

$$\alpha(M, A) = \sum_{i \geq 1} \frac{\mu(\pi(A)(i))}{i}, \tag{2.28}$$

where $\pi(A)$ is the measurable projection of $A \subset V$ into $\overline{V} := \text{lin}\,(L) \oplus T$ along S and $\pi(A)(i)$ are the sets in (2.26).

Before the proof let us write an identity that is a consequence of (1.1) and (1.2). Recall that $\text{lin}\,(L) \perp T$. For $C \subset \text{lin}\,(L) \oplus T$, bounded measurable set, we have

$$\mu(C) = \int_T \left(\int_Q |(C - y - x) \cap L| dx \right) dy = \int_P |(C - z) \cap L| dz, \tag{2.29}$$

where $P := Q \oplus T$. Hence, taking into account (2.3) we can write

$$\mu(C) = \int_{\varphi(C)} |(C - z) \cap L| dz . \tag{2.30}$$

Proof of Assertion 2.2. Denote $C := \pi(A)$. The definition (2.26) of $C(i)$ shows that

$$\{z \in P \quad \text{and} \quad (C(i) - z) \cap L \neq \emptyset\} \Rightarrow |(C(i) - z) \cap L| = i . \tag{2.31}$$

Using identity (2.30) this yields

$$\mu(C(i)) = \int_{\varphi(C(i))} |(C(i) - z) \cap L| dz = i\mu(\varphi(C(i))) . \tag{2.32}$$

(2.31) also yields

$$\varphi(C(i)) := \{z \in P : |(C(i) - z) \cap L| > 0\} = \{z \in P : |(C(i) - z) \cap L| = i\} , \tag{2.33}$$

consequently

$$\varphi(C(i)) \cap \varphi(C(j)) = \emptyset \quad \text{if} \quad i \neq j . \tag{2.34}$$

So we have

$$\varphi(C) = \bigcup_{i \geq 1} \varphi(C(i)) \quad \text{and} \quad \mu(\varphi(C)) = \sum_{i \geq 1} \mu(\varphi(C(i))) \tag{2.35}$$

and this gives (2.28) by (2.32) and (2.12).

The identity (2.30) implies another decomposition. Recall again $\lin(L) \perp T$. For $C \subset \lin(L) \oplus T$ denoting

$$\overline{C}(i) := \{z \in P : |(C - z) \cap L| = i\} , \quad i = 0, 1, \ldots \tag{2.36}$$

we have

$$\varphi(C) = \bigcup_{i \geq 1} \overline{C}(i) \quad \text{and} \quad \mu(\varphi(C)) = \sum_{i \geq 1} \mu(\overline{C}(i)) . \tag{2.37}$$

The relations (2.27), (2.33) and (2.34) show that $\varphi(C(i)) = \overline{C}(i), i \geq 1$, and by (2.32) we have

$$\mu(C) = \sum_{i \geq 1} i \cdot \mu(\overline{C}(i)) . \tag{2.38}$$

Let $\overline{C}(i) \neq \emptyset$. We claim that

$$|(C - C) \cap L| \geq 2i - 1. \tag{2.39}$$

Indeed, for $z \in \overline{C}(i)$ we have i points $c_1, c_2, \ldots, c_i \in C$ and $u_1, u_2, \ldots, u_i \in L$ such that $c_j = u_j + z, j = 1, 2, \ldots, i$. We can assume that $c_1 \prec c_2 \prec \ldots \prec c_i$, where "$\prec$" is the lexicographic ordering in R^n. So we have $c_1 - c_i \prec c_1 - c_{i-1} \prec \ldots \prec c_1 - c_2 \prec \theta \prec c_2 - c_1 \prec \ldots \prec c_i - c_1$ and these elements all belong to L, that proves (2.39).

Using (2.37) and (2.38) we get

$$|(C - C) \cap L| \geq 2\frac{\mu(C)}{\mu(\varphi(C))} - 1. \tag{2.40}$$

(2.40) implies

Assertion 2.3. Under the conditions of Assertion 2.2,

$$\alpha(M, A) \geq \frac{2\mu(\pi(A))}{1 + |(\pi(A) - \pi(A)) \cap L|}. \tag{2.41}$$

Inequality (2.40) is an extension of (1.3) to lower dimensional point lattices. (1.3) has also been extended to structures more general than R^n, [11].

Let us note that in (2.40) we assumed $\text{lin}\,(L) \perp T$ (this is the condition for (2.30) to hold) and that the $\pi(A)$ in (2.41) is the measurable projection of A into $\text{lin}\,L \oplus T$ along S, where M is in regular form, i.e., $\text{lin}\,(L), S$ and T are mutually orthogonal subspaces of V.

The last assertion of this section deals with the behaviour of $\alpha(M, \tau A)$ where τ is a regular linear transformation of V onto V.

Assertion 2.4. Let $M := L \oplus S \subset V$ and $T \subset V$ be such that $\text{lin}\,(L) \oplus S \oplus T = V, \text{lin}\,(L) \perp T$ and $(\text{lin}\,(L) \oplus T) \perp S$. Let τ be a regular linear transformation of V onto V such that

$$\tau(\text{lin}\,(L) \oplus T) = \text{lin}(L) \oplus T, \quad \tau S = S. \tag{2.42}$$

Denote by τ_0 the restriction of τ onto the subspace $\text{lin}\,(L) \oplus T$. Denote $M_0 := (\tau_0^{-1} L) \oplus S$, where τ_0^{-1} is the inverse transformation. Then

$$\alpha(M, \tau A) = |\det(\tau_0)|\alpha(M_0, A), \tag{2.43}$$

where $|\det(\tau_0)|$ is the absolute value of the determinant of τ_0.

This statement is not needed in what follows, we omit the proof. If $S = \{\theta\}$ then all conditions of the assertion are automatically satisfied, $\tau = \tau_0$, and the proof of this case is easy. The conditions for τ allow us to transform the general case to a discrete one. The conditions for τ are also automatically satisfied for the "transformation" $\tau x := tx$, where $t \geq 0, x \in V$. In this case τ_0^{-1} means $1/t$ and $|\det(\tau_0)| = t^{r+p}$, where $r = \dim(L)$ and $p = \dim(T)$. So (2.43) yields

$$\alpha(M, tA) = t^{r+p} \cdot \alpha(M_0, A) \tag{2.44}$$

where $M_0 := ((1/t)L) \oplus S$.

Taking in (2.44) $S := \{\theta\}, L := \{$ the set of integer points of a coordinate hyperplane $R^k\}$ and writing $\alpha(M, tA), \alpha(M_0, A)$ in the form (2.28), we get a result proved in [1].

3. The Main Theorem

We shall formulate and prove the main result mentioned in the Introduction for a broader class of sets than convex bodies. We call the set $B \subset E \subseteq R^n$, where E is a linear subspace, *star-shaped in E*, if there is a point $b \in E$ such that

$$\lambda(B - b) \subseteq B - b \quad \text{for all} \quad 0 \leq \lambda \leq 1. \tag{3.1}$$

Now we have

Theorem 3.1. Let $A \subset V \subseteq R^n$ be a bounded measurable set and $M := L \oplus S \subset V$. Assume that $(A - z) \cap \text{lin}(M)$ is star-shaped in $\text{lin}(M)$ for all $z \in V$ where it is non-empty. Then

$$\alpha(M, \lambda A) \leq \lambda^p \alpha(M, A) \quad \text{for all} \quad 0 \leq \lambda \leq 1, \tag{3.2}$$

where $p := \dim(V) - \dim(M)$.

Proof. Let $T \subset V$ be such that $\text{lin}(M) \perp T$ and $\text{lin}(M) \oplus T = V$, so $p = \dim(T)$. Denote $V' = \text{lin}(M)$. By (2.13) we have for all $0 \leq \lambda \leq 1$

$$\alpha(M, \lambda A) = \int_T \alpha'(M, (\lambda A - y) \cap V') dy. \tag{3.3}$$

If $\alpha(M, \lambda A) = 0$ then (3.2) holds trivially.

Similarly, (3.2) is trivial if $\lambda = 0$. Let $0 < \lambda$ and $\alpha(M, \lambda A) > 0$. Let $y = \lambda v, v \in T$. Changing the variable in the last integral we get

$$\alpha(M, \lambda A) = \lambda^p \int_T \alpha'(M, (\lambda A - \lambda v) \cap V')dv. \tag{3.4}$$

Let $v \in T$ be such that

$$\alpha'(M, (\lambda A - \lambda v) \cap V') > 0. \tag{3.5}$$

It is clear that $(\lambda A - \lambda v) \cap V' \neq \emptyset$, consequently $(A - v) \cap V' \neq \emptyset$. The star-shapedness of $(A - v) \cap V'$ implies that there is a $b(v) \in V'$ such that

$$\lambda(((A - v) \cap V') - b(v)) \subseteq ((A - v) \cap V') - b(v). \tag{3.6}$$

The set on the left-hand side of (3.6) is clearly equal to

$$((\lambda A - \lambda v) \cap V') - \lambda b(v), \tag{3.7}$$

which shows, using properties (2.5), (2.6) of α', that

$$\alpha'(M, (\lambda A - \lambda v) \cap V') \leq \alpha'(M, (A - v) \cap V'). \tag{3.8}$$

Integrating both sides of (3.8) over T and taking into account (3.3) (for λ and $\lambda = 1$), we get (3.2).

Corollary 3.2. Under the conditions of Theorem 3.1, the function

$$\alpha(M, tA) \cdot t^{-p} \tag{3.9}$$

is non-decreasing in $t, t \geq 0$.

Proof. Let $0 < t_1 < t_2$ and denote $B = t_2 A$. One can easily see that $(B - w) \cap V'$ is star-shaped in V' if $(A - z) \cap V'$ was. Hence

$$\alpha(M, \lambda B) \leq \lambda^p \alpha(M, B) \quad \text{for all} \quad 0 \leq \lambda \leq 1. \tag{3.10}$$

Taking $\lambda = t_1/t_2$ in (3.10) we get the monotonicity of (3.9).

Corollary 3.3. Let $K \subset V \subseteq R^n$ be a convex set and $M := L \oplus S \subset V$. Then the function

$$\alpha(M, tK) \cdot t^{-p} \tag{3.11}$$

is non-decreasing in $t, t \geq 0$, where $p := \dim(V) - \dim(M)$.

Proof. The set $(K - z) \cap V'$ is convex, hence star-shaped in V'.

We shall use Corollary 3.3, because we need the monotonicity of $\alpha(M, tA) \cdot t^{-p}$ for different $M - s$, hence the star-shapedness of $(A - z) \cap \lin(M)$ in $\lin(M)$ for different $M - s$ has to be assured. The latter condition is clearly true for all M if A is convex. The following example shows that the star shapedness of $(A - z) \cap \lin(M)$ in $\lin(M)$ is in some sense a necessary condition for (3.2) to hold. Namely, let $B \subset R^2$ be the convex hull of the points $(0,0), (0,-2), (1,-1), (2,1)$ and $(2,2)$, and let $D := B \cup -B$. Let $M := \text{int}((1,0)) \subset R^2$. The sets $(D - z) \cap \lin(M)$ are not star-shaped in $\lin(M)$ and easy computations show that

$$\alpha(M, (3/4)D) = 2.5 > \frac{3}{4}\alpha(M, D) = 2.25. \tag{3.12}$$

Denote the open kernel of the D by D_0. The set D_0 was used by Rogers (see, [5], p. 80) as an example of a set in R^2 which is star-shaped in R^2 and for which the successive minima theorem is not true. Namely, for this set we have

$$1 < \mu(D_0) \cdot \nu_1 \cdot \nu_2, \tag{3.13}$$

where ν_1, ν_2 are the successive minima of $D_0 - D_0$ with respect to the lattice of integer points in R^2.

4. Point Lattices

In this section we deal with $M - s$ such that $S = \{\theta\}$, that is, with $M = L$, where $L \subset V \subseteq R^n$ is a point lattice.

Lemma 4.1. Let $A \subset V \subseteq R^n$ be a bounded measurable set. Let $L_1 \subseteq L_2 \subset V$ be two point lattices. Then

$$\alpha(L_2, A) \leq \alpha(L_1, A). \tag{4.1}$$

If

$$(A - A) \cap L_1 = (A - A) \cap L_2 \tag{4.2}$$

then

$$\alpha(L_1, A) = \alpha(L_2, A). \tag{4.3}$$

Before turning to the proof, let us note that

$$(C - C) \cap L = D(L, C) \cup \{\theta\}, \tag{4.4}$$

where $D(L, C)$ is the set (2.23) (this is a simple consequence of the definition of $D(L, C)$).

Proof. The proof follows from the identity (2.28) (where now $S = \{\theta\}$). Let $L_1(a), A_1(i)$ and $L_2(a), A_2(i)$ be the sets (2.25) and (2.26) for L_1, L_2 and A. Hence

$$\alpha(L_1, A) = \sum_{i \geq 1} \frac{\mu(A_1(i))}{i}, \ \alpha(L_2, A) = \sum_{j \geq 1} \frac{\mu(A_2(j))}{j}. \tag{4.5}$$

The condition $L_1 \subset L_2$ implies that

$$|L_1(a)| \leq |L_2(a)|, \quad a \in A, \tag{4.6}$$

and from this we easily conclude that

$$i > j \Rightarrow A_1(i) \cap A_2(j) = \emptyset. \tag{4.7}$$

The $A_1(i)$ and $A_2(j)$ are disjunct decompositions of A, i.e.,

$$A = \bigcup_{i \geq 1} A_1(i) = \bigcup_{j \geq 1} A_2(j), \tag{4.8}$$

consequently

$$\alpha(L_1, A) = \sum_{i \geq 1} \sum_{j \geq 1} \frac{\mu(A_1(i) \cap A_2(j))}{i}, \tag{4.9}$$

$$\alpha(L_2, A) = \sum_{i \geq 1} \sum_{j \geq 1} \frac{\mu(A_1(i) \cap A_2(j))}{j}. \tag{4.10}$$

Taking into account (4.7) we can write

$$\begin{aligned} \alpha(L_1, A) &= \sum_{i \geq 1} \sum_{i \leq j} \frac{\mu(A_1(i) \cap A_2(j))}{i} \\ &\geq \sum_{i \geq 1} \sum_{i \leq j} \frac{\mu(A_1(i) \cap A_2(j))}{j}. \end{aligned} \tag{4.11}$$

But the right-hand side of (4.11) is equal to $\alpha(L_2, A)$. If condition (4.2) is satisfied, then by (4.4) we get that $D(L_1, A) = D(L_2, A)$, hence $L_1(a) = L_2(a)$ and $A_1(i) = A_2(i)$, so $\alpha(L_1, A) = \alpha(L_2, A)$.

Let $V \subseteq R^n$ be a linear subspace, $L_1 \subseteq L_2 \subseteq \ldots \subseteq L_q \subset V$ be point lattices and $A \subset V$ be a bounded measurable set. By (4.1) we have

$$\alpha(L_q, tA) \le \alpha(L_{q-1}, tA) \le \ldots \le \alpha(L_1, tA), \quad t \ge 0. \tag{4.12}$$

Denote for $j = 1, 2, \ldots, q - 1$

$$\Omega_j(A) := \{\omega \ge 0 : \alpha(L_q, tA) = \alpha(L_j, tA) \quad \text{for all} \quad 0 \le t \le \omega\}. \tag{4.13}$$

Clearly

$$\Omega_1(A) \subseteq \Omega_2(A) \subseteq \ldots \subseteq \Omega_{q-1}(A). \tag{4.14}$$

$\Omega_1(A)$ is non-empty, because for a sufficiently small ω we have $\alpha(L_j, tA) = \mu(A)$ for all j and all $0 \le t \le \omega$. (2.27) and (2.28) imply that $\alpha(L_j, A) = 0$ if and only if $\mu(A) = 0$, hence to avoid trivialities we shall always assume $\mu(A) > 0$.

Theorem 4.2. Let $K \subset V \subseteq R^n$ be a bounded convex set such that $\mu(K) > 0$ and $L_1 \subseteq L_2 \subseteq \ldots \subseteq L_q \subset V$ be point lattices. Let $\omega_0 = 0 \le \omega_1 \le \omega_2 \le \ldots \le \omega_{q-1} \le \omega_q := +\infty$ be a sequence such that

$$\omega_j \in \Omega_j(K), \quad j = 1, 2, \ldots, q - 1. \tag{4.15}$$

Then

$$\alpha(L_q, tK) \begin{cases} = \alpha(L_1, tK) & \text{if} \ \ 0 \le t \le \omega_1, \\[2mm] \ge t^{k_2} \cdot \omega_1^{-k_2} \cdot \alpha(L_1, \omega_1 K) & \text{if} \ \omega_1 \le t \le \omega_2, \\[2mm] \ge t^{k_{j+1}} \cdot \omega_j^{(k_j - k_{j+1})} \ldots \omega_2^{(k_2 - k_3)} \cdot \omega_1^{-k_2} \cdot \alpha(L_1, \omega_1 K) \end{cases} \tag{4.16}$$

$$\text{if} \qquad \omega_j \le t \le \omega_{j+1}, \\ j = 2, \ldots, q - 1,$$

$$\text{where } k_j := \dim(V) - \dim(L_j), \qquad j = 1, 2, \ldots, q.$$

Proof. (4.16) is a simple consequence of Corollary 3.3 and the definition of $\Omega_j(K)$. For $0 \le t \le \omega_1, \alpha(L_q, tK) = \alpha(L_1, tK)$ by the definition of $\Omega_1(K)$. Assume that we have proved (4.16) for $t = \omega_j$ and let $\omega_j \le t \le \omega_{j+1}$. By the definition of $\Omega_{j+1}(K)$ we have

$$t^{-k_{j+1}} \alpha(L_q, tK) = t^{-k_{j+1}} \alpha(L_{j+1}, tK) \tag{4.17}$$

and using the monotonicity of the right-hand side of (4.17) we get

$$t^{-k_{j+1}}\alpha(L_q, tK) \geq \omega_j^{-k_{j+1}}\alpha(L_{j+1}, \omega_j K). \tag{4.18}$$

But

$$\alpha(L_{j+1}, \omega_j K) = \alpha(L_q, \omega_j K) \tag{4.19}$$

and applying (4.16) to $\alpha(L_q, \omega_j K)$ (that is assumed to be proved) we get (4.16) for $\omega_j \leq t \leq \omega_{j+1}$.

Theoretically, Theorem 4.2 is true for any finite nested sequence of point lattices. However, if, say, $\dim(L_j) = \dim(L_{j+1})$, then $\omega_j^{(k_j - k_{j+1})} = \omega_j^0 = 1$, hence ω_j "disappears" from (4.16) in the sense that the same lower estimation holds for $\omega_{j-1} \leq t \leq \omega_j$ and for $\omega_j \leq t \leq \omega_{j+1}$, so the inequality can be formulated without ω_j, for $\omega_{j-1} \leq t \leq \omega_{j+1}$. This means that (4.16) is "effective" for such sequences $L_1 \subseteq \ldots \subseteq L_q$ only, when $0 \leq \dim(L_1) < \dim(L_2) < \ldots < \dim(L_q)$, implying that $q \leq \dim(V) + 1$.

Let K and L_i be as in Theorem 4.2. It is clear that if

$$(tK - tK) \cap L_q = (tK - tK) \cap L_j \tag{4.20}$$

for some $t > 0$, then

$$(t'K - t'K) \cap L_q = (t'K - t'K) \cap L_j \tag{4.21}$$

for all

$$0 < t' \leq t.$$

Consequently, if

$$(\beta_j \cdot (K - K)) \cap L_q = (\beta_j \cdot (K - K)) \cap L_j \tag{4.22}$$

for some $\beta_j \geq 0$, then

$$\beta_j \in \Omega_j(K). \tag{4.23}$$

So we have

Corollary 4.3. Let K and L_j be as in Theorem 4.2. Let $0 = \beta_0 \leq \beta_1 \leq \beta_2 \leq \ldots \leq \beta_{q-1} \leq \beta_q := +\infty$ be such that (4.22) holds. Then (4.16) is true with $\omega_j := \beta_j, j = 0, 1, 2, \ldots, q$.

Lemma 4.4. Let $A \subset V \subseteq R^n$ be a bounded measurable set and $L \subset V$ be a point lattice. If

$$\dim((A - A) \cap L) = k, \tag{4.24}$$

then there is a point lattice $L' \subseteq L$ such that $\dim(L') = k$ and

$$(A - A) \cap L' = (A - A) \cap L. \tag{4.25}$$

Before the proof let us recall the concept of a primitive subset of L ([3], [5]). Let $L \subset R^n$ be an r-dimensional point lattice. The linearly independent elements $u_1, \ldots, u_k \in L, k \leq r$, are called a "primitive subset" of L if

$$\left\{ \sum_{i=1}^{k} \lambda_i u_i : 0 \leq \lambda_i < 1, \ i = 1, \ldots, k \right\} \cap L = \{\theta\}. \tag{4.26}$$

For such a subset

$$\mathrm{int}(u_1, \ldots, u_k) = \mathrm{lin}(u_1, \ldots, u_k) \cap L, \tag{4.27}$$

i.e., $(u_1, \ldots, u_k)$ is a basis of a new k-dimensional point lattice $\mathrm{lin}(u_1, \ldots, u_k) \cap L$.

Indeed, let $v \in \mathrm{lin}(u_1, \ldots, u_k) \cap L$. Then

$$v = \sum_{i=1}^{k} \alpha_i u_i = \sum_{i=1}^{k} \{\alpha_i\} u_i + \sum_{i=1}^{k} v_i u_i$$

where $0 \leq \{\alpha_i\} < 1$ and v_i are integers. This is possible only if $\{\alpha_i\} = 0$ for all i, because $(u_1, \ldots, u_k)$ is primitive, and this implies (4.27).

Proof of Lemma 4.4. Denote $B = (A - A) \cap L$. Let us call the subset $u_1, \ldots, u_k \in B$ of linearly independent elements primitive with respect to B if

$$\left\{ \sum_{i=1}^{k} \lambda_i u_i : 0 \leq \lambda_i < 1, \quad i = 1, 2, \ldots, k \right\} \cap B = \{\theta\}. \tag{4.28}$$

Let $(u_1, \ldots, u_k)$ be primitive with respect to B.

If $(u_1, \ldots, u_k)$ is primitive (with respect to L), i.e., if (4.26) holds, then denoting $L' = \mathrm{int}(u_1, \ldots, u_k)$, all elements of B belong to L' and this implies (4.25). Denote $C := \{ \sum_{i=1}^{k} \lambda_i u_i : 0 \leq \lambda_i \leq 1, i = 1, 2, \ldots, k \}$.

If $(u_1, \ldots, u_k)$ is not primitive (with respect to L) then take the set $C \cap L$. We can prove easily by induction on k, that there is a set $u'_1, \ldots, u'_k \in C \cap L$ of linearly independent vectors that is primitive (with respect to L).

Then $L' = \text{int}\,(u'_1, \ldots, u'_k)$ is the wanted point lattice. The existence of $u_1, \ldots, u_k \in B$ so that (4.28) is true, can also be proved by induction on k.

Remark 4.5. Taking $L' = \text{lin}\,((A - A) \cap L) \cap L$ we get a discrete subgroup of L of dimension k and the statement of the lemma also follows from the well-known fact in the geometry of numbers that any discrete subgroup is a point lattice.

Corollary 4.6. Let $K \subset V \subseteq R^n$ be a bounded convex set such that $\mu(K) > 0$ and $L \subset V$ be a point lattice. Let $0 \le \delta_1 \le \delta_2 \le \ldots \le \delta_{q-1} \le \delta_q := +\infty$ and $k_1 \ge k_2 \ge \ldots \ge k_{q-1} \ge k_q = \dim(V) - \dim(L)$ be such that

$$\dim\,((\delta_j(K - K)) \cap L) = \dim(V) - k_j\,, j = 1, 2, \ldots, q - 1\,. \tag{4.29}$$

Then

$$\alpha(L, tK) \begin{cases} = \alpha(L', tK) \quad \text{if} \quad 0 \le t \le \delta_1\,, \\[2mm] \ge t^{k_2} \cdot \delta_1^{-k_2} \cdot \alpha(L', \delta_1 K) \quad \text{if} \quad \delta_1 \le t \le \delta_2\,, \\[2mm] \ge t^{k_{j+1}} \cdot \delta_j^{(k_j - k_{j+1})} \ldots \delta_2^{(k_2 - k_3)} \cdot \delta_1^{-k_2} \cdot \alpha(L', \delta_1 K) \end{cases} \tag{4.30}$$

$$\text{if } \delta_j \le t \le \delta_{j+1}\,,$$
$$j = 2, 3, \ldots, q - 1\,,$$

where $L' \subset L$ is a point lattice such that $\dim(L') = \dim(V) - k_1$ and $(\delta_1(K - K)) \cap L' = (\delta_1(K - K)) \cap L$.

Proof. By Lemma 4.4 there is a point lattice $L_{q-1} \subset L$ such that $\dim(L_{q-1}) = \dim(V) - k_{q-1}$ and

$$(\delta_{q-1}(K - K)) \cap L_{q-1} = (\delta_{q-1}(K - K)) \cap L\,. \tag{4.31}$$

By (4.21), (4.31) implies

$$(\delta_{q-2}(K - K)) \cap L_{q-1} = (\delta_{q-2}(K - K)) \cap L\,, \tag{4.32}$$

hence

$$\dim\left((\delta_{q-2}(K-K))\cap L_{q-1}\right) = \dim(V) - k_{q-2}. \qquad (4.33)$$

Using the lemma again we get $L_{q-2} \subset L_{q-1}$ such that $\dim(L_{q-2}) = \dim(V) - k_{q-2}$ and

$$(\delta_{q-2}(K-K))\cap L_{q-2} = (\delta_{q-2}(K-K))\cap L_{q-1} = (\delta_{q-2}(K-K))\cap L \qquad (4.34)$$

and so on. In the last step we get a point lattice $L' := L_1$. The numbers δ_j satisfy the conditions (4.22) (with $L_q = L$), hence the Corollary 4.3 yields (4.30).

Let $\Lambda \subset V$ be a point lattice of $\dim(\Lambda) = \dim(V)$, $K \subset V$ be a convex set such that $\mu(K) > 0$. Denote by ν_j the successive minima of $K - K$, i.e.,

$$\nu_j := \inf\{t \geq 0 : \dim\left((t(K-K))\cap\Lambda\right) \geq j\}, j = 1, 2, \dots, m := \dim(V). \qquad (4.35)$$

Then we have

Corollary 4.7.

$$\alpha(\Lambda, tK) \begin{cases} = t^m \mu(K) & \text{if} \quad 0 \leq t \leq \nu_1 \\ \geq t^{m-j}\left(\prod_{i=1}^{j}\nu_i\right)\mu(K) & \text{if} \quad \nu_j \leq t \leq \nu_{j+1} \end{cases} \qquad (4.36)$$
$$j = 1, 2, \dots, m, \nu_{m+1} := +\infty.$$

Proof. Assume first that

$$0 < \nu_1 < \nu_2 < \dots < \nu_m. \qquad (4.37)$$

Then for sufficiently small $\varepsilon > 0$ denoting $\bar{\nu}_j := \nu_j - \varepsilon$ we have $\nu_{j-1} < \bar{\nu}_j < \nu_j$, consequently by the definition of ν_j

$$\dim\left((\bar{\nu}_j(K-K))\cap\Lambda\right) = j - 1, j = 1, 2, \dots, m. \qquad (4.38)$$

It is clear that $\bar{\nu}_1 < \bar{\nu}_2 < \dots < \bar{\nu}_m$.

Applying Corollary 4.6 and taking into account that $\alpha(\{\theta\}, A) = \mu(A)$, we get

$$\alpha(\Lambda, tK) \begin{cases} = t^m \mu(K) & \text{if } 0 \le t \le \overline{\nu}_j \\ \ge t^{m-j} \left(\prod_{i=1}^{j} \overline{\nu}_i \right) \mu(K) & \text{if } \overline{\nu}_j \le t \le \overline{\nu}_{j+1} \\ & j = 1, 2, \dots, m. \end{cases} \tag{4.39}$$

If, say, $\nu_{j-1} < \nu_j = \nu_{j+1} < \nu_{j+2}$, then denoting again $\overline{\nu}_j = \nu_j - \varepsilon$, $\dim\left((\overline{\nu}_j(K - K)) \cap \Lambda\right) = \dim\left((\overline{\nu}_{j+1}(K - K)) \cap \Lambda\right) = j - 1$, and by the remark after the proof of Theorem 4.2, (4.39) holds again. Letting tend $\varepsilon \to 0+$ we get from (4.39) the inequality (4.36).

The result (4.36) is an extension of (1.5) to non-symmetric convex sets. It is clear that if $K = -K$ then $K - K = 2K$ and $\nu_i = \lambda_i/2$, where λ_i are the successive minima of K. The inequality (1.6) is a special case of (4.36) for $j = n$.

5. General M

As we have seen in Sec. 3 the main results, Theorem 3.1 and its corollaries, hold for the general $M := L \oplus S$, i.e., where S is not necessarily the $\{\theta\}$.

The question is what results of the previous section can be extended to this general case. The main difficulty of doing this is that in general the relation $\alpha(M_2, A) \le \alpha(M_1, A)$ is not true if $M_1 \subset M_2$, as is shown by the following examples. Let $e_1 := (1, 0), e_2 := (0, 1), a := (1, 1)$ be vectors in R^2, let $V = R^2$ and $L_1 := \text{int}(e_1, e_2), L_2 := \text{int}(e_1), L_1' := \text{int}(a), L_1'' := \text{int}(e_2), S_2 := \text{lin}(e_2)$. Let $b := \left(-\frac{1}{\sqrt{2}}, \frac{1}{\sqrt{2}}\right)$ and denote

$$A := \left\{ \lambda_1 a + \lambda_2\left(\frac{1}{2}b\right) : 0 \le \lambda_1, \lambda_2 \le 1 \right\},$$

$$B := \left\{ \lambda_1 a + \lambda_2 b : 0 \le \lambda_1, \lambda_2 \le 1 \right\},$$

$$C := \left\{ \lambda_1 e_1 + \lambda_2\left(\frac{1}{2}e_2\right) : 0 \le \lambda_1, \lambda_2 \le 1 \right\}$$

and

$$D := \left\{ \lambda_1\left(\frac{3}{2}e_1\right) + \lambda_2 e_2) : 0 \le \lambda_1, \lambda_2 \le 1 \right\}.$$

Denoting $M_2 := L_2 \oplus S_2$ we have: $L_1 \subset M_2$ and $\alpha(L_1, E) \leq \alpha(M_2, E)$ for all $E \subset R^2$; $L'_1 \subset M_2, \alpha(M_2, B) = 1 < \alpha(L'_1, B) = \sqrt{2}$ and $\alpha(L'_1, A) = \frac{\sqrt{2}}{2} < \alpha(M_2, A) = 1$; $L''_1 \subset M_2, \alpha(M_2, D) = 1 < \alpha(L''_1, D) = \frac{3}{2}$ and $\alpha(L''_1, C) = \frac{1}{2} < \alpha(M_2, C) = 1$. The reason for these irregularities is that $S_1 = \{\theta\} \neq S_2$.

Nevertheless, Lemma 4.1 can be extended to cases when $S_1 = S_2$.

Lemma 5.1. Let $A \subset V \subseteq R^n$ be a bounded measurable set. Let $M_1 := L_1 \oplus S, M_2 := L_2 \oplus S \subset V$ be of regular forms and such that $L_1 \subseteq L_2$. Then

$$\alpha(M_2, A) \leq \alpha(M_1, A). \tag{5.1}$$

If

$$(\pi(A) - \pi(A)) \cap L_1 = (\pi(A) - \pi(A)) \cap L_2, \tag{5.2}$$

then

$$\alpha(M_2, A) = \alpha(M_1, A), \tag{5.3}$$

where $\pi(A)$ is the measurable projection of A into $\text{lin}(L_1) \oplus T_1$ along $S(T_1$ is the subspace determined by regular form of $M_1)$.

Proof. Let T_1, T_2 be the subspaces determined by regular forms of M_1 and M_2, i.e., $\text{lin}(L_1) \perp T_1 \perp S, \text{lin}(L_2) \perp T_2 \perp S$ and $\text{lin}(L_1) \oplus T_1 \oplus S = \text{lin}(L_2) \oplus T_2 \oplus S = V$. These imply that $\text{lin}(L_1) \oplus T_1 = \text{lin}(L_2) \oplus T_2$, hence $\pi_1(A) = \pi_2(A)$, where π_1 and π_2 are the projections in $\alpha(M_1, A)$ and $\alpha(M_2, A)$. This in turn implies that in the identities (2.28), applied to M_1 and M_2, the sets $\pi_1(A)$ and $\pi_2(A)$ are equal, denote them by $\pi(A)$, so we can apply Lemma 4.1 to $\pi(A), L_1$ and L_2 in the space $V_1 := \text{lin}(L_1) \oplus T_1 = \text{lin}(L_2) \oplus T_2$.

Let $M_j := L_j \oplus S \subset V$, be in regular forms, $j = 1, 2, \ldots, q$, such that $L_1 \subseteq L_2 \subseteq \ldots \subseteq L_q$, and $A \subset V$ be a bounded measurable set of positive measure. Denote for $j = 1, 2, \ldots, q - 1$,

$$\overline{\Omega}_j(A) := \{\omega \geq 0 : \alpha(M_q, tA) = \alpha(M_j, tA) \text{ for all } 0 \leq t \leq \omega\} \tag{5.4}$$

Now we have the following extension of Theorem 4.2.

Theorem 5.2. Let M_j be as in the definition (5.4) of $\Omega_j(A)$ and let

$K \subset V$ be a bounded convex set with $\mu(K) > 0$. Let $\omega_0 = 0 \leq \omega_1 \leq \ldots \leq \omega_{q-1} \leq \omega_q := +\infty$ be a sequence such that

$$\omega_j \in \overline{\Omega}_j(K), \quad j = 1, 2, \ldots, q-1. \tag{5.5}$$

Then

$$\alpha(M_q, tK) \begin{cases} = \alpha(M_1, tK) & \text{if } 0 \leq t \leq \omega_1, \\ \geq t^{k_2} \cdot \omega_1^{-k_2} \cdot \alpha(M_1, \omega_1 K) & \text{if } \omega_1 \leq t \leq \omega_2, \\ \geq t^{k_{j+1}} \cdot \omega_j^{(k_j - k_{j+1})} \ldots \omega_2^{(k_2 - k_3)} \cdot \omega_1^{-k_2} \cdot \alpha(M_1, \omega_1 K) \end{cases} \tag{5.6}$$

$$\text{if } \omega_j \leq t \leq \omega_{j+1},$$
$$j = 2, \ldots, q-1,$$

where $k_j := \dim(V) - \dim(M_j), j = 1, 2, \ldots, q$.

Proof. Analogous to that of Theorem 4.2.

Also Lemma 4.4 can be extended.

Lemma 5.3. Let $A \subset V \subseteq R^n$ be a bounded measurable set and $M := L \oplus S \subset V$ be in regular form. Let $\pi(A)$ be the measurable projection of A into $\operatorname{lin}(L) \oplus T$ along S. If

$$\dim((\pi(A) - \pi(A)) \cap L) = k, \tag{5.7}$$

then there is a point lattice $L' \subseteq L$ such that $\dim(L') = k$ and

$$(\pi(A) - \pi(A)) \cap L' = (\pi(A) - \pi(A)) \cap L. \tag{5.8}$$

Using the above statements, all corollaries of the previous section can be extended appropriately to this more general case.

6. Concluding Remarks

Although the main results of this paper hold for general M, the "applications" of them, proved in Sec. 4 for discrete M and extended in Sec. 5 to more general M, do not go too "far" beyond the discrete case. We have some preliminary results also when $M_1 = L_1 \oplus S_1 \subset M_2 = L_2 \oplus S_2 \subset M_3 = L_3 \oplus S_3 \subset V$ and $S_1 \neq S_2 \neq S_3$, but they are so complicated that it is "unpublicable". Given a point lattice $\Lambda \subset R^n$ of full dimension and a bounded set $A \subset R^n$, Hadwiger [6] called the family $\mathcal{A} := \{(A + u)\}_{u \in \Lambda}$ "set-lattice" ("figuren-gitter").

We are dealing with the families $\overline{\mathcal{A}} := \{(A + u)\}_{u \in M}$ where $M = L \oplus S = \{(S + u)\}_{u \in L}$ is itself a set-lattice with a lower dimensional point lattice L and an unbounded set S. As M is a closed subgroup of R^n, the family $\overline{\mathcal{A}}$ might be called a "set-group". The set-lattice M is an extension of the point lattice in the sense that, while a point lattice is a collection of the point θ (the null-dimensional linear subspace) and its "periodic" translations, M is a collection of the linear subspace S and its periodic translations.

The main Theorem 3.1 seems to convey interesting new information into the problematics of successive minima. The whole of Sec. 4 shows that it lies in the depth of successive minima-type statements. This is also seen from (3.12), that seems to be the proper reason why the Rogers counterexample works.

The $\alpha(M, A)$ and the "theory" of this quantity described in this paper has an interesting "discrete" version. Namely, taking a point lattice $\Lambda \subset R^n$ instead of V and a point lattice $L \subset \Lambda$ instead of M, one can define the "measure of covering" Λ by the family $\{(B+u)\}_{u \in L}$, where $B \subset \Lambda$ is a finite set (the definition is analogous to that of $\alpha(M, A)$). This approach yielded some interesting improvements of some "classical" results concerning so called "star-numbers" of set-lattices, [12].

References

1. R. P. Bambah, A. Woods, and H. Zassenhaus, *Three proofs of Minkowski's second inequality in geometry of numbers*, J. Austral Math. Soc. **5** (1965) 453–462.

2. N. Bourbaki, *Elements de Mathematique. Livre VI. Integration*, Hermann, Paris, 1965 (Russian translation, Nauka, Moscow, 1967).

3. J. W. S. Cassels, *An Introduction to Geometry of Numbers*, Springer, Berlin-Heidelberg, 1959.

4. N. Dunford, and J. T. Schwartz, *Linear Operators, Part I: General Theory*, Interscience, New York, 1966.

5. P. M. Gruber, and C. G. Lekkerkerker, *Geometry of Numbers*, North-Holland, Amsterdam-New York, 1987.

6. H. Hadwiger, *Über Mittelwerte im Figurengitter*, Com. Math. Helv. **11** (1938) 221–233.

7. H. Hadwiger, *Vorlesungen über Inhalt, Oberfläche und Isoperimetrie*, Springer Berlin-Heidelberg, 1957.

8. P. Halmos, *Measure Theory*, Van Nostrand, New York, 1950.

9. E. Hewitt, and K. A. Ross, *Abstract Harmonic Analysis. Vol. I.*, Springer, Berlin-Heidelberg, 1963.

10. B. Uhrin, *Some useful estimations in geometry of numbers*, Period. Math. Hungar. **11** (1980) 95–103.

11. B. Uhrin, *Some remarks on the number of lattice points in difference sets*, Proc. A. Haar Memorial Conf., eds. K. Tandori and J. Szabados, Budapest, 1985, Coll. Math. Soc. J. Bolyai, Vol. 49, North-Holland, Amsterdam, 1987, pp. 929–937.

12. B. Uhrin, *On the star-number of a set-lattice*, MTA SZTAKI Közlemények, Budapest, 39/1987, pp. 229–257.

13. A. Weil, *Basic Number Theory*, Springer, Berlin-Heidelberg, 1967.

B. Uhrin
Computer and Automation Institute
Hungarian Academy of Sciences
1518 Budapest, p. f. 63
Hungary

A NON-ARCHIMEDEAN NUMBER FIELD AND ITS APPLICATIONS IN MODERN PHYSICS

Wang Shu-Tang

This is a further study of the work published in [1,2]. A non-archimedean number field called "generalized number system" (abbreviated as (GNS)) was initiated in [2]. This number field itself is a further development of a mathematical tool which was used by the author in solving the ω_μ-metrisation problem when studying general topology. This article contains further examinations on this field. It contains an application of this field to rest mass problem of the photon, and results on the representations of δ function as well as Heaviside's jump function, etc. Some analysis theoretical aspects are also included, which contain the extensions of the classical elementary functions e^x and $\ln x$. Finally, an extension of Taylor's formula to the present situation is also given.

As is well known, in the long history of Science development, natural science rules and its deductions are frequently expressed and treated by mathematical methods. For instance, classical mathematics (including statistics) can be used to deal with Newton's mechanics, thermodynamics, as well as electrodynamics and so on. In the mean time, while mathematical methods are being successfully applied in almost all science fields, in the recent near 40 years, theoretical physicists are frequently afflicted with divergence difficulties in calculation when they are studying quantum mechanical problems. According to the view point of the author such difficulties arise mainly because the modern physical world is multi-level (macroscopic, microscopic, etc.) in its character, while classical mathematics still retains the single level character in which $+\infty$ and $-\infty$ are usually considered in our macroscopic world as ideal objects and are treated by quite a different method compared with operations of ordinary real numbers. Non-standard analysis invented in the early 60's by Robinson [3] broke through the limitation of classical concept of real numbers in which infinitesimal and infinitely large numbers are legally introduced by logical method and used in dealing with many classical topics such as analysis, topology and some classical mechanics problems, etc. Many arguments are much simplified and become more intuitive by using non-standard methods to traditional ones. But Robinson's theory is based heavily on results of mathematical logic where infinitesimal and infinitely large numbers (being laid in different levels) have no exact and concrete expressions by using ordinary real numbers. This situation makes non-standard method not very convenient in computations and it involves more conceptional character. In the year 1963, the present author established a non-Archimedean number field, the so called "generalized number system" (abbreviated as (GNS)). He also used it to deal with the problem, giving the δ function more natural expressions. The main idea of (GNS) emerged from the author's further investigation into a mathematical tool which he used it to establish his ω_μ-metrisation theorem in general topology [1]. It should be emphasized that this field can be constructed from the real number field R by using pure algebraic method only, free from mathematical logic and is explicitly expressed through the real number field R. Unfortunately, such a number field was not published until 1979. Then, the author began to know about Robinson's work on non-standard analysis [3] only since 1978, from other chinese scholars, when the paper on generalized numbers was submitted for publication to a chinese top periodical "Scientia Sinicca". This paper was published in a 1979 issue of that periodical. A revised English version of that paper was published later in a Japanese periodical "Tsukuba J. Math", 1985.

In the present paper, we will give an account of this field and investigate some of its applications in theoretical physics. New results are included. In Sec. 1, some basis definitions are given. Section 2 contains some applications of (GNS) to modern physics and related topics. First, we discuss the rest mass problem of photon. Then more natural explicit expressions of some singular functions occuring frequently in the quantum field theory, such as Dirac's δ function, Heaviside's jump function, etc., are given. Finally, Sec. 3 contains some further analysis of theoretical aspects including extensions of the classical elementary functions e^x and $\ln x$. Basic properties of the later functions are discussed and an extension of classical Taylor's expansion formula will also be given.

1. Some Basic Definitions

1.1. *Generalized Number System (GNS)*

The number field which we are going to establish (called generalized number field) consists of the so-called generalized numbers, where each generalized number is a sequence of reals in which different terms representing quantities lie in different levels. We will give right now its exact description, as follows:

Let $x = (\ldots, x_{-m}, \ldots, x_0, \ldots, x_m, \ldots)$ be a sequence of real numbers where m is a natural integer, x_m's are real numbers and there are only a finite number of m's satisfying $x_{-m} \neq 0$. Define the following operations

1) Addition, Subtraction. If there is another $y = (\ldots, y_{-m}, \ldots, y_0, \ldots, y_m, \ldots)$, then define $x \pm y = (\ldots, x_{-m} \pm y_{-m}, \ldots, x_0 \pm y_0, , \ldots, x_m \pm y_m, \ldots)$.

2) Order. The order relation between x and y is taken to be lexicographical, i.e., $x < y$ means that there is an integer k_0 such that if $k < k_0$ then $x_k = y_k$, but $x_{k_0} < y_{k_0}$. Recall that $x_k(y_k)$ denotes the k-th term of $x(y)$.

3) Let c be a real number. Then define $cx = (\ldots, cx_{-m}, \ldots, cx_0, \ldots, cx_m, \ldots)$. If we further introduce the "unite element" of k-th level: $1_{(k)} = (\ldots, 0, \ldots, 0, 1, 0, \ldots)$, where 1 occupies the k-th term only and other terms are all 0, then each of the above given generalized number x can be expressed as $x = \sum_k x_k \times 1_{(k)}$.

4) Multiplication. Let x and y be given above, then the product $x \times y$ (or xy) is defined by

$$x \times y = \sum_m \left(\sum_{k+l=m} x_k \times y_l \right) \times 1_{(m)}$$

5) Division. It can be shown without difficulty that, under the above definition 4), if $y \neq 0$ (here $0 = (\ldots, 0, \ldots)$), then there exists a unique z (for each given x) satisfying $x = y \cdot z$. Then define $z = x/y$.

It is easy to see from the above definitions that the set $E = \{x : x = (\ldots, x_{-m}, \ldots, x_0, \ldots, x_m, \ldots)\}$ with the above operations is a field. It is called the generalized number field abbreviated as (GNS). Each member of E is a generalized number.

If we put $I_{(k)} = \{x : x = x_k \times 1_{(k)}, x_k \text{ a real number }\}$ and if the integers k, h satisfy $k < h$, then it is obvious that the numbers contained in $I_{(h)}$ are infinitely small with respect to $I_{(k)}$ and conversely non-zero numbers in $I_{(k)}$ are infinitely large compared with numbers in $I_{(h)}$. It is also obvious that the real number field R is isomorphic with $I_{(0)}$ and represents the quantities measurable within macroscopic world. (GNS) contains R in it as a subfield lying in a special level.

Remark. D. Laugwitz invented the field of generalized power series in 1968 [4]. (GNS) is a subfield of Laugwitz's field from the view point of algebraic structure. Roughly speaking, the later allows the suffixes to take values from real numbers which are not confined in integers. This makes the radical operations within it to be permissible. The author is aware of Laugwitz's work [4,5,6] only since 1978. In the following, investigations are confined within the scope of (GNS). In rare cases, the more elaborate Laugwitz's field will be needed in order that the discussion (when dealing with rest mass problem of photon) is more rigorous.

1.2. *Derivatives*

In the following context, by generalized functions we mean single valued mappings from (GNS) into (or onto) itself. Suppose $y = f(x) = (\ldots, y_{-m}, \ldots, y_0, \ldots, y_m, \ldots)$ is such a function, then for each $k, y_k = y_k(\ldots, x_i, \ldots)$ is a real valued function of infinitely many real variables. As usual let $\Delta x, \Delta y$, etc. denote variations, where $\Delta y = f(x + \Delta x) - f(x)$. Then $\Delta x = (\ldots, (\Delta x)_i, \ldots), \Delta y = (\ldots, (\Delta y)_i, \ldots)$, where $(\Delta y)_i = (f(x + \Delta x) - f(x))_i$ and without explicit mention, suffix always indicates corresponding term of a generalized number (constant or variable).

Definition 1.2.1. If there is a generalized number, denoted by $f'(a)$ (here a is arbitrary), such that for each generalized number $\varepsilon > 0$, there exists genralized $\delta > 0$, and the inequality

$$\left| \frac{\Delta y}{\Delta x} - f'(a) \right| < \varepsilon$$

holds for all $|\Delta x| < \delta$. Then $f'(a)$ is called the derivative of $y = f(x)$ of the point a where $|z|$ denotes z or $-z$ according to $z > 0$ or $z < 0. |0| = 0$.

Since each term of y depends on infinitely many numbers of reals, i.e., $y_k = y_k(\ldots, x_i, \ldots)$, some other derivatives are also useful. We intend to introduce them here. In order to do this, some other notations are needed. Let x and y be such that $x = (\ldots, 0, \ldots, 0, x_i, x_{i+1}, \ldots)$ and $y = (\ldots, 0, \ldots, 0, y_i, y_{i+1}, \ldots)$, then $x \ll y$ means $x_k \le y_k$ for all k.

Definition 1.2.2. Let $y = f(x), a$ be given as above. Suppose for each integer s there is a real number K_s such that when ε is real, $\varepsilon > 0$ and is arbitrarily given, then there exists $\delta = (\ldots, 0, \ldots, 0, \delta_i, \delta_{i+1}, \ldots)$, where $\delta_k > 0 (k \ge i)$ and satisfies $\left| \left(\frac{\Delta y}{\Delta x} \right)_s - K_s \right| < \varepsilon$ provided that $|\Delta x| \ll \delta$. Then the number $K = (\ldots, K_s, \ldots)$ is called the weak derivative of $y = f(x)$ at the point a.

Definition 1.2.3. If in the above Definition 1.2.2. we take $i = 0$, then the resulting derivative is called the restricted weak derivative.

Besides the above, the notion of (m, n) derivative was introduced and was used in [2]. In [2], we are concerned only with special terms of a generalized number.

1.3. *Integration*

In [2], as an extension of Lebesgue integral, the (GNL) integral and (G) integral have been introduced. By using them, a natural and explicit expression of Dirac's δ function was supplied. The reader is referred to paper [2] for details. We do not intend to write out the complicated integration procedure here. It should be noted right now that the integrand $f(x)$ defining the δ function has such properties: (a) The support of $f(x)$ is the set $E_1 = \{x : x_{-m} = 0, \text{ for } m > 0\}$, i.e., $\{x : f(x) \ne 0\}$ is the set E_1; (b) $f(x)$ depends only on the 1-th term x_1.

Definition 1.3.1. Let $y = f(x)$ be a generalized function such that (GNL) $\int f(x)dx$ exists. Suppose the support of $f(x)$ is the set $E^* = \{x : x_k = 0 \text{ for } k < h\}$, where h is an integer, and suppose that $f(x)$ depends on x_h only. Then the indefinite integral of $y = f(x)$ is defined as follows:

$$\int_a^{x^*} f(z)dz = \begin{cases} (\text{GNL}) \int \tilde{f}(z)dz + \sum_{i=1}^{\infty} (f(x_h^*) \times x_{h+i}^*) \times 1_{(h+i)}, \text{ where} \\ \qquad x^* = (\ldots, 0, \ldots, 0, x_h^*, x_{h+1}^*, \ldots); \\ (\text{GNL}) \int f(z)dz, \text{ for other } x^* \text{ and when } x_{h-1}^* > 0. \\ 0, \quad \text{otherwise, i.e., when } x_{h-1}^* < 0. \end{cases}$$

where a is a fixed point $a = (\dots, a_{-m}, \dots, a_0, \dots, a_m, \dots)$, e.g. $a = (\dots, 0, \dots)$ and $\tilde{f}(x)$ is defined by

$$\tilde{f}(x) = \begin{cases} f(x), & \text{for} \quad a \leq x \leq x^*, \\ \\ 0, & \text{otherwise}. \end{cases}$$

Roughly speaking, the integral involves two parts and since $f(x)$ is not depending upon $x_{h+1}, x_{h+2}, \dots$, hence the second part takes the infinite summation given above.

2. Some Applications in Modern Physics and Related Topics

2.1. *The Rest Mass of Photon*

According to A. Einstein's theory of relativity, the inertial mass m of a motional partical relates its rest mass m_0 as

$$m = \frac{m_0}{\sqrt{1 - \beta^2}} \quad, \quad \beta = v/c. \tag{1}$$

As for the photon, since its velocity $v = c$, the rest mass should be $m_0 = 0$. Otherwise one would get $m = \infty$ contradicting the observational fact. But injustice involved in calculation still remains, even one takes $m_0 = 0$ in formula (1). This is because in classical mathematics that $m = 0/0$ is impossible. Now it is clear from the view point of (GNS) that this difficulty occurs mainly because a (generalized) number may be zero in macroscopic world, while, at the same time, not zero in microscopic world. We can give the accurate calculation as follows. Suppose the rest mass m_0 of photon is $m_0 = (\dots, 0, \dots, \underset{\underset{\text{0-th term}}{\uparrow}}{0}, m_1, \dots)$, considering m in formula (1) is a finite real number, then the velocity may be supposed to be $v = (\dots, 0, \dots, 0, \underset{\underset{\text{0-th term}}{\uparrow}}{c}, 0, v_2, v_3, \dots)$. From (1)

$$m^2(c^2 - v^2) = c^2 m_0^2, \tag{2}$$

substitute the above v and m_0 into (2), and by comparing the leading term (in the present case it is the coefficient of $1_{(2)}$), one obtains $(-2c) \times m^2 \times v_2 = m_1^2$. Hence the relation between v_2 and m_1 can be expressed by the equation

$$v_2 = -\frac{m_1^2}{2cm^2}. \tag{3}$$

Remark. One might argue on the existence of square root contained in (1) within (GNS). In order to make the discussion more rigorous, one should further enlarge (GNS) by adding some new elements into it. One simple way which may be used is by allowing suffixes to take all real numbers (i.e., not restricted to integers) in the definition of generalized numbers. This is just the Laugwitz's field [4], in which radical operations are permissible. Under this situation, the electromagnetic mass m should be a number in that field with leading term m unchanged, and m_0, v given above both remain unchanged.

2.2. *Dirac's δ Function and Heaviside's Function*

Using (GNS) theory and (GNL) integral, the δ function can be expressed by an explicit form [2] as

$$g(x) = \begin{cases} g^*(x_1) \times 1_{(-1)}, & \text{for } x = (\ldots, 0, \ldots, 0, x_1, x_2, \ldots) \\ \\ 0, & \text{otherwise}, \end{cases} \tag{4}$$

where $g^*(t)$ is a sufficiently smooth real function on $(-\infty, \infty)$, whose support set is bounded and $\int_{-\infty}^{\infty} g^*(t)dt = 1$. Recall the term x_1 is a real variable.

Another singular function frequently used in quantum field theory which seems to be more strange is the well-known Heaviside's function. Its definition is

$$h(x) = \begin{cases} 0, \text{ for } x < 0 \\ \\ 1, \text{ for } x > 0, \end{cases} \tag{5}$$

and with condition $h'(x) = \delta(x)$.

From (4), it is obvious that $g(x)$ satisfies conditions listed in Definition 1.3.1. Let the indefinite integral of $g(x)$ be $h(x)$, then $h(x)$ satisfies (5). It is also clear that $h'(x) = g(x)$ follows from Definition 1.3.1 and Definition 1.2.1. So we have

Theorem 2.2.1. If $g(x)$ is a δ function defined by (4), then the indefinite integral $h(x) = \int_0^x g(z)dz$ is the Heaviside's jump function.

Remark. Using (GNS), a more simple representation of the Heaviside's function can be constructed (we do not intend to do it here). But

$h(x)$ of Theorem 2.2.1 seems more natural.

3. Further Analysis Topics

3.1. *Exponential and Logarithmic Functions*

Let $\{y_n\}, y_n = f_n(x)$ and $n \in N$, be a sequence of generalized functions. How to define the convergence $y_n \to y$? Recalling that each function y_n contains an infinite number of terms: $y_n = (\ldots, y_{-m}^{(n)}, \ldots, y_0^{(n)}, \ldots, y_m^{(n)}, \ldots)$ one can define $y_n \to y$ (here observe that superscript n means the n-th generalized function not the n-th term of a generalized number) as: for each k, the k-th term satisfies $y_k^{(n)} \to y_k$. In the following context we are interested and confine our discussions in the case that each term $y_k^{(n)}, y_k (k \in N$ or $-k \in N)$ depends only on a finite number of x_n's, say $x_0, x_1, \ldots, x_{n_k}$. Under such assumption, for each $k, y_k^{(n)} \to y_k (n \to \infty)$ means pointwise convergence of a sequence of functions from R^{n_k} to R. When the meaning of $y_k \to y$ is define as such, we will call it the convergence in weak sense, or simply weak convergence.

Now, introduce the following formal expression

$$e^x = 1 + x + \frac{1}{2!}x^2 + \ldots + \frac{1}{n!}x^n + \ldots, \tag{6}$$

where $x = (\ldots, 0, \ldots, 0, x_0, x_1, x_2, \ldots)$, i.e., the restriction of consideration is made: when $m > 0$, then $x_{-m} = 0$. This restriction simplifies our discussion in the following context very much.

Theorem 3.1.1. The series defined at the right hand side of the equality of (6) is convergent in the weak sense.

Proof. Put $x = \sum_{n=0}^{\infty} x_n \times 1_{(n)}$ and let $\xi_{k,n}$ be the k-th term of x^n. Then

$$\xi_{k,n} = \sum_{m_1 + \ldots + m_n = k} x_{m_1} \times x_{m_2} \times \ldots \times x_{m_n},$$

where m_s's are non-negative integers. We evaluate $|\xi_{k,n}|$ as

$$
\begin{aligned}
|\xi_{k,n}| &\leq \sum_{m_1+\ldots+m_n=k} |x_{m_1}| \times \ldots \times |x_{m_n}| \\
&\leq \left(\sum_{m_1=0}^{k} |x_{m_1}| \right) \times \ldots \times \left(\sum_{m_n=0}^{k} |x_{m_n}| \right) \\
&\leq [(k+1) \times \max\{|x_s| : 0 \leq s \leq k\}]^n \,.
\end{aligned}
\tag{7}
$$

The k-th term of the right hand side of the series of (6) is

$$
\xi_k = \sum_{n=0}^{\infty} \frac{1}{n!} \xi_{k,n} \,,
\tag{8}
$$

so, by (7), for each k we have the following estimation

$$
\begin{aligned}
|\xi_k| &\leq \sum_{n=0}^{\infty} \frac{1}{n!} [(k+1) \times \max\{|x_s| : 0 \leq s \leq k\}]^n \\
&= \exp\left\{ (k+1) \max\{|x_s| : 0 \leq s \leq k\} \right\} < \infty,
\end{aligned}
\tag{9}
$$

and shows that the series on the right hand side of (6) is weak convergent, that completes the proof.

The above function e^x defined by (6) is called the exponential function.

Theorem 3.1.2. For arbitrary x, y, the formula $e^{x+y} = e^x e^y$ holds. For $m \in N$, $x_{-m} = y_{-m} = 0$.

Proof. We should prove that for each k the following

$$
(e^{x+y})_k = \sum_{l+m=k} (e^x)_l \times (e^y)_m \,.
\tag{10}
$$

Recall that, in (10), $(e^{x+y})_k$ means the k-th term of e^{x+y}, the meanings of $(e^x)_l$, etc., are the same. For our aims, we consider the corresponding two polynomials $X = x_0 + x_1 t + \ldots + x_k t^k$, $Y = y_0 + y_1 t + \ldots + y_k t^k$, where x_s and y_s $(0 \leq s \leq k)$ are the s-th terms of x and y respectively. Put $e^X (= 1 + X + \frac{1}{2!}X^2 + \ldots) = \sum_{n=0}^{\infty} a_n t^n$, which is a real power series of indeterminate t. It is not difficult to verify that $a_n = (e^x)_n$, where

$(e^x)_n$ is the n-th term of the generalized number e^x. In fact, if we let $1_{(1)}$ correspond to t, then from definitions of operations between generalized numbers, $1_{(n)}$ should correspond to t^n. Comparing coefficient of t^n in e^X with the coefficient of $1_{(n)}$ in $e^x (= 1 + x + \frac{1}{2!}x^2 + \ldots)$, one gets the above assertion immediately. The same method shows that if we put $e^Y = \sum_{n=0}^{\infty} b_n t^n (e^{X+Y} = \sum_{n=0}^{\infty} c_n t^n)$, then $b_n = (e^y)_n (c_n = (e^{x+y})_n)$. But from the known equality $e^{X+Y} = (e^X) \times (e^Y)$, one can infer that $c_k = \sum_{l+m=k} a_l \cdot b_m$. From this and the preceding discussions, the desired (10) will follow at once.

Theorem 3.1.3. $(e^x)' = e^x$, where derivative is in the restricted weak sense.

Proof. In our discussion, derivations involved always mean in the restricted weak sense.

1) At first we prove $(x^n)' = nx^{n-1}$, where $n \in N$ and the k-th term $x_k = 0$ for $k < 0$. In fact this can be easily derived from $(x + \Delta x)^n - x^n = nx^{n-1}(\Delta x) + (\Delta x)^2 p(x, \Delta x)$. $p(x, \Delta x)$ is a polynomial in x and Δx.

2) Put $f(x) = e^x$. Then from (6) and the easily verified identity (a, b are generalized numbers): $a^n - b^n = (a - b)(a^{n-1} + a^{n-2}b + \ldots + b^{n-1})$, we have

$$\frac{f(x + \Delta x) - f(x)}{\Delta x} = 1 + \sum_{n=2}^{\infty} \frac{1}{n!}[x^{n-1} + x^{n-2}(x + \Delta x) + \ldots + (x + \Delta x)^{n-1}].$$

(11)

If we put $\xi_n = x^{n-1} + x^{n-2}(x + \Delta x) + \ldots + (x + \Delta x)^{n-1}$, and let $\xi_{k,n}$ be its k-th term, then the following estimation can be deduced:

$$|\xi_{n,k}| \le n[(k+1)K_k]^{n-1},$$

(12)

where

$$K_k = \max \left\{ \max_{s \le k}\{|x_s|\}, \max_{s \le k}\{|x_s + (\Delta x)_s|\}, \right.$$

(13)

in (13), as usual, $(\Delta x)_s$ denote the s-th term of Δx. In fact, to prove (12) it is only necessary to prove the following

$$|(x^i(x + \Delta x)^{n-i-1})_k| \le (k+1)^{n-1}K_k^{n-1}.$$

(14)

As usual, the subscript k indicates the k-th term of a generalized number. Here, we prove (14) only in case $i = n - 1$, i.e., $n - i - 1 = 0$, as other cases can be handled in the same way. Now (14) follows from the following deduction:

$$\left| (x^{n-1})_k \right| = \left| \sum_{m_1 + \ldots + m_{n-1} = k} x_{m_1} \times x_{m_2} \times \ldots \times x_{m_{n-1}} \right|$$

$$\leq \left(\sum_{m_1 = 0}^{k} |x_{m_1}| \right) \times \ldots \times \left(\sum_{m_{n-1} = 0}^{k} |x_{m_{n-1}}| \right)$$

$$\leq [(k+1)K_k]^{n-1} .$$

From (11) and (12), observing that $\xi_{n,k} \to n(x^{n-1})_k$ as $\Delta_x \to 0$ $(n, k$ fixed), it follows that $(e^x)' = e^x$ immediately. Thus the proof is completed.

Theorem 3.1.3. If $x \neq y$ then $e^x \neq e^y$.

Proof. Choose k_0 which satisfies: $x_{k_0} \neq y_{k_0}$ and if $k < k_0$ then $x_k = y_k$. Then $x = z + x', y = z + y'$, where $z = x_0 \times 1_{(0)} + \ldots + x_{k_0-1} \times 1_{(k_0-1)}$. It is clear that $x' \neq y'$. Now, we should prove that $e^{x'} \neq e^{y'}$ (from Theorem 3.1.1. $e^x = e^z \cdot e^{x'}, e^y = e^z \cdot e^{y'}$). In fact, from the power expressions

$$\begin{cases} e^{x'} & = 1 + x' + \frac{1}{2!}x'^2 + \ldots \\ e^{y'} & = 1 + y' + \frac{1}{2!}y'^2 + \ldots , \end{cases} \tag{15}$$

one infers that, if $k_0 > 0$, then the k_0-th term of $e^{x'}$ and $e^{y'}$ are x_{k_0} and y_{k_0} respectively and hence $e^{x'} \neq e^{y'}$. Otherwise, when $k_0 = 0$, then 0-th terms of them will be $e^{x'_0}$ and $e^{y'_0}$ respectively, which also imply $e^{x'} \neq e^{y'}$.

Before we define the Logarithmic function, the following theorem should be proven at first.

Theorem 3.1.4. For a fixed generalized number x, the equation $e^y = x$ has a solution y iff $x_0 > 0$. In this case the solution is unique.

Proof. Comparing corresponding terms of the equation

$$e^y = 1 + y + \ldots + \frac{1}{n!}y^n + \ldots = x, \tag{16}$$

one derives the following system of real equations

$$\begin{cases} 1 + y_0 + \frac{1}{2!}y_0^2 + \ldots + \frac{1}{n!}y_0^n + \ldots = x_0 \\ y_1 + y_0 y_1 + \ldots + \frac{1}{(n-1)!}y_0^{n-1}y_1 + \ldots = x_1 \\ \ldots\ldots\ldots\ldots\ldots\ldots\ldots\ldots\ldots\ldots\ldots\ldots\ldots\ldots\ldots\ldots \\ y_k + y_0 y_k + \ldots + \frac{1}{(n-1)!}y_0^{n-1}y_k + \ldots + \sum_k = x_k \\ \ldots\ldots\ldots\ldots\ldots\ldots\ldots\ldots\ldots\ldots\ldots\ldots\ldots\ldots\ldots\ldots\ldots\ldots\ldots, \end{cases} \qquad (17)$$

where $\sum_k$ is a polynomial of variables $y_0, y_1, \ldots, y_{k-1}$ and $\sum_1 = \sum_0 = 0\,(k \in N)$.

Rewrite (17) as

$$\begin{cases} e^{y_0} = x_0 \\ y_1 e^{y_0} = x_1 \\ \ldots\ldots\ldots\ldots \\ y_k e^{y_0} + \sum_k = x_k \\ \ldots\ldots\ldots\ldots\ldots \end{cases} \qquad (18)$$

Now it is clear, from the first equation of (18), that $e^{y_0} = x_0$ has a solution y_0 iff $x_0 > 0$. Furthermore, when $x_0 > 0$, one can decide from (18) $y_0, y_1, \ldots$, successively. This proves the first part of the theorem. The second part of the theorem is also clear.

A consequence easily seen from the above argument is that when x is real then y is purely real too. This is because in this case $x_1 = x_2 = \ldots = 0$, hence from (18) $y_1 = 0, \sum_2 = 0, y_2 = 0, \sum_3 = 0 \ldots$ will be obtained successively.

From the above theorem, the (uniquely defined) inverse of $e^y = x$ will be denoted as $y = \ln x$, which is called the logarithmic function. It is obvious that its domain $D(\ln x) = \{x : x_0 > 0\}$. In a similar way, elementary functions in calculus such as $\sin x$, $\cos x$, etc. can all be extended to the (GNS) case.

Theorem 3.1.5. The formula $(\ln x)' = \frac{1}{x}$ exists.

Proof. From $e^y = x$, if one substitutes $y = \ln x$, then an identity will be obtained. Then the desired formula will be proven through derivation of both sides of this identity. While the chain rule of derivation is needed in the above, we do not intend to discuss here.

3.2. *Extension of Taylor's Expansion Formula*

In this section, without explicitly stated, all derivatives occuring in the text will be in the restricted weak sense.

Theorem 3.2.1. If u and v are derivable generalized functions, then their product $u \cdot v$ is also derivable and $(u \cdot v)' = u' \cdot v + u \cdot v'$.

Proof. The proof is routine and is omitted.

Another restriction which the function under consideration possesses is: Suppose $y = f(x) = (\ldots, y_{-m}, \ldots, y_0, \ldots, y_m, \ldots)$ then, by assumption, each $y_k = y_k(x_0, x_1, \ldots, x_{n_k})$ depends upon a finite number of x_n's. For instance, for the sufficiently smooth real function $f_0(x_0)$, the naturally induced function mentioned in [2] is as such:

$$
\begin{aligned}
f(x) = &f_0(x_0) + f_0'(x_0) \times [x_1 \times 1_{(1)} + \ldots + x_n \times 1_{(n)} + \ldots] + \ldots \\
&+ f_0^{(n)}(x_0)/n![x_1 \times 1_{(1)} + \ldots + x_n \times 1_{(n)} + \ldots]^n + \ldots.
\end{aligned}
\tag{19}
$$

In (19), each k-th term $(f(x))_k$ depends upon the finite number of real variables $x_0, x_1, \ldots, x_k$ only.

Definition 3.2.1. For $g(x)$, if there is a derivable (in the restricted weak sense of course) generalized function $f(x)$ such that $f'(x) = g(x)$, then $f(x)$ is called a (weak sense) primitive function of $g(x)$, and $f(x) - f(a)$ is the (weak) indefinite integral which is denoted by ${}^*\!\int_a^x g(z)dz$, i.e., ${}^*\!\int_a^x g(z)dz = f(x) - f(a)$.

As in the classical case, the integration by part theorem is valid in the present case.

Theorem 3.2.2. If both $u(x)$ and $v(x)$ are derivable and if ${}^*\!\int_a^x v(y)u'(y)dy$ exists, then ${}^*\!\int_a^x u(y)v'(y)$ also exists and the following formula is true:

$$
{}^*\!\int_a^x u(y)v'(y)dy = u(y)v(y)|_a^x - {}^*\!\int_a^x u'(y)v(y)dy.
\tag{20}
$$

Proof. It is easy to verify that $(u \cdot v)' = u' \cdot v + u \cdot v'$. Taking derivatives of functions from both sides of Eq. (20), we can see that the theorem will be proven if one can show that $f'(x) \equiv 0$ implies $f(x) = \text{const}$. We are now going to give the later statement a proof. Suppose $f'(x) \equiv 0$, where, by the assumption stated before, $(f(x))_{-m} = 0$ when $m > 0$, and if $y = f(x) = (\ldots, 0, \ldots, 0, y_0, y_1, \ldots)$ then each y_k depends upon $x_0, x_1, \ldots, x_{n_k}$ only.

(a) $y_0 = c_0$ (constant).

Let $y_0 = y_0(x_0, \ldots, x_{n_0})$, we shall show that y_0 does not depend on x_{n_0} which will imply $y_0 = \text{const}$.

In fact, if we take $\Delta x = (\Delta x)_{n_0} \times 1_{(n_0)}$, i.e., Δx is a number lying in the n_0-th level of (GNS). Then the leading term of $\frac{\Delta y}{\Delta x}$ is

$$\eta_{(-n_0)} = \left[\frac{y_0(x_0, \ldots, x_{n_0-1}, x_{n_0} + (\Delta x)_{n_0}) - y_0(x_0, \ldots, x_{n_0})}{(\Delta x)_{n_0}} \right] \times 1_{(-n_0)},$$

from $f' \equiv 0, \eta_{(-n_0)} \to 0$ as $\Delta x \to 0$. Therefore

$$\frac{\partial y_0}{\partial x_{n_0}} \equiv 0. \tag{21}$$

From (21), one gets $y_0 = \text{const}$.

(b) Suppose it has been proved that $y_0 = \text{const}, y_1 = \text{const}, \ldots, y_{k-1} = \text{const}$. Then $y_k = \text{const}$. can be proven in the same way, with y_0 replaced by y_k in the preceding deduction. So, by induction the proof of $y = \text{const}$. is completed.

Theorem 3.2.3. If $f(x)$ has $(n+1)$-order derivatives at point a. Then the following formula is true:

$$f(x) = f(a) + f'(a)(x - a) + \ldots + \frac{f^{(n)}(a)}{n!}(x - a)^n$$
$$+ \frac{1}{N!} {}^*\!\!\int_a^x f^{(n+1)}(z)(x - z)^n dz. \tag{22$_n$}$$

Proof. Firstly, we have

$$f(x) = f(a) + {}^*\!\!\int_a^x f'(z)dz, \tag{23}$$

as from the definition of ${}^*\!\!\int_a^x f'(z)dz$, which was given in the beginning of this section.

Secondly, let $F(z) = f'(z)(x - z)$, then $F(z)$ is a restricted weak derivable, provided that $f'(z)$ is. Put $v(z) = f'(z)$ and $u(z) = x - z$, by (20) one would have

$$\int_a^{*x} f(z)dz = f'(a)(x - a) + \int_a^{*x} f''(z)(x - z)dz \,. \tag{24}$$

Substitute (24) into (23), then one gets $(22)_1$. The general case $(22)_n$ can be proved by induction. Routine details will be omitted.

For every number $z = (\dots, z_{-m}, \dots, z_0, \dots, z_m, \dots)$ introduce its "norm" as $\|z\| = \sup_k\{|z_k|\}$. The ordinary relations such as $\| - x\| = \|x\| \geq 0, \|x\| = 0$ iff $x = 0$, and $\|x + y\| \leq \|x\| + \|y\|$ are true in the present case.

The following is an extension of the classical Taylor's expansion theorem.

Theorem 3.2.4. Let $y = f(x)$ be the above mentioned generalized function, which has restricted weak derivatives of all orders. Suppose there is a real constant M such that, for all $n \in N, \|f^{(x)}(z)\| \leq M$. Then, in weak convergence sense, $\frac{1}{n!} \int_a^x f^{(n+1)}(z)(x - z)^n dz \to 0$ as $n \to \infty$ where x is fixed. Hence, under above conditions, we have the following formula:

$$f(x) = f(a) + f'(a)(x - a) + \dots + \frac{f^{(n)}(a)}{n!}(x - a)^n + \dots . \tag{25}$$

Proof. Let x^* be arbitrarily fixed such that $\|x^*\| \leq M$ and, for each negative $-m, (x^*)_{-m} = 0$.

(a) Estimation of the l-th term $((x - x^*)^n)_l$ of $(x - x^*)^n$. Let ζ_l denotes this term. Observe that ζ_l depends only upon $(x - x^*)_0, (x - x^*)_1, \dots, (x - x^*)_l$, subscripts indicate corresponding terms of $(x - x^*)$, then from

$$\zeta_l = [(x - x^*)^n]_l = \sum_{\substack{m_1 + \dots + m_n = l \\ m_i \geq 0}} [(x - x^*)_{m_1} \times \dots \times (x - x^*)_{m_n}],$$

we have

$$|\zeta_l| \leq [(l + 1) \cdot K_l]^n \,, \tag{26}$$

where

$$K_l = \max\{|(x - x^*)_s| : 0 \leq s \leq l\} \,. \tag{27}$$

(b) From the hypothesis of the theorem, if $f^{(n+1)}(x) = (\ldots, 0, \ldots, 0, \zeta_0, \zeta_1, \ldots)$ then each $|\zeta_n| \leq M$.

(c) Let η_k denote the k-th term of $\frac{1}{n!} f^{(n+1)}(x)(x - x^*)^n$. Then it is clear that

$$(n!)\eta_k \equiv \sum_{l+m=k} \zeta_l \times \zeta_m, \tag{28}$$

therefore

$$|\zeta_k| \leq \frac{1}{n!} \sum_{l+m=k} |\zeta_k| \cdot |\zeta_m| \leq \frac{1}{n!} \times M \times (k+1) \times [(k+1)K_k]^n. \tag{29}$$

From (29), if $n \to \infty$, then $\eta_k \to 0$.

(d) Let $\Theta_S = \Theta_S(x_0, \ldots, x_{n_s})$ denote the s-th term of $\frac{1}{n!} {}^*\!\int_a^x f^{(x+1)}(z)$ $(z - x^*)^n dz$. Since the integrand is restricted weak derivable, then one can prove that

$$\begin{cases} \frac{\partial \Theta_s}{\partial x_0} = \eta_s, \\[4pt] \frac{\partial \Theta_s}{\partial x_1} = \eta_{s-1}, \\[4pt] \quad \cdots\cdots\cdots\cdots \\[4pt] \frac{\partial \Theta_s}{\partial x_{n_s}} = \eta_{s-n_s}. \end{cases} \tag{30}$$

In fact, in the present case $\frac{1}{n!} {}^*\!\int_a^x f^{(n+1)}(z)(x^* - z)^n dz = (\ldots, \Theta_s, \ldots) = F(x)$. Then (30) can be derived from considering the limits of $\frac{\Delta F}{\Delta x}$ when $\Delta x \to 0$ and putting $\Delta x = (\Delta x)_0 \times 1_{(0)}, \Delta x = (\Delta x)_1 \times 1_{(1)}, \ldots, \Delta x = (\Delta x)_{n_s} \times 1_{(n_s)}$ respectively.

Finally, from (30), the above (c) and by making use of the classical differential mean value theorem concerning real functions, one gets $\frac{1}{n!} {}^*\!\int_a^x f^{(n+1)}(z)(x^* - z)^n dz \to 0$ provided $n \to \infty$, where x and x^* be arbitrarily fixed. In particular, if we take $x^* = x$, then the theorem is proved.

References

1. Wang Shu-Tang, *Remarks on ω_μ-additive spaces*, Fundamenta Mathematicae **55** (1964) 101–112.
2. Wang Shu-Tang, *Generalized number system and its applications* (I), Tsukuba J. of Math. (2) **9** (1985) 203–215.
3. A. Robinson, *Non-Standard Analysis*, North-Holland Publ. Co., 1974.
4. D. Laugwitz, *Eine Nichtarchimidishe Erweiterung Anordeneter Körper*, Math. Nachr. **37** (1968) 225–236.

5. D. Laugwitz, *Anwendungen Unendlichkleiner Zahlen*, I, Jour. für die reine und Angewandte Mat. **207** (1961) 53–60.
6. D. Laugwitz, *Anwendungen Unendlichkleiner Zahlen*, II, ibidem **208** (1961) 22–34.

Wang Shu-Tang
Mathematics Department
Northwest University
Xi'an, Shaanxi
People's Republic of China

THE DIFFERENTIAL GEOMETRY OF TWO TYPES OF ALMOST CONTACT METRIC SUBMERSIONS

Bill Watson

Riemannian submersions between two almost contact metric manifolds and between an almost contact metric manifold and an almost Hermitian one are defined and their properties elucidated. Many examples are given. The restrictions occassioned on O'Neill's configuration tensors. T and A, by assuming such different almost contact metric structures on the total space as K-contact and almost cosymplectic are derived. Resulting relationships on the (φ-) holomorphic sectional and bisectional curvatures are calculated. Many of these submersions are found to commute with the Laplacian and, when the associated manifolds are compact, inequalities between the k-th Betti numbers of the manifold result.

0. Background

Riemannian submersions have proved to be a useful tool in differential geometry, particularly in the construction of new Einstein metrics, a topic of considerable interest to modern physics as well as mathematics, and of new spaces of positive sectional curvature. The fundamental properties of Riemannian submersions were elucidated in foundational articles by A. Gray [G 3] and B. O'Neill [O'N] after several special initial results [R, He, Wo] and one paper [Y–I] which outlined their properties in local tensorial form. In the next decade, N. Wallach [W] utilized the sectional curvature increasing properties of Riemannian submersions to construct new manifolds of positive sectional curvature and R. Escobales, in his thesis [E], characterized certain well-known submersions (e.g., the Hopf mapping, $\pi : S^3 \to S^2$). J. Vilms [V] completely determined the properties of those Riemannian submersions which are totally geodesic mappings.

Mappings between manifolds which commute with either the codifferential or the Laplacian operator on r-forms were shown to be Riemannian submersions (Wa 1, Wa 2, G–I] which possess certain geometric properties (minimal fibers and, for $r \geq 2$, completely integrable horizontal distribution). The implications of this commutation property for the cohomological invariants of the total and base manifolds of a Riemannian submersion via the Hodge theory led the author to consider manifolds with G-structure for which minimally immersed invariant submanifolds are easily obtained (e.g., Kähler). Thus, almost Hermitian submersions [Wa 3] and the interesting subclass of almost semi-Kähler submersions [W–V] were defined and elucidated.

The present author also developed a setting for studying solitons in Yang-Mills gauge field theory based on the intertwining of several types of G, G' Riemannian submersions [Wa 4]. In particular, this construction includes a Boothby-Wang fibration of an odd-dimensional sphere over a Kähler manifold, an example of an almost contact metric submersion of type II which will be defined in this report.

After finishing this paper, we became aware of the recently published article "Almost contact metric submersions" by D. Chinea (Rend. Circ. Mat. Palermo, **34** (1985), 89–104) which also examines the fundamental properties of our type I submersions. Several additional almost contact metric structures (e.g., quasi-trans-Sasakian) are treated there.

1. Almost Hermitian and Almost Contact Metric Structures

All manifolds considered in this report are smooth (C^∞), paracompact, complete and connected. All mappings and tensor fields are smooth. An *almost Hermitian manifold* is a triple (M, g, J) where (M, g) is a Riemannian manifold and:

(i) J is a tensor field of type $(1, 1)$ on M satisfying $J \circ J = -\text{id}$,

(ii) g is an almost Hermitian structure of M; i.e., $g(JX, JY) = g(X, Y)$, for all $X, Y \in \mathfrak{X}(M)$, the Lie algebra of vector fields on M.

An almost Hermitian manifold is necessarily orientable and of even dimension. J bigrades the differential forms on M in the usual way. The *Kähler 2-form* of the almost Hermitian manifold (M^{2m}, g, J) is the differential 2-form, ϕ, of bidegree $(1, 1)$ defined by $\phi(X, Y) = g(X, JY)$. We say that the almost complex structure J is *integrable* if the Nijenhuis tensor S, given by

$$S(X,Y) = [X,Y] + J[JX,Y] + J[X,JY] - [JX,JY]$$

vanishes. The famous theorem of Newlander-Nirenberg [N-N] states that J is integrable if and only if (M^{2m}, J) is actually a complex manifold and J is the associated complex structure.

Several structures on the almost Hermitian manifold (M, g, J) are possible by imposing restrictions on the covariant derivative, $\nabla\Phi$, of the Kähler form of M. For instance, if $\nabla\Phi = 0$, then M is said to be *Kähler*. If $d\Phi = 0$, then M is *almost Kähler*. For definitions of other structures, see [G 2]. Such almost Hermitian manifolds have been extensively studied by Gray [G 1], [G 2], Gray and Hervella [G–H], Kotō[Ko], Yano [Y] and many others.

An *almost contact metric manifold* is a quintuple (M, ϕ, ξ, η, g), where (M, g) is a Riemannian manifold and

    (i) ξ is a distinguished unit vector field on M,

    (ii) η is the l-form on M which is g-dual to ξ,

    (iii) ϕ is a tensor field of type (1, 1) on M satisfying

$$\phi \circ \phi = -\mathrm{id} + \eta \otimes \xi,$$

    (iv) the Riemannian structure, g, satisfies

$$g(\phi X, \phi Y) = g(X, Y) - \eta(X)\eta(Y).$$

We denote the class of almost contact metric manifolds by $\mathcal{ACM}$. An almost contact metric manifold is necessarily orientable and of odd dimension. The basic reference is Blair's monograph [B]. The *fundamental 2-form*, Φ, of the almost contact metric manifold $(M^{2n+1}, \phi, \xi, \eta, g)$ is the differential 2-form given by $\Phi(X, Y) = g(X, \phi Y)$. As with the almost Hermitian manifolds, we shall define various sub-classes of the class of almost contact metric manifolds by imposing restrictions on the covariant derivative, $\nabla\Phi$, of Φ. In analogy again with the almost Hermitian situation, we say that (M, ϕ, ξ, η, g) is *normal* (and of class $\mathcal{N}$) if the obvious induced almost complex structure of $M \times \mathrm{I\!R}'$ is integrable. One may show that M is normal if and only if the tensor field

$$N^{(1)}(X, Y) = [\phi, \phi](X, Y) + 2d\eta(X, Y)\xi$$

vanishes identically (see [B], p. 47). It will also prove useful to define

$$N^{(2)}(X, Y) = (\mathcal{L}_{\phi X}\eta)Y - (\mathcal{L}_{\phi Y}\eta)X,$$

whose vanishing is implied by that of $N^{(1)}$. Several important properties of almost contact metric structures are collected in:

Proposition 1.1. Let $(M^{2n+1}, \phi, \xi, \eta, g)$ be an almost contact metric manifold. Then,

(a) $\eta \circ \phi = 0$,

(b) $\phi(\xi) = 0$,

(c) $2g((\Delta_X \phi)Y, Z) = 3d\Phi(X, \phi Y, \phi Z) - 3d\Phi(X, Y, Z)$
$$+ g(N^{(1)}(Y, Z), \phi X) + N^{(2)}(Y, Z)\eta(X)$$
$$+ 2d\eta(\phi Y, X)\eta(Z) - 2d\eta(\phi Z, X)\eta(Y),$$

(d) $N^{(2)}(X, Y) = d\eta(\phi X, Y) + d\eta(X, \phi Y)$.

Proof. See Blair [B].

We now gather together the definitions of several almost contact metric structures which willl be studied in this report.

$(M^{2n+1}, \phi, \xi, \eta, g)$ is said to be:

(i)   *Contact* (Class $\mathcal{C}$) if $\Phi = d\eta$,

(ii)   *K-contact* ($\mathcal{KC}$) if $\Phi = d\eta$ and ξ is a Killing vector,

(iii)   *Nearly Sasakian* ($\mathcal{NS}$) if
$$(\nabla_X \phi)Y + (\nabla_Y \phi)X = 2g(X, Y)\xi - \eta(X)Y - \eta(Y)X,$$

(iv)   *Quasi-Sasakian* ($\mathcal{QS}$) if $d\Phi = 0$ and M is normal,

(v)   *Sasakian* ($\mathcal{S}$) if $\Phi = d_\eta$ and M is normal,

(vi)   *Nearly Cosymplectic* ($\mathcal{NCoS}$) if $(\nabla_X \phi)X = 0$,

(vii)   *Closely Cosymplectic* ($\mathcal{CCoS}$) if $d\eta = 0$ and $(\nabla_X \phi)X = 0$,

(viii)   *Almost Cosymplectic* ($\mathcal{ACoS}$) if $d\Phi = d\eta = 0$,

(ix)   *Cosymplectic* ($\mathcal{CoS}$) if $d\Phi = d\eta = 0$ and M is normal.

These classes are related by the following lattice in which each inclusion is strict (the dotted lines indicate the addition of the normality property):

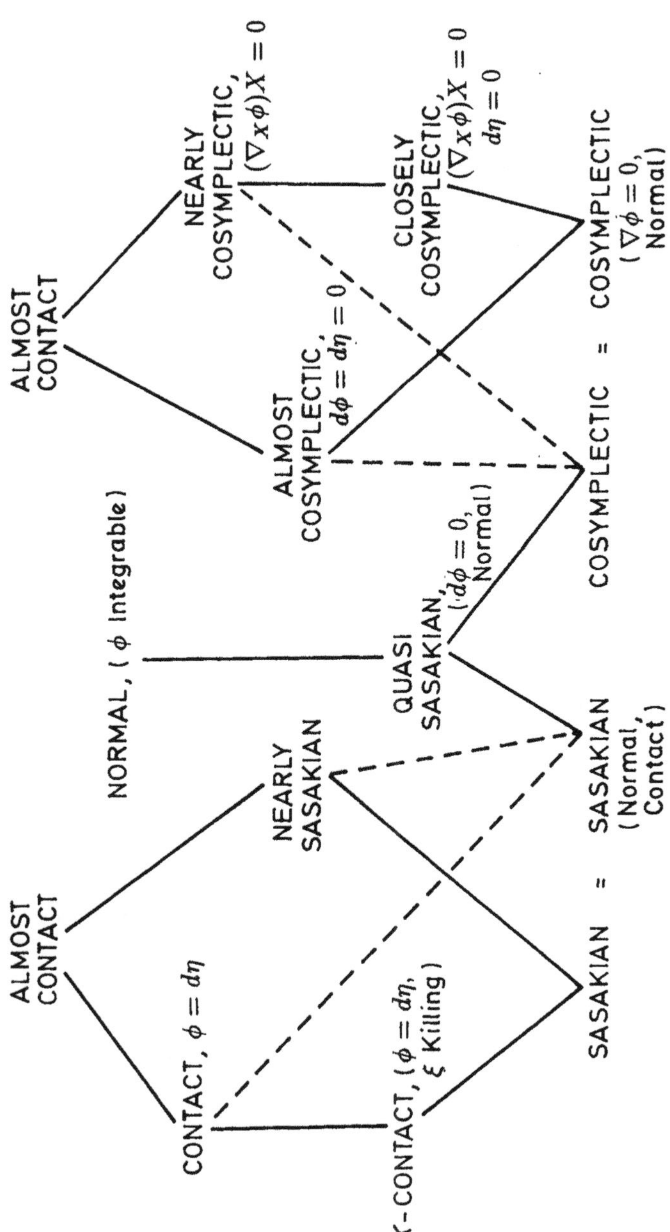

Fig. 1.1. Lattice of important contact metric structures.

Theorem 1.2.

(a) If M is contact, then

$$2g((\nabla_E \phi)F, G) = g(N^{(1)}(F, G), \phi E) + 2d\eta(\phi F, E)\eta(G) - 2d\eta(\phi G, E)\eta(F),$$

(b) If M is quasi-Sasakian, then,

$$g((\nabla_E \phi)F, G) = d\eta(\phi F, E)\eta(G) - d\eta(\phi G, E)\eta(F),$$

(c) If M is Sasakian, then,

$$(\nabla_E \phi)F = g(E, F)\xi - \eta(F)E,$$

(d) If M is almost cosymplectic, then,

$$2g((\nabla_E \phi)F, G) = g([\phi, \phi](F, G), \phi E),$$

(e) If M is cosymplectic, then,

$$\nabla_E \phi = 0.$$

Gray and Hervella [G–H] decomposed the space of trilinear covariant tensors on an almost complex vector space which satisfy the same basic identities as the Kähler 2-form. $U(n)$ has a natural representation on this tensor space which has four irreducible components producing sixteen invariant subspaces. Each of these subspaces corresponds to a different class of almost Hermitian manifolds; e.g., the $\{0\}$-subspace is the class of Kähler manifolds. Some classes are not well studied; e.g., the G_1 and G_2 manifolds of Vidal and Hervella [V–H]. Further results on submanifolds and fibrations of almost Hermitian manifolds in these classes are to be found in [G 1] and [Wa 3].

D. Janssens and L. Vanhecke [J–V] demonstrated a similar decomposition of the space of trilinear covariant tensors on an almost contact vector space, V^{2n+1}, which satisfy the same basic identities as the covariant derivative of the fundamental 2-form of an almost contact metric manifold. The natural representation of $U(n) \times \{1\}$ on this tensor space has 12 irreducible components and thus 4096 invariant subspaces. For instance, the $\{0\}$-subspace corresponds to the class of cosymplectic manifolds. The correspondence between the invariant subspaces and some of the well-known classes of almost contact metric manifolds is unfortunately not direct.

In small dimensions, the lattice of structures is diminished. In much the same way that every orientable Riemannian 2-manifold carries a Kähler structure, every orientable compact 3-manifold can be given a contact structure [Mar].

2. Almost Contact Metric Submersions

Riemannian submersions have been studied extensively by Gray [G 3] and O' Neill [O' N] to whom the reader is referred for the fundamental definitions and identities. In particular, we shall use the notation $\mathcal{H}$ (resp. $\mathcal{V}$) for the horizontal (resp. the vertical) distribution in the tangent bundle of the total space of a Riemannian submersion, $F \to M \xrightarrow{\pi} N$. The same letters will denote the projection mappings. We denote horizontal vector fields on M by X, Y and Z, vertical vector fields by U, V and W, and general vector fields by E and F. We shall make extensive use of O' Neill's T and A configuration tensors [O' N].

Let $(M^{2m+1}, \phi, \xi, \eta, g)$ and $(N^{2n+1}, \phi', \xi', \eta', g')$ be almost contact metric manifolds. Let $\pi : M \to N$ be a Riemannian submersion which satisfies:

(i) $\pi_* \phi E = \phi' \pi_* E$, and

(ii) $\pi_* \xi = \xi'$.

Then π is said to be an *almost contact metric submersion of type I*. If M is of structure type $\mathcal{P}$, we shall often call such a fibration, a "$\mathcal{P}$-submersion of type I."

In order that the fibre submanifolds of such a fibration inherit directly some of the differential geometric structure from the total space, M, it is essential that the vertical distribution be invariant under the action of the structure tensor, ϕ. In fact,

Proposition 2.1. Let $\pi : (M, \phi, \xi, \eta, g) \to (N, \phi', \xi', \eta', g')$ be an almost contact metric submersion of type I. Then,

(a) $\pi^* \eta' = \eta$,

(b) $\pi^* \Phi' = \Phi$,

(c) $\phi\{\mathcal{V}\} \subseteq \mathcal{V}$,

(d) $\phi\{\mathcal{H}\} \subseteq \mathcal{H}$,

(e) $\xi \in \mathcal{H}$, and

(f) $\eta(\mathcal{V}) = 0$.

Proof. (a) $\pi^*\eta'(X_*) = \eta'(\pi_*X) = g'(\xi', \pi_*X) = g'(\pi_*\xi, \pi_*X) = g(\xi, X) = \eta(X)$. (b) is proved in a similar fashion. (c) is obvious from $\pi_*\phi E = \phi'\pi_*E$. To see (d), note that $g(V, \phi X) = -g(\phi V, X) = 0$. For (e), decompose ξ as $\xi = \xi_1 + \xi_2$ with $\xi_1 = \mathcal{V}\xi$ and $\xi_2 = \mathcal{H}\xi$. Now $\eta(\xi) = 1 = \eta(\xi_1) + \eta(\xi_2)$. But $\eta(\xi_2) = g(\xi_2, \xi_2) = g'(\pi_*\xi_2, \pi_*\xi_2) = g'(\xi', \xi') = 1$. Thus, $\eta(\xi_1) = \|\xi_1\|^2 = 0$. (f) is immediate from this last assertion.

In order to define an almost complex structure, $\hat{J}$, on the fibres, F, of our almost contact metric submersion of type I, we need only take for $\hat{J}$ the restriction of ϕ to the vertical distribution; i.e., $\hat{J}V = \mathcal{V}\phi V$. By assertion (c) of Proposition 2.1, the sections of the tangent bundle of a fibre submanifold are $\hat{J}$-invariant. Moreover, $\hat{J}\hat{J}V = \phi\phi V = -V + \eta(V)\xi = -V$, because $\eta(\mathcal{V}) = 0$. For the same reasons, $\hat{g}(\hat{J}U, \hat{J}V) = \hat{g}(U, V)$ and the fibres $(F, \hat{J}, \hat{g})$ are almost Hermitian manifolds of dimension $2(m - n)$.

Propsition 2.2. Let $\pi : M \to N$ be an almost contact metric submersion of type I. Then,

     (a) $\hat{\Phi}(U, V) = \Phi(U, V)$,
     (b) $\hat{S}(U, V) = N^{(1)}(U, V)$,
     (c) $\pi^*N^{(1)'}(X, Y) = N^{(1)}(X, Y)$, and
     (d) $\mathcal{H}(\nabla_X\phi)Y$ is the basic vector field associated to $(\nabla'_{X_*}\phi')Y_*$, for basic vector fields, X and Y on M.

Proof. (a) and (c) are obvious. To see (b), we show that $d\eta(U, V) = 0$. But $d\eta(U, V) = \frac{1}{2}\{U \cdot \eta(V) - V \cdot \eta(U) - \eta([U, V])\}$. Since the vertical distribution is integrable, $d\eta$ vanishes on $\mathcal{V}$. (d) follows from the third assertion of Lemma 1 of [O' N].

For almost contact metric submersions of type I, we essentially pulled-back the G-structure of the base space $(U(n) \times \{1\})$ and embedded it into that of the total space $(U(m) \times \{1\})$. Further properties of such G, G'-Riemannian submersions are developed in [Wa 4]. If we pull-back the $U(n)$-structure of an almost Hermitian manifold, N^{2n}, into that of an almost contact metric manifold with its $U(m) \times \{1\}$-structure, we allow the possibility of a new type (called type II) of Riemannian submersion.

Let $(M^{2m+1}, \phi, \xi, \eta, g)$ be an almost contact metric manifold and let π :

$M \to N$ be a Riemannian submersion onto the almost Hermitian manifold (N^{2n}, J', g') which satisfies:

$$J'\pi_* E = \pi_* \phi E .$$

Then π is an *almost contact metric submersion of type* II.

As noted before, we shall often speak of "$\mathcal{P}$-submersions of type II." We now investigate the ϕ-invariance of the vertical and horizontal distributions:

Proposition 2.3. Let $\pi : (M^{2m+1}, \phi, \xi, \eta, g) \to (N^{2n}, J', g')$ be an almost contact metric submersion of type II. Then,
- (a) $\pi^* \Phi' = \Phi$,
- (b) $\phi\{\mathcal{V}\} \subseteq \mathcal{V}$,
- (c) $\phi\{\mathcal{H}\} \subseteq \mathcal{H}$,
- (d) $\xi \in \mathcal{V}$, and
- (e) $\eta(\mathcal{H}) = 0$.

Proof. $J'\pi_* \xi = \pi_* \phi\xi = 0$. Thus, $\pi_* \xi = 0$, proving (d). (a) is similarly obtained. Let $V \in \mathcal{V}$. Then $\pi_* \phi V = J'\pi_* V = 0$, so $\phi V \in \mathcal{V}$. Now take $X \in \mathcal{H}$. We have $g(V, \phi X) = -g(\phi V, X) = 0$ and $\phi X \in \mathcal{H}$. The last assertion follows from $\eta(X) = g(\xi, X)$ and the fact that ξ is vertical.

The fibres of an almost contact metric submersion of type II inherit a natural almost contact metric structure from the total space by restriction of the tensors ϕ, ξ and η. Recall that ξ is a vertical vector field.

Proposition 2.4. Let $\pi : M \to N$ be an almost contact metric submersion of type II. Then,
- (a) $d\eta(X, Y) = -\frac{1}{2}\eta([X, Y]) = -\eta(A_X Y)$,
- (b) $\pi^* S'(X, Y) = N^{(1)}(X, Y)$, and
- (c) $\mathcal{H}(\nabla_X \phi)Y$ is the basic vector field associated to $(\nabla'_{X_*} J')Y_*$ for basic vector fields, X and Y, on M.

Proof. (a) is clear from the standard expression for the exterior differential of a 1-form and the fact that $A_X Y = \frac{1}{2}\mathcal{V}[X, Y]$. Since ξ is vertical, $\pi^* S'(X, Y) = N^{(1)}(X, Y)$. Finally, (c) follows from [O'N].

We recall that Vilms [V] showed that a Riemannian submersion for which both T and A vanish (a totally geodesic submersion) is covered by a Riemannian projection mapping.

3. Transference of Structures and Examples

An almost Hermitian submersion, $\pi : M \to N$, whose total space is Hermitian (resp., quasi-Kähler, almost Kähler, nearly Kähler or Kähler) transfers that structure to the fibres and the base space. The case in which M is almost semi-Kähler ($\delta\Phi = 0$) is exceptional and is treated in [W–V]. We now turn our attention to the analogous question for almost contact submersions.

Theorem 3.1. Let $\pi : (M^{2m+1}, \phi, \xi, \eta, g) \to (N^{2n+1}, \phi', \xi', \eta', g')$ be an almost contact metric submersion of type I with fibres $(F^{2(m-n)}, \hat{J}, \hat{g})$,

If M is:	then N is:	and F is:
(a) Normal	Normal	Hermitian
(b) Quasi-Sasakian	Quasi-Sasakian	Kähler
(c) Nearly cosymplectic	Nearly cosymplectic	Nearly Kähler
(d) Closely cosymplectic	Closely cosymplectic	Nearly Kähler
(e) Almost cosymplectic	Almost cosymplectic	Almost Kähler
(f) Cosymplectic	Cosymplectic	Kähler.

Proof. The proof is straightforward using the fact that π^* commutes with d.

Observe that when M is contact (including K-contact and Sasakian), it cannot be the total space of an almost contact metric submersion of type I with non-trivial (positive dimensional) fibres because $d\eta = 0$ on F implies that the non-trivial 2-form Φ_F vanishes. We shall see later that the only almost contact metric submersions of type I with M, nearly Sasakian are trivial; i.e., Riemannian covering mappings.

The analogue of Theorem 3.1 for submersions of type II is:

Theorem 3.2. Let $\pi : (M^{2m+1}, \phi, \xi, \eta, g) \to (N^{2n}, J', g')$ be an almost contact metric submersion of type II with fibres, $(F^{2(m-n)+1}, \hat{\phi}, \hat{\xi}, \hat{\eta}, \hat{g})$

If M is:	then N is:	and F is:
(a) Normal	Hermitian	Normal
(b) Quasi-Sasakian	Kähler	Quasi-Sasakian
(c) Nearly Sasakian	Nearly Kähler	Nearly Sasakian
(d) Contact	Almost Kähler	Contact
(e) K-contact	Almost Kähler	K-contact
(f) Sasakian	Kähler	Sasakian
(g) Nearly cosymplectic	Nearly Kähler	Nearly cosymplectic
(h) Closely cosymplectic	Nearly Kähler	Closely cosymplectic
(i) Almost cosymplectic	Almost Kähler	Almost cosymplectic
(j) Cosymplectic	Kähler	Cosymplectic

Proof. The transference to the fiber submanifolds is clear. Some work is required to see the structure on the base manifold, N. For instance, let M be nearly Sasakian. Then, for basic vector fields, X and Y, we have

$$(\nabla_X \phi)Y + (\nabla_Y \phi)X = g(X,Y)\xi - \eta(X)Y - \eta(Y)X \,.$$

Taking the horizontal parts of both sides yields:

$$\mathcal{H}(\nabla_X \phi)Y + \mathcal{H}(\nabla_Y \phi)X = 0 \,,$$

the left-hand side of which is uniquely associated to $(\nabla'_{X_*} J')Y_* + (\nabla'_{Y_*} J')X_*$. The vanishing of this sum implies that N is nearly Kähler.

W. Boothby and H. C. Wang [B–W] proved that a compact regular contact metric manifold, M, *always* fibres as a circle bundle, $S^1 \to M \to N \cong M/S^1$, in which the base space is a symplectic manifold whose fundamental 2-form, Φ', is integral. Moreover, the fundamental 1-form, η, of M defines a connection form on this principal bundle whose curvature form is the pull-back of Φ'. Y. Hatekeyama [Ha] showed that a compact regular contact metric space is K-contact and is Sasakian if and only if the base space of the induced Boothby-Wang fibration is Kähler. Combining Hatekeyama's result with the integrality of the fundamental 2-form of the base space of the Boothy-Wang fibre space, we see that N must be Hodge when and only when the total space is compact regular Sasakian. A specific example of such an almost contact metric submersion of type II is the Hopf fibration, $S^1 \to S^{2n+1} \xrightarrow{\pi} P_n(\mathbb{C})$.

An application of the Boothy-Wang fibration to a model of Yang-Mills gauge field theory has been described [Wa 4] in which an almost contact metric submersion with 3-structure [Wa 5] is constructed (up to certain topological obstructions) over an almost quaternionic manifold (in particular, the compactification of the realization of spacetime); one of the regular vector fields from the regular 3-structure then fibres the total space of the 3-submersion as a Boothby-Wang fibre space. The resulting almost Kähler base space then fibres as the total space of an almost Hermitian almost contact metric submersion [Wa 6] over the original almost quaternionic space.

A. Morimoto [Mor] produced a fibration of a compact normal almost contact manifold whose distinguished vector field, ξ, is regular. The construction is similar to that of Boothby and Wang. The Morimoto fibration can be made into a normal almost contact metric submersion of type II in the obvious way. Again, η is a connection 1-form whose differential is the pullback of the Kähler 2-form of the base space.

D. Blair and D. K. Showers [B–S] showed how to fibre a compact closely cosymplectic manifold whose distinguished vector field, ξ, is regular as a principal circle bundle over a nearly Kähler manifold. If the total space of the Blair-Showers fibration is cosymplectic, then the base space of this type II almost contact metric submersion is Kähler.

Y. Tashiro [Tas] studied almost contact metric structures induced onto the tangent sphere bundle, $S(N)$, of a Riemannian manifold, N, from the natural almost Kähler structure on the tangent bundle, $T(N)$. He found the induced structure on $S(N)$ to be contact, in general, and to be K-contact if and only if the Riemannian base manifold is of positive Riemannian constant sectional curvature; in which case, $S(N)$ is Sasakian. If we now take N to be almost Hermitian, then it is possible to show that the natural tangent sphere bundle projection mapping is a type II almost contact metric submersion. In this case, $S(N)$ is ϕ-normal when N is J-normal.

We again use Tashiro's sphere bundle construction to produce examples of almost contact metric submersions of type I. Let $(M^{2m+1}, \phi, \xi, \eta, g)$ be a normal contact metric manifold with $d\eta = 0$ (e.g., cosymplectic). The procedure of Tanno [Tn 1] gives an almost complex structure, J, on $T(M)$, the tangent bundle of M, which depends only on the almost contact structure of M, i.e., the tangent bundle projection mapping commutes with ϕ and J. Choosing the Sasaki [Sa] metric for $T(M)$ makes the projection mapping a Riemannian submersion. With respect to the Sasaki metric and the Tanno almost complex structure, $T(M)$ is an almost Kähler manifold. Now, as before, we give the tangent unit sphere bundle, $S(M)$, the induced contact metric structure it obtains as a hypersurface of an almost Kähler manifold. In this way, the projection mapping, $\pi : S(M) \rightarrow M$, of the tangent unit sphere bundle is an almost contact metric submersion of type I.

4. The Geometry of the Fibres and the Integrability of the Horizontal Distribution

The imposition of the more restrictive almost contact metric structure onto the total space of a Riemannian submersion places further restrictions on the embedding of the fibre submanifolds and may determine the integrability of the horizontal distribution. In order to study these effects, we recall the *O'Neill configuration tensors* [O'N] of type $(1, 2)$. Let $\pi : M \rightarrow N$ be a Riemannian submersion and E and F, vector fields on M. Then,

$$T_E F = \mathcal{H}\nabla_{\mathcal{V}E}(\mathcal{V}F) + \mathcal{V}\nabla_{\mathcal{V}E}(\mathcal{H}F), \quad \text{and}$$
$$A_E F = \mathcal{V}\nabla_{\mathcal{H}E}(\mathcal{H}F) + \mathcal{H}\nabla_{\mathcal{H}E}(\mathcal{V}F).$$

The basic properties of the T and A tensors are developed in [O'N] and [G 3] for Riemannian submersions and in [Wa 3] for almost Hermitian submersions. (We take this opportunity to remark that Theorem 4.1 of [Wa 3] can be strengthened to state that the horizontal distribution of a quasi-Kähler submersion is completely integrable.) The tensor T, on vertical vector fields, is essentially the second fundamental form of the immersion of the fibres. In particular, if T vanishes, then the fibres are totally geodesically immersed, and if tr$T$ (essentially the mean curvature vector field, H) vanishes, then the fibres are minimally immersed. The tensor A, on horizontal vector fields, is half the vertical component of $[X, Y]$. Thus, A vanishes if and only if the horizontal distribution is completely integrable.

We begin our analysis of the geometry of the fibre submanifolds and of the horizontal distribution by considering type I submersions. Recall that the total space of a type I submersion cannot have a contact structure, thereby eliminating contact, K-contact and Sasakian structures on the total space from our study. Furthermore, nearly Sasakian total spaces are associated only with covering spaces; we therefore refrain from analyzing them. We begin with the nearly cosymplectic case.

Theorem 4.1. Let $F \to M \to N$ be an almost contact metric submersion of type I with M, nearly cosymplectic. Then,

(a) $T_U(\hat{J}V) + T_{\hat{J}U}V = 2\phi T_U V$,
(b) $T_V(\hat{J}U) = \phi T_V U$,
(c) $T_{\hat{J}U}X = \hat{J}T_U X - 2T_U(\phi X)$,
(d) $T_U \xi = 0$,
(e) $A_X(\phi Y) = A_{\phi X}Y$,
(f) $A_X(\phi Y) = \frac{1}{3}\hat{J}A_X Y$,
(g) $A_X \xi = 0$,
(h) $A_X(\phi X) = 0$,
(i) $A_X(\hat{J}U) + A_{\phi X}U = 2\phi A_X U$, for basic X, and
(j) $A_X(\hat{J}V) = 3\phi A_X V$.

Proof. On a nearly cosymplectic manifold, $(\nabla_U \phi)U = 0 = \nabla_U(\phi U) - \phi\nabla_U U$. Taking horizontal projections gives (b). Then the standard polarization trick yields (a). Straightforward use of the skew-symmetry property of T in $g(T_{\hat{J}U}X, V)$ produces (c). $T_U \xi = 0$ because ξ is a Killing vector on a nearly cosymplectic manifold [B]. A similar analysis on $(\nabla_X \phi)X = 0$ yields the identities on A.

Theorem 4.2. Let $F \to M \to N$ be a nearly cosymplectic submersion of type I. The fibre submanifolds are minimal in M.

Proof. Let $\{E_i, \hat{J}E_i\}$ be a local $\hat{J}$-basis for the vector fields on the fibres. Recall that the mean curvature vector field is given by

$$H = \sum_{i=1}^{m-n} \{T_{E_i}E_i + T_{\hat{J}E_i}(\hat{J}E_i)\}.$$

Now

$$T_{\hat{J}E_i}(\hat{J}E_i) = \phi T_{\hat{J}E_i}E_i = \phi\phi T_{E_i}E_i = -T_{E_i}E_i + \eta(T_{E_i}E_i)\xi = -T_{E_i}E_i$$

Therefore, $H = 0$.

Theorem 4.3. Let $F \to M \to N$ be an almost contact metric submersion of type I with M, almost cosymplectic. Then,

    (a) $A_X(\phi Y) = \hat{J} A_X Y$,
    (b) $A_X(\hat{J} V) = \phi A_X V$,
    (c) $A_X \xi = 0$, and
    (d) $A_\xi \xi = 0$.

Proof. All of the claims will be established when we show (a). Since $d\Phi = 0$ and $d\eta = 0$ on an almost cosymplectic manifold, the fundamental relation (Proposition 1.1c) becomes:

$$g((\nabla_E \phi)F, G) = \frac{1}{2} g([\phi, \phi](F, G), \phi E).$$

Thus, for Y, basic, we have

$$g(A_X(\phi Y) - \hat{J} A_X Y, V) = \frac{1}{2} g([\phi, \phi](Y, V), \phi X).$$

Since $[Y, V]$ is vertical, the right hand side vanishes, yielding (a).

Theorem 4.4. Let $F \to M \to N$ be an almost contact metric submersion of type I with M, cosymplectic. Then, $A = 0$.

Proof. From $(\nabla_X \phi)Y = 0$, we immediately conclude that $A_X(\phi Y) = \hat{J} A_X Y$. By comparison with the proof of Theorem 4.1f, the theorem follows.

Theorem 4.5. Let $F \to M \to N$ be an almost contact metric submersion of type I with M, quasi-Sasakian. Then,

    (a) $T_U(\hat{J} V) = T_U V$,
    (b) $T_U(\phi X) = \hat{J} T_U X$,
    (c) $T_{\phi U} X = -\hat{J} T_U X$,
    (d) $T_U \xi = 0$,
    (e) $A_X(\phi Y) = \hat{J} A_X Y$,
    (f) $A_X(\phi X) = 0$,
    (g) $A_X \xi = 0$,
    (h) $A_X(\hat{J} U) = A_X U$, and
    (i) $A_{\phi X} U = -A_X U$.

Proof. Since M is quasi-Sasakian, $d\Phi = 0$, $N^{(1)} = 0$, and $N^{(2)} = 0$.

Therefore the standard relation (Proposition 1.1c) on $\nabla\phi$ becomes:

$$g((\nabla_E\phi)F, G) = d\eta(\phi F, E)\eta(G) - d\eta(\phi G, E)\eta(F) .$$

Because η vanishes on the vertical distribution as does $d\eta$ (see the proof of Propositon 2.2), we have $g(T_U(\hat{J}V) - \phi T_U V, Z) = 0$, establishing (a). (b) and (c) follow immediately. $T_U\xi = 0$ is a corollary of (b). Returning to the first equation, for X basic,

$$g(A_X(\phi Y) - \hat{J}A_X Y, U) = d\eta(\phi Y, X)\eta(U) - d\eta(\phi U, X)\eta(Y) .$$

Since $N^{(2)}$ vanishes, $d\eta(U, \phi X) = -d\eta(\phi U, X)$. Then,

$$d\eta(U, \phi X) = \frac{1}{2}\{U \cdot \eta(\phi X) - (\phi X) \cdot \eta(U) - \eta([U, \phi X])\} ,$$

which vanishes because $[U, \phi X]$ is vertical. Thus (e) is shown. (f) and (g) follow directly from this, while (h) and (i) are demonstrated by means of the skew-symmetry of A.

Theorem 4.6. Let $F \to M \to N$ be an almost contact metric submersion of type I with M, quasi-Sasakian. Then the fibre submanifolds are minimal.

Proof. Let $\{E_i, \hat{J}E_i\}$ be a local orthonormal basis for the vector fields on the fibres, F. Theorem 4.4a easily implies that $T_{E_i}(E_i) + T_{\hat{J}E_i}(\hat{J}E_i) = 0$. Therefore, $H = 0$.

Using the O'Neill T tensor, we now show that almost contact metric submersions of type I with M, nearly Sasakian, are Riemannian covering maps.

Theorem 4.7. Let $F^{2(m-n)} \to M^{2m+1} \xrightarrow{\pi} N^{2n+1}$ be an almost contact metric submersion of type I with M, nearly Sasakian. Then π is a Riemannian covering mapping.

Proof. It suffices to show that $m = n$ [Wo]. Since M is nearly Sasakian and η vanishes on the vertical distribution, we have

$$(\nabla_U\phi)U + (\nabla_U\phi)U = 2\|U\|^2\xi ,$$

from the defining relation of the nearly Sasakian structure. Taking the horizontal projection gives:

$$T_U(\hat{J}U) - T_U U = \|U\|^2 \xi \,.$$

Let $\{E_i, \hat{J}E_i\}$ be a local orthonormal basis for the vector fields on the fibre manifolds, F. Then,

$$T_{E_i}(\hat{J}E_i) - T_{E_i}(E_i) = \xi \,.$$

Thus,

$$-\phi H = 2(m-n)\xi \,.$$

But ξ cannot be a ϕ-image. Thus $m = n$.

In summary then, the horizontal distribution of a type I almost contact metric submersion whose total space is cosymplectic is integrable. Those whose total space is nearly Sasakian are covering mappings. Type I fibre spaces whose total space is quasi-Sasakian, almost cosymplectic, cosymplectic, nearly cosymplectic or closely cosymplectic have minimal fibres. There can be no type I fibre spaces whose total space is contact (including K-contact and Sasakian).

Theorem 4.8. Let $F \to M \to N$ be an almost contact metric submersion of type II with M contact. Then, for X and Y basic,
  (a) $T_{\phi V}\xi = \phi T_V \xi$,
  (b) $T_\xi(\phi X) = \phi T_\xi X$,
  (c) $A_X(\phi Y) = \phi A_X Y + g(X,Y)\xi$,
  (d) $A_X(\phi X) = \|X\|^2 \xi$,
  (e) $A_{\phi X}\xi = \phi A_X \xi - X$, and
  (f) $A_X(\phi V) = \phi A_X V - \eta(V)X$.

Proof. Since the fundamental relation (Proposition 1.1c) on the covariant derivative of ϕ is

$$g((\nabla_E \phi)F, G) = \frac{1}{2}g(N^{(1)}(F,G), \phi E) + d\eta(\phi F, E)\eta(G) - d\eta(\phi G, E)\eta(F),$$

on a contact manifold, we have:

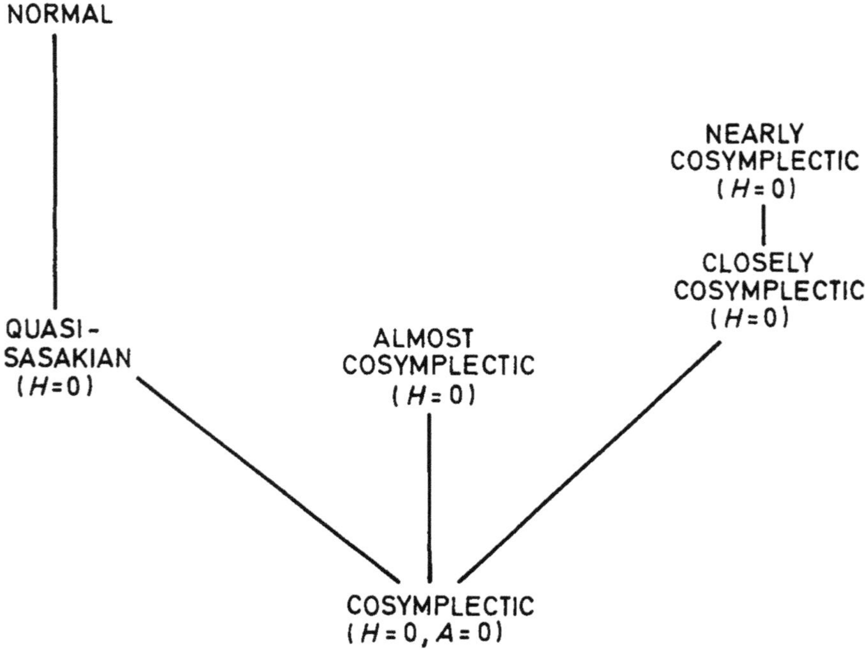

Fig. 4.1. Nontrivial almost contact metric submersions of type I with minimal fibres and/or integrable horizontal distribution.

$$g(T_U(\phi V) - \phi T_U V, X) = \frac{1}{2}g(N^{(1)}(V, X), \phi U) - d\eta(\phi X, U)\eta(V).$$

But $d\eta(\phi X, U) = 0 = d\eta(V, X)$. Therefore,

$$g(T_U(\phi V) - \phi T_U V, X) = \frac{1}{2}g([\phi, \phi](V, X), \phi U).$$

For $U = \xi$, relation (a) is immediate. (b) follows from a similar calculation. In an analogous manner,

$$g(A_X(\phi Y) - \phi A_X Y, U) = \frac{1}{2}g([\phi, \phi](Y, U), \phi X) + g(X, Y)\eta(U).$$

Since Y is basic, $[Y, U]$ is vertical and, therefore, $[\phi, \phi](Y, U)$ is vertical too. Thus, $A_X(\phi Y) - \phi A_X Y = g(X, Y)\xi$. The other assertions are proven straightforwardly.

Theorem 4.9. Let $F \to M \to N$ be an almost contact metric submersion of type II with M, K-contact. Then,

(a) $T_U \xi = 0$, and
(b) $T_\xi X = 0$.

Proof. The characteristic vector field, ξ, on a K-contact manifold is Killing, which is equivalent to $\nabla_E \xi = -\phi E$.

Theorem 4.10. Let $F \to M \to N$ be an almost contact metric submersion of type II with M, Sasakian. Then,

(a) $T_U(\phi V) = \phi T_U V$,
(b) $T_U(\phi X) = \phi T_U X$,
(c) $T_{\phi V} X = -\phi T_V X$, and
(d) $A_{\phi X} V = \phi A_X V$.

Proof. The assertions follow easily from the Sasakian identity:

$$(\nabla_E \phi)F = g(E, F)\xi - \eta(F)E .$$

For example, taking vertical parts of $\nabla_U(\phi X) - \phi \nabla_U X = g(X, U)\xi - \eta(X)V = 0$, yields relation (b).

Theorem 4.11. Let $F \to M \to N$ be an almost contact metric submersions of type II with M, Sasakian. The fibre submanifolds are minimally immersed submanifolds of M.

Proof. As usual, $T_U(\phi V) = \phi T_U V$ implies $T_U U + T_{\phi U}(\phi U) = 0$. Let $\{E_i, \phi E_i, \xi\}$ be a ϕ-basis locally for the vector fields on F. Since $T_\xi \xi = 0, H = 0$.

Theorem 4.12. Let $F \to M \to N$ be an almost contact metric submersion of type II with M, nearly Sasakian. Then,

(a) $T_U(\phi V) + T_{\phi U} V = 2\phi T_U V$,
(b) $T_V(\phi V) = \phi T_V V$,
(c) $T_\xi \xi = 0$,
(d) $A_X(\phi Y) - A_{\phi X} Y = 2g(X, Y)\xi$, and
(e) $A_X(\phi X) = \|X\|^2 \xi$.

Proof. We show assertions (d) and (e), leaving the others to the reader. Recall that the defining relation for a nearly Sasakian structure is:

$$(\nabla_E\phi)F + (\nabla_F\phi)E = 2g(E,F)\xi - \eta(E)F - \eta(F)E.$$

Taking the vertical projection on each side yields:

$$A_X(\phi Y) - \phi A_X Y + A_Y(\phi X) - \phi A_Y X = 2g(X,Y)\xi.$$

Thus,

$$A_X(\phi Y) - A_{\phi X}Y = 2g(X,Y)\xi.$$

Theorem 4.13. Let $F \to M \to N$ be an almost contact metric submersion of type II with M, nearly Sasakian. The fibre submanifolds are minimal.

Proof. Utilizing assertions (b) and (c) of Theorem 4.11, the proof is identical to that of Theorem 4.10.

Theorem 4.14. Let $F \to M \to N$ be an almost contact metric submersion of type II with M, quasi-Sasakian. Then,
  (a) $T_\xi(\phi V) = \phi T_\xi V$,
  (b) $T_\xi \xi = 0$,
  (c) $T_U(\phi X) = \phi T_U X + \frac{1}{2}\eta([X,\phi U])\xi$,
  (d) $T_\xi(\phi X) = \phi T_\xi X$,
  (e) $A_X(\phi Y) = \phi A_X Y + d\eta(X,\varphi Y)\xi$,
  (f) $A_X(\phi X) = d\eta(X,\varphi X)\xi$, and
  (g) $A_X(\phi U) = \phi A_X U$ for basic X.

Proof. The fundamental relation in Proposition 1.10 takes the following form for quasi-Sasakian manifolds ($d\Phi = 0, N^{(1)} = 0$):

$$g((\nabla_E\phi)F, G) = d\eta(\phi F, E)\eta(G) - d\eta(\phi G, E)\eta(F).$$

As an example of the calculations, we show (g). Now,

$$g((\nabla_U\phi)X, Y) = d\eta(\phi X, U)\eta(Y) - d\eta(\phi Y, U)\eta(X).$$

But η vanishes on the horizontal distribution and, for X basic,

$$\mathcal{H}(\nabla_U\phi)X = A_{\varphi X}U - \phi A_X U.$$

Theorem 4.15. The fibre submanifolds of an almost contact metric submersion of type II with M, quasi-Sasakian, are minimal.

Proof. Let $\{E_i, \phi E_i, \xi\}$ be an orthonormal local ϕ-basis for the vector fields on F. From the expression for $\nabla \phi$ given in the proof of Theorem 4.13,

$$g(T_V(\phi V) - \phi T_V V, X) = d\eta(X, \phi V)\eta(V).$$

Since E_i is orthogonal to ξ, we have $\eta(E_i) = 0$, from which

$$T_{E_i}(E_i) + T_{\phi E_i}(\phi E_i) = 0$$

follows quickly. Combining this with $T_\xi \xi = 0$, we obtain $H = 0$.

Theorem 4.16. Let $F \to M \to N$ be an almost contact metric submersion of type II with M, almost cosymplectic. Then,

(a) $A_X(\phi Y) = \phi A_X Y$,
(b) $A_X(\phi X) = 0$,
(c) $A_X(\phi V) = \phi A_X V$, and
(d) $A_X \xi = 0$.

Proof. The assertion of this theorem follow from the usual calculations utilizing the almost cosymplectic defining identities: $d\Phi = 0$ and $d\eta = 0$. No apparently useful relations on the O'Neill tensor, T, emerge.

Theorem 4.17. Let $F \to M \to N$ be an almost contact metric submersion of type II with M, cosymplectic. Then,

(a) $T_U(\phi V) = \phi T_U V$,
(b) $T_U(\phi X) = \phi T_U X$,
(c) $T_{\phi V} X = -T_V(\phi X)$,
(d) $T_U \xi = 0$, and
(e) $A = 0$.

Proof. The relations on T are clear from $\nabla_X \phi = 0$. The same relation on the covariant derivative of ϕ implies:

$$A_X(\phi Y) = A_{\phi X} Y \quad \text{as well as} \quad A_X(\phi V) = \phi A_X V.$$

Now, letting X be basic,

$$g(A_{\phi X}Y, V) = g(A_X(\phi Y), V) = -g(\phi Y, A_X V) = -g(\phi Y, \mathcal{H}\nabla_V X)$$
$$= g(Y, \mathcal{H}\nabla_V(\phi X)) = g(Y, A_{\phi X}V) = -g(A_{\phi X}Y, V).$$

Therefore, $A = 0$.

Theorem 4.18. Let $F \to M \to N$ be an almost contact metric submersion of type II with M, cosymplectic. The fibre submanifolds determined by the submersion are minimally immersed.

Proof. Obvious from Theorem 4.16a and 4.16d.

Theorem 4.19. Let $F \to M \to N$ be an almost contact metric submersion of type II with M, nearly cosymplecitc. Then,
    (a) $T_U(\phi V) + T_{\phi U}V = 2\phi T_U V$,
    (b) $T_U(\phi U) = \phi T_U U$,
    (c) $T_\xi \xi = 0$,
    (d) $A_X(\phi Y) = A_{\phi X}Y$,
    (e) $A_X(\phi Y) = \frac{1}{3}\phi A_X Y$,
    (f) $A_{\phi X}V = -\phi A_X V$,
    (g) $A_X(\phi V) = 3\phi A_X V$, and
    (h) $A_X \xi = 0$.

Proof. A typical calculation will establish (g). $(\nabla_E \phi)E = 0$ implies $(\nabla_X \phi)U + (\nabla_U \phi)X = 0$ which, in turn, yields, for basic $X, \nabla_X(\phi U) - \phi\nabla_X U + \nabla_U(\phi X) - \phi\nabla_U X = 0$. Taking horizontal projections gives $A_X(\phi U) - \phi A_X U + A_{\phi X}U - \phi A_X U = 0$ from which $A_X(\phi U) = 2\phi A_X U - A_{\phi X}U$. A short additional calculation shows that $A_X(\phi U) = 3\phi A_X U$.

Theorem 4.20. Let $F \to M \to N$ be an almost contact metric submersion of type II with M, nearly cosymplectic. The fibres submanifolds of the submersion are minimally immersed.

Proof. It is easy to derive from relation (b) of the previous theorem that:
$$T_U U + T_{\phi U}(\phi U) = -\eta(U)\phi T_{\phi U}\xi.$$

Let $\{E_i, \phi E_i, \xi\}$ be a ϕ-basis for the vector fields on F which is orthonormal. Then, as in the proof of Theorem 4.15, $\eta(E_i) = 0$. Since $T_\xi \xi = 0$, we have $H = 0$.

5. Curvature Relations

It is well known that a Riemannian submersion is Riemannian sectional curvature increasing on horizontal 2-planes [O'N]. The addition of a G-structure, say $U(n)$, with some restriction, say quasi-Kähler, makes the resulting almost Hermitian submersion, in this case, holomorphic sectional curvature *preserving* on horizontal holomorphic 2-planes [Wa 3]. Other, sometimes surprising, relations are possible.

We shall investigate the two sectional curvature tensors of primary interest on an almost contact metric manifold, $(M^{2n+1}, \phi, \xi, \eta, g)$. Let E and F be vector fields on M. The *ϕ-bisectional holomorphic curvature* tensor, B_ϕ, is given by:

$$B_\phi(E, F) = \|E\|^{-2} \|F\|^{-2} g(R(E, \phi E)F, \phi F).$$

When $F = E$ and $E \perp \xi$, we obtain the *ϕ-holomorphic sectional curvature* tensor:

$$H_\phi(E) = \|E\|^{-4} g(R(E, \phi E)E, \phi E).$$

These curvature tensors have been studied by others, e.g., [Mos], [Og], [Tn 3], and their relationships to the underlying topology of the almost contact metric manifold documented.

We let $B_J(E, F)$ and $H_J(E)$ denote the usual bisectional and sectional holomorphic curvature tensors, respectively, on an almost Hermitian manifold, (N, J, g).

Theorem 5.1. Let $F \to M \to N$ be an almost contact metric submersion of type I. Then the sectional curvature tensors have the following forms:

(a) $B_\phi(U, V) = \hat{B}_J(U, V) + \|U\|^{-2} \|V\|^{-2} \{ g(T_U(\phi V), T_{\phi U} V) - g(T_U V, T_{\phi U}(\phi V)) \},$

(b) $B_\phi(X, V) = \|X\|^{-2} \|V\|^{-2} \{ G((\nabla_V A)_X(\phi X), \hat{J}V) - g(A_X(\phi V), A_{\phi X} V) + g(A_X V, A_{\phi X}(\phi V)) - g((\nabla_{\phi V} A)_X(\phi X), V) + g(T_{\phi V} X, T_V(\phi X)) - g(T_V X, T_{\phi V}(\phi X)) \},$

B. Watson

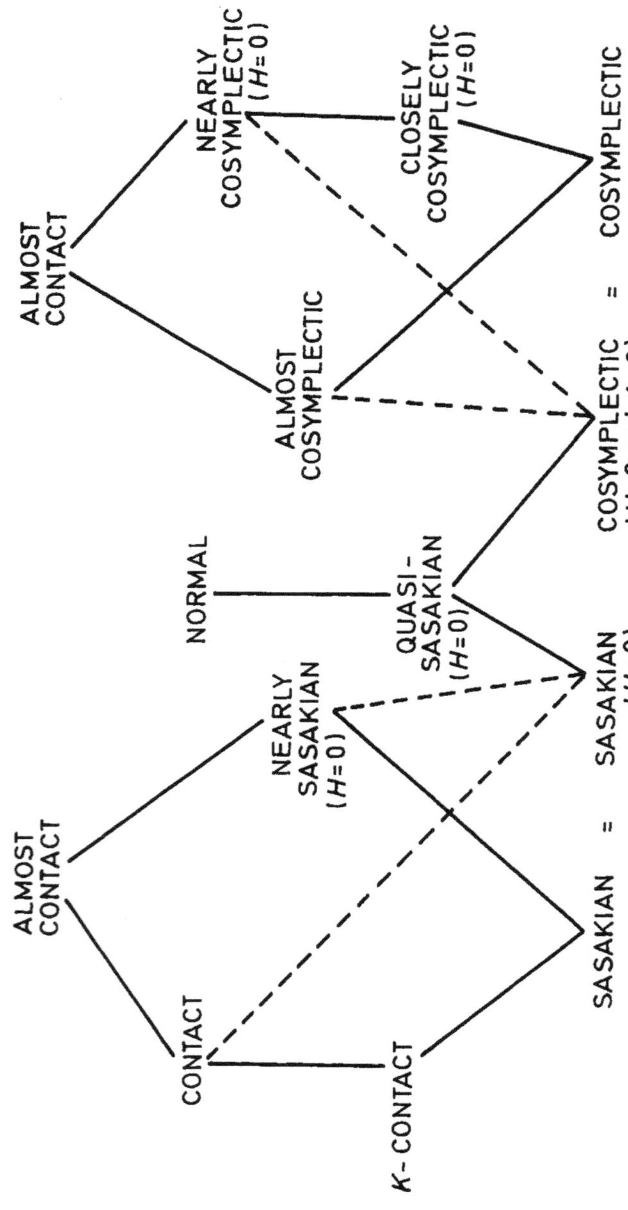

Fig. 4.2. Non-trivial almost contact metric submersions of type II with integrable horizontal distribution and/or minimal fibres.

(c) $B_\phi(X,Y) = B'_{\phi'}(X_*,Y_*) - \|X\|^{-2}\|Y\|^{-2}\{2g(A_X(\phi X), A_Y(\phi Y))$
$\qquad\qquad - g(A_{\phi X}Y, A_X(\phi Y)) - g(A_X Y, A_{\phi X}(\phi Y))\}$,

(d) $H_\phi(V) = \hat{H}_j(V) + \|V\|^{-4}\{\|T_V(\phi V)\|^2 - g(T_V V, T_{\phi V}(\phi V))\}$, and

(e) $H_\phi(X) = H'_{\phi'}(X_*) - 3\|X\|^{-4}\|A_X(\phi X)\|^2$.

We begin our study of the effects of certain almost contact metric structures on these curvature tensors with type I submersions in which the total space, M, is nearly cosymplectic.

Theorem 5.2. Let $F \to M \to N$ be an almost contact metric submersion of type I with M, nearly cosymplectic. Then,

$$B_\phi(X,Y) = B'_{\phi'}(X_*,Y_*) - \frac{2}{3}\|X\|^{-2}\|Y\|^{-2}\|A_X Y\|^2 .$$

Proof. On a nearly cosymplectic type I submersion, we have $A_X(\phi Y)$ $= \frac{1}{3}\hat{J}A_X Y$.

Corollary 5.2.1. Let $F \to M \to N$ be a nearly cosymplectic submersion of type I. Then,

$$B_\phi(X,Y) \leq B'_{\phi'}(X_*,Y_*)$$

and equality is obtained for all horizontal X and Y if and only if the horizontal distribution is completely integrable.

Theorem 5.3. Let $F \to M \to N$ be a nearly cosymplectic type I submersion. Then,

a) $H_\phi(V) = \hat{H}_j(V) + 2\|V\|^{-4}\|T_V V\|^2$, and

b) $H_\phi(X) = H'_{\phi'}(X_*)$.

Theorem 5.4. Let $F \to M \to N$ be an almost contact metric submersion of type I with M, almost cosymplectic. Then,

$$B_\phi(X,Y) = B'_{\phi'}(X_*,Y_*) + \|X\|^{-2}\|Y\|^{-2}\|A_X Y\| .$$

Theorem 5.5. Let $F \to M \to N$ be an almost contact metric submersion of type I with M, almost cosymplectic. Then,

$$H_\phi(X) = H'_{\phi'}(X_*) .$$

Theorem 5.6. Let $F \to M \to N$ be an almost contact metric submersion of type I with M, quasi-Sasakian. Then,

    (a) $B_\phi(U, V) = \hat{B}_{\hat{j}}(U, V) + 2\|U\|^{-2}\|V\|^{-2}\|T_U V\|^2$,

    (b) $B_\phi(X, V) = -2\|X\|^{-2}\|V\|^{-2}\{\|T_V X\|^2 - \|A_X V\|^2\}$, and

    (c) $B_\phi(X, Y) = B'_{\phi'}(X_*, Y_*) - 2\|X\|^{-2}\|Y\|^{-2}\|A_X Y\|^2$.

Proof. The usual calculations except for the vanishing of the terms contain A. Consider $g((\nabla_V A)_X(\phi X), \phi V)$. Now, for X basic,

$$
\begin{aligned}
(\nabla_V A)_X(\phi X) &= \nabla_V(A_X(\phi X)) - A_{\nabla_V X}(\phi X) - A_X(\nabla_V(\phi X)) \\
&= -A_{\mathcal{H}\nabla_V X}(\phi X) - A_X(\mathcal{H}\nabla_V(\phi X)) - A_X(\mathcal{V}\nabla_V(\phi X)) \\
&= -A_{A_X V}(\phi X) - A_X A_{\phi X} V - A_X T_V(\phi X) \\
&= A_{\phi X} A_X V - A_X A_{\phi X} V - \phi A_X T_V X = -\phi A_X T_V X \,.
\end{aligned}
$$

Similarly, $(\nabla_{\phi V} A)_X(\phi X) = -A_X T_V X$ because $\eta(A_X T_V X) = g(A_X T_V X, \xi) = 0$. Thus, $g((\nabla_V A)_X(\phi X), \phi V) = g((\nabla_{\phi V} A)_X(\phi X), V)$.

Corollary 5.6.1. Let $F \to M \to N$ be a quasi-Sasakian submersion of type I. Then,

$$
B_\phi(X, Y) \leq B'_{\phi'}(X_*, Y_*)
$$

and equality is obtained for all horizontal X and Y if and only if the horizontal distribution is completely integrable.

Theorem 5.7. Let $F \to M \to N$ be an almost contact metric submersion of type I with M, quasi-Sasakian. Then,

    (a) $H_\phi(V) = \hat{H}_{\hat{j}}(V) + 2\|V\|^{-4}\|T_V V\|^{-2}$, and

    (b) $H_\phi(X) = H'_{\phi'}(X_*)$.

Theorem 5.8. Let $F \to M \to N$ be an almost contact metric submersion of type II. Then the sectional curvature tensors take the following forms:

    (a) $B_\phi(U, V) = \hat{B}_{\hat{j}}(U, V) + \|U\|^{-2}\|V\|^{-2}\{g(T_U(\phi V), T_{\phi U} V) - g(T_U V, T_{\phi U}(\phi V))\}$,

    (b) $B_\phi(X, V) = \|X\|^{-2}\|V\|^{-2}\{g((\nabla_V A)_X(\phi X), \phi V) - g(A_X(\phi V), A_{\phi X} V) + g(A_X V, A_{\phi X}(\phi V)) - g((\nabla_{\phi V} A)_X(\phi X), V) + g(T_{\phi V} X, T_V(\phi X)) - g(T_V X, T_{\phi V}(\phi X))\}$,

(c) $B_\phi(X,Y) = B'_{J'}(X_*,Y_*) - \|X\|^{-2}\|Y\|^{-2}\{2g(A_X(\phi X), A_Y(\phi Y))$
$\qquad\qquad - g(A_{\phi X}Y, A_X(\phi Y)) - g(A_X Y, A_{\phi X}(\phi Y))\}$,

(d) $H_\phi(V) = \hat{H}_{\hat\phi}(V) + \|V\|^{-4}\{\|T_V(\phi V)\|^2 - g(T_V V, T_{\phi V}(\phi V))\}$, and

(e) $H_\phi(X) = H'_{J'}(X_*) - 3\|X\|^{-4}\|A_X(\phi X)\|^2$.

Theorem 5.9. Let $F \to M \to N$ be an almost contact metric submersion of type II with M, contact. Then,

$$B_\phi(X,Y) = B'_{J'}(X_*,Y_*) - 2 - \|X\|^{-2}\|Y\|^{-2}g(X,Y)^2 \,.$$

Proof. Since $A_X(\phi X) = \|X\|^2\xi$, we have $g(A_X(\phi X), A_Y(\phi Y)) = \|X\|^2\|Y\|^2$. On the other hand, $g(A_X Y, A_{\phi X}(\phi Y)) = g(A_X Y, \phi A_{\phi X}Y) + g(X,Y)g(A_X Y, \xi) = -g(\phi A_X Y, A_{\phi X}Y) + \eta(A_X Y)^2 = -\|A_X Y\|^2 + \eta(A_X Y)^2$. Similarly, $g(A_{\phi X}Y, A_x(\phi Y)) = \|A_X Y\|^2 - g(X,Y)^2 - \eta(A_X Y)^2$. Hence, $B_\phi(X,Y) = B'_{J'}(X_*,Y_*) - 2 - \|X\|^2\|Y\|^2 g(X,Y)^2$.

Corollary 5.9.1. Let $F \to M \to N$ be a contact submersion of type II. Then,

$$B_\phi(X,Y) < B'_{J'}(X_*,Y_*)\,.$$

Theorem 5.10. Let $F \to M \to N$ be an almost contact metric submersion of type II with M, contact. Then,

$$H_\phi(X) = H'_{J'}(X_*) - 3\,.$$

Corollary 5.10.1. Let $F \to M \to N$ be a contact submersion of type II. Then,

$$H_\phi(X) < H'_{J'}(X_*)\,.$$

Theorem 5.11. Let $F \to M \to N$ be an almost contact metric submersion of type II with M, Sasakian. Then,

a) $B_\phi(U,V) = \hat{B}_{\hat\phi}(U,V) + 2\|U\|^{-2}\|V\|^{-2}\|T_U V\|^2$, and

b) $B_\phi(X,V) = 2\|X\|^{-2}\|V\|^{-2}\{\eta(V)^2\|X\|^2 - \|T_V X\|^2 - \|A_X V\|^2\}$.

Proof. (a) is straightforward using the fact that $T_U(\phi V) = \phi T_U V$ on a Sasakian submersion of type II. For (b), we emulate the proof of Theorem 5.6 using $A_{\phi X} V = \phi A_X V$ and $A_X(\phi V) = \phi A_X V - \eta(V)X$ to show that $g((\nabla_V A)_X(\phi X), \phi V) = g((\nabla_{\phi V} A)_X(\phi X), V)$. Then we are able to demonstrate that $g(A_X V, A_{\phi X}(\phi V)) - g(A_X(\phi V), A_{\phi X} V) = -2\|A_X V\|^2 + 2\|X\|^2 \eta(V)^2$ and that $g(T_{\phi V} X, T_V(\phi X)) - g(T_V X, T_{\phi V}(\phi X)) = -2\|T_V X\|^2$.

Theorem 5.12. Let $F \to M \to N$ be an almost contact metric submersion of type II with M, Sasakian. Then,

$$H_\phi(V) = \hat{H}_{\hat{\phi}}(V) + 2\|V\|^{-4}\|T_V V\|^2 .$$

Corollary 5.12.1. Let $F \to M \to N$ be an almost contact metric submersion of type II with M, Sasakian. Then,

$$H_\phi(V) \geq \hat{H}_{\hat{\phi}}(V) .$$

Theorem 5.13. Let $F \to M \to N$ be an almost contact metric submersion of type II with M, nearly Sasakian. Then,
    a) $H_\phi(V) = \hat{H}_{\hat{\phi}}(V) + 2\|V\|^{-4}\|T_V V\|^2$,
    b) $H_\phi(X) = H'_{J'}(X_*) - 3$.

Proof. Assertion (a) is easy from $T_V(\phi V) = \phi T_V V$ on nearly Sasakian submersions of type II. (b) follows from the fact that $A_X(\phi X) = \|X\|^2 \xi$.

Corollary 5.13.1. Let $F \to M \to N$ be an almost contact metric submersion of type II with M, nearly Sasakian. Then,
    a) $H_\phi(V) \geq \hat{H}_{\hat{\phi}}(V)$, and
    b) $H_\phi(V) \leq H'_{J'}(X_*)$.

Theorem 5.14. Let $F \to M \to N$ be an almost contact metric submersion of type II with M, almost cosymplectic. Then,

$$B_\phi(X, Y) = B'_{J'}(X_*, Y_*) - 2\|X\|^{-2}\|Y\|^{-2}\|A_X Y\|^2 .$$

Proof. On an almost cosymplectic submersion of type II,

$$g(A_{\phi X} Y, A_X(\phi Y)) = -\|A_X Y\|^2 + \eta(A_X Y)^2 \,.$$

But $A_X \xi = 0$ implies that $\eta(A_X Y) = 0$. Similarly, $g(A_X Y, A_{\phi X}(\phi Y)) = -\|A_X Y\|^2$.

Corollary 5.14.1. Let $F \to M \to N$ be an almost cosymplectic submersion of type II. Then,

$$B_\phi(X, Y) \le B'_{J'}(X_*, Y_*)\,,$$

and equality is obtained for all horizontal X and Y if and only if the horizontal distribution is completely integrable.

Theorem 5.15. Let $F \to M \to N$ be an almost contact metric submersion of type II with M, almost cosymplectic. Then,

$$H_\phi(X) = H'_{J'}(X_*)\,.$$

Theorem 5.16. Let $F \to M \to N$ be an almost contact metric submersion of type II with M, cosymplectic. Then,
    (a) $B_\phi(U, V) = \hat{B}_{\hat\phi}(U, V) + 2\|U\|^{-2}\|V\|^{-2}\|T_U V\|^2$,
    (b) $B_\phi(X, V) = -2\|T_V X\|^2\|X\|^{-2}\|V\|^{-2}$,
    (c) $B_\phi(X, Y) = B'_{J'}(X_*, Y_*)$,
    (d) $H_\phi(V) = \hat{H}_{\hat\phi}(V) + 2\|V\|^{-4}\|T_V V\|^2$, and
    (e) $H_\phi(X) = H'_{J'}(X_*)$.

Proof. A direct computation using $T_U(\phi V) = \phi T_U V$ and the vanishing of the horizontal integrability tensor, A.

Corollary 5.16.1. Let $F \to M \to N$ be an almost contact metric submersion of type II with M, cosymplectic. Then,

$$B_\phi(U, V) \ge \hat{B}_{\hat\phi}(U, V)$$

and equality is obtained for all vertical U and V if and only if the fibres are totally geodesically immersed if and only if M is covered by a Riemannian product manifold, V, one of whose factors is Kähler and the other cosymplectic.

Theorem 5.17. Let $F \to M \to N$ be an almost contact metric submersion of type II with M, nearly cosymplectic. Then,

(a) $B_\phi(X, Y) = B'_{J'}(X_*, Y_*) - \frac{8}{9}\|X\|^{-2}\|Y\|^{-2}\|A_X Y\|^2$,

(b) $H_\phi(V) = \hat{H}_{\hat{\phi}}(V) + 2\|V\|^{-4}\|T_V V\|^2$, and

(c) $H_\phi(X) = H'_{J'}(X_*)$.

Proof. In the calculation of $B_\phi(X, Y)$, we take advantage of the facts that $A_X(\phi X) = 0$ and $A_{\phi X} Y = A_X(\phi Y)$ for a nearly cosymplectic submersion of type II in order to reduce the analysis to the last term:

$$g(A_X Y, A_{\phi X}(\phi Y)) = -\|A_X Y\|^2 + \eta(A_X Y)^2 .$$

But $A_X \xi = 0$ implies $\eta(A_X Y) = 0$, as before. Therefore the calculation in brackets in Theorem 5.8c becomes:

$$\left\{ 0 - \frac{1}{9}\|A_X Y\|^2 + \|A_X Y\|^2 \right\} .$$

The other assertions follow immediately.

Corollary 5.17.1. Let $F \to M \to N$ be an almost contact metric submersion of type II with M, nearly cosymplectic. Then,

$$B_\phi(X, Y) \le B'_{J'}(X_*, Y_*)$$

and equality is obtained for all horizontal X and Y if and only if the horizontal distribution is completely integrable.

6. Cohomology

In [Wa 2], the author proved several relations on the cohomology of the manifolds associated to a Riemannian submersion. These were obtained as corollaries of the following:

Theorem 6.1. Let $\pi : M \to N$ be a smooth surjective mapping between compact Riemannian manifolds. Then,

(a) π^* satisfies $\pi^*\delta_N\omega = \delta_M\pi^*\omega$, for all $\omega \in \Lambda^1(N)$ if and only if π is a Riemannian submersion with minimal fibres [Wa 2],

(b) π^* satisfies $\pi^*\Delta_N\omega = \Delta_M\pi^*\omega$, for all $\omega \in \Lambda^r(N), r \geq 2$, if and only if π is a Riemannian submersion with minimal fibres and completely integrable horizontal distribution [G–I].

A Riemannian submersion, $\pi : M \to N$, induces a linear isometry,

$$\pi^* : \Lambda^r(N) \to \Lambda^r(M),$$

and, if $*$ commutes with δ or Δ, this restricts to an isometry,

$$\pi^* : \mathcal{H}^r(N) \to \mathcal{H}^r(M),$$

of the finite dimensional real vector space of harmonic r-forms on N into the same space on M. Thus, when π^* commutes with δ or Δ on the r-forms of N, we have,

$$b_r(N) \leq b_r(M).$$

Theorem 6.2. Let $F \to M^{2m+1} \to N^{2n+1}$ be an almost contact metric submersion of type I, with M compact. If M possesses one of the following structures, then the fibre submanifolds are minimal:

(a) quasi-Sasakian,

(b) almost cosymplectic,

(c) nearly cosymplectic,

(d) closely cosymplectic, or

(e) cosymplectic.

Theorem 6.3. Let $F \to M^{2m+1} \to N^{2n}$ be an almost contact metric submersion of type II with M, compact. If M possesses one of the following structures, then the fibre submanifolds are minimal:

(a) nearly Sasakian,

(b) Sasakian,

(c) quasi-Sasakian,

(d) nearly cosymplectic,

(e) closely cosymplectic, or

(f) cosymplectic.

Theorem 6.4. Let $F \to M \to N$ be an almost contact metric submersion identified in Theorem 6.2 or Theorem 6.3. Then,

$$b_1(N) \leq b_1(M).$$

A specific instance of Theorem 6.4 is the well-known result of Tanno [Tn 2]:

Theorem 6.5. Let $S^1 \to M^{2m+1} \to N^{2n+1}$ be a Boothby-Wang fibration of the compact regular contact manifold, M. Then,

$$b_1(N) = b_1(M).$$

Other relevant results on the Betti numbers of compact contact metric manifolds have been obtained by S. Goldberg [Go], S. Tachibana [Tac], and others.

Theorem 6.6. Let $F \to M \to N^{2n+1}$ be an almost contact metric submersion of type I with M, cosymplectic and compact. Then, for $r = 0, 1, \ldots, 2n + 1$,

$$b_r(N) \leq b_r(M).$$

Proof. Since both $H = 0$ and $A = 0$, the submersion mapping commutes with the Laplacian on r-forms of N for all available r.

We remark that it is easy to see [B–G] that all Betti numbers of a compact cosymplectic manifold are positive.

Theorem 6.7. Let $F \to M \to N$ be an almost contact, metric submersion of type II, with M, cosymplectic and compact. Then, for $r = 0, 1, \ldots, 2n$,

$$b_r(N) \leq b_r(M).$$

Bibliography

[B]   D. E. Blair, *Contact Manifolds in Riemannian Geometry*, Springer-Verlag, New York, 1976.

[B-G]   D. E. Blair and S. I. Goldberg, *Topology of almost contact manifolds*, J. Diff. Geom. **1** (1967) 347–354.

[B-S]   D. E. Blair and D. K. Showers, *Almost contact manifolds with Killing structure tensors*, II, J. Diff. Geom. **9** (1974) 577–582.

[B-W]   W. M. Boothby and H. C. Wang, *On contact manifolds*, Ann. Math. **68** (1958) 721–734.

[E]   R. H. Escobales, *Riemannian submersions with totally geodesic fibers*, J. Diff. Geom. **10** (1975) 253–276.

[Go]   S. I. Goldberg, *On the topology of compact contact manifolds*, Tôhoku Math. J. **20** (1968) 106–110.

[G-I]   S. I. Goldberg and T. Ishihara, *Riemannian submersions commuting with the Laplacian*, J. Diff. Geom. **13** (1978) 139–144.

[G 1]   A. Gray, *Minimal varieties and almost Hermitian submanifolds*, Michigan Math. J. **12** (1965) 273–287.

[G 2]   A. Gray, *Some examples of almost Hermitian manifolds*, Illinois J. Math. **10** (1966) 353–366.

[G 3]   A. Gray, *Pseudo-Riemannian almost product manifolds and submersions*, J. Math. Mech. **16** (1967) 715–738.

[G-H]   A. Gray and L. Hervella, *The sixteen classes of almost Hermitian manifolds*, Ann. Mat. Pura Appl. **123** (1980) 35–58.

[Ha]   Y. Hatekeyama, *Some notes on differentiable manifolds with almost contact structures*, Tôhoku Math. J. **15** (1963) 176–181.

[He]   R. Hermann, *A sufficient condition that a mapping of Riemannian manifolds be a fibre bundle*, Proc. Amer. Math. Soc. **11** (1960) 236–242.

[J-V]   D. Janssens and L. Vanhecke, *Almost contact structures and curvature tensors*, Kodai Math. J. **4** (1981) 1–27.

[Ko]   S. Koto, *Some theorems on almost Kählerian spaces*, J. Math. Soc. Japan **12** (1960) 422–433.

[Mar]   J. Martinet, *Formes de contact sur les variétiés de dimension 3*, Proc. Liverpool Singularities Symp., II, Springer-Verlag, New York, 1971, pp. 142–163.

[Mor]   A. Morimoto, *On normal almost contact structures*, J. Math. Soc. Japan **15** (1965) 420–436.

[Mos]   E. M. Moskal, *Contact manifolds of positive curvature*, Ph. D. thesis Univ. of Illinois, Urbana, 1966.

[N-N] A. Newlander and L. Nirenberg, *Complex analytic coordinates in almost complex manifolds*, Ann. Math. **65** (1957) 391–404.

[Og] K. Ogiue, *On almost contact manifolds admitting axiom of planes or axiom of free mobility*, Kodai Math. Sem. Reps. **16** (1964) 223–232.

[O'N] B. O'Neill, *The fundamental equations of a submersion*, Michigan Math. J. **13** (1966) 459–469.

[R] B. Reinhart, *Foliated manifolds with bundle-like metrics*, Ann. Math. **69** (1959) 119–132.

[Sa] S. Sasaki, *On the differential geometry of tangent bundles of Riemannian manifolds*, Tôhoku Math. J. **10** (1958) 338–354.

[Tac] S. Tachibana, *On harmonic tensors in compact Sasakian spaces*, Tôhoku Math. J. **17** (1965) 271–284.

[Tn 1] S. Tanno, *Almost complex structures in bundle spaces over almost contact manifolds*, J. Math. Soc. Japan **17** (1965) 166–186.

[Tn 2] S. Tanno, *The topology of contact Riemannian manifolds*, Illinois J. Math. **12** (1968) 700–717.

[Tn 3] S. Tanno, *Sasakian manifolds with constact ϕ-holomorphic sectional curvature*, Tôhoku Math. J. **21** (1969) 501–507.

[Tas] Y. Tashiro, *On contact structures of tangent sphere bundles*, Tôhoku Math. J. **21** (1969) 117–143.

[V-H] E. Vidal and L. Hervella, *New classes of almost Hermitian manifolds*, Tensor (N.S.) **33** (1979) 293–299.

[V] J. Vilms, *Totally geodesic maps*, J. Diff. Geom. **4** (1970) 73–79.

[W] N. Wallach, *Compact homogeneous Riemannian manifolds of strictly positive curvature*, Ann. Math. **96** (1972) 277–295.

[Wa 1] B. Watson, *Manifold maps commuting with the Laplacian*, J. Diff. Geom. **5** (1971) 85–95.

[Wa 2] B. Watson, *δ-commuting maps and Betti numbers*, Tôhoku Math. J. **27** (1975) 137–152.

[Wa 3] B. Watson, *Almost Hermitian submersions*, J. Diff. Geom. **11** (1976) 147–165.

[Wa 4] B. Watson, *G, G' — Riemannian submersions and solutions to the Yang-Mills equations of general relativity*, in Global Analysis — Analysis on Manifolds, ed. G. Rassias, Teubner Press, Leipzig, 1983, pp. 324–349.

[Wa 5] B. Watson, *Almost contact metric 3-submersions*, Intl. J. Math. Math. Sci. **7** (1984) 667–688.

[W-V]   B. Watson and L. Vanhecke, *The structure equation of an almost semi-Kähler submersion*, Houston Math. J. **5** (1979) 295–305.

[Wo]   J. Wolf, *Differentiable fibre spaces and mappings compatible with Riemannian metrics*, Michigan Math. J. **11** (1964) 65–70.

[Y]   K. Yano, *Differential Geometry on Complex and Almost Complex Spaces*, Pergamon Press, Oxford, 1955.

[Y-I]   K. Yano and S. Ishihara, *Fibred spaces with invariant Riemannian metrics*, Kodai Math. Sem. Reps. **19** (1967) 317–360.

Bill Watson
Department of Mathematics
St. Johns's University
Jamaica, NY 11439
USA

THE MATH. HERITAGE OF C.F. GAUSS (pp. 862-871)
edited by George M. Rassias
©1991 World Scientific Publ. Co. Singapore

HYPERSURFACE WITH CONSTANT MEAN CURVATURE IN H^{n+1}

Wu Bao-Qiang

In this paper, we prove that a complete hypersurface with constant mean curvature and non-negative Ricci curvature in hyperbolic space form is umbilical.

It is well known that Nomizu, K. and Smyth, B. proved that [1] if M is a non-negatively curved hypersurface with constant mean curvature in the Euclidean space, or the Euclidean sphere, then it is the standard sphere or the product immersion of two spheres. In [2], S. T. Yau further proved that if M is a compact hypersurface with constant mean curvature and non-negative Ricci curvature in the Euclidean space, then M is umbilical. A natural question is that if the theorem holds for hyperbolic space form and even in a weaker topological restriction. In this paper, we shall answer the question and shall prove:

Theorem 1. A complete hypersurface with constant mean curvature and non-negative Ricci curvature in hyperbolic space form is umbilical.

Preliminary

Let H^{n+1} denote the connected hyperbolic space form with constant sectional curvature -1. Let M be its complete hypersurface with constant mean curvature and non-negative Ricci curvature.

We choose a local field of orthonormal frames $e_1,\ldots,e_{n+1}$ in H^{n+1} such that, restricted to M, the vectors $e_1,\ldots,e_n$ are tangent to M and e_{n+1} is normal to M. Unless stated otherwise, we use the following convention on the ranges of indices:

$$A,B,C,D = 1,\ldots,n+1$$
$$i,j,k,l,\ldots = 1,\ldots,n$$

and we shall agree that repeated indices are summed over the respective ranges. With respect to the frame field of H^{n+1} chosen above, let $\omega^1,\ldots,\omega^{n+1}$ be the field of dual frames.

First of all, we have [3]

$$K^A{}_{BCD} = -\delta_{AC}\delta_{BD} + \delta_{AD}\delta_{BC}. \tag{1}$$
$$R^i{}_{jkl} = -\delta_{ik}\delta_{jl} + \delta_{il}\delta_{jk} + h_{ik}h_{jl} - h_{il}h_{jk} \tag{2}$$

where $h_{ik}e_{n+1} = B(e_i,e_k)$, B is the second fundamental form with values in the normal bundle.

Equation (4) below and the hypothesis say constant mean curvature nH of M is not zero, without loss of generality, suppose

$$\sum h_{ii} = nH > 0. \tag{3}$$

In the case,

$$R^i{}_k = -(n-1)\delta_{ik} + nHh_{ik} - \sum h_{ij}h_{jk}. \tag{4}$$
$$R = \sum R^i{}_i = -n(n-1) + n^2H^2 - \|B\|^2 \tag{5}$$

where $\|B\|^2$ is the square of the length of the second fundamental form, i.e., $\|B\|^2 = \sum h_{ij}h_{ij}$.

The Codazzi equation shows

$$h_{ijk} = h_{ikj} \tag{6}$$

where, we define h_{ijk} and h_{ijkl} respectively by

$$\sum h_{ijl}\omega^l = d\,h_{ij} - \sum h_{il}\omega^l_j - \sum h_{lj}\omega^l_i. \tag{7}$$
$$\sum h_{ijkl}\omega^l = d\,h_{ijk} - \sum h_{ljk}\omega^l_i - \sum h_{ilk}\omega^l_j - \sum h_{ijl}\omega^l_k. \tag{8}$$

From (1), (2), (4), (5) and (6), we obtain [3], [4]

$$\Delta h_{ij} = nH\delta_{ij} - nh_{ij} + nH\sum h_{il}h_{lj} - \|B\|^2 h_{ij} \qquad (9)$$

where Δ is the Laplace operator.

Proof of Theorem 1.

Put $\cup M = \{(p,v) \in TM | \langle v,v \rangle = 1\}$, where TM is the tangent bundle over M and $\langle , \rangle$ the Riemannian interior product, and $\cup M_p = \{v \in T_pM | \langle v,v \rangle = 1\}$. We define $\varphi_p : \cup M_p \approx S^{n-1} \to \mathbb{R}$ as follows

$$\varphi_p(v) = \langle B(v,v), B(v,v) \rangle(p). \qquad (10)$$

Clearly, $\varphi_p \in C^\infty(\cup M_p, \mathbb{R})$. We define a function $\Phi : M \to \mathbb{R}^+ = \{a \in \mathbb{R} \mid a \geq 0\}$ by

$$\Phi(p) = \max_{v \in \cup M_p} \varphi(v). \qquad (11)$$

A point p is called a regular point if there exists a neighborhood U of p and a smooth section s in the unit tangent vector bundle $\cup U$ on U such that for any $q \in U$,

$$\Phi(q) = \langle B(\nu_q, \nu_q), B(\nu_q, \nu_q) \rangle$$

where $s|_q = \nu_q$. Otherwise, the point p is called singular point. Suppose

$$\Omega = \{p \in M | p \text{ is a regular point}\} ,$$
$$\Omega' = M \backslash \Omega . \qquad (12)$$

At a regular point p, we have

Lemma 1. There exist a coordinate neighborhood U of p and a locally orthonormal frame field on U such that for any $q \in U$, we have

$$\Phi(q) = h_{11}(q) > 0 \qquad (13)$$
$$h_{1i}(q) = 0 \ (i \geq 2) \qquad (14)$$
$$h_{11}(q) \geq h_{ii}(q) \quad (i \geq 2) \qquad (15)$$

especially at p,

$$h_{ij}(p) = 0 \quad (i \neq j) . \qquad (16)$$

Proof. We may choose a neighborhood U of p and an orthonormal frames field $\{e_1, \ldots, e_n\}$ such that e_1 is mentioned above, $\Phi(q) = h_{11}(q)$. For any $\nu_q = (\cos\theta\, e_1 + \sin\theta\, e_i)_q (i \geq 2)$, from (11), we get

$$0 = d_\theta \varphi_q(\nu_q)|_{\theta=0}$$
$$0 \geq d^2{}_{\theta\theta} \varphi_q(\nu_q)|_{\theta=0}\,.$$

A straightforward computation gives

$$h_{11}(q) h_{1i}(q) = 0\,. \tag{17}$$
$$h_{11}(q) \geq h_{ii}(q)\,. \tag{18}$$

(3), (11), (17) and (18) together imply (13) and (14).

From (13) and (14), we known that e_1 is a characteristic vector of the shape operator A. Furthermore, we can choose $\{e_2, \ldots, e_n\}$ such that they are also the characteristic vectors of the shape operator at p. Hence

$$A e_i = h_i e_i \quad (i = 1, \ldots, n) \tag{19}$$

i.e., (16) holds and for the sake of unity, we write $h_1 = \Phi(p)$.

The following computation is taken at p unless otherwise stated. Substituting (19) into (9),

$$\begin{aligned}
\Delta h_{11} &= nH - n h_1 + n H h_1^2 - h_1 \sum h_i^2 \\
&= \sum (h_i - h_1) + h_1^2 \sum h_i - h_1 \sum h_i^2 \\
&= \sum (h_1 - h_i)(h_1 h_i - 1)\,.
\end{aligned} \tag{20}$$

(19) and (3) give

$$\begin{aligned}
0 \leq R^i{}_i &= -(n-1) + h_i(nH - h_i) \\
&\leq -(n-1) + (n-1) h_i h_1 \\
&= (n-1)(h_1 h_i - 1)\,.
\end{aligned} \tag{21}$$

(20) and (21) together imply

$$\Delta h_{11}(p) \geq \frac{1}{n-1} \sum (h_1 - h_i) R^i{}_i \geq 0\,. \tag{22}$$

For distinguishing the difference between the Laplace of the components of the tensor B and one of the function Φ, we give the following lemma

Lemma 2.
$$\Delta\Phi(p) \geq \Delta h_{11}(p). \tag{23}$$

Proof. Since (8), we have

$$
\begin{aligned}
h_{11kl} &= h_{11k,l} - \sum h_{j1k}\Gamma_{1l}^j - \sum h_{1jk}\Gamma_{1l}^j - \sum h_{11j}\Gamma_{kl}^j \\
&= h_{11k,l} - \sum h_{11j}\Gamma_{kl}^j - 2\sum h_{1jk}\Gamma_{1l}^j
\end{aligned}
\tag{24}
$$

where $\omega_j^i = \sum \Gamma_{jk}^i \omega^k$, $dh_{ijk} = \sum h_{ijk,l}\omega^l$. Similarly, let $dh_{ij} = \sum h_{ij,l}\omega^l$. (7) and (14) give

$$\sum h_{11l}\omega^l = \sum h_{11,l}\omega^l, \quad \text{i.e.,} \quad h_{11l} = h_{11,l} \tag{25}$$

(7) and (14) together imply

$$
\begin{aligned}
h_{1jk} &= h_{1j,k} - \sum h_{1l}\Gamma_{jk}^l - \sum h_{1j}\Gamma_{1k}^l \\
&= \Gamma_{1k}^j(h_1 - h_j) \quad (j \geq 2).
\end{aligned}
\tag{26}
$$

Hence, by (24), (26), (25) and (15), we obtain

$$
\begin{aligned}
\Delta\Phi(p) &= \sum h_{11k,k} - \sum h_{11j}\Gamma_{kk}^j \\
&= \sum h_{11kk} + 2\sum h_{1jk}\Gamma_{1k}^j \\
&= \Delta h_{11}(p) + 2\sum (\Gamma_{1k}^j)^2(h_1 - h_j) \geq \Delta h_{11}(p).
\end{aligned}
\tag{27}
$$

As to singular point, we have [5].

Lemma 3. The n-dim measure of Ω' is zero and $\Omega' \cap D$ is not dense in D if D is any connected domain in M.

If $\Phi|_\Omega = \text{const}$, from the Lemma 3 above and continuity of Φ, we know that $\Phi \equiv \text{const}$, then (21), (27) and (22) become equalities. Moreover, (21) shows that $h_i \equiv \text{const} = \Phi$. Hence M is umbilical.

If $\Phi|_\Omega \neq \text{const}$, we suppose that there is $p \in \Omega$ such that $\Phi(p) = A > 0$ and $A < \sup \Phi$. We define a function $\mathscr{G}_k(x)$ as follows

$$\mathscr{G}_k(x) = \frac{\Phi(x)}{(\rho(x) + 2)^k} \quad (0 < k < 1), \tag{28}$$

where $\rho(x)$ denote the geodesic distance between p and x. Since $\mathscr{G}_k(p) = 2^{-k} A > 0, \lim_{\rho(x) \to \infty} \mathscr{G}_k(x) = 0$, there will exist $x_k \in M$ such that $\mathscr{G}_k(x_k) = \max \mathscr{G}_k(x)$. Whether $x_k \in \Omega$ or $x_k \in \Omega'$, from Lemma 3, there exists a small open set $U_k \in \Omega$ such that $x_k \in \partial \bar{U}_k$. Take a small enough neighborhood U_k' of x_k and one V_k' of p such that $V_k' \subset \Omega$ and $U_k' \cap V_k' = \phi$. Then a smooth vector field e_1 on U_k can be smoothly extended to U_k'. We define

$$\widetilde{\Phi}_k(x) = \begin{cases} h_{11}(x) & x \in U_k', \\ \Phi(x) & x \in V_k', \\ 0 & x \bar{\in} U_k' \cap V_k'. \end{cases} \tag{29}$$

(29), (11) and continuity of Φ give

$$\widetilde{\Phi}_k(x) \leq \Phi(x), \quad \widetilde{\Phi}_k|_{\bar{U}_k \cap U_k'} = \Phi|_{\bar{U}_k \cap U_k'},$$
$$\widetilde{\Phi}_k|_{V_k'} = \Phi|_{V_k'}. \tag{30}$$

In addition,

$$\widetilde{\Phi}_k(x_k) = \Phi(x_k), \quad \widetilde{\Phi}_k(p) = \Phi(p). \tag{31}$$

Constricting U_k' amd V_k' slightly, we get U_k'' and V_k'' respectively. In view of the standard theorem on intercept functions, we can define a smooth function $h_k : M \to \mathbb{R}$ as follows

$$h_k(x) = \begin{cases} 1, & x \in U_k'' \cup V_k'', \\ 0, & x \bar{\in} U_k' \cup V_k'. \end{cases} \tag{32}$$

By (32), (29) and (30), we define $\Phi_k' : M \to \mathbb{R}$

$$\Phi_k'(x) = \widetilde{\Phi}_k(x) \cdot h_k(x) \leq \widetilde{\Phi}_k(x) \leq \Phi(x). \tag{33}$$

Because of (33), (31) and (32), we obtain

$$\Phi_k'(x_k) = \Phi(x_k), \quad \Phi_k'(p) = \Phi(p) \tag{34}$$

and Φ'_k is a smooth function on M. We define

$$\mathscr{G}'_k(x) = \frac{\Phi'_k(x)}{(\rho(x)+2)^k} \le \mathscr{G}_k(x)\,. \tag{35}$$

$\mathscr{G}'_k(x)$ can achieve its maximum at some point as $\mathscr{G}_k(x)$. (34), (35) and (32) together imply

$$\mathscr{G}'_k(x_k) = \mathscr{G}_k(x_k) = \max \mathscr{G}_k(x) \ge \max \mathscr{G}'_k(x) \tag{36}$$

hence $\mathscr{G}'_k(x_k) = \max \mathscr{G}'_k(x)$. A direct computation shows that at x_k,

$$\frac{\nabla \Phi'_k}{(\rho+2)^k} - \frac{k\Phi'_k \nabla\rho}{(\rho+2)^{k+1}} = 0 \tag{37}$$

$$\frac{\Delta \Phi'_k}{(\rho+2)^k} - \frac{2k\nabla\Phi'_k \cdot \nabla\rho}{(\rho+2)^{k+1}} - \frac{k\Phi'_k \cdot \Delta\rho}{(\rho+2)^{k+1}} + \frac{k(k+1)\Phi'_k}{(\rho+2)^{k+2}} \le 0\,. \tag{38}$$

From (37) and (38), noting $0 < k < 1$, we obtain

$$\Delta\Phi'_k \le \frac{2k\nabla\Phi'_k \cdot \nabla\rho}{\rho+2} + \frac{k\Phi'_k \cdot \Delta\rho}{\rho+2} - \frac{k(k+1)\Phi'_k}{(\rho+2)^2}$$

$$= \frac{k(k-1)\Phi'_k}{(\rho+2)^2} + \frac{k\Phi'_k \Delta\rho}{\rho+2} \tag{39}$$

$$< \frac{k\Phi'_k \cdot \Delta\rho}{\rho+2}\,.$$

As to $\Delta\rho$, because, the distance function ρ is almost differentiable everywhere and a Lipschitz function, hence in the sense of distribution we have [7]

Lemma 4. Let M be a complete Riemannian manifold with negative Ricci curvature, then

$$\Delta\rho \le (n-1)\rho^{-1}\,. \tag{40}$$

Combining (39) with (40), we get

$$\Delta\Phi'_k < \frac{k(n-1)\Phi'_k}{\rho(\rho+2)}\,. \tag{41}$$

On the other hand, let us now prove

Lemma 5.

$$\sup \Phi = \lim_{k \to 0} \Phi'_k(x_k) . \tag{42}$$

Proof. If this were not true, there would exist $\delta > 0$ and $x \in M$ such that

$$\Phi(x) > \Phi'_k(x_k) + \delta \tag{43}$$

for k small enough.

If $\rho(x_k) \to \infty$ when $k \to 0$, we have

$$\mathscr{G}_k(x) = \frac{\Phi(x)}{(\rho(x) + 2)^k} > \frac{\Phi'_k(x_k)}{(\rho(x_k) + 2)^k} = \mathscr{G}'_k(x_k)$$

when $\rho(x_k) > \rho(x)$. Noting (36), this contradicts the definition of x_k.

If, for some subsequence of k, $\{x_k\}$ converges to a point x_0, then we find

$$\lim_{k \to 0} [\rho(x) + 2]^k = \lim_{k \to 0} [\rho(x_k) + 2]^k = \lim_{k \to 0} [\rho(x_0) + 2]^k = 1 .$$

Since (43), it is clear that when k is small enough, $\mathscr{G}_k(x) > \mathscr{G}'_k(x_k) = \mathscr{G}_k(x_k)$, which is again a contradiction.

From Lemma 3, for any k we can always choose a sequence $\{x_{kl}\}$ such that $x_{kl} \in U''_k \cap \Omega$ and $\lim_{l \to \infty} x_{kl} = x_k$. Because $\Phi'_k \in C^\infty(M)$, (29), (42), (23) and (22) together give

$$\begin{aligned}
\Delta \Phi'_k(x_k) &= \lim_{l \to \infty} \Delta \Phi'_k(x_{kl}) \\
&= \lim_{l \to \infty} \Delta \Phi(x_{kl}) \\
&\geq \frac{1}{n-1} \lim_{l \to \infty} \sum (h_1 - h_i) R^i{}_i(x_{kl}) \\
&= \frac{1}{n-1} \sum (h_1 - h_i) R^i{}_i(x_k) \geq 0 .
\end{aligned} \tag{44}$$

If $\rho(x_k) \to \infty$ when $k \to 0$, then (42), (44) and (41) imply

$$\lim_{k \to 0} \Delta \Phi'_k(x_k) = \lim_{k \to 0} \sum (h_1 - h_i) R^i{}_i(x_k) = 0$$

and (44) (hence (21)) become equalities. Therefore, we have

$$\lim_{k \to 0} h_i(x_k) = \lim_{k \to 0} h_1(x_k) = \lim_{k \to 0} \Phi'_k(x_k) = \sup \Phi \, .$$

Namely, $H = \sup \Phi$, so M is umbilical.

If, for some subsequence of k, $\{x_k\}$ converges to point x_0, then, for the subsequence of k, from (41) and (44) again

$$\begin{aligned}
\lim_{k \to 0} \frac{k(n-1)\Phi'_k(x_k)}{\rho(\rho+2)} &\geq \lim_{k \to 0} \Delta \Phi'(x_k) \\
&\geq \frac{1}{n-1} \lim_{k \to 0} \sum (h_1 - h_i) R^i{}_i(x_k) \geq 0 \, .
\end{aligned}$$
(45)

Because ρ is a Lipschitz function and (42) holds for the subsequence of k, then

$$0 < \frac{1}{2}(\sup \Phi - A) < |\Phi(x_k) - |\Phi(p)| \leq L \cdot \rho(p, x_k)$$
(46)

for some k small enough, where $L = \text{const} > 0$ from (45) and (46), we claim that M is umbilical by same reason as above. Hence Theorem 1 holds.

Remark 1. Because there are only two types of umbilical hypersurfaces in hyperbolic space with non-negative sectional curvature, sphere and horosphere [8], then Theorem 1 may be rewritten.

Theorem 1'. A complete hypersurface of hyperbolic space form with constant mean curvature and non-negative Ricci curvature is a sphere or horosphere.

Corollary. A compact (complete and non-compact) hypersurface of hyperbolic space form with constant mean curvature and non-negative Ricci curvature is a sphere (horosphere).

Remark 2. By an extrinsic sphere, we mean a totally umbilical submanifold with nonzero parallel mean curvature vector [9], hence we have

Theorem 1''. A complete hypersurface of hyperbolic space form with constant mean curvature and non-negative Ricci curvature is an extrinsic sphere.

As for minimal hypersurface, in the weaker condition on the Ricci curvature, we have

Theorem 2. A minimal hypersurface of hyperbolic space form with the Ricci curvature, which is not less than $-(n-1)$, is totally geodesic.

It is easy from (4).

References

1. K. Nomizu, and B. Smyth, *A formula of Simons' type and hypersurfaces of constant mean curvature*, J. Diff. Geom. **3** (1969) 367–377.
2. S. T. Yau, *Submanifolds with constant mean curvature* (II), Amer. Jour. Math. **97** (1975) 76–100.
3. S. S. Chern, M. Do Carmo and S. Kobayashi, *Minimal submanifold of a sphere with second fundamental form of constant length*, Functional Analysis and Related Fields, Springer (1970) 59–75.
4. B. Y. Chen, *Geometry of Submanifolds*, Marcel Dekker Inc., 1973.
5. Wu B-Q, *Complete totally real submanifolds of CP^n with parallel mean curvature vector*, Acta Math. Sinica, to appear.
6. S. T. Yau, *Harmonic functions on complete Riemannian manifolds*, Comm. Pure. Appl. Math. **28** (1975) 201–228.
7. S. T. Yau and R. Schoen, *Diffferential Geometry*, Scientific Publishing House, China, 1988.
8. T. E. Cecil and P. J. Ryan, *Tight and Tant Immersions of Manifolds*, Pitman Advanced Publishing Program, 1985.
9. B. Y. Chen, *Geometry of Submanifolds and its Applications*, Science Univ. of Tokyo, 1981.

Wu Bao-Qiang
Xuzhou Teachers College
221009 Xuzhou
People's Republic of China

THE MATH. HERITAGE OF C.F. GAUSS (pp. 872-881)
edited by George M. Rassias
©1991 World Scientific Publ. Co. Singapore

THE POINCARÉ DENSITY

Shinji Yamashita

As is well known, the notion of non-Euclidean geometry goes back to Gauss, and a systematic study of its metric was done by Poincaré using a model in the unit disk. Let f be holomorphic and locally univalent in a domain Ω in the complex plane with at least two boundary points. Let ρ_Ω be the density of the Poincaré metric $\rho_\Omega(z)|dz|$ in Ω. Then, f is univalent in Ω with convex image $f(\Omega)$ if and only if $\rho_\Omega^{-1}|f'|$ is superharmonic in Ω. In particular, Ω is convex if and only if ρ_Ω^{-1} is superharmonic in Ω. Suppose that Ω is simply connected. Is there a point $z \in \Omega$ such that $\rho(z) < \rho(w)$ for all $w \in \Omega \backslash \{z\}$? Roughly speaking, the answer is "yes" if Ω is close to "bounded" and close to "convex" at the same time.

1. Introduction

A subdomain Ω of the complex plane $\mathbb{C} = \{|z| < +\infty\}$ with at least two boundary points has the Poincaré metric $\rho_\Omega(z)|dz|$. The density function ρ_Ω is defined in Ω so that

$$(\rho_\Omega \circ \varphi)(z)(1 - |z|^2)|\varphi'(z)| = 1, \quad z \in D,$$

for a universal covering projection φ from $D = \{|z| < 1\}$ onto Ω. The definition is independent of the specified choice of φ. In particular, $\rho_D^{-1}(z)$ $(\equiv 1/\rho_D(z)) = 1 - |z|^2$; see [1, Chapter 1].

Let $LS(\Omega)$ be the family of functions f holomorphic with f' never vanishing in Ω. Let $S(\Omega)$ be the family of f holomorphic and univalent in Ω, and let $CS(\Omega)$ be the family of $f \in S(\Omega)$ with $f(\Omega)$ convex. Then, $CS(\Omega)$ is nonempty if and only if Ω is simply connected.

Theorem 1. Let $f \in LS(\Omega)$. Then $f \in CS(\Omega)$ if and only if $\rho_\Omega^{-1}|f'|$

is superharmonic in Ω.

Set $f_0(z) \equiv z$. Then, Ω is convex if and only if $f_0 \in \mathrm{CS}\,(\Omega)$. Therefore, a domain Ω is convex if and only if ρ_Ω^{-1} is superharmonic in Ω. Thus, an analytic property of ρ_Ω is equivalent to a topological property of Ω. Recall that the Gauss curvature identity $\rho_\Omega^{-2}\Delta \log \rho_\Omega \equiv 4$ shows the superharmonicity of $-\log \rho_\Omega$ in a general Ω. If Ω is convex, then $\rho_\Omega^{-1} = \exp(-\log \rho_\Omega)$ is superharmonic, further.

Suppose that $\Omega \neq \mathbb{C}$ is a simply connected domain. Our next subject is to find $z_0 \in \Omega$ such that

$$\rho_\Omega(z_0) < \rho_\Omega(z) \quad \text{for each} \quad z \in \Omega\backslash\{z_0\}. \tag{1.1}$$

If z_0 exists, then it is unique. There are examples of Ω for which (1.1) is false for each $z_0 \in \Omega$; see U and B in (IV) in Sec. 3. To set a further condition we write for $f \in \mathrm{LS}\,(\Omega)$,

$$\lambda(f) = f''/f' \quad \text{and} \quad \sigma(f) = \lambda(f)' - 2^{-1}\lambda(f)^2\,.$$

Let Ω_1 and Ω_2 be simply connected, proper subdomains of $\mathbb{C}$ and let $C(\Omega_1,\Omega_2)$ be the family of conformal homeomorphisms from Ω_1 onto Ω_2. The quantity

$$\delta(\Omega) = \sup_{z \in \Omega} \rho_\Omega(z)^{-2}|\sigma(f)(z)|\,,$$

where $f \in C(\Omega, D)$, is called the distance of Ω from a disk; $\delta(\Omega)$ is independent of the particular choice of f; see [4, p. 61]. It is easy to observe that [4, (1.11), p. 55]

$$\delta(\Omega) = \sup_{z \in D}(1 - |z|^2)^2|\sigma(f)(z)| \quad \text{for each} \quad f \in C(D,\Omega)\,.$$

We call Γ_z a geodesic path starting at $z \in \Omega$ if there exists $f \in C(D,\Omega)$ such that $\Gamma_z = f(L)$ and $f(0) = z$, where $L = \{0 \leq x < 1\}$ is the radius of D. For each pair $z, w \in \Omega, z \neq w$. We have one and only one Γ_z with $w \in \Gamma_z$. Let $\partial\Omega$ be the boundary of Ω in $\mathbb{C}$. We symbolically write

$$\lim_{z \to \partial\Omega} \rho_\Omega(z) = +\infty\,,$$

if, given $\varepsilon > 0$ we may find a bounded and closed set $K \subset \Omega$ such that $\rho_\Omega(z) > \varepsilon$ for each $z \in \Omega \backslash K$.

Theorem 2. Let $\Omega \neq \mathbb{C}$ be a simply connected domain in $\mathbb{C}$. Suppose that

$$\delta(\Omega) \leq 2 \tag{1.2}$$

and

$$\lim_{z \to \partial\Omega} \rho_\Omega(z) = +\infty. \tag{1.3}$$

Then (1.1) holds. More precisely, there exists $z_0 \in \Omega$ such that $\rho_\Omega(z)$ strictly increasingly tends to $+\infty$ as $z \to \partial\Omega$ along each Γ_{z_0}.

The last sentence means that, for f in the definition of $\Gamma_{z_0}, \rho_\Omega(f(x))$ is a strictly increasing function of $x \in L$; (1.3) shows that $\rho_\Omega(f(x)) \to +\infty$ as $x \to 1$.

2. Proof of Theorem 1

Only if.

Since Ω is simply connected, $\varphi \in C(D, \Omega)$. Then $g = f \circ \varphi \in \mathrm{CS}\,(D)$, so that, for each fixed $z \in D$, the following function of $\zeta \in D$ with the Taylor expansion

$$\frac{g\left(\frac{\zeta+z}{1+\bar{z}\zeta}\right) - g(z)}{(1 - |z|^2)g'(z)} = \zeta + a(g, z)\zeta^2 + \dots$$

is a member of $\mathrm{CS}(D)$, where

$$a(g, z) = -\bar{z} + 2^{-1}(1 - |z|^2)\lambda(g)(z), \quad z \in D. \tag{2.1}$$

The coefficient theorem [2, Corollary, p. 45] now yields that

$$|a(g, z)| \leq 1, \quad z \in D. \tag{2.2}$$

Setting $\Phi(z) = (1 - |z|^2)|g'(z)|, z \in D$, taking the logarithms of both sides, and partially differentiating them by z $(\partial/\partial z = 2^{-1}(\partial/\partial x - i\partial/\partial y))$ we have

$$(\Phi_z/\Phi)(z) = (1 - |z|^2)^{-1} a(g, z), \quad z \in D. \qquad (2.3)$$

Partially differentiating both sides of (2.3) by $\bar{z}$ ($\partial/\partial\bar{z} = 2^{-1}(\partial/\partial x + i\partial/\partial y)$), we further have

$$(\Phi\Phi_{z\bar{z}} - \Phi_z\Phi_{\bar{z}})\Phi^{-2} = -(1 - |z|^2)^{-2},$$

which, together with

$$\Delta\Phi = 4\Phi_{z\bar{z}} \quad \text{and} \quad |\Phi_z|^2 = \Phi_z\Phi_{\bar{z}},$$

shows that

$$4^{-1}(1 - |z|^2)^2(\Phi^{-1}\Delta\Phi)(z) = (1 - |z|^2)^2(|\Phi_z|/\Phi)^2(z) - 1, \, z \in D. \qquad (2.4)$$

It follows from (2.2), (2.3) and (2.4) that $\Delta\Phi \le 0$, or Φ is superharmonic in D. Letting γ be the inverse map of φ we now observe that $\rho_\Omega^{-1}|f'| = \Phi \circ \gamma$ is superharmonic in Ω.

If.

For a moment we dot not know that $\varphi : D \to \Omega$ is univalent. We set $h = f \circ \varphi$ and $H(z) = (1 - |z|^2)|h'(z)|, z \in D$. Then, $H = (\rho_\Omega^{-1} \cdot |f'|) \circ \varphi$ is superharmonic in D, whence $\Delta H \le 0$. We can replace g and Φ by h and H, respectively, in the displays (2.3) and (2.4); the resultings are denoted by $(2.k')$, $k = 3, 4$. Then, $(2.4')$ and $(2.3')$ show that $|a(h, z)| \le 1, z \in D$, or,

$$|-2|z|^2/(1 - |z|^2) + z\lambda(h)(z)| \le 2|z|/(1 - |z|^2),$$

whence

$$1 + \mathrm{Re}\,\{z\lambda(h)(z)\} \ge (1 - |z|)^2/(1 - |z|^2) > 0 \quad \text{for} \quad z \in D.$$

It then follows from the familiar criterion [2, Theorem 2.11, p. 42] that $h \in \mathrm{CS}(D)$. Now, if $w_1 \ne w_2$ in Ω, then we have $z_1 \ne z_2$ such that $w_k = \varphi(z_k), k = 1, 2$. Therefore, $f(w_1) = h(z_1) \ne h(z_2) = f(w_2)$. We have thus proved that $f \in S(\Omega)$ and $f(\Omega) = h(D)$ is convex.

3. Preliminary Remarks for Theorem 2

Some remarks will be given on the conditions (1.2) and (1.3) in Theorem 2.

(I) The condition (1.2) holds if and only if

$$\sup_{z \in D}(1 - |z|^2)^2 |\sigma(f)(z)| \leq 2 \qquad (1.2^*)$$

for a (hence for each) $f \in C(D, \Omega)$. F. W. Gehring and C. Pommerenke [3, Theorem 1] showed that if (1.2^*) is true, then $\partial\Omega$ itself or $\partial\Omega \cup \{\infty\}$ is a Jordan curve in $\mathbb{C}\cup\{\infty\}$ or Ω is the image of the band $B = \{-\pi/2 < \mathrm{Im}\, z < \pi/2\}$ by a Möbius transformation (or simply saying: Ω is a Möbius image of B).

(II) If Ω is the Möbius image of a convex domain, then (1.2) holds [4, Theorem 2.1, p. 63].

(III) The examples of nonconvex Ω's for which (1.2) is valid are the following. (III-1) The Möbius image $\{1/(z - \pi i/2); z \in B\}$ of B. (III-2) Given a number $k, 0 < k \leq 1/3$, we consider $f_n(z) = (z^{-n} + kz^n)^{-1/n} = z(1+kz^{2n})^{-1/n}$ in D, where n is a natural number. The function $F_n(z)$, defined on $\mathbb{C}\cup\{\infty\}$ by $F_n(z) = f_n(z)$ if $z \in D$ and $F_n(z) = [z^{-n}+k(\bar{z})^{-n}]^{-1/n}$ if $z \in (\mathbb{C} \setminus D) \cup \{\infty\}$, is a quasiconformal mapping of $\mathbb{C}\cup\{\infty\}$ onto itself. A computation shows that the complex dilatation $\mu_n(z)$ is 0 in D and $k(\bar{z}/z)^{-n-1}$ in $\mathbb{C}\setminus\overline{D}$, so that, for $\Omega_n = f_n(D)$, we have $\delta(\Omega_n) \leq 6\|\mu_n\|_\infty = 6k \leq 2$; see [4, Theorem 3.2, p. 72 and pp. 74–75]. For $n \geq 2$, the boundary $\partial\Omega_n$ waves $2n$ times between the circles $|z| = (1 \pm k)^{-1/n}$, so that Ω_n looks like a (monstrous) starfish with $2n$ legs. In particular, Ω_n is not convex for $n \geq 2$.

(IV) There are convex Ω's for which (1.3) is false. For the half plane $U = \{\mathrm{Im}\, z > 0\}$ we observe that $\rho_U(x + iy) = 1/(2y) \to 0$ as $y \to +\infty$. For the band B we have $\rho_B(z) = \{2\cos(\mathrm{Im}\, z)\}^{-1}$, so that ρ_B is constant on each line parallel to the real axis. In particular, we cannot drop condition (1.3) in Theorem 2.

(V) If the area of Ω is finite further, then (1.3) holds. Actually, for $f \in C(D, \Omega)$ and for $1/2 < |\zeta| < 1, z = f(\zeta)$, the subharmonicity of $|f'|^2$ in D

yields

$$(1 - |\zeta|^2)^{-2} \rho_\Omega(z)^{-2} = |f'(\zeta)|^2$$

$$\leq \{\pi(1 - |\zeta|)^2\}^{-1} \iint_{|w-\zeta|<1-|\zeta|} |f'(w)|^2 du dv , (w = u + iv).$$

Thus,

$$\rho_\Omega(z)^{-2} \leq 4\pi^{-1} \iint_{2|\zeta|-1<|w|<1} |f'(w)|^2 du dv .$$

The right-hand side tends to zero as $|\zeta| \to 1$, or $z \to \partial\Omega$.

(VI) If Ω is bounded and convex, then this is the case in (V), so that Theorem 2 is applicable to Ω. The same conclusion is true of Ω_n for each n in (III). Note that a convex domain in $\mathbb{C}$ is of finite area if and only if it is bounded.

(VII) An elementary but tedious computation starting from

$$\rho(z) = \rho_\Omega(z) = |(1 - |\zeta|^2)f'(\zeta)|^{-1} , z = f(\zeta), f \in C(D,\Omega), \qquad (3.1)$$

together with (1.2^*), shows that (1.2) is equivalent to

$$\delta(\Omega) = 2 \sup_{z \in \Omega} |\rho(z)^{-3}\rho_{zz}(z) - 2\rho(z)^{-4}\rho_z(z)^2| \leq 2 .$$

Therefore, (1.2) is actually the restriction in terms of ρ_Ω.

4. Proof of Theorem 2

We write $\rho = \rho_\Omega$. It follows from (1.3), together with the continuity of ρ, that there exists $z_0 \in \Omega$ such that $\rho(z_0) \leq \rho(z)$ for all $z \in \Omega$. There exists $f \in C(D,\Omega)$ such that $f(0) = z_0$ and $\Gamma_{z_0} = f(L)$. Then, (3.1) holds. Set $\zeta = \beta(w) = (e^w - 1)/(e^w + 1), w \in B$, and consider the function

$$F(w) = |(1 - \beta(w)^2)f'(\beta(w))|^{-1} = |(1 - \zeta^2)f'(\zeta)|^{-1} , w \in B . \qquad (4.1)$$

Then $F(u)$ is of C^∞ for the real variable u, and further if $0 \leq u < +\infty$, then $F(u) = \rho(f(\xi)), \xi = \beta(u) \in L$. By the property of ρ, $F(0) \leq F(u), 0 \leq u < +\infty$, so that $F_u(0) \geq 0$ must follow. Taking the logarithms of both sides of (4.1), then partially differentiating them by w and paying attention to $2\beta'(w) = 1 - \zeta^2$, we have

$$F_w(w)/F(w) = 2^{-1}\zeta - 4^{-1}(1 - \zeta^2)\lambda(f)(\zeta). \tag{4.2}$$

Further partial differentiations to (4.2) yield

$$F_{w\bar{w}}(w)/F(w) - (|F_w(w)|/F(w))^2 = 0 \tag{4.3}$$

and

$$F_{ww}(w)/F(w) - (F_w(w)/F(w))^2$$
$$= 8^{-1}\{2-(1 - \zeta^2)^2\sigma(f)(\zeta)\} - (F_w(w)/F(w))^2, \ w \in B. \tag{4.4}$$

It then follows from (4.3) and (4.4) that

$$4F_{uu}(w)/F(w) = 8(F_{w\bar{w}}(w)/F(w)) + 8\,\mathrm{Re}\,(F_{ww}(w)/F(w))$$
$$= 8(|F_w(w)|/F(w))^2 + (2 - \mathrm{Re}\,\{(1 - \zeta^2)^2\sigma(f)(\zeta)\}), w \in B. \tag{4.5}$$

It then follows from (4.5) and (1.2*) that $F_{uu}(u) \geqq 0$ for $0 \leqq u < +\infty$, so that $F_u(u)$ is nondecreasing there. Since $F_u(0) \geqq 0$, it follows that $F_u(u) \geqq 0$ for $0 \leqq u < +\infty$. Suppose that there exists $u_0, 0 < u_0 < +\infty$, such that $F_u(u_0) = 0$. Then $F_u(u) \equiv 0$ in $\Sigma = \{0 \leqq u < u_0\}$, so that $F_{uu}(u) \equiv 0$ in Σ. It then follows from (4.5) that $F_w(u) \equiv 0$ on $\Sigma \subset B$, so that (4.2) shows that the holomorphic function $\lambda(f)(\zeta) = 2\zeta/(1 - \zeta^2)$ on $\beta(\Sigma)$. The unicity theorem now shows that $\lambda(f)(\zeta) \equiv 2\zeta/(1 - \zeta^2)$ in the whole D. Then there exist $a \neq 0$ and b such that $f(\zeta) = a \log\{(1 + \zeta)/(1 - \zeta)\} + b, \zeta \in D$. Thus Ω is mapped onto B by the obvious manner and we easily see that (1.3) is false (see (IV) and (5.1) in Sec. 5). We thus have $F_u(u) > 0$ for $0 < u < +\infty$, so that $F(u)$ is strictly increasing for $0 \leqq u < +\infty$. This completes the proof of Theorem 2.

If we assume (1.2) only, then a local strict minimum of ρ_Ω is its global strict minimum in Ω. More precisely, (1.1) holds if (1.3) in Theorem 3 is replaced by the following:

There exist $z_0 \in \Omega$ and $\epsilon > 0$ such that $\rho_\Omega(z_0) < \rho_\Omega(z)$ for each $z \in \{0 < |z - z_0| < \epsilon\} \subset \Omega$. $\tag{1.3*}$

The proof is the same.

5. How to Find z_0

The proof of Theorem 2 suggests a hint to find z_0 of (1.1) under conditions (1.2) and (1.3). Set

$$G(\zeta) = \rho_\Omega(z)^{-1} = (1 - |\zeta|^2)|g'(\zeta)|\,,$$

where $g \in C(D, \Omega)$, $z = g(\zeta)$ and $\zeta \in D$. To find the candidates of the point where ρ_Ω assumes the minimum is nothing but to find ζ with $G_\zeta(\zeta) = 0$ because this is equivalent to $G_\xi(\zeta) = G_\eta(\zeta) = 0, \zeta = \xi + i\eta$. We show that the equation $G_\zeta(\zeta) = 0$ has the only one root ζ_0 for which $z_0 = g(\zeta_0)$. Suppose that there are $\zeta_1 \neq \zeta_2$ for which $G_\zeta(\zeta_k) = 0, k = 1, 2$. Let $T \in C(D, D)$ satisfy $T(0) = \zeta_1$ and $T(r_2) = \zeta_2, 0 < r_2 < 1$. We then consider F of (4.1) for $f = g \circ T \in C(D, \Omega)$. It then follows that, for $-\infty < u < +\infty, F(u) = V(\beta(u))^{-1}, V = G \circ T$, so that

$$F_u(u) = (d/du)F(u) = -V(\beta(u))^{-2}Q(u)\,,$$

where

$$Q(u) = \beta'(u)\{G_\xi(\psi(u))\mathrm{Re}\, T'(\beta(u)) + G_\eta(\psi(u))\mathrm{Im}\, T'(\beta(u))\}$$

with $\psi = T \circ \varphi$. We obtain $0 < u_2 < +\infty$ such that $r_2 = \beta(u_2)$, so that $F_u(0) = F_u(u_2) = 0$. We can apply the argument in the proof of Theorem 2 to arrive at the contradiction that (1.3) is violated. Therefore, $z_0 = g(\zeta_0)$, where ζ_0 is the only one root of $G_\zeta(\zeta) = 0$, or equivalently, the root of

$$G_\zeta(\zeta)/G(\zeta) = -\bar\zeta/(1 - |\zeta|^2) + 2^{-1}\lambda(g)(\zeta) = 0\,.$$

Since

$$\rho_{\Omega_1}(z) = |f'(z)|\rho_{\Omega_2}(f(z)) \tag{5.1}$$

for $f \in C(\Omega_1, \Omega_2), z \in \Omega_1$, it is easy to find z_0 for some specified Ω under conditions (1.2) and (1.3).

We suppose that Ω is symmetric with respect to a straight line Λ. Then, z_0 must lie on $\Lambda \cap \Omega$. Actually we have only to search z_0 on $\Lambda \cap \Omega$ such that $\rho_\Omega(z_0) \leq \rho_\Omega(z)$ for all $z \in \Lambda \cap \Omega$. For the proof we may assume, without loss of generality, that Λ is the real axis. To prove that

$$\rho_\Omega(z) = \rho_\Omega(\bar z) \quad \text{for each} \quad z \in \Omega, \tag{5.2}$$

we choose $f \in C(D, \Omega)$ with $z = f(\zeta), \zeta \in D$, fixed. Set $g(w) = \overline{f(\bar{w})}, w \in D$. Then, $g \in C(D, \Omega), \bar{z} = g(\bar{\zeta})$ and $g'(w) = \overline{f'(\bar{w})}, w \in D$. It now follows from (5.1) that

$$\rho_\Omega(z) = |f'(\zeta)|^{-1}\rho_D(\zeta) = |g'(\bar{\zeta})|^{-1}\rho_D(\bar{\zeta}) = \rho_\Omega(\bar{z}).$$

Suppose that z_0 is not in $\Lambda \cap \Omega$. Then, (5.2) for $z = z_0$ yields a contradiction to (1.1).

Suppose that there exists a real Θ such that $e^{i\Theta} \neq 1$ and $\Omega = \{e^{i\Theta}z; z \in \Omega\}$. It then follows from (5.1) that $\rho_\Omega(e^{i\Theta}z) = \rho_\Omega(z)$ for all $z \in \Omega$. Therefore $z_0 = 0$ is the requested. As a specified case we can choose $\Theta = \pi/n$ for Ω_n in (II) in Sec. 3, so that 0 is the requested for $\Omega_n (n \geq 1)$.

If Ω is symmetric with respect to a point z_1, then Ω has z_1 as its center of gravity, and, of course, $z_0 = z_1$. The situation is different if Ω is not point-symmetric. For example, we consider the sector

$$S(\alpha) = \{z; 0 < |z| < 1 \quad \text{and} \quad |\arg z| < \alpha\}$$

for $0 < \alpha < \pi$. With the aid of (5.1) and $f \in C(S(\alpha), U)$, where

$$f(z) = \left(\frac{1 + iz^\beta}{1 - iz^\beta}\right)^2, \quad \beta = \pi/(2\alpha),$$

we have for $0 < x < 1$,

$$\rho_{S(\alpha)}(x) = |f'(x)|/\{2\operatorname{Im} f(x)\} = \left(\frac{\beta}{2}\right)\frac{1 + x^{2\beta}}{x\{1 - x^{2\beta}\}},$$

which assumes the minimum at

$$x_0(\alpha) = (\{1 + (\pi/\alpha)^2\}^{1/2} - (\pi/\alpha))^{\alpha/\pi}.$$

Therefore, $z_0 = x_0(\alpha)$ for $S(\alpha)$. On the other hand, the center of gravity of $S(\alpha)$ is

$$x_G(\alpha) = \frac{2\sin\alpha}{3\alpha}.$$

Thus, $x_0(\alpha) \neq x_G(\alpha)$ for $\alpha = \pi/2, \pi/3$, etc. It would be of interest to observe that

$$x_0(\alpha) \to \sqrt{2} - 1 \quad \text{and} \quad x_G(\alpha) \to 0 \quad \text{as} \quad \alpha \to \pi,$$

while

$$x_0(\alpha) \to 1 \quad \text{and} \quad x_G(\alpha) \to 2/3 \quad \text{as} \quad \alpha \to 0.$$

Since $x_0(\pi/n) > x_G(\pi/n)$ for $n = 2, 5$ and $x_0(\pi/n) < x_G(\pi/n)$ for $n = 3, 4$, it follows that there are two values α_1 and α_2 such that

$$x_0(\alpha_k) = x_G(\alpha_k), \quad k = 1, 2, \quad \text{and} \quad \pi/5 < \alpha_1 < \pi/4, \quad \pi/3 < \alpha_2 < \pi/2.$$

References

1. L. V. Ahlfors, *Conformal Invariants; Topics in Geometric Function Theory*, McGraw-Hill, New York, 1973.
2. P. L. Duren, *Univalent Functions*, Springer, New York, 1983.
3. F. W. Gehring and C. Pommerenke, *On the Nehari univalence criterion and quasicircles*, Comm. Math. Helv. **59** (1984) 226–242.
4. O. Lehto, *Univalent Functions and Teichmüller Spaces*, Springer, New York, 1987.

Shinji Yamashita
Department of Mathematics
Tokyo Metropolitan University
Fukasawa, Setagaya, Tokyo 158
Japan

THE MATH. HERITAGE OF C.F. GAUSS (pp. 882-886)
edited by George M. Rassias
©1991 World Scientific Publ. Co. Singapore

SOPHIE GERMAIN PRIMES

Samuel Yates

Sophie Germain was born in Paris, France, in 1776: the year that the United States of American declared its independence from the British monarchy, one year before the birth of Gauss, and thirteen years before the storming of the Bastille. This paper discusses her interest in Gauss and in his work in number theory; her own worthwhile contributions to number theory using foundations laid by Gauss; and the work of others, including this writer, with respect to pairs of primes known as Sophie Germain primes.

Justifiably proud of his accomplishments in the theory of numbers and personally pleased with his creativity and imagination, Carl Friedrich Gauss was understandably eager for applications of his ideas and recognition by others. If another mathematician — advanced or novice — wrote to him showing an interest in and comprehension of his published works, he welcomed the correspondent with open arms and a warm heart. On one such occasion, he wrote to a friend, "Recently I have had the joy of receiving a letter from a young Parisian geometer, Le Blanc, who is familiarizing himself enthusiastically with the higher arithmetic and gives proof that he has penetrated deeply into my *Disquisitiones Arithmeticae*" [4,p.67]. In terms more commonly in use today, a *geometer* is a pure mathematician, and *higher arithmetic* is number theory. Two more words in this quoted sentence deserve comment. Though Gauss alluded correctly to Le Blanc as "young", Le Blanc was a year older than Gauss; and though he referred to Le Blanc as "he", Le Blanc was a woman.

For this last inaccuracy, Gauss cannot be blamed. M. Le Blanc was a pseudonym used by a talented person, Mlle. Sophie Germain — creative and imaginative like Gauss himself — who had to contend with current social barriers in the paths of intelligent women who had normal inclinations to study, to perform, and to be recognized in philosophy, science, and mathematics. Nevertheless, she did indeed write philosophical works, interpret and explain scientific phenomena dealing with elasticity, and solve significant mathematical problems in number theory.

Despite sharp criticism by members of the eliticist French Academy of Science, they awarded her special prizes for her work in elasticity and gave her reserved seating at their public meetings, although they did not confer membership on her [1,2].

Her achievements in number theory, however, were considered quite noteworthy by her contemporaries. Gauss wrote, after he learned her identity, "A taste for the abstract sciences in general and, above all, the mysteries of numbers, is excessively rare. It is not a subject which strikes everyone; the enchanting charms of ths sublime science reveal themselves only to those who have the courage to go deeply into it. But when a person of the sex which, according to our customs and prejudices, must encounter infinitely more difficulties than men to familiarize herself with these thorny researches, succeeds nevertheless in surmounting these obstacles and penetrating the most obscure of them, then without doubt she must have the noblest courage, quite extraordinary talents, and a superior genius" [5,p.61].

Sophie Germain apparently never met Gauss personally, but, aside from their mathematical correspondence, their concern for each other was touching. When French military units occupied the German region where Gauss lived, Mlle. Germain, whose family had influential connections, asked the French general in charge to look into the well-being of Gauss [10]. The young officer who came to see Gauss told him that it was at the request of Sophie Germain, a woman whose name was unfamiliar to Gauss. It was not until a few months later that he learned that his benefactress and M. Le Blanc were the same person. In 1810 the Institut de France awarded Gauss a prize for his work in astronomy, and he refused to accept the money because it came from France. Sophie Germain arranged for part of the prize money to be used for "a pendulum clock which was sent to Gauss and used in his room for the remainder of his life." [4,p.93] After her death of cancer at the age of fifty five, Gauss, upon his return to the University at Göttingen to participate in a ceremony at which honorary doctorates were conferred, openly regretted that Sophie Germain was no longer alive to receive one of them. [4,p.93]

Sophie Germain's Influence on Mathematics

Sophie Germain proved that if P and Q are primes such that $Q = 2P + 1$, there are no integers x, y, and z, other than zero and multiples of P, such that $x^P + y^P = z^P$ [11,p.261]. This proof eliminates a large set of candidates for the contradiction of the conjecture commonly called Fermat's last theorem.

Prime P is usually called a Sophie Germain prime when there exists a prime Q that is 1 more than twice P; but we shall henceforth refer to both P and Q as Sophie Germains or S.G.'s. The largest presently known S.G. pair is $39051 \times 2^{6001} - 1$ and $39051 \times 2^{6002} - 1$. Discovered by Wilfrid Keller of Hamburg in 1986, each of these primes has 1,812 digits [8].

Euler showed that if P is congruent to 3(mod 4), Q divides the Mersenne number $2^P - 1$. The converse is also true [7]. This is a simple way of determining that each of the members of a particular large set of Mersenne numbers is composite without a considerable amount of time-consuming computation. For example, 3539 and 7079 are a pair of S.G.'s, and $3539 \equiv 3 \pmod 4$. Then Mersenne number $2^{3539} - 1$ is composite, being divisible by 7079.

S.G.'s and chains of S.G.'s have merited attention in the study of repunits and repetends, as observed by Yates [14]. A repunit $R(N)$ is a number written as N 1's and its value is $(10^N - 1)/9$. If prime number D divides $R(N)$ and divides no smaller repunit, D is said to be a primitive prime divisor of $R(N)$. The number of digits in the period or repetend (the smallest repeating string) of the decimal representation of a proper fraction whose denominator is D is N. For example, 7 and 13 are prime factors of 111111, which is $R(6)$. They are primitive prime divisors of $R(6)$ because neither 7 nor 13 divides any smaller repunit. Decimally, 1/7 is $.\overline{142857}$, and 1/13 is $.\overline{076923}$. Each of these decimals has a period whose length is exactly 6 digits. The prime factors of 111111 are 3, 7, 11, 13, and 37, but 3, 11, and 37 are not primitive divisors of $R(6)$ because each of them divides a smaller repunit. Therefore, 7 and 13 are the only primes that are denominators of fractions whose decimal representations have six-digit period lengths.

We can demonstrate that every repunit has at least one primitive divisor. Consequently, every positive integer is equal to the period length of a finite number of primes. If repunit $R(N)$ has only one primitive prime divisor, only that one prime can be the denominator of a proper fraction whose decimal representation is N digits in length [12]. Among the repunits that have this property — albeit not the only ones — are the prime repunits, the only known examples of which are $R(2), R(19), R(23), R(317)$, and $R(1031)$. All other repunits smaller than $R(145000)$ have been proved to be composite [3].

Fermat's theorem states that if B is not a multiple of prime M, then M divides $B^{M-1}-1$. For example, when $M = 13$ and $B = 10$, 13 divides $10^{12}-1$. We noted earlier that 13 divides $R(6)$, which is $(10^6 - 1)/9$. So, in this case, M divides $B^{(M-1)/2} - 1$, because 13 divides $10^6 - 1$. A prime M that divides $B^{(M-1)/2}-1$ and no expression of this form with a smaller exponent is said to be a half-period prime. If M divides no smaller expression of this form than one with exponent $M-1$, we say that B is a primitive root (mod M), using Gauss's terminology [6], and that B is a full period prime. Thus, 13 is a half period prime and 7 is a full period prime.

Approximately three eighths of all primes are full period primes, and approximately one third of all primes are half period primes [13,14]. The set of repetends for a full period prime M is a single set of cyclic permutations of a string of $M - 1$ digits. For example, $1/7 = .\overline{142857}, 2/7 = .\overline{285714}, 3/7 = .\overline{428571}, 4/7 = .\overline{571428}, 5/7 = .\overline{714285}$, and $6/7 = .\overline{857142}$. The set of repetends for a half period prime M is partitioned into two distinct subsets, each being a set of cyclic permutations of $(M - 1)/2$ digits. In the case of prime 13, $1/13 = .\overline{076923}$, and $2/13 = .\overline{153846}$. There are six proper fractions with denominator 13 which have repetends that are cyclic permutations of 076923, and there are six with 153846.

Using Gauss's quadratic residues [6], we can show that if S.G. prime Q is congruent to 7, 19, or 23(mod 40), then Q is a primitive prime divisor of repunit $R(Q - 1)$, and it is therefore a full period prime with period length $Q-1$. For example, 99023 and 198047 are an S.G. pair, and $198047 \equiv 7(\text{mod } 40)$. Then 198047 is a full period prime and a primitive prime divisor of $R(198046)$. The corresponding 198,046 repetends are cyclic permutations of the same string of 198,046 digits. Also, if S.G. prime Q is congruent to 3, 27, or 39(mod 40), then Q is a half period prime with a period length of P. The S.G. prime pair 7841 and 15683 are examples; the period length of 15683 is 7841.

Some congruence subclasses are heavily laden with Sophie Germain primes. Among the 556 primes less than 200,000 which are congruent to 23(mod 120), there are 533 full period primes, mostly S.G. Q primes.

There exist sequences or chains of Sophie Germain primes in which each prime after the first is 1 greater than twice the one before it. Examples of such ordered sets are (2,5,11,23,47) and (89,179,359,719,1439,2879). We have shown that no such sequence can be infinite, and that the maximum possible length of a sequence whose initial prime is an odd P is $P - 1$ terms. There is only one sequence with as many as eight terms when the initial prime is less than 20,000,000 [14,p.116]. Löh has found larger chains, including several with 12 terms and with initial values that are 12 to 15 digits long [9].

We are indebted to Carl Friedrich Gauss for inspiring and encouraging the brilliant Mlle. Germain, and to Sophie Germain herself for significant number theory contributions which serve as a basis for further discoveries.

References

1. L. L. Bucciarelli and N. Dworsky, *Sophie Germain: An Essay in the History of the Theory of Elasticity*, Reidel, Holland, 1980.
2. J. W. Dauben, *Review of book by Bucciarelli and Dworsky*, American Mathematical Monthly **92**(1985) 64–70.
3. H. Dubner, Letter to S. Yates, 3 September 1989.
4. G. W. Dunnington, *Carl Friedrich Gauss: Titan of Science*, Hafner, New York, 1955.
5. H. M. Edwards, *Fermat's Last Theorem*, Springer-Verlag, New York, 1977.
6. C. F. Gauss, *Disquisitiones Arithmeticae*, Yale, Connecticut, 1966.
7. G. H. Hardy and E. M. Wright, *An Introduction to the Theory of Numbers*, Oxford, England, 1965.
8. W. Keller, Letter to S. Yates, 14 March 1988.
9. G. Löh, *Long chains of nearly doubled primes*, Mathematics of Computation (1989).
10. L M. Osen, *Women in Mathematics*, M.I.T., Massachusetts, 1974.
11. P. Ribenboim, *The Book of Prime Number Records*, Springer-Verlag, New York , 1988.
12. S. Yates, *Period lengths of exactly one or two primes*, Journal of Recreational Mathematics **18**(1985–86) 22–24.
13. S. Yates, *Prime Period Lengths*, Samuel Yates, New Jersey, 1975.
14. S. Yates, *Repunits and Repetend*, Samuel Yates, Florida, 1982.

Samuel Yates
157 Capri D, Kings Point
Delray Beach, Florida 33484
USA

THE MATH. HERITAGE OF C.F. GAUSS (pp. 887-899)
edited by George M. Rassias
©1991 World Scientific Publ. Co. Singapore

ON SAMPLING THEOREMS AND THE GAUSS-JACOBI MECHANICAL QUADRATURE

Ahmed I. Zayed

Sampling theory, which plays a vital role in modern communication theory, is an offspring of the theory of interpolation. C. F. Gauss was one of the greatest mathemticians who contributed to interpolation and approximation theory. In this article we discuss some of Gauss' major contributions to this theory.

Although this article is expository in nature, it also contains some new results. The first part of this article discusses some of Gauss' major contributions to interpolation theory, in particular, the Gauss-Jacobi mechanical quadrature formula, and also gives an overview of some of the main results in sampling theory, such as the Whittaker-Shannon-Kotel'nikov sampling theorem, Lagrange-type interpolation, and Kramer's sampling theorem. From these sampling theorems an analogue of the Gauss-Jacobi mechanical quadrature formula is obtained for some entire functions. In the second part of the article, some new results concerning the relationship between Lagrange-type interpolations and Kramer's sampling theorem are given and from which another generalization of the Gauss-Jacobi mechanical quadrature formula is obtained.

Keywords and phrases : sampling theorems, Lagrange interpolations, the Gauss-Jacobi mechanical quadrature

1. Introduction

The problem of reconstructing a function from its values at a given set of points, whether finite or infinite, is of great practical importance. This problem is closely related to the problem of uniquely determining a function from its values on a subset of its domain. The solution, of course, depends on the nature of the function at hand. While a polynomial of

---

AMS subject classification (1985): 41A05, 94A05

degree n is completely determined by its values at any $n+1$ distinct points, a continuous function is determined by its values on any dense subset of its domain, and an analytic function is determined by its values on any infinite sequence of points which, together with its unique limit point, belongs to the domain of analyticity of the function.

To see some of the practical applications of these problems, let us recall that communication engineers are very often interested in what they call "signals" and in the problem of reconstructing these signals from their sampled values. For mathematicians a signal means a function $f(t)$, usually but not necessarily real-valued, defined for every real t. For example, $f(t)$ may represent the voltage difference at time t between two points in an electrical circuit, or it may represent a speaking voice with its continuous variation of pitch and energy. Therefore, the engineering problem of reconstructing a signal from its sampled values is the same as the problem of reconstructing a function from its values at a given set of points. Clearly, the uniqueness of the solution is a highly desirable property. This problem lies at the heart of "Interpolation Theory", which was founded on the work of some of the greatest mathematicians of all time such as Newton, Lagrange, Legendre, Jacobi, and Gauss.

There are several interpolation and approximation formulae in the literature that bear the name of Gauss, such as the Newton-Gauss interpolation series ([5], [20], and [7] p. 47) and the Gauss-Jacobi mechanical quadrature theorem. The latter is a theorem on approximate integration, but before we state it, let us recall that if $w(x) > 0$ is a weight function defined on some finite closed interval $[a, b]$, then it is known ([18], p. 26) that there exists a system of orthonormal polynomials $\{P_n(x)\}_{n=0}^{\infty}$ with respect to $w(x)$, i.e., $\int_a^b P_n(x)P_m(x)w(x)dx = \delta_{nm}$, such that $P_n(x)$ is a polynomial of exact degree n and in which the coefficient of x^n is positive.

The Gauss-Jacobi mechanical quadrature theorem: Let $a < x_1 < x_2 < \ldots < x_n < b$ denote the zeros of $P_n(x), n \geq 1$, then there exist real numbers $\eta_1, \ldots, \eta_n$ such that

$$\int_a^b f(x)w(x)dx = \sum_{k=1}^{n} \eta_k f(x_k), \qquad (1.1)$$

whenever $f(x)$ is an arbitrary polynomial of degree $2n - 1$. The weight function $w(x)$ and the integer n uniquely determine these numbers η_k. In fact, $\eta_k = \int_a^b S_k(x)w(x)dx$,

$$S_k(x) = \frac{G(x)}{(x - x_k)G'(x_k)} \quad \text{and} \quad G(x) = \prod_{k=1}^{n}(x - x_k). \qquad (1.2)$$

The functions $S_k(x)$ are called the fundamental polynomials of the Lagrange interpolation corresponding to the set $\{x_1, \ldots, x_n\}$, see [18, p. 47].

Another well-known interpolation problem that Gauss was interested in was the problem of finding a function $f(x)$ that would agree with $n!$ at $x = n$ ([8], p. 230). Its solution, of course, is given by the Gamma function, i.e., $f(x) = \Gamma(x+1)$. A less well-known interpolation formula due to Gauss is the Gauss interpolation formula for trigonometric polynomials ([4], p. 38) which may be stated as follows: given $2n + 1$ points $-\pi < x_0 < x_1 < \ldots < x_{2n} < \pi$, construct a trigonometric polynomial $T(x)$ of degree n, i.e., a linear combination of $1, \cos x, \ldots, \cos nx, \sin x, \ldots, \sin nx$ such that $T(x_k) = w_k$ for some prescribed values $w_k, k = 0, 1, \ldots, 2n$. It is easy to verify that the solution may be given by $T(x) = \sum_{k=0}^{2n} w_k S_k(x)$ where

$$S_k(x) = \prod_{\substack{i=0 \\ i \neq k}}^{2n} \sin \frac{1}{2}(x - x_i) \bigg/ \prod_{\substack{i=0 \\ i \neq k}}^{2n} \sin \frac{1}{2}(x_k - x_i).$$

Most of these classical polynomial interpolation formulae such as Newton's, Lagrange's and Gauss' are inadequate to reconstruct signals from their sampled values since signals are not, in general, polynomials or even of polynomial growth. This is a consequence of the physical requirement that a signal should have finite energy E, where E is given by $E = \int_{-\infty}^{\infty} |f(t)|^2 dt$.

We note that signals possess Fourier transforms

$$F(w) = \frac{1}{\sqrt{2\pi}} \int_{-\infty}^{\infty} f(t)e^{itw} dt, \qquad (1.3)$$

which themselves have finite energy in view of Parseval's equality

$$\int_{-\infty}^{\infty} |F(w)|^2 dw = \int_{-\infty}^{\infty} |f(t)|^2 dt < \infty.$$

The function $F(w)$ is known as the amplitude spectrum of the signal $f(t)$, cf. [17].

The most important and common kinds of signals are the band-limited signals. These are signals whose amplitude spectra vanish outside some finite closed interval. If, for example, $F(w)$ vanishes outside $[-W, W], W > 0$, we shall say that $f(t)$ is band-limited to the band $[-W, W]$ or $f(t)$ is of band-width W. Thus, if $f(t)$ is of band-width W, then from the inversion formula for the Fourier transform, it can be represented by

$$f(t) = \frac{1}{\sqrt{2\pi}} \int_{-W}^{W} F(w)e^{-itw}\, dw\,, \quad F \in L^2(-W, W)\,. \qquad (1.4)$$

It is a known fact ([15], [12]) that in this case $f(t)$ is an entire function of exponential type at most W.

The fundamental result in reconstructing band-limited signals from their sampled values is the Whittaker-Shannon-Kotel'nikov (WSK) sampling theorem ([1], [7]), which states that every signal $f(t)$ that is band-limited to $[-\pi W, \pi W]$ for some $W > 0$, i.e., f contains no frequencies higher than πW, can be completely reconstructed from its sampled values $f(\frac{k}{W}), k = 0, \pm 1, \pm 2, \ldots$, and the reconstruction is given by the interpolating formula

$$f(t) = \sum_{k=-\infty}^{\infty} f\left(\frac{k}{W}\right) \frac{\sin \pi(Wt - k)}{\pi(Wt - k)} = \sum_{k=-\infty}^{\infty} f(t_k) S_k(t) \qquad (1.5)$$

where $t_k = \frac{k}{W}$ and

$$S_k(t) = \frac{\sin \pi(Wt - k)}{\pi(Wt - k)}\,. \qquad (1.6)$$

This formula was originally proved by Whittaker ([21], [22]) from the viewpoint of interpolation theory, but its significance was recognized by Shannon [16] and Kotel'nikov [11], who used it in communication theory. This formula also has a very interesting history for which we refer the reader to ([1], [2], [7], and [9]).

This theorem has been generalized in different ways. In one direction the equidistantly-spaced sampling points $t_k = \frac{k}{W}, k = 0, \pm 1, \pm 2, \ldots$ are replaced by non-equidistantly spaced ones with more general sampling functions $S_k^{**}(t)$ than $S_k(t)$ given by (1.6).

The main result in this direction, which is due to Paley and Wiener ([15], p. 114, [12]), states that if $\{t_k\}_{k=-\infty}^{\infty}$ is a sequence of real numbers such that $\sup_k |t_k - \frac{k}{W}| < \frac{1}{4W}$, and

$$G(t) = (t - t_0) \prod_{k=1}^{\infty} \left(1 - \frac{t}{t_k}\right)\left(1 - \frac{t}{t_{-k}}\right),$$

then for any entire function $f(t)$ of exponential type at most πW which is square integrable on the real line, i.e., $\int_{-\infty}^{\infty} |f(t)|^2 dt < \infty$, we have the sampling representation

$$f(t) = \sum_{k=-\infty}^{\infty} f(t_k) S_k^{**}(t) \tag{1.7}$$

where

$$S_k^{**}(t) = \frac{G(t)}{(t - t_k)G'(t_k)}. \tag{1.8}$$

When $t_k = \frac{k}{W} = -t_{-k}, k = 0, 1, 2, \ldots, G(t)$ becomes $\frac{\sin \pi W t}{\pi W}$ and (1.7), (1.8) reduce to (1.5), (1.6). The sampling series (1.7) can be regarded as an extension of the Lagrange interpolation formula to band-limited signals and this is the reason we call it a Lagrange-type interpolation series.

For the sake of consistency with the Lagrange interpolation formula, the sampling functions $S_k^{**}(t)$ are sometimes called the fundamental functions associated with the sampling expansion.

In another direction, Weiss [19] and Kramer [10] generalized the Whittaker-Shannon-Kotel'nikov sampling theorem by replacing the kernel function e^{iwt} in (1.4) by a more general one. Although Weiss announced his results first, he never published the proof. But it was Kramer who actually published the complete work two years later and that is why this result is now known in the literature as Kramer's sampling theorem. Kramer's sampling theorem is stated as follows:

Let $K(x,t) \in L^2(I)$ for each $t \in \mathbb{R}, I = [a,b]$ being some finite closed interval, and let $E = \{t_k\}_{k \in \mathbb{Z}}$ be a countable sequence of reals such that $\{K(x,t_k)\}_{k \in \mathbb{Z}}$ forms a complete orthogonal set of functions in $L^2(I)$. If

$$F(t) = \int_I K(x,t)f(x)dx \, (t \in \mathbb{R}) \tag{1.9}$$

for some $f \in L^2(I)$, then F admits the sampling representation

$$F(t) = \sum_{k=-\infty}^{\infty} F(t_k) S_k^{*}(t), \tag{1.10}$$

where

$$S_k^*(t) = \frac{\int_I K(x,t)\overline{K(x,t_k)}dx}{\int_I |K(x,t_k)|^2 dx}\,.\tag{1.11}$$

The kernel $K(x,t)$ is usuall chosen to be a solution of a self-adjoint boundary-value problem (see e.g., Weiss [19], Kramer [10], Campbell [3], Haddad, Yao and Thomas [6]).

In recent papers [23] and [24], we showed that these two generalizations of the Whittaker-Shannon-Kotel'nikov sampling theorem, that is the Lagrange-type interpolation series given by (1.7) and the Kramer's sampling series given by (1.10), are related.

More precisely, we showed that any function that has a sampling expansion in the scope of Kramer's theorem also has a Lagrange-type interpolation expansion, provided that the kernel $K(x,t)$ associated with Kramer's theorem arises from a regular Sturm-Liouville boundary-value problem [23]. It has also been shown that the results remain valid in cases where the kernel $K(x,t)$ arises from some singular Sturm-Liouville boundary-value problems [24]. One of the main results in [24] can be stated as follows:

Theorem 1. Consider the following regular Sturm-Liouville problem

$$y'' - q(x)y = -\lambda y = -t^2 y\,, -\infty < a \le x \le b < \infty\tag{1.12}$$

$$y(a)\cos\alpha + y'(a)\sin\alpha = 0\tag{1.13}$$

$$y(b)\cos\beta + y'(b)\sin\beta = 0\tag{1.14}$$

where $q(x)$ is continuous on (a,b) and tends to finite limits as $x \to a$ and $x \to b$. Let $\phi(x,\lambda)$ and $\chi(x,\lambda)$ be the solutions of (1.12) such that

$$\phi(a,\lambda) = \sin\alpha\,, \phi'(a,\lambda) = -\cos\alpha$$

$$\chi(b,\lambda) = \sin\beta\,, \chi'(b,\lambda) = -\cos\beta\,.$$

Let $f(x) \in L^2(a,b)$,

$$F(\lambda) = \int_a^b f(x)\phi(x,\lambda)dx\tag{1.15}$$

and

$$F^*(\lambda) = \int_a^b f(x)\chi(x,\lambda)dx\,.\tag{1.16}$$

Then, $F(\lambda)$ and $F^*(\lambda)$ are entire functions of order $1/2$ and type η with

$0 \leq \eta \leq b - a$ that admit the following sampling representations:

$$F(\lambda) = \sum_{n=0}^{\infty} F(\lambda_n) \frac{G(\lambda)}{(\lambda - \lambda_n)G'(\lambda_n)}, \tag{1.17}$$

$$F^*(\lambda) = \sum_{n=0}^{\infty} F^*(\lambda_n) \frac{G(\lambda)}{(\lambda - \lambda_n)G'(\lambda_n)}, \tag{1.18}$$

where $\{\lambda_n\}_{n=0}^{\infty}$ are the eigenvalues of the problem (1.12)–(1.14) and $G(\lambda)$ is the Wronskian $W(\phi, \chi) = \phi(x, \lambda)\chi'(x, \lambda) - \phi'(x, \lambda)\chi(x, \lambda)$ of the two functions ϕ and χ which, without loss of generality, may be written in the form

$$G(\lambda) = \begin{cases} \prod_{n=0}^{\infty} \left(1 - \frac{\lambda}{\lambda_n}\right) & \text{if none of the eigenvalues is zero} \\ \lambda \prod_{n=1}^{\infty} \left(1 - \frac{\lambda}{\lambda_n}\right) & \text{if one of the eigenvalues say } \lambda_0 = 0. \end{cases}$$
$$\tag{1.19}$$

The two series (1.17) and (1.18) converge uniformly on any compact subsets of the complex-plane and they are related via the relation $F^*(\lambda_n) = k_n F(\lambda_n)$ for all n, where $k_n = \frac{\phi(x, \lambda_n)}{\chi(x, \lambda_n)}$ which is neither zero nor ∞.

Observe that the sampling points are the eigenvalues of problem (1.12)–(1.14) and the fundamental functions $S_k(\lambda) = \frac{G(\lambda)}{(\lambda - \lambda_k)G'(\lambda_k)}$ are of the same form as those in (1.8).

One of the interesting features of this theorem is that it sheds some light on possible connections between interpolation theory and the theory of differential equations, in particular, the theory of boundary-value problems.

We also remark that upon formally multiplying (1.17) by $w(\lambda)$ and integrating, one obtains

$$\int_c^d F(\lambda)w(\lambda)d\lambda = \sum_{n=0}^{\infty} \eta_n F(\lambda_n), \tag{1.20}$$

where

$$\eta_n = \int_c^d S_k(\lambda)w(\lambda)d\lambda,$$

and

$$S_k(\lambda) = \frac{G(\lambda)}{(\lambda - \lambda_n)G'(\lambda_n)}.$$

This may be viewed as an analogue of the Gauss-Jacobi mechanical quadrature formula given by (1.1), (1.2) for some entire functions of exponential type.

In (1.1) and (1.20) if we set $w(\lambda) = 1$, we obtain

$$\int_c^d f(x)dx = \sum_{k=1}^n \eta_k f(x_k)\,,$$

where

$$\eta_k = \int_c^d S_k(x)dx\,, S_k(x) = \frac{G(x)}{(x - x_k)G'(x_k)}\,,\ G(x) = \prod_{k=1}^n (x - x_k)\,,$$

and

$$\int_c^d F(\lambda)d\lambda = \sum_{k=0}^\infty \eta_k F(\lambda_k)\,,$$

where

$$\eta_k = \int_c^d S_k(\lambda)d\lambda\,,\ S_k(\lambda) = \frac{G(\lambda)}{(\lambda - \lambda_k)G'(\lambda_k)}\,,$$

$G(\lambda)$ is given by (1.19).

Since the solution of problem (1.12)–(1.14) is usually a function of $\sqrt{\lambda}$, it is more convenient to use the variable $t = \sqrt{\lambda}$. Thus, let $\lambda = t^2, \lambda_n = t_{\pm n}^2, t_{-n} = -t_n$ and set $\tilde{G}(t) = G(\lambda), \tilde{\phi}(x,t) = \phi(x,\lambda)$ and $\tilde{F}(t) = F(\lambda)$.

It is easy to verify that (cf. [24], Cor. 2.1)

$$\int_c^d \tilde{F}(t)dt = \sum_{k=0}^\infty \eta_k \tilde{F}(t_k)\,, \tag{1.21}$$

where

$$\eta_k = \int_c^d \frac{\tilde{G}(t)(2t_k)}{(t^2 - t_k^2)\tilde{G}'(t_k)}dt\,.$$

As an example, let us consider the boundary-value problem

$$y'' = -t^2 y \qquad 0 \le x \le \pi$$
$$y(0) = 0 = y(\pi)\,.$$

It is easy to see that $\tilde{\phi}(x,t) = \frac{\sin xt}{t}$, the eigenvalues are $\lambda_k = k^2, k = 1, 2, \ldots, \tilde{G}(t) = \frac{\sin \pi t}{\pi t}$, hence $\tilde{G}'(t_k) = \frac{(-1)^k}{k}$. If we set $f(x) = 1$ in (1.15), we have

$$\tilde{F}(t) = \int_0^\pi \frac{\sin xt}{t}dx = \frac{1 - \cos \pi t}{t^2}$$

hence,

$$\tilde{F}(k) = \begin{cases} \frac{2}{k^2} & \text{if } k \text{ is odd} \\ 0 & \text{if } k \text{ is even}. \end{cases}$$

Therefore, (1.21) for $c = 0$ and $d = \infty$, gives

$$\int_0^\infty \left(\frac{1 - \cos \pi t}{t^2}\right) dt = \sum_{k=1}^\infty \frac{2}{(2k-1)^2} \int_0^\infty \frac{2(2k-1)^2(-1)^{2k-1}\sin \pi t}{\pi t(t^2 - (2k-1)^2)} dt$$

$$= \sum_{k=1}^\infty \frac{4}{\pi}(-1)^{2k-1} \int_0^\infty \frac{\sin \pi t}{t(t^2 - (2k-1)^2)} dt = 4 \sum_{k=1}^\infty \frac{1}{(2k-1)^2} = \frac{\pi^2}{2}.$$

Since Kramer's sampling theorem is known to hold in the case where the kernel $K(x,t)$ arises from self-adjoint boundary-value problems associated with nth order differential operators, it is natural to ask if the results of [23], [24], in particular, Theorem 1 cited above can also be extended to this case as well.

One of the aims of this article is to answer this question and to show that the extension is indeed possible. Since some of the proofs are rather technical, we shall only mention the main results in the following section and publish the complete version of the work somewhere else.

2. The Main Result

Consider the regular differential expression $L(y)$ defined by

$$L(y) = p_0(x)y^{(m)} + p_1(x)y^{(m-1)} + \ldots + p_m(x)y, \quad -\infty < a \le x \le b < \infty$$

where $\frac{1}{p_0}, p_1, \ldots, p_m$ are real-valued continuous functions on $[a,b]$. It is known ([13], p. 8) that $L(y)$ is self-adjoint if and only if m is even, say $m = 2n$, and $L(y)$ can be put in the form

$$L(y) = (-1)^n (p_0 y^{(n)})^{(n)} + (-1)^{n-1}(p_1 y^{(n-1)})^{(n-1)} + \ldots + p_n y.$$

By the quasi-derivatives of a function y related to the expression $L(y)$, we mean the functions $y^{[1]}, y^{[2]}, \ldots, y^{[2n]}$ defined by the formulae

$$y^{[0]} = y \text{ and } y^{[k]} = \frac{d^k y}{dx^k} \qquad \text{for } k = 1, 2, \ldots, n-1$$

$$y^{[n]} = p_0 \frac{d^n y}{dx^n}$$

$$y^{[n+k]} = p_k \frac{d^{n-k} y}{dx^{n-k}} - \frac{d}{dx}(y^{[n+k-1]}) \quad \text{for } k = 1, 2, \ldots, n.$$

Let $U_j(y), j = 1, 2, \ldots, 2n$ be $2n$ linear forms in the $4n$ quantities $y(a), y^{[1]}(a), \ldots, y^{[2n-1]}(a), y(b), y^{[1]}(b), \ldots, y^{[2n-1]}(b)$, i.e.,

$$U_j(y) = \sum_{k=1}^{2n} \alpha_{j,k} y^{[k-1]}(a) + \sum_{k=1}^{2n} \beta_{j,k} y^{[k-1]}(b), \; j = 1, \ldots, 2n.$$

Now consider the boundary-value problem

$$L(y) = \lambda y \tag{2.1}$$

$$U_j(y) = 0, \; j = 1, 2, \ldots, 2n. \tag{2.2}$$

It is known ([14], p. 77) that this problem is self-adjoint if and only if

$$\sum_{v=1}^{n} \alpha_{j,v} \bar{\alpha}_{k,2n-v+1} - \sum_{v=1}^{n} \alpha_{j,2n-v+1} \bar{\alpha}_{k,v}$$

$$= \sum_{v=1}^{n} \beta_{j,v} \bar{\beta}_{k,2n-v+1} - \sum_{v=1}^{n} \beta_{j,2n-v+1} \bar{\beta}_{k,v} .$$

Thus, from now on we shall assume that this condition always holds.

Let $A = (\alpha_{j,k}), B = (\beta_{j,k}); j, k = 1, 2, \ldots, 2n$. We also shall assume, without loss of generality, that A is nonsingular. Let

$$\tilde{U}_1(y) = \sum_{k=1}^{2n} \alpha_{1,k} y^{[k-1]}(a) \text{ and } \tilde{\tilde{U}}_1(y) = \sum_{k=1}^{2n} \beta_{1,k} y^{[k-1]}(b),$$

hence $U_1(y) = \tilde{U}_1(y) + \tilde{\tilde{U}}_1(y)$. Under the assumption that the problem

$$L(y) = \lambda y$$

$$\tilde{U}_1(y) = c, \; \tilde{\tilde{U}}_1(y) = -c, c \neq 0$$

$$U_j(y) = 0, j = 2, 3, \ldots, 2n - 1$$

has a unique solution $\phi(x, \lambda)$ for each λ that is a simple eigenvalue of (2.1)–(2.2), $\phi(x, \lambda)$ becomes also a solution of the problem

$$L(y) = \lambda y$$

$$U_j(y) = 0, \; j = 1, \ldots, 2n - 1.$$

Now let us define the function $G(\lambda)$ by

$$G(\lambda) = U_{2n}(\phi) = \sum_{k=1}^{2n} \{\alpha_{2n,k} \phi^{[k-1]}(a, \lambda) + \beta_{2n,k} \phi^{[k-1]}(b, \lambda)\} .$$

Then, it is easy to see that $G(\lambda)$ is an entire function in λ and that $\tilde{\lambda}$ is a zero of $G(\lambda)$ if and only if it is an eigenvalue of (2.1)–(2.2).

Now we can state our main result which is a generalization of Theorem 1.

Theorem 2. Let $\{\lambda_n\}_{n=0}^{\infty}$ be the eigenvalues of the boundary-value problem (2.1)–(2.2) and assume that they are all simple. Let

$$F(\lambda) = \int_a^b f(x)\phi(x,\lambda)dx \,, \ f \in L^2(a,b)\,.$$

Then, $F(\lambda)$ is an entire function which admits the following sampling expansion

$$F(\lambda) = \sum_{n=0}^{\infty} F(\lambda_n)\frac{G(\lambda)}{(\lambda - \lambda_n)G'(\lambda_n)}\,.$$

An analogue of (1.20) can now be obtained.

Corollary 1. Under the assumptions of Theorem 2, we have

$$\int_c^d F(\lambda)d\lambda = \sum_{n=0}^{\infty} \eta_n F(\lambda_n)\,,$$

where

$$\eta_n = \int_c^d \frac{G(\lambda)}{(\lambda - \lambda_n)G'(\lambda_n)}d\lambda\,.$$

References

1. P. Butzer, *A survey of the Whittaker-Shannon sampling theorem and some of its extensions*, J. Math. Res. Exposition **3** (1983) 185–212.
2. P. Butzer, W. Splettsöβer and R. Stens, *The sampling theorem and linear predictions in signal analysis*, Jahresber. Deutsch. Math. - Verein. **90** (1988) 1–70.
3. L. Campbell, *A comparison of the sampling theorems of Kramer and Whittaker*, J. SIAM **12** (1964) 117–130.
4. P. Davis, *Interpolation and Approximation*, Dover Publ., New York, 1975.
5. C. Gauss, *Carl Friedrich Gauss Werke*, Band 8, Königl. Gesellschaft Wiss. Gottingen, Teubner, Leipzig, 1900.

6. A. Haddad, K. Yao and J. Thomas, *General methods for the derivation of sampling theorems*, IEEE Trans. Inf. Theory, **IT-13** (1967) 227–230.

7. J. Higgins, *Five short stories about the cardinal series*, Bull. Amer. Math. Soc. (1) **12** (1985) 45–89.

8. E. Hille, *Analytic Function Theory*, Vol. 1, Chelsea Publ., New York, 1976.

9. A. Jerri, *The Shannon sampling theorem - its various extensions and applications: A tutorial review*, Proc. IEEE 65, 11 (1977) 1565–1596.

10. H. Kramer, *A generalized sampling theorem*, J. Math. Phys. **38** (1959) 68–72.

11. V. Kotel'nikov, *On the carrying capacity of the 'ether' and wire in telecommunications*, Material for the First All-Union Conference on Questions of Communication, Izd. Red. Upr. Svyazi RKKA, Moscow (Russian).

12. N. Levinson, *Gap and Density Theorems*, Amer. Math. Soc. Colloq. Publ., vol. 26, Providence, R.I., 1940.

13. M. Naimark, *Linear Differential Operators*, Vol. 1, *Elementary Theory of Linear Differential Operators*, George Harrap & Co. Ltd., London, U.K., 1967.

14. M. Naimark, *Linear Differential Operators*, Vol. II, *Linear Differential Operators in Hilbert Space*, George Harrap & Co. Ltd., London, U.K., 1968.

15. R. Paley and N. Wiener, *Fourier Transforms in the Complex Domain*, Colloq. Publ., vol. 19, Amer. Math. Soc., Providence, R.I., 1934.

16. C. Shannon, *Communication in the presence of noise*, Proc. IRE, **37** (1949) 10–21.

17. D. Slepian, *Some comments on Fourier analysis, uncertainty and modeling*, SIAM Review (3) **25** (1973) 379–393.

18. G. Szegö, *Orthogonal Polynomials*, Colloq. Pub. vol. 23, Amer. Math. Soc., Providence, R.I., 1939.

19. P. Weiss, *Sampling theorems associated with Sturm-Liouville systems*, Bull. Amer. Math. soc. **63** (1957) 242.

20. J. Whittaker, *Interpolatory Function Theory*, Cambridge Univ. Press, Cambridge, England, 1935.

21. J. Whittaker, *On the Fourier theory of the cardinal function*, Proc. Edinburgh Math. Soc. 1 (1929 b) 169–176.

22. J. Whittaker, *On the cardinal function of interpolation theory*, Proc. Edinburgh Math. Soc. 1 (1929 a) 41–46.

23. A. Zayed, G. Hinsen and P. Butzer, *On Lagrange interpolation and Kramer-type sampling theorems associated with Sturm-Liouville problems*, SIAM J. Appl. Math. (3) **50** (1990), 893–909.

24. A. Zayed, *On Kramer's sampling theorem associated with general Sturm-Liouville problems and Lagrange interpolation*, to appear in SIAM J. Appl. Math.

Ahmed I. Zayed
Mathematics Department
California Polytechnic State University
San Luis Obispo
CA 93407
USA

Current Address:
Mathematics Department
University of Central Florida
Orlando, FL 32816
U.S.A.

Author Index

www.ingramcontent.com/pod-product-compliance
Ingram Content Group UK Ltd.
Pitfield, Milton Keynes, MK11 3LW, UK
UKHW021824150726
7214IPUK00017B/304